Geologic Column (Mesozoic and Cenozoic Eras)		
Cenozoic Era	**Quaternary Period**	
	Holocene Epoch	12,000
	Pleistocene	1.6 Ma
	Tertiary	
	Pliocene	
	Piacenzian Stage	3,4
	Zanclian	5.2
	Miocene	
	Messinian	6.7
	Tortonian	10.4
	Serravallian	14.2
	Langhian	16.3
	Burdigalian	21.5
	Aquitanian	23.3
	Oligocene	
	Chattain	29.3
	Rupelian	35.4
	Eocene	(34)
	Priabonian	38.6
	Bartonian	42.1
	Lutetian	50.0
	Ypresian	56.5
	Paleocene	
	Thanetian	60.5
	Danian	65.0
Mesozoic Era	**Cretaceous**	
	Senonian	
	Maastrichtian	74.0
	Campanian	83.0
	Santonian	86.6
	Coniacian	88.5
	Gallic	
	Turonian	90.4
	Cenomanian	97.0
	Albian	112.0
	Aptian	124.5
	Barremian	131.8
	Neocomian	
	Hauterivian	135.0
	Valanginian	140.7
	Berriasian	145.6
	Jurassic	
	Triassic	

LATE CRETACEOUS AND CENOZOIC HISTORY OF NORTH AMERICAN VEGETATION

LATE CRETACEOUS AND CENOZOIC HISTORY OF NORTH AMERICAN VEGETATION

NORTH OF MEXICO

ALAN GRAHAM

New York Oxford

Oxford University Press

1999

Oxford University Press

Oxford New York
Athens Auckland Bangkok Bogotá Buenos Aires Calcutta
Cape Town Chennai Dar es Salaam Delhi Florence Hong Kong Istanbul
Karachi Kuala Lumpur Madrid Melbourne Mexico City Mumbai
Nairobi Paris São Paulo Singapore Taipei Tokyo Toronto Warsaw

and associated companies in
Berlin Ibadan

Published by Oxford University Press, Inc.
198 Madison Avenue, New York, New York 10016

Oxford is a registered trademark of Oxford University Press

Graham, Alan.
Late Cretaceous and Cenozoic history of North American vegetation, north of Mexico / by Alan Graham.
p. cm.
Includes bibliographical references and indexes.
ISBN 0-19-511342-X
1. Paleobotany—Cretaceous. 2. Paleobotany—Cenozoic.
3. Plants, Fossil—North America. I. Title.
QE924.G73 1998
561'.197—dc21 97-5713

1 3 5 7 9 8 6 4 2

Printed in the United States of America
on acid-free paper

Contents

Prologue

The vegetation of North America comprises 17,000–20,000 species of vascular plants or about 7% of the world's flora. As treated here, these species are arranged into seven plant formations: tundra, coniferous forest, deciduous forest, grassland, shrubland/chaparral–woodland–savanna, desert, and tropical. The current version of these formations is primarily a product of environmental change and biotic response during the past 70 million years (m.y.) of the Late Cretaceous Period and Cenozoic Era. The first unequivocal angiosperms appear in the Valanginian/Hauterivian [140–135 millions of years ago (Ma); Crane, 1989, 1993; Crane et al., 1995; Taylor and Hickey, 1996], and by the Late Cretaceous they had diversified and radiated to become the dominant group of terrestrial vascular plants. The Late Cretaceous also marks the time when most plant fossils can be assigned to modern forms at least to the level of family, facilitating the reconstruction of vegetation types, past environments, migrations, and lineages on the basis of paleopalynological (plant microfossil) and paleobotanical (plant megafossil) evidence. The development of vegetation is a continuum, but for the purposes of this history the Late Cretaceous is a convenient starting point. The region under consideration is North America north of Mexico, paralleling the geographic coverage of the Flora of North America project (Flora of North America Editorial Committee, 1993).

OUTLINE

The outline used in tracing this history begins with the composition and arrangement of the modern vegetation presented in Chapter 1. This sets the goal or end point of the survey.

In Chapter 2 some of the physical factors are discussed that have determined terrestrial paleoenvironments and influenced the course of vegetational history over the past 70 m.y. This material is included because the intent of the survey is not just to describe stages in the development of North American vegetation through time, but also to outline the principal causes for the changes. A broad overview of these factors is provided by Crowley and North (1991), and their general effect on global biotas is summarized by Zubakov and Borzenkova (1990). There is an excitement attendant to the realization that we now have the basic methodologies and are approaching the time when the myriad processes involved can be integrated into a fuller outline of the planet's physical and biological history. The factors are numerous and interactive, however, and until the development of several large-scale projects and sophisticated data-gathering devices, an explanation of the dynamics of environmental change and biotic response was impossible. These include the Climate Long-Range Investigation and Mapping Program (CLIMAP 1976, 1981, 1984; which is concerned with reconstruction of the last glacial and interglacial climate); the Cooperative Holocene Mapping Project (COHMAP, 1988; which is concerned with reconstruction of northern hemisphere climates for selected time intervals of the Holocene), Past Global Changes (PAGES, a project of the International Geosphere-Biosphere Programme, see Workshop Report Series 96–3, 4, 1996), the Greenland Ice Core Project (GRIP), the Greenland Ice Sheet Program II (GISP2), the Deep Sea Drilling Project (DSDP), and its successor the Ocean Drilling Program (ODP); satellites and Geographic Information Systems (GIS); and supercomputers. A convenient way to approach these relationships is through computer modeling, such as the General Circulation Model (GCM) or Community Climate Model (CCM) used at the National Center for Atmospheric Research (NCAR), the Geophysical Fluid Dynamics Laboratory at the National Oceanographic and Atmospheric Administration (GFDL/NOAA), the Goddard Institute for

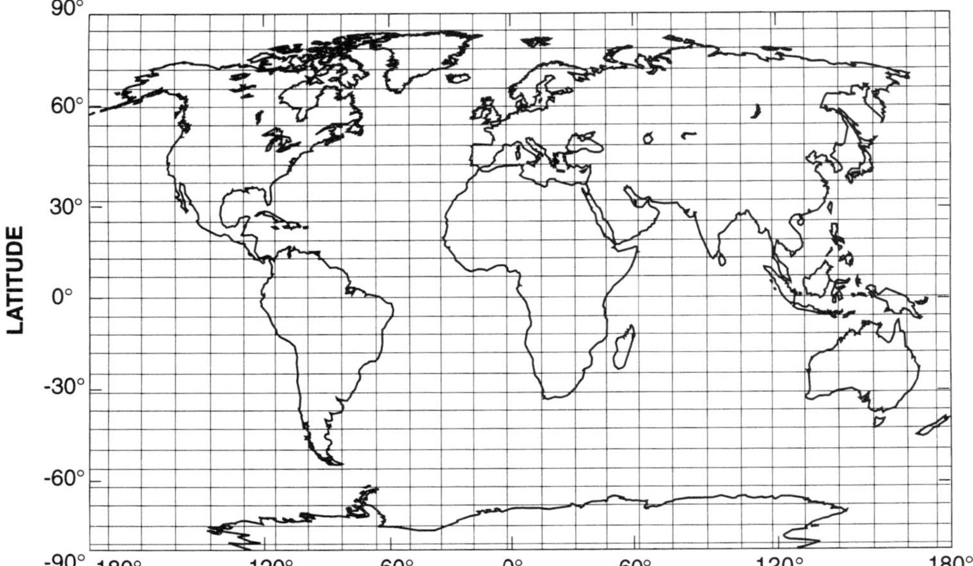

Figure 1. Global climate model grid at 7.8° × 10° resolution. Reprinted from Hansen et al. (1993) with the permission of the National Geographic Society.

Space Studies at the National Aeronautics and Space Administration (GISS/NASA), and elsewhere. For those not familiar with the mathematics and philosophical bases of computer modeling, some guidelines are useful for assessing the information. General reviews are given in Crowley and North (1991), Gates (1982), Hannon and Ruth (1994), Hansen et al. (1993), Imbrie (1982), and Schneider and Dickinson (1974). More detailed reviews of modeling theory are provided by Schlesinger (1988), Sellers et al. (1997), Trenberth (1992), and Washington and Parkinson (1986).

It is important to recognize that modeling presently involves developing, testing, and refining programs that will eventually approximate real-world situations. For the moment, however, the programs prescribe certain boundary conditions (e.g., the position and configuration of continents at 100 Ma), and hold others quasi-static at modern values, do not incorporate them, or parameterize (estimate) them. Then a change called an altered boundary condition or forcing mechanism is introduced, such as the progressive uplift of a mountain system or an extensive plateau, and the effects are noted, usually on one to a few factors. These are called sensitivity tests, in contrast to simulation models that attempt to reconstruct an actual past climate. Examples of the latter are rare because all boundary conditions must be specified (e.g., ice, albedo, insolation, CO_2 concentration, cloud cover, sea-surface temperatures), and much of this information is unknown for past times. The purpose of a sensitivity test may be to determine the effect of regional orogeny on atmospheric circulation patterns and subsequent precipitation and temperature regimes. Often it is found that the results of the simulation (model performance) correspond to real-world conditions (validation data or ground truth) based on paleopalynological or paleobotanical evidence (e.g., development of arid vegetation to the lee of rising mountains). In other instances the simulated effect may run counter to known conditions, prescribing instead high rainfall in areas where arid vegetation existed. The first result verifies the accuracy of the simulation because it replicates a condition known to have existed. The second result shows that the model needs improvement.

Model units consist of a series of grids, each representing areas several hundred kilometers in length and width (e.g., 7.8° latitude × 10° longitude; Fig. P.1) and a few kilometers in height. The CCM used at NCAR to model North American climates has a grid resolution of 4.4° × 7.5°. The geography and continental configuration is idealized and is commonly called "wonderland" (Fig. P.2). The spatial resolution or grid constraints of the second simulation noted above may have allowed the edge of the North American Cretaceous and early Cenozoic epicontinental sea to extend too far west on the idealized continent, resulting in simulated rainfall in an area that, in reality, was dry. This testing of model prediction on known past climates is called hindcasting and it gives an estimate of the model's accuracy for eventual forecasting. Paleontology provides the principal data base for hindcasting.

The impressive strength of sensitivity tests is that they can be used to specify and to some extent quantify the role of individual factors that determine climate. One reason it is not yet possible to completely model climate, however, is because each of these factors interacts to amplify, dampen, or cancel out the effect of the others through an immensely complex system of feedbacks. On a scale of geologic time, the position and orientation of the Earth relative to the sun is changing, solar luminosity varies, mountains are rising and eroding, seaways are opening and closing, atmospheric and oceanic circulation patterns are fluctuating, continents are drifting, volcanoes are erupting,

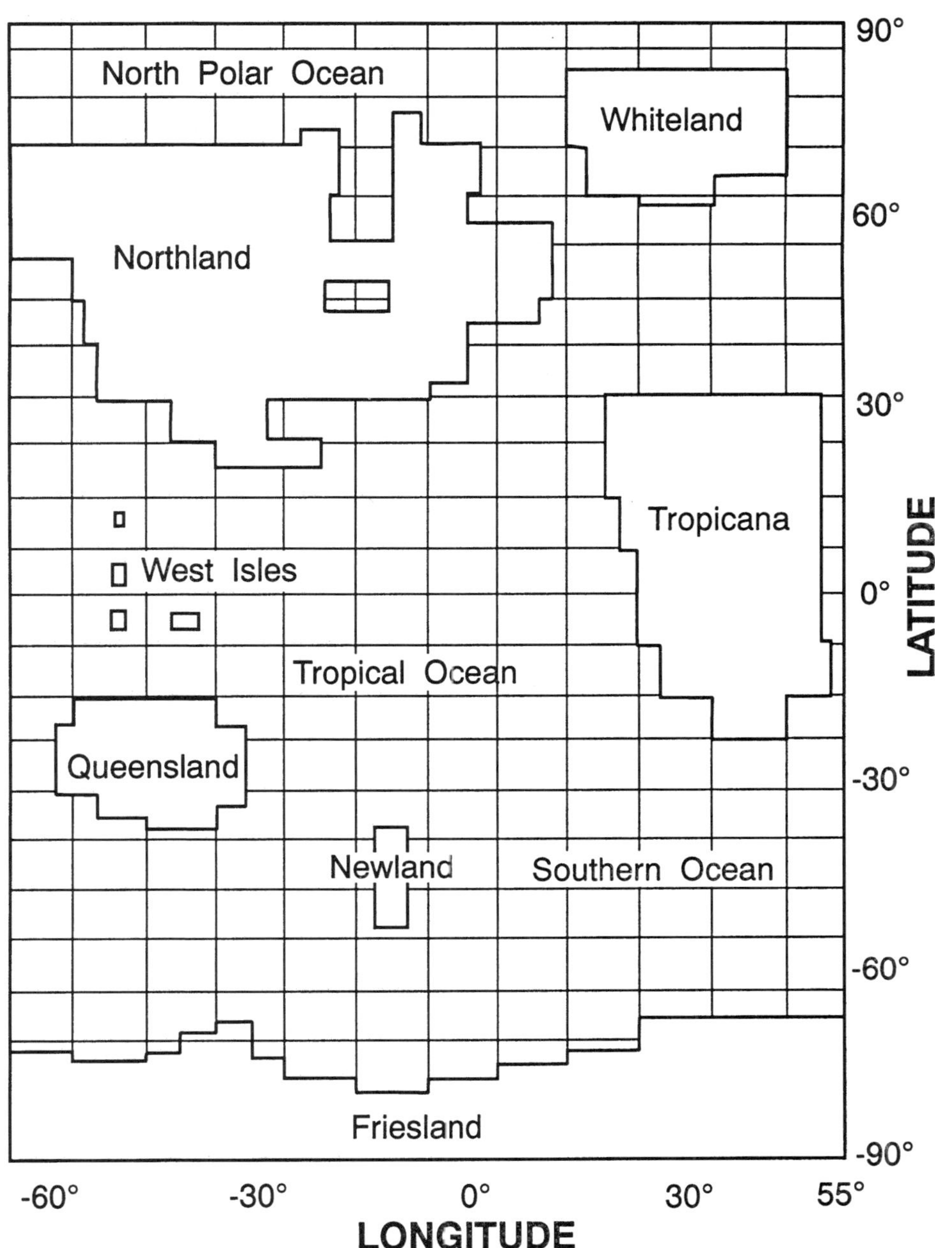

Figure 2. Wonderland model geography. Reprinted from Hansen et al. (1993) with the permission of the National Geographic Society.

atmospheric and oceanic CO_2 and marine salinity concentrations are changing, El Niños occur, chaos microvibrates the climate system, glaciers and vegetation types are expanding and contracting, albedoes are changing, catastrophes are occurring, and organisms are evolving. The result is that climatic–biotic interactions are more complex than our present ability to model them. Nonetheless, modeling through the use of supercomputers, including parallel computing with Connection Machines (64,000 simultaneous calculations, 10 billion calculations/s), teraflops (computing on the trillionfold/s level), or new photon-based (rather than electric current-based) microcircuitry allowing computers to utilize the speed of light, is the only approach that offers even the possibility of integrating these factors into a quantified statement of long-term environmental change. This is especially true if full ocean and atmospheric interactions are to be coupled in simulation models (Bye, 1996).

Chapter 3 discusses results of context information from other disciplines that is essential for reconstructing past communities and environments. Two of these, paleotemperature and sea-level changes, are also forcing mechanisms (Chapter 2), but because the records are typically presented as curves that include all of Late Cretaceous and Cenozoic time, they serve as especially convenient frameworks for viewing vegetational history. For this reason, paleotemperature and sea-level changes are emphasized separately in Chapter 3.

A temporal context is provided by radiometric dating. In addition to the well-known methods based on carbon 14

(^{14}C) and potassium/argon (K/Ar), there are about 15 other measurements that allow an absolute age to be placed on a wide variety of sediments and structures, even from minute samples such as individual seeds or teeth (Saarnisto, 1988; Williams et al., 1993, appendix, table A.1). Among these are rubidium/strontium (Rb/Sr), uranium/thorium/lead (U/Th/Pb), fission-track dating (Wagner and Van den Haute, 1992), thermoluminescence dating (Aitken, 1985), electron spin resonance (ESP) dating, amino acid racemization analysis, and argon 39/argon 40 ($^{39}Ar/^{40}Ar$, including single-crystal laser probe method; Chen et al., 1996). The application of radiometric techniques facilitates correlation between floras in reconstructing regional vegetation and allows fossil floras from different regions to be placed in a proper vertical sequence to track environmental, distributional, and lineage changes through time. They also allow correlation between terrestrial strata and marine strata that are typically subdivided into nannofossil and planktic foraminiferal zones. These zones, in turn, may be aligned with the magnetic polarity time scale (sequence of pole reversals), providing a precise chronological framework for terrestrial and marine fossiliferous sediments that lack minerals suitable for radiometric dating. Seismic profiles through subsurface strata are patterns of sound reflection that vary with erosion–deposition sequences and from which rock types and environments of deposition can be inferred. These permit a further comparison between terrestrial and marine events (Shackleton and Opdyke, 1973). The methods are described in numerous publications, including recent ones by Dalrymple (1991, especially chapter 2), Easterbrook (1988), Goodfriend (1992), McDougall and Harrison (1988), and Roth and Poty (1989). It should be noted that dates obtained by the ^{14}C technique and presented in the older literature did not take into account the changing amounts of radiocarbon in the atmosphere over time, and they are ~(approximately) 2 ky (thousand years) too young prior to about 12,500 years ago and 250–500 years in maximum error in later times. For example, the warming period at 14,450 ± 250 years before present (B.P.) is dated at ~12,500 years by the older ^{14}C measurements. The ^{14}C chronometer has not been calibrated for dates younger than 40 ka. Unless stated otherwise, the dates for this interval cited in the text are in ^{14}C years. Similarly, many of the early dates obtained by the K/Ar method (Evernden and James, 1964), which was used to calculate the age of Mesozoic and Tertiary rocks, were from contaminated sediments and require the conversion of old to new constants (Dalrymple, 1979). Other samples came from sites adjacent to the ones preserving the flora or fauna, and some of these were later found to be significantly different in age. New and recalibrated dates are being published, and they are the ones mentioned in the text.

The framework for describing events in Earth history is the geologic time scale. This scale continues to be refined; several versions are available that provide the sequence of eras, periods, epochs, and stages, as well as estimates of their span in absolute number of years (e.g., Berggren et al., 1995; Haq and van Eysinga, 1987; Harland et al., 1989). The compilations by Berggren et al. (1995) and Harland et al. (1989) show the age and stratigraphic placement of formations and fossil assemblages mentioned in this text. In addition to the formal time units shown on the generalized geologic column in the frontispiece and in the more detailed scales shown in Figs. 5.1, 5.19, 6.13, 6.14, 7.1, and 7.2, other terms encountered in the literature are Cainozoic (Cenozoic Era), Paleogene or Early Tertiary (Paleocene through Oligocene), Neogene or Late Tertiary (Miocene and Pliocene), ice ages (Pleistocene), and Holocene (postglacial or Recent; time after ~12 Kya).

Another important context is faunal history, which reflects prevailing vegetation types and general environmental conditions. Two other contexts are conceptual in nature, but they nonetheless influence our reading of vegetational history from the fossil record. These are the older and much-debated concept of geofloras and the newly emerging concept of a boreotropical flora.

Other contexts are important to paleoenvironmental reconstructions, but they are beyond the scope of the present text. One is the study of land forms and sedimentology (Ziegler et al., 1987). Tillites, fluvioglacial deposits, ice-rafted debris, rock striations, and geomorphic features such as moranes, V-shaped valleys, kettle-hole lakes, aretes, cols, and cirques all tell of glacial conditions. In contrast, filled mud cracks, increasing amounts of eolian (wind blown) dust, and evaporites such as gypsum and salt crystals reveal subhumid environments and reduced vegetation cover. Peat and swamp sediments, which may be subsequently indurated into lignite and coal, form most extensively under humid conditions and high water tables in temperate environments and in near-shore coastal marine to brackish waters in tropical environments where biomass production exceeds physical, microbial, and oxidative removal. Bauxites (from Les Baux in Provence, France) are produced from the weathering of aluminum-containing rocks under tropical conditions. The characterization of the late Middle Eocene Green River Flora of Colorado and Utah as a seasonally arid subtropical shrubland/chaparral–woodland–savanna is supported by the presence of seasonal evaporites and deeply oxidized redbeds. Another valuable contribution of sedimentology to paleopalynological and paleobotanical studies is facies (sediment-type) analysis. The generalized nature of some fossils (e.g., many fern spores of the Blechnaceae, Polypodiaceae, Pteridaceae complex) or their age (pre-Eocene) precludes assigning the taxa to the proper association and habitat. However, some of these fossils are consistently found in facies known to be deposited under specific depositional settings. For example, in Paleocene deposits from the Bighorn Basin of Wyoming, the fern spore *Deltoidiospora* is most abundant in sediments representing pond and swamp environments (Farley, 1989; Farley and Traverse, 1990). Facies analysis can help reveal the diversity of habitats within a region and allows some generalized fossil types to be placed in the proper paleocommunity (Farley, 1990). Even so, it is usually only the extremes of climate that are revealed, while within that range there are few clear-cut indications of climate available from sedimentological analysis alone. Readers interested in this subject are referred to the standard texts in sedimentology as they relate to basin analysis and paleogeography (e.g., Miall, 1984).

Another intent of this text, in addition to providing cause and effect and context information, is to describe the impressive array of methodologies available specifically for the study of fossil floras and to note the strengths and limitations inherent in each approach. This material is presented in Chapter 4. The principal disciplines are paleopalynology, which for terrestrial vegetation includes the study of fossil pollen, spores, and phytoliths; and paleobotany, which includes the study of plant megafossils such as leaves, cuticles, cones, flowers, fruits, seeds, and wood. An important use of the information is in reconstructing paleoclimates, and two approaches are available for this purpose.

The modern analog or nearest living relative (NLR) method involves establishing the composition of fossil floras by relating various elements to the most morphologically similar living species. The assemblages are then arranged in a horizontal sequence to reconstruct regional and global vegetation and paleoenvironmental conditions at any one point in time and in a vertical sequence to determine paleoenvironmental trends, vegetational change, evolutionary patterns, and the origin of biogeographic relationships on the basis of the ecology and distribution of presumed modern analogs.

Another approach is foliar physiognomy: the study of the various leaf size categories, margin types (leaf margin analysis, LMA), texture variations, and special features (e.g., the presence or absence of drip tips) found in different climatic regimes. In the modern flora the percent representation of these features varies in a general way with rainfall and especially temperature. The relationship affords an opportunity to reconstruct paleoclimates independent of taxonomic identification and the assumed ecological equivalency of fossil and modern species. Methods are being developed by which qualitative observations and measurements of physiognomic characters can be treated statistically through multivariate correspondence analysis and polynomial regressions as part of an important new program called the Climate-Leaf Analysis Multivariate Program (CLAMP; Wolfe, 1993). The compilation of data for statistical analysis and the construction of matrices for computer analyses can expand and enhance the precision of observations, standardize the information gathered, and provide a reproducible method of reconstructing paleoclimates from fossil leaf assemblages. Ideally, the use of microfossils and megafossils, and among megafossils the application of the modern analog and foliar physiognomy approaches, is most effective for paleovegetation analysis. Where only microfossils or megafossils are available or where the time interval is beyond the use of modern analogs (viz., earlier than about the Middle Eocene), reconstructions are more tentative and depend upon an objective recognition of the strengths and weaknesses of each methodology and the available context information.

No single method is adequate for unraveling the complex history of North American vegetation and for tracing terrestrial paleoenvironments through 70 m.y. of time. In earlier days it was commonplace for investigators to pursue their speciality independent of others working in the same discipline and to be mostly indifferent to results from other lines of inquiry. This frequently led to the existence of various "schools" (literally and philosophically) wherefrom picturesque figures boisterously defended their collections and theories against competitors and new ideas. They often went to great lengths to establish priority, accumulate authorship of new taxa, increase the holdings of type specimens for their museums, or to "do a number" on a colleague who was viewed as particularly obstreperous. Paleontologists from competing institutions have been known to write hurriedly scribbled descriptions of new species on pieces of paper in the field and give them to couriers who would rush them into town to be published in the next edition of the local newspaper. In the early 1900s the North American paleontology establishment reacted to Wegener's unsettling challenge to the permanence of continents and ocean basins with a voice that was strong, virtually united, and ultimately wrong. At the American Association of Petroleum Geologists' 1926 symposium on continental drift, University of Chicago geologist Rollin T. Chamberlain ranted whether we can call geology a science when it is "possible for a theory such as this to run wild?" (Sullivan, 1974, p. 11). Paleobotanist Edward W. Berry of Johns Hopkins University characterized Wegener's method, and by implication Wegener himself, as not scientific, but taking "the familiar course of an initial idea, a selective search through the literature for corroborative evidence, ignoring most of the facts that are opposed to the idea, and ending in a state of auto-intoxication in which the subjective idea comes to be considered as an objective fact" (Sullivan, 1974, p. 12).

It would be naive to assume that such attitudes are exclusively a thing of the past. In addition to valid scientific debate, the paleontological literature contains its share of phenomenon chasing, tenacious adherence to previously expressed opinions, and point–counterpoint arguments that focus as much on discrediting as on objectively assessing the strengths and weaknesses of a theory. For beginning students and the general reader, this can result in a confusing melange of opinions expressed as proclamations; overemphasized strengths of a particular methodology; and oversimplistic characterizations of alternative views, approaches, or theories. Founding practitioners can be prone to unrestrained advocacy, while the establishment can be rather unreceptive to challenge and innovation. Nonetheless, a hallmark of most present-day investigations is an effort to place results derived from specialized disciplines into the broadest possible context based on all available information. New interdisciplinary coalitions are being forged, and these are to the benefit of each specialized approach.

With the goal of the survey defined as explaining the origin and development of the existing plant formations of North America and discussions of cause and effect, context, and methodologies as background, it is possible to consider this history in a more integrated way. The record of North American vegetation is discussed in Chapters 5–8 for the interval between ~70 Ma and the present. Four stages emerge from three major climatic changes that produced effects that are evident in the plant fossil record. Prominent among these are global decreases in temperature and associated extinction–diversification–migration events in the biotic realm, which occurred in the Middle Eocene, the Middle Miocene, and the Quaternary, marking the beginning of extensive northern hemisphere glaciations. These time segments provide a convenient framework for considering the development of North American vegetation between the Late Cretaceous and the Early Eocene (70–50 Ma, Chapter 5), the Middle Eocene and the Early Miocene (50–16.3 Ma, Chapter 6), the Middle Miocene and the Pliocene (16.3–1.6 Ma, Chapter 7), and during the Quaternary (1.6 Ma to the present, Chapter 8). Emphasis is placed on terrestrial paleoenvironments and on the time of appearance and subsequent development of the major plant formations and associations.

Chapter 9 traces the origins of current biogeographic relationships of the North American flora, primarily with the Mediterranean region, the dry regions of South America, eastern Asia, and eastern Mexico. Until recently this could be done through only traditional paleopalynological and paleobotanical methodologies, estimates of land–sea relationships through time, and intuitive interpretations based on the relationships, distributions, and modes of dispersal of modern taxa. Now it is possible to add new molecular techniques involving allozyme–isozyme analysis, genetic distance calculations, and the unsettled issue of the existence and use of molecular clocks. Patterns of distribution are being analyzed through emphasis on the movement of land (vicariance biogeography), as well as through the movement of organisms (dispersal biogeography). A union is being attempted between historical geology, biogeography, phylogenetics, and cladistics into a synergistic approach in which results are summarized in the form of area cladograms. This is also an unsettled methodology, but it offers the potential of discussing biogeography in ways other than the traditional narrative format. These innovations constitute a whole new world of research from that practiced by Linnaeus in the middle 1700s or Asa Gray in the middle 1800s and their early attempts to define and explain the origin of geographic affinities of the North American Flora.

AUDIENCES

The audiences to which this text is addressed are those who share an interest in the environments and events that have shaped North American vegetation and who have a general background in the biological and geological sciences. For the specialist, species lists are provided for the major fossil floras and a list of technical papers is included after each chapter. For the general reader, terms are defined, specialized units of measurements are explained, and widely used common names for familiar plants are given as they are encountered in the text. At the end of each chapter a supplemental bibliography of General Readings is also provided. These include relevant articles in *American Scientist*, *Natural History*, *Scientific American*, *Smithsonian Magazine*, and other sources, as well as reviews in *BioScience*, *Nature*, and *Science*.

Early versions of the book were also used in a course on North American vegetation for upper level undergraduates and graduate students. For this purpose, the presentation has been designed as a course or seminar text, or as a reference to be used in conjunction with lectures, student presentations, and outside readings. In the planning stages it was anticipated that the subject would be restricted to vegetational history and that a few representative floras would be used to describe North American plant communities at various times during the Late Cretaceous and Cenozoic. After using the material in the course, which included students with an array of academic backgrounds, it became clear that a more expansive treatment would serve better. Many geology students have few or no courses in botany, while many biologists take only the introductory courses in geology. Thus, a broader self-contained survey that included causes, context, and methodology was deemed most practical. Three decades of teaching have convinced me that subjects take on interest in proportion to the background brought to the course by the students and/or provided by the instructor, the demonstrated importance or "sowhatness" of the subject, and the effort spent in presenting and learning the subject. Attempting to provide a minimum of background and establish the relevancy of the subject for those mostly having one-time contact with the topic results in a somewhat encyclopedic text, but the decision was a deliberate and carefully considered one based on teaching experience.

A perception that has undoubtedly influenced my writing is that an increasing number of students seem most comfortable as passive recipients of information. This is in contrast to learning endeavors that involve active participation. One of my most challenging and ultimately satisfying educational experiences as a beginning student was reading several texts that were important in biology and geology, but that were outside my immediate speciality and for which I had only a minimum background. These included Carl Swanson's *Cytology and Cytogenetics*, G. Ledyard Stebbins, Jr.'s *Variation and Evolution in Plants*, and later Verne Grant's *Plant Speciation*, Robert Sokal and Peter Sneath's *Principles of Numerical Taxonomy*, and Alan Shaw's *Time in Stratigraphy*. I eventually struggled to a general understanding and a lasting appreciation of this material through persistence and a great deal of help from others. Although the present survey is intended to provide adequate background for the subject matter discussed, dictionaries, glossaries, additional texts, libraries, peers, and teachers in related subjects do exist and are sources of additional explanatory information. A more full, lasting, and satisfying understanding of the topic can be gained by assuming the role of an active participant and using the text as part of a cooperative venture.

Terms of nomenclature are explained at the end of the Prologue and are repeated the first time the symbols are used. Paleolatitudes and the former positions of continents follow Scotese and Sager (1989) and Smith et al. (1981). The species lists given for the principal floras discussed in the text, together with mention of abundant or otherwise important members of allied floras, follow the spellings, family assignments, and nomenclature of the original author, augmented by recent additions to the flora. A brief description of families and the general worldwide distribution of extant families and genera can be found in Willis (1985); the scientific and common names, range, and diagnostic features of North American trees are given in Elias (1980); distributions and general habitats for extant North American plant taxa are in the *Flora of North America* (Flora of North America Editorial Committee, 1993 et. seq. 1); botanical terms are explained in Allaby (1992); a list of the common and scientific names of many extinct and extant mammals of North America is given in Graham and Mead (1987, appendix I) and in the FAUNMAP Working Group (1996, footnote 5); and definitions of geologic terms are given in Allaby and Allaby (1990) and Jackson (1997). Figures, tables, chapters, and appendices cited in upper case (e.g., Fig. 1) refer to material in this text, while lower case (fig. 1) refers to material in other publications.

ACKNOWLEDGMENTS

The genesis of this book was an invitation from Nancy Morin, Missouri Botanical Garden (presently the American Association of Botanical Gardens and Arboreta), and Luc Brouillet, University of Montreal, to prepare a review of the Late Cretaceous and Tertiary vegetation of North America for the *Flora of North America* (Graham, 1993). Many of the illustrations and computer-designed graphics were prepared by Cathy Brown, William Megenhardt, Larry Rubens, Diane Sperko, and Samuel Tussing of the AV Services at Kent State University. Kirk Jensen and MaryBeth Branigan at Oxford University Press have skillfully guided the book to completion.

A number of individuals have read chapters or portions thereof, or provided discussions, advice, photographs, permission to use published and unpublished material, and other kinds of valuable assistance and cooperation. These are Theodore Barkely, Kathleen Blase, Larry Bliss, Gerard Bond, Robert Berner, Thure Cerling, Peter Clark, Rachael Craig, Peter Crane, Aureal Cross, Margaret Davis, Alastair Dawson, Hazel Delcourt, Paul Delcourt, David Dilcher, Deborah Elliott-Fisk, David Engebretson, Patrick Fields, Norman Frederiksen, John Freudenstein, Bruce Graham, Kathryn Gregory, Eric Grimm, Niels Gundestrup, Jan Hardenbol, Leo Hickey, Hsioh-Yu Hou, Gordon Jacoby, Herbert Meyer, Karl Niklas, James MacMahon, Steven Manchester, Rolf Mathewes, John McAndrews, Richard Miller, John Oldow, Dolores Piperno, Donald Prothero, William Ruddiman, Sammantha Ruiz, Samuel Savin, Thomas Schmidlin, Linda Shane, Ruth Stockey, Charles Stockton, Ralph Taggart, Alfred Traverse, Bruce Tiffney, Thomas Van Devender, Terry Vaughan, S. David Webb, Elisabeth Wheeler, Peter Wigand, M. A. J. Williams, Scott Wing, Jack Wolfe, and Zhengyi Wu. Parts of the text were read by several anonymous reviewers and their suggestions and encouragement were most helpful. Introductory chapters were reviewed by biology graduate student Beverly Brown, and the entire manuscript was read by Shirley A. Graham. Valued assistance in the tedious tasks of proofreading and checking literature references was capably provided by Sonja Kaderly and Earl Landree. The efforts of these colleagues and students are deeply appreciated. Copyright holders granting permission to use their material are acknowledged in the figure legends.

The information summarized here is based on the work of many individuals. They include early workers who laid the foundation for Late Cretaceous and Cenozoic paleopalynological and paleobotanical studies in North America. A new generation of paleopalynologists and paleobotanists are now refining older concepts and providing new insights through the innovative application of modern methodologies. Their contributions are important not only in their own right, but also in providing an essential part of the data base required for large-scale environmental modeling projects and for assessing the real-world accuracy of the modeled results. In addition, much of what we theorize about future environmental and evolutionary change must be assessed against the documented events and trends of the past. Reseachers in paleontology, including the disciplines of paleopalynology and paleobotany, provide a significant part of this documentation and their efforts are becoming increasingly recognized as necessary for a proper understanding of the modern Earth and its biota and for anticipating trends that are likely to shape its future.

References

Aitken, M. J. 1985. Thermoluminescence Dating. Academic Press, New York.

Allaby, A. and M. Allaby (eds.). 1990. The Concise Oxford Dictionary of Earth Sciences. Oxford University Press, Oxford, U.K.

Allaby, M. (ed.). 1992. The Concise Oxford Dictionary of Botany. Oxford University Press, Oxford, U.K.

Berggren, W. A., D. V. Kent, M.-P. Aubry, and J. Hardenbol (eds.). 1995. Geochronology, Time Scales and Global Stratigraphic Correlation. Society of Sedimentary Geology, Tulsa, OK. Special Publ. 54.

Bye, J. A. T. 1996. Coupling ocean–atmosphere models. Earth-Sci. Rev. 40: 149–162.

Chen, Y., P. E. Smith, N. M. Evensen, D. York, and K. R. Lajoie. 1996. The edge of time: dating young volcanic ash layers with the ^{40}Ar–^{39}Ar laser probe. Science 274: 1176–1178.

CLIMAP Project Members. 1976. The surface of the ice-age Earth. Science 191: 1131–1137.

———. 1981. Seasonal reconstruction of the Earth's surface at the last glacial maximum. Geological Society of America, Boulder, CO. Map Chart Series MC-36.

———. 1984. The last interglacial ocean. Quatern. Res. 21: 123–224.

COHMAP Members. 1988. Climatic changes of the last 18,000 years: observations and model simulations. Science 241: 1043–1052.

Crane, P. R. 1989. Paleobotanical evidence on the early radiation of nonmagnoliid dicotyledons. Plant Syst. Evol. 162: 165–191.

———. 1993. Time for the angiosperms. Nature 366: 631–632.

———, E. M. Friis, and K. R. Pedersen. 1995. The origin and early diversification of angiosperms. Nature 374: 27–33.

Crowley, T. J. and G. R. North. 1991. Paleoclimatology. Oxford Monographs on Geology and Geophysics. Oxford University Press, Oxford, U.K. Monograph 16.

Dalrymple, G. B. 1979. Critical tables for conversion of K-Ar ages from old to new constants. Geology 7: 558–560.

———. 1991. The Age of the Earth. Stanford University Press, Stanford, CA.

Easterbrook, D. J. (ed.). 1988. Dating Quaternary Sediments. Geological Society of America, Boulder, CO. Special Paper 227.

Elias, T. S. 1980. The Complete Trees of North America. Van Nostrand Reinhold, New York.

Evernden, J. F. and G. T. James. 1964. Potassium-argon dates and the Tertiary floras of North America. Am. J. Sci. 262: 945–974.

Farley, M. B. 1989. Palynological facies fossils in nonmarine environments in the Paleogene of the Bighorn Basin. Palaios 4: 565–573.

———. 1990. Vegetation distribution across the early Eocene depositional landscape from palynological analysis. Palaeogeogr., Palaeoclimatol., Palaeoecol. 79: 11–27.

——— and A. Traverse. 1990. Usefulness of palynomorph concentrations in distinguishing Paleogene depositional environments in Wyoming (U.S.A.). Rev. Palaeobot. Palynol. 64: 325–329.

FAUNMAP Working Group. 1996. Spatial response of mammals to Late Quaternary environmental fluctuations. Science 272: 1601–1606.

Flora of North America Editorial Committee (eds.). 1993. Flora of North America North of Mexico. Oxford University Press, Oxford, U.K.

Gates, W. L. 1982. Paleoclimatic modeling—a review with reference to problems and prospects for the pre-Pleistocene. In: Geophysics Study Committee, Climate in Earth History. National Academy Press, Washington, D.C., pp. 26–42.

Goodfriend, G. A. 1992. Rapid racemization of aspartic acid in mollusc shells and potential for dating over recent centuries. Nature 257: 399–401.

Graham, A. 1993. History of vegetation: Cretaceous (Maastrichtian)–Tertiary. In: Flora of North America Editorial Committee (eds.), Flora of North America North of Mexico. Oxford University Press, Oxford, U.K. Vol. 1, pp. 57–70.

Graham, R. W. and J. I. Mead. 1987. Environmental fluctuations and evolution of mammalian faunas during the last deglaciation in North America. In: W. F. Ruddiman and H. E. Wright, Jr. (eds.), North America and Adjacent Oceans during the Last Deglaciation. The Geology of North America, DNAG (Decade of North American Geology). Geological Society of America, Boulder, CO. Vol. K-3, pp. 371–402.

Hannon, B. M. and M. Ruth. 1994. Dynamic Modeling. Springer-Verlag, Berlin.

Hansen, J., A. Lacis, R. Ruedy, M. Sato, and H. Wilson. 1993. How sensitive is the world's climate? Natl. Geogr. Res. Explor. 9: 142–158.

Haq, B. U. and F. W. B. van Eysinga. 1987. Geological Time Table. Elsevier Science Publishers, Amsterdam. Wall chart.

Harland, W. B., R. L. Armstrong, A. V. Cox, L. E. Craig, A. G. Smith, and D. G. Smith. 1989. A Geologic Time Scale 1989. Cambridge University Press, Cambridge, U.K.

Imbrie, J. 1982. The role of prediction in paleoclimatology. In: Geophysics Study Committee, Climate in Earth History. National Academy Press, Washington, D.C. pp. 21–25.

Jackson, J. A. (ed.). 1997. Glossary of Geology, Amer. Geol. Inst., Falls Church, VA.

McDougall, I. and T. M. Harrison. 1988. Geochronology and Thermochronology by the $^{40}Ar/^{39}Ar$ Method. Oxford University Press, Oxford, U.K.

Miall, A. D. 1984. Principles of Sedimentary Basin Analysis. Springer-Verlag, Berlin.

Roth, E. and B. Poty (eds.). 1989. Nuclear Methods of Dating. Kluwer Academic Publishers, Dordrecht.

Saarnisto, M. 1988. Time-scales and dating. In: B. Huntley and T. Webb III (eds.), Vegetation History. Kluwer Academic Publishers, Dordrecht. pp. 77–112.

Schlesinger, M. E. (ed.). 1988. Physically-Based Modelling and Simulation of Climate and Climatic Change. Kluwer Academic Publishers, Dordrecht.

Schneider, S. H. and R. E. Dickinson. 1974. Climate modeling. Rev. Geophys. Space Phys. 12: 447–493.

Scotese, C. R. and W. W. Sager (eds.). 1989. Mesozoic and Cenozoic Plate Reconstructions. Elsevier Science Publishers, Amsterdam.

Sellers, P. J. et al. (+ 12 authors). 1997. Modeling the exchanges of energy, water, and carbon between continents and the atmosphere. Science 275: 502–509.

Shackleton, N. J. and N. D. Opdyke. 1973. Oxygen isotope and paleomagnetic stratigraphy of equatorial Pacific core V28–238: oxygen isotope temperatures and ice volumes on a 10^5 and 10^6 year scale. Quatern. Res. 3: 39–55.

Smith, A. G., A. M. Hurley, and J. C. Briden. 1981. Phanerozoic Paleocontinental World Maps. Cambridge University Press, Cambridge, U.K.

Sullivan, W. 1974. Continents in Motion, the New Earth Debate. McGraw-Hill, New York.

Taylor, D. W. and L. J. Hickey (eds.). 1996. Flowering Plant Origin, Evolution and Phylogeny. Chapman & Hall, New York.

Trenberth, K. E. (ed.). 1992. Climate System Modeling. Cambridge University Press, Cambridge, U.K.

Wagner, G. A. and P. Van den Haute. 1992. Fission-Track Dating. Enke, Stuttgart, and Kluwer, Norwell, MA.

Washington, W. M. and C. L. Parkinson. 1986. An Introduction to Three-Dimensional Climate Modeling. University Science Books, Mill Valley, CA.

Williams, M. A. J., D. L. Dunkerley, P. De Deckker, A. P. Kershaw, and T. Stokes. 1993. Quaternary Environments. Edward Arnold, London.

Willis, J. C. 1985. A Dictionary of the Flowering Plants and Ferns (revised by H. K. Airy Shaw). 8th (student) edition. Cambridge University Press, Cambridge, U.K.

Wolfe, J. A. 1993. A method of obtaining climatic parameters from leaf assemblages. U.S. Geol. Surv. Bull. 2040: 1–71.

Ziegler, A. M. et al. (+ 6 authors). 1987. Coal, climate and terrestrial productivity: the present and Early Cretaceous compared. In: A. C. Scott (ed.), Coal and Coal-Bearing Strata: Recent Advances. Geological Society of America, Boulder, CO. Special Publ. 32, pp. 25–49.

Zubakov, V. A. and I. I. Borzenkova. 1990. Global Palaeoclimate of the Late Cenozoic. Developments in Palaeontology and Stratigraphy, 12. Elsevier Science Publishers, Amsterdam.

General Readings

Barron, E. J. 1992. Lessons from past climates. Nature 360: 533.

———. 1995. Climate models: how reliable are their predictions? Consequences 1: 17–27.

Bell, G. 1992. Ultracomputers: a teraflop before its time. Science 256: 64.

Berry, W. B. N. 1987. Growth of a Prehistoric Time Scale, Based on Organic Evolution. Blackwell Science Publishers, London. 202 pp.

Broecker, W. 1997. Future directions of paleoclimate research. Quatern. Sci. Rev. 16: 821–825.

Bradley, D. 1993. Will future computers be all wet? Science 259: 890–892.

Cipra, B. 1996. Microprocessors deliver teraflops. Science 271: 598.

Culotta, E. 1993. Is the geological past a key to the (near) future? Science 259: 906–908.

Emeagwali, P. 1991. The ways of counting. Michigan Today 23: 1–4.

Hanson, B. and D. Voss. 1992. Large scale measurements. Science 256: 289.

Kerr, R. A. 1994. Climate modeling's fudge factor comes under fire. Science 265: 1528.

———. 1997. Climate-evolution links weaken. Science 276: 1968. [with reference to mammal evolution].

Pool, R. 1992. Massively parallel machines usher in next level of computing power. Science 256: 50–51.

Trenberth, K. E. 1997. The use and abuse of climate models. Nature 386: 131–133.

Vrba, E. S., G. H. Denton, T. C. Partridge, and L. H. Burckle (eds.). 1995. Paleoclimate and Evolution, with Emphasis on Human Origins. Yale University Press, New Haven, CT.

Waldrop, M. M. 1993. Computing at the speed of light. Science 259: 456.

References to Frontispiece

Anderson, D. L., T. Tanimoto, and Y.-S. Zhang. 1992. Plate tectonics and hot spots: the third dimension. Science 256: 1645–1651.

Bloxham, J. and D. Gubbins. 1989. The evolution of the Earth's magnetic field. Sci. Am. 261: 68–75.

Buffett, B. A., H. E. Huppert, J. R. Lister, and A. W. Woods. 1992. Analytical model for solidification of the Earth's core. Nature 356: 329–331.

Constable, C. 1993. About turn for reversals. Nature 361: 305–306.

Davies, G. F. 1992. Plates and plumes: dynamos of the Earth's mantle. Science 257: 493–494.

Duba, A. 1992. Earth's core not so hot. Nature 359: 197–198.

Fukao, Y. 1992. Seismic tomogram of the Earth's mantle: geodynamic implications. Science 258: 625–630.

Hoffman, K. A. 1988. Ancient magnetic reversals: clues to the geodynamo. Sci. Am. 258: 76–83.

Jacobs, J. A. 1992. The Earth's core in a nutshell. Nature 356: 286–287.

Jeanloz, R. 1993. The mantle in sharper focus. Nature 365: 110–111.

Kerr, R. A. 1991. Deep rocks stir the mantle pot. Science 252: 783.

———. 1992. Having it both ways in the mantle. Science 258: 1576–1578.

Knittle, E. and R. Jeanloz. 1991. Earth's core–mantle boundary: results of experiments at high pressures and temperatures. Science 251: 1438–1443.

Langmuir, C. H. 1993. Deep thoughts on the mantle. Nature 364: 191–192.

Menzies, M. A. (ed.). 1990. Continental Mantle. Clarendon Press, Oxford, U.K.

Sautter, V., S. E. Haggerty, and S. Field. 1991. Ultradeep (>300 kilometers) ultramafic xenoliths: petrological evidence from the transition zone. Science 252: 827–830.

Vidale, J. E. and H. M. Benz. 1992. Upper-mantle seismic discontinuities and the thermal structure of subduction zones. Nature 356: 678–683.

Nomenclature

AP	arboreal (tree) pollen
AU	Astronomical unit (mean distance between the Earth and sun; 149,600,000 km, 93,000,000 mi)
B.P.	before the present
CCM	Community Climate Model
chenoam	pollen of the families Chenopodiaceae–Amaranthaceae, which cannot be distinguished
CLAMP	Climate-Leaf Analysis Multivariate Program
CLIMAP	Climate Long-Range Investigation and Mapping Program
COHMAP	Cooperative Holocene Mapping Project
D-O	Dansgaard-Oeschger event
DSDP	Deep Sea Drilling Project
EBM	energy balance model
ENSO	El Niño-Southern Oscillation Event
ESP	electron spin resonance (dating)
GCM	General Circulation Model
GFDL	Geophysical Fluid Dynamics Laboratory
GENESIS	global environmental and ecological simulation of interactive systems
GIS	Geographic Information System
GISP2	Greenland Ice Sheet Program II
GISS	Goddard Institute for Space Studies
GLAS	Goddard Laboratory for Atmospheric Sciences
GPTS	global polarity time scale
GRIP	Greenland Ice Core Project
Gt	gigaton (billions of metric tons)
HM	half mountains (mode in computer simulations)
ITCZ	intertropical convergence zone
J	joules (energy of 1 amp of current passed through a resistance of 1 ohm for 1 s)
K	thousand
Ky	thousand years
Kya	thousands of years ago
K–T	Cretaceous–Tertiary
LMA	leaf margin analysis
LO	last occurrence
M	mountains (mode in computer simulation)
Ma	millions of years ago
m.y.	million years
MART	mean annual range of temperature
MAT	mean annual temperature
mb	millibar (1/1000 of a bar, or 1000 dynes/cm^2)
MEDD	Middle Eocene Diversity Decline
MJ	megajoules
mt	metric tons
m.y.	million years (duration in time)
NADW	North Atlantic deep water
NAP	nonarboreal (herb) pollen
NASA	National Aeronautics and Space Administration
NALMA	North American Land Mammal Age
NCAR	National Center for Atmospheric Research
NEO	near Earth object
NLR	nearest living relative
NM	no mountains (mode in computer simulations)
NOAA	National Oceanographic and Atmospheric Administration
ODP	Ocean Drilling Program
PAGES	Past Global Changes
PAR	pollen accumulation rate
Pg	petagram (10^{15} g = 1 gt)
pl	paleolatitude
ppmv	parts per million by volume
RSL	relative sea level
SADW	south Atlantic deep water
SCLF	single crystal laser fusion
SCOR	scientific committee on oceanic research

SEM	scanning electron microscopy
SST	sea-surface temperature
TAMS	tandem accelerator mass spectrometry
TCT	pollen of the families Taxodiaceae–Cupressaceae–Taxaceae, which cannot be distinguished
TIMS	thermal ionization mass spectrometry
TOMS	total ozone mapping spectrometer
VSP	vertical seismic profile
WBC	western boundary current
~	approximately
δ	change in (e.g., $\delta^{18}O$)

LATE CRETACEOUS AND CENOZOIC HISTORY OF NORTH AMERICAN VEGETATION

ONE

Setting the Goal: Modern Vegetation of North America

Composition and Arrangement of Principal Plant Formations

Vegetation is the plant cover of a region, which usually refers to the potential natural vegetation prior to any intensive human disturbance. The description of vegetation for an extensive area involves the recognition and characterization of units called formations, which are named with reference to composition (e.g., coniferous), aspect of habit (deciduous), distribution (western North America), and climate, either directly (tropical) or indirectly (tundra). Further subdivisions are termed associations or series, such as the beech–maple association or series within the deciduous forest formation.

Formations and associations constitute a convenient organizational framework for considering the development of vegetation through Late Cretaceous and Cenozoic time. For this purpose seven extant plant formations are recognized for North America (Fig. 1.1, Table 1.1): (1) tundra, (2) coniferous forest, (3) deciduous forest, (4) grassland, (5) shrubland/chaparral–woodland–savanna, (6) desert, and (7) elements of a tropical formation. Several summaries are available for the modern vegetation of North America, including Barbour and Billings (1988), Barbour and Christensen (1993, vegetation map fig. 5.1), Küchler (1964), and Vankat (1979).[1] The following discussions are based primarily on these surveys.

TUNDRA FORMATION

Tundra (Fig. 1.2) is a treeless vegetation dominated by shrubs and herbs, and it is characteristic of the cold climates of polar regions (Arctic tundra) and high-altitude regions (alpine tundra). In the Arctic tundra a few isolated trees or small stands may occur locally, such as *Picea glauca* (white spruce), but these are always in protected habitats. The Arctic region experiences nearly continuous darkness in midwinter, and nearly continuous daylight in midsummer. There is a short growing season of only 6–24 weeks; this accounts, in part, for the fact that 98% of all Arctic tundra plants are perennials (Vankat, 1979). Strong winds are another feature of the Arctic landscape, often exceeding 65 km/h for 24 h or more. They likely account for the frequency of rosettes, persistent dead leaves, and the cushion growth form, in the center of which wind velocities may be reduced by 90%. The harsh growing conditions also result in leaves of the microphyllous size class being comparable to those of desert plants. Vegetative reproduction and self-pollination is common, and phenotypic plasticity is high among Arctic tundra plants.

Intensely cold and dry air is maintained by a high-pressure system over the Yukon Territory and the western part of the Northwest Territories. The importance of this system in determining the distribution of plant formations in the north is made evident by comparing the annual variation in its position to the limits of the Arctic tundra and boreal forest. The location of the system varies from a line in winter that parallels the southern boundary of the boreal coniferous forest to a line in the summer that generally parallels the northern boreal coniferous forest–tundra boundary (Bliss, 1988).

Another feature of the Arctic tundra environment is permafrost, defined as frozen soil in which temperatures persist below 0°C for 2 or more years. Permafrost extends down 400–600 m in the Arctic region, and it occurs southward in increasingly discontinuous patches into the boreal forest. There is generally permafrost thawing in summer to a depth of 20–60 cm and to 100–200 cm along rivers.

The southern boundary of the Arctic tundra is defined by the limit of permafrost and it follows the contact with the boreal forest, extending farthest south to ~55°N along the east coast of North America because of the cold Labrador Current. The warmer Japan current flows along the northwest coast of the continent (tundra to ~60°N).

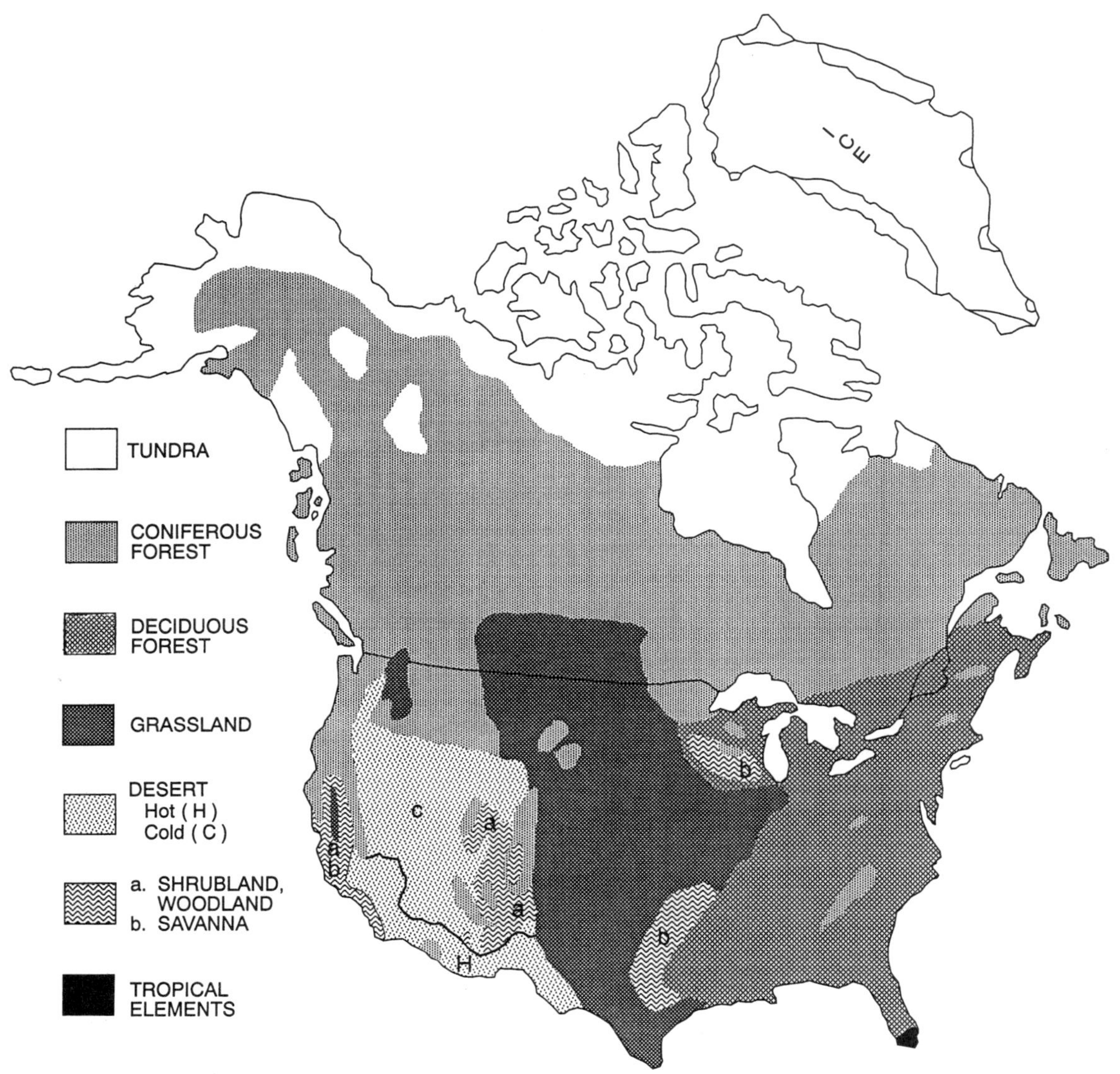

Figure 1.1. Vegetation diagram of North America.

Biodiversity is relatively low in the Arctic tundra; the North American Arctic consists of ~700 species of vascular plants. Distributions, however, are broad and ~50% of the vascular plant species are circumboreal. The development of this formation involved invasion of many preadapted plants from adjacent alpine (northern and central Rocky Mountains), coastal, bog, and marsh communities (especially from montane central Asia), rather than invasion by a few taxa that subsequently diversified. Evidence for this are the relatively few genera restricted to tundra habitats. Rather, most species of tundra plants belong to genera well represented in other types of environments. The taxonomy of tundra plants suggests a recent origin for the formation because of the few genera and the large number of subspecies and varieties that compose tundra vegetation.

Geographically, the Arctic region is divided into a high Arctic zone, including the high Arctic polar desert and the high Arctic semidesert, and a low Arctic zone, which includes most of the tundra (Table 1.2). The dividing line is at ~72° N latitude. Tundra covers ~4 million km^2 or ~19% of North America north of Mexico.

The high Arctic polar desert receives less precipitation than the warm deserts in the southwestern United States; moisture is available only about 3 months per year. Snow cover ranges from 20 to 50 cm, and it typically melts by late July to early August (Bliss, 1988). The low angle of the sun results in low intensity radiation and cold temperatures, which are usually below 12°C in midsummer. These environmental factors are reflected in leaf physiognomy; the leaf area of plants of the high Arctic are generally less than that for plants of the low Arctic. Vegetation cover ranges from 0% in areas of permanent snow, to typically 1–5% in ice-free areas with gravelly out-wash soils, to slightly more in protected areas with better developed soils. Vast areas of the high Arctic polar desert, including the Queen Elizabeth Islands, are virtually devoid of plant cover. Only rosette (e.g., *Draba*), cushion (*Saxifraga*), and mat-forming (*Puccinellia*) species are able to withstand the

Table 1.1. Summary of North American vegetation types.

- *Tundra formation*
  - High Arctic associations
    - Graminoid–moss
    - Dwarf-shrub heath
    - Cushion plant–cryptogam polar semidesert
    - Cryptogam–herb
    - Graminoid steppe
    - Herb barrens
  - Low Arctic associations
    - Forest tundra (transitional)
    - Tall shrub
    - Low shrub
    - Dwarf-shrub heath
    - Cottongrass–dwarf-shrub heath
    - Graminoid–moss
    - Cushion plant–herb–cryptogam polar semidesert
  - Alpine
- *Coniferous forest formation*
  - Boreal coniferous forest (taiga) association
    - Closed forest
    - Lichen woodland
    - Forest–tundra–ecotone (e.g., aspen-parkland association)
    - Shrublands (frequently included in Arctic tundra)
    - Bogs
  - Montane coniferous forest association
    - Appalachian montane coniferous forest
    - Western montane coniferous forest
      - Rocky Mountains region
        - Riparian and canyon forest
        - Short conifer woodland
        - *Pinus ponderosa* woodland
        - Madrean pine–oak woodland
        - *Pseudotsuga menziesii* forest
        - Cascadian forests
        - Montane seral forests
        - Subalpine white pine forests
        - Treeline vegetation
        - Meadows and parks
      - Sierra Nevada region zone forests
        - Montane (transition)
        - Upper montane (Canadian)
        - Subalpine (Hudsonian)
      - Pacific Northwest region
        - *Pseudotsuga menziesii–Tsuga heterophylla* forest
        - *Picea sitchensis–Tsuga heterophylla* forest
        - *Sequoia sempervirens* forest
        - Klamath Mountains mixed evergreen forest
        - Sierran-type mixed conifer forest
        - Subalpine forests and parklands
          - *Abies amabilis–Tsuga heterophylla* forest
          - *Abies magnifica* var. *shastensis* forest
          - *Tsuga mertensiana* forest
          - Subalpine parkland
- *Deciduous forest formation associations*
  - Mixed mesophytic
  - Beech–maple
  - Maple–basswood
  - Lake states forest
  - Oak–chestnut
  - Oak–hickory
  - Southern mixed hardwood
  - Pine woods
  - Flood–plain forest
- *Grassland formation associations*
  - Tallgrass prairie
  - Mixed grass
  - Shortgrass
  - Desert grassland
  - Palouse Prairie
  - California grassland
- *Shrubland/chaparral–woodland–savanna formation associations*
  - Piñon–juniper (and other juniper) woodland
  - Scrub oak–mountain mahogany shrubland
  - Chaparral
  - Broad–sclerophyll forest
  - California woodland
    - Blue oak woodland
    - Southern oak woodland
    - Mixed evergreen forest
  - Savanna
- *Desert formation associations*
  - Great Basin
  - Mojave Desert
  - Sonoran Desert
  - Chihuahuan Desert
- Tropical and subtropical elements

Based on Barbour and Billings (1988), Barbour and Christensen (1993), and Vankat (1979).

Figure 1.2. Tundra formation—Graminoid–moss association (*Carex stans*, *Eriophorum angustifolium*), Devon Island, Northwest Territories. Photograph courtesy of Larry Bliss.

Table 1.2. Comparison of environmental and biotic characteristics of low and high Arctic in North America.

Characteristics	Low Arctic	High Arctic
Length of growing season (months)	3–4	1.5–2.5
Mean July temperature (°C)/ppt (mm)	8–12/35.6	3–6/20.8
Frobisher Bay, NWT, 64°	3.9/53	
Baker Lake, NWT, 65°	10.8/36	
Tuktoyaktuk, NWT, 70°	10.3/22	
Kotzebue, AK, 67°	12.4/44	
Umiat, AK, 69°	11.7/24	
Anamagssalik, Greenland, 65°	7.4/35	
Godthaab, Greenland, 64°	7.6/59	
Umanak, Greenland, 71°	7.8/12	
Barter Island, AK, 70°	4.6/18	
Barrow, AK, 71°	3.8/22	
Cambridge Bay, NWT, 69°	8.1/22	
Sacks Harbour, NWT, 72°	5.5/18	
Resolute, NWT, 75°	4.3/26	
Isachsen, NWT, 79°	3.3/21	
Eureka, NWT, 80°	5.3/13	
Alert, NWT, 80°	3.9/18	
Scoresbysund, Greenland, 70°	4.7/38	
Nord, Greenland, 81°	4.2/12	
Botanical/vegetational		
Total plant cover (%)		
Tundra	80–100	80–100
Polar semidesert	20–80	20–80
Polar desert	1–5	1–5
Vascular plant flora (species)	700	350
Bryophytes	Common	Abundant
Lichens (common growth form)	Fruticose, foliose	Crustose, foliose
Vascular (common growth form)	Woody, graminoid,	Graminoid, cushion, rosette

Precipitation (ppt). Adapted from Bliss (1988).

extreme climate. In one survey of six islands, only 17 species of vascular plants and 14 species of cryptogams were recorded. The vascular plant cover averaged 1.8%, the cryptogams 0.7%, and 98% of the ice-free area was bare soil. The high Arctic semidesert is often included within the polar desert. It covers ~25% of the northern islands and 50–55% of the southern islands. The vegetation consists of cushion plant–cryptogam and cryptogam–herb species. Compared to the low Arctic tundra, the high Arctic tundra generally consists of herbaceous rather than woody species; there is much less plant cover; and lichens and mosses are much more prominent (up to 95% of the ground cover in some areas) with graminoid–moss tundra also extensively developed.

The low Arctic region extends from the high Arctic zone south to the northern limits of the boreal forest. It is characterized by a snow cover of 1–5 m or more that usually does not melt until about mid-June. The plant cover is nearly continuous, consisting of tall shrubs (2–5 m) to dwarf shrub heath (5–20 cm), with abundant grasses, sedges, lichens, and mosses. There is a general increase southward in total plant cover, plant height, and woody vascular plant species and a decrease in the prominence of lichens and mosses, although they are still a conspicuous and important component of the vegetation. The low Arctic tundra is more diverse in composition than that of the high Arctic and will be used to describe the vegetation in more detail.

Six principal types of low Arctic tundra are recognized (Bliss, 1988). The tall shrub association has *Salix alaxensis* (willow) as a common member, and also includes *S. glauca* spp. *richardsonii*, *S. pulchra*, and *Alnus crispa* (alder). Species of *Betula* (birch) grow along river terraces, stream banks, steep slopes, and lake shores where they form the tallest component of the vegetation (2–5 m in height).

The low shrub association consists of plants generally between 40 and 60 cm in height. The prominent ones are *Betula nana* ssp. *exilis*, and *Salix glauca* spp. *richardsonii*, with a ground cover of *Carex lugens* (sedge), *C. bigelowii*, *Eriophorum vaginatum* (cottongrass), *Vaccinium uliginosum* (heath), *V. vitis-idaea* spp. *minus*, *Empetrum nigrum* ssp. *hermaphroditum*, *Ledum palustre* spp. *decumbens*, *Arctostaphylos alpina*, *A. rubra*, and *Rubus chamaemorus*. The common lichens are *Cetraria nivalis*, *C. cucullata*, *Cladonia gracilis*, *C. mitis*, *C. rangiferina*, and *Thamnolia vermicularis*. The mosses include *Aulacomnium turgidum*, *Hylocomium splendens*, and *Polytrichum juniperinum*.

The dwarf-shrub heath association is usually found in relatively small parcels of a few hundred square meters. In North America it occurs scattered from Alaska to Hudson Bay and into Greenland on well-drained soils of river terraces, slopes, and uplands where winter snows are at least 20–30 cm deep. The vegetation includes *Ledum palustre* spp. *decumbens*, *Vaccinium uliginosum*, *V. vitis-idaea*, *Empetrum nigrum* spp. *hermaphroditum*, *Loiseleuria procumbens*, *Rhododendron lapponicum*, *R. hamschaticum*, *Cassiope tetragona*, *Betula nana* spp. *exilis*, *B. glandulosa*, deformed *Picea mariana* (black spruce), dwarf *Salix reticulata* and *S. herbacea*, *Dryas integrifolia*, *Phyllodoce caerulea*, *Carex bigelowii*, *C. misandra*, *Luzula nivalis*, and various lichens and mosses.

The cottongrass–dwarf shrub heath association is common along rolling uplands between mountains and along the wet coastal plain of Alaska, the Yukon Territory, and the Mackenzie River delta. The conspicuous element in this association is the cottongrass *Eriophorum vaginatum* spp. *vaginatum* in the west and *E. vaginatum* spp. *spissum* in the east. Heath shrubs include *Vaccinium vitis-idaea* spp. *minus*, *V. uliginosum* spp. *alpinum*, *Ledum palustre* spp. *decumbens*, and *Empetrum nigrum* spp. *hermaphroditum*. Other shrubs are *Betula nana* spp. *exilis* and *Salix pulchra*. The ground cover consists of two common sedges, *Carex bigelowii* and *C. lugens*; the mosses *Dicranum elongatum*, *Aulacomnium turgidum*, *A. palustre*, *Rhacomitrium lanuginosum*, *Hylocomium splendens*, *Tomenthypnum nitens*, and several species of *Sphagnum*; and lichens such as *Cetraria cucullata*, *C. nivalis*, *Cladonia rangiferina*, *C. mitis*, *C. sylvatica*, *Dactylina arctica*, and *Thamnolia vermicularis*.

The graminoid–moss association is a treeless wetland with grasses and sedges as the prominent elements. The association is common along foothills and coastal regions in Alaska. The sedges are *Carex aquatilis*, *C. rariflora*, *C. rotundata*, *C. membranaceae*, *Eriophorum angustifolium*, *E. scheuchzeri*, and *E. russeolum*; grasses are *Arctagrostis latifolia*, *Dupontia fisheri*, *Alopecurus alpinus*, and *Arctophila fulva*; mosses are *Hylocomium splendens*, *Tomenthypnum nitens*, and species of *Aulacomnium*, *Ditrichum*, *Calliergon*, *Drepanocladus*, and *Sphagnum*. Other associates include *Menyanthes trifoliata*, *Equisetum variegatum*, *Arctophila fulva*, *Potentilla palustris*, and *Hippuris vulgaris*. This type of tundra extends eastward, but it is less common in Greenland. In more temperate regions grasses often form salt marshes and true grasslands, but these associations are rare in the Arctic.

The cushion plant–herb–cryptogam polar semidesert association occurs on wind-swept slopes and ridges, which are often dominated by cushions or mats of *Dryas integrifolia* and *D. octopetala*. These species may be a component of fellfields, which are rocky areas in cold climates with shrubs growing between the rocks and boulders. They are most common in the extensive areas of barren rock and gravel of the eastern Arctic. Other cushion plants are *Silene acaulis*, *Saxifraga oppositifolia*, and *S. tricuspidata* associated with the lichens *Alectoria nitudula*, *A. ochroleuca*, *Cetraria cucullata*, and *C. nivalis*, and the moss *Rhacomitrium lanuginosum*.

Alpine tundra occurs in the Cascade Mountains, the Sierra Nevada, the Rocky Mountains, and the Appalachian Mountains of central New England northward at elevations above ~1200 m. Compared to the low Arctic zone, alpine regions above timberline have more uniform day and night periods throughout the year, light intensity is greater, and there is greater fluctuation in temperatures. For example, temperatures may rise to 40°C at the soil surface during summer days and may drop by 20°C in a few seconds under passing clouds. The soil is generally drier in an alpine tundra due to faster drainage on the steep slopes. The substrate is often rocky, and the plants grow among the rocks and boulders to form fellfields. Most alpine tundras include wet meadows dominated by sedges and grasses and dry meadows that are drained by the later part of the summer. Tall shrub communities of *Salix* may occur along streams.

Many alpine tundra plants are also found in the Arctic tundra, suggesting a common origin. For example, 45% of the plants in an alpine tundra from the Rocky Mountains in Montana and 75% in an Appalachian Mountain tundra in New Hampshire also occur in the Arctic tundra (Vankat, 1979).

CONIFEROUS FOREST FORMATION

Boreal Coniferous Forest or Taiga

One of the defining and most obvious features of the boreal coniferous forest is the predominance of a relatively few genera of gymnosperms (e.g., *Abies*, *Larix*, *Picea*, *Thuja*, a few species of *Pinus*). Gymnosperms are also dominant along the Gulf Coast of the southeastern United States and on the higher mountain slopes of the American west. In my view there is probably a historical component to the explanation for the abundance of gymnosperms in these regions. Each of the areas is characterized by physiological aridity, even though precipitation may be moderate or even high. In the boreal coniferous forest, soil water is frozen for 8–9 months of the year. Along the Gulf Coast of the southeastern United States the soils are predominantly deep sand where water percolates through rapidly; at mid- to high elevations in the west, surface runoff from the mountain slopes reduces opportunity for absorption. The origin and early diversification of modern gymnosperms was primarily during the climatically tumultuous times of the Permian, which included widespread aridity as reflected by evaporites and extensive, deeply oxidized redbeds. As a consequence, gymnosperms possess a number of features that allow them to cope with climatic and physiological aridity and to occupy underexploited ecological niches such as regions of frozen soil, deep sand, and steep slopes. Adaptive morphological features include sunken stomata, thick cuticles, circular leaves (or circular in the aggregate fasicles) that present minimum surface area for water loss, and convoluted mesophyll cells that provide increased surface area to maintain adequate physiological activity in compensation for reduced leaf volume. Physiological adaptations include tannins stored in the cell vacuoles that constitute a colloid system. As water is lost from the leaves, a point is reached where the vacuole contents change from a liquid to a gel, further reducing water loss. One environment that renders plants with these adaptations at a competitive advantage is in the region just below the low Arctic tundra.

The boreal coniferous forest (Fig. 1.3; Shugart et al., 1992) covers ~28% of North America north of Mexico and extends from the low Arctic tundra to transition zones with various vegetation types to the south.[2] In eastern North America it intergrades with the hemlock and white pine–northern hardwood forest association (lakes state forest) of the deciduous forest, while in the midcontinent region there is a transition through a deciduous forest phase into the central prairie. In the west it merges with the Rocky Mountains and northwest coastal coniferous forest, sharing such species as *Picea glauca*, *Pinus contorta* (lodgepole pine), and *Populus tremuloides* (quaking aspen). The ecotone with the low Arctic tundra is comparatively sharp and narrow in the east and west, but it is up to 235 km wide in central Canada (Elliott-Fisk, 1988).

Environments are generally similar and grade northward into those of the low Arctic tundra: low temperatures, low precipitation, and a short growing season of 75–120 days (Table 1.3). The evergreen habit of the dominant trees is a response to the short growing season. Temperatures range from a summer maximum of 15°C to a winter minimum of –35°C. The distribution is determined by a complex of factors, but temperature is significant. In general, the northern limit of the boreal coniferous forest follows the 13°C July isotherm, while the southern limit corresponds to about the 18°C July isotherm (Elliott-Fisk, 1988). Rainfall within the formation ranges from about 240–500 mm, but water is frozen for 8–9 months of the year.

The area was extensively glaciated in the east by the Laurentide ice sheet and in the west by the Cordilleran ice sheet, resulting in a flat topography with numerous lakes. Typical succession on these lakes is from open water to encroaching *Sphagnum* mat, shrubs, trees, and finally to a filled basin covered with boreal forest trees. The accumulating organic material produces poorly oxygenated, acidic environments that reduce microbial activity and favor preservation of pollen, spores, seeds, wood, and cuticles, as well as diatoms, phytoliths, insects and other invertebrates, and even small mammals. As a consequence, a great deal is known about the late-glacial and postglacial history of the boreal forest from pollen, spore, and other analyses.

Fire is an important factor in maintaining the coniferous forest (Crutzen and Goldammer, 1993; Johnson, 1992). Many tree species regenerate quickly after fire by sprouting, and fire is important in seed dispersal of serotinous-cone species. These cones are sealed by resin that melts during a fire, releasing the seeds to a cleared, more nutrient-rich, less competitive environment. In the past, fire was a frequent and natural component of the coniferous forest environment, resulting from volcanic activity, light-

Figure 1.3. Coniferous forest formation—boreal coniferous forest (taiga) association (*Abies balsamea*, *Larix laricina*, *Picea glauca*, *P. mariana*), Chippewa National Forest, Minnesota. Photograph courtesy of the National Agricultural Library, Forest Service Photo Collection.

Table 1.3. Climatological data for northern boreal forest sites in Alaska and western, central, and eastern Canada.

Parameter	Fairbanks, Alaska	Doll Creek, Yukon	Ennadai Lake, NWT	Napaktok Bay, Labr.
Mean daily temp. (°C)				
(January)	−24	−35	−31.5	−20
(July)	15	14	13	7.5
Mean annual temp. (°C)	−3	−10	−9.3	−5
Mean annual precipitation (mm)	290	240	290	500
Mean growing season length (days)	100	120	100	75
Mean annual net radiation (kly/yr)*	22	20	19	18

Adapted from Elliot-Fisk (1988).
* Kilolangley/year.

Figure 1.4. Coniferous forest formation—boreal coniferous forest association (*Pinus banksiana*), Chippewa National Forest, Minnesota. Photograph courtesy of the National Agricultural Library, Forest Service Photo Collection.

ning, and in more recent times from the hunting and agricultural practices of the AmerIndian. The ground was cleared of debris and the fires scorched only the lower, older trunks of the trees, facilitating seed dispersal and sprouting. In recent times management policy has been largely to prevent fires, with the result that underbrush accumulates in great abundance. Fires now reach the crowns and frequently cause much more extensive and long-term damage. Another modern factor significantly affecting the range of the coniferous forest is acid rain, defined as precipitation with a pH of less than 5.6. There are areas in the northeastern United States where the pH of rain water is as low as 3.0.

The trees of the boreal coniferous forest are comparatively small, usually under 15 m in height. At its northern limit approaching the low Arctic tundra, the trees are reduced to ~4 m in height or to even smaller shrubby forms, the canopy becomes more open, and the forest breaks up into small stands or individual trees.

The closed forest phase of the boreal coniferous forest (Fig. 1.3) has the lowest species diversity of any North American forest vegetation, with only nine principal tree genera; only the tundra is lower in number of species. This is likely due to disturbances during the Quaternary and to the youthfulness of much of the landscape. *Picea glauca* is the most widespread tree and is common on acidic, alluvial soils along river valleys. *Picea mariana* is the most frequently occurring tree in the boreal forest. It is common on upland, basic, shallow soils or bedrock, and on poorly drained, acidic soils surrounding *Sphagnum* bogs. Other trees include *Abies balsamea* (balsam fir), *Larix laricina* (tamarack, larch), *Pinus banksiana* (jack pine; Fig. 1.4), *P. contorta*, *Thuja occidentalis* (white cedar), *Betula papyrifera* (paper birch), *Populus balsamifera* (balsam poplar), *P. tremuloides*, and coastal species such as *Tsuga heterophylla* and *Chamaecyparis nootkatensis* (Alaskan cedar and yellow cypress). Quaking aspen is dominant in the aspen-parkland association that forms one transition between the boreal coniferous forest and the northern part of the grassland formation. The aspen-parkland association is often seral (temporal), as evidenced by the frequent uniform size of the trees in individual stands. Shrubs include *Alnus crispa* (green alder), *Betula glandulosa* (dwarf birch), *Salix* sp., and various Ericaceae (heaths), although these are mostly part of the low Arctic tundra. There is an understory of lichens (*Cladonia* or reindeer lichen), mosses (*Hylocomium splendens*, *Pleurozium schreberi*, and especially *Sphagnum* sp.), ferns (*Osmunda*), and lycopods (*Lycopodium*). The northern region of the forest is abundant in lichens and is often treated as a separate phase called the lichen woodland. It includes *Picea mariana*, *P. glauca*, and the lichens *Stereocaulon paschale* in the west and *Cladonia stellaris* in the east.

The bog flora of the boreal coniferous forest includes *Sphagnum* and trees and shrubs on or surrounding the basin, depending on the state of bog maturity and the extent of basin filling. The common trees are *Picea mariana* and *Larix laricina* that are associated with orchids, heaths, composites, sedges, and grasses. In southern Alaska *Tsuga heterophylla*, *Thuja plicata* (western arborvitae), *Chamaecyparis nootkatensis*, and *Pinus contorta* generally replace *Picea mariana*.

Montane Coniferous Forest

Appalachian Montane Coniferous Forest

This forest is a continuation of the boreal coniferous forest formation, and it is generally similar in composition from north to south (Greller, 1988). Toward the south it occurs at progressively higher elevations: above 150 m in Maine, 762 m in the Green Mountains of Vermont and in the White Mountains of New Hampshire, 1280 m in the Catskill Mountains of southeastern New York state, and 1524 m in the Great Smoky Mountains of Tennessee and North Carolina [e.g., Mt. Mitchell, 2037 m (6684 ft), and Clingmans Dome, 2025 m (6643 ft), the highest peaks east of the Rocky Mountains]. The average temperature is ~5°C for the coldest month and ~15°C for the warmest month. Annual precipitation is 1500 mm in the northern part of the range and 2250 mm in the southern part. Thus, the Appalachian montane coniferous forest occupies regions that are warmer and wetter than the boreal coniferous forest.

The dominant trees are *Abies* at the higher elevations with *Picea* intermixed at lower elevations. The most important fir in the north is *A. balsamea*; in the south it is replaced by *A. fraseri* (fraser fir), a closely related species that may have evolved via founder effect mechanisms in relatively recent times. The principal spruce throughout is *Picea rubens* (red spruce). Associated with the coniferous element is *Betula papyrifera* in the north, *B. lutea* (yellow birch) throughout the range, and *Rhododendron* (azalea, rhododendron) on exposed slopes in the south.

Western Montane Coniferous Forest

Rocky Mountains Region The Rocky Mountains are the most continental of the western mountain chains. The prevailing westerlies provide the highest annual precipitation on the western slopes, ranging from 400 mm in the south to 1500 mm in the north. The coniferous forest of the Rocky Mountains extends from northern Alberta to New Mexico, with outliers at higher elevations from central Utah to western South Dakota (e.g., *Picea glauca* in the Black Hills). To the west, coniferous forests of the Sierra Nevada and the Cascade Mountains and Coast Ranges of the Pacific Northwest are generally recognized as separate floristic provinces. In the north the Rocky Mountains coniferous forest blends into the boreal coniferous forest, and in the south at lower elevations it frequently borders woodland vegetation of *Pinus edulis* (piñon pine) and various species of *Juniperus* (juniper). It also extends along the Sierra Madre Oriental and Sierra Madre Occidental to the Transvolcanic Belt of Mexico (the Madrean region of Axelrod and Raven, 1985), giving the formation a latitudinal range in the New World of from ~65° N to ~19° N latitude.

Zonation is one of the most evident and widely recognized aspects of the Rocky Mountains vegetation, especially since the definitive work of Merriam (1890) describing six altitudinally arranged life zones. For the region of northern Arizona and adjacent areas these zones, as classically described, are the lower Sonoran life zone (a southern desert scrub, to ~200 m elevation), upper Sonoran life zone (desert grassland to 1100 m) and piñon and juniper woodland (to 1700 m), transition life zone (ponderosa pine forest to 2100 m), Canadian life zone (douglas fir forest to 3000 m), Hudsonian life zone (spruce and fir forest to 3500 m), and Arctic and Alpine life zone (alpine tundra to 3800 m; Lowe, 1980). Three of these (transition, Canadian, and Hudsonian) are characterized by coniferous forest. This life zone nomenclature has been modified in recent times, especially in regions beyond the deserts where the altitudinal belts are commonly referred to as foothill (transition), montane (Canadian), subalpine (Hudsonian), and alpine (Arctic and Alpine).

Peet (1988) notes that, "Given the length of the cordillera, the surprise is not that latitudinal variation exists in Rocky Mountain forests but that major vegetation types, often representing elevational zones, remain essentially constant over great distances." Elevation, moisture gradients, and soil combine to produce the recognizable units that do exist, and Peet (1988) recognizes 11 vegetation types.

Riparian and canyon forest in the drier lowlands consist of *Populus trichocarpa*, *P. sargentii*, *P. fremontii*, *P. angustifolia* (cottonwoods), and *Salix* sp. With increasing elevation these species are replaced by *P. angustifolia*, *Alnus tenuifolia*, and *Betula occidentalis*; to the south bordering the Madrean region *Platanus wrightii* (sycamore), *Juglans major* (walnut), *Fraxinus velutina* (ash), and *Alnus oblongifolia* become the common deciduous species.

A short conifer woodland of *Juniperus monosperma* (one-seed juniper) and *Pinus edulis* forms a transition between the drier plains vegetation and the montane coniferous forest of the southern Rocky Mountains. At higher altitudes from Mexico to southern British Columbia, the lower montane zone is characterized by a *Pinus ponderosa* woodland, although overgrazing has caused species of *Quercus* (oak) and *Cercocarpus* (mountain mahogany) to replace ponderosa pine in many areas. The effect of overgrazing on ponderosa pine may be indirect in that when grasses are reduced, fewer fires control the competing shrubby vegetation. The southern margin of the ponderosa pine association grades into the Madrean pine–oak woodland of the southwestern United States and Mexico.

Near the upper elevational limits of the ponderosa pine, the *Pseudotsuga menziesii* forest forms a transition to the upper montane zone. Peet (1988) recognizes a Cascadian forest and a montane seral forest. The latter is a consequence of cyclic and unpredictable disturbances that are characteristic of the elevated and tectonically active landscape. Even though the communities are theoretically seral in nature, the frequency of disturbances such as fire, wind, insects, disease, browsing, avalanches, landslides, extreme weather, volcanism, and human activity make these com-

Figure 1.5. Coniferous forest formation—montane coniferous forest association (treeline vegetation; *Abies lasiocarpa, Picea engelmanii*), Colorado.

munities a constant, albiet shifting, feature of the Rocky Mountains vegetation. Two common members of the seral forest are *Populus tremuloides*, the most widespread tree in North America, and *Pinus contorta*.

Members of the white pine group (*Pinus* subgenus *Haploxylon*) are able to withstand the low temperatures and strong winds of high-altitude ridges and exposed slopes, so they form subalpine white pine forests. The important species are *P. albicaulis* (whitebark pine) in the north, *P. aristata* (bristlecone pine) in the south, and *P. flexilis* (limber pine) throughout most of the region. *Pinus longaeva* (Great Basin bristlecone pine) replaces bristlecone pine on the high peaks of the Great Basin. Treeline vegetation (Fig. 1.5) typically consists of *Picea engelmannii* (Engelmann spruce) and *Abies lasiocarpa* (subalpine fir); members of the white pine group grow on exposed sites.

Open areas of mostly grasses and sedges occur scattered through the coniferous forest of the Rocky Mountains. The smaller ones are called meadows and the larger ones parks. The presence of meadows and parks at lower elevations (e.g., in valleys like Estes Park, CO) is thought to result from a combination of soil texture (typically fine alluvial soils), drought, and exposure on south-facing slopes and steep hillsides. At high elevations, saturated peaty soils, low temperatures, and high winds are considered to be important factors. At midaltitude these communities include *Carex utriculata, C. aquatilis*, and *Calamagrostis canadensis*; at higher altitudes *Eleocharis pauciflora, Caltha leptosepala, Deschampsis caespitosa, Carex illota, Podagrostis humilis, Phleum alpinum, Poa reflexa, Erigeron peregrinus*, and *Bistorta bistortoides* are common components. Occasional boggy areas include *Sphagnum* and the ericads *Vaccinium myrtillus, Kalmia polifolia*, and *Gaultheria humifusa*.

Sierra Nevada Region Altitudinal zonation is evident among the plant communities of the Sierra Nevada, but the proximity to the Pacific Ocean causes the ecotones to be higher. For example, in the southern Sierra Nevada alpine tundra begins at ~3500 m, while in the Cascade Mountains and mountains further north it begins at ~2000 m. In the south the area is characterized by cool wet winters and dry warm summers (Mediterranean climate), and 80–85% of the precipitation falls during the winter months. Farther north winter precipitation constitutes 60–65% of the annual total, mostly as snow, which may reach 20 m annually in the subalpine zone. Along the western slopes of the Sierra Nevada rainfall averages 250–350 mm at low elevations and 1250 mm in the upper montane and subalpine zones.

The montane (transition) zone forest of the Sierra Nevada (Fig. 1.6) typically begins at ~1500 m and extends to ~2500 m. It includes *Pinus ponderosa* as the dominant element, with *P. jeffreyi* (jeffrey pine) growing at higher elevations; both extend into the upper montane zone. *Pinus lambertiana* (sugar pine) is present in the central Sierra Nevada,

Figure 1.6. Coniferous forest formation—montane coniferous forest association (Sierra Nevada region, montane zone forest; *Abies concolor*), California. Photograph courtesy of the National Agricultural Library, Forest Service Photo Collection.

and *Sequoiadendron giganteum* (giant sequoia; Fig. 1.7) is dominant in scattered groves in the central and southern Sierra Nevada at elevations from 825 to 2680 m. Other coniferous elements, often representing extensions from montane communities, are *Pseudotsuga menziesii*, *Abies concolor* var. *lowiana* (California white fir), and *Calocedrus decurrens* (= *Libocedrus decurrens*, incense cedar). Associated angiosperm trees are *Quercus kelloggii* (black oak), *Q. chrysolepis* (canyon live oak), *Cornus nuttallii* (dogwood), *Acer macrophyllum* (bigleaf maple), and *Cercocarpus ledifolius*; associated shrubs are *Arctostaphylos*, *Ceanothus*, *Chamaebatia*, *Castanopsis*, *Lithocarpus*, *Prunus*, *Ribes*, *Symphoricarpus*, and *Vaccinium* (Barbour, 1988).

The upper montane (Canadian) zone forest extends from 2500 to 3000 m and is characterized by *Pinus contorta* var. *murrayana*, which is occasionally replaced at higher altitudes on mesic sites by *Abies magnifica* (red fir). Seral stands of *Populus tremuloides* form local aspen parklands (Barbour, 1988). Understory shrubs are not common, but *Arctostaphlos*, *Ceanothus*, *Castanopsis*, and *Ribes* can be found.

The subalpine (Hudsonian) zone forest extends typically from 3000 to 3500 m. These forests differ from those in the Rocky Mountains in that they are not characterized by *Picea* and *Abies*. Rather, the dominant species are *Pinus albicaulis*, *P. flexilis*, *Tsuga mertensiana* (mountain hemlock), *Pinus contorta* var. *murrayana*, *P. monticola* (western white pine), and *P. balfouriana* spp. *austrina* (foxtail pine). Associates include *Populus tremuloides* var. *aurea*, *Pinus jeffreyi*, *P. monophylla*, *Abies concolor* var. *lowiana*, *A. magnifica*, *Juniperus occidentalis* spp. *australis*, and occasional prostrate shrubs of *Arctostaphylos*, *Artemisia*, *Holodiscus*, *Phyllodoce*, *Salix*, *Ribes*, and *Vaccinium*.

Along the colder and drier eastern slopes, altitudinal zones are not as distinct. The upper montane zone (2500–2900 m in the northern Sierra Nevada, 2800–3400 m in the south) includes *Abies magnifica*, *A. concolor* var. *lowiana*, and *Pinus contorta*; the lower montane zone (2000–2500 m in the north, 2600–2800 m in the south) has open stands of *P. jeffreyi* as the prominent species. The lower limits of the coniferous forest merge into the *Artemisia tridentata* (sagebrush) association of the Basin and Range Province.

Pacific Northwest Region This region consists of the Pacific coast coniferous forest: the Coast Ranges of midcoastal California; the Cascade Mountains of coastal Oregon, Washington, and British Columbia; the Olympic Mountains of northwest Washington; and the British Columbia Coast Ranges. These forests form a relatively narrow belt along the western coast of North America from about central California to Alaska. They generally extend less than 150 km inland and range from sea level to several thousand meters in elevation. They parallel the coniferous forests of the Sierra Nevada and are bordered on the north by the boreal forest and Rocky Mountains coniferous forest and on the south by woodland and scrubland communities.

Figure 1.7. Coniferous forest formation—montane coniferous forest association (Sierra Nevada region; *Sequoiadendron giganteum*), California. Photograph courtesy of the National Agricultural Library, Forest Service Photo Collection.

Relatively mild climates are maintained by proximity to the Pacific Ocean, and temperatures are mostly above −15°C even in the northernmost portions. Rainfall ranges from 650 mm along the eastern slopes in the south to 4000 mm in northwestern Washington, producing temperate rainforest vegetation on the Olympic Peninsula. In the south 75% of the precipitation falls in the winter, and fog is an important source of moisture; precipitation is more uniformly distributed in the north.

Franklin (1988) notes that one characteristic of the Pacific Northwest coniferous forest is the dominance of numerous coniferous species that are the largest and longest lived representatives of their genera. These include *Pseudotsuga menziesii*, *Tsuga heterophylla*, *Thuja plicata*, *Picea sitchensis*, and *Sequoia sempervirens* (Fig. 1.8). Where these forests merge with those of the Sierra Nevada, shared species include *Tsuga mertensiana* and *Picea glauca*. Species in the Cascade Mountains and in the northern Rocky Mountains coniferous forest are *Tsuga heterophylla*, *Thuja plicata*, and *Taxus brevifolia* (Pacific yew). The latter was earlier a source of taxol used in the treatment of some cancers, and damage to the trees from harvesting the bark threatened the species and the source of the drug. Taxol was later obtained from the leaves and twigs of other species (*T. baccata*, Europe; *T. wallichiana*, Himalayas) and is now being synthesized (Suffness, 1995).

The distribution of several important Pacific Northwest coniferous elements is being rapidly modified by lumbering. Less than 5% of the coast redwoods remain uncut, and only about one-half of these are protected in parks. The cutting of douglas fir in Washington and Oregon accounts for over one-third of the annual lumber production in the United States, making douglas fir the most important commercial timber tree in North America (Vankat, 1979).

Franklin (1988) recognizes six major forest types within the Pacific Northwest coniferous vegetation. The principal one is the *Pseudotsuga menziesii* (seral)–*Tsuga heterophylla* (climax) forest. It ranges from sea level to about 1000 m elevation and includes associates such as *Thuja plicata*, *Abies grandis* (grand fir), *Picea sitchensis*, *Pinus monticola*, *P. lambertiana*, *Calocedrus decurrens*, *Abies amabilis* (Pacific silver fir), and *Taxus brevifolia*. The minor hardwood element includes *Alnus rubra* (red alder), *Acer macrophyllum*, *Prunus emarginata* (bitter cherry); *Populus trichocarpa* (black cottonwood) and *Fraxinus latifolia* (Oregon ash) along waterways; and *Arbutus menziesii* (Pacific madrone), *Castanopsis chrysophylla* (golden chinkapin), and *Quercus garryana* (Oregon white oak) on drier sites. Common shrubs are *Gaultheria shallon*, *Holodiscus discolor*, *Berberis nervosa*, and *Rhododendron macrophyllum*.

The *Picea sitchensis*–*Tsuga heterophylla* forest grows under distinctly coastal, maritime climates from northern California to Alaska. Associates include *Pinus contorta*, *Pseudotsuga menziesii*, *Thuja plicata*, *Abies grandis*, *A. amabilis*, *Acer macrophyllum*, *Alnus rubra* (on former logging sites); *Sequoia sempervirens* in the south; *Tsuga mertensiana* and *Chamaecyparis nootkatensis* in the north;

Figure 1.8. Coniferous forest formation—montane coniferous forest association (Pacific Northwest region; *Sequoia sempervirens* forest), Del Norte County, California. Photograph courtesy of the National Agricultural Library, Forest Service Photo Collection.

and *Pinus contorta* in swamps and dunes. The shrub layer consists of *Acer circinatum*, *Rubus spectabilis*, *Vaccinium parvifolium*, *V. alaskaense*, and *Menziesia ferruginea*.

The *Sequoia sempervirens* forest (Fig. 1.8) occurs along a zone only ~16 km wide from northern coastal California to southern Oregon.[3] Associated with the coast redwood are *Pseudotsuga menziesii* and *Lithocarpus densiflorus* (tan oak); *Tsuga heterophylla*, *Umbellularia californica* (California bay), and *Alnus rubra* (moist sites); *Picea sitchensis* (coastal); and *Arbutus menziesii*, *Calocedrus decurrens*, and *Pinus attenuata* (drier sites). Common shrubs are *Gaultheria shallon*, *Rhododendron macrophyllum*, and *Vaccinium ovatum*.

In northwestern California and southwestern Oregon the Klamath Mountains mixed evergreen forest is a mosaic of vegetation derived from the surrounding areas. The common trees are *Pseudotsuga menziesii*, mixed with *Lithocarpus densiflorus*, *Quercus chrysolepis*, *Arbutus menziesii*, and *Castanopsis chrysophylla*.

Coniferous elements from the Sierra Nevada extend into the southern Pacific Northwest forests and form the Sierran-type mixed conifer forest (Franklin, 1988). Represented are *Pseudotsuga mensiesii*, *Pinus lambertiana*, *P. ponderosa*, *Abies concolor*, and *Calocedrus decurrens*. At higher altitudes subalpine forests and parklands develop, within which various phases can be recognized such as the *Abies amabilis–Tsuga heterophylla* forest, the *Abies magnifica* var. *shastensis* forest, the *Tsuga mertensiana* forest, and the subalpine parkland.

As noted earlier, the montane coniferous forest associations are a comparatively uniform vegetation characteristic of mid- to high elevations in North America. Similarities in composition are evident between the Appalachian region and the western mountains and especially within the western cordilleras. Associations are recognized, in part, on the basis of the geographic separation of the several mountain systems and on altitudinal zonation within individual systems. Where physiography allows a particular climate to extend across topographic barriers and connect otherwise isolated regions, coniferous forest elements in these disjunct areas are often shared. For example, Pacific air can penetrate the coastal sierras at places along the Cascade Mountains and bring heavy rains and cold temperatures to the western slopes of the Rocky Mountains near the Columbia Plateau and adjacent Canada. At such places the Pacific coastal and Rocky Mountains coniferous forests share species that elsewhere are used to distinguish the associations. Cascadian elements in this region of the Rocky Mountains include *Tsuga heterophylla*, *Thuja plicata*, *Abies grandis*, *Taxus brevifolia*, *Tsuga mertensiana*, *Larix lyallii*, *Menziesia ferruginea*, *Oplopanax horridum*, *Philadelphus lewisii*, *Rhododendron albiflorum*, *Sorbus sitchensis*, *Vaccinium globulare*, *V. membranaceum*, and *Xerophyllum tenax* (Peet, 1988). The relationship between the Cascade and Rocky Mountains coniferous forests is one example of the variable composition of associations. Clearly any change in climate or topography would result in greater opportunity for intermingling of the dominant and associated species between regions currently separated geographically, and there is no guarantee that after reshuffling the principal tree component and associated vegetation would remain the same. In fact, results from Quaternary pollen analysis indicate it is unlikely. This raises the question of how constant paleocommunities remain in composition through time, which has implications for scenarios attempting to trace the history of major vegetation units. The fluid composition of various closely related coniferous forest associations seem to necessitate an equally flexible component to any paleoecological concept in order to explain the history of the North American vegetation. These ideas will be considered in Chapter 3, but similarities between the Cascade and Rocky Mountains coniferous forests in parts of their range provide an introduction to the temporal and dynamic nature of plant associations.

DECIDUOUS FOREST FORMATION

The eastern deciduous forest formation covers ~11% of the continent, and it is the most diverse and species-rich component of the North American vegetation. Whereas the boreal forest is characterized by about nine principal tree genera, at least 67 canopy and subcanopy trees and shrubs combine in various ways to constitute the associations of the deciduous forest. The canopy averages ~30 m in height, although Greller (1988) quotes old accounts of dense virgin forests in Ohio with oaks, hickories, pines, and other trees considerably larger than in any modern second-growth forests. The formation is bounded on the west by grasslands from western Minnesota to eastern Texas where lower moisture and, formerly, fires by AmerIndians limited its distribution. To the north it is bordered by the boreal forest from Minnesota to northern Maine where it becomes limited primarily by low temperatures, lower light intensity, and a shorter growing season. These and other factors combine to render deciduous plants at a competitive disadvantage to evergreen plants such as conifers. Deciduous components blend with boreal coniferous species in the upper midwest to form a transitional vegetation called the lake states forest. Elements of the deciduous forest are also found in western North America where they are most common as riparian vegetation along river valleys and floodplains (e.g., *Salix*, *Alnus*, *Populus*).

Precipitation is ~1250 mm along the Atlantic coast, 850 mm along the forest–grassland boundary, 750 mm in the Great Lakes region, and 1500 mm along the Gulf Coast. Average January temperatures range from −6.7°C in the north to 4.4°C in the south; average temperatures in July are 21.1 and 26.7°C, respectively. Under these temperature regimes the growing season is 120 days in the north and 250 days in the south (Vankat, 1979).

Several systems are available for classifying the associations of the deciduous forest formation (Braun, 1950; Greller, 1988; Vankat, 1979). Following Vankat (1979), eight associations are recognized. The mixed mesophytic association occurs on moist, well-drained sites in the central parts of the formation in the unglaciated Appalachian Plateau and Cumberland Mountains of eastern Tennessee and Kentucky. This is the richest and most diverse association and includes as dominants *Fagus grandifolia* (American beech), *Liriodendron tulipifera* (tulip tree), *Tilia* (basswood, several species), *Acer saccharum* (sugar maple), *Aesculus octandra* (buckeye), *Quercus rubra* (red oak), *Q. alba* (white oak), *Tsuga canadensis* (eastern hemlock), plus 30 or more others that commonly form more local associations.

The beech–maple association (Fig. 1.9) is restricted primarily to glaciated areas of Ohio, western Indiana, and southern Michigan. The dominants are *Fagus grandifolia* and *Acer saccharum*, and species of *Fraxinus* (ash) and *Ulmus americana* (American elm) are common associates.

The maple–basswood association occupies the smallest area, occurring principally in east-central and southeastern Minnesota, southwestern Wisconsin, and northeastern Iowa. The boundaries are not clear and it has been suggested that this is a consequence of former burning practices by the AmerIndians. The dominants are *Acer saccharum* and *Tilia americana* (replacing *Fagus grandifolia*), and various species of oak are subdominants.

The lake states forest association is a hemlock–white pine–northern hardwood association and, as noted previously, it is transitional between the deciduous forest and the boreal coniferous forest. It is distributed from western Ontario and Minnesota, across southern Canada and New England, to the Maritime Provinces of eastern Canada. At different places the association may be dominated by coniferous species, deciduous species, or a combination of both. Widespread members include *Pinus strobus* (white pine), *P. resinosa* (red pine), and *Betula lutea*. Species with boreal affinities are *Betula papyrifera*, *Populus tremuloides*, and *Pinus banksiana*. Deciduous forest elements are *Acer rubrum* (red maple), *A. saccharum*, *Fagus americana*, *Tilia americana*, and *Tsuga canadensis*. In stands dominated by conifers the principal species are *Pinus strobus*, *P. resinosa*, and *P. banksiana* on drier sites; the boreal species *Picea glauca*, *Picea mariana*, and *Larix laricina* are in and around scattered bogs. To the east *Picea rubens* and *Tsuga* sp. are common. Stands of white pine were formerly much more extensive, but they have been intensively lumbered since about 1623, initially to provide masts for British sailing ships, and have been replaced mostly by hardwoods.

The oak–chestnut association no longer exists in its original form because of the virtual elimination of mature trees of *Castanea dentata* (American chestnut) by the ascomycete fungus *Endothia parasitica*. The parasite was introduced from China in about 1900 and was first reported on American trees in New York City in 1904 (Greller, 1988; Vankat, 1979). By 1920, 50% of the chestnut trees in the north had been destroyed, and by 1930 50% were gone from southern forests. *Castanea dentata* is now primarily a scattered understory tree reproducing by vegetative methods, and it has been replaced mostly by oaks (*Quercus borealis* var. *maxima*, *Q. prinis*, and *Q. alba*).

The most widespread association in the eastern deciduous forest formation is the oak–hickory association, extending from New Jersey south along the piedmont to the Gulf states, west to central Texas, and north to Minnesota.

Figure 1.9. Deciduous forest formation—beech–maple association (*Fagus grandifolia, Acer saccharum*), Nicolet National Forest, Wisconsin. Photograph courtesy of the National Agricultural Library, Forest Service Photo Collection.

At its western edge it grades into the grassland formation as a savanna. The dominant oaks in the south and southwest are *Quercus stellata* (post oak) and *Q. marilandica* (blackjack oak); in the west are *Quercus alba* (white oak), *Q. rubra*, *Q. velutina*, and *Q. macrocarpa* (bur oak). The hickories are *Carya cordiformis* (bitternut hickory), *C. tomentosa* (mockernut hickory), *C. ovalis* (red hickory), and *C. ovata* (shagbark hickory). The region of greatest species diversity is in the low hills of the Ozark Plateau and Ouachita Mountains of central Arkansas. In the east and south and the piedmont region of Georgia, pines become an important component. The settlement of Georgia's rich piedmont began in 1773, and by 1840, 87% of the area was under cultivation. During the Civil War ~10% of the land was abandoned, followed by another 30% during the agricultural depression of the late 1880s, and 35% following spread of the boll weevil in the 1920s (Vankat, 1979). Most of the abandoned land was regenerated by pine. In the south the prominent species are *Pinus taeda* (loblolly pine) and *P. echinata*, while in the north *P. rigida* (pitch pine) and *P. virginiana* (Virginia pine) are the dominants. The pine phase is fire dependent, and without fire hardwoods replace them in most areas of the oak–hickory association.

The southern mixed hardwood association is scattered along the coastal plain from North Carolina to Texas, and various phases are influenced by the level of the water table and the water-retention capacity of the soil. It is a multidominant assemblage and individual stands may have from five to nine prominent tree species. The more common ones in well-drained areas with rich soils are *Fagus americana*, *Quercus alba*, *Q. virginiana* (live oak), *Q. laurifolia* (laurel oak), and *Magnolia grandiflora* (evergreen magnolia).

Pine woods are locally developed on deep sandy soils of the coastal plain and include the northern pine barrens that extend south to near the Delaware Bay (Christensen, 1988). *Pinus rigida* is the dominant, mixed with *P. echinata*, *Comptonia peregrina*, *Kalmia latifolia*, *Gaylussacia baccata*, *G. frondosa*, *G. dumosa*, *Quercus ilicifolia*, *Q. prinoides*, *Ilex glabra*, and *Clethra alnifolia*. To the south is the sandhill pine forest of *Pinus palustris* (longleaf pine) and the understory grass *Aristida stricta* (wire grass). Associated vegetation defining three phases of the forest includes *Quercus laevis* (turkey oak), *Nyssa sylvatica* (black gum), *Diospyros virginiana* (persimmon), *Lyonia mariana* (staggerbush), and *Gaylussacia dumosa* (dwarf huckleberry)—the pine–turkey oak sandridge phase; *Quercus marilandica*, *Q. margaretta* (sandhill post oak), *Q. indica* (bluejack oak), and *Liquidambar styraciflua* (sweet gum)—

Figure 1.10. Deciduous forest formation—flood-plain forest association (*Taxodium distichum*), Louisiana. Photograph courtesy of the National Agricultural Library, Forest Service Photo Collection.

the fall-line sandhill phase; and *Q. laevis*, *Q. marilandica*, *Q. indica*, *Q. rubra*, *Pinus elliottii* (slash pine), and *Diospyros virginiana*—the Florida sandhill phase.

Another pineland along the southeastern coastal plain is the sand pine scrub of *Pinus clausa* (sand pine) and associated vegetation of *Serenoa repens* (saw palmetto), *S. etonia* (scrub palmetto), *Quercus germinata*, *Q. myrtifolia*, *Q. virginiana*, *Q. inopina*, *Carya floridana*, *Lyonia ferruginea*, and *Ceratonia ericoides*.

The flood-plain forest association (Fig. 1.10) is a riparian vegetation. A southern phase is characterized by *Taxodium distichum* (bald cypress), *Chamaecyparis thyoides* (southern white cedar), *Pinus glabra* (spruce pine), *Nyssa* (tupelos, *N. sylvatica*, *N. aquatica*, *N. ogeche*), *Quercus lyrata* (overcup oak), *Q. laurifolia*, *Q. michauxii* (basket oak), *Q. falcata* (cherry bark oak), *Q. phellos* (willow oak), *Carya aquatica* (water hickory), *Ilex verticillata* (winterberry), *I. opaca* (American holly), *Gleditsia aquatica* (water locust), *Asimina triloba* (pawpaw), *Lindera benzoin* (spicebush), *Sabal palmetto* (palmetto), *Itea virginica*, *Styrax americanum*, *Cornus foemina*, *Acer rubrum*, *Ulmus americana*, *Fraxinus caroliniana*, *F. profunda*, *F. pennsylvanica*, *Magnolia virginiana* (sweet bay), *Persea borbonia* (red bay), *Cyrilla racemiflora*, *Leucothoe racemosa*, *Lyonia lucida*, *Clethra alnifolia*, and *Liquidambar styraciflua* (sweet gum). In the north, floodplain habitats are occupied by species of *Populus*, *Salix*, *Ulmus*, and *Fraxinus*.

The deciduous forest formation is found in several disjunct regions throughout the northern hemisphere, in addition to outliers in the American west. The forest is especially well developed in eastern Asia, and a phase intermixed with warm temperate to tropical elements occurs along the eastern escarpment of the Mexican Plateau. It was also extensive in central and western Europe during the Tertiary, disappearing there as a result of climatic changes in the Late Tertiary and the continental glaciations of the Pleistocene. One of the goals of Cenozoic vegetational history studies is to better understand the origin of such disjunct floristic relationships (Chapter 9); these attempts are facilitated by the fact that almost all of the tree genera, many of the shrubs, and some of the herbs mentioned in the above discussion are represented in the fossil record.

GRASSLAND FORMATION

When French explorers moved south from eastern Canada into the central plains, they encountered a vast expanse of grassland never before witnessed by western Europeans, and applied the closest French word for a community dominated by grasses and forbs—*prairie* (meadow). The grassland is the most extensive of the North American plant formations, originally covering an estimated 265,987,700 ha of the 770 million ha in the United States (~30%, Sims, 1988; ~21% of North America north of Mexico, Barbour and Christensen, 1993). It extends from Indiana south along the western edge of the deciduous forest to the Front Ranges of the Rocky Mountains and from southern Canada through northern Mexico. The grassland exists, in part, as a result of insufficient precipitation during the spring and summer growing season to support trees. Precipitation decreases from east to west and from north to south. Annual rainfall is ~1000 mm in the tallgrass prairie of the northeast and ~200 mm in the desert grasslands of the southwest. Annual temperatures range from 3°C in mountain grasslands to 15°C in the desert grasslands. One of the characteristics of the North American grassland climate is the distinctly seasonal rainfall and the extremes in temperature. In fact, the range of climatic parameters in the grassland is greater than for any other North American plant formation (Sims, 1988). In the central part of the continent temperatures may be –35°C in the winter and reach 45°C in the summer. The growing season is from 120 days in the north to 200 days in the south and extends virtually year-round in the southwestern United States and northern Mexico. In the California grasslands rainfall is 150–500 mm and falls mostly during the winter months. Fire is an important, if not essential, factor in the maintenance of the modern North American grassland (Collins and Wallace, 1990; Wright and Bailey, 1982). The front of one fire early in this century covered over 200 km in a single day (Vankat, 1979).

The grasslands include ~600 genera and 7500 species of plants (Sims, 1988). About 95% are perennial, and the Leguminosae–Fabaceae and Compositae–Asteraceae contribute the largest number of associated forbs. Another characteristic of the grassland formation is the dramatic change in composition that often takes place throughout the growing season. In one prairie in Wisconsin, 17 new species came into flower each week from April through October (Vankat, 1979).

Six associations are recognized within the grassland formation: the tallgrass prairie, mixed grass prairie, and shortgrass prairie in the central part of the continent; the desert grassland of the arid southwest in Arizona, New Mexico, Texas, and northern Mexico; the Palouse Prairie in parts of Washington, Oregon, and Idaho; and the California grassland in the Central Valley of California between the Coast Ranges and Sierra Nevada.

Grasses of the tallgrass prairie association may reach 3 m in height near the end of the growing season. This is the most fertile of the associations and, because of its agricultural value for growing wheat and corn, little of the native tallgrass prairie remains today (Samson and Knopf, 1994). The tallgrass prairie is the most mesic of the grassland associations and is best developed along the margin of the deciduous forest. A lobe extending eastward into Indiana and Illinois is known as the Prairie Peninsula (Stuckey and Reese, 1981; Transeau, 1935). The principal grasses are *Andropogon gerardi* (big bluestem), *Sorghastrum nutans* (Indian grass), *Panicum virgatum* (switchgrass), *Elymus canadensis* (Canada dry grass), and *Spartina pectinata* (prairie cordgrass, mainly in wet, poorly drained areas). In the drier uplands are *Andropogon scoparius* (little bluestem), *Stipa spartea* (needlegrass), *Sporobolus heterolepis* (prairie dropseed), *Bouteloua curtipendula* (sideoats grama), and *Koeleria cristata* (junegrass). An elongated isolated lobe called the blackland prairie (*Andropogon*, *Stipa*) runs from north of Dallas, to south near Austin, to San Antonio (Küchler, 1964).

The mixed grass association (Fig. 1.11) lies between the tallgrass and shortgrass association. The short phase of the association includes *Bouteloua gracilis* (blue grama), *B. hirsuta* (hairy grama), and *Buchloë dactyloides* (buffalo grass). The taller phase consists of *Andropogon scoparius* and *Stipa comata* (needle and thread grass).

The westernmost portion of the central grasslands, a physiographic area known as the High Plains, is occupied mostly by the shortgrass association that extends into Mexico as far as Jalisco (22° N). Rainfall is less than in the tallgrass and mixed grass associations and the plants typically reach only 20–50 cm in height. The dominants are *Bouteloua gracilis* and *Buchloë dactyloides*, associated with *Stipa comata*, *Aristida longiseta* (wire grass), *Agropyron smithii* (western wheat grass), *Sporobolus cryptandus*, *Muhlenbergia torreyi*, *Koeleria cristata*, and *Hillaria jamesii*. Before the advent of more rational farming practices, wind erosion of the soil was extensive and created the great dust bowl storms of the 1930s. In March 1935, 37,000 metric tons of soil per cubic kilometer of air was recorded above Wichita, Kansas, that came from an area over 400 km away. In some areas vegetation cover during this period was reduced by 98% (Vankat, 1979).

The desert grassland association occupies plateaus in the arid southwest region of west Texas, southern New Mexico, Arizona between the desert and piñon–juniper woodlands, and into Mexico (McClaran and Van Devender, 1995). The prominent grasses are *Bouteloua eriopoda* (grama grass), *Hilaria belangeri* (tobasa grasses), *Oryzopsis hymenoides*, *Muhlenbergia porteri*, and *Aristida* sp. However, drying climates and recent overgrazing have caused shrubs to invade from adjacent desert vegetation, and much of the area is occupied by *Prosopis glandulosa* (mesquite), *Larrea tridentata* (creosote bush), *Acacia* (acacias), *Yucca* (yucca, Spanish bayonet), and *Opuntia* (prickly pear).

The California grassland association in the agriculturally rich Central Valley now includes mostly introduced annual grasses and other plants; between 50 and 90% of the present vegetation cover in uncultivated areas is comprised of exotic species. Cool-season perennials have been replaced by cool-season annuals, partially because of the grazing of perennials in preference to the lower quality forage of many of the annuals. The original dominant was *Stipa pulchra* (needlegrass), which has now been replaced by such grasses as

Figure 1.11. Grassland formation—mixed grass association (*Bouteloua gracilis*), Colorado. Photograph courtesy of the National Agricultural Library, Forest Service Photo Collection.

Avena fatua (weedoats), *Hordeum murinum* (mouse barley), *Bromus* sp. (brome grasses), and *Festuca* sp. (fescues).

The Palouse Prairie association developed as a result of the rainshadow created by uplift of the Cascade Mountains. Originally the dominants were *Agropyron spicatum* (bluebunch wheat grass), *Festuca idahoensis* (Idaho fescue), *Elymus condensatus*, and *Poa secunda* (sandburg bluegrass). In the late 1800s the annual *Bromus tectorum* (downey chess) was introduced and it is now the dominant, along with *Artemisia tridentata* (sagebrush) that has come in with human disturbance.

SHRUBLAND/CHAPARRAL–WOODLAND–SAVANNA FORMATION

Mediterranean and Madrean Scrublands and Woodlands

The vegetation characteristic of lower elevations and drier parts of the western intermountain basins, valleys, slopes, and plateaus consists of three phases based on the spacing or density of the tree or shrub layer. Shrubland/chaparral or thickets are composed of a dense, mostly continuous layer of shrubs and small shrubby trees. Woodlands (Fig. 1.12) are vegetation composed principally of trees in which the crowns do not touch. Savanna is a vegetation in which the tree or shrub cover is typically less than 30% and a dense herbaceous layer usually dominated by grasses. All of these developed primarily in response to drier conditions created to the lee of the rising Sierra Nevada and Cascade Mountains. They cover ~1% of North America north of Mexico.

The scrub oak–mountain mahogany shrubland occurs from southern Wyoming and northern Utah to Mexico at elevations between 1250 and 1750 m and is typically found as a band between the coniferous forest and desert vegetation. The composition of the association varies considerably. The most wide-ranging species is *Cercocarpus ledifolius* (curl-leaf mountain mahogany). The greatest diversity is found in the vicinity of the southern Rocky Mountains where the dominants are *Quercus turbinella*

Figure 1.12. Shrubland/chaparral–woodland–savanna formation—juniper (*Juniperus occidentalis*) woodland, Central Oregon. Photograph courtesy of The National Archives.

(scrub oak), *Q. emoryi*, *Q. dumosa*, and *Q. chrysolepis* (an eastern extension of a California oak group south of the Colorado Plateau). In the north the species are *Quercus gambellii* (Gambell's oak), *Q. undulata* (a hybrid between *Q. gambelii* and several southern evergreen live oaks; Tucker, 1961), *Cercocarpus breviflorus*, *C. betuloides*, and *C. montanus*. Other shrubs include *Artemisia tridentata* var. *vaseyana*, *Acer grandidentatum*, *Purshia tridentata*, *Rhus triloba*, *Rhamnus corcea*, *Fallugia paradoxa*, *Cowania mexicana*, *Arctostaphylos pungens*, *A. pringlei*, *Ceanothus greggii*, *Garrya flavescens*, *G. wrightii*, *Eriodictyon angustifolium*, and species of *Amelanchier*, *Symphoricarpus*, and *Berberis* (West, 1988).

Chaparral is a version of the shrubland found mostly in California from southern Oregon (Rogue River watershed, 43° N latitude) to northern Baja California (Sierra San Pedro Martir, 30° N latitude; Keeley and Keeley, 1988). Chaparral and related shrub communities also extend along the Sierra Madre Occidental to southern Durango, Mexico, as summer-wet Madrean vegetation. It is a dense, sclerophyllous vegetation most common in California between the foothills of the Sierra Nevada and the Pacific coast from sea level to ~2000 m elevation. The shrub layer is 1–4 m tall and is frequently so dense that it is virtually impenetrable. The word chaps is derived from chaparral, denoting pieces of leather tied across the legs for protection while riding through this vegetation. The area is characterized by the cool wet winters and hot dry summers typical of a Mediterranean climate (Arroyo et al., 1995). Precipitation ranges from 200 to 1000 mm, and 80–90% falls from November to April. The mean winter minimum temperature is between 0 and 10°C, while the summer maximum reaches 40°C. If summer fires are frequent, the vegetation trends toward grassland. If they are less frequent, the development of oak woodland is favored (Vankat, 1979).

The composition of the California chaparral is variable from one locality to another and may include from one to as many as 20 dominants. Keeley and Keeley (1988) present a table listing 63 of the common species. The most widely distributed member is *Adenostoma fasiculatum* (chamise), frequently associated with *Salvia mellifera*, *S. alpina* (California coastal sage), *Eriogonum fasciculatum*, and species of *Arctostaphylos* (manzanitas) and *Ceonothus*. The range of the California chaparral is expanding as a result of overgrazing, mining, burning, logging, and construction.

The broad-sclerophyll forest has a range that mostly corresponds to that of the California chaparral. It occurs on more mesic sites and has a tree layer that is from 5 to 10 m in height compared to 1–4 m for the chaparral. It is also more common than chaparral in the north where annual rainfall is greater (1000 mm) and less common in the south where rainfall within its range is ~350 mm. Mean annual temperatures are below freezing in the winter to over 30°C in the summer. The broad-sclerophyll forest extends to ~1500 m elevation where it is commonly bordered by the lower montane coniferous forest. The most prominent species are *Quercus dumosa* (scrub oak), *Q. chrysolepis* (canyon live oak), *Q. agrifolia* (coast live oak), *Q. wislizenii* (interior live oak), *Arbutus menziesii* (madrone), *Pinus sabiniana* (digger pine), *P. coulteri* (Coulter pine), *Umbellularia californica*, *Aesculus californica* (California buckeye), *Heteromeles arbutifolia* (chaparral holly), *Prunus lilcifoia* (chaparral cherry), *Cercocarpus betuloides*, and *Malosma* (*Rhus*) *laurina* (laurel sumac).

The California woodland vegetation is found in the interior valleys of California from near Los Angeles to northern California at elevations between 150 and 1000 m. The forest consists of trees from 5 to 10 m in height with an open canopy covering 30–50% of the ground. The dominants are *Quercus douglasii* (blue oak), *Q. lobata* (California white oak), and *Q. agrifolia* (coast live oak).

The piñon–juniper woodland is best developed in the Great Basin and in the drier parts of the southern Rocky Mountains. It extends to the eastern slope of the Sierra Nevada and from Arizona and New Mexico through southern and eastern Colorado, Utah, and Nevada, to southern Idaho (Miller and Wigand, 1994; West, 1988). The association has a discontinuous distribution along the high mesas, foothills, and lower elevations of large mountain systems and a continuous range over many of the smaller mountains. In most areas the piñon–juniper woodland occurs at elevations between 1500 and 2250 m. Upslope it is frequently bordered by ponderosa and other pine forests and at lower elevations by *Artemisia* shrubland, desert, or grassland. It is an extensive association covering ~24 million ha in the western United States and ~19 million ha in the intermountain region between the Sierra Nevada–Cascade Mountains and the Rocky Mountains. Rainfall is between 250 and 500 mm, and mean annual summer temperatures are ~20°C with highs to 30°C. The structure and composition of this woodland is comparatively simple and includes as the principal dominants *Pinus edulis*, *P. monophylla*, *P. cembroides*, *Juniperus erythrocarpa* (redberry juniper), *J. occidentalis* (western juniper), *J. scopulorum* (Rocky Mountains juniper), *J. monosperma*, *J. deppeana* (alligator juniper), and *J. osteosperma* (Utah juniper).

The savanna in North America is a fire-dependent, primarily grassland–deciduous forest boundary vegetation. It also includes open pinelands in the southeastern United States, areas of the central valley of Oregon, and the more open parts of the California woodland vegetation. Before human disturbance, in the east it extended as a 75–175 km wide belt from Minnesota to Texas. In composition it is a mixture of grassland and oak–hickory association species, including *Quercus macrocarpa* in the north and *Q. stellata* and *Q. marilandica* in the south. Further south in Texas it borders the blackland prairie, and *Juniperus* and *Prosopis* are common associates (Küchler, 1964). Fire is important to the perpetuation of the savanna; because fire is now controlled in many areas, woodland or forest vegetation has invaded areas once occupied by oak savanna.

DESERT FORMATION

Deserts are areas receiving less than ~120 mm of available precipitation per year. Most North American deserts receive slightly more and are only semiarid. The annual rainfall is characteristically variable in amounts, and it is unevenly distributed throughout the year. Temperature regimes differ among the different kinds of deserts (warm deserts versus cold deserts), but at least some season of the year has high temperatures. Another feature of the desert is high winds that, combined with low and unpredictable rainfall, high temperatures, and high light intensities, produce the most biologically significant facet of desert climates: a low precipitation/evaporation ratio. Even during the rainy period there is usually rapid runoff or rapid percolation of water through the coarse soils, particularly in summer rain deserts (e.g., Chihuahuan). These areas receive rain in the form of brief and intense thunderstorms, in contrast to winter rain deserts (e.g., Mojave) where precipitation periods are longer and less intense (MacMahon, 1988). As a consequence, there has been evolutionary selection and elaboration of characters and processes that compensate for water stress, and these constitute the syndrome of features typically associated with desert plants: low stature, wide spacing, alleopathy, deep penetrating or extensive shallow root systems, coriaceous and microphyllous leaves, pubescense, replacement of leaves by spines, presence of leaves only during the rainy season, succulence, pleating of stems to accommodate rapid changes in volume with sudden water availability, slow growth rate, water-soluble growth-inhibiting substances in the seed coat, rapid reproductive cycle of annuals, and others. With these adaptations many desert plants withstand ambient temperatures in excess of 55°C where, for example, pubescence can reduce plant surface temperatures by 13°C (Vankat, 1979). The effectiveness of these modifications is apparent from an observation cited in MacMahon (1988) that water loss from a plot of bare soil and one containing *Larrea tridentata* was virtually the same.

There are four deserts on the North American continent (~5% of North America north of Mexico; Fig. 1.13, Table 1.4), in addition to the polar desert of the Arctic. These are the Great Basin, Mojave, Sonoran, and Chihuahuan deserts. The Great Basin is a cold desert receiving ~60% of its precipitation as snow, and it is often included in the shrubland vegetation of the Basin and Range Province. The others are warm deserts in the southwestern United States and adjacent Mex-

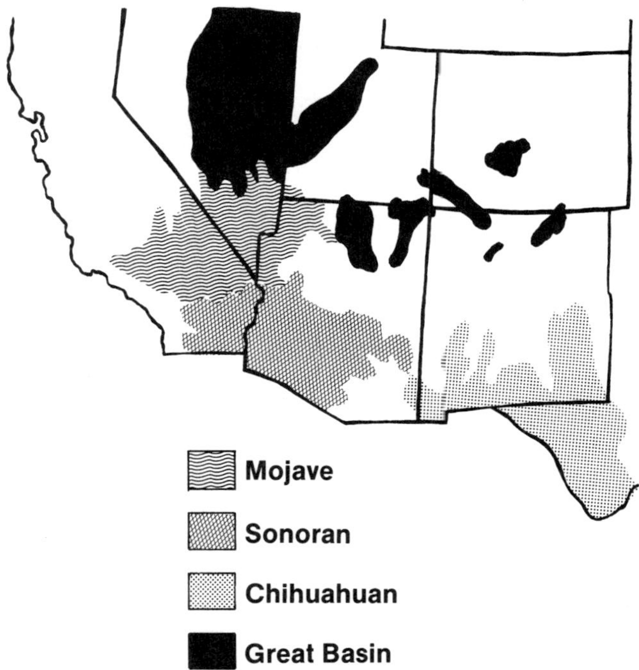

Figure 1.13. Distribution of North American deserts. Reprinted from MacMahon (1988) with the permission of Cambridge University Press.

ico, receiving most of their precipitation in the form of summer or winter rains (MacMahon and Wagner, 1985).

The Great Basin Desert association corresponds in distribution to the Great Basin physiographic province, occurring mostly in the zone between the Sierra Nevada–Cascade Mountains and the Rocky Mountains at elevations between 1200 and 2500 m (Osmond and Pitelka, 1990). It is a result of the rainshadow created by the Sierra Nevada and Cascade Mountains. The Great Basin Desert is also typical of broad valley floors that receive precipitation from runoff that slowly evaporates, producing salt-rich playa lakes. Annual rainfall is between 100 and 250 mm. Winter temperatures usually do not go above freezing, and summer frosts occur at night in the valleys, but during the summer day the temperatures may reach above 40°C. At higher elevations the association grades into piñon–juniper woodland. Cacti and other succulents typical of the warm deserts are less common in the Great Basin Desert.

Areas of low soil salinity support stands of *Artemisia tridentata*, the most widespread vegetation type in the intermountain lowlands. In soils of high salinity *Atriplex confertifolia* (shadscale) is the conspicuous element. Common associates are *Artemisia spinescens*, *Atriplex nuttallii* (Nuttall saltbush), *Atriplex spinosa*, *Salicornia utahensis*, *Suaeda fruticosa*, *S. torreyana*, *Sarcobatus vermiculatus* (greasewood), *Eurotia lantana* (winter fat), and several perennial grasses such as *Festuca idahoensis*, *Agropyron (Pseudoroegneria) spicatum*, *A. (Pascopyrum) smithii*, *Elymus lanceolatus*, *Stipa thungergia*, *S. arida*, and *S. speciosa*. Overgrazing has increased the area covered by sagebrush, and the annual grass *Bromus tectorum* has become widespread along with *Chrysothamnus* (rabbit bush), *Ephedra* (Morman tea), and *Tetradymia* (horse brush). The term steppe or woody steppe is primarily a European term for temperate shrub vegetation with plentiful short grasses often growing on dry seasonally cold plains and mountain slopes. One difference between savanna–grassland and steppe is that the latter is summer dry.

Table 1.4. Approximate areas of North American deserts.

Unit	Area (km^2)	Portion of NA Desert Area (%)
Great Basin Desert	409,000	32.0
Sonoran Desert	275,000	21.5
Mojave Desert	140,000	11.0
Chihuahuan Desert	453,000	35.5
Total desert	1,277,000	100
Warm deserts	868,000	68.0
Cold deserts	409,000	32.0
Basin and range province	1,717,000	—

Adapted from MacMahon (1988).

Figure 1.14. Desert Formation—Sonoran Desert association (*Carnegiea gigantea*, *Larrea tridentata*).

The Mojave Desert association is the smallest of the North American deserts, extending from southern California through southern Nevada to western Arizona. Elevation is from -86 m in Death Valley to 1200 m, and precipitation ranges from 125 mm in the western portions to 50 mm in the east. Winter temperatures reach 15°C during the day and are below freezing at night, while summer temperatures are from 30° to 40°C in the basins to 18° to 30°C on mountain summits.

The Mojave Desert is often considered a transition between the Great Basin Desert to the north and the Sonoran Desert to the south. However, MacMahon (1988) notes that ~25% of the total species and 80% of the ~250 annuals are endemic. The dominants are *Larrea tridentata* (creosote bush) and *Franseria* (*Ambrosia*) *dumosa* (bur sage), which cover ~75% of the region. In the north *Artemisia tridentata* and *Atriplex confertifolia* come in at the higher elevations, and in the west *Yucca schidigera* (Mojave yucca) and *Y. brevifolia* (Joshua tree) are characteristic members of the vegetation. In the Mojave Desert, like the Great Basin Desert, cacti are present but are not as common as in the Chihuahuan Desert.

The Sonoran Desert association (Fig. 1.14) is the most varied of the North American deserts in terms of temperature, rainfall, and composition. It extends from southern California and southern Arizona into Baja California and northwestern mainland Mexico. Most of the area consists of plains and moderate mountainside slopes called bajadas, which formed by the coalescence of alluvial fans. It is the lowest of the North American deserts, occupying sites generally between sea level and 900 m, but mostly below 600 m. The Sonoran Desert is not a rainshadow desert, but it is maintained primarily by a persistent high-pressure system. Precipitation is variable and irregular. Rainfall ranges from near 0 mm in the west to 35 mm in the east, to as much as 700 mm on some low mountain peaks. In the summer temperatures reach a maximum of 40°C, winter highs are around 20°C, and frosts are rare. This makes the Sonoran Desert the hottest and most temperature variable of the North American deserts; on bare rock daily temperatures may fluctuate by 60°C within a 24-h period.

These variations in climate also make this desert the most diverse in composition (Shreve and Wiggins, 1964). The region around Tucson, Arizona has many small trees and shrubs related to those of the Sinaloan (Mexico) thorn forest. In the arid lowlands *Franseria* (*Ambrosia*) *dumosa* and *Larrea tridentata* are the most common species, but each has a varied associated vegetation. These include *Ephedra trifurca*, several species of *Dalea*, *Sapium biloculare*, *Condalia lycioides*, *Prosopis glandulosa*, and *Encelia farinosa* (brittlebush). At higher elevations *Cerdidium microphyllum* (palo verde), *Fouquieria splendens* (ocotillo), *Acacia greggii*, and *Carnegiea gigantea* (saguaro, Fig. 1.14) are conspicuous elements.

The Chihuahuan Desert association occurs from southern New Mexico and western Texas into north-central Mexico. In elevation it extends from ~400 m along the Rio Grande to 1500 m on the central Mexican Plateau. The landscape consists mostly of a plain dissected by valleys and numerous small sierras. Precipitation (150–400 mm) is slightly higher than in the other deserts, and most rain falls in the summer between June and September. Summer temperatures average 5–10°C lower than in the Sonoran Desert; winter temperatures are also lower, and night frosts

Figure 1.15. Warm-temperate to subtropical vegetation, Everglades National Park, Florida. Photograph courtesy of Sammantha Ruiz, National Park Service, Everglades National Park.

are more common. The dominants are *Flourensia cernua* (tarbush), *Acacia schaffneri*, *Mimosa biuncifera*, *Larrea tridentata*, *Fouquieria splendens*, *Agave lechuguilla* (lechuguilla), *Koeberlinia spinosa* (allthorn), *Hechtia scariosa*, *Leucophyllum frutescens*, *Prosopis* sp., *Yucca filifera*, *Y. carnerosana*, and *Agave salmiana*. Cacti and succulents are abundant and form the distinctive element of the Chihuahuan Desert. They include *Myrtillocactus geometrizans*, *Opuntia robusta*, *O. leucotricha*, *O. streptacantha*, *O. cantabrigiensis*, *O. phaeacantha*, *O. leptocaulis*, and *Echinocactus horizonthalonius*.

TROPICAL ELEMENTS

A tropical formation does not occur in the region covered in this survey. Rather, elements of tropical vegetation extend various distances into peninsular Florida and along the southern Atlantic and Gulf Coast. The most distinctive are the mangroves (black, *Avicennia nitida*; white, *Laguncularia racemosa*; and red, *Rhizophora mangle*) often associated with *Conocarpus erecta* (buttonwood; Myers and Ewel, 1990; Odum and McIvor, 1990). With approximately 80% of the Florida coastline commercially developed, the most extensive remnants of this community are found in the protected habitats of the Everglades National Park (Fig. 1.15) and the J. N. "Ding" Darling National Wildlife Refuge on Sanibel Island (Fig. 1.16).

SUMMARY

These seven formations constitute the modern vegetation of North America. They are the product of environmental–biological interactions over mostly the past 70 m.y., and in composition and distribution they represent only the latest of many previous and future versions. To better understand this vegetation, it is necessary to know some of the factors that have shaped its history.

References

Arroyo, M. T. K., P. H. Zedler, and M. D. Fox (eds.). 1995. Ecology and Biogeography of Mediterranean Ecosystems in Chile, California and Australia. Springer-Verlag, Berlin.

Axelrod, D. I. and P. H. Raven. 1985. Origins of the cordilleran flora. J. Biogeog. 12: 21–47.

Figure 1.16. Tropical elements—J. N. "Ding" Darling National Wildlife Refuge, Sanibel Island, FL (red mangrove, *Rhizophora mangle*). Photograph courtesy of Kathleen Blase.

Barbour, M. G. 1988. Californian upland forests and woodlands. In: M. G. Barbour and W. D. Billings (eds.), North American Terrestrial Vegetation. Cambridge University Press, Cambridge, U.K. pp. 131–164.

——— and W. D. Billings (eds.). 1988. North American Terrestrial Vegetation. Cambridge University Press, Cambridge, U.K.

——— and N. L. Christensen. 1993. Vegetation of North America north of Mexico. In: Flora of North America Editorial Committee (eds.), Flora of North America North of Mexico. Oxford University Press, Oxford, U.K. Vol. 1, pp. 97–131.

Bliss, L. C. 1988. Arctic tundra and polar desert biome. In: M. G. Barbour and W. D. Billings (eds.), North American Terrestrial Vegetation. Cambridge University Press, Cambridge, U.K. pp. 1–32.

Braun, E. L. 1950. Deciduous Forests of Eastern North America. Blakiston Co., Philadelphia, PA.

Christensen, N. L. 1988. Vegetation of the southeastern coastal plain. In: M. G. Barbour and W. D. Billings (eds.), North American Terrestrial Vegetation. Cambridge University Press, Cambridge, U.K. pp. 317–363.

Collins, S. L. and L. L. Wallace. 1990. Fire in North American Tallgrass Prairies. University of Oklahoma Press, Norman, OK.

Crutzen, P. J. and J. G. Goldammer. 1993. Fire in the Environment. Wiley, New York.

Elliott-Fisk, D. L. 1988. The boreal forest. In: M. G. Barbour and W. D. Billings (eds.), North American Terrestrial Vegetation. Cambridge University Press, Cambridge, U.K. pp. 33–62.

Franklin, J. F. 1988. Pacific Northwest forests. In: M. G. Barbour and W. D. Billings (eds.), North American Terrestrial Vegetation. Cambridge University Press, Cambridge, U.K. pp. 103–130.

Goward, S. N., C. J. Tucker, and D. G. Dye. 1985. North American vegetation patterns observed with the NOAA-7 advanced very high resolution radiometer. Vegetatio 64: 3–14.

Greller, A. M. 1988. Deciduous forest. In: M. G. Barbour and W. D. Billings (eds.), North American Terrestrial Vegetation. Cambridge University Press, Cambridge, U.K. pp. 287–316.

Johnson, E. A. 1992. Fire and Vegetation Dynamics, Studies from the North American Boreal Forest. Cambridge University Press, Cambridge, U.K.

Keeley, J. E. and S. C. Keeley. 1988. Chaparral. In: M. G. Barbour and W. D. Billlings (eds.), North American Terrestrial Vegetation. Cambridge University Press, Cambridge, U.K. pp. 165–207.

Küchler, A. W. 1964. Manual to Accompany the Map Potential Natural Vegetation of the Conterminous United States. American Geographic Society, New York. Special Publ. 36.

Lowe, C. H. 1980. Arizona's Natural Environment. University of Arizona Press, Tucson, AZ.

MacMahon, J. A. 1988. Warm deserts. In: M. G. Barbour and W. D. Billlings (eds.), North American Terrestrial Vegetation. Cambridge University Press, Cambridge, U.K. pp. 231–264.

——— and F. H. Wagner. 1985. The Mojave, Sonoran and Chihuahuan Deserts of North America. In: M. Evenari, I. Noy-Meir, and D. W. Goodall (eds.), Hot Deserts and Arid Shrublands, A. Elsevier Science Publishers, Amsterdam. pp. 105–202.

McClaran, M. P. and T. R. Van Devender (eds.). 1995. The Desert Grassland. University of Arizona Press, Tucson, AZ.

Merriam, C. H. 1890. Results of a biological survey of the San Francisco Mountain region and desert of the Little Colorado, Arizona. U.S.D.A., N. Am. Fauna 3: 1–136.

Miller, R. F. and P. E. Wigand. 1994. Holocene changes in semiarid pinyon–juniper woodlands. BioScience 44: 465–474.

Myers, R. L. and J. J. Ewel (eds.). 1990. Ecosystems of Florida. University of Central Florida Press, Orlando, FL.

Odum, W. E. and C. C. McIvor. 1990. Mangroves. In: R. L. Myers and J. J. Ewel (eds.), Ecosystems of Florida. University of Central Florida Press, Orlando, FL. pp. 517–548.

Osmond, C. B. and L. F. Pitelka (eds.). 1990. Plant Biology of the Basin and Range. Ecological Studies 80. Springer-Verlag, Berlin.

Peet, R. K. 1988. Forests of the Rocky Mountains. In: M. G. Barbour and W. D. Billings (eds.), North American Terrestrial Vegetation. Cambridge University Press, Cambridge, U.K. pp. 63–101.

Samson, F. and F. Knopf. 1994. Prairie conservation in North America. BioScience 44: 418–421.

Shreve, F. and I. L. Wiggins. 1964. Vegetation and Flora of the Sonoran Desert. Stanford University Press, Stanford, CA.

Shugart, H. H., R. Leemans, and G. B. Bonan (eds.). 1992. A Systems Analysis of the Global Boreal Forest. Cambridge University Press, Cambridge, U.K.

Sims, P. L. 1988. Grasslands. In: M. G. Barbour and W. D. Billings (eds.), North American Terrestrial Vegetation. Cambridge University Press, Cambridge, U.K. pp. 265–286.

Spicer, R. A. and J. L. Chapman. 1990. Climate change and the evolution of high-latitude terrestrial vegetation and floras. Trends Ecol. Evol. 5: 279–284.

Stuckey, R. L. and K. J. Reese (eds.). 1981. The Prairie Peninsula—in the "shadow" of Transeau. Proc. of the 6th North American Prairie Conference, Ohio State University, Columbus, OH, 1978. Ohio Biological Survey, Columbus, OH.

Suffness, M. (ed.). 1995. Taxol, Science and Applications. CRC Press, Boca Raton, FL.

Transeau, E. N. 1935. The prairie peninsula. Ecology 16: 423–437.

Tucker, J. M. 1961. Studies in the *Quercus undulata* complex. I. A preliminary statement. Am. J. Bot. 48: 202–208.

Vankat, J. L. 1979. The Natural Vegetation of North America, An Introduction. Wiley, New York.

West, N. E. 1988. Intermountain deserts, shrub steppes, and woodlands. In: M. G. Barbour and W. D. Billings (eds.), North American Terrestrial Vegetation. Cambridge University Press, Cambridge, U.K. pp. 209–230.

Wright, H. A. and A. W. Bailey. 1982. Fire Ecology: United States and Southern Canada. Wiley, New York.

General Readings

Abrams, M. D. 1992. Fire and the development of oak forests. BioScience 42: 346–353.

Chapin, F. S. III and C. Körner (eds.). 1995. Arctic and Alpine Biodiversity: Patterns, Causes, and Ecosystem Consequences. Springer-Verlag, Berlin.

Cohn, J. P. 1996. The Sonoran Desert. BioScience 46: 84–87.

Gower, S. T. and J. H. Richards. 1990. Larches: deciduous conifers in an evergreen world. BioScience 40: 818–826.

Gregory, S. V., F. J. Swanson, W. A. McKee, and K. W. Cummins. 1991. An ecosystem perspective of riparian zones. BioScience 41: 540–551.

Joern, A. and K. Keeler. 1994. The Changing Prairie. Oxford University Press, Oxford U.K.

Kareiva, P. 1996. Diversity and sustainability on the prairie. Nature 379: 673.

Mlot, C. 1990. Restoring the prairie. BioScience 40: 804–809.

Solbrig, O. T., E. Medina, and J. F. Silva (eds.). 1996. Biodiversity and Savanna Ecosystem Processes: A Global Perspective. Springer-Verlag, Berlin.

Van Cleve, K. et al. (+ 5 authors). 1983. Taiga ecosystems in interior Alaska. BioScience 33: 39–44.

Whelan, R. J. 1995. The Ecology of Fire. Cambridge University Press, Cambridge, U.K.

Notes

1. For patterns of vegetation evident from the Advanced Very High Resolution Radiometer, a sensor on U.S. National Oceanographic and Atmospheric Administration meteorological satellites, see Goward et al. (1985).

2. The most northerly forest in the world is a small patch of conifers near 72° 30' N near Ary-Mas in northern Russia (Spicer and Chapman, 1990).

3. Among the prominent organizations effective in the preservation of these forests and other North American biotas are the following:

The Nature Conservancy
1815 North Lynn Street
Arlington, VA 22209

Save-The-Redwoods League
114 Sansome Street
San Francisco, CA 94104

Sierra Club
730 Polk Street
San Francisco, CA 94109

CHAPTER TWO

Cause and Effect

Factors Influencing the Composition and Distribution of North American Plant Formations through Late Cretaceous and Cenozoic Time

The arrangement of vegetation over the landscape is a product of interactions between the environment, the ecological characteristics of individual organisms, barriers, dispersal potential, epidemic disease, anthropogenic influences, and the partially serendipitous factor of propagule availability. Within the complex of environmental factors, several are of special importance in tracing the history of North American plant communities. They include climate; plate tectonics as a mechanism for orogeny, volcanism, land bridges, and terranes; and catastrophes. Each have numerous interacting subcomponents, feedbacks, and amplifiers, and although constraints of format make it necessary to discuss these separately and sequentially, they are interconnected and pertubation of one affects the entire system.

Diagrams summarizing these factors are presented at the end of the following sections. The diagrams are not intended as models for, indeed, the single factor of climate could be expanded into a component so vastly complex that it would be counterproductive to a general summary. Similarly, the hydrological cycle, which involves the largest movement of any substance on Earth, cannot be fully treated because a "systems" view of its role in influencing climate is not available (Chahine, 1992) and the roles of water vapor (a greenhouse gas) and cloud cover are just now being quantified (Cess et al., 1995; Ramanathan et al., 1995). Rather, the diagrams illustrate some of the factors and relationships discussed in the text and serve as a reminder of the complex interactive nature of physical and biotic events.

CLIMATE

Plants are limited in their ecological amplitude. Several important corollaries follow from this observation; one of the most fundamental is that changes in climate cause extinctions, promote evolution, and alter the range and habitats of organisms. Because climate plays a central role in the arrangement of modern communities (Gates, 1993; Kareiva et al., 1993; Woodward, 1987) and in the development and distribution of past assemblages (Brenchley, 1984; Crowley and North, 1991; Hecht, 1985a), reference to some elements of general climatology is necessary for understanding the diversification, radiation, and reshuffling of North American paleocommunities during the Late Cretaceous and Cenozoic.

Some General Aspects of Climate

Climate is the long-term weather of a region; it consists of a plexus of dynamic, interrelated elements that are collectively responsible for precipitation, temperature, and atmospheric and oceanic circulation[1] (Barry and Chorley, 1992). Variations in these elements determine the different regional climates, the seasonal changes in climate within a region, and fluctuations in climate over time. The driving force behind regional differences and temporal changes is insolation, which is the amount of solar radiation or heat reaching the Earth's surface.

The spatial distribution of heat varies because of the curvature of the Earth: a given amount of heat is dispersed over a greater surface area at the poles than at the equator (Fig. 2.1). In addition, solar rays impinge on the Earth at a more oblique angle near the poles, resulting in greater scattering. At the equator the amount of solar energy received at the top of the atmosphere is 311,520 cal/cm^2/year, while near the poles it is less than half this amount (130,300 cal/cm^2/year; List, 1951, table 133). Consequently, any change in insolation has a magnified effect on polar climates compared to equatorial climates, as evidenced by the distribution and extent of glaciers.

Annual changes, or seasons, are a result of the tilt of the

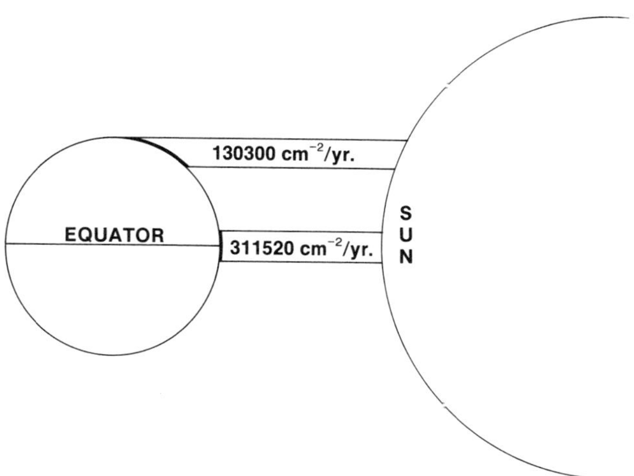

Figure 2.1. Effect of the Earth's curvature on the distribution of equal amounts of heat from the sun at the equator and at the poles, as represented by the widths of the two sets of lines and the length of the dark bars, respectively. Any change in incident radiation will have a magnified effect at the poles compared to the equatorial regions.

Earth during its orbit around the sun (Fig. 2.2). The tilt is presently 23.5° from the orbital plane, so that 1 day of the year, June 21 (the northern hemisphere summer), the sun's rays are at the northernmost point where they strike the Earth perpendicularly (the summer solstice, Fig. 2.2A). This latitude is marked by the Tropic of Cancer at 23.5° N, and it crosses North America at a point just south of Matehuala, Mexico, westward toward Mazatlan. On December 21 (the winter solstice), at a point in the orbit opposite the June 21 position (Fig. 2.2B), it is the southern pole that is inclined toward the sun, and its rays reach the southernmost point where they strike the Earth perpendicularly (the southern hemisphere summer). This latitude is marked by the Tropic of Capricorn at 23.5° S, and it crosses South America between Rio de Janeiro and São Paulo westward toward Antofagast in northern Chile. The term solstice refers to the two times of the year when the perpendicular rays of the sun reach these most distant latitudes and momentarily have no apparent northward or southward motion. The related event *equinox* (Latin for equal night) refers to the two times of the year when the sun crosses the equator (September 22 and March 21, Fig. 2.2C,D). There are 12 h of day and night at the equator all year; at the equinox in the midlatitudes there are about 15 h of day in the summer and 9 h in the winter, and at the solstices in the high latitudes about half of the days have nearly 24 h of light and about half have nearly 24 h of darkness. Clearly, any change in the tilt of the Earth's axis would affect not only insolation and consequently the various elements of climate, but also influence the fundamental physiological process of photoperiod and induce changes in both the composition and distribution of plant formations. As will be seen later, the tilt of the Earth has changed over time, and that has played an important role in the history of the Earth's biota.

The annual progression of the direct rays of the sun over the Earth's surface produces a differential warming of the atmosphere that contributes to atmospheric circulation. The equatorial region is the warmest latitudinal zone because it is at or near a right angle to the sun's rays throughout the year. This warm equatorial air rises, cools, and

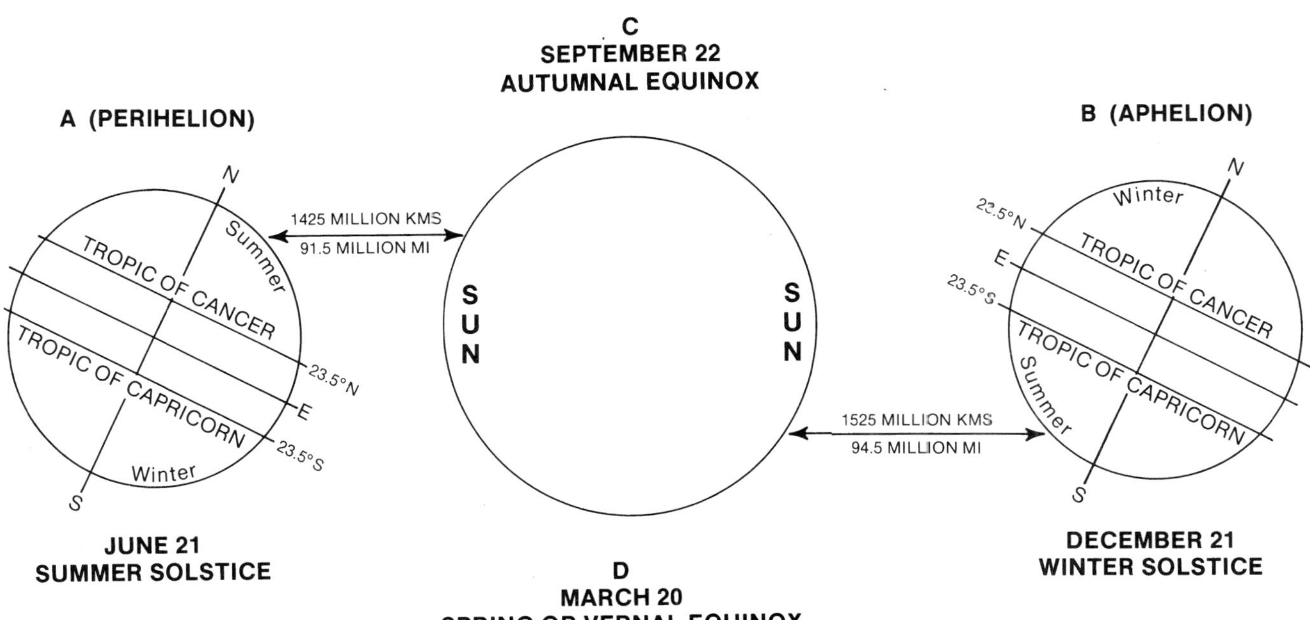

Figure 2.2. Position and present orientation of the Earth relative to the sun at the summer and winter solstices. Presently, the perihelion is January 4 and the aphelion is July 5. The solstices do not correspond to the perihelion and aphelion.

loses its moisture; hence, the equatorial zone has high temperatures, large amounts of annual rainfall, and tropical vegetation. At about 30° N and 30° S latitudes, across northern Mexico and southernmost Brazil, respectively, the cooled dry air begins to descend, warming as it approaches the Earth's surface, and creates the high-pressure systems that characterize these latitudes. The result is low precipitation, high evapotranspiration potential, and desert vegetation. This north–south circulation pattern of the equatorial and midlatitudes is known as the Hadley regime after British lawyer George Hadley, who in 1735 wrote an early work on the northeast trade winds. The Sonoran Desert exists, in part, because it is located beneath the descending arm of the Hadley convection cell.

At about 60° N and 60° S latitudes, from the southern coast of Alaska across to mid-Hudson Bay and through the Drake Passage between South America and Antarctica, respectively, the reheated air begins to rise again and at the poles descends to contribute to the climatic environments of the polar deserts. However, Hadley regimes are weaker outside the equatorial zone and give way to a different primary form of atmospheric circulation. At the mid- to high latitudes the circulation has less the form of a convection cell, but rather it occurs as two to five superimposed east-trending waves called the Rossby regime (after Swedish-American meteorologist Carl Rossby). The separation between the Hadley and Rossby regimes is around 30° N and 30° S latitudes. The Rossby regime comprises waves of long amplitude upon which are superimposed shorter waves (Trewartha and Horn, 1980, fig. 5.3). The lows and highs shown on daily weather maps in the high northern and southern latitudes generally correspond to the troughs and ridges of the Rossby regime. Within these westerly winds (viz., from the west) there are zones of faster flowing air. At an altitude of ~12 km the average flow is 50–100 km/h (30–60 mi/h), but portions have air speeds of 150–300 km/h (90–180 mi/h). These cores of faster flowing air are the polar and subtropical jet streams. Cyclones (zones of low pressure with associated precipitation) and fronts (zones of weather instability and steeply gradient temperature at the interface between highs and lows) develop beneath the jet streams and are carried eastward by the westerly flow of air. The jet streams shift south in the winter, bringing cold Arctic air into the midlatitudes, and north in the summer. For example, a subtropical high near 40° N in July shifts to near 25° N in January. Consequently, any factor that alters the nature and course of atmospheric circulation and the seasonal meandering of the jet streams will affect regional climates and vegetation. As will be seen later, orogeny (mountain building) and ice sheets of substantial height are two such factors; they have also played an important role in the history of the Earth's climate and biota.

In addition to the vertical patterns of wind distribution, there are horizontal patterns created by gravity, rotation of the Earth, frictional drag between the air and the Earth's surface, and atmospheric pressure, which is the weight of air per unit area [14.7 lb/in.^{2}, or 1013.3 mb (millibars) at sea level]. Atmospheric pressure is the most fundamental because it is the factor that initially sets air into motion. Air flows from zones of high pressure at about 30° N and 30° S latitudes into zones of low pressure at the equator and at about 60° N and 60° S latitudes (Fig. 2.3). Isobars connect areas of similar atmospheric pressure, and the present distribution of winter and summer highs and lows as outlined by isobars is shown in Figs. 2.4 and 2.5.

The direction of the wind is affected at the surface by the Coriolis Effect (after French physicist and mathematician Gaspard de Coriolis who described it in 1844), a deflecting force generated by the spinning of the Earth. This force deflects winds circulating around highs and lows either to the east or west to create the easterlies, westerlies, and trade winds in the idealized pattern shown in Fig. 2.3B. Within regions of constant high pressure, air movement is relatively static, forming a zone known as the Horse Latitudes; these are two belts located over the oceans (at about 30–35° N and S latitudes) that have high barometric pressure, calms, and light variable winds (Fig. 2.3A).

In the Pacific, systems of high and low pressure (the southeast and northeast trade winds) meet along a zone of generally low pressure and cloudiness just north of the equator called the intertropical convergence zone (ITCZ, Fig. 2.3B). Its position is seasonally between 10° N (August and September) and 3° N (February and March). The location of the ITCZ is one factor reflecting global climatic processes, as evidenced by events associated with a shift of only a few degrees latitude.

El Niño–Southern Oscillation Events

Although the relationship is not well understood, a southward displacement of the ITCZ to near or just below the equator is an early signal for the development of an El Niño–Southern Oscillation (ENSO) event (Diaz and Markgraf, 1992; Philander, 1983). The effects of the Southern Oscillation component are reciprocal between the Pacific and Indian Oceans (e.g., a change in wind velocity that produces higher sea levels in one region corresponds to lower sea levels in the other, hence, oscillation). An El Niño is the eastern Pacific and coastal South American component of an ENSO event that begins with an overall warming and consequent atmospheric disturbance in the equatorial Pacific. [For a color illustration based on satellite thermography, see Holden (1992)]. Technically, it is defined as the appearance of anomalously warm water along the coast of Ecuador and Peru as far south as Lima (12° S), during which a normalized sea surface temperature (SST) anomaly exceeding 1 standard deviation occurs for at least four consecutive months at three or more of the following five coastal stations: Talara, Puerto Chicama, Chimbote, Isla Don Martin, and Callao [Quinn et al., 1987; Scientific

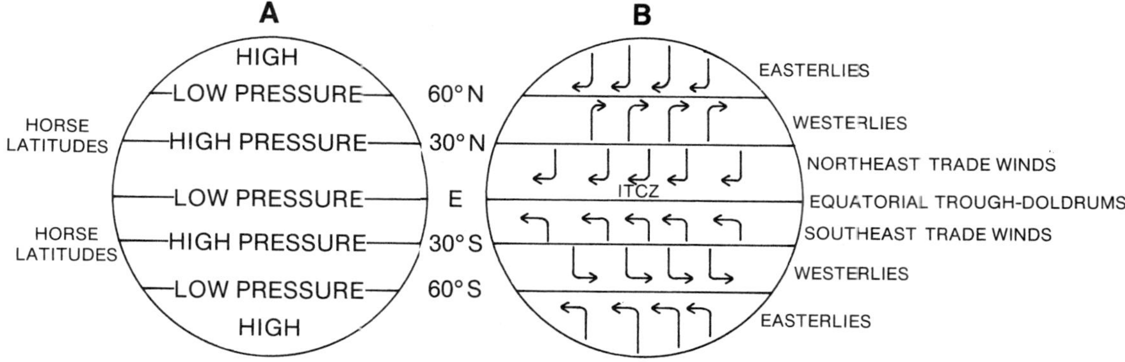

Figure 2.3. Schematic distribution of idealized high- and low-pressure systems and corresponding wind directions.

Committee on Oceanic Research (SCOR), 1983]. The name was coined by Peruvian fishermen because the event generally occurs around the Christmas season. The occurrence of El Niños varies, but they commonly appear every 3–7 years. They begin in the northern hemisphere spring, last from 1–2 years, and alternate with normal or cooler than normal episodes called La Niñas. There is an initial warming of equatorial eastern Pacific waters from 1 to 2°C to 5°C or more above average that displaces the track of the subtropical jet stream and weakens the tropical easterly trade winds off the western coast of South America. This allows warm ocean currents to flow more strongly eastward, disrupting the cold Humboldt Current, increasing atmospheric moisture, and producing climates not only atypical for western South America but also for distant parts of the world, including North America. Among the climatic effects of a typical ENSO event are droughts in Australia, Indonesia, and southern Africa, and the bleaching of coral reefs due to a westerly shift in wind direction that lowers sea level in the west and raises it in the east (Glynn and de Weerdt, 1991). There are excessive rains along the west coast of South America and in the southwestern United States (viz., the Texas to southern California floods of early 1992, flooding in the midwest in 1993), unusually cold winters in northeastern North America, and unusually dry warm winters in the midcontinent region and the west. For example, during non–El Niño years the snowfall in Cleveland, Ohio, is ~100 in./yr, while during an El Niño, it is half that amount. The 1982–1983 event (Glynn, 1990; Table 2.1), one of the most intense oceanographic and cli-

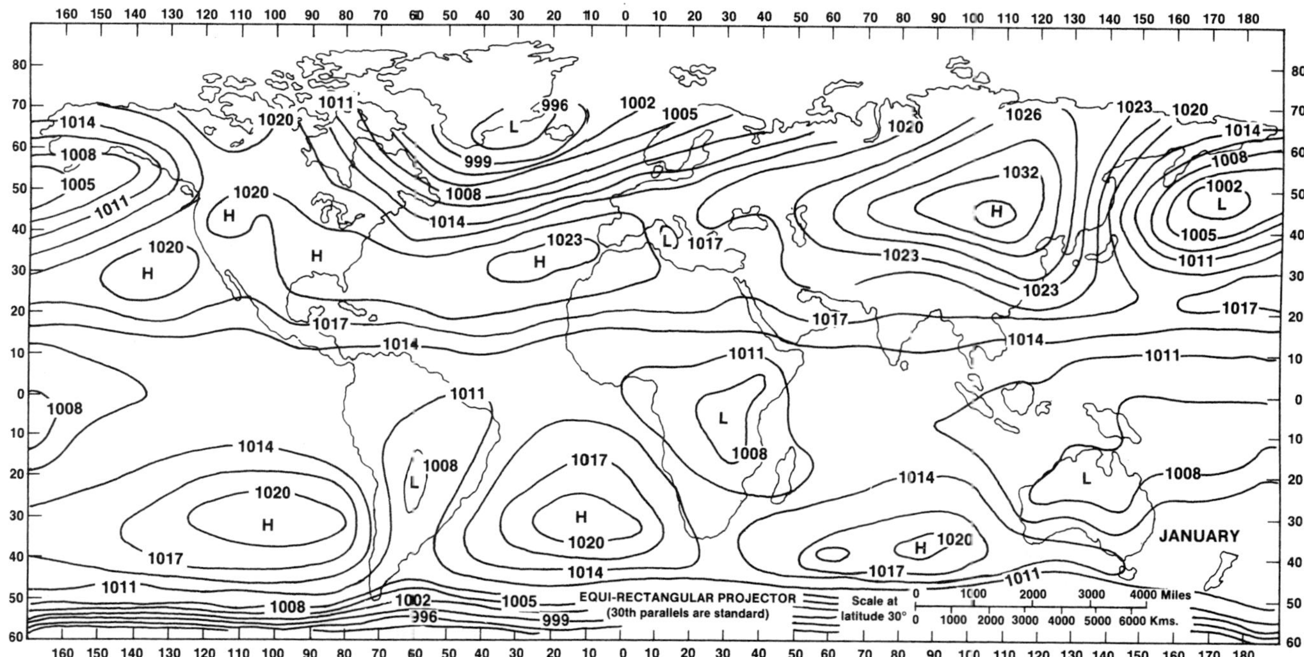

Figure 2.4. Distribution of January high- and low-pressure systems. Adapted from Trewartha and Horn (1980) and based on Mintz and Dean (1952).

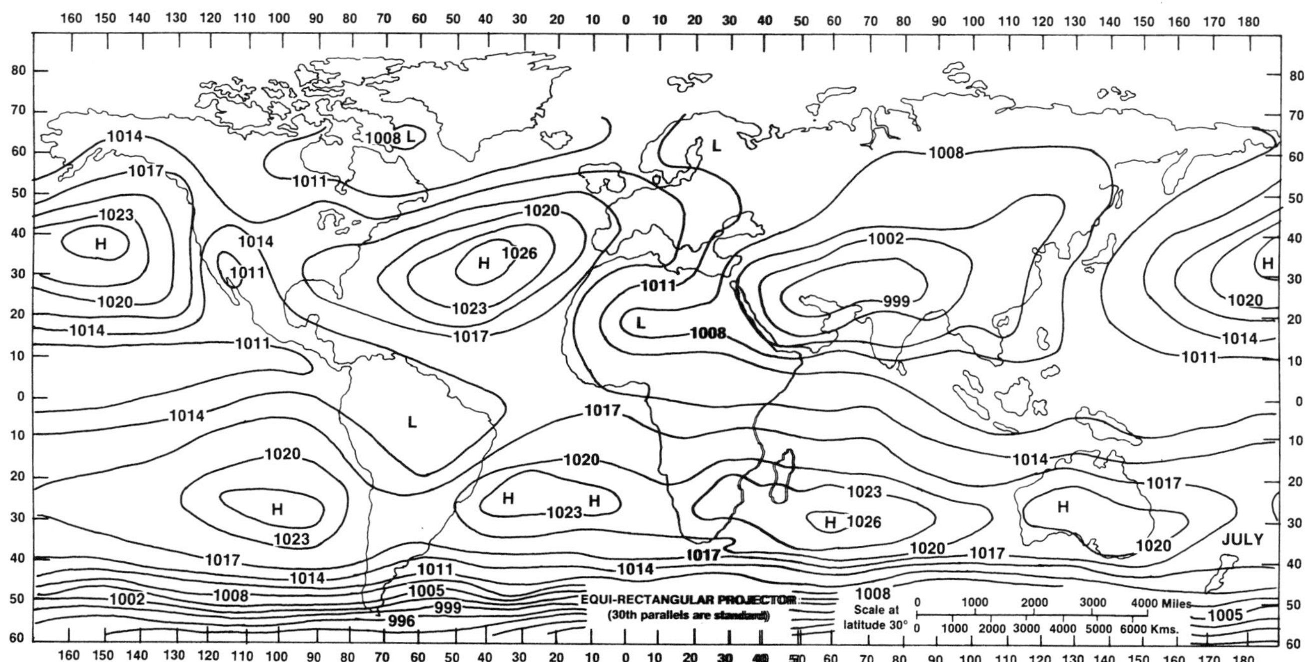

Figure 2.5. Distribution of July high- and low-pressure systems. Adapted from Trewartha and Horn (1980) and based on Mintz and Dean (1952).

matic disturbances in the Pacific in the past 100 years, was anomalous because the effects were evident in May rather than in the fall and occurred in the central and western Pacific rather than along the west coast of South America. Rainfall increased sixfold to 2540 mm (100 in.) over an 8-month period, converting the coastal desert of Ecuador and northern Peru into a lake-dotted grassland. The El Niño of 1997–1998 is predicted to be comparable in intensity to the 1982–1983 event. A strong El Niño produces a global warming of ~0.1–0.2°C. The cause(s) of ENSO events are not known, but there is evidence to suggest a relationship with variations in the rate of the Earth's rotation that may produce cyclic changes in the atmospheric and oceanic distribution of heat and moisture (Hide and Dickey, 1991). The economic impact of an El Niño is evident by the extensive media coverage, such as CNN's Financial Market Reports on the Web during September–October, 1997: "Markets weather El Niño—onset of phenomenon blows more volatility into commodities trading; El Niño hits nickle producer; El Niño hits coffee crop; Making El Niño into dinero—Small firms still profit from storm pattern." The importance of El Niños to short-term weather is further witnessed by the current 40-nation effort to establish an international research center to forecast and predict the impact of El Niños on climate (Macilwain, 1995).

Eventually it may be possible to detect El Niños in times preceding historical records. Long-chain organic molecules such as alkenones enter the sedimentary record as biochemical fossils. In coccolithophorid algae such as *Emiliania huxleyi*, the degree of alkenone unsaturation has been shown to be temperature dependent (increasing unsaturation as the water cools). If under favorable conditions alkenone patterns are preserved in sediments for long periods of time, ocean temperature changes could be resolved on a near-annual basis (Kennedy and Brassell, 1992). The irregularity (chaos) in the appearance and duration of El Niños is also being investigated (Jin et al., 1994; Tziperman et al., 1994), but presently it cannot be explained.

Table 2.1. Chronology of strong (S) to very strong (VS) ENSO events in nineteenth and twentieth centuries.

1803–1804 (S+)	1877–1878 (VS)	1925–1926 (VS)
1814 (S)	1884 (S+)	1932 (S)
1828 (VS)	1891 (VS)	1940–1941 (S)
1844–1845 (S+)	1899–1900 (S)	1957–1958 (S)
1864 (S)	1911–1912 (S)	1972–1973 (S)
1871 (S+)	1917 (S)	1982–1983 (VS)

Adapted from Quinn et al. (1978, 1987); see also Mass and Portman (1989).

THE IDEALIZED PATTERNS of temperature, rainfall, and wind direction are modified locally by three geographic features of the Earth's surface: topography, large bodies of water, and large masses of land. Large-scale topographic effects are induced by the presence and orientation of mountain systems and plateaus. The windward (west) side of the western mountains of North America receives moisture from the westerlies off the Pacific Ocean (orographic precipitation), while a rainshadow is created to the east (leeward) side. Rainshadows occur to the east of the Sierra Nevada, contributing to the development of the arid Basin and Range Province vegetation, and to the east of the Rocky Mountains, contributing to the development of the High Plains and Great Plains woodland and grassland vegetation. In addition to these regional effects, topography on a smaller scale is important to local climates. The eastern suburbs of Cleveland, Ohio (e.g., Lorain), receive ~1000 mm (40 in.) of snow each year. Fifty miles to the east at Chardon, the annual snowfall is 2690 mm (106 in.) and the seasonal maximum is 4100 mm (161 in.; Schmidlin, 1989). Winds circulating clockwise around Canadian high-pressure systems bring cold Arctic air across Lake Erie, which is shallow and normally does not freeze until February. The lake provides a source of moisture and a warmer surface than the frozen land. Snow is generated when the warmed, moisture-laden winds off the lake cross back onto colder land on the southern shore. This system combines with an elevational difference of 230 m (690 ft) to produce up to 4 times more snow in Chardon than in Lorain, which are at the same latitude and only 50 mi apart.

Another local effect of topography, in combination with other factors, is the generation of the Santa Ana winds of southern California. The winds develop around a high-pressure anticyclone system, flow over the Sierra Nevada, and gain heat on the western side as they subside. Additional heat is gained from crossing the warm lower elevations. The velocity of the winds increases during the fall and winter months when clockwise circulating winds to the north converge either with counterclockwise circulating winds to the south or with less intense highs. These pulses of strong winds gain additional speed as they are funneled through the canyons of the Sierra Nevada. The result is a strong, hot, dry, dusty winter wind that contributes significantly to the extent and intensity of forest fires in southern California, to high evapotranspiration, and to the maintenance of arid shrubland/chaparral–woodland vegetation. Prior to significant elevations in the Sierra Nevada (~10 Ma), these seasonal winds did not develop, which was one of the altered boundary conditions that sustained more mesic vegetation in the region.

Large bodies of water buffer surrounding terrestrial habitats against extreme annual temperature fluctuations. Oceans receive about 80% of the total solar radiation reaching the Earth's surface. Marine waters have a specific heat of 0.93 cal/g/°C, compared to that of land with 0.19–0.60 cal/g/°C, and yield heat more slowly than land. Thus, oceans serve as vast reservoirs of heat that maintain climatic stability (Berggren 1982). The mean annual temperature (MAT) of marine waters ranges from ~0°C at the ocean bottom to ~28°C at the equatorial surface, and it usually varies annually by only a few degrees. In contrast, MAT on land ranges from approximately −40 to 30°C and can vary annually by 50°C. The extensive epicontinental sea on the North American continent during the Late Cretaceous was a significant factor in producing the equable maritime climates and low thermal gradients characteristic of that period.

In addition, oceans and large lakes are responsible for sea and land breezes that can modify local climate over short distances. At sunrise land surfaces warm faster than adjacent bodies of water and a rising flow of air is generated. Cool air from over the water is drawn inland and may decrease average daytime temperatures by several degrees. For example, the daily mean maximum July temperature at Milwaukee, Wisconsin, on the south shore of Lake Michigan is 25°C (77°F) while at Madison, which is 125 km (75 mi) inland, it is 28°C (82°F).

Large masses of land alter the idealized global zonation of climate by setting the broad pattern of ocean circulation and by producing extremes of temperature and precipitation characteristic of interior continental climates. The withdrawal of the epicontinental sea during the Cenozoic promoted greater seasonal temperature extremes (continentality) in the interior of North America.

Another consequence of wind patterns induced by insolation regimes are ocean surface currents (Fig. 2.6). High- and low-pressure systems, prevailing winds, the position and configuration of the continents, and the spinning motion of the Earth produce gyres—great circulating systems of surface ocean water (Pedolosky, 1990). A factor further contributing to the motion of marine water is the difference in salinity in different parts of the ocean. There is less oceanic rainfall under the descending branch of the Hadley cell, with consequent higher salinity and greater evaporation from marginal seas (Price et al., 1993). The center of the gyres marks the position of semipermanent, seasonally variable, high-pressure systems; the spinning of the Earth produces the familiar clockwise circulation in the northern hemisphere and the counterclockwise motion in the southern hemisphere about these columns of air, and these motions are transferred to the surface of ocean waters. The southern limits of the northern hemisphere gyres are defined by the seasonally migrating ITCZ, the northern limits by the high latitude easterlies, and the east–west limits by the location and configuration of the continents (Berggren, 1982). The circulating motion of the gyres pulls warm water from the equatorial region and cold water from the poles, contributing to the global distribution of heat, and is responsible for many features of climate, and especially noticeable along continental coastlines. Where cold currents flow along the coast, such as the Humboldt or Peru Current along the western coast of South America, the cool dry winds that blow onto the land extract moisture from the atmosphere and contribute to the formation of the dry climate and vegetation of the Chilean Atacama Desert. Studies of diatomites from western South America show that upwellings developed at ~39 Ma, indicating that a proto-Humboldt Current was in existence by that time. Upwellings are an upward movement of cold, nutrient-rich bottom waters, and they appear during periods of increased circulation of surface waters. These increases occur with the opening and closing of land bridges, such as the Drake Passage between South America and Antarctica,

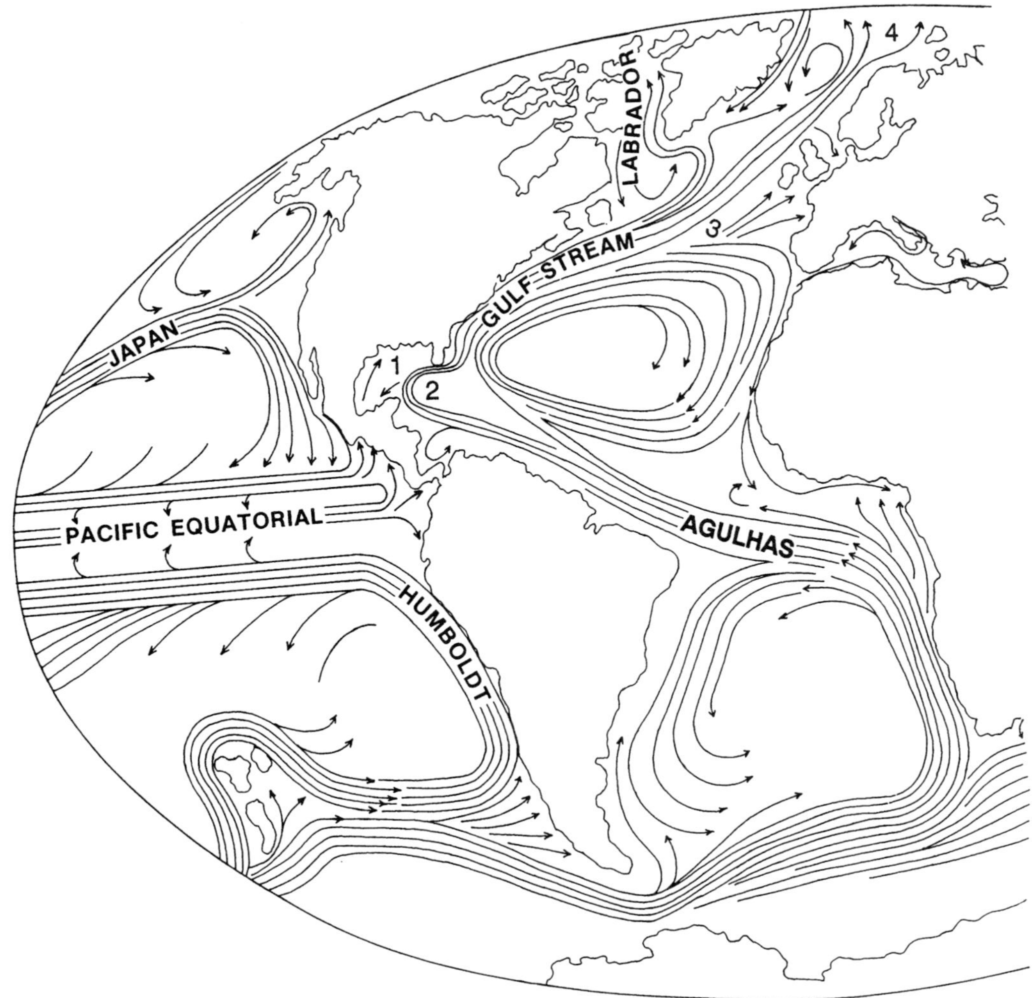

Figure 2.6. Principal ocean currents affecting North American climates: (1) East Gulf loop, (2) Florida current, (3) North Atlantic drift, (4) east Greenland current.

and across the Isthmus of Panama. Upwellings produce blooms of marine organisms in response to increased nutrient levels, and these blooms are preserved in ocean sediments. The South American diatomites correspond to a drop in sea level at ~40 Ma, which probably represents early glaciation on Antarctica and the initiation of cold South Atlantic Deep Water (SADW) production.

Where the current is warm, such as the Gulf–Agulhas current flowing into the western North Atlantic Ocean, the effect on local climates is quite different. About 30 million m^3 of seawater are transported through the Straits of Florida into the North Atlantic each second (annual mean). The Agulhas Current of the Indian Ocean flows westward around the tip of Africa and feeds considerable amounts of warm water into the Gulf Stream (Fig. 2.6; Bird, 1991). The average temperature of this water is 17.4°C. These Western Boundary Currents (WBCs) displace climatic zones poleward by 5–10°. The presence of tropical zooxanthellate corals (those with dinoflagellate symbionts) in Bermuda (latitude 33° N) is one biogeographic consequence of the current. Farther north, westerly winds blowing across these waters are warmed; by the time they reach the coast of England [average cold month temperature of 5.5°C (41°F)], this sea to air temperature exchange produces a maritime climate very different from the continental climate in Labrador [average cold month temperature of −25°C (−13°F)], which is at the same latitude (51.5° N). The North Atlantic current also receives an influx of highly saline water from the Mediterranean which increases the density of the water. The cooled, saline-dense current begins to sink in the Labrador and Nordic Seas to form North Atlantic Deep Water (NADW) that flows southward at depths greater than 1000 m, eventually reaching the east coast of South America. Compared to the 17.4°C temperature of the north-flowing surface water, the returning south-flowing water has an estimated average temperature of 5.4°C (Covey and Barron, 1988). At the tip of South America it trends east along the Antarctic coast, then north to join the warm equatorial Agulhas Current that feeds back into the Gulf Stream. This conveyor belt system has a period of ~1600 years. Intermittent NADW production began in the early Oligocene, suggesting cooler climates and further glaciation on Antarctica, and reached a pattern typical of modern times between 10 and 7 Ma. The Isthmus of Panama began closing about this time with the formation of a submerged sill that deflected and increased the flow of the Gulf Stream. The Isthmus was essentially closed by ~3.5 Ma, at which time upwellings increased along the western Atlantic coasts. In the present configuration of the continents, a part of the Agulhas Current is deflected northward by the projection of Brazil (Fig. 2.6) and it augments heat transport by the warm Gulf Stream. A slightly eastward position of South America, as during the Tertiary, would cause a greater southern deflection, which would alter ocean circulation and the global distribution of heat. Seaways and land bridges across the North Atlantic Ocean and North Pacific Ocean, through Panama, and between South America and Antarctica were altered during the Cenozoic. These changes had a significant effect on ocean circulation, climate, and the history of the Earth's biota.

In the North Pacific less heat transport occurs, and the effect on regional temperatures is quite different from that in the North Atlantic. Bodö, Norway, has a January average temperature of −2°C and a July average of 14°C. In contrast, the January and July values at Nome, Alaska, at a similar latitude but off the Pacific Ocean are −15 and 10°C, respectively.

The temperature, precipitation, wind, and ocean currents generated by insolation and modified by the position and configuration of continents and their topography produce the climates typical of different regions of North America. Within each component of climate the parameters most important to the composition and distribution of plant formations are the extremes, as in temperature and precipitation, because it is the extremes that interact with the ecological tolerances of individual organisms to determine potential distributions. The MATs of Des Moines, Iowa, and Oakland, California, are ~11 and ~13°C, respectively; but the annual range at Oakland is 9.5°C, and at Des Moines it is 30.4°C, and the two regions support completely different potential natural vegetation. Average sea-level temperatures for January and July in North America are shown in Figs. 2.7 and 2.8, respectively, and precipitation for the same months are shown in Figs. 2.9 and 2.10, respectively. The values reflect the extremes in yearly variations with which all organisms in a given region must contend and be physiologically capable of surviving. It is interesting to note that present mean seasonal extremes are greater than the MAT difference between the glacial and interglacial phases of the Pleistocene (Crowley and North, 1991). The interaction of climatic extremes and ecological tolerances, modified by other more local factors of the environment and depending on propagule availability, produce patterns in the distribution of plant species. The particular patterns determined by present conditions are shown on various vegetation maps (Barbour and Billings, 1988; Barbour and Christensen, 1993; Küchler, 1964) and on the summary diagram in Fig. 1.1.

It is evident from the preceding discussion that these patterns are tenuous and that a change in any one of the numerous factors that determine climate will, in turn, influence the composition and arrangement of plant formations and associations. It may seem that the changes evident over geologic time could be attributed primarily to known fluctuations in the amount of radiation emanating from the sun because insolation is the source of energy determining the Earth's climate. At a 150 million km (93 million mi) mean distance from the sun, a surface oriented perpendicular to the sun receives ~2.0 cal/cm^2/min. This value is known as the solar "constant." In fact, there have been fluctuations in the amount of heat coming from the sun, and this may have been a factor in determining patterns of temperature during the Cenozoic (Haigh, 1996; Hoyt and Schatten, 1997; Kerr, 1996a, General Readings). However, changes in solar heat alone do not sufficiently explain the Earth's climatic history; there is presently no evidence to document that it has fluctuated in a fashion corresponding to the patterns of extinctions, evolutionary peaks, or the migrational history of the Earth's biota. Periodically, solar flares of ultraviolet radiation are trapped in the Earth's magnetic field and are concentrated above the north and south magnetic poles. At an elevation of about 100 km (60 mi), these charged particles collide with atmospheric gases to produce the aurora borealis and aurora australis. Although such variations in solar radiation are visually spectacular, they do not appear to cause any significant changes in global climate. [For a possible correlation between sun spot activity and short-term climate fluctuations, see Flam (1993), Friis–Christensen and Lassen (1991), Kerr (1995, 1996b)]. As noted by Covey (1996), "It would be foolish to assume that historical solar luminosity changes are insignificant. We should not, however, make the opposite error and jump to the conclusion that they explain all—or even most—of the climate variations of the past few centuries." If fluctuations in solar radiation alone do not correspond in detail to the history of the Earth's climate and biota, then other factors affecting insolation must be involved, such as variations in the position and orientation of the Earth relative to the sun. These variations have occurred over time, and they are a driving force behind global climatic change (Berger et al., 1984).

Milankovitch Variations

The Serbian astronomer and mathematician Milutin Milankovitch knew from previous studies (e.g., Croll, 1875) that cyclic changes in insolation occurred as a result of variations in the distance and orientation of the Earth relative to the sun. The variations derive from fluctuations in the gravitational pull on the Earth's equatorial bulge by the moon, sun, and planets in the solar system, notably Jupiter and Saturn, depending on their positions and distances

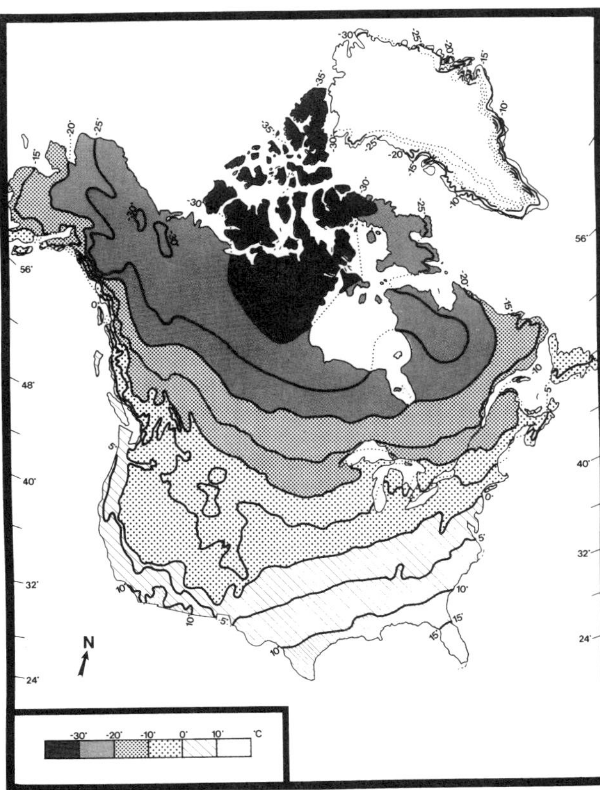

Figure 2.7. Mean January temperatures for North America. Based on the Climatic Atlas of North and Central America—Maps of Mean Temperature and Precipitation (World Meteorological Organization, 1979). Scale 1:10,000,000.

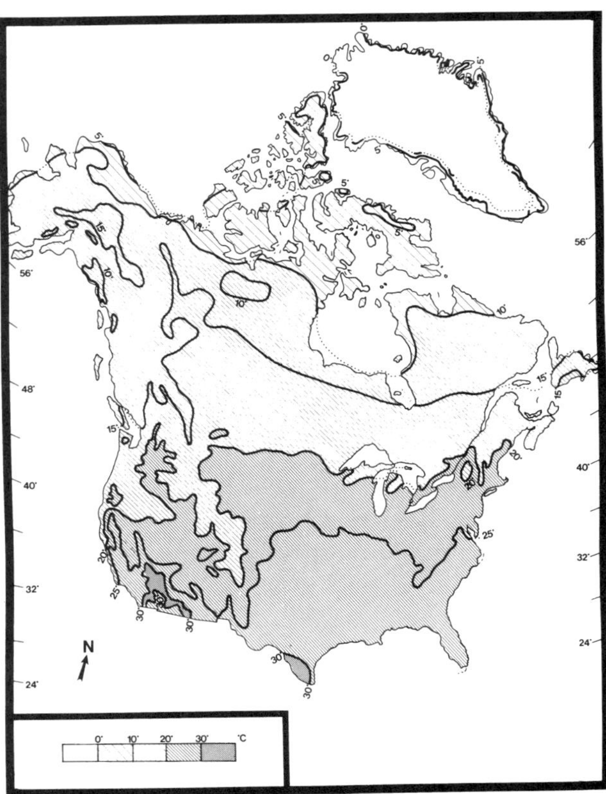

Figure 2.8. Mean July temperatures for North America. Based on the Climatic Atlas of North and Central America—Maps of Mean Temperature and Precipitation (World Meteorological Organization, 1979). Scale 1:10,000,000.

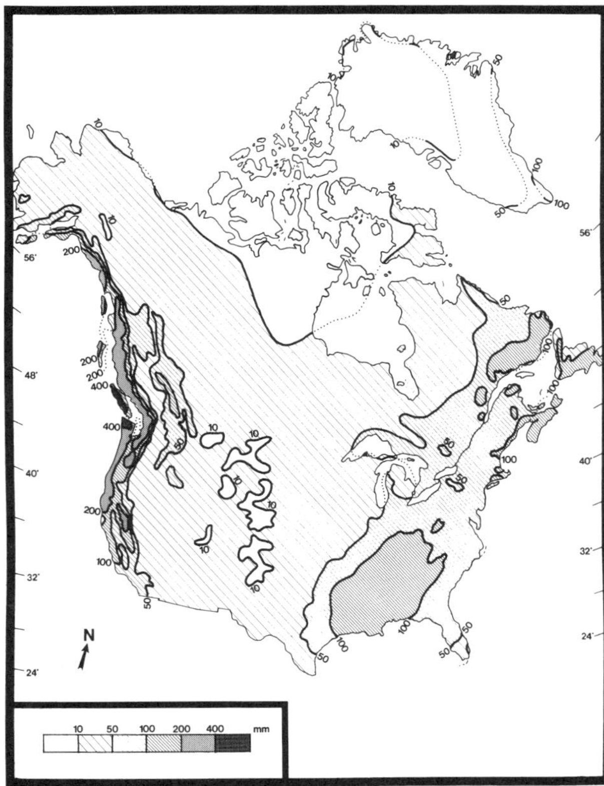

Figure 2.9. Mean January precipitation for North America. Based on the Climatic Atlas of North and Central America—Maps of Mean Temperature and Precipitation (World Meteorological Organization, 1979). Scale 1:10,000,000.

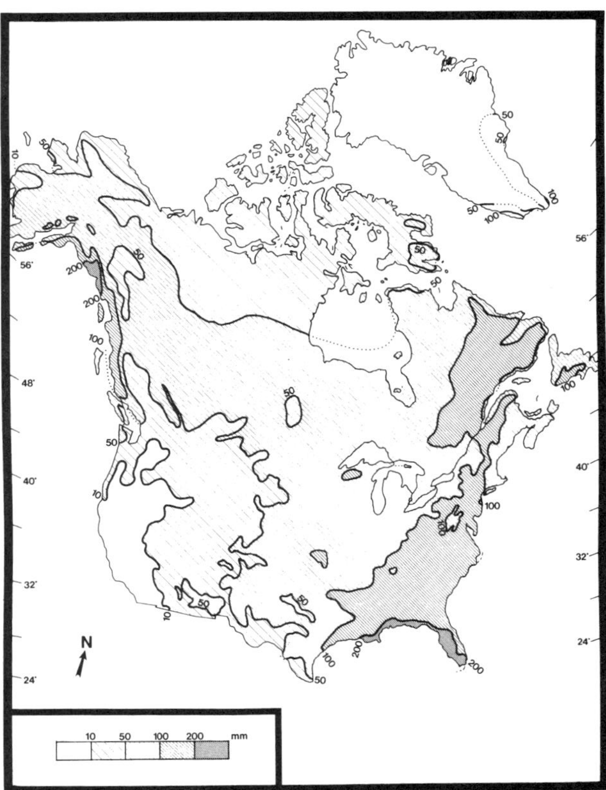

Figure 2.10. Mean July precipitation for North America. Based on the Climatic Atlas of North and Central America—Maps of Mean Temperature and Precipitation (World Meteorological Organization, 1979). Scale 1:10,000,000.

relative to the Earth. The spatial relationship between the Earth and sun varies in three ways. One is in the tilt of the Earth's axis or obliquity. The tilt is presently at 23.5° (Fig. 2.2), but it varies between 22.1° and 24.5°. For example, at 9 Kya the tilt was 24.2°. Obliquity affects the distribution of heat over the Earth's surface, particularly the insolation differential between the high and low latitudes (large effects at the poles, less toward the equator), and the intensity of the seasons. An increase in the tilt results in greater seasonal contrast (warmer summers and colder winters). For the northern hemisphere the maximum difference can be ~14°C (+7°C in July and −7°C in January). The period of the tilt variation is ~41,000 years.

The second variation is in the eccentricity of the elliptical orbit (Clemens and Tiedemann, 1997). If the Earth's orbit around the sun were a perfect circle, the eccentricity would be zero; if one dimension were twice as long as the other, the eccentricity would be 0.866. At present it is 0.017, and it varies between 0.005 and 0.06. The perihelion is the point where the Earth is nearest the sun, which is now reached on January 3 (Fig. 2.2); the aphelion is the point where it is furthest away, which is now reached on July 4. About 9 Kya the perihelion was July 30. Presently the Earth receives ~6% more heat at perihelion than at aphelion, while at maximum eccentricity the difference is 27%. Changes in eccentricity alternately intensify the seasons in one hemisphere and moderate them in the other. The period of the eccentricity variation is ~100 ka, and there is a secondary period at ~400 ka.

The third variation is in the circular movement or precession of the polar axes. As the Earth spins there is a slight movement in the direction toward which the polar axes point, so that over time they would inscribe a circle if projected onto a distant background. The torque exerted on the Earth by the moon and planets accounts for about two-thirds of the precession rate and that exerted by the sun accounts for the remaining one-third. At present the north pole points toward a star that is called the pole star or Polaris (the end star in the handle of the Little Dipper). In about 12 Kya it will point toward Vega. This precession is a superimposed, second-order factor that reinforces the intensity of the seasons, lessening it in one hemisphere (presently the northern hemisphere, with comparatively mild winters) and increasing it in the other (presently the southern hemisphere, with comparatively cold winters). In contrast to the tilt variable, the effect of changes in precession is small at the poles (~0.5%) and larger toward the equator. It takes ~23 Ky for the Earth's axis to inscribe a full imaginary circle as it precesses, and a secondary cycle takes ~19 Ky.

The three kinds of variations just described are euphemistically referred to as tilt, stretch, and wobble. Milankovitch (1920, 1930, 1941) calculated their individual and combined effects on insolation for each latitude over time. He concluded that when periods of low solar radiation coincide with minimal seasonal ranges in temperature (viz., greater summer accumulation of winter snow), glacial conditions develop. The combined effects of the three variations can vary summer insolation in the northern hemisphere by a maximum of ~20%. His calculations predicted glaciations every 100 Ky. When high insolation is matched with maximum seasonal ranges in temperature (viz., greater summer melting of winter snow), the glacial periods should end by these calculations for 10–12 Ky. Both cycles have been confirmed by a variety of marine and ice-core evidence (Imbrie et al., 1992, 1993; Schwarzacher, 1993; see following section). Milankovitch's original calculations have been reviewed by Berger (1980).

At different times in Earth history and at different latitudes each of the variations have dominated solar-controlled climatic patterns. Before 2.4 Ma the dominant cycles were 23 Ky and 19 Ky precessions, while after 2.4 Ma the 41 Ky obliquity cycle dominated. During the interval between 800 and 620 Kya, climatic systems in the high latitudes underwent a change from primarily 41 Ky obliquity forcing to the primarily 100 Ky eccentricity forcing of the present (Fischer, 1982; Shackleton et al., 1990).

In the years following Milankovitch's work there was limited acceptance of the theory for two principal reasons. One was that the scale of the theory, both temporal and spatial, made it impossible to test directly or to establish a cause and effect relationship with climatic changes and the Pleistocene glaciations. With the technology of the day it was also impossible to determine whether other cyclic events evident in the geologic record, such as repeating cycles of sedimentation, corresponded to periods predicted by the Milankovitch calculations. Supporting evidence was accumulating, however (Broecker and van Donk, 1970), and it is somewhat ironic that strongest support for the theory, which was originally proposed to explain continental glaciations, should ultimately come from the sea.

$^{18}O/^{16}O$ Ratios

In 1947 Harold Urey suggested that the relative amounts of oxygen isotopes in the mineral walls of marine organisms were a function of the ocean water temperature and more ^{18}O was taken up as the water cooled (see also Epstein et al., 1951; Urey et al., 1951). Various marine organisms such as molluscs, Coccolithophorae, and foraminifera synthesize their calcium carbonate walls from molecules available in the ocean water; Emiliani (1955) demonstrated that the shell chemistry of modern foraminifera does indeed reflect the oxygen isotope ratio of the ambient water. (One molecule of seawater per 500 has the ^{18}O molecule.) The shells become incorporated into bottom sediments accumulating at depths less than ~4 km, the calcite compensation depth, and they preserve a record of fluctuating $^{18}O/^{16}O$ ratios. At greater depths the shells are dissolved in the unsaturated water. In 1966 the research vessel Challenger began to retrieve cores from all the ocean basins, and in 1968 the cores formed the basis for the DSDP. An account of these historic and scientifically important voyages is provided by Hsü (1992), and the project continued as the ODP.

The cores were analyzed for a variety of climatic indicators, including ^{18}O and ^{16}O. The waters of the world's oceans mix on a time scale of ~1600 years (ventilation rate), so their isotopic composition is relatively uniform and results can be correlated among the different ocean basins (Hecht, 1985b). A 30-year trend in the ^{18}O composition of modern precipitation and surface air temperature confirmed the relationship (Rozanski et al., 1992), and the isotope ratios were calibrated to real temperature values by comparison with modern analogs and through aquaria experiments. A factor that complicates the derivation of temperature estimates from the $^{18}O/^{16}O$ data is that the ratio also varies with salinity and ice volume (Chapter 3). Nevertheless, sequences of cool–warm intervals are clearly evident in the ocean cores, and when the timing of magnetic pole reversals and extrapolations were used to date the cycles, they revealed periodicities of 23, 41, and 105 Ky (Hays et al., 1976; see also Shackleton et al., 1990); the 100 Ky cycle was the primary one (Fig. 2.11).

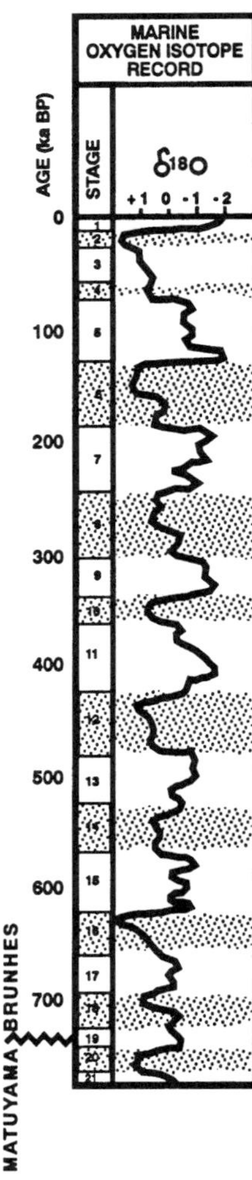

Figure 2.11. The marine oxygen isotope record through the past 800,000 years. Stippling indicates periods of cooling temperatures. Matuyama–Brunhes marks a reversal in the Earth's magnetic pole. Adapted from Williams et al. (1993) and based on Imbrie et al. (1984). Reprinted with the permission of M. A. J. Williams and Kluwer Academic Publishers.

In addition to DSDP and ODP core segments with patterns of high versus low ^{18}O concentrations, support for the Milankovitch theory has come from several other sources. An early source was from patterns of sea-level change preserved as erosion terraces on emergent *Acropora palmata* coral reefs in Barbados. The reefs grow at the water surface so that a rise in sea level, as during an interglacial, results in new growth to keep pace with rising ocean waters. The top of an ancient reef thus marks a high stand in sea level and, by inference, a period of warm climate. When the sea level is lowered, as during glacial periods and cold climates, a terrace is eroded into the coral mass. Eighteen major terraces are preserved on the island. By using uranium series dating techniques, Broecker et al. (1968) and Mesolella et al. (1969) reported high stands of the sea at 82, 103, and 122 Kya that were roughly in agreement with the tilt and precessional orbital variations calculated by Milankovitch for the lower latitudes (45° N at 88, 110, and 132 Kya; the "Barbados sea-level model"; Ku et al., 1990).

Changes in North Atlantic Ocean currents also provided a clue to the timing of glacial events. Ruddiman and McIntyre (1976, 1984) traced the course of the Gulf Stream by analyzing cores for shifts in the distribution of temperature-sensitive marine species. During the interglacials the Gulf Stream flowed northeast toward Great Britain, while during glacial times it flowed more easterly toward Spain. The fixed point of the gatelike swings was near Cape Hatteras, North Carolina. Comparison of the swings with the magnetic time scale revealed a cycle of 100 Kya.

There are also indications of similar cyclic events in much older deposits (Berger et al., 1984, 1992). The Newark Basin in northern New Jersey is a remnant of a series of filled rift valleys about 200 Ma old that developed in the Late Triassic as North America separated from Africa. Individual units about 6 m thick are recognizable on the basis of sediment type, texture, fossils, and other organic matter content. Van Houten (1964) suggested these cycles might correspond to periods of climatic change induced by orbital variations, but in the absence of accurate age determinations it was not possible to establish a close association. Recently, Olsen (1986) and Olsen and Kent (1996) applied radiometric techniques to the sediments and found ages of 25, 44, 100, and 400 Ky, which compare to orbital periods of 23, 41, 100, and 400 Ky. Fourier analysis of Mesozoic climate proxy data and DSDP cores show similarities between rates of sedimentation, accumulation of calcareous marine microfossils, and orbital cycles in Middle Cretaceous deposits in Italy (100 Ma; 100 and 400 Ky cycle), Late Triassic deposits near Lyme Regis in southern coastal England (200 Ma; 40 Ky cycle), and the Late Cretaceous of the South Atlantic (DSDP site 516F, 30° S; precessional cycle of ~23 Ky; Park et al., 1993). Other examples are given by Arthur and Garrison (1986) and Berger et al. (1984).

Although the Milankovitch theory is now generally accepted, the correspondence between cyclic climatic events and periods of the orbital cycles has been challenged.

Winograd et al. (1988) studied ^{18}O variations in calcite deposited from ground water on the walls of a submerged open fault called Devils Hole near Ash Meadows, Nevada, which is 120 km northwest of Las Vegas. The deposits preserve a continuous 250 Ky record of climatic change in the Great Basin between the Middle and Late Pleistocene (~310–50 Kya). The general sequence of events is similar to that in ocean sediments and ice cores, but the timing is different. The dates from Devils Hole suggest that the previous interglacial [Termination II in the terminology of Broecker and van Donk (1970)] began about 147 Kya or earlier; other records, in agreement with the timing of the orbital cycles, place the date at about 127 Kya. An earlier interglacial was dated at 272 Kya, while the marine and ice-core date is 244 Kya.

In the midst of these discussions, additional support for the Milankovitch theory came from new dates obtained from the coral reefs that corrected for inaccuracies due to diagenesis (Gallup et al., 1994). Dating of Termination II from Barbados coral by a refined alpha particle counting technique (Rea et al., 1985) yielded an approximate age of 126 Ky; a new method based on thermal ionization mass spectrometry (TIMS) yielded an age between 122 and 130 Ky. Both dates are in agreement with those from ice cores and marine sediments for Termination II and are consistent with the timing of the Milankovitch cycles. Spectral analysis of layers of eolian (wind blown) dust particles in a DSDP core from the central North Pacific revealed fluctuations associated with the comparatively moist interglacial and drier glacial climates, and these are also consistent with dates from the marine realm for Termination II. Subsequent peaks in the Barbados coral curve were dated at 23, 41, and 104 Kya. Supporters of the Milankovitch variations have suggested that dates from Devils Hole are probably inaccurate because of contamination by ^{230}Th (Edwards and Gallup, 1993).

Longer and better dated cores from Devils Hole later provided a 500 Ky record (between 560 and 60 Kya; Winograd et al., 1992), and uranium-thorium dates of coral reefs from New Guinea (135 Ky) and in the Bahamas (132 Ky) suggest an end to glacial cycles earlier than predicted by the Milankovitch model. Imbrie et al. (1993) believed there were flaws in the dates and interpretation of the Devils Hole material because calculations show that direct application of the Devils Hole chronology to ocean cores would result in implausible rates of marine sedimentation (see response, Ludwig et al., 1993). U-Th dates of marine sediments for the last two interglacial periods (120–127 and 189–190 Ky) are in accord with variations in the Earth's orbit as a fundamental cause of Quarternary climatic change (Slowey et al., 1996).

The latest contribution to these discussions has been application of U_{235} /Protactinium$_{231}$ dating to the Devils Hole carbonates that confirms the original date for warming at ~140 Kya (Edwards et al., 1997). The view is now emerging that because the Barbados coral data and sea-level changes are in agreement with the Milankovitch cycles, they are the principal cause for global-scale climatic changes during the Quaternary. Confirmation of the Devil Hole dates are now suggested to reflect the local climatic cycles of the southwestern United States.

In addition to these problems, some other challenging questions have not been fully answered. In particular, it is unclear why orbital variations should produce extensive glacial–interglacial cycles during the last ~2.4–1.6 m.y. but not earlier. It is possible that overriding high CO_2 concentrations in the Mesozoic and Paleogene had waned by the Neogene to the point that the Milankovitch variations could be expressed as a primary forcing mechanism of climate. Another problem is that a change in the shape of the orbit (100 Ky cycle) alters insolation by about 0.5%, while tilt and precession can change it by as much as 14–27%. This raises the question of why major climatic cycles should follow the pattern of least heat fluctuation. Clearly some amplification of the 100 Ky cycle is needed, and explanations are now being formulated. They include patterned overlap between equinox and perihelion (Crowley et al., 1992), CO_2 concentration (see later section), variations in the frequency of the obliquity cycle (Liu, 1992), and possibly variations in the accretation rate of interplanetary dust (~10 K tons of extraterrestrial material fall on the Earth each year). Variation in this amount may result from changes in the inclination of the orbital plane that follow a 100 Ky cycle (see papers mentioned in Brownlee, 1995).

With regard to the change from one dominant forcing mechanism to another, Imbrie et al. (1993) suggest that ice sheets, resulting from precession and obliquity forcing, upon exceeding a critical size act as negative feedback that imposes the slower eccentricity cycle of change.

Another problem is that northern and southern hemisphere glaciations are nearly synchronous (Schneider and Little, 1997), but because of the Earth's tilt they should alternate. This may be due partly to the facts that Antarctica is surrounded by a circumpolar ocean current and it is a smaller continent unjoined to other land masses (less continentality). Thus, it may have a higher threshold for the formation of glaciers. Models are being developed that explore the role of cooler equatorial oceans in both amplifying orbitally-induced changes in seasonality and in the interhemispheric correlation of glacial events (Bard et al., 1997; Beck et al., 1997; Curry and Oppo, 1997; Webb et al., 1997). Finally, climates can change abruptly, within a few to several hundred years, as shown by analyses of ice cores from Greenland; abrupt changes are not explained by the gradual and steady pace of the orbital cycles. It is clear that other processes that produce more modest and rapid shifts in climate must be superimposed on the Milankovitch variations. Witnessing the debates and the fine-tuning of significant unifying concepts imparts to each period of science its own uniqueness and special importance. Emerging data from the ice cores and refinements in the orbital theory of glaciations are among those that mark this period. It is a challenging and fascinating time to be contemplating Earth history and biotic response because existing technologies are being pushed to their limits, new methodologies are being developed, and broad answers to immensely complex questions are just being formulated.

Greenland Ice-Core Record

During the last deglaciation there was an abrupt cooling event in Europe between ~11 and 10 Kya called the Younger Dryas (Broecker et al., 1988; Severinghaus et al., 1998). Pollen and spore diagrams revealed that forests that had begun to recolonize western Europe were replaced by tundra vegetation, while in eastern North America the vegetation mostly continued a trend toward deciduous forest. Such rapid shifts in climate are now being confirmed through the European GRIP (Dansgaard et al., 1984, 1993; GRIP Members, 1993; Grootes et al., 1993; Johnsen et al., 1992; Paterson and Hammer, 1987; Taylor et al., 1993a,b), and the U.S. GISP2. Annual layers are present in the ice, and this allows precise dating of climatic events (Meese et al., 1994). The air trapped in bubbles in the ice (~10% of the ice volume) can be analyzed for CO_2, nitrogen, and methane, revealing changes in atmospheric chemistry (Science, 1993); water from the ice can be studied for oxygen-isotope composition, revealing changes in temperature. Electrical signals also provide a record of conductivity through the cores, which vary with the amount of wind-blown dust. The amount of dust increases during cool-dry-windy glacial intervals and reduces conductivity. The record can also be tied to tree-ring chronologies (LaMarche and Hirschboeck, 1984) and to volcanism (Zielinski et al., 1994; but see Fiedel, 1995; Southon and Brown, 1995).

The GRIP and GISP2 cores, taken only about 28 km apart, are similar down to ~2700 m (100 Kya), which includes the present interglacial, the last glacial, and part of the previous interglacial. Below that level, in the Sangamon (North America) or Eemian (Europe) interglacial (marine isotope stage 5e), they differ (Grootes et al., 1993; Taylor et al., 1993b); the differences have been attributed to thrust faulting and distortion by ice flow over uneven terranes. These effects may extend above the last interglacial boundary (Alley et al., 1995). Nonetheless, the records from the upper levels of both cores show that in the period from 40 to 8 Kya there were sudden changes of 5–10°C that sometimes lasted less than 5 years, and in the past 8 Ky changes of up to 10–12°C were recorded in a few decades and lasted as little as 70 years. During the Younger Dryas climates were dry, windy, and ~14±3°C colder than today, while at the end of that interval near the Holocene boundary (~10 Kya), it was wetter, less windy, and the average temperature increase was +1°C/decade or steeper than the most extreme forecasts for greenhouse warming during the next 100 years. These brief oscillations on the order of 100 to a few thousands of years evident in the ice cores are called Dansgaard-Oeschger (D-O) events (Bender et al., 1994; Broecker and Denton, 1989; see Figs. 8.5, 8.6). Events lasting longer than ~2 Ky are also evident in the Antarctic ice-core record, suggesting that they were of global extent (Mayewski et al., 1996). The last D-O cycle was the Younger Dryas (11–10 Kya), which was more evident in Europe than in North America, and its end marked the beginning of the Holocene. The principal transition took place over a remarkably brief 40 years (Taylor et al., 1997). Among the principal revelations of the Greenland ice-core record is that in addition to shifts between glacial and interglacial conditions at Milankovitch-time intervals, climates have switched rapidly between cold and warm states over the past 100 Ky (and probably throughout the Quaternary; Bond et al., 1997). Thus, additional climate-controlling processes must be operating, and one likely control is in the heat transport provided by ocean currents (Birchfield and Broecker, 1990; Broecker, 1997; Broecker et al., 1990; Lehman and Keigwin, 1992). The process is described as the salinity–density ocean current model or the thermohaline cell (the "conveyer belt").

Thermohaline Cell

The D-O events preserved in the Greenland ice cores cluster into packets of temperature declines, and during the most intensely cold phases a Heinrich event occurs followed by a prominent warming. Heinrich events (after Harmut Heinrich, 1988) are spaced at ~10 Ky (13–7 Ky) intervals and are numbered H-1 (top, beginning of Termination I, ^{14}C age ~14,500 years) through H-6 (~70 Kya; first phase of the last glacial; see Marine Geology, 1996, for a collection of papers on Heinrich events). They are evident in marine sediments as zones of ice-rafted terrestrial debris, which are associated with layers of depleted foraminifer and other marine fossils, very cold temperatures, and low surface salinity.[2] Fluctuations on the time scales of D-O and Heinrich events are not restricted to the high latitudes. Grimm et al. (1993) noted vegetation trends in Florida that correlate with the last five Heinrich events; moranes in Chile and rock varnish (manganese-rich coatings on rock) in the Great Basin appear to follow a similar pattern; cycles of abundance of the marine alga *Florisphera profunda* following Heinrich events 1-5 occur in the equatorial Atlantic (McIntyre and Molfino, 1996); and D-O events are evident in Antarctic ice cores (Vostok site, Dansgaard et al., 1993; Raynaud et al., 1993). The cycles also match the NADW variations to the south based on $^{13}C/^{12}C$ ratios in benthic foraminifera.

Although these climatic changes appear global in extent, the mechanisms are presently unclear. Bond et al. (1992, 1993) found that layers of marine debris discovered in the North Atlantic resulted from calving of numerous icebergs as they expanded and neared the coasts at intervals correlated with cold periods in the Greenland ice cores. One source of the debris was the Laurentide ice sheet via the Hudson Strait and the Gulf of St. Lawrence, and similar layers were derived from Iceland. The layers thin from west to east. There also were periods of increased iceberg calving at intervals of 2–3 Ky corresponding to longer D-O or mini-Heinrich events (Bond and Lotti, 1995). The abundances in marine planktic foraminifera and the layers of marine debris both varied in accordance with the terrestrial ice-core records for the past 90 Ky, and all three correlate well with the marine oxygen isotope record. This provided the long-sought link between temperature and the rapid changes observed in the ice cores and in the marine sediments. The question is, what causes the rapid temperature changes?

One of the fundamental processes influencing world climates is the poleward transfer of heat from equatorial regions to the high latitudes. This is accomplished by wind and ocean currents (Fig. 2.6). As discussed earlier, warm low-latitude waters flow northward via the Gulf Stream into the North Atlantic Ocean. The lower ambient temperature cools the water and the flow of the westerlies across to northern Europe increases evaporation, producing a cold, saline-dense, heavy water that sinks and flows southward (NADW). This pulls additional warm water from the south in a conveyer belt system at an average flow rate of 20 million m^3/s, or ~100 times that of the Amazon River. The prevailing westerlies convey substantial heat toward Europe. Clearly, any disruption of the conveyer belt would cool the North Atlantic region in general and western Europe in particular; the latter is confirmed by tree-ring analyses and by fossil pollen and spore studies of Younger Dryas age sediments (Björck et al., 1996). Rerouting of meltwaters from the Mississippi River to the St. Lawrence drainage with retreat of the ice margin has been proposed (Broecker et al., 1989; but see de Vernal et al., 1996). Bond et al. (1992) and Broecker (1994) suggest that the increased iceberg calving from the eastern North America Laurentide and Iceland ice sheets and the melting of sea ice dilutes the saline-dense North Atlantic water, preventing or reducing the rate at which it sinks to form NADW. This in turn disrupts thermohaline circulation (the "salt oscillator"). Models indicate that freshwater input of 0.06×10^6 m^3/S, about one-fourth of the discharge of the Amazon River, could shut down the conveyor (Manabe and Stouffer, 1995; Rahmstorf, 1995).

The mechanisms causing the periodicity in dilution are being explored through several models. One focuses on glacial mechanics. When continental glaciers first begin to form they are anchored (frozen) to the substrate and thicken without appreciable spreading. As they thicken, pressure on the base increases and the ice further serves as an insulation layer that traps geothermal heat. Distortion and melting of the base of the glacier causes the ice to flow and calve icebergs (freshwater) into the ocean. The amounts of water are estimated at half the volume of the present Greenland ice sheet in a few hundred years. Upon thinning, the glacier settles to the substrate, and the cycle repeats on a proposed ~7 Ky cycle. However, the discovery that the large Laurentide and small Iceland ice sheets both discharged ice and meltwaters into the North Atlantic Ocean on about the same schedule (Bond and Lotti, 1995) suggests that glacial mechanics were a result and not a cause of the climatic changes. Also, millennial-scale oscillations occur in the Holocene when there were no large northern ice sheets. The discovery that rapid climatic changes correlated with the North Atlantic variability occurred in the tropical Atlantic region (Hughen et al., 1996) suggests that global forcing mechanisms are involved, possibly solar variability.

The Milankovitch variations provide one forcing mechanism for long-term climatic change. The ice-core data and marine sedimentary record reveal the need for other mechanisms that explain sudden short-term reversals. Although not fully understood, it is becoming increasingly likely that such mechanisms operate by periodically disrupting oceanic heat transfer to the North Atlantic region through dilution by glacial meltwater (Johnson and Lauritzen, 1995; Keigwin et al., 1994; McManus et al., 1994; Paillard and Labeyrie, 1994; Rahmstorf, 1994).

CO_2 Concentrations

Much of the literature on atmospheric levels of CO_2 and other radiatively active or greenhouse gases [about 40, including reactive trace gases such as ammonia (NH_3), chlorofluorocarbons (CFCs), methane (CH_4), nitrous oxide (N_2O), ozone (O_3), and reduced sulfur (SO_2) compounds] deals with their effect on recent climate and human economies (Crowley, 1991; Dacey et al., 1994; Dowlatabadi and Morgan, 1993; Dunnette and O'Brien, 1992; Hansen et al., 1991; Kareiva et al., 1993; Kaufmann and Stern, 1997; Kelley and Wigley, 1992; Kiehl and Briegleb, 1993; Lutjeharms and Valentine, 1991; Majumdar et al., 1992; Peters and Lovejoy, 1992; Rind et al., 1992; Rubin et al., 1992; Schlesinger and Ramankutty, 1992; Schneider, 1994; Science, 1993, including cover illustration; Nature, 1992). For example, it is known that the CO_2 level rose from ~280 ppm in the preindustrial years of 1750–1800 to 315 in 1958, 338 in 1980, and 351 in 1988; it is expected to reach 600 ppm between 2030 and 2080 A.D. At present ~6–8 gigatons (Gt, billions of metric tons) of carbon are produced annually through anthropogenic activities, which are mainly industrial and automobile emissions and burning of the rain forest. About one-half of this amount escapes to the atmosphere; ~2 Gt are taken up by the ocean; and much of the remaining ~2 Gt is stored in the terrestrial biosphere in soil, peat, and plant (e.g., the northern biosphere, Keeling et al., 1996; deep-rooted grasses of the South American savannas, Fisher et al., 1994) and animal biomass. Processes and the net CO_2 balance from among the various components of the terrestrial realm are not well understood, as indicated by the designates "the missing sink" and "an enigmatic terrestrial reservoir" (Hanson, 1993; Oechel et al., 1993; Siegenthaler and Sarmiento, 1993;

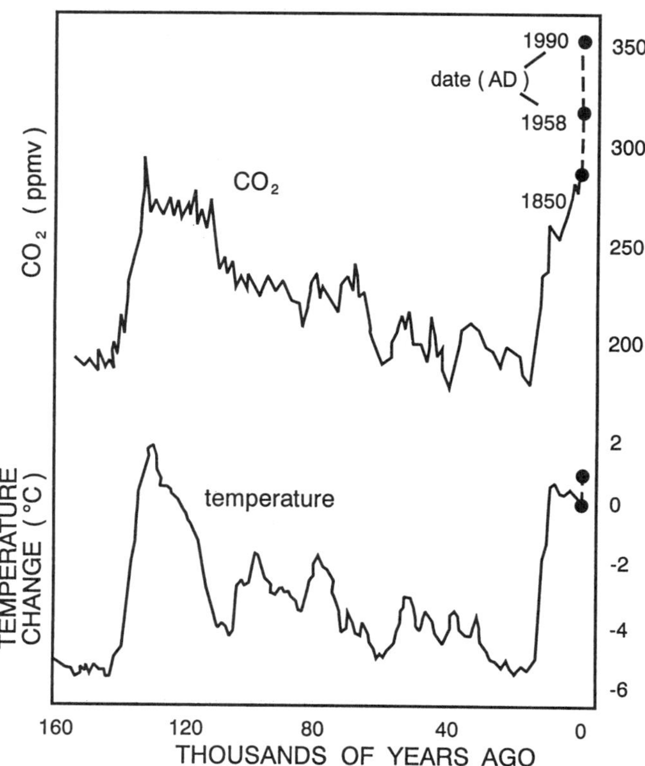

Figure 2.12. CO_2 and temperature records for the past 160 ka from Antarctic ice cores, and recent atmospheric measurements. Reprinted from Hansen et al. (1993) with the permission of the National Geographic Society.

Smith and Shugart, 1993). Measurements over a 55-day period in the wet and dry seasons from a tropical rain forest in Brazil showed the forest is a net absorber of CO_2 (8.5 ± 2.0 mol/m^2/year; Grace et al., 1995). In 1990 deforestation in the low latitudes emitted 1.6 ± 0.4 petagrams (Pg = 10^{15}/g = 1 Gt) of carbon, while forest expansion in the mid- to high latitudes incorporated 0.7 ± 0.2 Pg into the biomass, for a net input into the atmosphere of 0.9 ± 0.4 Pg (Dixon et al., 1994). Natural changes in vegetation cover (e.g., forest to desert) not only affect the albedo of the Earth's surface and the amount of dust in the atmosphere, but also the carrying capacity of CO_2 by the biosphere. CO_2 allows solar radiation of short wavelengths to penetrate the atmosphere, but it retards the reradiation of longer wave infrared heat energy. Modeling experiments show that a doubling of CO_2 concentration in the atmosphere raises the mean annual surface temperature by 2–5°C (Geophysics Study Committee, 1982; Wigley and Jones, 1981; 1.5–4.5°C, Cess et al., 1993; Houghton et al., 1990, 1992). Increased surface temperatures would have a significant impact on agricultural patterns and require major readjustments in local cultivation and management practices (Overpeck et al., 1991).

The amount of CO_2 in the atmosphere has also fluctuated over geologic time (Berner, 1991; Crowley, 1993), and within the past 15 Ky (Faure, 1990) to 160 Ky there is a clear covariation with atmospheric temperature (Fig. 2.12). The sources of these data for relatively recent times of ~100–300 Kya are many and varied. One is the measurement of isotopic carbon contributed by the atmosphere and by plant roots to form veins and nodules of soil carbonate. Plants tend to take up more of the lighter ^{12}C than the heavier ^{13}C, and the ratio can be used to distinguish between the plant-derived component from root cell respiration and the atmosphere-derived components in the paleosols. This approach shows covariation between high versus low atmospheric CO_2 concentration and warm versus cool temperatures (Cerling, 1991).

A related technique involves distinguishing between the ^{13}C in paleosol carbonate contributed by C3 and C4 plants (Cerling et al., 1993; Morgan et al., 1994; see Nature, 1994, for response and reply). In the photosynthesis of C3 plants, CO_2 is split and the carbon is taken up by ribulose diphosphate, which splits into two molecules of phosphoglyceric acid (Calvin cycle). The ^{13}C composition of C3 plants averages −26%. About 85% of flowering plants follow the C3 pathway. In C4 plants the CO_2 combines with the 3-C phosphoenol pyruvate to form 4-C malate or aspartic acid, which is transported to bundle sheath cells (Hatch-Slack cycle). There CO_2 is released and used in the Calvin cycle reactions. The ^{13}C composition of C4 plants averages −13%. The C4 pathway is derived from C3, and approximately one half of the grasses are C4 plants; these are mostly tropical and subtropical in distribution. The C4 photosynthesis is favored under low concentrations of CO_2, while higher concentrations favor the C3 pathway (Polley et al., 1993), and the isotopic carbon signature is preserved in plant debris from paleosols and in ungulate tooth enamel. Beginning in the Middle Miocene and especially by 7–6 Ma, plants with the C4 pathway began increasing in several parts of the world (Cerling et al., 1997; Quade et al., 1989), suggesting that a lower threshold of atmospheric CO_2 had been reached that, in turn, is consistent with evidence of progressively lowering temperature and associated dryness during the Neogene. An increase in atmospheric CO_2 between 9 and 7 Kya (postglacial climatic optimum) is recorded in a shift from C4-dominant grasses to C3-dominant shrubs on an alluvial fan system in New Mexico (Cole and Monger, 1994; but see Boutton et al., 1994). Similar trends for the past 100 Ma are preserved in carbon isotopes from marine phytoplankton (Freeman and Hayes, 1992). Another technique being developed is the measurement of ^{18}O in phosphate structures (e.g., fossil horse teeth 18.2–8.5 m.y. old from Nebraska; Bryant et al., 1994). Results show a depletion of ^{18}O during this interval that is consistent with the gradual trend toward lower temperatures and less evaporation of the ^{18}O.

Another source of information on ancient CO_2 concentrations is from the ^{13}C content of planktonic and benthic foraminifera (Shackleton et al., 1983). These studies demonstrate that the CO_2 concentration of the atmosphere was lower during the glacial ages than in the last interglacial, and during the last glacial maximum at 18 Kya it was ~30% lower than today. Atmospheric CO_2 was ~220 ppm at 18 Kya, and 260–300 ppm during the interglacials. It is estimated that the global content of carbon in the phytomass varied between 968 Gt at glacial maxima (18 Kya) and 2319 Gt at present (Adams et al., 1990). The $^{13}C/^{12}C$ ratios in mosses and sedges preserved in peat are also being used (White et al., 1994). These analyses provide a resolution of about a decade and show increases at 12,800 years B.P. (a warming episode in the North Atlantic), 10 Kya (end of the Younger Dryas cold period), and 4400 years B.P. (modern interglacial climates).

The most detailed data are just now being published, and they involve carbon isotopic analysis of air bubbles of CO_2 trapped in ice cores (Jouzel et al., 1993; Neftel et al., 1982). These are from the Antarctica Vostok ice core representing the past 500,000 years (Maier-Reimer et al., 1990), the GRIP Camp Century and Dye 3 cores from Greenland, and the GISP2 cores representing the past 150,000 years. These document a close relationship between CO_2 concentration and polar temperatures for the past 100,000 years.

Modeling experiments are also underway to determine the effect of CO_2 fluctuations on global climates in more ancient times. Experiments for the Cretaceous at 100–65 Ma, using several alternative seafloor spreading rate formulations, produce an atmospheric CO_2 concentration several times higher than at present (Berner, 1994; Berner et al., 1983; near sevenfold, Berger and Spitzy, 1988), corresponding to the warm temperatures of that period. In contrast, the cooling trend of the Middle to Late Cenozoic involves a general decline in CO_2 concentration to ~2.5-fold. A source of these data is air trapped in amber, including material from the Dominican Republic (Late Oligocene to Early Miocene), the Baltic (Oligocene to Eocene), and Manitoba (Senonian, 75–95 Ma; Berner and Landis, 1988; but see Technical Comments, Science, 1988).

It is becoming increasingly clear that changes in atmospheric CO_2 concentration have been a significant factor in determining paleoclimates. The causes for these changes is varied. As climates warm, land and water surfaces release more CO_2 as a positive feedback. This is one component of the CO_2–temperature relationship. Another derives from geologic processes. A convenient way of envisioning these processes is as a series of sinks (CO_2 utilizing reactions) and pumps (CO_2 generating reactions) in which each rate can be influenced by various geological and biological events. The balance between them is important as evidenced by the fact that if today's sinks were just 10% larger than the pumps, CO_2 would disappear from the air–ocean system in only 5 m.y.

Carbon dioxide is involved in the biological carbon cycle through several processes. During the buildup of complex organic molecules by plants and animals, it is stored as fixed carbon (a sink); in decay and in the erosion of buried organic material and kerogen, CO_2 along with ammonia and methane is produced (a pump). Although these tend to balance out over long periods of time, on shorter time scales there can be a significant lag period with a resulting effect on climate. For example, the carbon utilized in catabolic processes by organisms, which is washed into the sea by erosion of organic-containing coastal and shelf sediments and carried in by rivers, sinks at the relatively rapid rate of ~150 m/day. However, it takes 100–1000 years before the CO_2 released near the ocean floor by dissolution is returned to the surface because of the slow circulation of the deep ocean. Periodically the release can be delayed even longer by changes in surface temperatures, glacial meltwaters, land bridges, salinity concentrations, and other factors. It is recognized that the biological carbon cycle and factors that affect ocean circulation and thermal stratification have an effect on long-term climate through the release of CO_2, but at present these factors cannot be integrated and accurately quantified.

In contrast, the effect of fluctuations in the geochemical carbon cycle is clearer. In this cycle there can be significant changes in the rate of production and consumption of CO_2; the recycling period is long, and it varies over time. A principal cause for changes in atmospheric CO_2 concentration is variation in the rate of weathering and metamorphism of silicate rocks (Berner, 1994; Berner et al., 1983). The process is expressed by the following generalized reaction [note the utilization of CO_2 during weathering and later burial storage (a sink) and its release during metamorphism (a pump)]:

$$CO_2 + CaSiO_3 \text{ (e.g., wollastonite)} \underset{\text{metamorphism}}{\overset{\text{weathering}}{\rightleftarrows}} CaCO_3 + SiO_2$$

or

$$CO_2 + MgSiO_3 \text{ (e.g., enstatite)} \underset{\text{metamorphism}}{\overset{\text{weathering}}{\rightleftarrows}} MgCO_3 + SiO_2$$

Any change in the rate of weathering or metamorphism of silicate rocks will affect the atmospheric balance of CO_2. This is the basis for the contention by Raymo and Ruddiman (1992) and Molnar and England (1990) that uplift with consequent increased rates of geochemical weathering of the Tibetan Plateau, the Himalayas, and the Rocky Mountains during the past 40 m.y. caused a drawdown of CO_2 and was, therefore, a contributing factor to the resulting cooler climates (see also, Nature, 1993; Partridge et al., 1995).[3]

Another source of variation in CO_2 concentration is the process of degassing. CO_2 is released at Benioff zones and along midocean ridges by metamorphism and heating of the water. In addition, CO_2 can be generated from mineral springs and from continental and oceanic flood basalts. A rough estimate of the amount of CO_2 emitted to the atmosphere through degassing is ~0.09 x 10^{15} g of C/year as CO_2 (Arthur, 1982). Consequently, an increase in tectonic activity, as occurs during periods of major plate reorganization (viz., in the Cretaceous), will release considerable quantities of CO_2 to the atmosphere and contribute toward warming climates through greenhouse effect mechanisms. Seafloor spreading accounts for ~80% of the world's volcanic activity. Conversely, during times of tectonic quiescence, less CO_2 is released and there is a trend toward cooler climates as greater amounts of heat escape the troposphere. Sea level can also affect CO_2 concentration by influencing the amounts of silicate and organic material available for erosion (Chapter 3).

It is becoming clearer that CO_2 has played other significant but less direct and still poorly understood roles in paleoclimatic change (Keir, 1992). Some studies suggest that changes in CO_2 concentration may amplify the 100 Ky Milankovitch cycle. In discussing the Milakovitch variations it was noted that uncertainty still exists as to how the redistribution of relatively small amounts of heat from changes in eccentricity could produce the dramatic climatic effects of the past 1.6 m.y. The ice records for the past 40 Ky from Camp Century, Greenland, and Byrd Station and Dome C, Antarctica, that were studied by Neftel et al. (1982) yielded CO_2 concentrations of ~180–240 ppm at the last glacial maximum compared to 280 ppm for pre-Industrial Revolution times. As the most recent glaciers began retreating ~16 Kya, CO_2 concentration rapidly declined. This pattern in the ice-core record is paralleled by the marine isotope record from the eastern tropical Pacific that was extended back to 340 Kya by Shackleton et al. (1983). When changes in CO_2 concentration were compared with the Milankovitch variations, it was found that there was first a change in the orbital cycle, followed by a change in CO_2 concentration and then a change in climate. It was further documented that the greatest CO_2 concentrations matched the 100 Kya eccentricity period. The implication is that orbital variation may be a cause and CO_2 fluctuation an agent or amplifer of climatic change.

How these correlated changes in the Milankovitch variations, CO_2 concentrations, and glacial cycles are brought about is currently under study (Saltzman and Verbitsky, 1994). According to one scenario they involve the utilization of CO_2 by planktonic organisms to build their shells and tissues. In the deep ocean the remains sink to great depths where the carbonate dissolves and the organic matter is oxidized. The resulting carbon may be prevented from returning immediately to the surface by the especially steep temperature and salinity–density gradients that develop at high latitudes during glacial times. The sea is the primary carbon sink, and it presently stores an estimated 38 Gt of dissolved carbon (~65 times the amount in the atmosphere). Eventually the carbon does return to the ocean surface at polar latitudes where more is released than can be utilized by the plankton and it escapes back to the atmosphere as CO_2. What causes fluctuations in the amount and timing of CO_2 circulating between the deep ocean, the polar ocean surface, and the atmosphere has yet to be determined. However, Milankovitch-induced warming trends do raise surface water temperature and allow more CO_2 to escape. The full answer is still forthcoming, but changes in CO_2 concentration may amplify the eccentricity variant and the mechanism likely involves chemical and physical processes in the high latitude oceans (Oppo and Lehman, 1993).

PLATE TECTONICS

The patterns of climatic change induced by variations in the orbital cycle are modified on a local to regional scale by various consequences of plate tectonics (Gordon and Stein, 1992; Hill et al., 1992). These are described as telluric changes in climate; that is, they are derived from alterations in geography and topography that form the boundaries to the fluid spheres of water and atmosphere. The most dramatic are the separation, transport, and collision of continents over great distances. The separation of Australia from Antarctica, beginning ~50 Ma (early Eocene), brought its south-temperate biota into contact with tropical assemblages of southeast Asia as marked by Wallace's Line between Java and Borneo and Lombok and the Celebes.

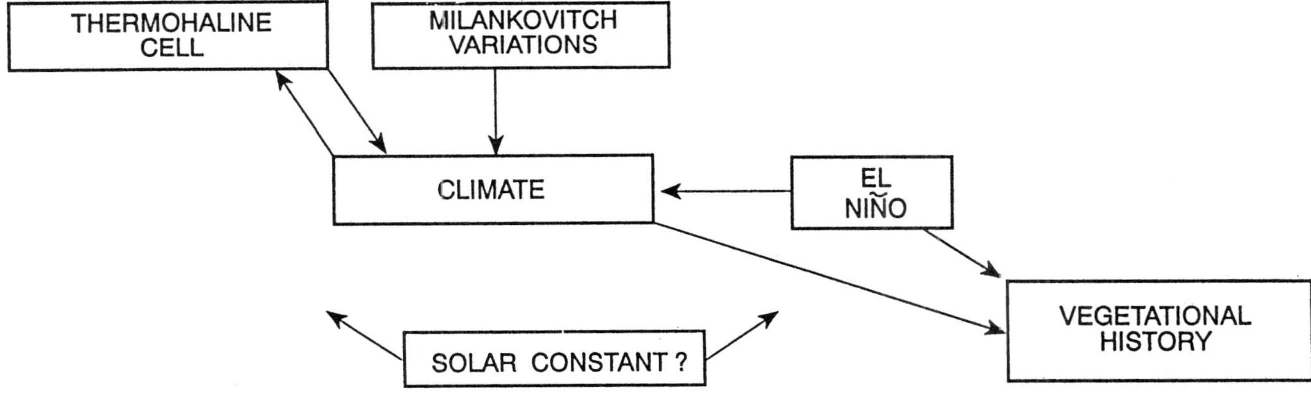

India separated from Africa by at least the Middle Cretaceous as is evident from oceanic basalts of that age between Africa and Madascagar or India. The island continent drifted northward from south-temperate, across equatorial-tropical, into north-temperate zones until it collided with island arcs located off south-central Asia in the Early Eocene (50.3 Ma) but possibly as early as the Late Cretaceous (Beck et al., 1995). Thus, the time of early uplift of the Himalayas and Tibetan Plateau resulting from the collision is unsettled, but they were certainly substantially elevated by the Miocene (~20 Ma; Harrison et al., 1992).

Although the movement of India was geographically removed a great distance from the North American continent, there were effects that illustrate the global interactions between major changes in the physical environment, world climate, and biotic history. Uplift of the Tibetan Plateau altered atmospheric circulation, and the northward drift of India diverted westward-flowing Pacific waters into the counterclockwise subtropical gyre that presently characterizes the southern Pacific Ocean. One consequence of that diversion was to provide a moisture source for the Antarctic continent that, along with the opening of the Drake Passage that thermally isolated the continent, contributed to the formation of glaciers there in the Oligocene. Tasmania remained attached to Antarctica until ~30 Ma, at which time deep-water circulation was established around Antarctica. The circum-Antarctica current transports ~200 million m^3 of water per second, which is probably the largest volume of any ocean current. This, in turn, chilled the southern oceans, affected ocean circulation, and reduced global sea levels. The Middle to Late Miocene drying trend evident from North American biotas was partially due to colder ocean waters and reduced evaporation resulting from the formation of glaciers on Antarctica.

North America experienced less longitudinal movement across climatic zones during the Cenozoic. Movement is estimated at an average rate of 2.2 cm/year during the past 10 m.y. (Anders, 1994), and the North American biota has developed more under the influence of cycles of orbital-induced climatic changes, augmented by other consequences of plate tectonic activity and catastrophes. In particular, the role of plate tectonics in the evolution of North American vegetation has been: to serve as the mechanism for uplift, tilting, extension, compression, transpression, and rotation of crustal blocks in the western part of the continent (orogeny); volcanism in western North America; fluctuating land barriers across Beringia to eastern Asia in conjunction with sea-level changes; the generation of similar connections and barriers across the North Atlantic to western Europe; and the accretion of fossil-bearing exotic or suspect terranes onto the western margin of the continent. Some of the results and chronologies of plate tectonic activity relevant to the development of the North American landscape are presented below and are summarized in Table 2.2. A physiographic outline of North America, an index of place names mentioned in the following text, and a landform map (Thelin and Pike, 1991), are given in Figs. 2.13–2.15, respectively.

Orogeny

A valuable context for interpreting North American vegetational history and a major factor in modifying global and regional climate is orogeny, the process of mountain building through uplift, folding, and thrusting of strata. In North America the two principal orogenic systems are the Appalachian Mountains and the western cordilleras including the Rocky Mountains, Sierra Nevada, Cascade Mountains, Coast Ranges, and other more local units.

The Appalachian Mountains extend 3000 km from central Alabama to Newfoundland and include outliers such as the Ouachita, Arbuckle, and Wichita Mountains in Arkansas and Oklahoma, and a terminating western fragment in the Marathon region of west Texas (King, 1977). To the north, the system continues across the North Atlantic into the British Isles and northwestern Europe as the Caledonian and Hercynian Mountains. The Appalachian Mountains had their origin in ancient megasutures resulting from collision of land masses that formed Pangaea and in the breakup of Pangaea to form the central and North Atlantic Ocean (Bally and Palmer, 1989; Sacks and Secor, 1990). The principal episode involving uplift of this sys-

Table 2.2. Summary of Cretaceous and Tertiary events in development of North American landscape mentioned in text.

145 Ma (Early Cretaceous)	Spreading evident in Labrador Sea
131 Ma (early Middle Cretaceous)	Spreading evident in Davis Strait and southern Baffin Bay
~112 Ma (Middle Cretaceous)	Crustal separation between Labrador Shelf and Greenland
90 Ma (late Middle Cretaceous)	Ingression of Arctic sea into Baffin Bay rift
90–70 Ma (Late Cretaceous)	Mixing of Arctic–Atlantic waters marking formation of Davis Strait (70 Ma); Cape Dyer (Baffin Isl.)–Disko Isl. (Greenland) separated by ~150 km (70 Ma); early formation of Idaho Batholith (90–70 Ma); Blacktail, Snowcrest, Madison Ranges (Montana; 70 Ma)
65 Ma (K–T boundary)	Asteroid impact
65–38 Ma (K–T boundary to Late Eocene)	Volcanism increases in Davis Strait and north of Ellesmere Isl. (Thulean volcanism), Cape Dyer–Disko Isl. separated by another 200 km (57–55 Ma); crustal separation between Greenland and NW Europe; Rocky Mts. uplifted to ~1 km (0.6 mi); fragmentation of the old Farallon plate to form the Juan de Fuca and other subplates (55 Ma); beginning of Proto-Aleutian arc (55–50 Ma); beginning of Bitteroot Dome (50–45 Ma); Uinta Range (45 Ma); formation of Great Basin topography through extension and differential collapse (45 Ma north, 35 Ma south); beginning of volcanism in the Cascades (44–38 Ma); major uplift of Aleutian arc (40 Ma)
33 Ma	Beginning of Tertiary volcanism in Sierra Nevada
30–29 Ma (Middle Oligocene)	Most Aleutian arc islands present (30 Ma); Pacific plate begins to impinge on western edge of North American continent
27 Ma (early Late Oligocene)	Subduction of Juan de Fuca plate activates Garibaldi–Pemberton volcanic belt (SW Br. Columbia; to 7 Ma; then 2.5 Ma to present)
26 Ma (middle Late Oligocene)	First Pacific ridge segments arrive off Baja California (28–26 Ma); San Andreas fault structure develops (29 Ma), on-land activation (22 Ma), modern single-fault configuration (12 Ma)
24 Ma (Oligo–Miocene)	Uplift of Colorado Plateau ~3 km (2 mi)
23 Ma (Early Miocene)	Beginning uplift and volcanism in California Coast Ranges
20–18 Ma (Early Miocene)	Extension in Sonoran Desert (to ~10 Ma); continued uplift of Rocky Mts.
18 Ma (Early Miocene)	Block faulting and differential uplift on Colorado Plateau (to 10 Ma), subsidence and erosion to present
17 Ma (Early Miocene)	Cascade volcanic belt ~100 km long; Columbia River basalts (17–6 Ma)
15 Ma (Middle Miocene)	Aleutian arc essentially established in modern configuration
14 Ma (Middle Miocene)	Major extension in Death Valley; Transverse Ranges (California; to Plio-Pleistocene); subsidence Iceland–Faeroe Ridge with interchange of Norwegian–Greenland Seas cold water with Atlantic Ocean warmer water
10–5 Ma (Late Miocene)	Rocky Mts. further uplifted; Cascade volcanic belt to present length and width (by 8–9 Ma); Alert Bay belt (Vancouver Isl.) formed (8–2.5 Ma)
6.5 Ma (Late Miocene)	Opening Gulf of California; Cascades and Coast Ranges begin to cast rainshadow
3.4 Ma (Middle Pliocene)	Latest and principal disruption of land through Beringia

tem was the Alleghenian at 300–250 Ma (Middle Pennsylvanian to Late Permian). The formation of the Appalachian Mountains was completed by the end of the Paleozoic, and they have since undergone 250 m.y. of erosion to produce relatively low elevations and rounded domelike structures. The highest peaks are Mt. Mitchell (2037 m, 6684 ft) and Clingmans Dome (2025 m, 6642 ft) in the Blue Ridge Mountains of North Carolina and Tennessee, respectively, and Mt. Washington (1916 m, 6288 ft) in the White Mountains of New Hampshire.

In contrast, the western cordilleras are younger (Late Cretaceous to present), and the modern physiographic diversity is considerably greater. The highest peaks include Mt. McKinley in the Alaska Range (6194 m, 20,320 ft), Mt. Logan in the St. Elias Range of the southwest Yukon Territories (6050 m, 19,840 ft), Mt. Whitney in the Sierra Nevada of California (4418 m, 14,494 ft), and several others over 3400 m (11,000 ft) in elevation. The history of these mountains has had a more direct influence on the development of modern North American vegetation than that of the older Appalachian Mountains.

In the Middle Jurassic (175 Ma) a foredeep developed at the present site of the Rocky Mountains, followed by foreland subsidence of about 3 km between 84 and 66 Ma and deformation and uplift of 1 km above sea level at 65 Ma (Bird, 1988). Deformation through thrustfaulting intensified in the late Paleocene (~55 Ma) and waned during the Eocene (~40 Ma). The period in the development of the Rocky Mountains between ~55 and 40 Ma constitutes the latter part of the Laramide orogeny [Mid–Late Cretaceous (Coniacian) to the Middle Eocene]. By the end of the Eocene, erosion had produced a landscape similar to the modern High Plains (Scott, 1975; but see below). According to the geologic data summarized by Ruddiman and Kutzbach (1989, 1990) and Ruddiman et al. (1989), the Rocky Mountains gained about half their present altitude between 10 and 5 Ma. Regional uplift began in the southern Rocky Mountains at ~20 Ma (Middle Miocene) and in-

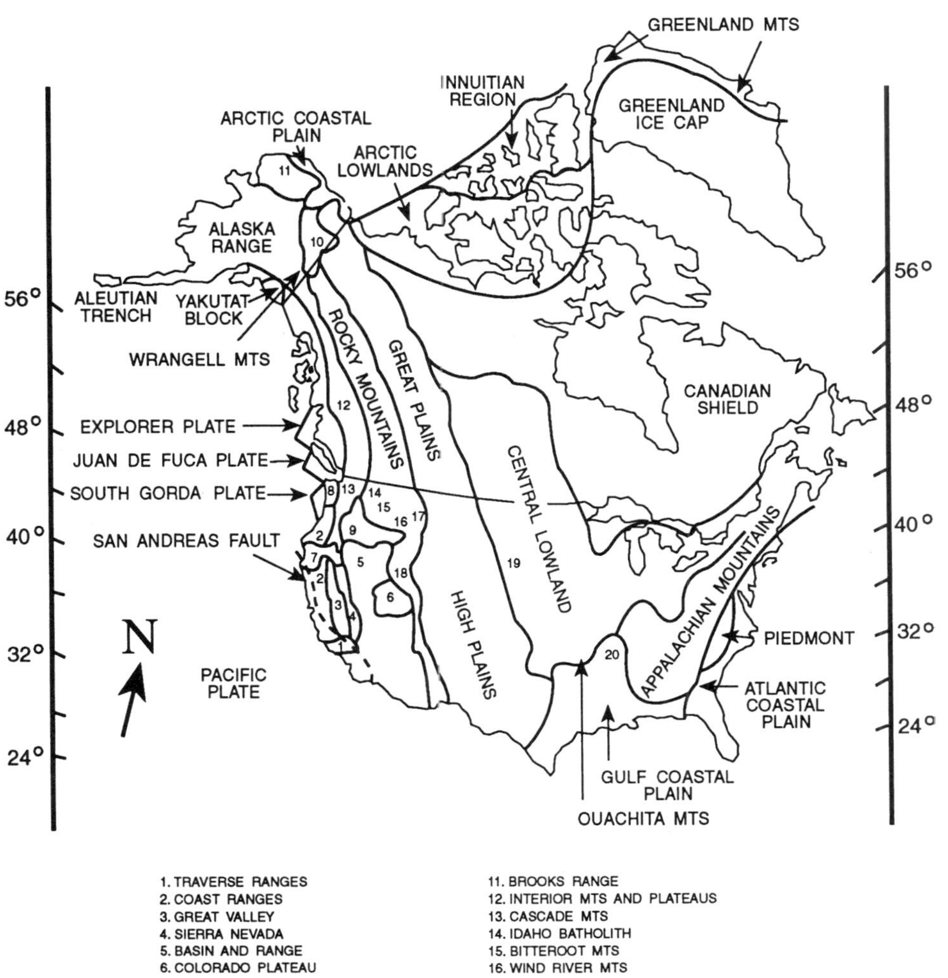

Figure 2.13. Physiographic diagram of North America with reference to features mentioned in the text.

tensified at ~12 Ma, and a particularly intense uplift occurred between 7 and 4 Ma. The Rocky Mountains orogeny is reflected by a change in sediments carried eastward onto the High Plains by the Arkansas and Platte Rivers. At ~40 Ma these sediments were mostly clay and silt, changing to sand at ~20 Ma, then to pebbles and cobbles between 10 and 5 Ma. Downcutting of the Arkansas, Platte, and Colorado Rivers also increased during this time. Deposition of eroded sediments and uplift along the flanks of the Rocky Mountains raised the western plains to form the High Plains. Downcutting by rivers indicates some uplift has occurred within the past 5 Ma (Eaton, 1986). Uplift of the Brooks Range along the North Slope of Alaska began in the Berriasian (~145 Ma), continued into the Aptian (124 Ma), and further intensified between the Albian (112 Ma) and Turonian (90 Ma; Mull, 1979, 1985). The Idaho Batholith formed between 90 and 70 Ma, the Blacktail–Snowcrest and Madison Ranges in Montana were formed in the latest Cretaceous, crustal shortening is evident in the British Columbia section of the southern Canadian Rockies in the Paleocene, the Uinta Range formed in the middle Eocene (~45 Ma), and the Bitteroot Dome was uplifted between 50 and 45 Ma.

The cause of the orogeny is complex and multifaceted, but it involved a plate of Pacific oceanic lithosphere (the Farallon plate, the ocean floor east of the mid-Pacific Rise; Fig. 2.16) underthrusting the western edge of the North American plate. The Farallon plate now has been mostly subducted beneath North America; however, it fragmented at ~55 Ma into a series of smaller plates and subplates, some of which are still impinging on the margins of the continent (e.g., the Gorda and Juan de Fuca plates along the coast of Washington and Oregon; Fig. 2.13; Clarke and Carver, 1992; Savage and Lisowski, 1991). The rate of subduction of the Farallon Plate during periods of intense tectonism is estimated at ~12 cm/year.

In sections from Wyoming the horizontal shortening of strata caused by uplift was ~5–10%. However, crustal thickness here increased from 33 to 55 km, which was far more than would have resulted from a shortening of 5–10%. The extra material may be accounted for in a model proposed by Bird (1988) in which the subduction track,

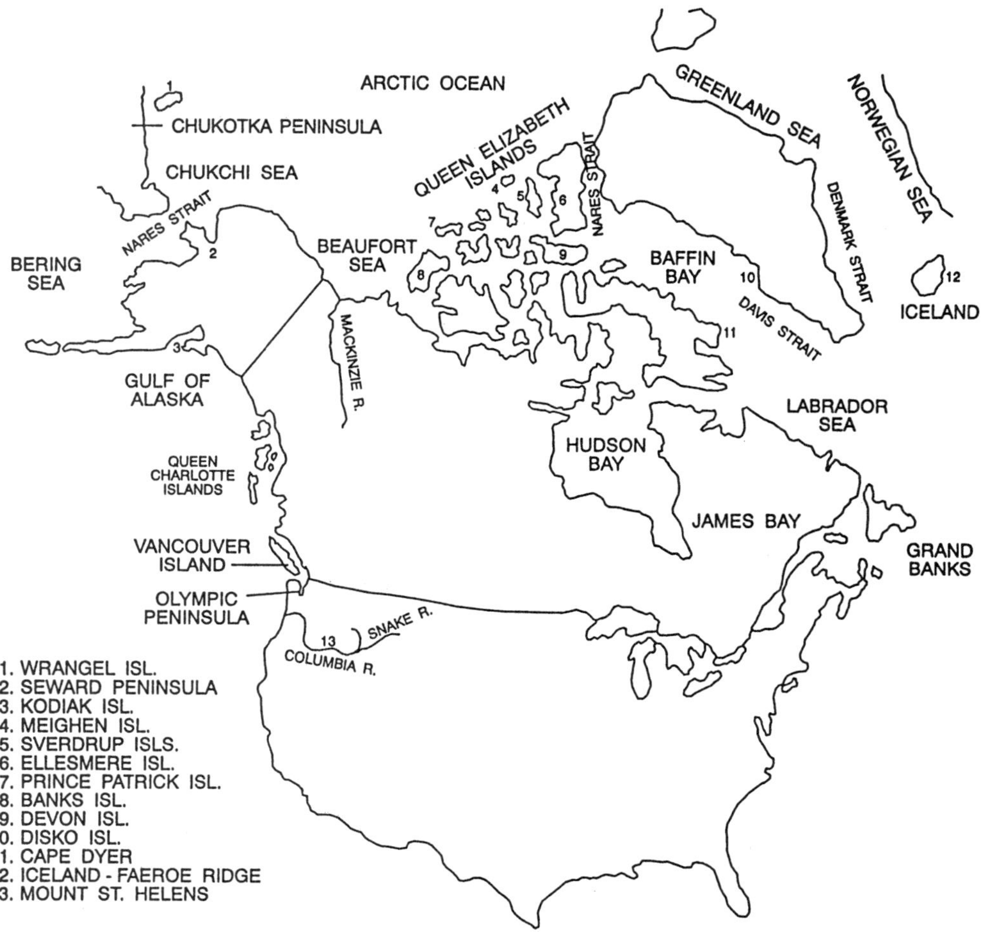

Figure 2.14. Index to place names mentioned in the text.

Figure 2.15, facing page. Landform map of conterminous United States (Thelin and Pike, 1991). U.S. Geological Survey terminology:
Laurentian Upland: 1. Superior Upland. Atlantic Plain: 2. Continental shelf (not on map); 3. Coastal Plain, a. Embayed section, b. Sea Island section, c. Floridian section, d. East Gulf Coastal Plain, e. Mississippi Alluvial Plain, f. West Gulf Coastal Plain.Appalachian Highlands: 4. Piedmont province, a. Piedmont Upland, b. Piedmont Lowlands; 5. Blue Ridge province, a. Northern section, b. Southern section; 6. Valley and Ridge province, a. Tennessee section, b. Middle section, c. Hudson Valley; 7. St. Lawrence Valley, a. Champlain section, b. Northern section (not on map); 8. Appalachian Plateaus, a. Mohawk section, b. Catskill section, c. Southern New York section, d. Allegheny section, e. Kanawha section, f. Cumberland Plateau section, g. Cumberland Mountain section; 9. New England Province, a. Seaboard Lowland section, b. New England Upland section, c. White Mountain section, d. Green Mountain section, e. Taconic section, 10. Adirondack province. Interior Plains: 11. Interior Low Plateaus, a. Highland Rim section, b. Lexington Plain, c. Nashville Basin; 12. Central Lowland, a. Eastern Lake section, b. Western Lake section, c. Wisconsin Driftless section, d. Till Plains, e. Dissected Till Plains, f. Osage Plains; 13. Great Plains province, a. Missouri Plateau (glaciated), b. Missouri Plateau (unglaciated), c. Black Hills, d. High Plains, e. Plains Border, f. Colorado Piedmont, g. Raton section, h. Pecos Valley, i. Edwards Plateau, k. Central Texas section. Interior Highlands: 14. Ozark Plateaus, a. Springfield–Salem plateaus, b. Boston "Mountains"; 15. Ouachita province, a. Arkansas Valley, b. Ouachita Mountains.Rocky Mountain System: 16. Southern Rocky Mountains, 17. Wyoming Basin, 18. Middle Rocky Mountains, 19. Northern Rocky Mountains. Intermontane Plateaus: 20. Columbia Plateau, a. Walla Walla Plateau, b. Blue Mountain section, c. Payette section, d. Snake River Plain, e. Harney section; 21. Colorado Plateaus, a. High Plateaus of Utah, b. Uinta Basin, c. Canyon Lands, d. Navajo section, e. Grand Canyon section, f. Datil section; 22. Basin and Range province, a. Great Basin, b. Sonoran Desert, c. Salton Trough, d. Mexican Highland, e. Sacramento section. Pacific Mountain System: 23. Cascade–Sierra Mountains, a. Northern Cascade Mountains, b. Middle Cascade Mountains, c. Southern Cascade Mountains, d. Sierra Nevada; 24. Pacific Border province, a. Puget Trough, b. Olympic Mountains, c. Oregon Coast Range, d. Klamath Moutains, e. California Trough, f. California Coast Ranges, g. Los Angeles Ranges; 25. Lower Californian province.

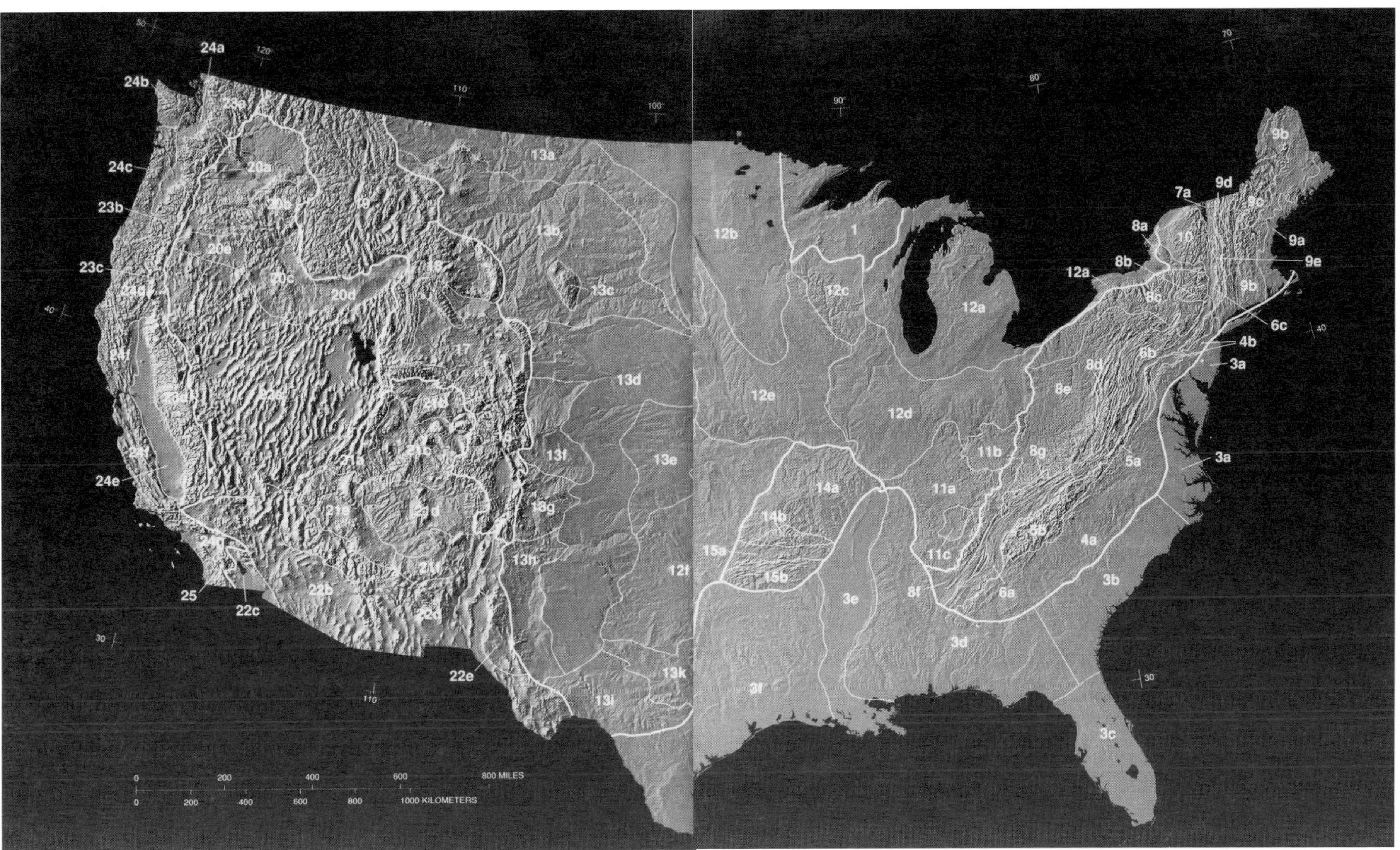

24a
24b
24c
23b
23c
24e
25
22c
22e
20a
20d
20e
20c
13a
13b
13c
13d
13e
13f
13h
13k
13i
12f
12a
12b
12c
12d
12e
1
14a
14b
15a
15b
3e
3f
8f
3d
3b
3c
3a
4a
4b
5a
5b
6a
6b
6c
8a
8b
8c
8d
8e
8g
9a
9b
9c
9d
9e
7a
10
11a
11b
11c
17
18
22b
0
200
400
600
800 MILES
800
1000 KILOMETERS

from southwest to northeast, is simulated as nearly horizontal, and crustal material is added (scraped) onto the western edge of the developing Rocky Mountains. Finite-element modeling techniques further reveal that the mantle layer of the North American lithosphere (at a depth of ~100 km) is stripped away in the subduction process and displaced eastward as far as the Black Hills.

One effect of the uplift was the development of a zone of extensional strain (a stretching and thinning of strata) immediately west of the uplifting Rocky Mountains. The thin crust, in combination with a reduced underlying mantle lithosphere, resulted in a subsurface zone of heat and structural weakness. Extension processes created sutures for extrusion of volcanics and facilitated differential collapse of the region (formation of grabens and half-grabens) to create the topography of the Basin and Range Province. The events primarily responsible for the modern configuration of this province began in the Middle Eocene (~45 Ma) in southern British Columbia and spread southeast to New Mexico by the Oligocene (35 Ma; Bird, 1988). A major extension in Death Valley occurred during the last 14 m.y. In the Sonoran Desert region, extension began in the Early Miocene (20–18 Ma), immediately followed by rapid uplift at a rate of ~7.2 mm/year. The area has been relatively inactive since 10 Ma. In general, the fast rate of regional extension was between 15 and 10 Ma.

The earliest dates for Tertiary volcanism in the Sierra Nevada cluster around 33 Ma and mark the beginning of that system. The Colorado Plateau, a tectonic block 45 km in thickness, was uplifted ~3 km beginning ~24 Ma. Extension forces developed in the Early Miocene (20–18 Ma), creating block faults, and portions were differentially uplifted by another kilometer with reference to the Basin and Range Province floor between 18 and 10 Ma for a total uplift of ~4 km. The present mean elevation is 2 km. The difference is likely due to erosion and subsidence resulting from further extension in the Great Basin in Pliocene and Quaternary times. Most of the net uplift has been since ~12–10 Ma.

With the Farallon plate subducted beneath North America, the present Pacific plate began to impinge on western North America in the Middle to Late Oligocene (~29–30 Ma; Fig. 2.16). In terms of plate mechanisms, the contact changed from oblique transpression (forces generated by oblique rather than direct contact between plate margins; Farallon plate) to a transform boundary (faults developing through shearing as one plate slides past another; Pacific

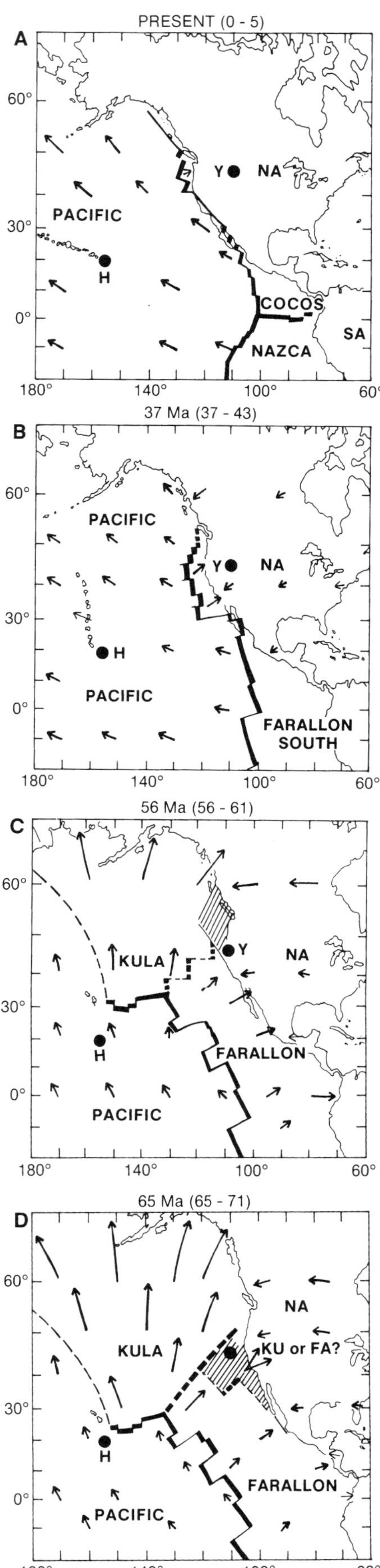

Figure 2.16. Position of the principal plates impinging on the North American continent during the four phases of the Late Cretaceous and Cenozoic: (A) the present (0–5), (B) 37 Ma (37–43), (C) 56 Ma (56–61), (D) 65 Ma (65–71). Y, Yellowstone hot spot. Reprinted from Winterer et al. (1989) with the permission of the Geological Society of America.

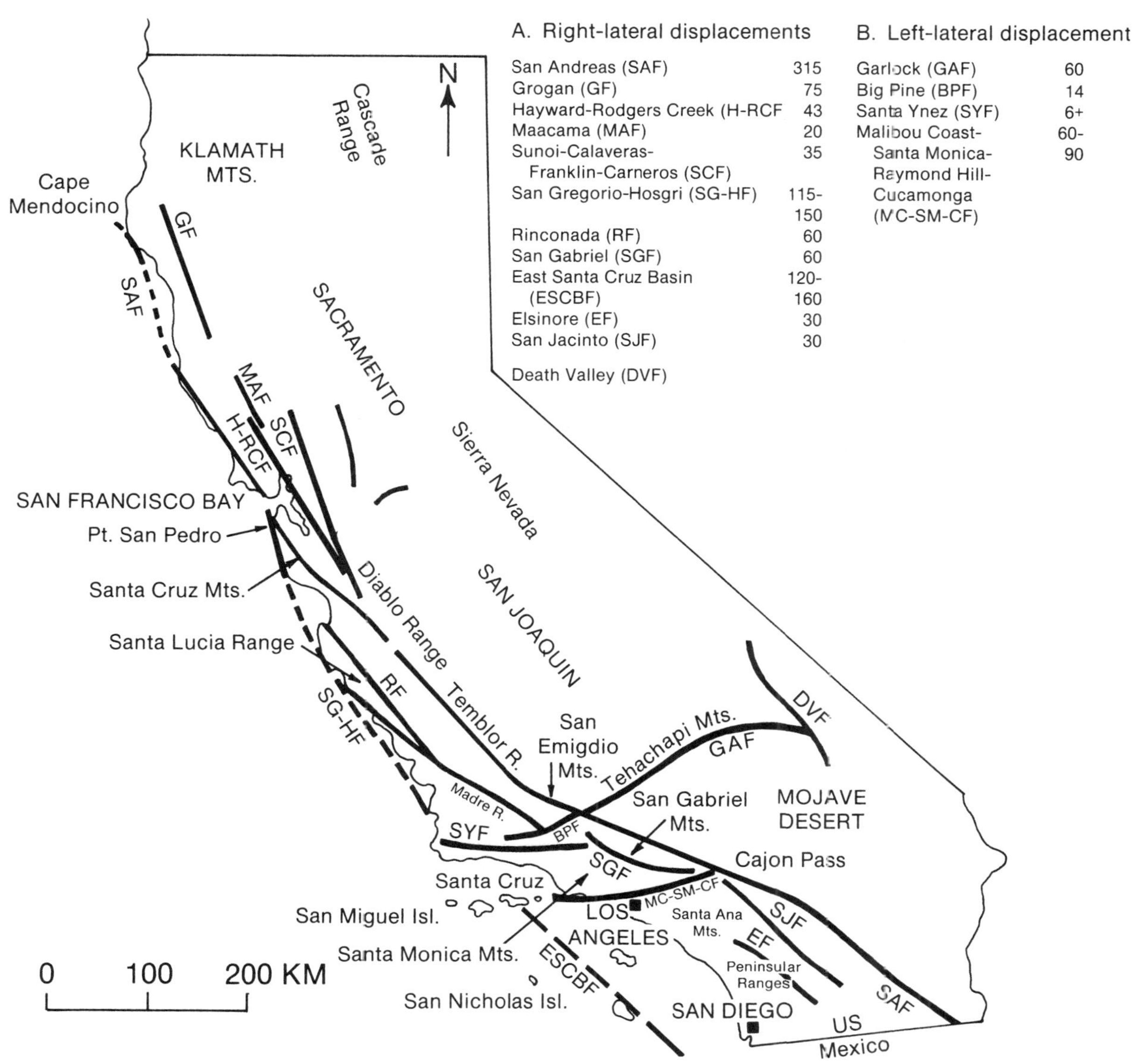

Figure 2.17. Distribution and displacement (in meters) of major fault systems in California. Reprinted from Oldow et al. (1989) with the permission of the Geological Society of America.

plate). The western edge of North America began to rise up and over the east Pacific rise (the old Pacific–Farallon spreading center, Fig. 2.16) by the Early Miocene (~17 Ma), and this continues to the present. In the Los Angeles earthquake of January 17, 1994, and the strong aftershocks through late March, the Sierra Nevada rose in places by ~380 cm (15 in.). The first Pacific ridge segments arrived off northern Baja California 28–26 Ma (Winterer et al., 1989). Among the consequences of the transform boundary mechanics was the initial development of the 1000 km long San Andreas fault system at ~29 Ma, and the first on-land manifestations were at 22 Ma. The modern, single major-fault configuration dates from ~12 Ma, and the system attained its present form ~5 Ma. The San Andreas fault system consists of *the* fault plus about 40 other associated ones that cover an area 200 km wide (Wallace, 1990).

Right-lateral slip now totals more than 315 km (Fig. 2.17), and there are some 8000 tremors a year along the fault. Uplift and volcanism in the California Coast Ranges began in the Oligocene (23 Ma), opening of the Gulf of California by sea-floor spreading began in the Miocene (Atwater, 1970), and transfer of Baja California to the Pacific plate began at 5–6 Ma. Baja California and areas of California west of the San Andreas fault are moving northward, relative to the rest of North America, at about 5 cm/year (Fig. 2.17). In 50 m.y. this will place the terrane off the coast of Alaska. The Traverse Ranges of California rotated clockwise and rose by overthrusting and transpression in Middle Miocene to Plio–Pleistocene times. About two-thirds of the elevation of the central Sierra Nevada was attained during the past 10 m.y. (Huber, 1981), and about two-thirds of the elevation of their west flank was attained since the

Mio–Pliocene (5–6 Ma), with 1 km or more of uplift during the past 3 m.y. The Sierra Nevada and the Cascade Mountains began to cast a rainshadow to the east during the latest Miocene and especially in the Pliocene. In the Pliocene there was an increased differentiation of vegetation east and west of the Sierra Nevada, and at high elevations there was a change from temperate hardwoods and evergreens to subalpine conifererous forest (Axelrod, 1962). Tectonic deformation continued into Late Holocene and modern times, as evidenced by earthquakes of 7.6–7.8 magnitude in the Humboldt Bay region of northern California in the past 1700 years [Clarke and Carver, 1992; see papers in Science (1992) regarding the Puget Sound region]. The central Coast Ranges of California were uplifted along block faults during the past 2–3 m.y. (Christensen, 1965).

The volcanic arc that constitutes the Aleutian Islands between western Alaska and Kamchatka developed in the Eocene (55–50 Ma) through capture of an accretionary wedge from the Cretaceous Kula plate. Subduction beneath the North American plate was particularly active in the Oligocene–Miocene and in the Pliocene–Quaternary. The mountain massif including Mt. McKinley in the Alaska Range underwent rapid uplift ~6 Ma (Fitzgerald et al., 1993). At this time rivers that flowed south into the Gulf of Alaska reversed to flow northward as at present. Volcanism associated with principal uplift of the Tibetan Plateau began at 13–14 Ma (Coleman and Hodges, 1995; Turner et al., 1993); much of that elevation, which affected atmospheric circulation patterns, was attained by ~14 Ma (Coleman and Hodges, 1995) and at ~8 Ma (Harrison et al., 1992). It should be noted that the Tibetan Plateau is the largest topographic feature on Earth and has an average elevation of 5 km and a surface of over 5 million km^2.

The uplift of the western North American mountains and Asian plateaus had a significant and immensely complex effect on the evolution of North American vegetation through the creation of orographic rainfall and rainshadow provinces, by altering atmospheric circulation, and by increasing erosion of silicate rocks. The latter process draws down CO_2, allowing more heat to escape, and results in a lowering of temperatures (see previous section on CO_2). It is ironic that even though an initial stimulus for the development of large computers was to predict weather and assess the impact of various factors on long-term climate, it is still not possible to predict or precisely model the effect of major changes in topography on global and regional climates. In fact, the permutations of such events are so complex that weather predictions, climatic trends, El Niños (Jin et al., 1994; Tziperman et al., 1994), fluid dynamics (atmospheric and oceanic circulation patterns), and motion of the inner planets are frequently used as examples of nonlinear relationships that are dealt with in the revitalized field of mathematics called chaos theory (Allen et al., 1993; Hanski et al., 1993; Lewin, 1992; Ruelle, 1991; Stewart, 1989; Waldrop, 1992). Chaotic or unforced fluctuations (Fig. 2.18) are superimposed on those derived from Milankovitch cycles, continentality, orogeny, CO_2 changes, and other processes, and may be a particularly important variable in ocean circulation and heat transport.

Nonetheless, it is generally recognized that uplift of major mountain systems and plateaus affects climate, and computer modeling is a convenient and powerful way to investigate single factors involved in climate–orogeny interactions (i.e., single-variable sensitivity tests). The CCM has been used to simulate large-scale uplift of the Himalayas and Tibetan Plateau and the Rocky Mountains and Colorado Plateau (Kutzbach et al., 1989; Manabe and Broccoli, 1990; Manabe and Terpstra, 1974; Ruddiman and Kutzbach, 1989, 1990; Ruddiman et al., 1989). In one sensitivity test using the CCM at NCAR, the presence or absence of mountains and plateaus was altered (M, mountains, simulating conditions at present; HM, half mountains, 10–3 Ma; NM, no mountains, with no exact paleophysiographic equivalent, but closest to conditions assumed to have prevailed in western North America prior to ~40 Ma). Other parameters were held constant [e.g., plate movement, isthmian connections (Beringia, North Atlantic, Panama, and the Drake Passage), CO_2 concentrations, Arctic sea ice, continentality, albedo effect of vegetation and snow cover, and ocean temperature]. This allowed the effect of orogeny on atmospheric circulation and climate to be isolated and more precisely defined. Such experiments are run in perpetual January and July modes to determine the effects on the extremes of annual climate.

In the models the general effect of mountain and plateau uplift in Asia and western North America was a redirection of the mean planetary waves in the northern hemisphere troposphere at middle and high latitudes (Ruddiman and Kutzbach, 1989). In the NM simulation (Fig. 2.19) a continuous westerly circumpolar jet stream with relatively small amplitude between troughs (low pressure) and ridges (high pressure) in the Rossby regime is present at latitudes north of 30° N. Patterns of annual precipitation are closely associated with these storm tracks, and they are uniformly distributed across the continent (west to east). This most closely approaches known conditions in western North America in the Late Cretaceous and earliest Cenozoic prior to extensive uplift. In the M experiment (Fig. 2.19) wind tracks change from nearly zonal (viz., more uniform climates, both regionally and seasonally) to the meandering, seasonal patterns of the present; the jet stream becomes discontinuous; the amplitude of the troughs and ridges increases; the mean positions of the troughs and ridges become more stationary (viz., a large stationary trough develops to the lee of the Rocky Mountains with greater regional differentiation of climate); wind flow, initially from the west, becomes northwesterly along the western coast of the United States by orographic deflection, which decreases the moisture content of the avected air; air flow becomes more northerly in the midwest; and a stronger southerly component develops along the east coast, bringing warm, moist summer winds into the region, as at present.

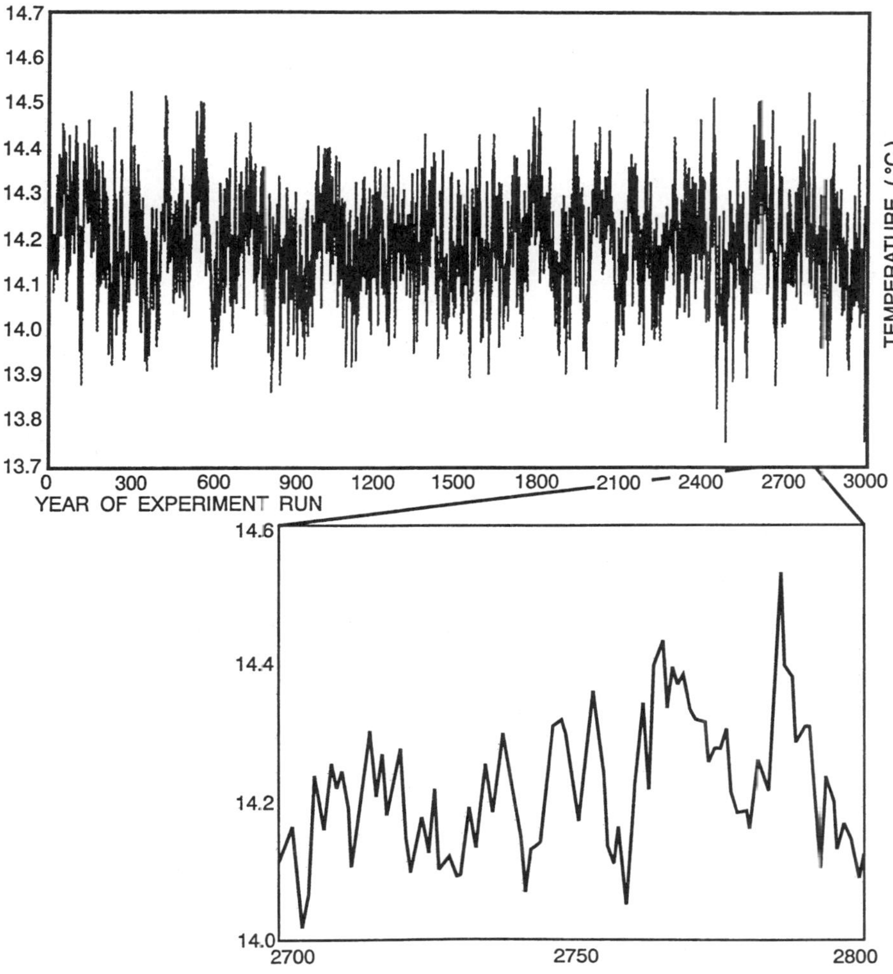

Figure 2.18. Unforced (chaotic) global temperature variations in a 3 ka run of the Global Climate Model (GCM) using the atmospheric composition of 1850 and a mixed-layer ocean of 250-m maximum depth. Reprinted from Hansen et al. (1993) with the permission of the National Geographic Society.

Climatic changes resulting from these alterations in atmospheric circulation are an increase in January precipitation in eastern and southern North America and a decrease in central North America (winter drying), an increase in January precipitation over the southern and southeastern United States and a decrease west of the Rocky Mountains, drier summers along the American west coast (transition to a Mediterranean climate) consistent with the gradual elimination of summer wet vegetation, and drier winters in the northern plains. Temperatures in the uplifted regions decrease by about 6°C (January cooling) because of rising elevation and a stronger northerly flow of wind patterns; colder winters and slightly cooler summers also develop in eastern and central North America, which is consistent with the elimination of much of the older (Eocene) subtropical vegetation; and there is little change in the far southeast, which is consistent with the persistence of elements of this vegetation into Late Cenozoic and modern times. A summary of these changes is given in Table 2.3.

Thus, the presence of mountains produces a broad spectrum of climatic effects within the immediate region and in distant parts of the world (Ruddiman, 1997). For example, among the effects of the uplift of the western North American and central Asian mountains and plateaus, as predicted by the CCM, are a weakening and western shift of an Icelandic low, where it merges with and reinforces a trough located off the east coast of the United States (Ruddiman and Kutzbach, 1989); a cooling of parts of the Arctic Ocean and east-central North America because of stronger northerly air flow; and a warming in Alaska because of a stronger southerly air flow (see below).

Not all of the predictions of the CCM are borne out by the geological record (Table 2.3), and refinements in the model continue to be made (e.g., Sloan and Barron, 1990). The present version calls for wetter summers and winters in the southern Rocky Mountains and the High Plains, contrary to a wealth of paleobotanical information documenting a drying trend. This inconsistency is likely a result of

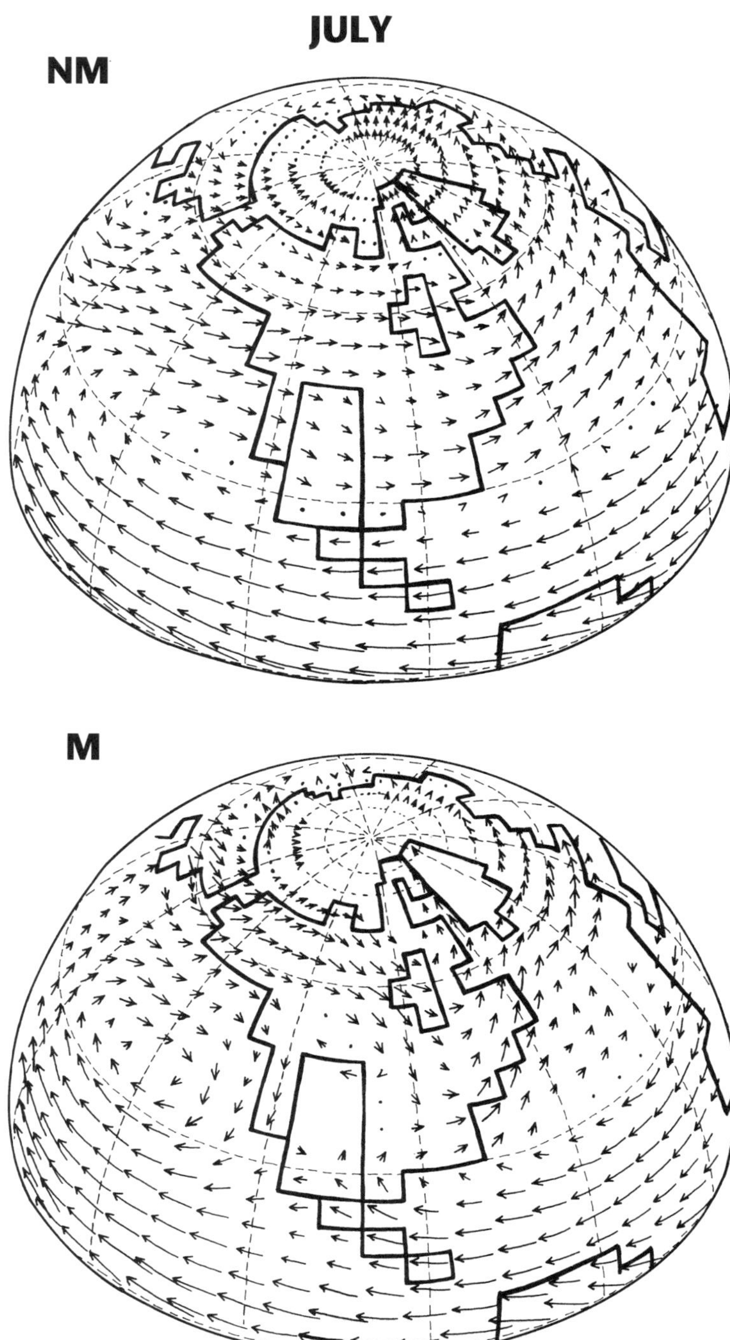

Figure 2.19. Computer simulation showing the effect of uplift of the Tibetan Plateau and Rocky Mountains on wind directions in July for the North American continent. NM, no mountains; M, mountains. Taken from Ruddiman and Kutzbach (1990) and reprinted with the permission of W. F. Ruddiman and the Royal Society of Edinburgh.

the relatively coarse grid constraints (Fig. P.1) that allow the Gulf of Mexico to extend about 5° W of its actual position, favoring transport of moisture to the High Plains and southern Rocky Mountains (Ruddiman and Kutzbach, 1989). The model also calls for warmer winters in interior Alaska, probably because it simulates broad plateaus more precisely than it resolves large and relatively narrow mountain systems like the Rocky Mountains into individual components having more local effects. Full-year integrations will also improve the precision of the model.

In addition to effects on climate, orogeny affects the course of development of regional vegetation in several other ways. Increased topographic diversity provides a greater number of habitats and increases the potential biotic inventory and community types for a region. Uplift and subsequent erosion creates new or disturbed habitats that are important to the establishment and perpetuation of hybrids. These interacting factors form a feedback system within the model, because with increased biotic and habitat diversity, the appearance and persistence of novel phenotypes accelerates, which further alters the composition of the communities and increases the inventory.

The uplift of mountains is one form of barrier that can divide the range of a species into geographically and, thus,

Table 2.3. Summary of simulated versus actual change in climate for North America based on CCM.

Region	Simulated Change	Data Vs. Model	Observed Change
W. coast NA	Drier summers	+	Drier summers
Am. N. plains	Drier winters (colder)	+	Drier winters (colder)
S. Rockies and S. plains	Wetter summers and winters	x	Drier at low elev.; wetter (?) at high elev.
E. NA	Colder winters and summers	+	Colder winters (and summers?)
SE U.S. and Gulf Coast	Wetter, little temp. change	+	Little temp. change
Alaska	Warmer winters, cooler summers	x	Colder, drier

Adapted from Ruddiman and Kutzback (1989).
(+) indicates agreement between simulated and actual change; (-) indicates disagreement.

reproductively isolated populations, resulting in vicariance speciation (Chapter 9). When substantial elevations are achieved, the peaks constitute isolated climatic islands; the arrangement of these peaks, in combination with the dispersal potential of the propagules, provide pathways for the migration of alpine, boreal, and temperate species during periods of climatic change. The range expansion of *Pinus banksiana*, *Abies*, and *Picea* into the southern Appalachian Mountains during the Late Quaternary was likely facilitated by increased target areas at high altitudes during periods of cooling climate. The enhanced probabilities afforded by topographic diversity for the establishment of isolated populations further promotes speciation through founder effect mechanisms, which are abrupt geographic and reproductive isolation of segments of a population through long-distance dispersal (Giddings et al., 1989). As noted, mountains create orographic rainfall–rainshadow systems and alter high-altitude atmospheric circulation, leading to the development of different kinds of vegetation on the windward and leeward sides.

The preceding scenario emphasizes mountain building and plateau uplift as agents of climatic change that, in turn, drive changes in vegetation through time. Molnar and England (1990), Burbank (1992), and Small and Anderson (1995) proposed an alternative model wherein climate is an agent for late Cenozoic mountain and plateau uplift (see also Caldeira et al., 1993; Raymo and Ruddiman, 1993; Volk, 1993). They suggest that climate-induced increases in mechanical erosion from Late Cenozoic cooling and accompanying increases in precipitation and storminess account for part of the uplift. Although an increase in the elevation of highlands by erosion may at first seem paradoxical, it is based on the principle of isostacy whereby depressed surfaces rebound as overlying material is removed. According to Raymo and Ruddiman (see also Prell and Kutzbach, 1992) the two models may not be incompatible:

> It is consistent with both hypotheses to argue that uplift of Tibet (and possibly elsewhere) is a plausible first cause of Cenozoic climatic changes (through circulation changes and weathering), but that climatic change (in particular, glacier activity) then caused additional erosion, exhumation and isostatic uplift in other regions . . . Thus, the arguments of Molnar and England can be viewed as a positive feedback mechanism, whereby uplift-induced erosion initiates global cooling, which then causes further glacial erosion and cooling worldwide. (1992, p. 122)

In this combined version there would be initial uplift, resulting in cooling via alteration of atmospheric circulation and a decrease in CO_2 concentration from erosion of silicate rocks, some consequent impoverishment in vegetation, followed by increased erosion, more uplift, additional climatic change, and further alteration in vegetation. Molnar and England (1990) believe that considerably greater uplift had been achieved in the Rocky Mountains earlier than 10–7 Ma suggested by data summarized by Ruddiman and Kutzbach (1989, 1990) and Ruddiman et al. (1989) and that the juvenile aspect of the terrane is due in large part to recent climate-induced erosion followed by isostatic uplift rather than to recent tectonic uplift. Current efforts at paleoelevation analysis discussed in Chapter 6 suggest that substantial elevations indeed may have been achieved in parts of western North America by the Eocene–Oligocene. In western Nevada elevations may have been higher than at present because of compression forces associated with the Laramide orogeny, followed by collapse from regional extension. This would allow the Late Cenozoic lowering of temperature to be given a more prominent role in explaining the biotic changes evident in the Middle Tertiary fossil record. The relative importance of uplift and global temperature change in influencing vegetational history is currently a subject of active discussion (Summerfield and Kirkbride, 1992; see also reply by Molnar and England; Axelrod, 1981). Although extremely complex, such hypotheses are probably approaching the real-world intricacies of orogenic–climatic–biotic interactions.

Disentangling the roles of orogeny and simultaneous solar-induced change in climate is important in interpreting vegetational history in western North America. In many areas the older fossil assemblages, with prominent warmth-requiring, broad-leaved, evergreen components, are followed by deciduous vegetation characteristic of cooler conditions. This involves uplift, providing cooler high-altitudinal zones, and global cooling. Associated geomorphic evidence, such as increased downcutting by streams, can be interpreted either as reflecting steeper gradients (orogeny) or as a greater volume of water and trans-

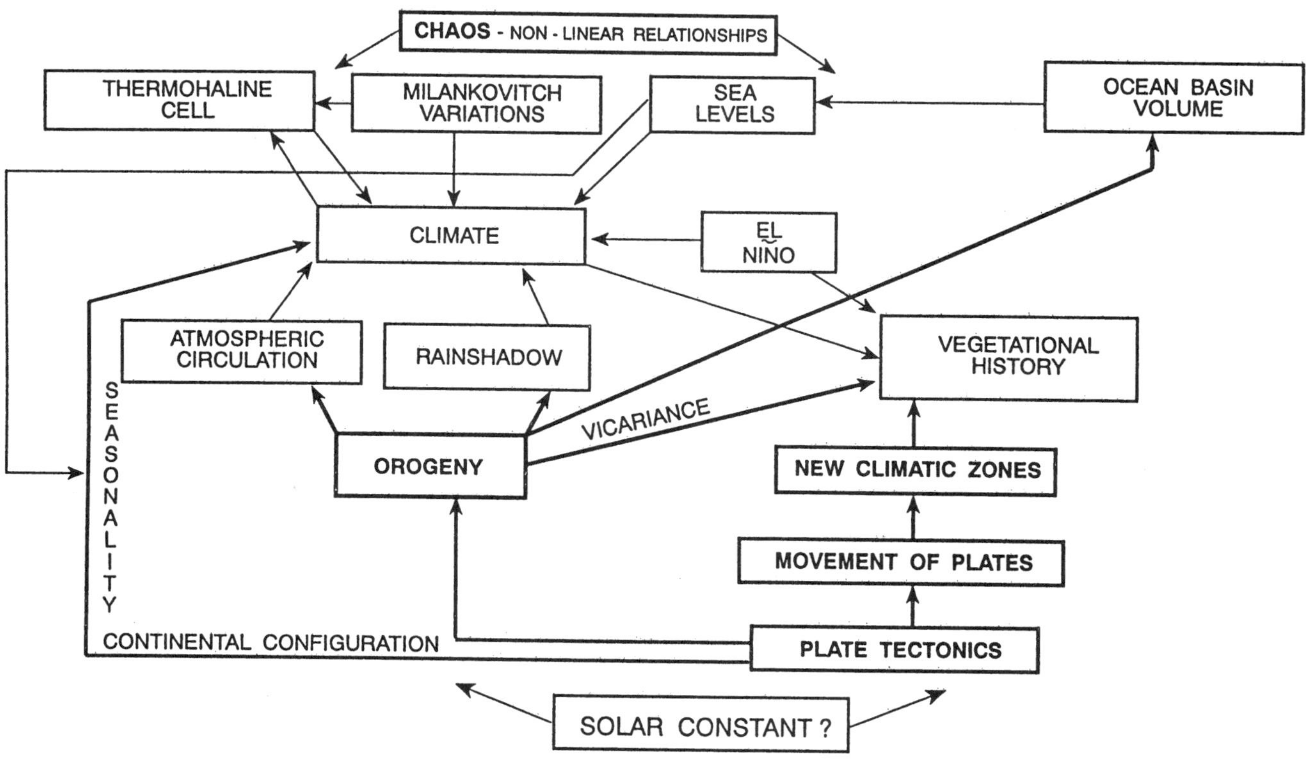

port load due to increased precipitation, and later due to meltwater from developing alpine glaciers (climatic change). If a real decrease in temperature of 6–9°C occurred during the past 15 m.y., using a lapse rate of ~6°C/km, this would give a decrease in estimated uplift of about 1500 m if the temperature decline was not taken into account. As an example, MacGinitie (1953), using the modern analog method, estimated the paleoelevation of the Late Eocene–Early Oligocene Florissant flora of central Colorado at less than 915 m. Meyer (1992), using foliar physiognomy, estimated MATs of lowland and upland fossil floras, and using an assumed lapse rate of 5.9°C/km, suggested a paleoelevation of 2450 m. The present elevation is ~2600 m. Thus, one paleobotanical analysis suggests uplift of about 1685 m since the Early Oligocene, while more recent estimates suggest virtually no net uplift.

Computer simulations have clearly documented the multifaceted effect of orogeny on atmospheric circulation and subsequent surface climates. As expected, however, the models also document that no single factor is sufficient to account for the climate changes detected by $^{18}O/^{16}O$ ratios in DSDP or ODP cores and evidenced by glaciations, sea-level curves, and alterations in the Earth's biota. It is estimated that deep-water temperature in the early Eocene was up to 12°C warmer than at present (Miller et al., 1987) and that deep and surface waters were 15°C warmer in the southern Pacific Ocean (Shackleton and Kennett, 1975). High temperatures also characterized land surfaces in the Early Eocene, then declined throughout the rest of the Cenozoic. Orogeny can account for only some, albeit a significant, part of this change: "[T]he amount of cooling remains far short of that required to explain long-term Cenozoic cooling. Additional factors, such as changes in atmospheric composition, seem to be required to explain global Cenozoic cooling" (Raymo and Ruddiman, 1992, p. 118) (Table 2.2, Diagram above).

Volcanism

> "[T]here was a thick darkness in all the land of Egypt for three days" (Exodus). [Eruption of Santorini (Thera), Aegean Sea, second millenium B.C. (Rampino et al., 1988)]

Another factor interacting with orbital-induced and orogenically influenced climatic trends is volcanism (Decker and Decker, 1991; Fiocco et al., 1997; Wood and Kienle, 1990). Volcanic activity frequently occurs in association with orogeny, is a direct consequence of plate tectonics, and is generated through one of three mechanisms. Volcanoes may form over hot spots, which are fissures in the Earth's mantle through which deep-seated magma emerges onto the surface as lava. This includes the type of volcanic activity evident at Yellowstone National Park (Anders, 1994) and in the Hawaiian Islands. Another type results in rift volcanoes that form where plates are diverging. These volcanoes may be submerged or they may emerge along midocean ridges as islands, such as Iceland. The third kind of volcanism develops along subduction zones where the edge of one plate dips under another along deep ocean trenches bordering the active margins of continents. Rocks at the leading edge of the subducting plate become molten at depths usually between 100 and 150 km and rise to the surface along fissures as arcs of volcanic activity called Benioff zones (after Hugo Benioff, California Institute of Technology). The arcs may be offshore such as those of the Antilles, Aleutians, Japanese, and Philippine Islands or inland such as those in western North and South America.

Volcanism in the Cascade Mountains began 44–38 Ma (Middle Eocene) in connection with subduction of the Pacific plate, and by 17 Ma (Early Miocene) a belt of volcanoes 100 km wide extended along nearly the entire western margin of the continent. By 8–9 Ma (Late Miocene) the arc had narrowed to its approximate present width and the southern margin of the Cascade Mountains had retreated northward to its present terminus at Mt. Lassen.

The Garibaldi–Pemberton volcanic belt in southwest British Columbia was activated with the subduction of the Juan de Fuca plate (a remnant of the old Farallon plate, Fig. 2.13) between 27 and 7 Ma, and the Alert Bay belt on Vancouver Island formed between 8 and 2.5 Ma. Further volcanic activity was evident in the Garibaldi belt between 2.5 Ma and the present through subduction of the Juan de Fuca and Explorer subplates.

To the north, the Colville foredeep formed on the north shore of Alaska in the Early Cretaceous (~135–140 Ma), which marks the beginning of the fold and thrust Brooks Range. Uplift was strong into the Aptian (113 Ma) and between the Albian and the Turonian (113–85 Ma). The proto-Aleutian arc began to develop 55–50 Ma as a result of subduction of the Pacific plate beneath the North American plate. Between 80 and 45 Ma subduction was rapid, estimated at 15–25 cm/year, and the volcanic arc formed as a result of jamming of the subduction zone, with accretion of Cretaceous strata between the trench and the continental margin (Oldow et al., 1989). Major uplift occurred at 40 Ma (Winterer et al., 1989), many of the islands had formed by 30 Ma, and the modern arc was essentially established after ~15 Ma. The highest elevations in the adjacent Alaska Range (e.g., Mt. McKinley, 6194 m, 20,230 ft) were attained within the past 5–6 m.y.

In recent times there has been increased opportunity to study volcanic eruptions with advanced technology (e.g., the total ozone mapping spectrometer, or TOMS, Nimbus-7 satellite) and to more precisely quantify the effects on global climate, particularly on temperature (Mass and Portman, 1989).

On March 20, 1980, a seismograph operated by the University of Washington recorded an earthquake of Richter Scale magnitude 4 on Mount St. Helens in the Cascade Mountains of Washington. On March 25, 47 earthquakes with magnitudes of 3 or more were registered within a 12-h period; on May 18, 1980, at 8:47 P.M., the first major volcanic eruption occurred in the conterminous United States since that of Lassen Peak in 1914–1917. The initial eruption was followed by other explosions on May 25, June 12, July 22, August 7, and October 16–18. The force of the explosions was 1.7×10^{18} joules (J, 1 J= the energy of 1 amp of current passed through a resistence of 1 ohm for 1 s). For comparison, a 1-megaton nuclear bomb generates 4.2×10^{15} J, so over a 9-day period, Mount St. Helens produced energy equivalent to 400 1-megaton bombs or the energy equivalent to 27,000 Hiroshima explosions at one every second for 9 h.

Mount St. Helens (Fig. 2.20), called Loowit (Lady of Fire) by the Pacific Northwest AmerIndians, erupted twice about 13 Kya and 20 times in the last 4500 years, at an average interval of once every 225 years. Over the 9 days of its most recent activity, ~1 km^3 of ash was discharged and ~2.7 km^3 of volcanic rock was displaced. Mount St. Helens is one of 15 active volcanoes in the Cascade Mountains from Lassen Peak in California to Mt. Garibaldi in British Columbia. Other volcanic eruptions equaling or surpassing Mount St. Helens include:

Asama (Japan)	1783
Laki (Iceland)	1783
Mayon (Philippines)	1814
Tambora (Indonesia)	1815
Krakatoa (Indonesia)	1883
Mt. Augustine (Alaska)	1883
Tarawera (New Zealand)	1886
Mt. Pélé, Soufriere (Antilles)	1902
Santa Maria (Guatemala)	1902
Ksudach (Kamchatka, Russia)	1907
Katmai (Alaska)	1912
Bezymianny (Kamchatka, Russia)	1956
Awu (Indonesia)	1966
Fuego (Philippines)	1974
Mt. Pinatubo (Philippines)	1991
Rabaul (Papua New Guinea)	1994

Some 400 active volcanoes are known to have erupted around the Pacific ring of fire in historic times, and the Earth presently has 37,000 km of subduction-zone volcanic chains. There are over 260 volcanoes, volcanic complexes, and volcanic fields in the western United States and Canada (Wood and Kienle, 1990). Long Valley caldera in California erupted 700 Kya, produced 500 km^3 of ash, and deposited 5 cm of ash across most of the central United States. The Toba eruption of Sumatra 70 Kya was tens of times larger than any historic eruption, and 2.2–4.4 billion tons of sulfuric acid fell on the Earth over a 6-year period. Mount Mazama exploded 7 Kya and formed Crater Lake in Oregon.

In the 1980 Mount St. Helens activity, ash columns reached an altitude of 18 km. Aerosol particles were carried to Spokane, Washington, 430 km to the northwest, where visibility was reduced to near darkness. Three days after the eruption the ash had crossed the North American continent, and aerosols still remain suspended in the atmosphere.

In 1982 El Chichón (1260 m, 4130 ft), located in the southern Mexican state of Chiapas near latitude 17° N, erupted between March 28 and April 4 (Parker and Brownscombe, 1983; Rampino et al., 1988; Tilling et al., 1984). Approximately 10–20 million metric tons (mt) of gaseous sulfur dioxide and other aerosols were ejected and sent

Figure 2.20. Eruption of Mount St. Helens, May 18, 1980. Photo by Austin Post; photograph number 17, U.S.G.S. Photographic Library.

more than 25 km into the atmosphere, and ~0.3 km^3 of pyroclastic material was discharged (Hoffer et al., 1982). This was nearly double the amount from Mount St. Helens. Within a few weeks a narrow band of ejecta had encircled the Earth, and within a year the stratosphere in most of the northern hemisphere contained aerosols produced by the eruption. The volcano was also active ~600, 1250, and 1700 years ago and earlier at intervals of ~600 years. The June 1991 eruption of Mt. Pinatubo in the Philippines ejected 30 mt of aerosols into the atmosphere (Brasseur and Granier, 1992; McCormick et al., 1995; Minnis et al., 1993) and produced the greatest ash clouds observed during the satellite era. Global cooling was estimated at 0.5°C for the following 2–4 years. Estimates are 100 mt for Tambora, and 50 mt for Krakatoa.

Ash tends to settle out within a few days; it is microdroplets of sulfuric acid, formed from ejected hydrogen sulfide and sulfur dioxide, that mostly intercept and scatter incoming insolation at ~25-km altitude (the Junge Layer; Hofmann and Rosen, 1983; Kotra et al., 1983). The stratosphere warms because the aerosols are more effective at reflecting the shortwave solar radiation than the longwave Earth-emitted radiation (Minnis et al., 1993). After the El Chichón eruption, temperatures in the stratosphere over the equatorial region rose by ~4°C. A warming of the upper atmosphere through aerosol concentration reduces the temperature cline, the strength of the Hadley Regime, and the force of the westerly trade winds. The Earth's surface (troposphere) cools because of reduced insolation, although there is considerable local and seasonal variation (Robock and Mao, 1992). The stratospheric aerosol cloud of El Chichón was about 100 times more dense than the ash-rich ejections of Mount St. Helens and will last for years. The previous El Chichón eruption about 600 years ago was several times the size of the latest one.

Continuing plate tectonic activity along the west coast

of North America is widely recognized through the devastating and highly publicized earthquakes that periodically affect the area. On April 25, 1992, a 7.1 earthquake occurred at Cape Mendocino, California, along the subduction zone between the Gorda and North American plates (Oppenheimer et al., 1993). Less well known is the frequency of smaller earthquakes. In the 4-day period of January 18–22, 1991, the U.S. Geological Survey at Menlo Park recorded 62 earthquakes in southern California registering 1 or more on the Richter Scale and 10 registering 2 or more, which is one measurable earthquake every 90 min. There is also evidence that residual spreading from the east Pacific rise (the remnant of the mid-Pacific ridge between the Pacific plate and the old Farallon plate) is still occurring near the southern terminus of the Juan de Fuca plate. At ~525 km west of the Oregon coast a string of 17 new lava mounds, constituting 50 million m^3 of new ocean floor, appeared between 1981 and 1987. Associated with the lava was the release of an estimated 100 million m^3 of water at 350°C. Such observations are important because they provide insight into the nature of plate tectonically induced events that shaped the evolution of western North America during the Cenozoic.

Model simulations focusing on volcanism, holding other factors constant, indicate that abrupt cooling of the lower atmosphere occurs during the first 2–3 months after a major eruption (Kelly and Sear, 1984) and that mean temperature drops by ~0.1–0.3°C for 1–3 years. A lowering of 0.5°C occurred after the eruption of Krakatoa, and 1816 was known as "the year without a summer" after the 1815 eruption of Tambora. The drop after El Chichón was predicted to be ~0.4°C, but, because of the simultaneous ENSO warming (Table 2.1), none was recorded. When the ENSO effects are compensated for in the model, a drop of several tenths of a degree results (Angell, 1988; Bradley, 1988; Mass and Portman, 1989). Models predicted a cooling of ~0.5°C by late 1992 after the 1991 eruption of Mt. Pinatubo. As of April 1992, a global cooling of ~0.4°C was evident when adjusted for the 1992 ENSO event (McCormick et al., 1995). The horizontal effects of an eruption are detectable almost immediately because of rapid west–east transport by directional winds. Effects to the north and south are delayed a few months because the aerosol cloud spreads more slowly by convection cell mechanisms. There seems to be little or no effect on atmospheric pressure or precipitation, even when large events are composited (Mass and Portman, 1989).

To better understand the effects of a few tenths of a degree change in MAT on climate, it may be noted that temperatures of the 1980s were generally high, 1990 was then the warmest on record, and the winter of 1991–1992 was the mildest ever recorded. Media coverage showed caravans of midwestern farmers carrying hay to drought-stricken farmers in the southeastern United States. The MAT for 1990 was 0.5°C above normal, and temperatures for the 1980s were 0.34°C above previous decades. The year 1991 might have been even warmer were it not for the eruption of Mt. Pinatubo.

The question relevant to Late Cretaceous and Cenozoic vegetational history is whether aerosols resulting from volcanic eruptions have any long-term effect on climate. The answer is still forthcoming because the sophisticated technology necessary to collect and analyze the data is new, observations are constrained by the infrequency and unpredictability of major eruptions, the response time of the ocean–atmosphere system is 4–5 years after volcanic pertubation, and many other predictable and unpredictable events are occurring simultaneously. As noted earlier, the El Chichón eruption occurred during the anomalous ENSO event of 1982, and the eruption of Mount Agung (Indonesia) in 1963 was followed by an ENSO event in 1963. Although a cause and effect relationship has been suggested (Handler, 1984), other analyses do not support such a hypothesis (Mass and Portman, 1989). Nonetheless, when ENSOs and eruptions occur simultaneously, it is more difficult to isolate the effects of volcanism alone on global climates. Such complex interactions continue in modern times as shown by the August 1997 eruptions of Montserrat in the West Indies and predictions that the simultaneous El Niño would be the most intense since the 1982–1983 event.

In addition to ENSO events and volcanism, the complex nature of factors determining climates is further illustrated by the fact that Late Cenozoic orogenies induced drying trends in western North America, which reduced surface vegetation and increased the amount of dust in the atmosphere (Rea et al., 1985). Dust particles between 1 and 10 microns (μm) in diameter may remain suspended indefinitely and are distributed globally in the upper troposphere (Li et al., 1996; Tegen et al., 1996). An increase in dust is recorded in oceanic sediments from the Pliocene onward (Janecek, 1985; Leinen and Heath, 1981). As noted by Ruddiman and Kutzbach (1989), the formation of local or regional deserts through orogeny may load the atmosphere with dust that produces a cooling effect on climate. Data for soil erosion in modern times show that impressive volumes can be involved. Sediment loss from forested regions presently amounts to 1.7 t/ha/year, compared to 18 t/ha/year for croplands in the United States. The effect of dust particles may be augmented by volcanism and mollified by ENSO events. Carbon dioxide released by volcanism tends to warm the atmosphere through a greenhouse effect mechanism, while the simultaneous emission of other aerosols lowers atmospheric temperatures.

Even with these complexities, there is evidence that aerosols from volcanic activity influence global climates to some degree by reducing insolation and altering atmospheric circulation patterns (Zielinski et al., 1994). High acid levels were recorded in ice layers from Greenland (Hammer et al., 1980), corresponding to eruptions of Krakatoa (1883), Tambora (1815), Laki (1783, Iceland), Eldgja (10th century, Iceland), and Kekla (1104, Iceland). The highest and most sustained level occurred between 1350 and 1700 A.D., correlating with a period of cold climate and glacial advance known as the Little Ice Age (Tilling et al., 1984). Also, the eruption of Toba 70 Kya corresponds to the early Wisconsin glaciation. However, the acidity spikes may have been augmented, to some extent, by tropospheric transport from volcanic activity in nearby Iceland (Hammer, 1984; Rampino et al., 1988); temperature records from the GRIPS2 ice core from Greenland now show that Toba aerosols had no long-term effect on climate.

The pace of volcanic activity is a reflection of the rate of subduction. For example, in Indonesia and Japan the annual rate of subduction is presently ~6–7 cm/year and volcanic eruptions are frequent (~one/year), while in western North America the slippage is 2–3 cm/year and eruptions are less frequent. In comparison, convergence rates between the Juan de Fuca and North American plates decreased fivefold between 30 Ma and the present, and this decrease correlates with a steady wane in volcanic intensity (Oldow et al., 1989). Recall that the rate of subduction of the Farallon plate has been as high as 12 cm/year and that subduction of the Pacific plate along the Aleutian Trench was 15–25 cm/year at 45 Ma.[5] During times when volcanic eruptions are more frequent and intense, they probably contribute to a greater lowering of temperatures for longer periods of time than the 3-year lowering of 0.3–0.4°C measured after the eruption of individual volcanoes in recent times, perhaps affecting temperatures for decades to a few hundred years (Mass and Portman, 1989). Although volcanism is not presently regarded as a primary forcing mechanism for long-term climatic change, depending on the timing, frequency, and coordination of volcanic events, the effect may be to trigger change in an unstable climate near a "bifurcation point" (Crowley and North, 1991) or to intensify or mollify trends controlled primarily by orbital variations, major orogenies, CO_2 concentration, albedo changes, continentality, and other factors.

On a local level, accounting for the effects of volcanism can be important in reconstructing vegetational history and paleoenvironments based on fossil floras (Cross and Taggart, 1982; Taggart and Cross, 1980). Thick deposits of volcanic ash alter soil characteristics and reset the successional process (del Moral and Wood, 1993). In the absence of adequate data on local or regional volcanism, the resulting changes in vegetation can be mistakenly attributed to fluctuating climate.

In addition, volcanic eruptions have had consequences related to other aspects of vegetational history. The mud (lahars) and lava flows associated with volcanism dam drainage systems, creating numerous lakes that serve as depositional basins for organic matter. In the case of Mount St. Helens, mud flows along Toutle Creek crested 9 m (27 ft) above any previously recorded flood level. Between 17 and 6 Ma, 120–150 volcanic eruptions produced the Columbia River basalts, covering an area of about 200,000 km^2 (somewhat larger than the state of Washington) with a maximum thickness of 2.5 km (Hooper, 1982). The total lava generated was over 700 billion m^3 within the first few days of the eruption. Similar flood basalts are known from the older South African Karroo Series (175–195 Ma), the South American Paraná Series (133 Ma), and the Indian Deccan Traps (65–69 Ma). Ash particles from these eruptions settled onto surfaces of the newly created lakes and formed a fine-grained matrix for the preservation of fossils, frequently with remarkable surface detail. The matrix may also contain fossil pollen and spores and other plant and animal microfossils. In addition, the ash charges the waters with silicates that promote the growth of diatoms, resulting in extensive deposits of diatomite, another fine-grained sediment conducive to the preservation of plant and animal fossils. Many areas in western North America are characterized by fossil-bearing shales of volcanic origin, such as the Ruby Basin flora of the Late Eocene or Early Oligocene age in southwestern Montana, or diatomite, such as the Middle to Late Miocene Trout Creek flora of southeastern Oregon. Frequently the deposits are associated with thick sequences of lava (Figs. 2.21, 2.22) that provide material for radiometric dating. Thus, an important consequence of volcanism in western North America has been to produce geologic conditions suitable for the preservation and dating of one of the most extensive sequences of Tertiary floras and faunas in the world (diagram p. 62).

Land Bridges

The role of land bridges in vegetational history is not simple. The most evident biological consequence is that at different times and to varying degrees they form a connecting link, filter, or barrier to the migration of terrestrial organisms and marine organisms. At the same time, however, they alter climate by affecting ocean circulation. The connections may be complete or partial, prevailing climates may vary over time, and while links are being forged in one place (e.g., the Isthmus of Panama), they can be broken elsewhere (e.g., across the Bering Straits). During the Late Mesozoic and Cenozoic the northern portions of North America were connected to other continents via two land bridges: one across the North Pacific to Asia and another across the North Atlantic to Europe.

Beringia

In August of 1728 Russian explorer Vitus Bering sailed through the strait between the Seward and Chukotka Peninsulas, demonstrating that the New World and Old World continents were indeed separate. Later soundings showed, however, that water in the northeastern half of the Bering Sea and portions of the Bering Strait and Chukchi Seas was less than 100 fathoms (600 ft) deep and mostly less than 100 ft. At the closest point North America is separated from Asia by only 80 km. The kinds of communities are generally similar throughout the region, and paleonto-

Figure 2.21. Lava flow of the Columbia Plateau along highway 95 near the Idaho–Oregon border.

logical studies have documented a similarity among terrestrial biotas on both sides of the strait for most of the Tertiary. The area has long been recognized as an integrated biotic and physiographic unit, and in 1937 Eric Hultén applied the name Beringia to the region (Fig. 2.14).

From the standpoint of geology and plate tectonics, the Cenozoic history of Beringia is complex. Numerous plates and plate fragments (Pacific, Kula, Juan de Fuca) impinged on the region, and exotic terranes consisting of rotated blocks, volcanic arc segments, and forearc basins were added to its margin. The fringing Aleutian Islands arose as a volcanic arc above the subduction zone between the Pacific and North American plates, beginning principally ~40 Ma; seamounts (guyots), originating over the Hawaiian Islands hot spot, continue to be subducted into the Aleutian trench. From the standpoint of physical geography the situation is simpler: the area was mostly above sea level for most of the Tertiary. The disruption of land connections between North America and Asia dates from ~3 Ma (Repenning and Brouwers, 1992), after which interchange was determined primarily by Quaternary glaciations, sea-level changes, and climate (Hopkins et al., 1982; West, 1996). Continuous land was available between 60 Kya and 25 Kya (oxygen isotope stage 3), and a sea-level fall of 121 m between 20 Kya and 18 Kya (late glacial; oxygen isotope stage 2) produced the maximum extent of the bridge during Late Pleistocene time (~1500 km wide). After ~11 Kya little exchange was possible (Hoffecker et al., 1993).

The North Atlantic

The history of land connections across the North Atlantic is essentially the history of the breakup of northern Laurasia through the mechanism of plate tectonics (Dawes and Kerr, 1982; Ziegler, 1988). Developing rift systems separated Greenland from northwestern Europe through the Denmark Strait and the Norwegian–Greenland Seas and from mainland North America through the Labrador Sea–Davis Strait–Baffin Bay region (Fig. 2.14). The separation involved north to south rifting in the North Atlantic and adjacent Arctic Ocean in the early Mesozoic, subsequent development of a spreading center in the central Atlantic, and eventual propagation of the northern terminus of the mid-Atlantic Ridge system into the Arctic region.

In the Early Jurassic (~180 Ma) there was evidence of crustal extension and differential subsidence of graben structures on the Grand Banks and in the East Newfoundland Basin as a result of rifting and wrench tectonics. In general, wrench deformation (e.g., in northern Baffin Bay) is a compensation for rifting in adjacent areas (e.g., in the

Figure 2.22. Ash deposits along highway 95 near the Idaho–Oregon border.

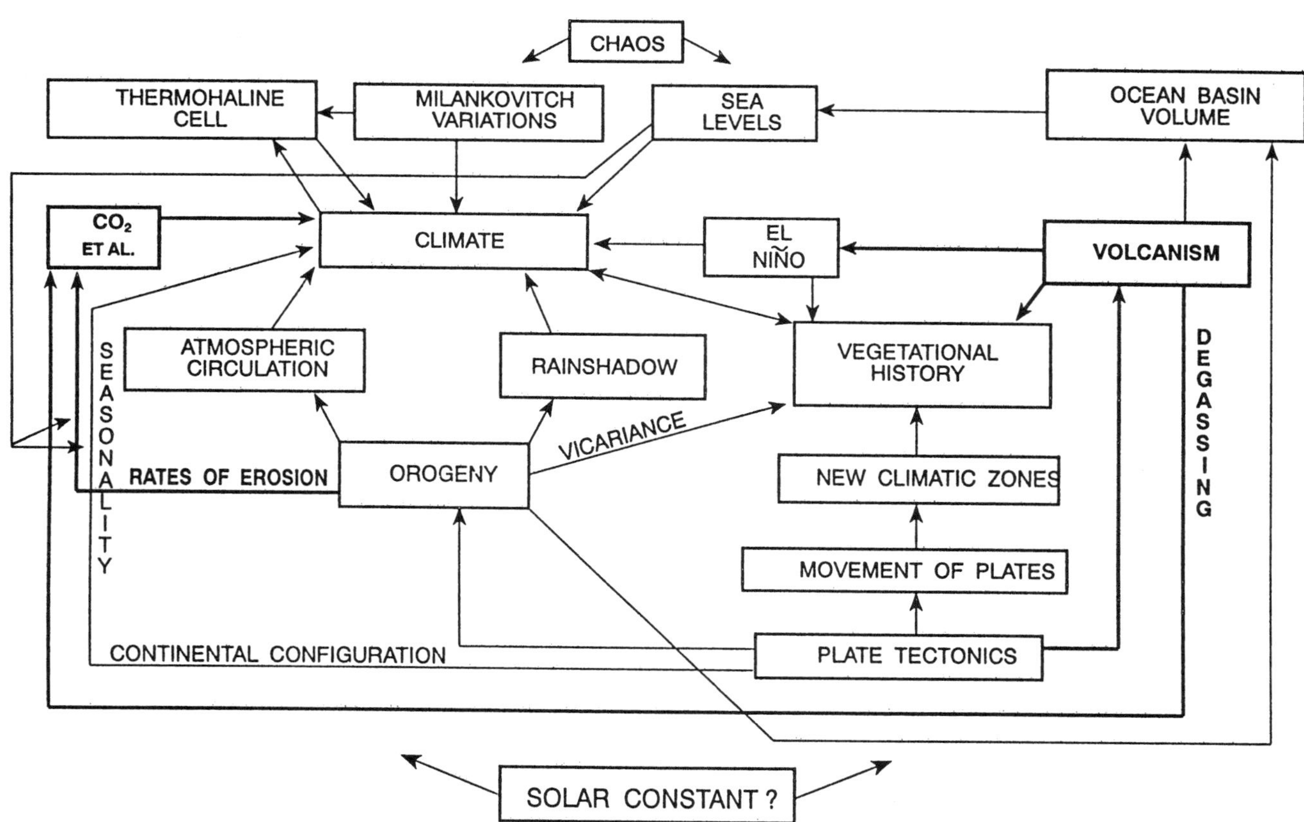

Labrador Sea). The Norwegian–Greenland Sea rift system was also active. However, tectonic activity was considerably less in both regions than in the preceding Triassic, and the Mid-Jurassic actually represents a period of waning influence for locally controlled rifting events. After the Mid-Jurassic the center of plate tectonic activity shifted to the central Atlantic as a continuation of the opening of the Atlantic Ocean from the south (Ziegler, 1988). At this time (180 Ma) actual crustal separation, as opposed to crustal extension, was evident at the central mid-Atlantic Ridge; by the Late Jurassic spreading was proceeding northward at the rate of ~3.4 cm/year, decreasing to ~2.3 cm/year in the Early Cretaceous. By the Neocomian, spreading had reached the southern part of the North Atlantic in the Labrador Sea, as evidenced by dikes (tabular bodies of igneous rock that cut across adjacent rocks) in southwestern Greenland and diatremes (tubular volcanic vents through flat, overlying rocks formed by explosive ejection of gas-charged magmas) in southeastern Labrador. Crustal extension in the East Newfoundland Basin is estimated at between 100 and 200 km, while to the north it is much less. By the Barremian (Early Cretaceous) rifting was also evident in the Davis Strait and in southern Baffin Bay. According to Ziegler (1988), during the Late Jurassic and Early Cretaceous the Labrador–Baffin Bay Rift was extending into the southern Canadian Archipelago at a rate of ~8 cm/year. Extension in the Norwegian–Greenland Sea Rift system also continued during the Late Jurassic and Early Cretaceous.

In the Aptian–Albian, actual crustal separation developed between the Labrador Shelf and Greenland. North Atlantic seafloor spreading extended rapidly into the Labrador Sea, accompanied by crustal extension in Baffin Bay induced by a counterclockwise rotation of Greenland relative to North America. In the Turonian the Arctic Sea invaded into the Baffin Bay Rift, and in the Maastrichtian there is evidence of a mixing of Arctic and Atlantic waters that marks the formation of the Davis Strait. Igneous activity in the Aptian–Albian through the Early Paleocene also occurred along the north coast of Greenland as a result of wrench movements between Svalbard and northeast Greenland, which compensated for continued extension along the Norwegian–Greenland Rift.

In the Late Paleocene to the earliest Eocene, volcanism increased in the Davis Strait and north of Ellesmere Island. This activity is known as the Thulean volcanism, and it produced basalts estimated at volumes equivalent to those of the Late Cretaceous Deccan basalts of India. An earlier extension in the Cretaceous had separated Cape Dyer on Baffin Island from Disko Island in west-central Greenland by ~150 km (Ziegler, 1988). During the latest Paleocene to Early Oligocene the distance increased by another 200 km through seafloor spreading. Spreading terminated in the south Labrador Sea in the Early Oligocene. Crustal separation occurred in Baffin Bay during the Late Paleocene and continued into the Early Oligocene. The period of Thulean volcanism also represents the final stage in rifting before initial crustal separation was achieved between Greenland and northwest Europe in the earliest Eocene (~56 Ma; chron 24; Eldholm et al., 1994). This brought to an end the early Norwegian–Greenland Rift system that had affected tectonics between Greenland and Europe for 275 Ma since the Late Carboniferous.

Subsidence of the Iceland–Faeroe Ridge in the Middle Miocene allowed interchange of cold waters from the Norwegian–Greenland Seas with warmer waters from the Atlantic Ocean. This event essentially established the present configuration of the North Atlantic.

Several aspects of Late Cretaceous and Cenozoic paleophysiography of the North Atlantic region are important from the standpoint of terrestrial biogeography. First, the separation between North America and western Europe progressed from south to north with propagation of the mid-Atlantic Ridge. In a very general sense, therefore, two land bridges can be recognized, particularly for the area between Greenland and Europe (McKenna, 1983). A southern Thulean route, at 45–50° N paleolatitude, extended across Greenland–Iceland into Scotland and afforded mostly continuous land surfaces through the end of the Cretaceous. Although the mixing of Arctic and North Atlantic waters through Baffin Bay and the Labrador Sea is first evident in the Maastrichtian (~66 Ma), no major physical barriers existed for most terrestrial species until the Middle Eocene. Fossil floras indicate that this southern portion of the North Atlantic land bridge served as a migration route for tropical and subtropical elements from the Late Cretaceous through about the early Middle Eocene.

The northern portion of the land bridge through Svalbard (northern Scandinavia to northern Greenland), at paleolatitude 55–60° N (to near 75° N; McKenna, 1972), is known as the DeGeer route and it remained intact longer and initial crustal separation was evident in the earliest Eocene (~56 Ma). The DeGeer route probably provided nearly continuous land surfaces until the Early Oligocene; because of its northerly position, it was occupied in the Late Cretaceous and Early Tertiary by more temperate and temperate to subtropical transitional assemblages. After that time the region was likely populated by progressively cool to cold temperate vegetation, and glacial conditions developed locally in the Miocene.

To some extent, reconstructing biogeographic events through the North Atlantic land bridge is conditioned by data from other regions. Recently it has been found that genera such as *Weigela* (Caprifoliaceae, presently extinct in North America) grew in the Arctic region during the Neogene (Matthews and Ovenden, 1990), while present-day North American genera such as *Dulichium* (Cyperaceae) and *Diervilla* (Caprifoliaceae) are known from the fossil record of Asia. Previously the North Atlantic connection had to accommodate all such interchange, but recent studies document their presence in the Beringian region during the Neogene. The Beringian pathway became increasingly more likely as the route of exchange of temperate elements as rifting in the North Atlantic continued through the Neogene. Thus, as a clearer picture emerges of Arctic vegetational history, it becomes unnecessary to overload the North Atlantic land bridge with migrations that were physically, climatically, or temporally unlikely.

Second, the interchange of most terrestrial organisms is not significantly affected during early phases of rifting in which only narrow barriers exist. There is usually a lag time between the inception of extension, crustal separation, and the appearance of distinct terrestrial biotic provinces; and there have been numerous intervening land surfaces across the North Atlantic Ocean throughout the Late Cretaceous and Cenozoic. Many terrestrial organisms continued to migrate through the region even after the appearance of channels, bays, straits, and narrow to moderately broad seas.

Third, the structural separations were overprinted with changes in sea level. Late Cretaceous sea levels are estimated to have been as much as 300 m higher than at present (Vail and Hardenbol, 1979), while during the maximum of the Late Cenozoic glaciations they were from an estimated 92 m to more than 100 m lower (Geophysics Study Committee, 1990; Savin and Douglas, 1985). There were other intervening fluctuations in land–sea relationships that were caused by a variety of factors, several of which are discussed by Ziegler (1988).

The interchange of terrestrial biotas during the Late Cretaceous and Cenozoic, like that through Beringia, was influenced but not determined solely by physiography and sea levels. Changes in climate from the warm, equitable conditions of the Late Cretaceous through the Early Eocene, in the transitional period of the Middle Eocene and Early Oligocene, to the glacial conditions of the Late Cenozoic were also important in defining the timing and the kinds of communities migrating across the North Atlantic land bridge. The biotic elements and communities populating the region in the Late Cretaceous and Tertiary are described in Chapters 5–8 (diagram p. 65).

Terranes

Another consequence of plate tectonics that is relevant to North American vegetational history in the Late Cretaceous and Cenozoic is the terrane: sediments; portions or entire volcanic island arcs, forearc basins (e.g., the Eel River Basin of offshore California), and seamounts; fragments of continents; and other suspect-exotic-allochthonous (transported) materials that have been carried by plate movement and accreted onto the active margins of distant land masses (Dewey et al., 1991; Hashimoto and Uyeda, 1983; Howell, 1989; Jones et al., 1983). One kind of terrane is crustal blocks, which are fault-bound units with a lithology, paleomagnetic configuration (remnant magnetism), and/or fossil content that is different from adjacent strata. An example is Wrangellia (from the Wrangell Mountains 400 km east of Anchorage; Jones et al., 1977; Stone et al., 1982). In the Triassic it was situated at or below the equator and reached the vicinity of the paleo-Oregon-British Columbia coast in the Middle to Late Cretaceous (70–100 Ma). Pieces were later carried farther north through nearly 24° of latitude by fault movement and are now found on the Queen Charlotte Islands, and in the Wrangell Mountains of southern Alaska. Present west-coastal Oregon includes other fragments sutured on after Wrangellia was in place. Support for the concept that parts of west-coastal North America are from distant regions comes from Vancouver Island. The remnant magnetism of mollusk-bearing rocks indicates that in the Late Cretaceous these rocks were located at ~Pl 25±3°N, or off Baja California, 3500 km to the south (Ward et al., 1997).

Sediments may also be added to coastal margins in the form of accretionary prisms when ocean-bottom material is scraped onto continental margins at V-shaped subduction zones. Accretionary prisms are found near Kodiak Island and in the Gulf of Alaska. Thus, much of the west coast of North America, from Baja California to Alaska, and extending inland up to 500 km, is a collage of some 100 exotic terranes added from Permian through Early Mesozoic time, and especially between 200 (Early Jurassic) and 50 Ma (Early Eocene; Churkin, 1983; Jones et al., 1983, fig. 1).

Terranes may be continental in size (e.g., India, which includes a complex of smaller terranes) or comparatively small, such as the individual basaltic seamounts that have been periodically added to the Oregon coast. The distances involved are from a few kilometers to thousands of kms for terranes appressed to the California coast near San Francisco. The orientation may change during transport and docking, as shown by the Channel Islands off southern California, which were rotated clockwise 70–90° during the Neogene.

The implication of exotic terranes for vegetational history is apparent. If fossil-bearing units have been transported significant distances, paleoenvironmental reconstructions are valid for the original locale of the terrane at the time the fossils were being deposited, not for the eventual docking site. Similarly, biogeographic patterns (geographic relationships, pathways of migration, hypothesized land bridges, and seaways) can be misconstructed if the fossils were not part of the regional flora. A case in point is the Miocene Mint Canyon flora of southern California (Axelrod, 1940, 1986, 1987) that was transported northward 200–300 km through some 2–3° of latitude (1° of latitude = 110.68 km or 68.7 mi). Another example is the Carmel flora (12–11 Ma) from the middle of the Monterey Formation of California that was transported from the west coast of Mexico (D. I. Axelrod, personal communication, 1996). During the Late Cretaceous the Salinia terrane (off the coast of southern California) was located ~2600 km to the south, and some Late Cretaceous volcanics north of San Francisco (Point Arena) were probably displaced ~2900 km between 90 and 48 Ma. The west coast of North America in

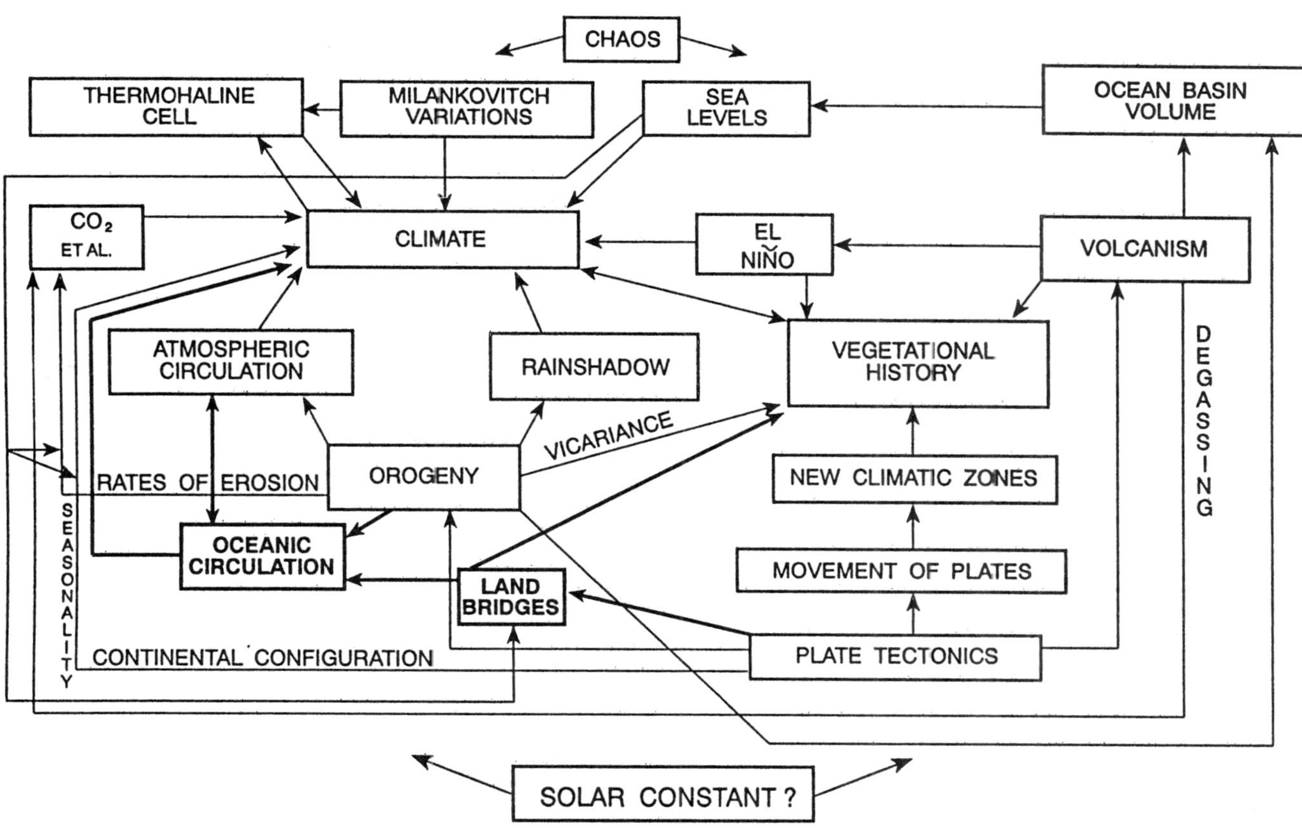

general, and Alaska in particular, includes terranes accreted onto the continental margin. Recognition of this fact is essential to the proper interpretation of North American fossil floras along the west coast and for the accurate reconstruction of paleoenvironments (diagram p. 66).

CATASTROPHES

Catastrophes are devastating events in geologic time that are unique or follow no apparent predictable pattern (Ager, 1993; Clube, 1989; Huggett, 1990; Sharpton and Ward, 1990). The effects may be global and influence the subsequent course of evolution for various plant and animal groups via extinctions and new selective pressures, including modification of world climates. The most prominent example is the bolide (asteroid or comet) impact at the end of the Cretaceous.[6] Other effects are more regional in scope. The late Wisconsin Missoula Floods devastated extensive areas in the state of Washington, modifying edaphic and landscape features and resetting successional trends back to the initial colonization of barren rock. Finally, there are catastrophes occurring at present that are part of the complex web of interacting factors influencing global climates. One of the most tragic and potentially dangerous to the welfare of human societies is the devastation of the world's tropical forests and other natural vegetation. About 2–5% of the Earth's land area burns each year, and this generates 5–25% of the radiative climatic forcing from human-produced greenhouse gas emissions (Levine, 1991; review by Overpeck, 1992). Plant transpiration and evaporation are responsible for ~65% of the total precipitation falling on land or ~70,000 km^3 of water per year. Response to forest clearing is rapid because water resides in the atmosphere only ~10 days before returning to Earth. With nearly two-thirds of this plant formation already destroyed, the process constitutes a kind of mindless experiment on the climatic consequences of a major alteration in the Earth's vegetation cover (Dunnette and O'Brien, 1992).

Asteroid Impact

Paleontological records show that periods of increased extinction and ecosystem collapse followed by recovery and appearance of new systems and lineages have occurred several times in the past (Fig. 2.23; Belton et al., 1992, 1994). There was a notable die-off in the Late Devonian known as the Frasnian–Famennian event and another at the end of the Permian. One of the most prominent was the terminal Cretaceous event (65 Ma) in which one-half of all genera and 75% of all marine species were estimated to have perished. Alvarez (1986) and Alvarez et al. (1980) proposed that an asteroid collided with the Earth at the end of the Cretaceous, sending large amounts of debris into

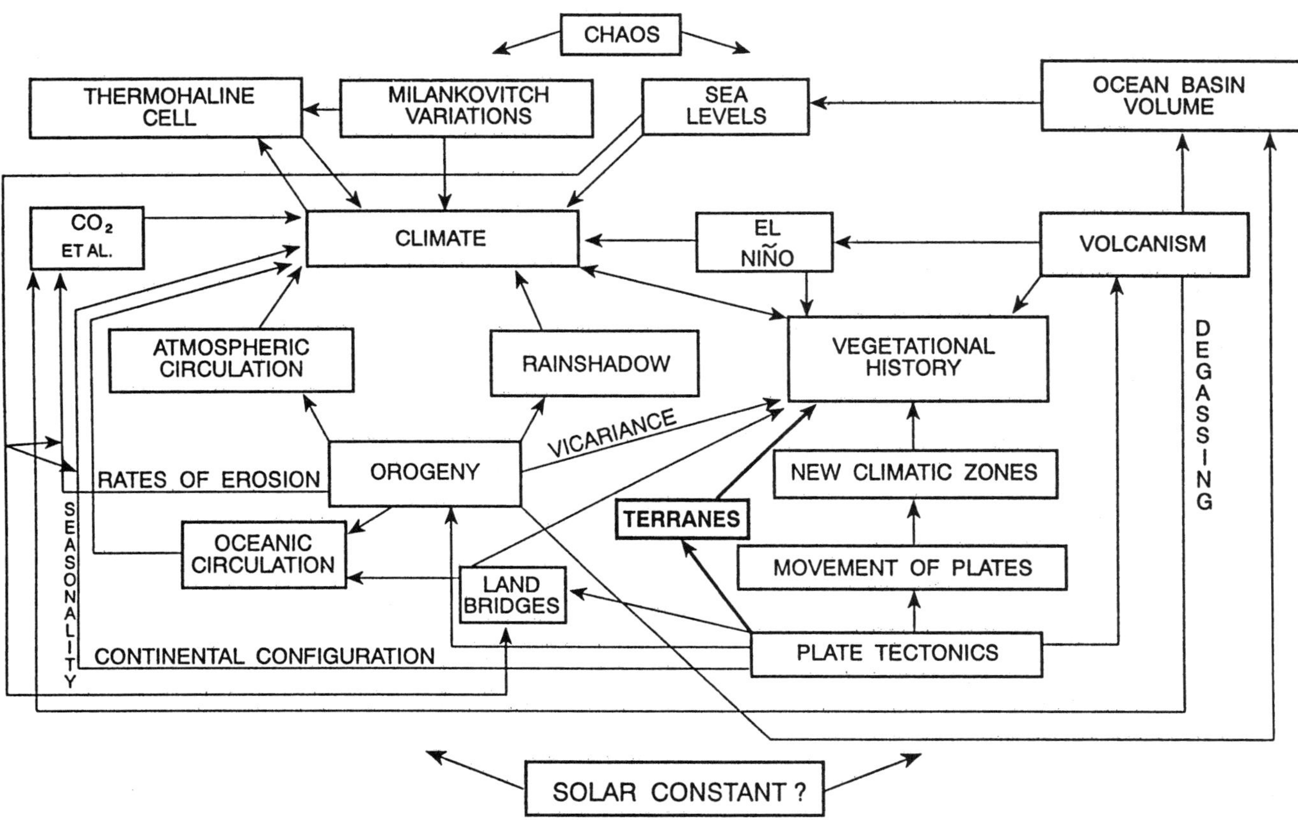

the stratosphere (Covey et al., 1994) and creating darkness and cooling that reduced the rate of photosynthesis. Other consequences were shock waves, acid rain (Retallack, 1996), and global wildfires. The size of the asteroid was estimated at 10±4 km in diameter, the amount of ejecta at 60 times the object's mass, and the duration of the effects at 3–5 years. The force of the impact was estimated at 10,000 times the world's nuclear arsenal. An attractive feature of the theory is that it provides a unique explanation for widespread and sudden extinction in various plant and animal groups that cannot be accounted for solely by changes in climate [Crowley and North, 1988, 1991 (chapter 9); Sheehan et al., 1991; see *Marine Micropaleontology* 29(2), 1996, for a series of papers on the El Kef blind test]. Even the simultaneous occurrence of several reinforcing events such as phases of the Milankovitch variations, ENSO events, and volcanism (e.g., contemporaneous eruption of the Late Cretaceous Deccan flood basalts) are inadequate, as well as improbable, to account for the magnitude and global extent of the extinctions. The oxygen isotope record reflects no major deviation from the gradual temperature decline for the 10–15 m.y. period bracketing the Cretaceous–Tertiary [K–T] boundary (Crowley and North, 1988).

An early basis for the theory was the discovery in 1978 of the platinum group or siderophile metal iridium (Ir) in anomalous concentrations at the K–T boundary. Iridium is depleted in the Earth's crust, but it is more abundant in chondritic meteorites that cross the Earth's orbit. Amounts of 20, 30, and 160 times the concentration in adjacent strata have been found in K–T boundary clays. The initial sites of these iridium anomalies were in Italy, New Zealand, and Denmark; but by 1985, 15 sites had been found in such widespread localities as Spain, the South Atlantic, the North Pacific, New Mexico, and Texas.

There were two principal objections to the original version of the theory. First, not all plants and animals were affected. Some animal groups, such as benthic foraminifera and some birds (Cooper and Penny, 1997) crossed the boundary with little or no change, and others that disappeared near the K–T boundary were already in decline prior to the impact. In particular, the rudists (extinct, shallow marine water bivalves with the habit of corals) disappeared about 1.5 m.y. before the event. Other bivalves, such as the inoceramids, also show a gradual decline prior to the end of the Cretaceous. Many terrestrial organisms, including several plant groups, crossed the K–T boundary without marked extinction. In contrast, planktic organisms, the non-avian dinosaurs (near the top of the food chain), and many ammonite species disappeared abruptly (Marshall and Ward, 1996). A later impact of a meteorite 3.4 km in diameter at 50.8 Ma on offshore Nova Scotia resulted in few extinctions (Jansa, 1993), and the consequences of the Pliocene (2.15 Ma) Eltanin impact 1,500 km southwest of Chile are still under study (Gersonde et al., 1997). These observations introduce a selective component to the mass extinction and require accommodation for cooling climates and the gradual retreat of the seas from

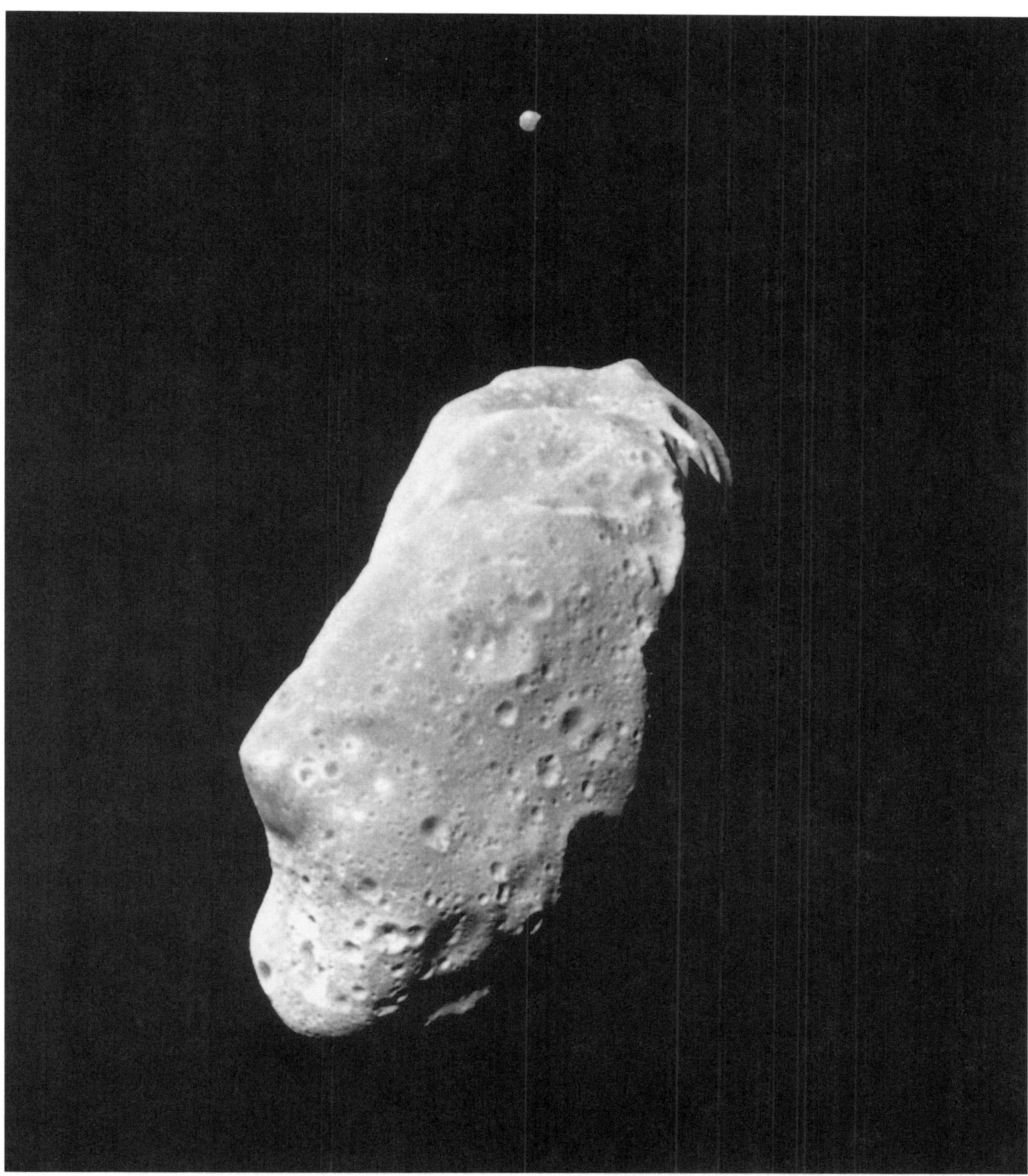

Figure 2.23. Asteroid 243 Ida and its newly discovered moon photographed from NASA's Galileo spacecraft, August 28, 1993, from a distance of ~10,870 km (6755 mi). The asteroid is ~56 km (35 mi) long. It is in the main asteroid belt between Mars and Jupiter and the 243rd discovered since the discovery of the first asteriod early in the 19th century. The photograph was taken as the spacecraft flew past Ida on its way to Jupiter where it went into orbit in December 1995. Photograph courtesy of the Jet Propulsion Laboratory, Pasadena, California. See also Belton et al. (1994).

the continental interiors and margins. They demonstrate that catastrophes, by definition, are superimposed on environmental trends and evolutionary processes already in progress and that no single event can be burdened with the responsibility of accounting for the history and extinction patterns of all groups. As noted by Clemens (quoted in Morell, 1993; Clemens and Nelms, 1993), it is equally important for the theory to account for gradual and preimpact extinctions and for survivals. These were inconsistent with the proposed 3–5 year period of global near darkness originally proposed. Evidence presented at the Snowbird, Utah, conferences in 1981 and 1988 (Sharpton and Ward, 1990) indicate that modern microfloras consume existing food reserves within 10–100 days, a fact that allowed scaling down of the dark period to 4–6 months. Extinctions at the K–T boundary have been verified from DSDP cores at ~44% of all genera and ~70% of the species in marine plankton (compared to 14% of freshwater genera and 20% of terrestrial genera).

A second problem was that no crater of the proper age and size was known. Initially it was speculated that the bolide landed in the ocean. Later it was suggested that the Chicxulub (Mayan, tail of the devil) structure (180-km diameter) buried ~2 km beneath the Yucatan Peninsula ($^{40}Ar/^{39}Ar$ age of 64.98±0.05 m.y.; Krogh et al., 1993; Swisher et al., 1992), the Manson crater (36-km diameter) buried under central Iowa (radiometric age originally reported at ~66 m.y.), and a site in the Pacific Ocean (2-km diameter, DSDP 577, 32° 26' N and 157° 43' E, Shatsky Rise; Robin et al., 1993) were multiple craters resulting from shattering as the bolide entered the Earth's atmosphere (Hildebrand and Boynton, 1990; Izett et al., 1991; Kring and Boynton, 1992; Maurrasse and Sen, 1991; Sharpton et al., 1992). The diameter of the Chicxulub crater was recently estimated at 300 km (Sharpton et al., 1993), but subsequent studies reaffirmed the 180-km value (Hildebrand et al., 1995). The diameter is important because impact energy is about the cube of crater size, so that the larger size would mean a force 10 times greater than the smaller size. Chicxulub samples contain shocked quartz (stishovite, quartz grains with striations that form under sudden intense pressure), and samples from around the Gulf–Caribbean region contain shocked quartz that splashed out from the impact site. These are from Haiti, near Brazos, Texas, and Mimbral north of Tampico, Mexico. The samples from the latter locality also contain ripple marks and debris from land vegetation that was washed into the ocean by an accompanying tsunami (tidal wave) estimated at a kilometer or more in height. Other evidence includes the presence of siderophile (iron-rich) elements such as chromium, rhodium, scandium, and titanium, and two amino acids that are rare in Earth material but common in meteorites (Zhao and Bada, 1989). Recently the isotopic composition of oxygen and the impact glasses strontium and neodymium from Chicxulub compared well with those from the Beloc Formation in Haiti (64.42±0.06 Ma; Dalrymple et al., 1993). The isotopic chemistry of the Manson sediments is different, which led Blum et al. (1993) to conclude that Chicxulub is the primary source crater for the K–T catastrophy and that the Manson melt rocks are unrelated to the event. A new date on the Manson impact structure (73.8±0.3 Ma; Izett et al., 1993), with splash (tsunami) debris in the coeval Crow Creek Member of the Pierre Shale of South Dakota, supports this view.

The plausibility of an asteroid colliding with the Earth in the past has been enhanced by reports of other collisions (Grieve, 1996), as in the Late Eocene (Chesapeake Bay, Poag, 1997; Popigai crater in Siberia, Bottomley et al., 1997), recent near misses, and near hits. The asteroid 1991 BA, which was about 5–10 m in diameter, came within 170,000 km of the Earth (Scott et al., 1991). If it had hit, the velocity of impact would have been 21.2 km/s and the explosive force would have been ~40 kilotons of TNT (3 times that of the Hiroshima bomb). A near hit occurred over Tunguska, Siberia, on June 30, 1908 (Chyba et al., 1993; Svetson, 1996), that felled trees over an area of 1000 km^2, and over one-third of the atmospheric ozone was destroyed by the explosion. Asteroid particles were found embedded in tree resin at the site. It is now estimated that the frequency of impacts of near Earth objects (NEOs) 50 m in diameter is about one per century. The break-up of comet Shoemaker 9 in the summer of 1994 provided spectacular documentation of a comet impact on Jupiter. These observations have engendered some chilling proposals to avert future collisions, including deflection by rocket-launched masses and fragmentation by nuclear explosions (see discussions in Ahrens and Harris, 1992; Sagan and Ostro, 1994).

Although various aspects of the theory are still being argued and refined, the consensus is that an impact did occur at the end of the Cretaceous. In tracing the Late Cretaceous and Cenozoic history of North American terrestrial vegetation and environments, it is thus necessary to monitor the record for effects that may be related to this type of catastrophy. Evergreen components were disappearing from mesothermal vegetation of the midnorthern latitudes at about this time, initiating a trend from evergreen to deciduous vegetation. Temperatures were also declining from Middle Cretaceous highs and may have reached a threshold that favored development of deciduous vegetation, but the trend could have been intensified by the bolide impact (Wolfe, 1987; Chapter 5).

Missoula Floods

In eastern Washington and northern Idaho on the Columbia Plateau, there is a vast area of erosional and depositional land forms called the channeled scablands (Fig. 2.24). These geomorphic features cover a region equivalent to the southeastern one-fourth of Washington or an area of ~45,000 km^2 (17,000 mi^2). Between 1923 and 1969 J Harlen Bretz[7] (e.g., Bretz et al., 1956) published a series of papers suggesting that this unique landscape was a result of colossal floods

Figure 2.24. The scablands of eastern Washington resulting from the Missoula floods. Photograph courtesy of Rachael Craig.

during the late Wisconsin glacial period (his Spokane Flood). Bretz's ideas were initially rejected for a variety of reasons. The event was unique and the scale incomprehensible, there was no known source for the huge amounts of water required, and no mechanism was proposed for the sudden release of the water. The early 1900s was also a time of rigid adherence to the gradualist principle, or uniformitarianism; processes acting today are those that have acted in the past to produce the structures and events evident in the geological record. Random, unprecedented events with no known cause evident from presently observable processes constituted catastrophism and were summarily rejected. A sharp distinction between these concepts is no longer recognized, and catastrophic events are accepted as occasionally superimposed on uniformitarian processes. As late as 1981 Raup noted that, "To consider crises as a legitimate and important part of the formative process in ecology and evolution is new to many and anathema to some." Derek Ager (1993) compared this view of geologic history to the life of a soldier: long periods of boredom and short periods of terror. Recent studies by Baker (1973), Craig (1987), and Waitt (1985) provided satisfactory explanations and mechanisms for the Missoula floods, which are now documented as the most extensive in recent Earth history. [See also Baker et al. 1993 regarding a superflooding event in Siberia.]

The Purcell Trench lobe of the Cordilleran ice sheet advanced from the north during the Fraser glaciation (25–10 Kya) and abutted against the Bitteroot Range in northern Idaho (Waitt, 1985, fig. 1). The ice lobe blocked the Clark Fork River, creating glacial Lake Missoula to the east of the ice dam. Flathead Lake in Montana is a remnant of glacial Lake Missoula. The volume of the lake was ~2500 km^3 or about 6 times the amount of water in Lake Erie (458 km^3).

When the lake reached a critical depth (~600 m), the ice dam became buoyant and the seal against the mountain was broken. The result was the release of vast amounts of waters that scoured the landscape, carved elongated parallel channels called coulees into the basalts of the Columbia Plateau, and back-flooded up the Columbia River and its tributaries. Ripple trains were formed with wavelengths of 150 m and amplitudes of 15 m; boulders 10 m in diameter were carried more than 3.5 km; and discharge rates reached 17 x 10^6 m^3/s, which is more than 10 times the combined flow of all the rivers of the world. These catastrophic floods are called *jökulhlaups* (Icelandic for outbursts) from similar events at Grimsvötn, Iceland. As the water level fell, the ice settled back into place and the cycle repeated. Waitt (1985) identified up to 60 of these hydraulic cycles between 15.3 and 12 Kya.

The Lake Missoula sediments are related stratigraphically to those of glacial Lake Bonneville to the south. Great

Salt Lake in Utah is a remnant. Radiocarbon dates show that the alluvial spillway damming the north end of Lake Bonneville broke by head cutting between 15 and 14 Kya, causing a sudden drop in the lake level of 115 m. The waters flowed through the Snake River Canyon to the Columbia River, and near Lewiston, Idaho, the deposits are merged with those of the Missoula floods. Evidence for cataclysmic flooding is also found in Box Canyon along the Big Lost River in east central Idaho (Rathburn, 1993). Discharge is estimated at 60,000 m^3/s, making it the third largest known jökulhlaups.

The periodic removal of vegetation over a combined area exceeding 45,000 km^2, followed by relatively continuous succession after 12 Kya, would be reflected in the composition of the regional pollen and spore rain onto adjacent bog and lake surfaces. Therefore, high resolution pollen and spore profiles derived from these sediments may reflect changes due to an exceptional and local environmental history superimposed upon regional climatic trends.

The Missoula floods further exemplify the intricacy and dynamic nature of factors controlling environmental change and influencing vegetational history. Embedded within the rhythmites of back-flooded valleys are three lenses of volcanic ash. Two are from eruptions of Mount St. Helens ~13 Kya, and one records the eruption of Mount Mazama 7 Kya. Thus, the effects of volcanism discussed earlier are added to the consequences of flooding, long-term climatic trends resulting from variations in the orbital cycles, fluctuations in the poleward transport of heat, CO_2 concentration changes, and other events. Collectively, these constitute some of the complex and interacting factors that determined late Wisconsin environments in the region and influenced the course of development of the plant formations.

Deforestation

> [O]n Tuesday, July 22d [1494], he departed for Jamaica . . . The sky, air, and climate were just the same as in other places; every afternoon there was a rain squall that lasted for about an hour. The admiral writes that he attributes this to the great forests of that land; he knew from experience that formerly this also occurred in the Canary, Madeira, and Azore Islands, but since the removal of forests that once covered those islands they do not have so much mist and rain as before. (the biography of Christopher Columbus, Colon, 1959)

From the end of the Cretaceous through the Middle Eocene, warm-temperate to tropical vegetation extended from equatorial regions to Beringia and the North Atlantic land bridge. Subsequent climatic trends toward cooler and seasonally dry climates have restricted the range of tropical forests in the northern hemisphere of the New World today to below 17° N latitude near southern Veracruz, Mexico. Many parts of the North American landscape once occupied by dense, evergreen forests now support deciduous forests, grasslands, and deserts. In earlier discussions, climate, augmented and interacting with orogeny, volcanism, various consequences of plate tectonic activity, and catastrophes have been emphasized as determining the composition and distribution of North American plant formations. It is worthwhile to note, however, that there is a feedback circuit whereby vegetation influences climate (Bonan et al., 1992; Foley et al., 1994; Kutzbach et al., 1996). This is accomplished through at least three mechanisms. First, various types of plant cover impart a different albedo (pattern of heat reflection from surfaces; Berger, 1982, table 3.2). Albedo is ~15 for forested land and ~30 for deserts (Table 3.1). Second, transpiration rates of water into the atmosphere from leaf surfaces differ among vegetation types; and third, the type and extent of vegetation cover affects the rate of erosion and, hence, dust and CO_2 concentration.

It is not possible to specifically quantify the climatic effects of changes in plant cover from dense forest to a mosaic of open savanna, shrubland, grassland, and desert vegetation. As with most large-scale multifactor events, however, computer simulation models can aid in assessing whether there is a potential effect, and if so, its theoretical extent and how the variables interrelate. Shukla and Mintz (1982) used the atmospheric GCM at Goddard Laboratory for Atmospheric Sciences (GLAS) to simulate two extremes—an Earth completely covered with vegetation and moist soils (wet-soil case) and one completely devoid of vegetation (dry-soil case). Ocean-surface temperatures, continental topography, and surface albedo were held at current conditions.

In the wet-soil case precipitation in North America was ~3–6 mm/day, and in the dry-soil case it was reduced to 1 mm/day or less. (Eastern North America had more precipitation from ocean water vapor.) The temperature simulations show that regions north of ~20° S latitude were 15–25°C warmer in the dry-soil case. The conversion of high polar landscapes from sparse tundra or bare soil to forest has a modeled effect comparable to a doubling of CO_2 concentration. From such data Shukla and Mintz (1982) conclude that

> Calculations with a numerical model of the atmosphere show that the global fields of rainfall, temperature, and motion strongly depend on the land-surface evapotranspiration. This confirms the long-held idea that the surface vegetation, which produces the evapotranspiration, is an important factor in the earth's climate.

Using a more recent modified version of the CCM at NCAR, McGuffie et al. (1995) also demonstrated a worldwide effect of tropical deforestation on climate. From the viewpoint of vegetational history this means that changes in the composition and distribution of North American plant formations through Cenozoic time are not simply a response to climatic trends and alterations in the physical environment. Rather, as environmental change alters the vegetation cover, the alteration in turn influences climate,

Figure 2.25. Destruction of vegetation and subsequent erosion north of Tehuantepec, Mexico.

Figure 2.26. Destruction of vegetation, Guatemala. Photograph courtesy of Bruce Graham.

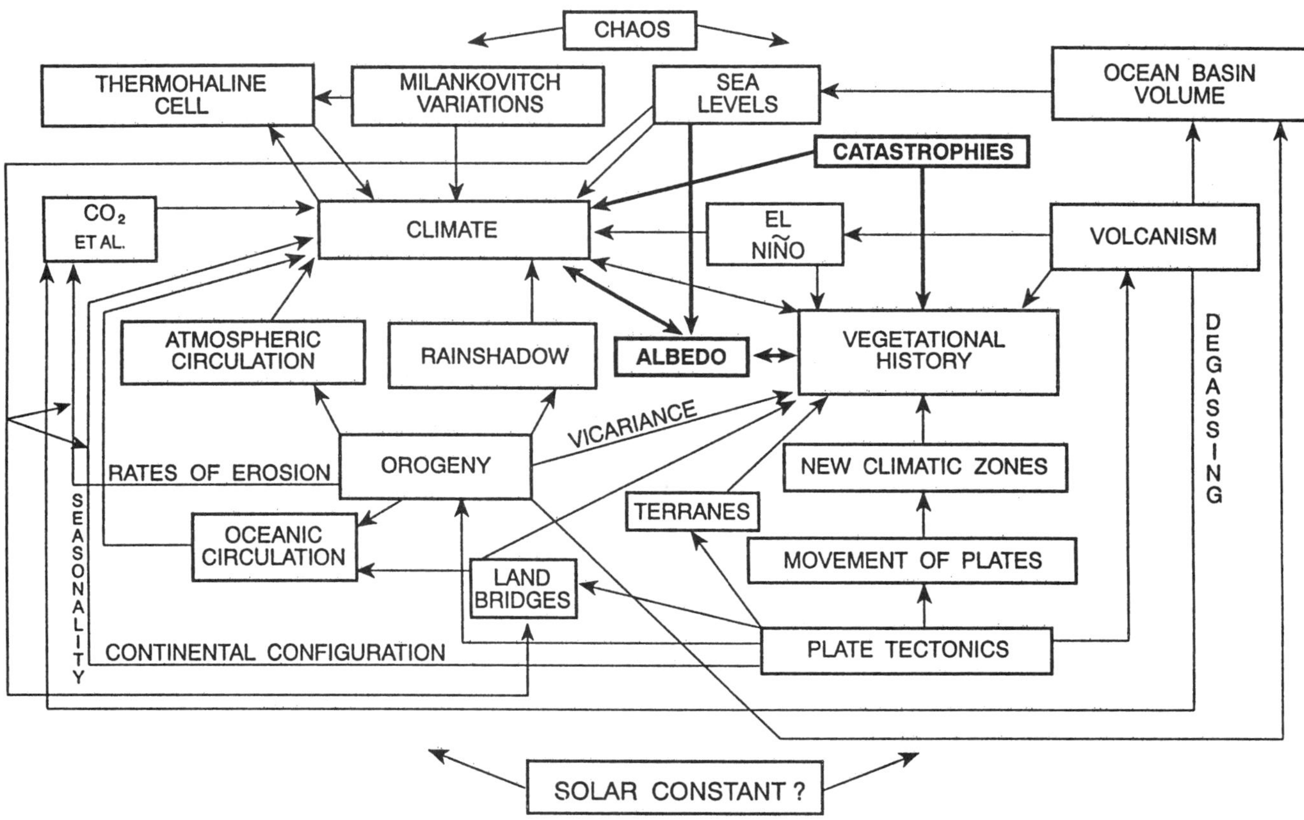

either accelerating or countering precipitation and temperature trends. The Food and Agriculture Organization (FAO, United Nations), estimated 10 years ago that 70,000 km^2 of rain forest were being destroyed each year at 100 acres every 3 min. The current rate is 100 acres/min, or an area equal to the size of a football field every second. This is in addition to 100,000 km^2 that are drastically altered each year, and other vegetation is being similarly impacted throughout warm-temperate to tropical America (Figs. 2.25, 2.26). Salati et al. (1983; Salati and Vose, 1984) estimate that half the current rainfall in the Amazon Basin is associated with forest cover and that clear-cutting may cause temperatures to rise by 5°C or more. Other aspects of deforestation and climate are discussed by Crutzen and Andreae (1990), Levine (1991), Sanford et al. (1985), Shukla et al. (1990), and Woodwell et al. (1983).

The many factors influencing vegetational history discussed in this chapter are summarized in the diagram on p. 72. In considering this illustration it is worth noting a comment by Crowley and North:

> The above studies indicate that it is unlikely that the long-term evolution in past climates will be explained by only one factor. One of the goals of paleoclimatology is to integrate the effects of all of the above changes and then compare the results with the long-term history mapped by geologists. A substantial amount of additional work is required to reach this goal. (1990)

References

Adams, J. M., H. Faure, L. Faure-Denard, J. M. McGlade, and F. I. Woodward. 1990. Increases in terrestrial carbon storage from the last glacial maximum to the present. Nature 348: 711–714.

Ager, D. 1993. The New Catastrophism. Cambridge University Press, Cambridge, U.K.

Ahrens, T. J. and A. W. Harris. 1992. Deflection and fragmentation of near-Earth asteroids. Nature 360: 429–433.

Allen, J. C., W. M. Schaffer, and D. Rosko. 1993. Chaos reduces species extinction by amplifying local population noise. Nature 364: 229–232.

Alley, R. B. et al. (+ 5 authors). 1995. Comparison of deep ice cores. Nature 373: 393–394.

Alvarez, L. W., W. Alvarez, F. Asaro, and H. V. Michel. 1980. Extraterrestrial cause for the Cretaceous–Tertiary extinction. Science 208: 1095–1108.

Alvarez, W. 1986. Toward a theory of impact crises. EOS (Trans. Am. Geophys. Union) 67: 649–658.

Anders, M. H. 1994. Constraints on North American plate velocity from the Yellowstone hotspot deformation field. Nature 369: 53–55.

Angell, J. K. 1988. Impact of El Niño on the delineation of tropospheric cooling due to volcanic eruptions. J. Geophys. Res. 93: 3697–3704.

Arthur, M. A. 1982. The carbon cycle—controls on atmospheric CO_2 and climate in the geologic past. In: Geophysics Study Committee, Climate in Earth History. National Academy Press, Washington, D.C. pp. 55–67.

———. and R. E. Garrison (eds.). 1986. Milankovitch cycles through geologic time. Paleoceanography 1: 369–586 [Special Section].

Atwater, T. 1970. Implications of plate tectonics for the Cenozoic tectonic evolution of western North America. Geol. Soc. Am. Bull. 81: 3513–3536.

Axelrod, D. I. 1940. The Mint Canyon flora of southern California: a preliminary statement. Am. J. Sci. 238: 577–585.

———. 1962. Post-Pliocene uplift of the Sierra Nevada, California. Geol. Soc. Am. Bull. 73: 183–198.

———. 1981. Altitudes of Tertiary forests estimated from paleotemperature. In: Proc. of the Symposium on the Qinghai-Xizang (Tibet) Plateau (Beijing, China). Science Press, Beijing, and Gordon & Breach Science Publishers, New York. Vol. 1, pp. 131–137.

———. 1986. Analysis of some palaeogeographic and palaeoecologic problems of palaeobotany. Palaeobotanist 35: 115–129.

———. 1987. The late Oligocene Creede flora, Colorado. Univ. Calif. Publ. Geol. Sci. 130: 1–235.

Baker, V. R. 1973. Paleohydrology and Sedimentology of Lake Missoula Flooding in Eastern Washington. Geological Society of America, Boulder, CO. Special Paper 144.

———, G. Benito, and A. N. Rudoy. 1993. Paleohydrology of late Pleistocene superflooding, Altay Mountains, Siberia. Science 259: 348–350.

Bally, A. W. and A. R. Palmer (eds.). 1989. The Geology of North America; An Overview. DNAG (Decade of North American Geology). Geological Society of America, Boulder, CO. Vol. A.

Barbour, M. G. and W. D. Billings (eds.). 1988. North American Terrestrial Vegetation. Cambridge University Press, Cambridge, U.K.

——— and N. L. Christensen. 1993. Vegetation. In: Flora of North America Editorial Committee (eds.), Flora of North America North of Mexico. Oxford University Press, Oxford, U.K. Vol. 1, pp. 97–131.

Bard, E., F. Rostek, and C. Sonzogni. 1997. Interhemispheric synchrony of the last deglaciation inferred from alkenone palaeothermometry. Nature 385: 707–710.

Barry, R. G. and R. J. Chorley. 1992. Atmosphere, Weather, and Climate. Routledge, London. 6th edition.

Beck, J. W., J. Récy, F. Taylor, R. L. Edwards, and G. Cabioch. 1997. Abrupt changes in Early Holocene tropical sea surface temperature derived from coral records. Nature 385: 705–707.

Beck, R. A. et al. (+ 13 authors). 1995. Stratigraphic evidence for an early collision between northwest India and Asia. Nature 373: 55–58.

Belton, M. J. S. et al. (+ 19 authors). 1992. Galileo encounter with 951 Gaspra: first pictures of an asteroid. Science 257: 1647–1652.

——— et al. (+ 19 authors). 1994. First images of asteroid 243 Ida. Science 265: 1543–1547.

Bender, M. et al. (+ 6 authors). 1994. Climate correlations between Greenland and Antarctica during the past 100,000 years. Nature 372: 663–666.

Berger, A. L. 1980. The Milankovitch astronomical theory of paleoclimates: a modern review. Vistas Astron. 24: 103–122.

———, J. Imbrie, J. Hays, G. Kukla, and B. Saltzman (eds.). 1984. Milankovitch and Climate. NATO ASI Series C: Mathematical and Physical Sciences. D. Reidel Publ. Co., Dordrecht. Vol. 126, Parts 1 and 2.

———, M. F. Loutre, and J. Laskar. 1992. Stability of the astronomical frequencies over the Earth's history for paleoclimate studies. Science 255: 560–566.

Berger, W. H. 1982. Climate steps in ocean history—lessons from the Pleistocene. In: Geophysics Study Committee (eds.), Climate in Earth History. National Academy Press, Washington, D.C. pp. 43–54.

——— and A. Spitzy. 1988. History of atmospheric CO_2: constraints from the deep-sea record. Paleoceanography 3: 401–411.

Berggren, W. A. 1982. Role of ocean gateways in climatic change. In: Geophysics Study Committee (eds.), Climate in Earth History. National Academy Press, Washington, D.C. pp. 118–125.

Berner, R. A. 1991. A model for atmospheric CO_2 over Phanerozoic time. Am. J. Sci. 291: 339–376.

———. 1994. 3GEOCARB II: a revised model of atmospheric CO_2 over Phanerozoic time. Am. J. Sci. 294: 56–91.

——— and G. P. Landis. 1988. Gas bubbles in fossil amber as possible indicators of the major gas composition of ancient air. Science 239: 1406–1409.

———, A. C. Lasaga, and R. M. Garrels. 1983. The carbonate-silicate geochemical cycle and its effect on atmospheric carbon dioxide over the past 100 million years. Am. J. Sci. 283: 641–683.

Bevis, M. et al. (+ 10 authors). 1995. Geodetic observations of very rapid convergence and back-arc extension at the Tonga Arc. Nature 374: 249–251.

Birchfield, G. E. and W. S. Broecker. 1990. A salt oscillator in the glacial Atlantic? 2. A "scale analysis" model. Paleoceanography 5: 835–843.

Bird, J. 1991. Supercomputer voyages to the southern seas. Science 254: 656–657.

Bird, P. 1988. Formation of the Rocky Mountains, western United States: a continuum computer model. Science 239: 1501–1507.

Björk, S. et al. (+ 10 authors). 1996. Synchronized terrestrial-atmospheric deglacial records around the North Atlantic. Science 274: 1155–1160.

Blum, J. D. et al. (+ 6 authors). 1993. Isotopic comparison of K/T boundary impact glass with melt rock from the Chicxulub and Manson impact structures. Nature 364: 325–327.

Bonan, G. B., D. Pollard, and S. L. Thompson. 1992. Effects of boreal forest vegetation on global climate. Nature 359: 716–718.

Bond, G. C. and R. Lotti. 1995. Iceberg discharges into the North Atlantic on millennial time scales during the last glaciation. Science 267: 1005–1010.

——— et al. (+ 13 authors). 1992. Evidence for massive discharges of icebergs into the North Atlantic Ocean during the last glacial period. Nature 360: 245–249.

——— et al. (+ 6 authors). 1993. Correlations between climate records from North Atlantic sediments and Greenland ice. Nature 365: 143–147.

——— et al. (+ 9 authors). 1997. A pervasive millennial-scale cycle in North Atlantic Holocene and glacial climates. Science 278: 1257–1266.

Bottomley, R., R. Grieve, D. York, and V. Masaitis. 1997. The age of the Popigai impact event and its relation to events at the Eocene/Oligocene boundary. Nature 388: 365–368.

Boutton, T. W., S. R. Archer, and L. C. Nordt. 1994. Scientific correspondence: Climate, CO_2 and plant abundance. Nature 372: 625–626.

Bradley, R. S. 1988. The explosive volcanic eruption signal in northern hemisphere continental temperature records. Climatic Change 12: 221–243.

Brass, G. W. et al. (+ 6 authors). 1982. Ocean circulation, plate tectonics, and climate. In: Geophysics Study Committee (eds.), Climate in Earth History. National Academy Press, Washington, D.C. pp. 83–89.

Brasseur, G. and C. Granier. 1992. Mount Pinatubo aerosols, chloroflurocarbons, and ozone depletion. Science 257: 1239–1242.

Brenchley, P. J. (ed.). 1984. Fossils and Climate. Wiley-Interscience, New York.

Bretz, J H., H. T. U. Smith, and G. F. Neff. 1956. Channeled scabland of Washington: new data and interpretations. Bull. Geol. Soc. Am. 67: 957–1049.

Broecker, W. S. 1994. Massive iceberg discharges as triggers for global climate change. Nature 372: 421–424.

———. 1997. Thermohaline circulation, the Achilles Heel of our climate system: will man-made CO_2 upset the current balance? Science 278: 1582–1588.

———, G. Bond, M. Klas, G. Bonani, and W. Wolfli. 1990. A salt oscillator in the glacial Atlantic? I. The concept. Paleoceanography 5: 469–477.

——— and G. H. Denton. 1989. The role of ocean–atmosphere reorganizations in glacial cycles. Geochim. et Cosmochim. Acta 53: 2465–2501.

——— and J. van Donk. 1970. Insolation changes, ice volumes, and the O^{18} record in deep-sea cores. Rev. Geophys. Space Phys. 8: 169–198.

——— et al. (+ 5 authors). 1968. Milankovitch hypothesis supported by precise dating of coral reefs and deep-sea sediments. Science 159: 297–300.

——— et al. (+ 6 authors). 1988. The chronology of the last deglaciation: implications to the cause of the Younger Dryas event. Paleoceanography 3: 1–19.

——— et al. (+ 6 authors). 1989. Routing of meltwater from the Laurentide ice sheet during the Younger Dryas cold episode. Nature 341: 318–321.

Brownlee, D. E. 1995. A driver of glaciation cycles? Nature 378: 558.

Bryant, J. D., B. Luz, and P. N. Froelich. 1994. Oxygen isotopic composition of fossil horse tooth phosphate as a record of continental paleoclimate. Palaeogeogr., Palaeoclimatol., Palaeoecol. 107: 303–316.

Burbank, D. W. 1992. Causes of recent Himalayan uplift deduced from depositional patterns in the Ganges Basin. Nature 357: 680–683.

Caldeira, K., M. A. Arthur, R. A. Berner, and A. C. Lasaga. 1993. Scientific correspondence: Cooling in the Late Cenozoic. Nature 361: 123–124.

Cerling, T. E. 1991. Carbon dioxide in the atmosphere: evidence from Cenozoic and Mesozoic paleosols. Am. J. Sci. 291: 377–400.

———, Y. Wang, and J. Quade. 1993. Expansion of C4 ecosystems as an indicator of global ecological change in the late Miocene. Nature 361: 344–345.

Cerling, T. E. et al. (+ 6 authors). 1997. Global vegetation change through the Miocene/Pliocene boundary. Nature 389: 153–158.

Cess, R. D. et al. (+ 29 authors). 1993. Uncertainties in carbon dioxide radiative forcing in atmospheric general circulation models. Science 262: 1252–1255.

——— et al. (+ 19 authors). 1995. Absorption of solar radiation by clouds: observations versus models. Science 267: 496–499.

Chahine, M. T. 1992. The hydrological cycle and its influence on climate. Nature 359: 373–380.

Christensen, M. N. 1965. Late Cenozoic deformation in the central Coast Ranges of California. Geol. Soc. Am. Bull. 76: 1105–1124.

Churkin, M., Jr. 1983. Tectonostratigraphic terranes of Alaska and northeastern USSR—a record of collision and accretion. In: M. Hashimoto and S. Uyeda (eds.), Accretion Tectonics in the Circum-Pacific Regions. Terra Scientific Publishing Company, Tokyo. pp. 37–42.

Chyba, C. F., P. J. Thomas, and K. J. Zahnle. 1993. The 1908 Tunguska explosion: atmospheric disruption of a stony asteroid. Nature 361: 40–44.

Clarke, S. H., Jr., and G. A. Carver. 1992. Late Holocene tectonics and paleoseismicity, southern Cascadia subduction zone. Science 255: 188–192.

Clemens, S. C. and R. Tiedemann. 1997. Eccentricity forcing of Pliocene-early Pliocene climate revealed in a marine oxygen-isotope record. Nature 385: 801–804.

Clemens, W. A. and L. G. Nelms. 1993. Paleoecological implications of Alaskan terrestrial vertebrate fauna in latest Cretaceous time at high paleolatitudes. Geology 21: 503–506.

Clube, S. V. M. (ed.). 1989. Catastrophes and Evolution. Cambridge University Press, Cambridge, U.K.

Cole, D. R. and H. C. Monger. 1994. Influence of atmospheric CO_2 on the decline of C4 plants during the last deglaciation. Nature 368: 533–536.

Coleman, M. and K. Hodges. 1995. Evidence for Tibetan Plateau uplift before 14 Myr ago from a new minimum age for east–west extension. Nature 374: 49–52.

Colon, F. 1959. The Life of Christopher Columbus by His Son Ferdinand (translated and annotated by B. Keen). Rutgers University Press, New Brunswick, NJ.

Cooper, A. and D. Penny. 1997. Mass survival of birds across the Cretaceous–Tertiary boundary: molecular evidence. Science 275: 1109–1113.

Covey, C. 1996. Sun–climate links. Science 272: 179.

——— and E. Barron. 1988. The role of ocean heat transport in climatic change. Earth-Sci. Rev. 24: 429–445.

———, S. L. Thompson, P. R. Weissman, and M. C. MacCracken. 1994. Global climatic effects of atmospheric dust from an asteroid or comet impact on Earth. Global Planet. Change 9: 263–273.

Craig, R. G. 1987. Dynamics of a Missoula flood. In: L. Meyer. and D. Nash (eds.), Eighteenth Annual Geomorphology Symposium: Catastrophic Flooding. Allen and Unwin, London. pp. 305–332.

Croll, J. 1875. Climate and time in their geological relations—a theory of secular changes of the Earth's climate. D. Appleton, New York.

Cross, A. T. and R. E. Taggart. 1982. Causes of short-term sequential changes in fossil plant assemblages: some considerations based on a Miocene flora of the northwest United States. Ann. Missouri Bot. Gard. 69: 676–734.

Crowley, T. J. 1991. Utilization of paleoclimate results to

validate projections of a future greenhouse warming. In: M. E. Schlesinger (ed.), Greenhouse-Gas-Induced Climatic Change: A Critical Appraisal of Simulations and Observations. Elsevier Science Publishers, Amsterdam. pp. 35–45.

———. 1993. Geological assessment of the greenhouse effect. Bull. Am. Meterol. Soc. 74: 2363–2373.

———, K.-Y. Kim, J. G. Mengel, and D. A. Short. 1992. Modeling 100,000-year climate fluctuations in pre-Pleistocene time series. Science 255: 705–707.

——— and G. R. North. 1988. Abrupt climate change and extinction events in Earth history. Science 240: 996–1002.

——— and ———. 1990. Modeling onset of glaciation. Ann. Glaciol. 14: 39–42.

——— and ———. 1991. Paleoclimatology. Oxford Monographs in Geology and Geophysics. Oxford University Press, Oxford, U.K. Monograph 16.

Crutzen, P. J. and M. O. Andreae. 1990. Biomass burning in the tropics: impact on atmospheric chemistry and biogeochemical cycles. Science 250: 1669–1677.

Curry, W. B. and D. W. Oppo. 1997. Synchronous, high-frequency oscillations in tropical sea surface temperatures and North Atlantic deep water production during the last glacial cycle. Paleoceanography 12: 1–14.

Dacey, J. W. H., B. G. Drake, and M. J. Klug. 1994. Stimulation of methane emission by carbon dioxide enrichment of marsh vegetation. Nature 370: 47–49.

Dalrymple, G. B., G. A. Izett, L. W. Snee, and J. D. Obradovich. 1993. $^{40}Ar/^{39}Ar$ age spectra and total-fusion ages of tektites from Cretaceous–Tertiary boundary sedimentary rocks in the Beloc Formation, Haiti. U.S. Geol. Survey Bull. 2065: 1–20.

Dansgaard, W. et al. (+ 5 authors). 1984. North Atlantic climatic oscillations revealed by deep Greenland ice cores. In: J. E. Hansen and T. Takahashi (eds.), Climate Processes and Climate Sensitivity. American Geophysical Union, Washington, D.C. Geophys. Monogr. 29, pp. 288–298.

——— et al. (+ 10 authors). 1993. Evidence for general instability of past climate from a 250-kyr ice-core record. Nature 364: 218–220.

Dawes, P. R. and J. W. Kerr (eds.). 1982. Nares Strait and the Drift of Greenland: A Conflict in Plate Tectonics. Meddelelser øm Grønland, Geoscience 8, Copenhagen.

Decker, R. W. and B. B. Decker. 1991. Mountains of Fire: the Nature of Volcanoes. Cambridge University Press, Cambridge, U.K.

del Moral, R. and D. M. Wood. 1993. Early primary succession on a barren volcanic plain at Mount St. Helens, Washington. Am. J. Bot. 80: 981–991.

de Vernal, A., C. Hillaire-Marcel, and G. Bilodeau. 1996. Reduced meltwater outflow from the Laurentide ice margin during the Younger Dryas. Nature 381: 774–777.

Dewey, J. F., I. G. Gass, G. B. Curry, N. B. W. Harris, and A. M. C. Sengör (eds.). 1991. Allochthonous Terranes. Cambridge University Press, Cambridge, U.K.

Diaz, H. F. and V. Markgraf (eds.). 1992. El Niño—Historical and Paleoclimatic Aspects of the Southern Oscillation. Cambridge University Press, Cambridge, U.K.

Dixon, R. K. et al. (+ 5 authors). 1994. Carbon pools and flux of global forest ecosystems. Science 263: 185–190.

Dowlatabadi, H. and M. G. Morgan. 1993. Integrated assessment of climate change. Science 259: 1813, 1932.

Dunnette, D. A. and R. J. O. O'Brien (eds.). 1992. The Science of Global Change, the Impact of Human Activities on the Environment. American Chemical Society, Washington, D.C.

Eaton, G. P. 1986. A tectonic redefinition of the southern Rocky Mountains. Tectonophysics 132: 163–193.

Edwards, R. L., H. Cheng, M. T. Murrell, and S. J. Goldstein. 1997. Protactinium-231 dating of carbonates by thermal ionization mass spectrometry; implications for Quarternary climate change. Science 276: 782–786.

——— and C. D. Gallup. 1993. Technical comments: Dating of the Devils Hole calcite vein. Science 259: 1626.

Eldholm, O., A. M. Myhre, and J. Thiede. 1994. Cenozoic tectomagmatic events in the North Atlantic: potential paleoenvironmental implications. In: M. C. Boulter and H. C. Fisher (eds.), Cenozoic Plants and Climates of the Arctic. NATO ASI Series 1: Global Environmental Change. Springer-Verlag, Berlin. Vol. 27, pp. 35–55.

Emiliani, C. 1955. Pleistocene temperatures. J. Geol. 63: 538–578.

Epstein, S., R. Buchsbaum, H. A. Lowenstam, and H. C. Urey. 1951. Carbonate-water isotopic temperature scale. Bull. Geol. Soc. Am. 62: 417–426.

Faure, H. 1990. Changes in the global continental reservoir of carbon. Palaeogeogr., Palaeoecol., Palaeoclimatol. 82: 47–52.

Fiedel, S. J. 1995. Technical comment: The GISP ice core record of volcanism since 7000 B.C. Science 267: 256.

Fiocco, G., D. Fu'a, and G. Visconti. 1997. The Mount Pinatubo eruption, effects on the atmosphere and climate. Springer-Verlag, Berlin.

Fischer, A. G. 1982. Long-term climatic oscillations recorded in stratigraphy. In: Geophysics Study Committee (eds.), Climate in Earth History. National Academy Press, Washington, D.C. pp. 97–104.

Fisher, M. J. et al. (+ 6 authors). 1994. Carbon storage by introduced deep-rooted grasses in the South American savannas. Nature 371: 236–238.

Fitzgerald, P. G., E. Stump, and T. F. Redfield. 1993. Late Cenozoic uplift of Denali and its relation to relative plate motion and fault morphology. Science 259: 497–499.

Flam, F. 1993. Is Earth's future climate written in the stars? Science 262: 1372–1373.

Flora of North America Editorial Committee (eds.). 1993. Flora of North America North of Mexico. Oxford University Press, Oxford, U.K.

Foley, J. A., J. E. Kutzbach, M. T. Coe, and S. Levis. 1994. Feedbacks between climate and boreal forests during the Holocene epoch. Nature 371: 52–54.

Freeman, K. H. and J. M. Hayes. 1992. Fractionation of carbon isotopes by phytoplankton and estimates of ancient CO_2 levels. Global Biogecchem. Cycles 6: 185–198.

Friis-Christensen, E. and K. Lassen. 1991. Length of the solar cycle: an indicator of solar activity associated with climate. Science 254: 698–700.

Gallup, C. D., R. L. Edwards, and R. G. Johnson. 1994. The timing of high sea levels over the past 200,000 years. Science 263: 796–800.

Gates, D. M. 1993. Climatic Change and its Biological Consequences. Sinauer Associates, Sunderland, MA.

Gates, W. L. 1982. Paleoclimatic modeling—a review with reference to problems and prospects for the pre-Pleistocene. In: Geophysics Study Committee (eds.), Climate in Earth History. National Academy Press, Washington, D.C. pp. 26–42.

Geophysics Study Committee (eds.). 1982. Climate in Earth History. National Academy Press, Washington, D.C.

———. 1990. Sea-Level Change. National Academy Press, Washington, D.C.

Gersonde, R. et al. (+ 12 authors). 1997. Geological record and reconstruction of the Late Pliocene impact of the Eltanin asteroid in the southern ocean. Nature 390: 357–363.

Giddings, L. V., K. Y. Kaneshiro, and W. W. Anderson (eds.). 1989. Genetics, Speciation, and the Founder Effect. Oxford University Press, Oxford, U.K.

Glynn, P. W. (ed.). 1990. Global Ecological Consequences of the 1982–83 El Niño–Southern Oscillation. Elsevier Oceanography Series. Elsevier Science Publishers, Amsterdam. Vol. 52.

——— and W. H. de Weerdt. 1991. Elimination of two reef-building hydrocorals following the 1982–83 El Niño warming event. Science 253: 69–71.

Gordon, R. G. and S. Stein. 1992. Global tectonics and space geodesy. Science 256: 333–342.

Grace, J. et al. (+ 11 authors). 1995. Carbon dioxide uptake by an undisturbed tropical rain forest in southwest Amazonia, 1992 to 1993. Science 270: 778–780.

Greenland Ice-Core Project (GRIP) Members. 1993. Climate instability during the last interglacial period recorded in the GRIP ice core. Nature 364: 203–207.

Grieve, R. A. F. 1996. Meteorite impact craters of North America. In: R. L. Bishop (ed.), Observer's Handbook 1997. Royal Astronomical Soc. Can., Toronto. pp. 185–188.

Grimm, E. C., G. L. Jacobson, Jr., W. A. Watts, B. Hansen, and K. A. Maasch. 1993. A 50,000-year record of climate oscillations from Florida and its temporal correlation with the Heinrich events. Science 261: 198–200.

Grootes, P. M., M. Stuiver, J. White, S. Johnsen, and J. Jouzel. 1993. Comparison of oxygen isotope records from the GISP2 and GRIP Greenland ice cores. Nature 366: 552–554.

Haigh, J. D. 1996. The impact of solar variability on climate. Science 272: 981–984.

Hammer, C. U. 1984. Traces of Icelandic eruptions in the Greenland ice sheet. Jökull 34: 51–65.

———, H. B. Clausen, and W. Dansgaard. 1980. Greenland ice sheet evidence of post-glacial volcanism and its climatic impact. Nature 288: 230–235.

Handler, P. 1984. Possible association of stratospheric aerosols and El Niño type events. Geophys. Res. Lett. 11: 1121–1124.

Hansen, J. et al. (+ 7 authors). 1991. Regional greenhouse climate effects. In: M. E. Schlesinger (ed.), Greenhouse-Gas-Induced Climatic Change: A Critical Appraisal of Simulations and Observations. Elsevier Science Publishers, Amsterdam. pp. 211–229.

———, A. Lacis, R. Ruedy, M. Sato, and H. Wilson. 1993. How sensitive is the world's climate? Natl. Geogr. Res. Explor. 9: 143–158.

Hanski, I., P. Turchin, E. Korpimäkl, and H. Henttonen. 1993. Population oscillations of boreal rodents: regulation by mustelid predators leads to chaos. Nature 364: 232–235.

Hanson, B. 1993. Editorial—evolution of atmospheres. Science 259: 875.

Harrison, T. M., P. Copeland, W. S. F. Kidd, and A. Yin. 1992. Raising Tibet. Science 255: 1663–1670.

Hashimoto, M. and S. Uyeda (eds.). 1983. Accretion Tectonics in the Circum-Pacific Regions. Terra Scientific Publishing Company, Tokyo.

Hays, J. D., J. Imbrie, and N. J. Shackleton. 1976. Variations in the Earth's orbit: pacemaker of the ice ages. Science 194: 1121–1132.

Hecht, A. D. (ed.). 1985a. Paleoclimate Analysis and Modeling. Wiley-Interscience, New York.

———. 1985b. Paleoclimatology: a retrospective of the past 20 years. In: A. D. Hecht (ed.), Paleoclimate Analysis and Modeling. Wiley-Interscience, New York. pp. 1–25.

Heinrich, H. 1988. Origin and consequences of cyclic ice rafting in the northeast Atlantic Ocean during the past 130,000 years. Quatern. Res. 29: 142–152.

Hide, R. and J. O. Dickey. 1991. Earth's variable rotation. Science 253: 629–637.

Hildebrand, A. R. and W. V. Boynton. 1990. Proximal Cretaceous–Tertiary boundary impact deposits in the Caribbean. Science 248: 843–847.

———, M. Pilkington, M. Connors, C. Ortiz-Aleman, and R. E. Chavez. 1995. Size and structure of the Chicxulub crater revealed by horizontal gravity gradients and cenotes. Nature 376: 415–417.

Hill, R. I., I. H. Campbell, G. F. Davies, and R. W. Griffiths. 1992. Mantle plumes and continental tectonics. Science 256: 186–193.

Hoffecker, J. F., W. R. Powers, and T. Goebel. 1993. The colonization of Beringia and the peopling of the New World. Science 259: 46–53.

Hoffer, J. M., F. Gomez, and P. Muela. 1982. Eruption of El Chichón volcano, Chiapas, Mexico, 28 March to 7 April 1982. Science 218: 1307–1308.

Hofmann, D. J. and J. M. Rosen. 1983. Sulfuric acid droplet formation and growth in the stratosphere after the 1982 eruption of El Chichón. Science 222: 325–327.

Holden, C. 1992. Energizing El Niño? Science 256: 741.

Hooper, P. R. 1982. The Columbia River basalts. Science 215: 1463–1468.

Hopkins, D. M., J. V. Matthews, Jr., C. E. Schweger, and S. B. Young (eds.). 1982. Paleoecology of Beringia. Academic Press, New York.

Houghton, J. T., B. A. Callander, and S. K. Varney. 1992. Climate Change 1992. IPCC (Intergovernmental Panel on Climate Change) Supplement. Cambridge University Press, Cambridge, U.K.

———, G. J. Jenkins, and J. J. Ephraums (eds.). 1990. Climate Change: The IPCC Scientific Assessment. Cambridge University Press, Cambridge, U.K.

Howell, D. G. 1989. Tectonics of Suspect Terranes, Mountain Building and Continental Growth. Chapman and Hall, New York.

Hoyt, D. V. and K. H. Schatten. 1997. The Role of the Sun in Climate Change. Oxford University Press, Oxford, U.K.

Hsü, K. 1992. Challenger at Sea, A Ship that Revolutionized Earth Science. Princeton University Press, Princeton, NJ.

Huber, N. K. 1981. Amount and Timing of Late Cenozoic Uplift and Tilt of the Central Sierra Nevada, California—Evidence from the Upper San Joaquin River Basin. U.S. Geological Survey, Denver, CO. Professional Paper 1197, pp. 1–28.

Huggett, R. J. 1990. Catastrophism: Systems of Earth History. Edward Arnold, London.

Hughen, K. A., J. T. Overpeck, L. C. Peterson, and S. Trumbore. 1996. Rapid climate changes in the tropical Atlantic region during the last deglaciation. Nature 380: 51–54.

Hultén, E. 1937. Outline of the History of Arctic and Boreal Biota during the Quaternary Period. Bokforlags Aktiebolaget Thule, Stockholm.

Imbrie, J., A. C. Mix, and D. G. Martinson. 1993. Milankovitch theory viewed from Devils Hole. Nature 363: 531–533.

——— et al. (+ 8 authors). 1984. The orbital theory of Pleistocene climate: support from a revised chronology of the marine ^{18}O record. In: A. L. Berger, J. Imbrie, J. Hays, G. Kukla, and B. Saltzman (eds.), Milankovitch and Climate: Understanding the Response to Astronomical Forcing. Reidel, Dordrecht. pp. 269–305.

——— et al. (+ 17 authors). 1992. On the structure and origin of major glaciation cycles 1. Linear responses to Milankovitch forcing. Paleoceanography 7: 701–738.

——— et al. (+ 18 authors). 1993. On the structure and origin of major glaciation cycles 2. The 100,000-year cycle. Paleoceanography 8: 699–735.

Izett, G. A., W. A. Cobban, J. D. Obradovich, and M. J. Kunk. 1993. The Manson impact structure: $^{40}Ar/^{39}Ar$ age and its distal impact ejecta in the Pierre Shale in southeastern South Dakota. Science 262: 729–732.

———, G. B. Dalrymple, and L. W. Snee. 1991. $^{40}Ar/^{39}Ar$ age of Cretaceous–Tertiary boundary tektites from Haiti. Science 252: 1539–1542.

Janecek, T. R. 1985. Eolian sedimentation in the northwestern Pacific Ocean: a preliminary examination of the data from Deep Sea Drill. Proj. sites 576 and 578. Initial Rep. Deep Sea Drilling Project 86: 589–603.

Jansa, L. F. 1993. Cometary impacts into ocean: their recognition and the threshold constraint for biological extinctions. Palaeogeogr., Palaeoclimatol., Palaeoecol. 104: 271–286.

Jin, F.-F., J. D. Neelin, and M. Ghil. 1994. El Niño on the Devil's staircase: annual subharmonic steps to chaos. Science 264: 70–72.

Johnsen, S. J. et al. (+ 9 authors). 1992. Irregular glacial interstadials recorded in a new Greenland ice core. Nature 359: 311–313.

Johnson, R. G., and S.-E. Lauritzen. 1995. Hudson Bay–Hudson Strait jökulhlaups and Heinrich events: a hypothesis. Palaeogeogr., Palaeoclimatol., Palaeoecol. 117: 123–137.

Jones, D. L., D. G. Howell, P. J. Coney, and J. W. H. Monger. 1983. Recognition, character, and analysis of tectonostratigraphic terranes in western North America. In: M. Hashimoto and S. Uyeda (eds.), Accretion Tectonics in the Circum-Pacific Regions. Terra Scientific Publishing Company, Tokyo. pp. 21–35.

———, N. J. Silberling, and J. Hillhouse. 1977. Wrangellia—a displaced terrane in northwestern North America. Can. J. Earth Sci. 14: 2565–2577.

Jouzel, J. et al. (+ 16 authors). 1993. Extending the Vostok ice-core record of palaeoclimate to the penultimate glacial period. Nature 364: 407–412.

Kareiva, P. M., J. G. Kingsolver, and R. B. Huey (eds.). 1993. Biotic Interactions and Global Change. Sinauer Associates, Sunderland, MA.

Kaufmann, R. K. and D. I. Stern. 1997. Evidence for human influence on climate from hemispheric temperature relations. Nature 388: 39–44.

Keeling, R. F., S. C. Piper, and M. Heimann. 1996. Global and hemispheric CO_2 sinks deduced from changes in atmospheric O_2 concentration. Nature 381: 218–221.

Keigwin, L. D., W. B. Curry, S. J. Lehman, and S. Johnsen. 1994. The role of the deep ocean in North Atlantic climate change between 70 and 130 Kyr ago. Nature 371: 323–326.

Keir, R. 1992. Packing away carbon isotopes. Nature 357: 445–446.

Kelly, P. M. and C. B. Sear. 1984. Climatic impact of explosive volcanic eruptions. Nature 311: 740–743.

——— and M. L. Wigley. 1992. Solar cycle length, greenhouse forcing and global climate. Nature 360: 328–330.

Kennedy, J. A. and S. C. Brassell. 1992. Molecular records of twentieth-century El Niño events in laminated sediments from the Santa Barbara Basin. Nature 357: 62–64.

Kerr, R. A. 1995. A fickle sun could be altering Earth's climate after all. Science 269: 633.

———. 1996a. A new dawn for sun–climate links? Science 271: 1360–1361.

———. 1996b. Ice rhythms: core reveals a plethora of climate cycles. Science 274: 499–500.

Kiehl, J. T. and B. P. Briegleb. 1993. The relative roles of sulfate aerosols and greenhouse gases in climate forcing. Science 260: 311–314.

King, P. B. 1977. The Evolution of North America. Princeton University Press, Princeton, NJ.

Kotra, J. P., D. L. Finnegan, W. H. Zoller, M. A. Hart, and J. L. Moyers. 1983. El Chichón: composition of plume gases and particles. Science 222: 1018–1021.

Kring, D. A. and W. V. Boynton. 1992. Petrogenesis of an augite-bearing melt rock in the Chicxulub structure and its relationship to K/T impact spherules in Haiti. Nature 358: 141–144. (See also discussion in Nature 362: 503–504 and 363: 503–504.)

Krogh, T. E., S. L. Kamo, V. L. Sharpton, L. E. Marin, and A. R. Hildebrand. 1993. U-Pb ages of single shocked zircons linking distal K/T ejecta to the Chicxulub crater. Nature 366: 731–734.

Ku, T.-L., M. Ivanovich, and S. Luo. 1990. U-series dating of last interglacial high sea stands: Barbados revisited. Quatern. Res. 33: 129–147.

Küchler, A. W. 1964. Manual to Accompany the Map Potential Natural Vegetation of the Conterminous United States. American Geographical Society, New York. Special Publ. 36.

Kutzbach, J., G. Bonan, J. Foley, and S. P. Harrison. 1996.

Vegetation and soil feedbacks on the response of the African monsoon to orbital forcing in the Early to Middle Holocene. Nature 384: 623–626.
———, P. J. Guetter, W. F. Ruddiman, and W. L. Prell. 1989. Sensitivity of climate to late Cenozoic uplift in southern Asia and the American west: numerical experiments. J. Geophys. Res. 94: 18393–18407.
LaMarche, V. C., Jr., and K. K. Hirschboeck. 1984. Frost rings in trees as records of major volcanic eruptions. Nature 307: 121–126.
Lehman, S. J. and L. D. Keigwin. 1992. Sudden changes in North Atlantic circulation during the last deglaciation. Nature 356: 757–762.
Leinen, M. and G. R. Heath. 1981. Sedimentary indicators of atmospheric activity in the northern hemisphere during the Cenozoic. Palaeogeogr., Palaeoclimatol., Palaeoecol. 36: 1–21.
Levine, J. S. (ed.). 1991. Global Biomass Burning: Atmospheric, Climatic, and Biospheric Implications. MIT Press, Cambridge, MA.
Lewin, R. 1992. Complexity: Life at the Edge of Chaos. MacMillan, New York.
Li, X., H. Maring, D. Savoie, K. Voss, and J. Prospero. 1996. Dominance of mineral dust in aerosol light-scattering in the North Atlantic trade winds. Nature 380: 416–419.
List, R. J. 1951. Smithsonian Meterological Tables, Smithsonian Miscellaneous Collections. Smithsonian Institution Press, Washington, D.C. Vol. 114, 6th revised edition.
Liu, H.-S. 1992. Frequency variations of the Earth's obliquity and the 100-kyr ice-age cycles. Nature 358: 397–399.
Ludwig, K. R., K. R. Simmons, I. J. Winograd, B. J. Szabo, and A. C. Riggs. 1993. Technical comments: dating of the Devils Hole calcite vein, response. Science 259: 1626–1627.
Lutjeharms, J. R. E. and H. R. Valentine. 1991. Sea-level changes: consequences for the southern hemisphere. Climatic Change 18: 317–337.
MacGinitie, H. D. 1953. Fossil Plants of the Florissant Beds, Colorado. Carnegie Institute of Washington, D.C. Contrib. Paleontol. Publ. 599.
Macilwain, C. 1995. El Niño centre is forecasting "first." Nature 378: 228.
Maier-Reimer, E., U. Mikolajewicz, and T. J. Crowley. 1990. Ocean general circulation model sensitivity experiment with an open Central American isthmus. Paleoceanography 5: 349–366.
Majumdar, S. K., L. S. Kalkstein, B. M. Yarnal, E. W. Miller, and L. M. Rosenfeld (eds.). 1992. Global Climate Change: Implications, Challenges and Mitigation Measures. Pennsylvania Academy of Science, Philadelphia, PA.
Manabe, S. and J. Broccoli. 1990. Mountains and arid climates of middle latitudes. Science 247: 192–195.
——— and R. J. Stouffer. 1995. Simulation of abrupt climate change induced by freshwater input to the North Atlantic Ocean. Nature 378: 165–167.
——— and T. B. Terpstra. 1974. The effects of mountains on the general circulation of the atmosphere as identified by numerical experiments. J. Atmos. Sci. 31: 3–42.
Marine Geology. 1996. 131 (April): whole issue.
Marshall, C. R. and P. D. Ward. 1996. Sudden and gradual molluscan extinctions in the latest Cretaceous of western European Tethys. Science 274: 1360–1363.
Mass, C. F. and D. A. Portman. 1989. Major volcanic eruptions and climate: a critical evaluation. J. Clim. 2: 566–593.
Matthews, J. V., Jr., and L. E. Ovenden. 1990. Late Tertiary plant macrofossils from localities in Arctic/subarctic North America: a review of the data. Arctic 43: 364–392.
Maurrasse, F. and G. Sen. 1991. Impacts, tsunamis, and the Haitian Cretaceous–Tertiary boundary layer. Science 252: 1690–1693.
Mayewski, P. A. et al. (+ 13 authors). 1996. Climate change during the last deglaciation in Antarctica. Science 272: 1636–1638.
McCormick, M. P., L. W. Thomason, and C. R. Trepte. 1995. Atmospheric effects of the Mt Pinatuba eruption. Nature 373: 399–404.
McGuffie, K., A. Henderson-Sellers, H. Zhang, T. B. Durbidge, and A. J. Pitman. 1995. Global climate sensitivity to tropical deforestation. Global Planet. Change 10: 97–128.
McIntyre, A., and B. Molfino. 1996. Forcing of Atlantic equatorial and subpolar millennial cycles by precession. Science 274: 1867–1870.
McKenna, M. C. 1972. Was Europe connected directly to North America prior to the middle Eocene? In: T. Dobzhansky, M. K. Hecht, and W. C. Steere (eds.), Evolutionary Biology. Appleton-Century-Crofts, New York. Vol. 6., pp. 179–189.
———. 1983. Holarctic landmass rearrangement, cosmic events, and Cenozoic terrestrial organisms. Ann. Missouri Bot. Garden 70: 459–489.
McManus, J. F. et al. (+ 5 authors). 1994. High-resolution climate records from the North Atlantic during the last interglacial. Nature 371: 326–329.
Meese, D. A. et al. (+ 8 authors). 1994. The accumulation record from the GISP2 core as an indicator of climate change throughout the Holocene. Science 266: 1680–1682.
Mesolella, K. J., R. K. Matthews, W. S. Broecker, and D. L. Thurber. 1969. The astronomical theory of climatic change: Barbados data. J. Geol. 77: 250–274.
Meyer, H. W. 1992. Lapse rates and other variables applied to estimating paleoaltitudes from fossil floras. Palaeogeogr., Palaeoclimatol., Palaeoecol. 99: 71–99.
Milankovitch, M. 1920. Théorie mathematique des phenomènes thermiques produits par la radiation solaire. Gauthier-Villars, Académie Yugoslave des Sciences et des Arts de Zagreb, Yugoslavia.
———. 1930. Mathematische klimalehre und astronomische théorie de klimschwankungen. In: W. Köppen and R. Geiger (eds.), Handbuch der Klimatologie, 1. Teil A. Borntraeger, Berlin.
———. 1941. Canon of Insolation and the Ice Age Problem. Köninglich Serbische Akadémie, Beogra, Serbia. [English translation of the Israel Program for Scientific Translation, published for the U.S. Dept. Commerce and the National Science Foundation].
Miller, K. G., R. G. Fairbanks, and G. S. Mountain. 1987. Tertiary oxygen isotope synthesis, sea level history, and continental margin erosion. Paleoceanography 2: 1–19.
Minnis, P. et al. (+ 6 authors). 1993. Radiative climate forc-

ing by the Mount Pinatubo eruption. Science 259: 1411–1415.

Mintz, Y. and G. Dean. 1952. The Observed Mean Field of Motion of the Atmosphere. Air Force Cambridge Research Center, Cambridge, U.K. Geophys. Res. Paper 17.

Molnar, P. and P. England. 1990. Late Cenozoic uplift of mountain ranges and global climate change: chicken or egg? Nature 346: 29–34.

Morell, V. 1993. How lethal was the K–T impact? Science 261: 1518–1519.

Morgan, M. E., J. D. Kingston, and B. D. Marino. 1994. Carbon isotopic evidence for the emergence of C4 plants in the Neogene from Pakistan and Kenya. Nature 367: 162–165.

Mull, C. G. 1979. Nanushuk Group deposition and late Mesozoic structural evolution of the central and western Brooks Range and Arctic Slope. In: T. S. Ahlbrandt (ed.), Preliminary Geologic, Petrologic, and Paleontologic Results of the Study of Nanushuk Group Rocks, North Slope, Alaska. U.S. Geological Survey, Denver, CO. Circular 794, pp. 5–13.

———. 1985. Cretaceous tectonics, depositional cycles, and the Nanushuk Group, Brooks Range and Arctic Slope, Alaska. U.S. Geol. Surv. Bull. 1614: 7–36.

Nature. 1992. 357 (May 28): whole issue.

Nature. 1993. 361: 123–124.

Nature. 1994. 371: 112–113.

Neftel, A., H. Oeschger, J. Schwander, B. Stauffer, and R. Zumbrunn. 1982. Ice core sample measurements give atmospheric CO_2 content during the past 40,000 yr. Nature 295: 220–223.

NRC Panel on Climatic Variation, U.S. Committee for GARP. 1975. Understanding Climatic Change—A Program for Action. National Academy Press, Washington, D.C.

Oechel, W. C. et al. (+ 5 authors). 1993. Recent change of Arctic tundra ecosystems from a net carbon dioxide sink to a source. Nature 361: 520–523.

Oldow, J. S., A. W. Bally, H. G. Avé Lallemant, and W. P. Leeman. 1989. Phanerozoic evolution of the North American Cordillera; United States and Canada. In: A. W. Bally and A. R. Palmer (eds.), The Geology of North America; An Overview. DNAG (Decade of North American Geology), The Geology of North America. Geological Society of America, Boulder, CO. Vol. A, pp. 139–232.

Olsen, P. E. 1986. A 40-million-year lake record of early Mesozoic orbital climatic forcing. Science 234: 842–848.

——— and D. V. Kent. 1996. Milankovitch climate forcing in the tropics of Pangaea during the Late Triassic. Palaeogeogr., Palaeoclimatol., Palaeoecol. 122: 1–26.

Oppenheimer, D. et al. (+ 19 authors). 1993. The Cape Mendocino, California earthquakes of April 1992: subduction at the triple junction. Science 261: 433–438.

Oppo, D. W. and S. J. Lehman. 1993. Mid-depth circulation of the subpolar North Atlantic during the last glacial maximum. Science 259: 1148–1152.

Oberbeck, V. R. and R. L. Mancinelli. 1994. Asteroid impacts, microbes, and the cooling of the atmosphere. BioScience 44: 173–177.

Overpeck, J. T. 1992. Review: Global biomass burning: atmospheric, climatic, and biospheric implications, J. S. Levine (ed.). Nature 356: 670.

———, P. J. Bartlein, and T. Webb III. 1991. Potential magnitude of future vegetation change in eastern North America: comparisons with the past. Science 254: 692–695.

Paillard, D. and L. Labeyrie. 1994. Role of the thermohaline circulation in the abrupt warming after Heinrich events. Nature 372: 162–164.

Park, J., S. L. D'Hondt, J. W. King, and C. Gibson. 1993. Late Cretaceous precessional cycles in double time: a warm-Earth Milankovitch response. Science 261: 1431–1434.

Parker, D. E. and J. L. Brownscombe. 1983. Stratospheric warming following the El Chichón volcanic eruption. Nature 301: 406–408.

Partridge, T. C., G. C. Bond, C. J. H. Hartnady, P. B. deMenocal, and W. F. Ruddiman. 1995. Climatic effects of late Neogene tectonism and volcanism. In: E. S. Vrba, G. H. Denton, T. C. Partridge, and L. H. Burckle (eds.), Paleoclimate and Evolution, with Emphasis on Human Origins. Yale University Press, New Haven, CT. pp. 8–23.

Paterson, W. S. B. and C. U. Hammer. 1987. Ice core and other glaciological data. In: W. F. Ruddiman and H. E. Wright, Jr. (eds.), North America and Adjacent Oceans during the Last Deglaciation. The Geology of North America, DNAG (Decade of North American Geology). Geological Society of America, Boulder CO. Vol. K-3, pp. 91–109.

Pedolosky, J. 1990. The dynamics of the oceanic subtropical gyres. Science 248: 316–322.

Peters, R. L. and T. E. Lovejoy (eds.). 1992. Global Warming and Biological Diversity. Yale University Press, New Haven, CT.

Philander, S. G. H. 1983. El Niño southern oscillation phenomena. Nature 302: 295–301.

Poag, C. W. 1997. The Chesapeake Bay bolide impact: a convulsive event in Atlantic coastal plain evolution. Sedimentary Geol. 108: 45–90.

Polley, H. W., H. B. Johnson, B. D. Marino, and H. S. Mayeux. 1993. Increase in C3 plant water-use efficiency and biomass over glacial to present CO_2 concentrations. Nature 361: 61–64.

Prell, W. L. and J. E. Kutzbach. 1992. Sensitivity of the Indian monsoon to forcing parameters and implications for its evolution. Nature 360: 647–652.

Price, J. F. et al. (+ 8 authors). 1993. Mediterranean outflow mixing and dynamics. Science 259: 1277–1282.

Quade, J., T. E. Cerling, and J. R. Bowman. 1989. Development of Asian monsoon revealed by marked ecological shift during the latest Miocene in northern Pakistan. Nature 342: 163–166.

Quinn, W. H., V. T. Neal, and S. E. Antunez de Mayolo. 1987. El Niño occurrences over the past four and a half centuries. J. Geophys. Res. 92: 14449–14461.

———, D. O. Zopf, K. S. Short, and R. T. W. Kuo Yang. 1978. Historical trends and statistics of the southern oscillation, El Niño, and Indonesian droughts. Fish. Bull. 76: 663–678.

Rahmstorf, S. 1994. Rapid climate transitions in a coupled ocean–atmosphere model. Nature 372: 82–85.

———. 1995. Bifurcations of the Atlantic thermohaline circulation in response to changes in the hydrological cycle. Nature 378: 145–149.

Ramanathan, V. et al. (+ 7 authors). 1995. Warm pool heat budget and shortwave cloud forcing: a missing physics? Science 267: 499–503.

Rampino, M. R., S. Self, and R. B. Stothers. 1988. Volcanic winters. Annu. Rev. Earth Planet. Sci. 16: 73–99.

Rathburn, S. L. 1993. Pleistocene cataclysmic flooding along the Big Lost River, east central Idaho. Geomorphology 8: 305–319.

Raup, D. M. 1981. Introduction: what is crisis? In: M. H. Nitecki (ed.), Biotic Crises in Ecological and Evolutionary Time. Academic Press, New York. pp. 1–12.

Raymo, M. E. and W. F. Ruddiman. 1992. Tectonic forcing of late Cenozoic climate. Nature 359: 117–122.

———. 1993. Scientific correspondence: Cooling in the late Cenozoic. Nature 361: 124.

Raynaud, D. et al. (+ 5 authors). 1993. The ice record of greenhouse gases. Science 259: 926–934.

Rea, D. K., M. Leinen, and T. R. Janecek. 1985. Geologic approach to the long-term history of atmospheric circulation. Science 227: 721–725.

Repenning, C. A. and E. M. Brouwers. 1992. Late Pliocene–early Pleistocene ecologic changes in the Arctic Ocean borderland. U.S. Geol. Surv. Bull. 2036: 1–37.

Retallack, G. J. 1996. Acid trauma at the Cretaceous–Tertiary boundary in eastern Montana. GSA Today 6: 1–7.

Rind, D., C. Rosenzweig, and R. Goldberg. 1992. Modelling the hydrological cycle in assessments of climatic change. Nature 358: 119–122.

Robin, E., L. Froget, C. Jéhanno, and R. Rocchia. 1993. Evidence for a K/T impact event in the Pacific Ocean. Nature 363: 615–617.

Robock, A. and J. Mao. 1992. Winter warming from large volcanic eruptions. Geophys. Res. Lett. 19: 2405–2408.

Rozanski, K., L. Araguás-Araguás, and R. Gonfiantini. 1992. Relation between long-term trends of oxygen-18 isotope composition of precipitation and climate. Science 258: 981–985.

Rubin, E. S. et al. (+ 6 authors). 1992. Realistic mitigation options for global warming. Science 257: 148–149; 261–266.

Ruddiman, W. F. 1997. Tectonic Uplift and Climatic Change. Plenum, New York.

Ruddiman, W. F. and J. E. Kutzbach. 1989. Forcing of late Cenozoic northern hemisphere climate by plateau uplift in southern Asia and the American west. J. Geophys. Res. 94: 18409–18427.

———. 1990. Late Cenozoic plateau uplift and climate change. Trans. R. Soc. Edinburgh: Earth Sci. 81: 301–314.

Ruddiman, W. F. and A. McIntyre. 1976. Northeast Atlantic paleoclimatic changes over the past 600,000 years. In: R. M. Cline and J. D. Hays (eds.), Investigation of Late Quaternary Paleoceanography and Paleoclimatology. Geological Society of America, Boulder, CO. Memoir 145, pp. 111–146.

———. 1984. Ice-age thermal response and climatic role of the surface Atlantic Ocean, 40° to 63° N. Geol. Soc. Am. Bull. 95: 381–396.

Ruddiman, W. F., W. L. Prell, and M. E. Raymo. 1989. Late Cenozoic uplift in southern Asia and the American west: rationale for general circulation modeling experiments. J. Geophys. Res. 94: 18379–18391.

Ruelle, D. 1991. Chance and Chaos. Princeton University Press, Princeton, NJ.

Sacks, P. E. and D. T. Secor, Jr. 1990. Kinematics of late Paleozoic continental collision between Laurentia and Gondwana. Science 250: 1702–1705.

Sagan, C. and S. J. Ostro. 1994. Dangers of asteroid deflection. Nature 368: 501.

Salati, E., T. E. Lovejoy, and P. B. Vose. 1983. Precipitation and water recycling in tropical rain forests with special reference to the Amazon Basin. Environmentalist 3: 67–72.

Salati, E. and P. B. Vose. 1984. Amazon Basin: a system in equilibrium. Science 225: 129–138.

Saltzman, B. and M. Verbitsky. 1994. CO_2 and glacial cycles. Nature 367: 419.

Sanford, R. L., Jr., J. Saldarriaga, K. E. Clark, C. Uhl, and R. Herrera. 1985. Amazon rain-forest fires. Science 277: 53–55.

Savage, J. C. and M. Lisowski. 1991. Strain measurements and the potential for a great subduction earthquake off the coast of Washington. Science 252: 101–103.

Savin, S. M. and R. G. Douglas. 1985. Sea level, climate, and the Central American land bridge. In: F. G. Stehli and S. D. Webb (eds.), The Great American Biotic Interchange. Plenum, New York. pp. 303–324.

Schlesinger, M. E. and N. Ramankutty. 1992. Implications for global warming of intercycle solar irradiance variations. Nature 360: 330–333.

Schmidlin, T. W. 1989. Climatic summary of snowfall and snow depth in the Ohio snowbelt at Chardon. Ohio J. Sci. 89: 101–108.

Schneider, R. and M. Little. 1997. Southern and northern hemisphere climatic changes: synchronous or not? PAGES 5:12.

Schneider, S. H. 1994. Detecting climatic change signals: are there any "fingerprints"? Science 263: 341–347.

Schwarzacher, W. 1993. Cyclostratigraphy and the Milankovitch Theory. Elsevier Science Publishers, Amsterdam.

Science. 1988. 241: 717–724.

Science. 1992. 258: 1611–1630.

Science. 1993. 259 (February 12): cover.

Scientific Committee on Oceanic Research (SCOR), Working Group 55. 1983. Prediction of "El Niño." SCOR Proc. (Paris) 19: 47–51.

Scott, G. R. 1975. Cenozoic surfaces and deposits in the southern Rocky Mountains. In: B. F. Curtis (ed.), Cenozoic History of the Southern Rocky Mountains. Geological Society of America, Boulder, CO. Memoir 144, pp. 227–248.

Scott, J. V., D. L. Rabinowitz, and B. G. Marsden. 1991. Near miss of the Earth by a small asteroid. Nature 354: 287–289.

Severinghaus, J. P., T. Sowers, E. J. Brook, R. B. Alley, and M. L. Bender. 1998. Timing of abrupt climate change at the end of the Younger Dryas interval from thermally fractionated gases in polar ice. Nature 391: 141–146.

Shackleton, N. J., A. Berger, and W. R. Peltier. 1990. An alternative astronomical calibration of the lower Pleistocene timescale based on ODP site 677. Trans. R. Soc. Edinburgh: Earth Sci. 81: 251–261.

Shackleton, N. J., M. A. Hall, J. Line, and C. Shuxi. 1983. Carbon isotope data in core V19-30 confirm reduced carbon dioxide concentration in the ice age atmosphere. Nature 306: 319–322.

Shackleton, N. J. and J. P. Kennett. 1975. Late Cenozoic oxygen and carbon isotopic changes at DSDP site 284: implications for glacial history of the northern hemisphere. Initial Rep. Deep Sea Drilling Project 29: 801–807.

Sharpton, V. L. and P. D. Ward (eds.). 1990. Global Catastrophies in Earth History. An Interdisciplinary Conference on Impacts, Volcanism, and Mass Mortality, Snowbird, Utah, 1988. Geological Society of America, Boulder, CO. Special Paper 247.

Sharpton, V. L. et al. (+ 5 authors). 1992. New links between the Chicxulub impact structure and the Cretaceous/Tertiary boundary. Nature 359: 819–821.

Sharpton, V. L. et al. (+ 9 authors). 1993. Chicxulub multiring impact basin: size and other characteristics derived from gravity analysis. Science 261: 1564–1567.

Sheehan, P. M., D. E. Fastovsky, R. G. Hoffmann, C. B. Berghaus, and D. L. Gabriel. 1991. Sudden extinctions of the dinosaurs: latest Cretaceous, upper Great Plains, U.S.A. Science 254: 835–839.

Shukla, J. and Y. Mintz. 1982. Influence of land-surface evapotranspiration on the Earth's climate. Science 215: 1498–1501.

Shukla, J., C. Nobre, and P. Sellers. 1990. Amazon deforestation and climatic change. Science 247: 1322–1325.

Siegenthaler, U. and J. L. Sarmiento. 1993. Atmospheric carbon dioxide and the ocean. Nature 365: 119–125.

Sloan, L. C. and E. J. Barron. 1990. "Equable" climates during Earth history? Geology 18: 489–492.

Slowey, N. C., G. M. Henderson, and W. B. Curry. 1996. Direct U-Th dating of marine sediments from the two most recent interglacial periods. Nature 383: 242–244.

Small, E. E. and R. S. Anderson. 1995. Geomorphically driven late Cenozoic rock uplift in the Sierra Nevada, California. Science 270: 277–280.

Smith, T. M. and H. H. Shugart. 1993. The transient response of terrestrial carbon storage to a perturbed climate. Nature 361: 523–526.

Southon, J. R. and T. A. Brown. 1995. Technical Comment: The GISP ice core record of volcanism since 7000 B.C. Science 267: 256–257.

Stewart, I. 1989. Does God Play Dice?: The Mathematics of Chaos. Blackwell, New York.

Stone, D. B., B. C. Panuska, and D. R. Packer. 1982. Paleolatitudes versus time for southern Alaska. J. Geophys. Res. 87: 3697–3707.

Summerfield, M. A. and M. P. Kirkbride. 1992. Climate and landscape response. Nature 355: 306. (See also Molnar and England reply.)

Svetsov, V. V. 1996. Total ablation of the debris from the 1908 Tunguska explosion. Nature 383: 697–699.

Swisher C. C., III (+ 11 authors). 1992. Coeval $^{40}Ar/^{39}Ar$ ages of 65.0 million years ago from Chicxulub Crater melt rock and Cretaceous–Tertiary boundary tektites. Science 257: 954–958.

Taggart, R. E. and A. T. Cross. 1980. Vegetation change in the Miocene Succor Creek flora of Oregon and Idaho: a case study in paleosuccession. In: D. L. Dilcher and T. N. Taylor (eds.), Biostratigraphy of Fossil Plants. Dowden, Hutchinson, & Ross, Stroudsburg, PA. pp. 185–210.

Taylor, K. C. et al. (+ 7 authors). 1993a. The "flickering switch" of late Pleistocene climate change. Nature 361: 432–436.

Taylor, K. C. et al. (+ 9 authors). 1993b Electrical conductivity measurements from the GISP2 and GRIP Greenland ice cores. Nature 366: 549–552.

Taylor, K. C. et al. (+ 12 authors). 1997. The Holocene–Younger Dryas transition recorded at Summit, Greenland. Science 278: 825–827.

Tegen, I., A. A. Lacis, and I. Fung. 1996. The influence on climate forcing of mineral aerosols from disturbed soils. Nature 380: 419–422.

Thelin, G. P. and R. J. Pike. 1991. Landforms of the conterminous United States—a digital shaded-relief portrayal. [Text] U.S. Geological Survey, Denver, CO. Miscel. Invest. Ser. Map I-2206.

Tilling, R. I. et al. (+ 5 authors). 1984. Holocene eruptive activity of El Chichón volcano, Chiapas, Mexico. Science 224: 747–749.

Trewartha, G. T. and L. H. Horn. 1980. An Introduction to Climate. McGraw-Hill, New York. 5th edition.

Turner, S., C. Hawkesworth, J. Liu, N. Rogers, S. Kelley, and P. van Calsteren. 1993. Timing of Tibetan uplift constrained by analysis of volcanic rocks. Nature 364: 50–54.

Tziperman, E., L. Stone, M. A. Cane, and H. Jarosh. 1994. El Niño chaos: overlapping of resonances between the seasonal cycle and the Pacific Ocean-atmosphere oscillator. Science 264: 72–74.

Urey, H. C. 1947. The thermodynamic properties of isotopic substances. J. Chem. Soc. 562–581.

———, H. A. Lowenstam, S. Epstein, and C. R. McKinney. 1951. Measurement of paleotemperatures and temperatures of the Upper Cretaceous of England, Denmark, and the southeastern United States. Bull. Geol. Soc. Am. 62: 399–416.

Vail, P. R. and J. Hardenbol. 1979. Sea-level changes during the Tertiary. Oceanus 22: 71–79.

Van Houten, F. B. 1964 Cyclic lacustrine sedimentation, Upper Triassic Lockatong Formation, central New Jersey and adjacent Pennsylvania. Bull. Kansas State Geol. Survey 169: 497–531.

Volk, T. 1993. Scientific correspondence: Cooling in the late Cenozoic. Nature 361: 123.

Waitt, R. B., Jr., 1985. Case for periodic, colossal jökulhlaups from Pleistocene Glacial Lake Missoula. Bull. Geol. Soc. Am. 96: 1271–1286.

Waldrop, M. M. 1992. Complexity: The Emerging Science at the Edge of Order and Chaos. Simon & Schuster/Viking, New York.

Wallace, R. E. (ed.). 1990. The San Andreas Fault System, California. U.S. Geological Survey, Denver, CO.

Ward, P. D., J. M. Hurtado, J. L. Kirschvink, and K. L. Verosub. 1997. Measurements of the Cretaceous paleolatitude of Vancouver Island: Consistent with the Baja–British Columbia Hypothesis. Science 277: 1642–1645.

Webb, R. S., D. H. Rind, S. J. Lehman, R. J. Healy, and D. Sigman. 1997. Influence of ocean heat transport on the climate of the last glacial maximum. Nature 385: 695–699.

West, F. H. (ed.). 1996. American Beginnings: the prehistory and palaeocology of Beringia. University of Chicago Press, Chicago.

White, J. W. C., P. Ciais, R. A. Figge, R. Kenny, and V. Markgraf. 1994. A high-resolution record of atmospheric CO_2 content from carbon isotopes in peat. Nature 367: 153–156.

Wigley, T. M. L. and P. D. Jones. 1981. Detecting CO_2-induced climatic change. Nature 292: 205–208.

Williams, M. A. J., D. L. Dunkerley, P. De Deckker, A. P. Kershaw, and T. Stokes. 1993. Quaternary Environments. Edward Arnold, London.

Winograd, I. J., B. J. Szabo, T. B. Coplen, and A. C. Riggs. 1988. A 250,000-year climatic record from Great Basin vein calcite: implications for Milankovitch theory. Science 242: 1275–1280.

Winograd, I. J. et al. (+ 7 authors). 1992. Continuous 500,000-year climate record from vein calcite in Devils Hole, Nevada. Science 258: 255–260.

Winterer, E. L., T. M. Atwater, and R. W. Decker. 1989. The northeast Pacific Ocean and Hawaii. In: A. W. Bally and A. R. Palmer (eds.), The Geology of North America; an Overview. DNAG (Decade of North American Geology). Geological Society of America, Boulder, CO. Vol. A, pp. 265–298.

Wolfe, J. A. 1987. Late Cretaceous–Cenozoic history of deciduousness and the terminal Cretaceous event. Paleobiology 13: 215–226.

Wood, C. A. and J. Kienle (eds.). 1990. Volcanoes of North America: United States and Canada. Cambridge University Press, Cambridge, U.K.

Woodward, F. I. 1987. Climate and Plant Distribution. Cambridge University Press, Cambridge, U.K.

Woodwell, G. M. (+ 6 authors). 1983. Global deforestation: contribution to atmospheric carbon dioxide. Science 222: 1081–1086.

World Meterological Organization [F. Steinhauser (Tech. Super.)]. 1979. Climatic Atlas of North and Central America—Maps of Mean Temperature and Precipitation. World Meteorological Organization, Geneva.

Zhao, M. and J. L. Bada. 1989. Extraterrestrial amino acids in Cretaceous/Tertiary boundary sediments at Stevns Klint, Denmark. Nature 339: 463–465.

Ziegler, P. A. 1988. Evolution of the Arctic–North Atlantic and the Western Tethys. American Association of Petroleum Geologists, Tulsa, OK. Mem. 43.

Zielinski, G. A. et al. (+ 8 authors). 1994. Record of volcanism since 7000 B.C. from the GISP2 Greenland ice core and implications for the volcano-climate system. Science 264: 948–952. (See responses by Fiedel, 1995; Southon and Brown, 1995.)

General Readings

Adams, J. 1992. Paleoseismology: a search for ancient earthquakes in Puget Sound. Science 258: 1592–1593.

Albarède, F. 1997. Isotopic tracers of past ocean circulation; turning lead into gold. Science 277: 908–909.

Alley, R. B., and M. L. Bender. 1998. Greenland ice cores: frozen in time. Sci. Am. 278: 80–85.

Alvarez, W. 1997. *T. rex* and the Crater of Doom. Princeton University Press, Princeton, NJ.

——— and F. Asaro. 1990. An extraterrestrial impact. Sci. Am. 263: 78–84.

Amato, I. 1993. Coming in loud and clear—via chaos. Science 261: 429.

Andreae, M. O. 1996. Raising dust in the greenhouse. Nature 380: 389–390.

Anonymous. 1994. El Niño and climate prediction. Report to the nation on our changing planet. Univ. Corp. Atmos. Res. Spring.

Appenzeller, T. 1993a. Searching for clues to ancient carbon dioxide. Science 259: 908–909.

———. 1993b. Ancient climate coolings are on thin ice. Science 262: 1818–1819.

Ball, P. 1992. Raising the roof of the world. Nature 357: 642.

Barnes-Svarney, P. 1996. Asteroid, Earth Destroyer or New Frontier? Plenum, New York.

Bender, M. 1996. A quickening on the uptake? Nature 381: 195–196.

Berner, R. A. and A. C. Lasaga. 1989. Modeling the geochemical carbon cycle. Sci. Am. 260: 74–81.

Bigg, G. R. 1996. The Oceans and Climate. Cambridge University Press, Cambridge, U.K.

Blum, J. D. 1993. Zircon can take the heat. Nature 366: 718.

Bond, G. C. 1995. Climate and the conveyor. Nature 377: 383–384.

Boulton, G. S. 1993. Two cores are better than one. Nature 366: 507–508.

Briggs, J. C. 1991. A Cretaceous–Tertiary mass extinction? BioScience 41: 619–624.

Broecker, W. S. 1987. The biggest chill. Natural History 96: 74–82.

———. 1992. Upset for Milankovitch theory. Nature 359: 779–780.

———. 1994. An unstable superconveyor. Nature 367: 414–415.

———. 1995. The Glacial World According to Wally. Lamont-Doherty Earth Observatory, Palisades, NY.

——— and G. H. Denton. 1990. What drives glacial cycles? Sci. Am. 262: 49–56.

Burke, I. C. (+ 5 authors). 1991. Regional analysis of the central Great Plains—sensitivity to climate variability. BioScience 41: 685–692.

Carroll, M. R. 1997. Volcanic sulphur in the balance. Nature 389: 543–544.

Charles, C. 1997. Cool tropical punch of the Ice Ages. Nature 385: 681–683.

Clifton, H. E. and R. H. Dott, Jr. 1994. Sedimentology of the K–T boundary. Science 264: 1829. (See also Keller, 1994; Kerr, 1994c.)

Committee on International Science's Task Force on Global Biodiversity. 1989. Loss of Biological Diversity: A Global Crisis Requiring International Solutions. A report to the National Science Foundation. National Science Foundation, Washington, D.C.

Cook, F. A., L. D. Brown, and J. E. Oliver. 1980. The southern Appalachians and the growth of continents. Sci. Am. 243: 156–168.

Courtillot, V. E. 1990. A volcanic eruption. Sci. Am. 263: 85–92.

Crowley, T. J. 1996. Remembrance of things past: greenhouse lessons from the geologic record. Consequences 2: 3–12.

Culotta, E. 1993. Is the geological past a key to the (near) future? Science 259: 906–908.

Dauber, P. M. and R. A. Muller. 1996. The Three Big Bangs: Comet Crashes, Exploding Stars, and the Creation of the Universe. Addison-Wesley, Reading, MA.

Davies, G. F. 1992. Plates and plumes: dynamos of the Earth's mantle. Science 257: 493–494.

Decker, R. W. and B. B. Decker. 1981. The eruptions of Mount St. Helens. Sci. Am. 244: 68–80.

———. 1991. Mountains of Fire: The Nature of Volcanoes. Cambridge University Press, Cambridge, U.K.

Deming, D. 1995. Climatic warming in North America: analysis of borehole temperatures. Science 268: 1576–1577.

Gibbons, A. 1997. Did birds sail through the K–T extinction with flying colors? Science 275: 1068.

Glantz, M. H. 1996. Currents of change. El Niño's impact on climate and society. Cambridge University Press, Cambridge, U.K.

Gordon, A. L. 1996. Communication between oceans. Nature 382: 399–400.

Gould, S. J. 1967. Is uniformitarianism useful? Geol. Ed. 15: 149–150.

———. 1978. The great scablands debate. Nat. Hist. 87: 12–18.

Holden, C. 1992a. Pack rats' liquid legacy. Science 255: 155.

———. 1992b. Cracking secrets in Greenland ice. Science 257: 161.

———. 1996. Environment award honors the big chill. Science 272: 201. [Tyler Prize for Environmental Achievement to Willi Dansgaard, Claude Lorius, and Hans Oeschger; see The Greenland Ice-Core Record section, this chapter].

Howell, D. G. 1985. Terranes. Sci. Am. 253: 116–125.

Huggett, R. J. 1991. Climate, Earth Processes and Earth History. Springer-Verlag, Berlin.

International Geosphere–Biosphere Programme. 1992. Global Change; Reducing Uncertainties. Royal Swedish Academy of Sciences, Stockholm.

Johnson, K. R. 1993. Extinctions at the antipodes. Nature 366: 511–512.

Jones, D. L., A. Cox, P. Coney, and M. Beck. 1982. The growth of western North America. Sci. Am. 247: 70–84.

Jordan, T. H. and J. B. Minster. 1988. Measuring crustal deformation in the American west. Sci. Am. 259: 48–58.

Jouzel, J. 1994. Ice cores north and south. Nature 372: 612–613.

Kanamori, H. and T. H. Heaton. 1996. The wake of a legendary earthquake. Nature 379: 203–204.

Keller, G. 1994. K–T boundary issues. Science 264: 641. (See also Clifton and Dott, 1994; Kerr, 1994c.)

Kerr, R. A. 1983. Fading El Niño broadening scientists' view. Science 221: 940–941.

———. 1984. Carbon dioxide and the control of ice ages. Science 223: 1053–1054.

———. 1986. Mapping orbital effects on climate. Science 234: 283–284.

———. 1987. Milankovitch climate cycles through the ages. Science 235: 973–974.

———. 1990a. Marking the ice ages in coral instead of mud. Science 248: 31–33.

———. 1990b. The ice age bones of contention. Science 248: 32.

———. 1990c. Commotion over Caribbean impacts. Science 250: 1081.

———. 1991a. El Niño winners and losers declared. Science 251: 1182.

———. 1991b. Big squeeze points to a big quake. Science 252: 28.

———. 1991c. Yucatan killer impact gaining support. Science 252: 377.

———. 1991d. The stately cycles of ancient climate. Science 252: 1254–1255.

———. 1991e. Did a volcano help kill off the dinosaurs? Science 252: 1496–1498.

———. 1991f. Hugh eruption may cool the globe. Science 252: 1780.

———. 1991g. Dancing with death at Unzen volcano. Science 253: 271.

———. 1991h. Could the sun be warming the climate? Science 254: 652–653.

———. 1991i. Ocean mud pins down a million years of time. Science 254: 802–803.

———. 1991j. Extinction potpourri: killers and victims. Science 254: 942–943.

———. 1992a. 1991: warmth, chill may follow. Science 255: 281.

———. 1992b. A successful forecast of an El Niño winter. Science 255: 402.

———. 1992c. Extinctions by a one-two punch?/Extinction with a whimper. Science 255: 160–161.

———. 1992d. Unmasking a shifty climate system. Science 255: 1508–1510.

———. 1992e. Another impact extinction? Science 256: 1280.

———. 1992f. Hugh impact tied to mass extinction. Science 257: 878–880.

———. 1992g. A revisionist timetable for the ice ages. Science 258: 220–221.

———. 1993a. Pinatubo global cooling on target. Science 259: 594.

———. 1993b. Galileo reveals a badly battered Ida. Science 262: 33.

———. 1993c. Second crater points to killer comets. Science 259: 1543.

———. 1993d. Volcanoes may warm locally while cooling globally. Science 260: 1232.

———. 1993e. No way to cool the ultimate greenhouse. Science 262: 648.

———. 1993f. New crater age undercuts killer comets. Science 262: 659.

———. 1993g. The whole world had a case of the Ice Age shivers. Science 262: 1972–1973.

———. 1993h. Even warm climates get the shivers. Science 261: 292.

———. 1993i. How ice age climate got the shakes. Science 260: 890-892.

———. 1994a. Milankovitch plays climate in double-time in the tropics. Science 263: 174–175.

———. 1994b. Did Pinatubo send climate-warming gases into a dither? Science 263: 1562.

———. 1994c. Testing an ancient impact's punch. Science 263: 1371–1372. (See also Clifton and Dott, 1994; Keller, 1994.)

———. 1994d. Did the tropical Pacific drive the world's warming? Science 266: 544–545.

———. 1994e. Baring the secrets of asteroid Ida. Science 266: 1322.

———. 1995a. Darker clouds promise brighter future for climate models. Science 267: 454.

———. 1995b. How far did the west wander? Science 268: 635–637.

———. 1995c. U.S. Climate tilts toward the greenhouse. Science 268: 363–364.

———. 1995d. Chilly ice-age tropics could signal climate sensitivity. Science 267: 961.

———. 1995e. Exonerating an ice sheet. Science 267: 27.

———. 1995f. The seasons of global change. Science 267: 27–28.

———. 1995g. Studies say-tentatively-that greenhouse warming is here. Science 268: 1567–1568.

———. 1996a. Millennial climate oscillation spied. Science 271: 146–147.

———. 1996b. Volcano–ice age link discounted. Science 272: 817.

———. 1996c. Ice bubbles confirm big chill. Science 272: 1584–1585.

———. 1996d. A shocking view of the Permo-Triassic. Science 274: 1080.

———. 1996e. New way to read the record suggests abrupt extinction. Science 274: 1303–1304.

———. 1997a. Cores document ancient catastrophe. Science 275: 1265.

———. 1997b. Why the West stands tall. Science 275: 1564–1565.

———. 1997c. Second clock supports orbital pacing of the Ice Ages. Science 276: 680–681.

———. 1997d. Moderating a tropical debate. Science 276: 1793.

———. 1997e. More signs of a far-traveled west. Science 277: 1608.

Krumenaker, L. 1994. In ancient climate, orbital chaos? Science 263: 323.

Kunzig, R. 1989. Ice cycles. Discover May: 74–79.

Lacis, A. A. and B. E. Carlson. 1992. Keeping the sun in proportion. Nature 360: 297.

Lean, J. and D. Rind. 1996. The sun and climate. Consequences 2: 27–36.

Lehman, S. 1993. Ice sheets, wayward winds and sea change. Nature 365: 108–110.

———. 1997. Sudden end of an interglacial. Nature 390: 117–119.

Lehodey, P., M. Bertignac, J. Hampton, A. Lewis, and J. Picant. 1997. El Niño southern oscillation and tuna in the western Pacific. Nature 389: 715–718.

Lewis, J. S. 1995. Rain of Iron and Ice: The Very Real Threat of Comet and Asteroid Bombardment. Addison-Wesley, Reading, MA.

Lorius, C., J. Jouzel, D. Raynaud, J. Hansen, and H. Le Treut. 1990. The ice-core record: climate sensitivity and future greenhouse warming. Nature 347: 139–145.

Macdonald, K. C. and P. J. Fox. 1990. The mid-ocean ridge. Sci. Am. 262: 72–79.

——— and B. P. Luyendyk. 1981. The crest of the east Pacific rise. Sci. Am. 244: 100–116.

Mahlam, J. D. 1997. Uncertainties in projections of human-caused climate warming. Science 278: 1416–1417.

Marvin, U. B. 1973. Continental Drift, the Evolution of a Concept. Smithsonian Institution Press, Washington, D.C.

Masood, E. 1997a. Climate panel forecasts way ahead. Nature 385: 7.

———. 1997b. El Niño forecast fails to convince sceptics. Nature 388: 108.

McPhaden, M. J. 1994. The eleven-year El Niño? Nature 370: 326–327.

Melosh, H. J. 1993. Tunguska comes down to Earth. Nature 361: 14–15.

———. 1995. Around and around we go. Nature 376: 386–387 [reviews estimates of the size of the Chicxulub impact crater].

Michael, A. 1992. Three's a crowd in California. Nature 357: 111–112.

Muller, R. A. and G. J. MacDonald. 1995. Glacial cycles and orbital inclination. Nature 377: 107–108 [discusses the possibility that "[T]he 100-kyr glacial cycle is caused, not by eccentricity, but by a previously ignored parameter: the orbital inclination, i, the tilt of the Earth's orbital plane."].

Nadis, S. 1995. Asteroid hazards stir up defense debate. Nature 375: 174.

Newton, P. 1995. Perspectives past and present. Nature 378: 12 [discusses importance of the terrestrial carbon sink].

Oreskes, N., K. Shrader-Frechette, and K. Belitz. 1994. Verification, validation, and confirmation of numerical models in the Earth sciences. Science 263: 641–646.

Post, W. M. et al. (+ 5 authors). 1990. The global carbon cycle. Am. Sci. 78: 310–326.

Prance, G. T. (ed.). 1986. Tropical Rain Forests and the World Atmosphere. Westview Press, Boulder, CO.

——— and T. E. Lovejoy (eds.). 1985. Key Environments—Amazonia. Pergamon Press, New York.

Rampino, M. R. and S. Self. 1984. The atmospheric effects of El Chichón. Sci. Am. 250: 48–57.

Robock, A. 1996. Stratospheric control of climate. Science 272: 972–973.

Ropelewski, C. F. 1992. Predicting El Niño events. Nature 356: 476–477.

Ruddiman, W. F. and J. E. Kutzbach. 1991. Plateau uplift and climatic change. Sci. Am. 264: 66–75.

Sandweiss, D. H., J. B. Richardson III, E. J. Reitz, H. B. Rollins, and K. A. Maasch. 1996. Geoarcheological evidence from Peru for a 5000 years B.P. onset of El Niño. Science 273: 1531–1533.

Schneider, S. H. 1987. Climate modeling. Sci. Am. 256: 72–80.

Searle, M. 1995. The rise and fall of Tibet. Nature 374: 17–18.

Simkin, T. 1994. Distant effects of volcanism—how big and how often? Science 264: 913–914.

Smit, J. 1997. The big splash. Nature 390: 340–341.

Stanley, S. M. 1984. Mass extinctions in the ocean. Sci. Am. 250: 64–72.

Stauffer, B. 1993. The Greenland ice core project. Science 260: 1766–1767.

Steel, D. 1991. Our asteroid-pelted planet. Nature 354: 265–267.

Stöffler, D. and P. Claeys. 1997. Earth rocked by combination punch. Nature 388: 331–332.

Stone, R. 1995. If the mercury soars, so may health hazards. Science 267: 957–958.

Sullivan, W. 1974. Continents in Motion, the New Earth Debate. McGraw-Hill, New York.

Tibbetts, J. 1996. Farming and fishing in the wake of El Niño. BioScience 46: 566–569.

Walker, G. 1993a. Breaking up is hard to do. Nature 363: 584–585.

———. 1993b. Back to the future. Nature 362: 110.

Weaver, A. J. 1993. The oceans and global warming. Nature 364: 192–193.

———. 1995. Driving the ocean conveyor. Nature 378: 135–136.

Webb, R. S. and J. T. Overpeck. 1993. Carbon reserves released? Nature 361: 497–498.

White, J. W. C. 1993. Don't touch that dial. Nature 364: 186.
White, R. S. and D. P. McKenzie. 1989. Volcanism at rifts. Sci. Am. 261: 44–55.
Whitehead, J. A. 1989. Giant ocean cataracts. Sci. Am. 260: 50–57.
Williams, N. 1997. El Niño slows greenhouse buildup? Science 278: 802.
Williams, S. N. 1995. Erupting neighbors—at last. Science 267: 340–341.
Zahn, R. 1994. Core correlations. Nature 371: 289–290.
Zahnle, K. 1996. Leaving no stone unburned. Nature 383: 674–675.
Zuber, M. T. 1994. Folding a jelly sandwich. Nature 371: 650.

Notes

1. For purposes of mathematical manipulation in determistic modeling, climate is defined as the statistics of one or more components of the climatic system over a specified domain and specified time interval, including the mean, variance, and other moments (Gates, 1982; NRC Panel on Climatic Variation, U.S. Committee for GARP, 1975).

2. For a discussion of the role of fluctuations in warm, highly saline marginal seas in driving the thermohaline circulation of oceans in more ancient times, see Brass et al. (1982).

3. For an interesting article on atmospheric temperatures in the Precambrian, the roles of CO_2 and weathering in reducing these temperatures, and the effects on the evolution of ancient biological systems, see Overbeck and Mancinelli (1994).

4. For a satellite photograph showing the principal physical features of North America, see Flora of North America Editorial Committee (1993, fig. 1.5).

5. The fastest present rate of subduction is along the Tonga Trench at 24 cm/year (Bevis et al., 1995).

6. The terminology used to described objects in space is complex and is often used inconsistently and without clear definition outside the field of astronomy. The suffix "-oid" designates an object while it is in space. A *meteoroid* is of silicate or metalic (Fe, Ni) composition, ranging in size from sand particles to less than 1 km, originating in the asteriod belt between Mars and Jupiter, and usually destroyed as it passes through Earth's atmosphere. When a meteoroid enters a planet's atmosphere it is designated a *meteor*, and if it is large enough it can strike Earth's surface and become a *meteorite*. A *comet* is a low-density body composed mostly of ice around a meteoroid nucleus, originating in the Oort cloud at the orbit of Neptune and extending out 100,000 AU (astronomical unites, the mean distance between the Earth and Sun; 149,600,000 km, 93,000,000 mi), and destroyed gradually by evaporation as it approaches a sun or a planet's atmosphere. A *bolide* is the brightest of the meteors (a fireball or firery meteor). An *asteroid* is a big meteor (or small planet) ranging in size from ~1000 km to 1 km in diameter.

7. Bretz's first name is the letter J; it is not an abbreviation, and hence there is no period.

THREE

Context

The interaction between vegetation and the environment over time is one of the most complex of the Earth's integrated systems. In addition to the direct methodologies of paleopalynology and paleobotany, there are other techniques that provide independent sources of information for interpreting this interaction. These include paleotemperature analysis, sea-level changes, and faunal history. The first two are also forcing mechanisms as discussed in Chapter 2, but for this survey the summary curves can also serve as convenient context information. Each is a vast subject with an extensive literature, and all are presently generating considerable discussion. For paleotemperature analysis, unsettled issues include the extent of temperature change in equatorial waters during the Early and Middle Tertiary, which would affect the poleward transport of heat by conveyer-belt mechanisms. Estimates range from surface waters as warm or warmer than the present to considerably cooler. For the Neogene, CLIMAP estimates based on the ecology of coccolithophores, diatoms, radiolarians, and especially foraminifera are that temperatures in the tropics did not cool significantly; modeling results, terrestrial paleontological evidence, and new Barbados coral data suggest they cooled by ~5°C. There is uncertainty as to when glaciations began on Antarctica; recent estimates range from the Early Eocene to late Middle Eocene to Middle Oligocene (45–35 Ma; Birkenmajer, 1990; Leg 119 Shipboard Scientific Party, 1988). This affects interpretation of ^{18}O values during the Paleogene because they could reflect temperature alone or could be due to ocean water temperature and ice volume changes. Another challenge is to unravel the extent to which benthic temperature records track insolation-induced changes in water temperature versus new thresholds in ocean bottom-water circulation.

Discussions of sea-level fluctuation are presently focused on their causes during the preglacial Early Cenozoic. In faunal history the timing of the North American Land Mammal Ages (NALMAs) or provincial ages are being revised. For vegetational history much of the older literature describes events in terms of geofloras, but this conceptual context, at minimum, requires substantial renovation, and the boreotropical hypothesis is emerging as an alternative for envisioning biotic events in the high northern latitudes. Even with these uncertainties, however, a hallmark of recent studies is the gradual emergence of consistency between biotic and environmental histories based on the various independent lines of inquiry.

GLOBAL MARINE PALEOTEMPERATURE CURVE

The history of the Earth's paleotemperature is provided by analyses of ^{16}O and its isotope ^{18}O in the mineral walls of marine invertebrates (Savin, 1977, Savin and Douglas, 1985; Savin and Woodruff, 1990). The calcium carbonate–water oxygen isotope geothermometer is the most widely applied quantitative tool for estimating ancient ocean temperatures (Savin, 1982).

As noted in Chapter 2, the technique is based on the theoretical considerations and early demonstration by Urey (1947), Urey et al. (1951), Epstein et al. (1951), and Emiliani (1955) that the ratio of ^{18}O and ^{16}O incorporated into the calcium carbonate walls of such organisms as Coccolithophorae, foraminifera, and molluscs is temperature dependent: more ^{18}O is taken up as the water cools. The early equations of Urey have been modified several times; one presented by Shackleton (1974) is

$$T\,(^\circ\mathrm{C}) = 16.9 - 4.38(\delta^c - \delta^w) + 0.10(\delta^c - \delta^w)^2,$$

where T (°C) is the temperature in degrees Celsius, δ^c is the $^{18}O/^{16}O$ ratio of the carbonate wall material relative to a laboratory standard,[1] and δ^w is the ^{18}O value of the seawater in which the wall material was formed. The values are expressed as deviations from the standard (e.g., −2% = 2 parts/mil less than the standard). It has now been determined that each 0.23% change in the $^{18}O/^{16}O$ ratio is equal to a temperature change of ~1°C. The $^{13}O/^{16}O$ ratio can be measured to ~0.1/mil^2, which corresponds to a precision of ~0.5°C. During ice-free times, analyses made on benthic foraminifera record ocean bottom-water temperatures, which are near freezing and do not change appreciably during glacial–interglacial cycles; those made on planktonic species record the more variable surface temperatures. DSDP and ODP cores have been studied from the Atlantic, Indian, and Pacific oceans, and from the northern and southern hemispheres, including the equatorial regions. The oldest widespread ocean sediments for which marine paleotemperatures can be calculated theoretically is ~100 Ma; beyond that time most of the older ocean floor has been subducted. In practice, modification of the wall calcite through time mostly limits the technique to material ~60 m.y. old and younger.

Accurate paleotemperature analysis requires that the calcium carbonate in the walls of the fossils not be altered by diagenesis (chemical or physical changes in the sediment during and after deposition but prior to consolidation). The various kinds of diagenesis that can change the primary chemistry of the wall include recrystallization; chemical alteration; dissolution as the shells sink to great depths; and encrustation, which may overprint or erase the original isotopic composition. The first three can usually be detected by careful visual inspection, and the last is evident through scanning electron microscopy (SEM) or cathodoluminescence microscopy. The presence of algal symbionts in some species of planktonic foraminifera may also cause a problem. As the biology of modern forms becomes better known, it is possible to avoid or deemphasize these species or to compensate for the $^{18}O/^{16}O$ imbalance caused by the algae.

Technically, it is now a comparatively straightforward procedure to obtain $^{18}O/^{16}O$ data from marine sediments, and several paleotemperature curves have been published for the Late Cretaceous and Tertiary (e.g., Miller et al., 1987; Prentice and Matthews, 1988; Savin, 1977). It is much more complicated, however, to assign temperature values to these ratios. These values can be calculated directly only if the proportion of ^{18}O and ^{16}O in the water has remained constant over time and if the species analyzed faithfully accumulate ^{18}O and ^{16}O in equilibrium with the surrounding water. If some factor alters the relative amounts of ^{18}O and ^{16}O in the source water from which the calcium carbonate walls are formed, it is more difficult to relate a particular ratio to a specific temperature.

Oxygen isotope ratios have changed over time because water containing the lighter ^{16}O is evaporated and transported more readily from the ocean surface than the heavier ^{18}O. During interglacial times only ~2% of the world's water is stored in ice and the ^{16}O is returned to the ocean by precipitation (~80% falls on the ocean surface), runoff, and groundwater seepage. A relatively constant balance is maintained in the marine environment through the circulation of ocean currents. During glacial times, however, the return of ^{16}O was delayed because much of the ^{16}O rich water was held in continental ice sheets (~10%), and the relative concentration of the heavier (higher atomic weight) ^{18}O in marine waters increased. At the maximum of the Pleistocene glaciations (~18 Kya), the mean ^{18}O value of seawater was +1.2–1.6/mil, during past ice-free times it was −1.08/mil, and at present it is −0.18/mil (Savin and Woodruff, 1990). The ratios are a result of both water temperature and ice volume changes, and at the last glacial–interglacial interval the amplitude was ~1.7%. Approximately 0.4% was due to cooling and ~1.3% was ice-volume effect (Shackleton and Duplessy, 1986). As long as there is consensus about the presence of glaciers during a particular period, the imbalance can be factored into the calculations to determine temperature. If there is disagreement about the presence of ice, and there is for the Paleogene, different temperature values will be read from the same ratios.

Disentangling the effects of these two independent variables is one of the challenges in paleotemperature research. For example, it was assumed previously that in the Late Cretaceous and until about the Middle Miocene (~14 Ma) in the Tertiary, the Earth was essentially ice free and that the calcium carbonate of marine waters was in near isotopic equilibrium with that of the wall material (Shackleton and Kennett, 1975a,b). Thus, changes in the ratio were attributed solely to fluctuating temperatures. For equatorial surface waters, the present ~28°C temperature was estimated to have been ~18°C during the Paleogene. However, recent evidence indicates that mountain glaciers may have been present in Antarctica in the Early Eocene (Birkenmajer, 1990), that limited ice sheets were present along the margin of Antarctica during the late Middle Eocene and Early Oligocene (Leg 119 Shipboard Scientific Party, 1988), and that more extensive ice developed later in the Oligocene (~30 Ma; Miller et al., 1987; Poore and Matthews, 1984). After that time, development of the Antarctic ice cap was well under way, and after the Middle Pliocene (~3.4 Ma) northern hemisphere glaciations became extensive (Shackleton and Opdyke, 1977). For the Paleogene this means that isotopic values of equatorial waters were due in part to changes in the $^{18}O/^{16}O$ equilibrium and that temperatures were warmer than was thought earlier.

Several methods can be used to compensate for the imbalance in oxygen isotopic equilibrium during glacial climates. This imbalance, and those derived from local to regional variations in evaporation and precipitation, are most pronounced near continental margins due to fluctuations in the amount of runoff from streams and glacial meltwaters. Therefore, one procedure is to base calculations on planktic foraminifera from the open oceans at low latitudes where they are less influenced by high-latitude temperature changes associated with glaciation. A second method is to adjust the isotope data from the fossils according to the temperature requirements of morphologically similar or identical species. Modern studies recognize that certain species of foraminifera (e.g., *Globigerinoides sacculifer*) tend to deposit $^{18}O/^{16}O$ in their walls in close equilibrium with seawater (Shackleton and Kennett, 1975a), and these species can be selected or emphasized in the analyses. This modern analog method provides reliable data for those parts of the column where several members of an assemblage are morphologically similar to extant species, as is frequently the case from the Middle Tertiary through the Quaternary. Similarities between modern and fossil forms diminish in the Middle and Early Tertiary, just when the unsettled timing of continental glaciations also becomes a problem. The third method is the possibility of determining the $^{18}O/^{16}O$ ratios for periods just before, during, and immediately after glacial periods and applying a linear regression[2] to remove these trends from the data subsets. From these and other approaches it is estimated that each 10-m lowering of glacial sea level is equivalent to a change of ~0.1% in ^{18}O (Moore et al., 1982). Because sea-level fluctuations can be detected in sediments deposited along the passive margins of continents (see following section), these data can be incorporated into estimates of paleotemperatures from ^{18}O values. As noted previously, one remaining concern is how to distinguish between changes in marine temperatures due to climatic forcing versus sudden threshold reorganization of ocean circulation.

New methodologies are being developed that will eventually provide additional data on paleotemperatures, as well as independent checks on values calculated from $^{18}O/^{16}O$ ratios. One is based on an index (U^k_{37}) derived from the ratio of diunsaturated to triunsaturated forms of long chain (C37–C39) alkenones presently known from sediments extending back 650,000 years. The index varies with the temperature of the ambient water (as the water cools there is a higher proportion of the diunsaturated molecule), and it has been calibrated to specific temperatures through laboratory culture of the coccolithophorid alga *Emiliani huxleyi* (Eglinton et al., 1992). A shift in the index of 0.1 equals a change in temperature of ~3 ± 0.8°C. The technique allows the calculation of recent paleotemperature changes on a much finer scale (100 years) than previously possible.

Another ratio is the $^{13}C/^{12}C$ in organic carbon from marine sediments. The $^{13}C/^{12}C$ ratios in organic C_{35} hopanes and C_{27} steranes are preserved in sulfur-bound compounds likely derived from bacterial plankton at the base of the photic zone (upper 100 m) in coastal waters (Schoell et al., 1994). In Miocene sediments from the Monterrey Formation at Naples Beach near Los Angeles, California, ^{13}C closely tracked ^{18}O signals in foraminiferal inorganic carbonate as the ocean waters cooled.

Another innovation is the use of calcium carbonate otoliths (ear bones) from fish, which are preserved in the fossil record. The calcium carbonate is laid down in thin daily increments and incorporates the oxygen isotope ratio of the surrounding water. By counting and analyzing these otolitic equivalents to growth rings, there is the potential for determining ocean temperature changes at closely spaced intervals. Isotope fluctuations are already evident in otoliths from 3.5 Ma in Florida and Idaho, and otoliths preserved in rocks elsewhere that are as old as 150 m.y. offer the possibility of extending the record still further.

It is also possible to check results derived from oxygen isotope measurements from calcium carbonate with measurements of strontium in the carbonate. Strontium can replace calcium because it is chemically similar, and the rate of replacement varies with water temperature. The earlier CLIMAP estimate of tropical ocean-surface temperature during the recent glacial maximum (18 Kya) suggested little or no cooling (2°C or less) while high-latitude temperatures cooled by ~5°C. This was inconsistent with interpretations made from many terrestrial assemblages and with modeling, which predicted a cooling at the low latitudes. Recent measurements of strontium in the carbonate from Barbados corals suggest that equatorial waters cooled by ~5°C (Guilderson et al., 1994). Further, the oxygen isotopic composition of seawater sampled during Leg 154 of the ODP (site 925, Ceara Rise, tropical Atlantic Ocean) suggests deep water was cooler by 4°C at the last glacial maximum (Schrag et al., 1996).

A general global marine paleotemperature curve for the Late Cretaceous and Cenozoic is presented in Fig. 3.1. The current temperature scale, based on ^{18}O values when ice is present, is given along the inner left margin of Fig. 3.1; estimated values for the ice-free Tertiary are given along the outer left margin.

The curve shows that during the Maastrichtian ~70 Ma, ocean deep-water temperatures were between 15 and 20°C, or about 12–15°C warmer than at present, and gradually declined to about 10–12°C warmer by the K–T boundary. Present benthic temperatures average about 1–2°C. In the debate as to whether climatic change or an asteroid impact was decisive in the extinction of the large dinosaurs 65 Ma, dinosaur diversity in the High Plains region did not decline in concert with the gradual lowering of temperatures shown in Fig. 3.1 but instead declined suddenly and more rapidly (Sheehan et al., 1991).

Temperatures began to rise during the Paleocene, and by the Early Eocene values were at or near the highest for all of Phanerozoic time. The estimated range is between 11 and 15°C for bottom-water temperatures and possibly as high as 21°C for surface waters at 52° S (Miller et al., 1987). The present ocean-surface temperatures at the equator are between about 20 and 28°C. The warming trend across the Paleocene–Eocene boundary is also preserved in patterns of paleosol carbonates (a sharp decrease in $^{13}C/^{12}C$ ratios) and in mammalian tooth enamel (Koch et al., 1992). The cause(s) for the amounts of carbon in the biosphere to vary with temperature is not presently known (P. L. Koch, personal communication, 1996). A punctuated decline fol-

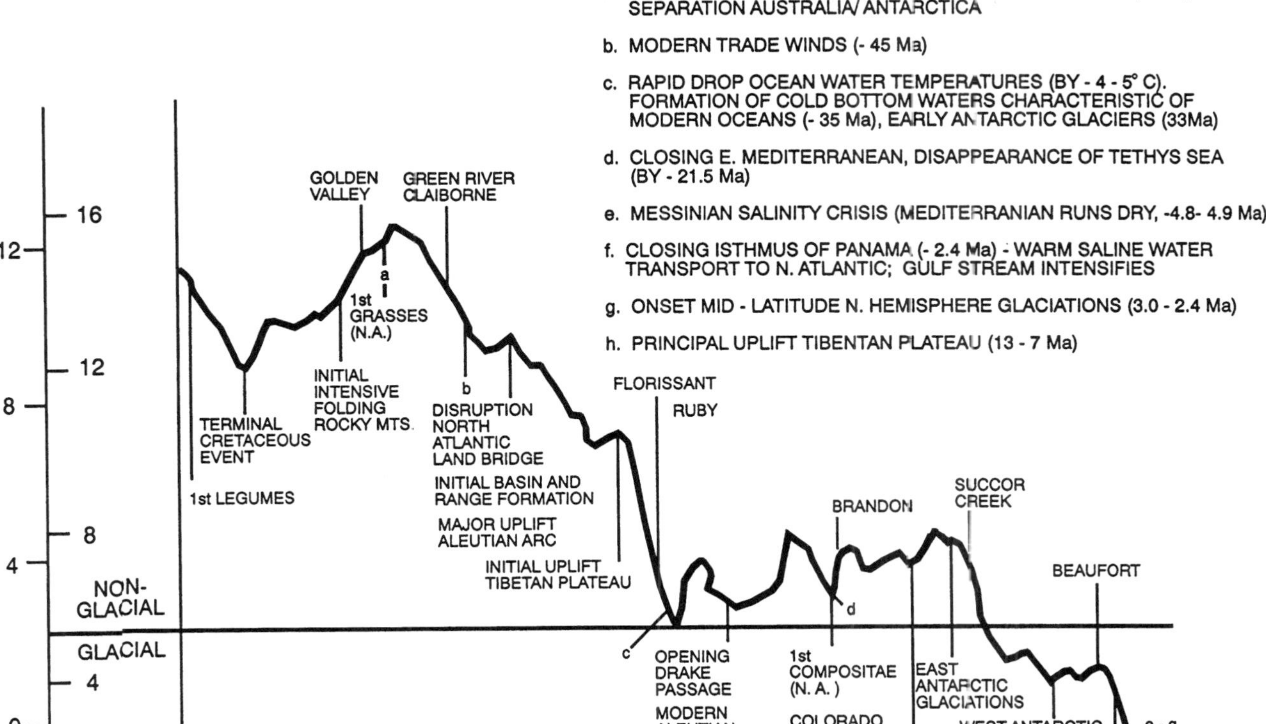

Figure 3.1. Composite benthic foraminiferal oxygen isotope record for Atlantic DSDP sites (based on Miller et al., 1987); major biological and geological events and selected North American fossil floras are superimposed. Separation of Australia from Antarctica, as expressed by shallow water circulation, is placed in the early Eocene (50–55 Ma), and deep water circulation is evident in the Early to Middle Oligocene (25–30 Ma). The placement of fossil floras along the curve provides a global paleotemperature context for assessing the paleoenvironments of the assemblages. Data from various sources.

lowed between the Middle Eocene (52–40 Ma) and the Late Eocene (40–38 Ma), to 8–10°C warmer. The difference between the Early to Middle Eocene high and Late Oligocene lows was ~10–11°C. Kennett and Shackleton (1976) proposed a cooling of 5°C during a period of about 100 Ky. Collectively, the ^{18}O values suggest a mostly ice-free world between the Late Cretaceous and the Early Oligocene. Between the Early Oligocene and the Middle Miocene temperatures fluctuated, but values at the beginning and end of this time span were about the same. Within this interval the circumpolar Antarctic current developed, the Antarctic ice sheet formed by the beginning of the Oligocene (35–30 Ma; there is a significant drop in sea level at ~30 Ma), and there was intensification of cold bottom-water flow. Glaciers likely appeared and disappeared during the Late Oligocene and Early Miocene (Miller et al., 1987). A second major decline in temperature occurred in the Middle Miocene (15–14 Ma). This decline marked the beginning of permanent ice cover on Antarctica and the initiation of glacial climates in the Arctic. Temperature gradients increased from low (equatorial) to high (polar) latitudes, and this may have initiated a more vigorous thermohaline circulation. A third abrupt lowering of temperatures is recorded between ~3 and 2.4 Ma, ushering in northern hemisphere glaciations. These glaciations were well underway by 1.6 Ma, a date used to mark the beginning of the Pleistocene Epoch, or Ice Age, which extends to the end of the last glacial advance at ~11 Kya.

For vegetational history, the critical question is to what extent the pattern of paleotemperature change evident in the marine environment applies to terrestrial habitats. Dorf (1964) attempted to reconstruct terrestrial paleoclimates for the Late Cretaceous and Tertiary of the western United States. The resulting curve (Fig. 3.2) did reflect some trends

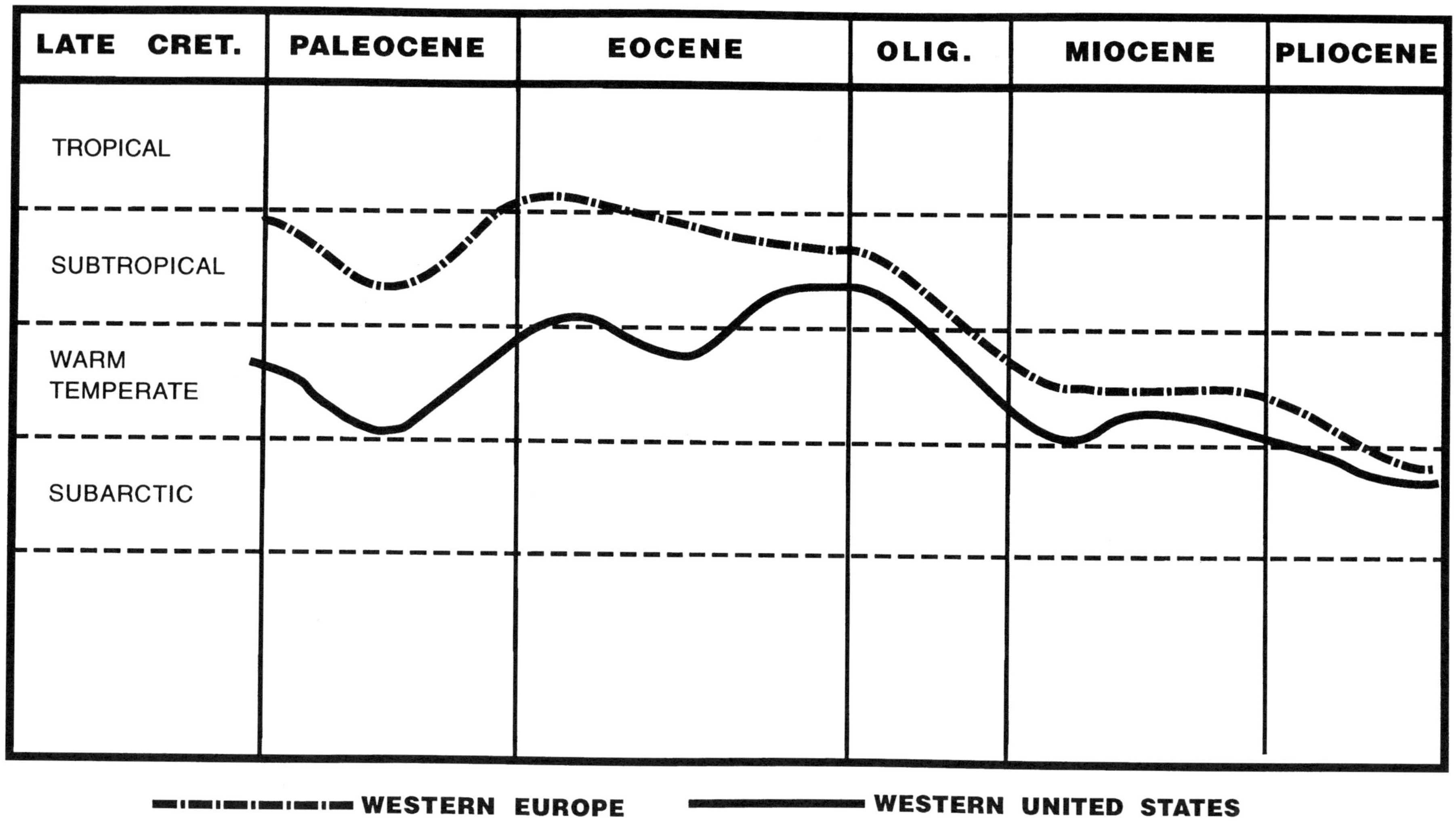

LATE CRET. | PALEOCENE | EOCENE | OLIG. | MIOCENE | PLIOCENE

TROPICAL

SUBTROPICAL

WARM TEMPERATE

SUBARCTIC

WESTERN EUROPE — WESTERN UNITED STATES

Figure 3.2. An early attempt at inferring climatic regimes for western Europe and the western United States from paleobotanical evidence. Reprinted from Dorf (1964) with the permission of Interscience Publishers.

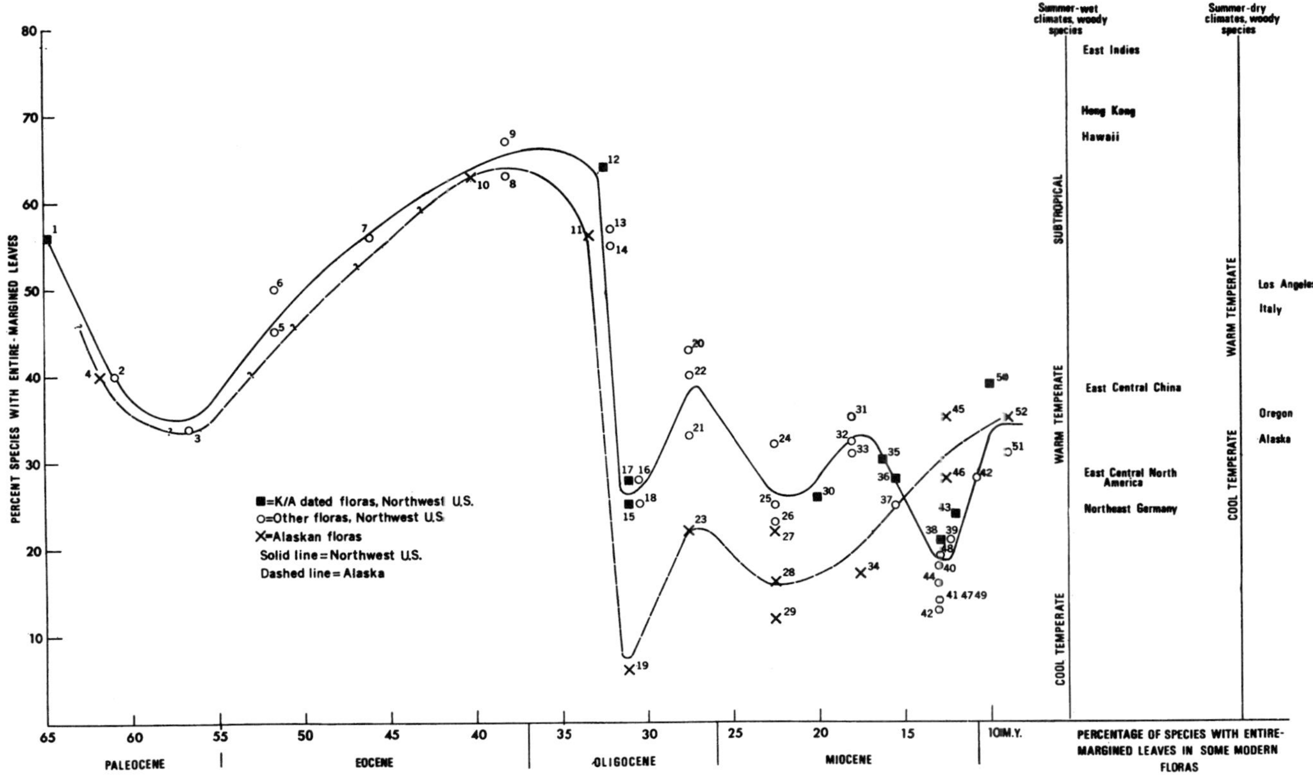

Figure 3.3. LMA curves from Tertiary floras of the northwestern United States and Alaska. Recent data place the peak of maximum warmth in the Late Paleocene and Early to Middle Eocene and the sharp drop in the Late Eocene. Adapted from Wolfe and Hopkins (1967); numbers along the curve refer to fossil floras mentioned in that publication.

subsequently identified from the marine environment. The warm conditions of the Paleogene, followed by the decline during the Eocene, and the climatic deterioration of Middle and Late Cenozoic times are broadly evident. Other features, such as the abruptness of the Middle Eocene and Middle Miocene climatic changes, temperature fluctuations in the Oligocene, and the timing of the events are not comparable. Overall the curve is not very similar to the one shown in Fig. 3.1. A curve similar to Dorf's (1964) is given by Tanai and Huzioka (1967) for Tertiary climatic changes in Japan. The highly generalized nature of these early reconstructions, based exclusively on the floristic composition (modern analog method) of comparatively few floras and mostly in the absence of radiometric dates, creates the impression that either marine and terrestrial environments had rather different climatic histories, the paleopalynological and paleobotanical record is not sufficiently sensitive to closely track global climatic changes, or regional conditions so overprint the response by terrestrial biotas to global changes that patterns derived from paleovegetation evidence reflect only local climates.

In 1967 Wolfe and Hopkins published an interpretation of Tertiary paleoclimates for northwestern United States and Alaska based on LMA. This important study included a larger number of floras from a more restricted region than the analyses of Dorf (1964), and several radiometric dates were available that provided more accurate information on the age and stratigraphic relationships among the assemblages. Their curve is shown in Fig. 3.3. [Other versions are given in Wolfe (1978) and Wolfe and Poore (1982)]. The Wolfe and Hopkins (1967) curve better resembles the later derived marine paleotemperature curve and suggests that temperature fluctuations evident from the marine environment reflect terrestrial paleoclimates more closely than indicated by previous studies.

A comparison was made later between marine paleotemperatures and terrestrial paleoclimates using pollen and spore floras from the Eocene of northwestern Europe (Hubbard and Boulter, 1983). Through the use of multivariate statistical analysis, three paleocommunities were identified from the total assemblage that were sufficiently close to modern forest analogs to allow estimates of summer maximum and winter minimum temperatures. There were warm periods at ~51–53 Ma (top of the London Clay and bottom of the Bracklesham beds), at ~46 Ma (Headon Beds), and for brief periods during the Oligocene (Bovey Tracey, Mochras,

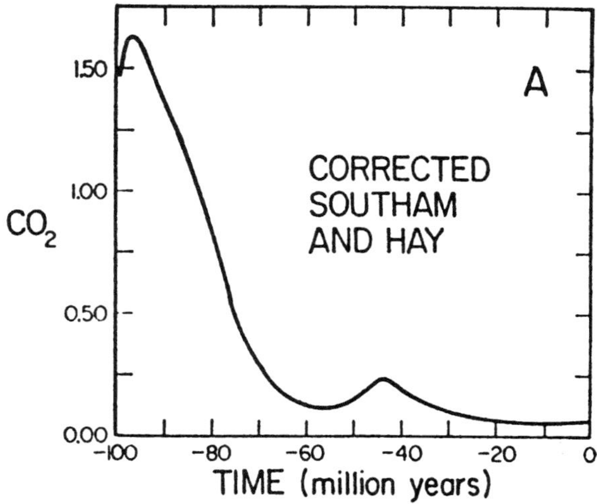

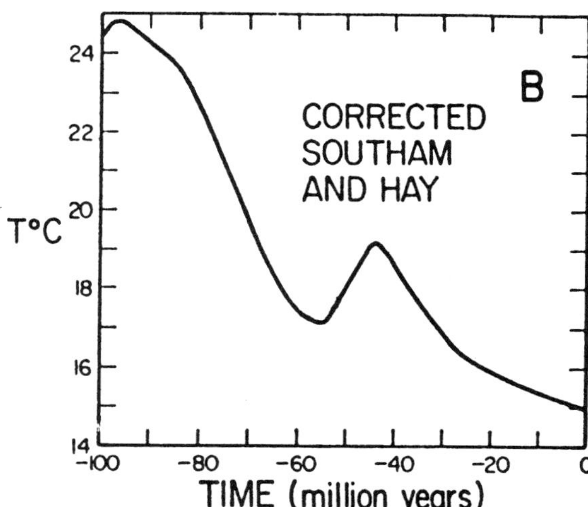

Figure 3.4. (A) Computer results for the mass of CO_2 (in 10^{18} mol) as a function of time using the land area curve of Barron et al. (1980) and the corrected spreading rate formulations of Southam and Hay (1977). (B) Computer results for worldwide mean annual air-surface temperature as a function of time corresponding to the situation for CO_2 depicted in (A). Adapted from Berner et al. (1983). [See Berner (1994) for revised CO_2 values.] Reprinted with the permission of R. A. Berner and the American Journal of Science.

and Lundy beds; 33 Ma); the Oligocene was generally cooler than the preceding Eocene. These results are similar to patterns shown on the curve in Fig. 3.1. When details of the diagrams based on paleovegetation evidence were further compared with $^{18}O/^{16}O$ data from Shackleton and Boersma (1981), the major trends were clearly recognizable in both records. Hubbard and Boulter (1983) concluded that there is considerable similarity between the marine and terrestrial plant record and that the Paleogene floras of northwestern Europe, where Cenozoic orogeny is not a complicating factor, were tracking genuine global climatic change. Terrestrial records from paleosol carbonates and mammalian tooth enamel also closely track fluctuations in the marine carbon isotope record (Koch et al., 1992).

An early indication of how results integrated from independent studies, including paleotemperature analysis, provide an internally consistent model of past environments was presented by Berner et al. (1983). Using a modification of the spreading center rates proposed by Southam and Hay (1977) and land–sea relationships as described by Barron et al. (1980), they calculated CO_2 values for Middle Cretaceous through Tertiary time (Fig. 3.4A; revised by Berner, 1994). When continental position, land–sea relationships, and CO_2 concentration were used to model expected air-surface paleotemperature patterns, a secondary temperature maximum was predicted by the model at ~55–40 Ma (Fig. 3.4B). When this curve was compared with those based on LMA (Wolfe and Hopkins, 1967) and planktonic carbonate values (Savin, 1977), the similarity in patterns was clearly evident (Fig. 3.5).

Some uncertainty remains as to the relative roles of orbitally induced temperature changes (Wolfe, 1971; Wolfe and Hopkins, 1967; Wolfe and Poore, 1982) and orogenic factors (Axelrod, 1981; Axelrod and Bailey, 1969) in the history of Tertiary vegetation in western North America. Problems in interpretation arise, in part, from the fact that while climates were changing, orogenic uplift and volcanic accumulations were simultaneously providing new higher altitudinal temperate zones, creating arid habitats to the lee of developing mountains, altering atmospheric circulation, and intensifying the rate of erosion of silicate rocks. Nonetheless, warm–cool periods are documented in the marine realm, and it is likely that the development of terrestrial vegetation was influenced by similar patterns (Wing et al., 1991). Within western North America in particular, the problem is to achieve a balance between the interwoven roles of orogeny and global climatic change. This makes even more valuable a resource such as marine paleotemperature analysis, which serves to better define one component of this multivariate system.

GLOBAL SEA-LEVEL CURVE

Sea levels are presently changing through a combination of human and other influences. The Gulf of Mexico is encroaching on the Mississippi River delta at a rate of ~1 mm/century because of channels, dikes, and canals that reduce the input of sediments to the delta. In contrast, Hudson Bay is rising by ~1 cm/century through isostatic rebound following deglaciation.

Sea levels have also changed significantly over the past 100 m.y., rising during this time between 1 and 2 mm/year. Vail and Hardenbol (1979) have estimated that the difference between the highest stand of the Late Cretaceous and present levels was over 300 m. If the polar ice caps of

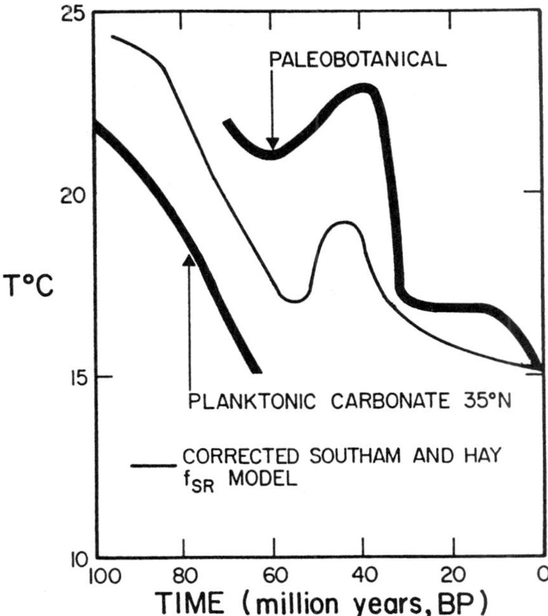

Figure 3.5. Plots of worldwide mean annual air-surface temperature versus time based on predictions from computer modeling (narrow line), compared with estimates from paleobotanical (Wolfe and Hopkins, 1967) and ^{18}O studies of planktonic foraminifera and nannofossils of Cretaceous sediments of the Shatsky Rise (North Pacific) as summarized by Savin (1977). Reprinted from Berner et al. (1983) with the permission of the American Journal of Science.

Antarctica (89% of present total continental ice) and Greenland (10%) were to melt, sea levels would rise by ~70 m. Melting of the Arctic ice sheet would have less effect because it is anchored in the Arctic Ocean and the result on water level would be analogous to the melting of ice in a glass of water. At the latest maximum of the Pleistocene glaciations at 18 Kya, sea level was more than 100 m lower than at present (Geophysics Study Committee, 1990; −121 ± 5, Fairbanks, 1989); in 100 years it probably will be ~0.5–1 m higher through a combination of thermal expansion of marine waters and glacial melting.

Changing sea levels have a variety of effects on climate and terrestrial plant communities (Warrick et al., 1993). During low stands, portions of the continental shelf are exposed and become vegetated. The Quaternary glaciations produced an increase of ~15% in land surface, and this allowed greater storage of carbon in the phytomass and in the adjacent sedimentary basins (Prentice and Fung, 1990). There is also greater erosion of exposed silicate rocks, and both processes constitute carbon sinks. At the same time vegetation buried from previous inundations is uncovered and decays, and CO_2 and methane are released. This process constitutes a carbon pump. In turn, higher sea levels cover portions of the continent, slowing erosion and the drawdown of CO_2.

When sea levels are high, low-lying interior regions of the continent are inundated. Marine waters have a specific heat of 0.93 cal/g/°C (compared to that of land with 0.60–0.19 cal/g/°C) and release heat more slowly. High sea levels result in extensive equable maritime climates, while low levels promote greater continentality or extremes in annual temperature variation. Model simulations have documented that in pre-Cretaceous times continental extent significantly affected climates in the interior (Yemane, 1993). Crowley et al. (1987), using a 2-dimensional energy balance model (EBM), found that land–sea configuration limited summer warming in the central portions of the large Gondwana continent and was important in the glaciations of the Late Paleozoic. An older model even predicted that the position and extent of the continents was primary (Barron and Washington, 1984). However, a newer global environmental and ecological simulation of interactive systems model (GENESIS) indicates a mean global cooling of only 0.2°C with Cretaceous geography, whereas a fourfold increase in atmospheric CO_2 raises surface temperature by 5.5°C. Although insolation and CO_2 concentration remain principal climate forcing mechanisms, seasonality must be factored in as lowering sea level drained the epicontinental sea from interior North America after the Cenomanian.

Albedo (reflectivity) is the ratio of the amount of solar radiation reflected by the Earth to the amount incident upon it (Thompson and Barron, 1981). Albedo, along with insolation, drives atmospheric and oceanic circulation; it fluctuates with sea level because the reflectivity of various land surfaces, vegetation, sea, and ice is different. A rise or fall by ~100 m would produce a change in albedo of 0.01, which would alter surface temperatures by ~1°C (Geophysics Study Committee, 1990). Barron et al. (1980) discussed the effect of land–sea distribution and eustatic fluctuations on surface albedos. They concluded that the most important factor in altering albedo over Late Cretaceous and Tertiary time was sea-level change. During the Quaternary, insolation patterns were further strongly influenced by ice sheet expansion (more reflective ice) and contraction (more absorptive vegetation) through feedback mechanisms. Albedo values for various surfaces are shown in Table 3.1.

During periods when inundation is extensive, as it was in the Late Cretaceous, terrestrial biotas may be separated into geographically and reproductively isolated populations, which promotes speciation by vicariance mechanisms. A lowering of sea level removes the barrier and promotes speciation through hybridization. These events are reflected in the fossil record in the form of floristic provinces that appear and disappear over time (Chapter 5). It is also noteworthy that 12 of the 13 NALMA boundaries, established in large part on the basis of immigrant taxa, coincide with drops in sea level (Opdyke, 1990; see following section on Faunas).

The record of sea-level change is preserved as repeating sequences of near-shore marine deposits (high stands), alternating with erosional surfaces (sequence boundaries) or with periods of nondeposition identifiable when planktonic foraminifera zones can be tied to a geochronometric scale (low stands). If no additional constraining geological information is available, the fluctuations are termed relative sea-level (RSL) changes because they could be the result of uplift and subsidence of the land, lowering or rising of the ocean surface, or both. A distinction between these two events can be made, however, when the sediments are deposited along the passive margins of continents. Passive margins are portions of the continental shelf and adjacent coastal plain that were relatively stable tectonically during the period under consideration, as, for example, the Atlantic and Gulf Coasts during the Cenozoic. This is in contrast to active margins, as along the west coast of North America during the Cenozoic. In the case of passive margins, the sequences of near-shore marine deposition, separated by erosional surfaces or periods of nondeposition (unconformities), can be attributed mostly to actual changes in the level of the sea. Such fluctuations are termed eustatic changes.

Table 3.1. Comparative albedo values for ocean and land surfaces (from Berger, 1982)[a].

Annual global average	14
Ocean	
low latitudes	4–7
mid-latitudes	4–19
high latitudes	6–50
Great Lakes	
minimum (summer)	6
maximum (winter)	55
Land	
desert	20–30
grasslands, coniferous and deciduous forests	15
wetlands	10
tropical rain forests	7
snow-covered land	35–82
Antarctic ice cap	85
Pack-ice in water	40–55

[a]Reflectivity in percent of incident light during noon

The method for constructing global sea-level curves was developed at Exxon by a group headed by Peter Vail. Acoustic signals are sent through underlying strata and are reflected in a pattern determined primarily by the density–porosity of the various rock layers. Solid rock returns a signal of amplitude, frequency, waveform, and velocity that is different from that of unconsolidated erosion surfaces. The result is a vertical seismic profile (VSP) that reflects periods of marine deposition (submergence) and erosion (emergence) at a given site.

Seismology has long been used in petroleum exploration as a means of detecting depositional settings and postdepositional histories favorable to the formation and accumulation of oil. The innovation of the Vail group, in applying otherwise standard seismic methodology, was to correlate among the numerous near-shore sedimentary profiles available from Exxon's worldwide operations. The original curves (Vail et al., 1977) were based on profiles from about 60 different regions. Within this global network, patterns that showed up at the same time along the passive margins of distant continents were attributed to global changes in sea level, while individual isolated patterns were attributed to local tectonics. The differentiation between local and global changes depends on accurate dating of the hiatuses on each passive margin, which is achieved through detailed biostratigraphic, magnetostratigraphic, and isotope stratigraphic studies (see Temporal Context Section).

The sea-level curve developed by Vail and coworkers is shown in Fig. 3.6. The discussions that followed publication of the curve centered primarily on three points. One was the cause of the changes. Glaciation is the most obvious mechanism for controlling eustatic changes in sea level (called glacioeustatic changes), and this was pro-

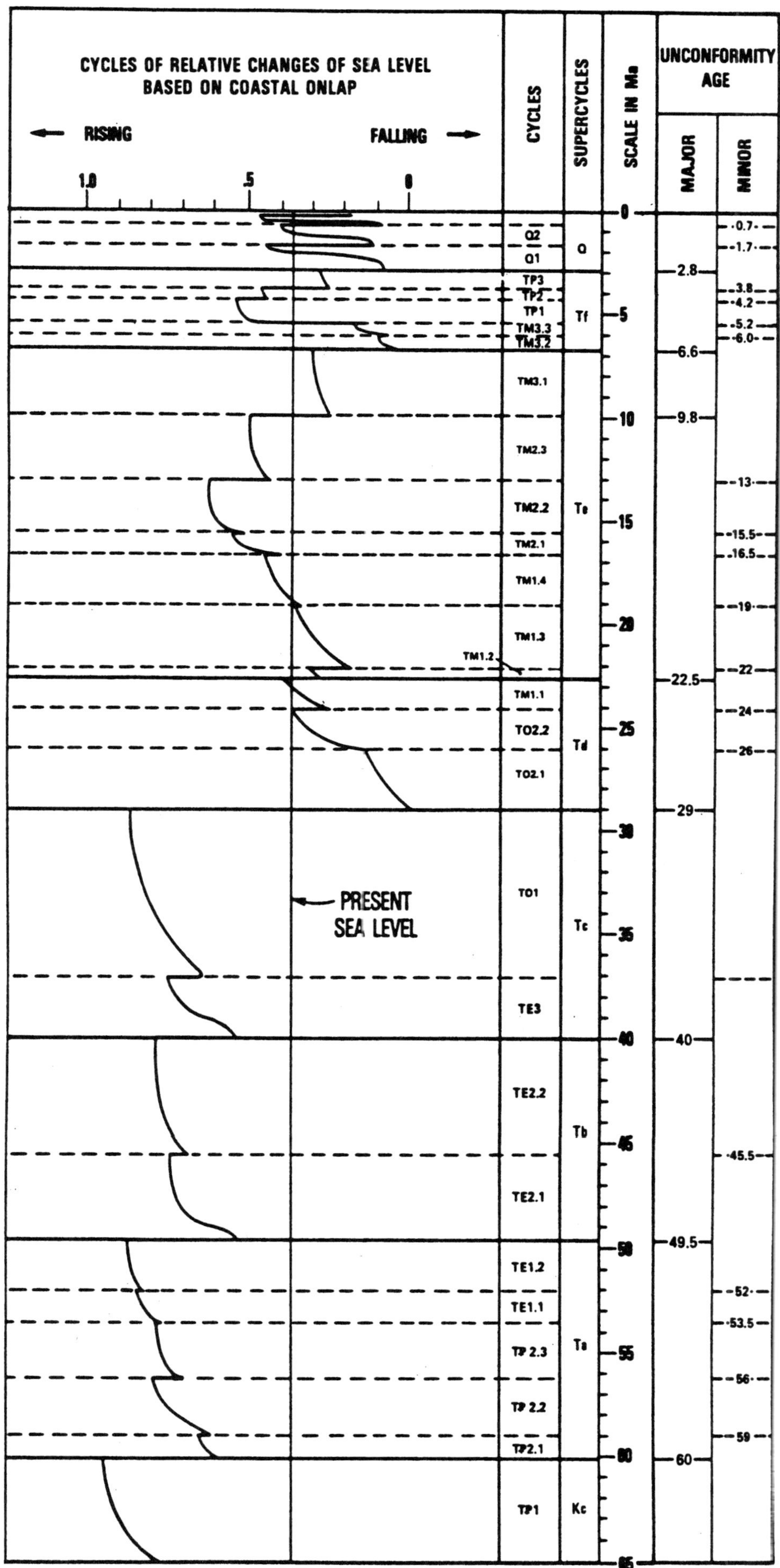

Figure 3.6. Global cycles of relative sea-level change during the Tertiary. Reprinted from Vail and Hardenbol (1979) with the permission of Oceanus.

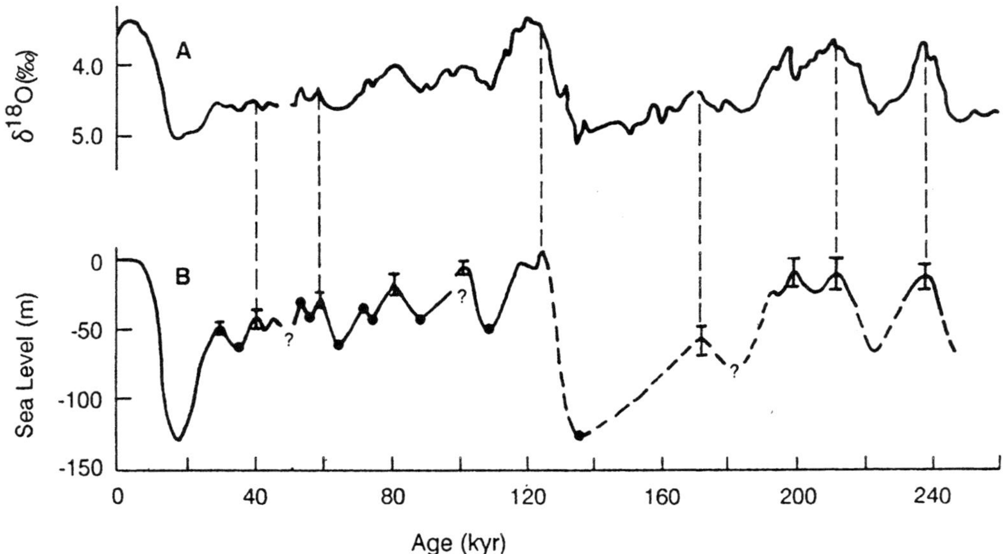

Figure 3.7. Correlation between (A) ^{18}O-measured cold-water intervals and (B) low–high stands of sea level. Based on Chappell and Shackleton (1986). Reprinted from the Geophysics Study Committee (1990) with the permission of the National Academy Press.

posed by the Vail group as the principal factor in causing the simultaneous worldwide appearance of erosion surfaces along the passive margins of continents. If so, a correlation between sea level and temperature as recorded by ^{18}O measurements would be expected, and this was the case (Chappell and Shackleton, 1986; Shackleton, 1987; Fig. 3.7). However, rapid falls identified near the end of the Mesozoic and in the Paleogene did not correspond to any known periods of major glacial advance. Thus, other factors must have been involved during preglacial times. One is a holdover from Cretaceous expansion of midocean ridge systems with an increase in midplate volcanism (Ager, 1981; Hays and Pitman, 1973; Pitman, 1978; Schlanger et al., 1981). Ridge expansion decreases ocean basin volume and displaces water onto the continental margin and into the interior. If the factors of ridge expansion and thermal expansion of the crust, upper mantle, and ocean water were operating alone, they could theoretically affect sea level by a ~175 m (see Harrison, 1990). Until recently it was not certain that changes in ridge volume covaried with the pattern of sea-level change closely enough to identify it as a significant factor. Gurnis (1990) further suggested that because the rate at which cold lithosphere is returned to the mantle along subduction zones has generally been ignored in the models, sea levels might actually rise or fall when increased spreading is considered alone. Improved modeling now reveals a better correlation between periods of continental flooding and increased plate velocities (Gurnis, 1993).

Other factors include outgassing of juvenile water (greatest near the beginning of rapid plate movement, as in the Jurassic and Cretaceous, and diminishing later); recycling of water through the crust and upper mantle (but this is mostly a balanced system); changes in the volume of groundwater (also generally balanced); redistribution of water among the ocean basins (e.g., the drying of the Mediterranean in the Late Miocene Messinian salinity crisis, 4.8–4.9 Ma; Hsü et al., 1977), with resulting alteration in ocean salinity; and the thermal contraction and expansion of water with changing paleotemperatures (steric changes). Sea level rises ~100 mm/1°C of temperature increase throughout the uppermost 500 m. If deep ocean waters were warmer by 10°C, as was the case during the Late Cretaceous and Early Tertiary, sea levels would be higher by ~10 m from this effect alone (Geophysics Study Committee, 1990). Recent measurements off the southern California coast show that the upper 100 m of water has warmed by 0.8°C in the past 42 years, resulting in a sea-level rise of 0.9 ± 0.2 mm/year, which is consistent with coastal tide gauge records (1–3 mm/year; Roemmich, 1992).

Other factors affecting past sea levels have been reviewed by Savin and Douglas (1985) and Southam and Hay (1981). Individually these account for fluctuations estimated from a few meters to negligible. Collectively, they reduce the amount of change that previously had to be attributed to unknown factors or to undetected glaciation for eustacy throughout the Late Cretaceous and Early Cenozoic.

Another concern was with the suddenness of the declines reflected in the jagged profile of the curve. These precipitous drops do not match the more gradual rates of decline evident in some Cretaceous and Tertiary deposits in the western and midcontinent regions of North America. However, current research on sedimentary processes along continental margins is beginning to produce models with smoother curves (Pitman, 1978). Transport of eroded material onto the continental shelf during a time of lowering sea level may be retarded across freshly exposed portions of the coastal plain until adequate slope has developed. This would give the impression from seismic profiles that shelf erosion had started later and, therefore, that the sea-level fall occurred over a shorter interval. Results from GRIP and GRIS2 further reveal that climates do switch from glacial to nonglacial modes abruptly, so that the sudden changes in sea level from the Late Eocene onward is not so perplexing.

The third concern was with the magnitude of the changes. The largest fall in all of Tertiary time was in the Late Oligocene (~30 Ma) when sea-level lowering was estimated at more than 200 m in 1–2 m.y. When it was assumed that no significant glaciers existed before the Middle Miocene, the fall was anomolous. It is now known that this period corresponds to an increase in ^{18}O values in benthic and in high southern latitude planktonic foraminifera and the development of Antarctic glaciers. The earlier estimate has been revised to just over 100 m (see Kerr, 1984). Other major declines at about 60, 49.5, 40, 22.5, 10.8, 6.6, and 4.2 Ma are not so clearly associated with glaciation. An observation now being incorporated into new versions of the curve (Haq et al., 1987) is that although the Atlantic and other continental margins are passive in the sense of not being active plate boundaries, subsidence is taking place. The U.S. continental margin is sinking at a rate of ~1–2 cm/1 Ky from continued cooling after the opening of the Atlantic Basin and from the accumulation of sediments (sediment loading; Watts, 1982). When tectonic subsidence at passive margins is taken into account, it reduces the magnitude of sea-level fall by about half for some parts of the curve. Revised versions of the curve by Haq et al. (1987) and Moore et al. (1987) show a gradual scaling down and smoothing out of the curve as new information becomes available on the interaction between thermal contraction along margins upon lithosphere cooling, the thermal contraction and expansion of ocean waters with temperature changes, sediment loading, midocean ridge crest volume and volcanism, and glaciation (Fig. 3.8). Miller et al. (1987) provide a summary that is useful as a working concept while mechanisms and details of the curve are being refined. They suggest that the principal mechanism for major eustatic fluctuations during the past 36 m.y. was ice-volume change and that for the Mesozoic, Paleocene, and parts of the Eocene the principal mechanism was the pace and aftermath of global seafloor spreading. In general, as more data accumulate, the compilations of Haq et al. (1987) and Vail et al. (1977) are being validated (Miller et al., 1996).

During the Late Cretaceous and Early Tertiary (Paleo-

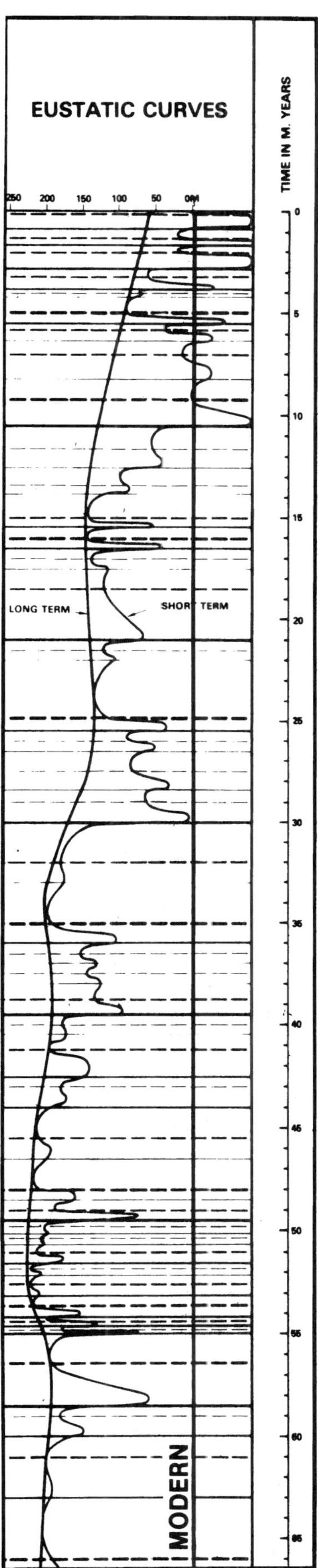

Figure 3.8. Cenozoic cycles of long-term and short-term sea-level change. Reprinted from Haq et al. (1987) with the permission of Jan Hardenbol, Peter Vail, and the American Association for the Advancement of Science.

cene, Eocene), large portions of the Atlantic and Gulf Coastal Plains were inundated up to the flanks of the Appalachian Mountains. Most of the continental interior, from the Gulf of Mexico to the Arctic Ocean, and from the Appalachian Mountains to the proto-Rocky Mountains, was periodically covered by shallow marine waters. This, in itself, would suggest equable maritime climates for emergent North America. When considered in conjunction with paleotemperatures for the same period (Fig. 3.1), a useful context emerges for estimating the terrestrial climate under which the vegetation grew. The major drop in sea level at ~30 Ma was preceded by a paleotemperature decline of ~5°C between the Late Eocene and earliest Oligocene. The climatic implications of lower sea level would be greater continentality in interior climates and changes in the Earth's albedo. Sea levels and paleotemperatures fluctuated but without a sustained trend between the Late Oligocene and the Middle Miocene (~30–15 Ma). Another major lowering of sea level occurred at ~10.8 Ma, preceded by a significant drop in paleotemperatures that probably marks the beginning of an expanded west Antarctic ice sheet and the early development of Arctic polar ice. After that time the sea-level curve closely tracks orbital-induced climatic changes (Matthews and Frohlich, 1991; Prentice and Matthews, 1991). Sea-level fluctuations that vary in concert with the Milankovitch cycles afford another insight into climatic change when ^{18}O data are sparse (Ye et al., 1993).

TEMPORAL CONTEXT

Most rocks undergoing erosion contain a form of iron called magnetite. This mineral settles out in an orientation aligned with the prevailing magnetic poles. As the magnetic poles reverse periodically, the magnetite tracks their polarity and position at the time of deposition. Past orientations like those of today are black on the charts presented in Fig. 5.1, and reversed polarities or anomalies are white. The intervals during which a normal (N) or reversed (R) polarity dominate are called chrons (C), and these are numbered in a vertical sequence to constitute a magnetostratigraphy of relative ages. The duration of a reversal is generally between 1000–6000 years and the most recent was the Brunhes–Matuyama reversal at ~780,000 years ago. Magnetochrons in continuous deep-sea sections have been tied to radiometric dates to form a magnetic polarity time scale of absolute ages. In some charts, units of the magnetostratigraphy are correlated with intervals designated P (e.g., P1b at chron C29-28, ~63 Ma). These are planktic foraminiferal zones. C or NP designates a calcareous or nannofossil zone and together they allow further comparison with environmental and biotic events in the marine realm (marine biostratigraphy). If seismic profiles are available, these can be used to establish another chronologic framework for faunal (NALMA) provincial ages and floristic history. In transgressive units, marine deposits may interfinger with fossil-bearing terrestrial sediments; the latter can then be tied to global events in other environments and comparisons can be made between the history of various biotic groups. Synchronous layers of volcanic debris settling onto marine, freshwater, and terrestrial surfaces can also provide key markers for dating and correlation (tephrachronology). These five techniques of isotopic stratigraphy, magnetostratigraphy, biostratigraphy, seismic stratigraphy, and tephrachronology establish a valuable temporal context and provide an impressive arsenal of techniques for organizing and comparing the records of various biological groups with geological–environmental events.

FAUNAS

The North American land mammal record is one of the most extensive and thoroughly studied in the world (Kurtén and Anderson, 1980; Savage and Russell, 1983; Webb, 1985, 1989; Webb et al., 1995; Woodburne, 1987). Many North American Late Cretaceous and Cenozoic floras occur intermingled with these faunas; because animal groups are mostly dependent upon the plant formations and associations for their habitat and food resources, a general correlation among the kinds of fauna, principal vegetation types, and environmental conditions is evident throughout the geologic record (e.g., Hutchison, 1982, fig. 5). An awareness of the faunal record, therefore, affords another context for interpreting vegetational history.

Fossil faunas provide an independent source of information on paleoenvironments. Westgate and Gee (1990) described a Middle Eocene assemblage from the Laredo Formation in southwest Texas containing the tropical Asian palm *Nipa*. The biological affinities of the other plants could not be established, or the taxa had broad ecological and geographic ranges. The associated fauna included *Carcharias* (shark), batoids (skates and rays), bony fishes (e.g., *Tarpon*), and Crocodylidae (Fig. 3.9). Although paleoenvironmental reconstructions based on plant material alone would have been equivocal, the associated fauna strengthened the interpretation that Middle Eocene climates in the region, and the ecological parameters of the Eocene nipas, were warm temperate to tropical. Faunas reflecting similar warm environments have been reported from Upper Cretaceous and Lower Tertiary deposits in the northern Great Plains (turtles, crocodilians, champsosaurs; Hutchison, 1982) and from Lower to Middle Eocene deposits from Ellesmere Island within the Arctic Circle (alligatorlike crocodilians, turtles, perissodactyl mammals; Dawson et al., 1976; McKenna, 1980; West and Dawson, 1978). The lower temperature limit of modern *Alligator* is ~4.4°C. This indicates that warm-temperate to tropical climates extended far into the high latitudes, at least along maritime coastal areas (including the margins of the epicontinental sea), during the Late Cretaceous, Paleocene,

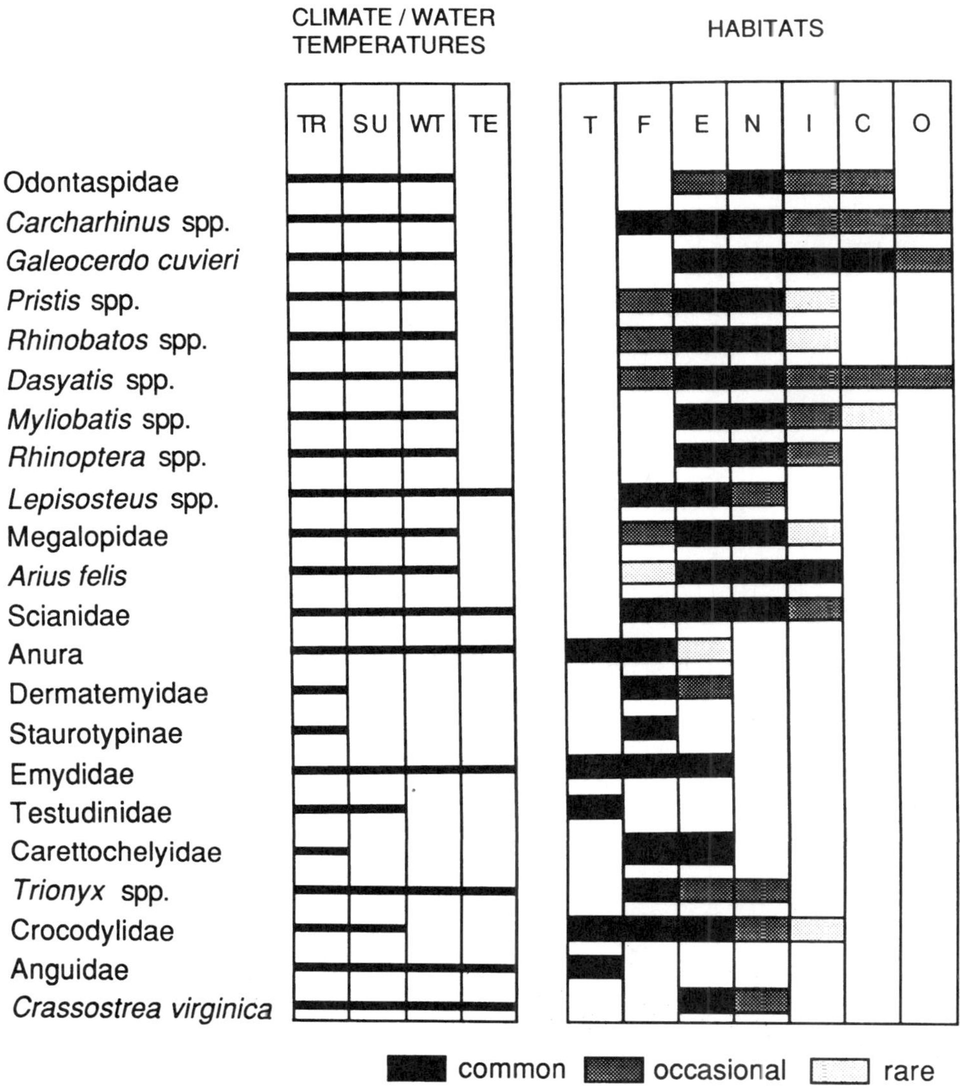

Figure 3.9. Habitats of modern relatives of paleoecologically useful species in the Casa Blanca fauna (Laredo Formation, Middle Eocene, Webb County, Texas). TR, tropical; SU, subtropical; WT, warm temperate; TE, temperate; T, terrestrial; F, fresh water; E, estuarine; N, near shore; I, infralittoral; C, circalittoral; O, oceanic. Reprinted from Westgate and Gee (1990) with the permission of Elsevier Science-NL.

and Early to Middle Eocene; interior portions of the continents may have been more seasonal. Such reconstructions, in turn, are consistent with the low physiographic relief of the period (Chapter 2), widespread epicontinental seas as evidenced from the global sea-level curve (Figs. 3.6, 3.8), high CO_2 concentrations, and warm conditions reflected by the oxygen isotope based paleotemperature curve (Fig. 3.1). In the Late Tertiary and Quaternary cooling climates are reflected in the large size attained by certain mammals. According to Bergmann's Rule, body size increases with decreasing temperatures. With larger size, volume and the capacity for heat production both increase more than skin surface and the capacity for heat loss. This trend in body size parallels the climatic deterioration after the Middle Miocene.

Fossil faunas provide an indication of the type and structure of vegetation prevailing in a region during a particular span of time. The increasing diversity and abundance of horses, camelids, and other browsing and grazing ungulates (hooved mammals) in North America beginning in the Late Eocene parallels the spread of seasonally dry woodland, savanna, and, eventually, grasslands. These low-density communities developed under dry conditions, and the grassland formation characterizes the interior lowlands where only moderate physiographic relief existed for much of the Tertiary. The physical conditions under which these vegetation types grew were not as conducive to the preservation of extensive and diverse floras as were those of the temperate deciduous forests. Vegetation surrounding the numerous lakes formed through volcanic activity in western North America during the Tertiary (Chapter 2) or growing along the shores of a gradually retreating epicontinental sea, as with the tropical to warm-temperate Eocene Mississippi Embayment floras, was more likely to be preserved. The presence of faunas typically associated with drier open forests, savanna, or grasslands can be especially useful in reconstructing the kind of plant formations and associations present when direct paleobotanical evidence is meager. Janis (1984) emphasized the usefulness of the different kinds of fossil ungulate communities in estimating the structure of vegetation growing under seasonally dry or cold climates.

As would be expected with any complex system, the faunal and plant evidence is not inevitably consistent through all of Late Cretaceous and Cenozoic time. There are instances where faunal assemblages and paleobotanical evidence from the same formation suggest different vegetation types and paleoenvironments. As mentioned earlier, such situations have often provided fodder for entertaining controversies that mostly make for good theater and bad science. The modern approach is to view these apparent incongruities as opportunities to refine existing concepts and to provide a better understanding of past biotic–environmental interactions.

As an example, Neogene deposits associated with the Columbia Plateau region preserve an extensive sequence of floras that contain a rich assemblage of deciduous forest elements now found in eastern North America, western North America, and temperate Asia. Extant and extinct species of *Acer*, *Alnus*, *Betula*, *Carya*, *Castanea*, *Cornus*, *Corylus*, *Diospyros*, *Engelhardia*, *Fagus*, *Fraxinus*, *Ginkgo*, *Glyptostrobus*, *Ilex*, *Juglans*, *Liquidambar*, *Magnolia*, *Nyssa*, *Ostrya*, *Platanus*, *Pterocarya*, *Populus*, *Quercus*, *Salix*, *Sassafras*, *Tilia*, and *Ulmus* were present in a region that today is mostly covered at low to middle elevations by an *Artemisia* (sagebrush) cold desert and riparian vegetation. The physiognomy of the vegetation suggested by the modern analogs is a dense hardwood forest. In contrast, many of the associated fossil mammalian faunas are rich in browsing and grazing ungulates more typical of open forests, savannas, and grasslands. The paleoclimatic implications of a landscape dominated by hardwood forest, versus one covered by grassland or savanna that supports herds of grazing mammals, are quite different and require either a plausible explanation or reinterpretation of one of the data sets. As it turns out, an explanation is available.

As noted in Chapter 2, one reason for the extensive sequence of fossil floras and faunas in western North America was the presence of conditions favorable to fossilization created by extensive volcanic activity. The lava flows blocked drainage systems and created numerous lakes that served as depositional basins, while ejection of large quantities of ash provided the matrix for exceptional preservation in fine-grained shales, tuffs (water-deposited ash), and diatomite. The extent of volcanic activity in the region is reflected by the sedimentary sequence shown in Figs. 3.10 and 3.11 from the Late Miocene Trout Creek locality in southeastern Oregon. The light bands in Fig. 3.10 are deposits of diatomite, and the grey ones are compressed layers of ash. Diatomite is a rapidly accumulating sediment, and many of the ash bands were also probably deposited in tens to a few hundreds of years. The approximately 20 layers of ash evident in the 55-ft section are testimony to frequent volcanism.

Taggart and Cross (1990) and Taggart et al. (1982) used evidence for volcanic activity from the nearby, and slightly older, Succor Creek flora to develop a model for explaining the association of abundant browsers and grazers with fossil floras that reflect dense forests. Fine-resolution sampling from zones immediately above the ash layers provided pollen and spore profiles that revealed successional stages (seres) of grasses and other herbs that were replaced by climax forest communities. These intervals were brief, lasting only a few to several hundred years, and would go undetected in normal megafossil sampling procedures. The overall regional plant fossil record would therefore suggest that forests dominated the landscape during virtually all of the Miocene. However, frequent volcanic activity provided a shifting mosaic of open, short-lived communities that were sufficient to sustain mobile herds of grazing (*Merychippus*, *Dromomeryx*) and woodland-browsing (*Anchitherium*, *Hypohippus*, *Ticholeptus*) horses and other ungulates. During times of diminishing food supply following volcanic eruptions, herds would likely congregate around lake margins, enhancing their representation in the faunal record of the region (e.g., Shotwell, 1968). The differences are resolved in a model that emphasizes short-term successional, as well as long-term climatic, changes.

Fossil faunas can provide an estimate of the age and stratigraphic relationships between strata when radiometric dates from plant-bearing strata are absent or inconclusive. The scale is the NALMA or provincial ages (Opdyke, 1990; Webb, 1985; Fig. 3.12; see Chapter 6 Context Summary). Each age is characterized by a fauna that can provide information on the paleoenvironments and expected associated vegetation types at specific times during the Late Cretaceous and Cenozoic. The NALMAs can be further plotted with reference to the paleotemperature curve constructed from the marine oxygen-isotope record (Fig. 3.13). It should be noted that ages for the subdivisions of the Paleogene are being revised (Berggren et al., 1992, 1995; D. R. Prothero, personal communication, 1996), and some may be older by ~2 m.y. and two polarity chrons (see Chapter 6). Some selected fossil floras are included in Fig. 3.13 as examples of the context that can be established for interpreting paleovegetation.

The fossil faunal record can often identify or confirm pathways of migration and allow more precise reconstructions of land connections than would be possible from plant evidence alone. Most plants and many animals can move across narrow to moderate barriers, but the interchange of large mammals more precisely defines the time

Figure 3.10. Exposure of the Trout Creek diatomite (Late Miocene), southeastern Oregon; ash layers represent volcanic eruptions. Adapted from Graham (1965).

and place of complete land connections. The pattern of new immigrations into North America from Asia via Beringia, from Europe via the De Geer and Thulean routes, and from South America through the Isthmus of Panama are preserved in pulses of affinities with each area. The timing and paleoenvironmental implications of these exchanges were reviewed by Webb (1985).

The Late Cretaceous Lancian NALMA is represented by the fauna of the Lance Formation in Wyoming and by the Hell Creek fauna of eastern Montana. The assemblages consist of Multituberculata (an extinct group of Mesozoic and Paleogene rodentlike organisms), Marsupialia (marsupials), and Eutheria (placentals). The depositional settings were swamps and waterways bordering the Cretaceous epicontinental sea. Primary geographic affinities are with Asia and Europe; there is no evidence of direct interchange with South America after the Maastrichtian and for most of the Tertiary. This reconstruction is consistent with the paleophysiography of the North Pacific and North Atlantic region during this period; connections with South America did not become established until late in the Pliocene.

The Paleocene (Puercan, Torrejonian, and Tiffanian land-mammal ages) witnessed extensive adaptive radiation of the placentals (12–13 species, double that of the Lancian), the marsupials declined from 13 to two species, and the dinosaurs disappeared completely (Webb, 1985). Three orders of large herbivores were present in the Middle to Late Paleocene: Pantodonta, Taeniodonta, and Dinocerata. The latter two were browsers, and all three orders probably inhabited local open woodlands (Webb, 1989). Late Cretaceous and Paleocene faunas have been described from sites ranging from Texas to Alberta, and their composition generally reflects widespread tropical to warm-temperate faunas distributed latitudinally along low thermal gradients.

The notable zoogeographic feature of the Wasatchian (Early Eocene) is the especially close affinity with European faunas. About one-half of the known land-mammal genera of North America were also present in Europe (mostly a consequence of migration from North America to Europe). This represents the strongest mammalian affinity that ever existed between the continents and suggests the North Atlantic land bridge was in full operation during the Early Eocene, both in terms of physical connection and climates suitable for the interchange of warm-temperate to subtropical forest-inhabiting mammals. The faunal evidence is, in turn, consistent with geological (McKenna, 1975; Chapter 2) and paleobotanical reconstructions (Tiffney, 1985) for the region.

In the Late Eocene, similarities between North American and European mammalian faunas declined but were maintained with Asia, a scenario compatible with the respective geologic history of the two land bridges. Also, a drop in sea level occurred at ~40 Ma in the Late Eocene (Duchesnian NALMA), which maintained the land connection between Asia and North America. Toward the end of the Eocene the temperature decline intensified a trend toward faunas occupying seasonally dry to more arid habitats (e.g., camelids), and Webb (1989) believes "that woodland savanna had become the predominant biome in North America, displacing the subtropical forests that had prevailed for the first 17 million years of Tertiary history." To a degree this trend is consistent with the composition and leaf physiognomy evidence from the late Middle Eocene Green River flora (46 Ma) of Colorado and Utah (MacGinitie, 1969) in which Leguminosae, Sapindaceae, Anacar-

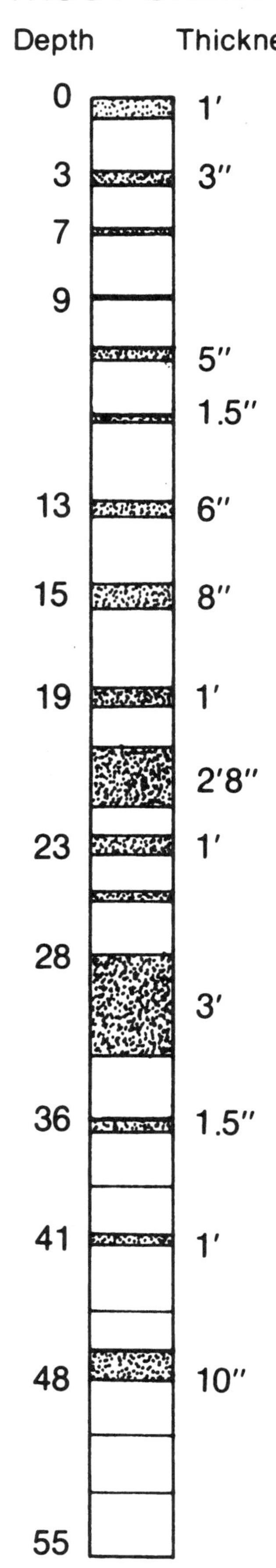

Figure 3.11. Diagram of the Trout Creek diatomite section showing number and extent of ash layers (stippled). Adapted from Graham (1988).

diaceae, species with smaller leaves, early grasses, and evaporitic and oxidized redbed deposits are evident.

In Chron 13R (37 Ma) there was another lowering of sea level, and at about 31–29 Ma (early Late Oligocene) a major fall of up to 200 m occurred (Figs. 3.6, 3.8) concomitant with a moderate decline in global temperatures (Fig. 3.1). Extensive land connections were present through Beringia. In the midcontinent region of North America floral diversity decreased, and LMA from the Florissant flora of Colorado reveal a decline in entire-margined leaves. Hutchison (1982) found a contemporaneous decline in aquatic reptile diversity. The faunal history of the Oligocene in Europe has been called the "Grande Coupure" (the great turnover) because of differences between the Early and Middle Tertiary assemblages. This was due primarily to the decline of immigrants from North America with disruption of the North Atlantic land bridge. New waves of immigrants arrived in North America from Asia through Beringia during the Chadronian. The introductions resulted in a change of ~60% in the North American land-mammal fauna and included, in particular, rodents (heteromyids, geomyids, castorids, sciurids, cylindrodontids, cricetids) and larger mammals (Canidae, Felidae, Mustelidae, Tapiridae, Rhinocerotidae, Anthracotheriidae, Tayassuidae). There was a pronounced shift from small mammal and arboreal forms, including many primates, to increased representation of terrestrial rodents and large herding ungulates of savanna and open-forest habitats (*Mesohippus*, *Leptomeryx*, *Poebrotherium*, *Meyrcoidodon*). These faunas persisted generally throughout the Oligocene into Hemingfordian time (Miocene; Figs. 3.12, 3.13). New introductions from Asia slowed in the Barstovian and Clarendonian.

Brown and Heske (1990) present interesting results on the effect that new introductions and exclusions (emigration, extinction) of rodents can have on the development of vegetation types. When three species of kangaroo rats (*Dipodomys*) were excluded from a Chihuahuan Desert shrub habitat in southeastern Arizona, there was a threefold increase in the density of tall perennial and annual grasses—a trend from shrubland to grassland. These data suggest that the introduction and persistence of new rodent and large mammal populations during the Cenozoic involved not just a passive range expansion into favorable vegetation types already determined by physical (land connections) and climatic factors, but that once introduced, their presence and later absence likely influenced the history of shrublands, grasslands, and savannas.

Miocene immigrants included cats (*Pseudaelurus*), bears (*Ursavus*, *Hemicyon*), beaver (*Anchitheriomys*, *Castor*), flying squirrel (*Blackia*), and the proboscideans (*Miomastodon*, *Gomphotherium*). These mostly reflect a mosaic of forest, open forest, and savanna vegetation across many parts of the continent. There was an increase in the diversity of browsing and grazing herbivores, including about 12 genera of horses (three browsers, nine grazers) and nearly as many genera of camels. The number of native un-

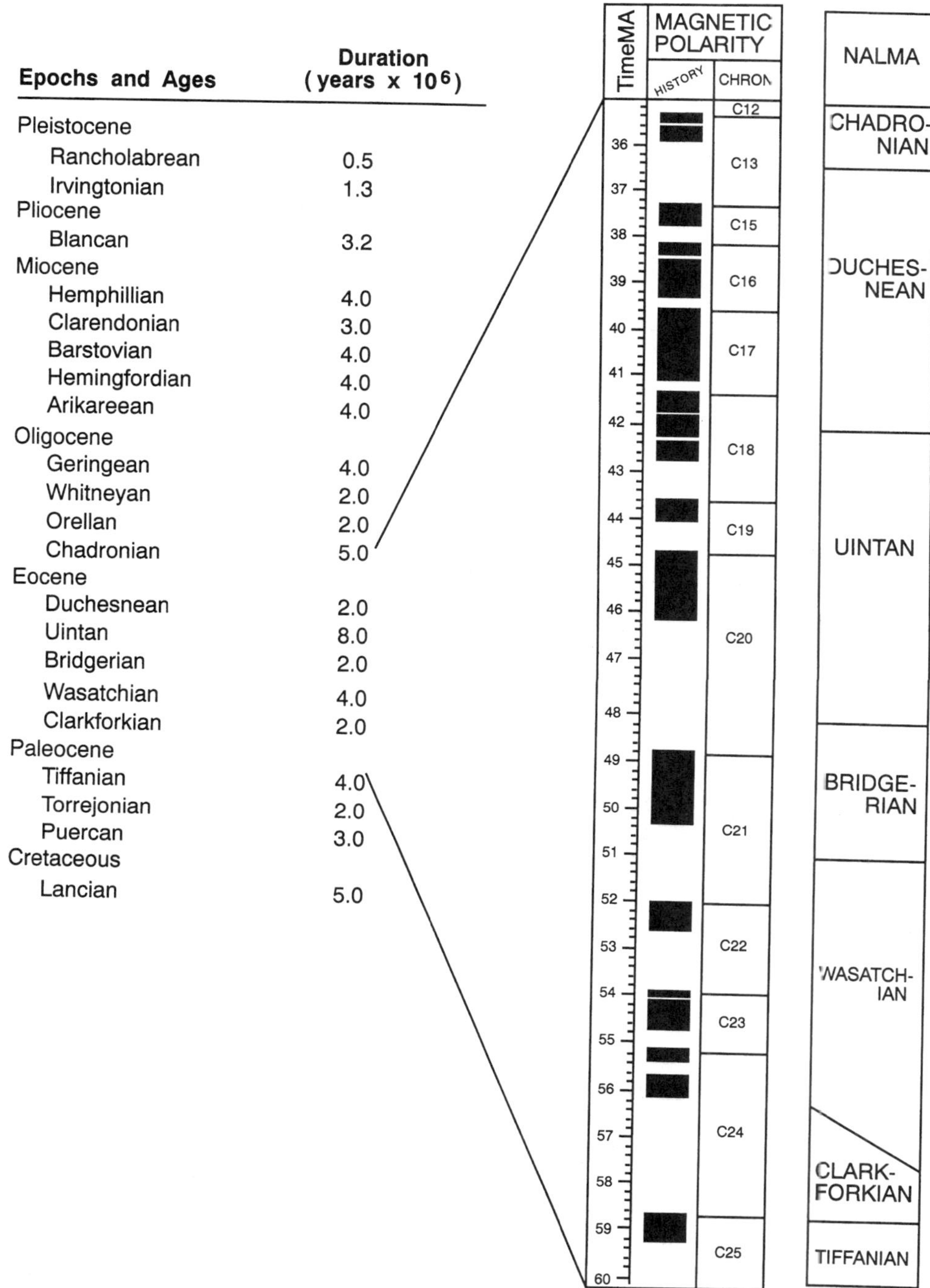

Epochs and Ages	Duration (years x 10^6)
Pleistocene	
Rancholabrean	0.5
Irvingtonian	1.3
Pliocene	
Blancan	3.2
Miocene	
Hemphillian	4.0
Clarendonian	3.0
Barstovian	4.0
Hemingfordian	4.0
Arikareean	4.0
Oligocene	
Geringean	4.0
Whitneyan	2.0
Orellan	2.0
Chadronian	5.0
Eocene	
Duchesnean	2.0
Uintan	8.0
Bridgerian	2.0
Wasatchian	4.0
Clarkforkian	2.0
Paleocene	
Tiffanian	4.0
Torrejonian	2.0
Puercan	3.0
Cretaceous	
Lancian	5.0

Figure 3.12. NALMAs and magnetic polarity time scale illustrating faunal context with which fossil floras can be compared. Compiled from Webb (1985) and Opdyke (1990). Note that recent age revisions (Berggren et al., 1992, 1995; Prothero, 1994; D. R. Prothero, personal communication, 1996; Prothero and Swisher, 1992) for the Paleogene are affecting NALMA provincial ages by ~+2 Ma (Chapter 6). Reprinted with the permission of S. David Webb, Plenum Publishing Corporation, and The University of Chicago Press (Journal of Geology), The University of Chicago.

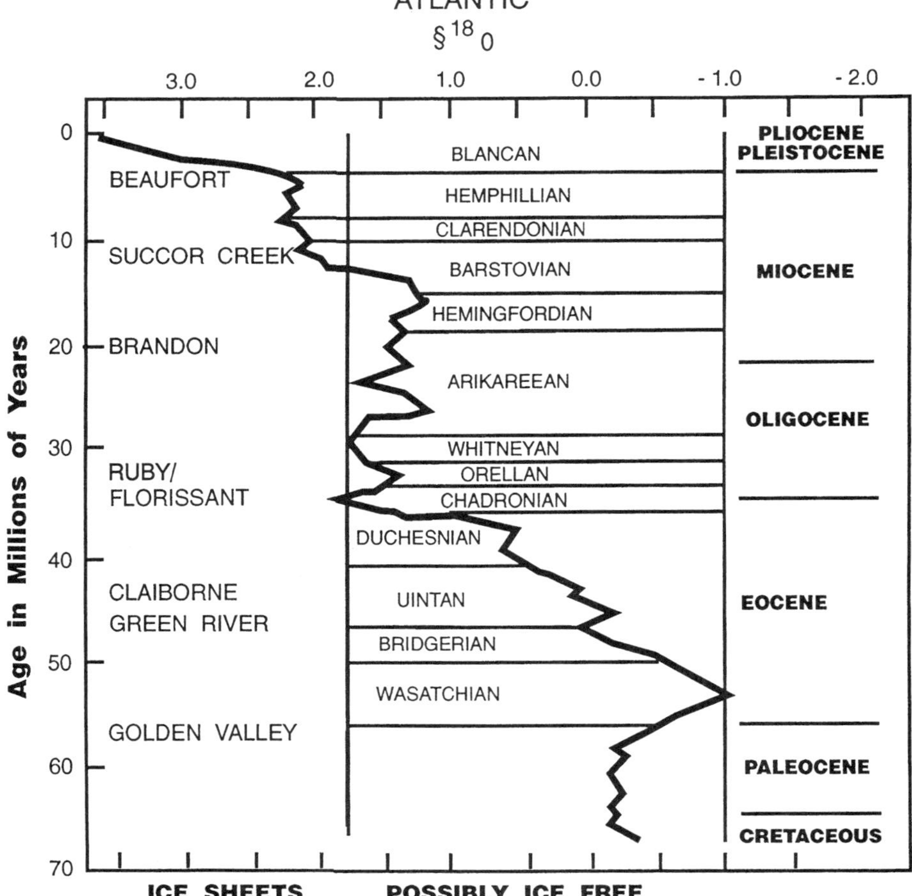

Figure 3.13. Composite Cenozoic ^{18}O record for benthic foraminifera compared to NALMA; selected fossil floras are included as examples of the faunal context available for interpreting vegetational history. Adapted from Opdyke 1990. (See note in caption to Figure 3.12.) Reprinted with the permission of The University of Chicago Press (Journal of Geology), The University of Chicago.

gulate genera doubled about every 3–5 m.y. from the Hemingfordian to the end of the Clarendonian. The reason for the increase is attributed to vegetational changes from woodland savanna to grassland savanna in many parts of North America (Webb, 1989). Paleobotanical evidence for the Late Eocene and later Cenozoic reflects forest and woodland vegetation; the faunal record suggests more open forests, savannas, and grasslands. The Miocene biota in parts of central North America has been compared to that of the modern African savannas (Webb, 1977); but as Janis (1984) points out, there are significant differences between the vegetation and ungulate communities in the two regions. Regardless of quantitative considerations, however, it is clear that the environmental and vegetational history patterns reflected by the flora and fauna are generally consistent.

Another major faunal turnover occurred in the Blancan, and more than two-thirds of the first mammals to appear were new introductions. This coincided in time with the Messinian (Mediterranean) salinity crisis, expansion of the great African savannas, and the divergence there of the ape–human lineage. The new forms in North America included a large number of Asian microtid rodents, *Petaurista* (a large, subtropical flying squirrel), *Parailurus* (an extinct relative of the lesser panda), *Ursus* (brown bear), *Cervus* (elk), *Homotherium* (a hyaenid), and *Brachypotherium* (a rhinocerotid that later gave rise to the amphibious North American genus *Teleoceras*). A distinctive feature of the new assemblage was an increased contingency from South America, including the sloth (*Glossotherium*), armadillos (*Dasypus*, *Holmesina*, *Glyptotherium*), and porcupine (*Erethizon*). This is consistent with estimates for uplift of the Isthmian region that formed a continuous land bridge by ~3.5 Ma (Fig. 3.14).

In the latest Miocene the trend toward open forests, savannas, and grasslands continued; but the development of even colder and drier conditions reduced the rich savanna ungulate fauna in North America. In the Pleistocene the formation of Arctic and alpine tundras, coalescence and spread of gymnosperms and associated elements into the extensive boreal coniferous forest or taiga, the appearance of deserts, the pronounced altitudinal zonation of communities within the western cordilleras, and severe winters in the midcontinent region resulted in a provincialization of the North American fauna and increased differentiation of regional communities (Webb, 1989). An overall trend during the Pleistocene was the replacement of large herbivores by smaller herbivores (Webb, 1969). The extinction of the

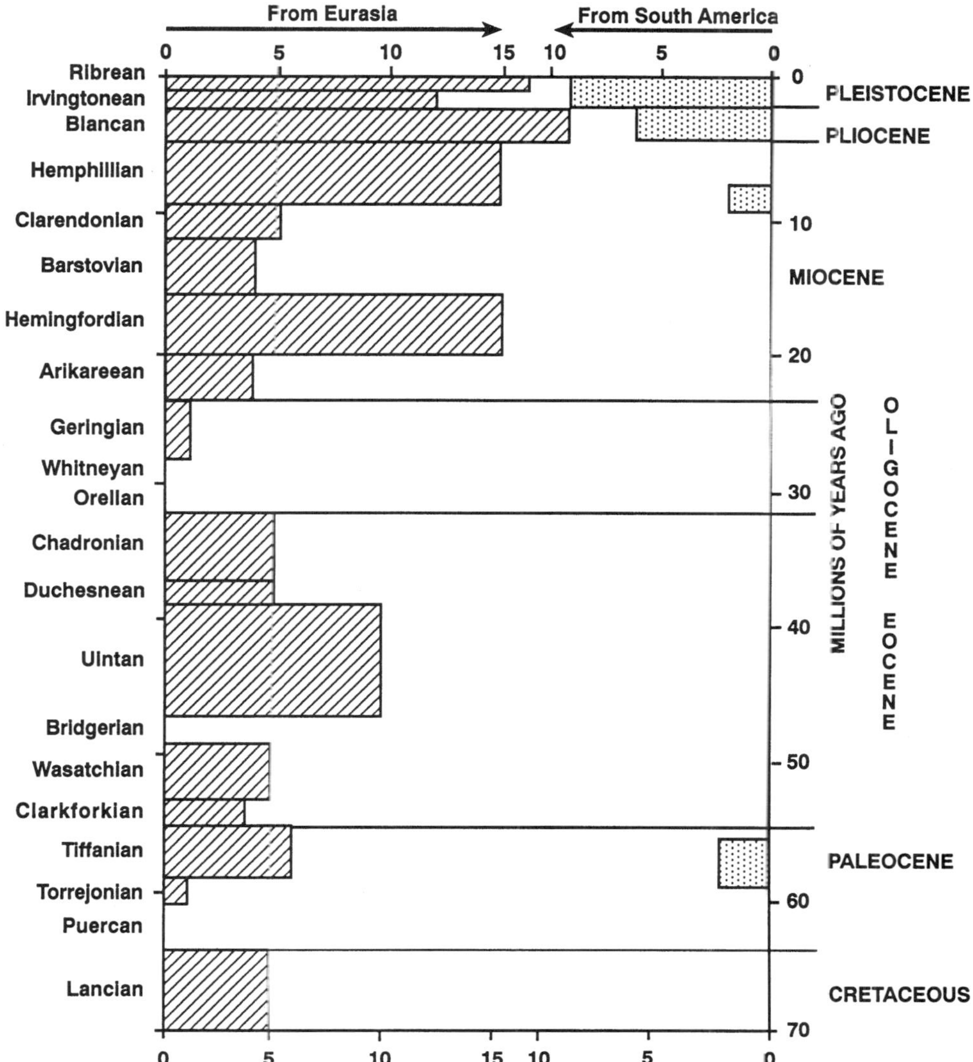

Figure 3.14. Immigrant land-mammal genera in North America during the past 70 m.y. Reprinted from Webb (1985) with the permission of S. David Webb and Plenum Publishing Corporation. (See note in caption to Figure 3.12.)

megafauna as a result of rapid climatic changes near the end of the Pleistocene was accelerated by Asian hunters who had crossed the Bering land bridge into North America (Barnosky, 1989; Martin and Wright, 1967).

It is clear from this brief summary that a knowledge of faunal history can provide valuable insights and supportive data for estimating the kinds of vegetation and environments existing in North America in the Late Cretaceous and Cenozoic. It can also help interpolate critical stratigraphic control including relative and absolute age data. Finally, the faunal record can provide information relevant to certain conceptual questions that are important in how we envision the history of plant communities.

Paleoenvironmental reconstructions for the Cenozoic are based, in part, on the present occurrence of presumed modern analogs. Environmental conditions such as rainfall and temperature regimes and altitudes are compiled, individual members of the assemblage are sorted out into paleocommunities according to the associations and ecological characteristics of their counterparts in the modern biota, preliminary environmental reconstructions are made, and the models are compared and refined with reference to the ancillary sources of data. If the modern ranges of some elements do not reflect their full ecological amplitude but reflect only that part into which they have moved since the last climatic change or barrier removal, this will be a weak point in the reconstruction. Often the consequences of these individual cases are mitigated by considering the ecological requirements of the entire assemblage; but if the species is distinctive, conspicuous, and/or numerically abundant, or if its significance is overemphasized for some reason, artifically narrow limits may be placed on the assumed paleoenvironment. Where an organism does occur is not always the same as where it can occur, and this concept is important in tracing environmental change and biotic history.

A case in point is provided by the migratory history of the edentate *Dasypus novemcinctus* (nine-banded armadillo; Humphrey, 1974). This is a South American genus that first appeared in North America in the Late Blancan mammal age after uplift of the isthmian land bridge ~3.5 Ma (Coates et al., 1992; Graham, 1992). It was first reported in northern Mexico and south Texas in 1854 (climatic data for Brownsville: January average temperature 59°F, annual rainfall 32.9 in.; United States Department of Agriculture Yearbook, 1941). By 1974 it had moved into Colorado, Kansas, and Missouri (climatic data for St. Louis: January average temperature 32.9°F, annual rainfall 36.67 in.). Although a warming climate is clearly a factor in the continued expansion northward, removal of barriers and migration rate (~4–10 km/year) have been important in determining the range of *Dasypus* since its introduction into North America. Paleoenvironmental reconstructions based on assumed ecological parameters from fossil and much of the pre-1854 distribution, compared to the present, would be different and a result of the fact that the range of *Dasypus* did not then reflect its full ecological amplitude. The northernmost limit of *Alligator* presently conforms to the 10°C isotherm; but when historical records are taken into account, the 4.4°C isotherm is a better estimate (Hutchison, 1982; Markwick, 1994). These and other examples from the faunal record reinforce emerging data from the plant record, especially from Quaternary pollen and spore studies, that many species are capable of living under a wider range of ecological conditions than their present distribution implies.

Another conceptual view is that communities are fluid in composition over time and that the particular versions witnessed today are only a few among many possible combinations. [For a similar view based on insect communities, see Elias (1991); for mammals, see the FAUNMAP Working Group (1996)]. The species comprising an association at any given point in time is a reflection of ecological compatibility; the opportunity to associate as determined by the availability and dispersal–survival potential of propagules in relation to existing barriers; the presence of soils suitable for each component; epidemic disease; time; and evolutionary alteration in the ecological parameters of individual lineages. The latter is frequently alluded to in the literature; but widespread fluctuations in the ecological requirements of a species through Cenozoic time, unaccompanied by any morphological change that would lead to its recognition as a novel taxon, are easier to assume than document (Elias, 1991). The fossil record generally reflects ecologically (rather than taxonomically) defined associations that are recognizable through time, but which frequently do involve the continued association of some of the dominants. My impression is that the definition and dynamic nature of plant paleocommunities is due more to ecological compatibility; the vicissitudes of propagule availability and dispersal potential; the existence, kinds, and extent of barriers; and the length of time since the last environmental reshuffling than to rapid widespread changes in the ecological requirements of numerous taxa accompanied by morphological stasis. A similar view is expresed by Hubbard and Boulter (1983), who note "that in certain family and generic groupings there has been no gross overall change in climatic and ecological preferences." Nonetheless, plant and animal communities are, within limits, temporal assemblages that undergo changes in composition through time. A contrasting view places more emphasis on the movement of plant communities as recognizable blocks in response to environmental change. No treatment of the Late Cretaceous and Tertiary history of North American vegetation would be complete without some comment on the turbulent history of geofloras.

GEOFLORA CONCEPT

The genesis of this idea was the application of the name arctotertiary flora by Engler (1879/1882) to an extant vegetation "Distinguished by numerous conifers and the numerous genera of [deciduous angiosperm] trees and shrubs that now dominate in North America or extratropical Eastern Asia and in Europe" (quoted from Mai, 1991).

Ralph W. Chaney (Museum of Paleontology, University of California, Berkeley) began his studies on the Tertiary floras of the western United States early in this century (e.g., Chaney, 1920). [For a synopsis of his life and work see Andrews (1980) and Gray and Axelrod (1971)]. By the 1950s he had recognized that Oligocene and Miocene floras there typically contained a strong temperate deciduous angiosperm and coniferous element, in contrast to older floras with a more tropical aspect, and younger ones with a strong subhumid (dry) component. Chaney added a time dimension to extant vegetation types with the term geoflora: "a group of plants which has maintained itself with only minor changes in composition for several epochs or periods of earth history, during which time its distribution has been profoundly altered . . ." (1959, p. 12).

Three geofloras were proposed for North America. In the Paleogene the Neotropical–Tertiary Geoflora extended into the Mississippi Embayment region, as represented by the Middle Eocene Claiborne flora. In the west the Clarno flora preserves an assemblage with tropical elements that reached northern Oregon. Individual elements extended even further north, occupying progressively more coastal sites in maritime environments. The Neotropical–Tertiary Geoflora reached its maximum northern extent during the Early Eocene, after which time it became progressively constricted toward equatorial regions in response to cooling and seasonally dry climates. The present northern limit of tropical vegetation in the sense of a tropical rain forest in North America is in southeastern Mexico in southern Veracruz state (17° N), while individual elements such as *Rhi-*

zophora (red mangrove) extend northward along the Florida coasts.

The Arcto–Tertiary Geoflora was described as a temperate deciduous angiosperm forest that presently characterizes much of eastern North America and parts of eastern Asia. The name implies a place (the Arctic) and a time of origin (the Tertiary) and, after the temperature decline of the late Eocene, it was viewed as extending its range progressively southward. The Arcto–Tertiary Geoflora was also identified in Early Tertiary high-latitude floras from Asia and Europe. It was considered to be a nearly continuous band of temperate vegetation encircling the middle latitudes of the northern hemisphere during the Middle Tertiary.

The Madro–Tertiary Geoflora (Axelrod, 1958) is a vegetation of dry to arid environments. Although early lineages and pre-adapted elements were present in the Paleogene, this geoflora became recognizable as a distinct vegetation type during the Mio–Pliocene. This was in response to reduced rainfall created by uplift of the coastal cordilleras and the Sierra Nevada and, to the north, by colder climates (viz., the cold deserts of much of the Basin and Range Province). From the perspective of geofloras, the history of North American vegetation was seen as a shifting mosaic of these three communities over the landscape in response to climatic and physiographic changes.

Much of the discussion about geofloras has centered on the Arcto–Tertiary Geoflora. Its early history was confused by an erroneous age assignment of plant-bearing strata in the Arctic. The Swiss paleobotanist and entomologist Oswald Heer (Andrews, 1980) was familiar with the European Neogene floras that had a strong temperate component. When floras from the Arctic region were discovered with temperate representatives (Heer, 1869), they were assigned a similar age; and an "Arctic Miocene" came into existence with floras that included broad-leaved deciduous elements. These floras are now known to be Paleocene and Eocene in age. The origin of the Arcto–Tertiary Geoflora was pushed back to the Late Cretaceous and Paleogene in the form of "blurred preliminary prints" (Chaney, 1967).

Although not emphasized in early considerations, these high-latitude temperate deciduous Paleogene floras are associated with tropical elements. Hollick (1936) described an assemblage from Berg Lake, Alaska. He identified *Populus*, *Juglans*, *Planera*, *Ulmus*, *Rhamnus*, *Cornus*, *Rhododendron*, and *Fraxinus*, as well as *Artocarpidium*, *Mohrodendron*, *Magnolia*, *Cinnamomum*, *Persea*, *Malpoenna*, *Terminalia*, and *Semecarpus*. The flora has since been revised to include *Dryopteris*, *Platycarya*, *Alnus*, *Knema*, *Myristica*, *Cinnamomophyllum*, *Luvunga*, *Malanorrhoea*, *Celastrus*, *Parashorea*, and *Alangium*. LMA has shown that 71% of Hollick's "species" have entire-margined leaves (Wolfe, 1977), and a number of paleotropical genera are now recognized (e.g., *Parashorea*). Fan palms are reported from the Paleocene of Alaska (57° and 61.5° N; Hollick, 1936) and from the Paleocene of Greenland (70° N; Koch, 1963). Assessment of the geoflora concept partly depends on whether the tropical elements are viewed as intermingled with temperate ones (based on present-day habitats of similar taxa) to constitute an assemblage unique in taxonomy and ecology, mostly separated in coastal and lowland habitats from inland and upland temperate elements, or part of a suspect terrane. This is unsettled. The current trend is toward interpreting the tropical elements as largely a coastal fringe community that was transported northward, but only moderate distances, during the Paleogene. This would leave intact a version of a deciduous forest formation that was not anomalously mixed extensively with tropical species. At a minimum, however, modification of the Arcto–Tertiary Geoflora would have to include an associated paratropical rain forest with Old World components. Valid criticisms of the original version of the geoflora concept are that it was based on an erroneous stratigraphy that affected estimates of its time of origin, presented too simplistic a view of vegetation types in the Arctic during the Paleogene, and underestimated the importance of Old World tropical components.

The principal concern, however, centers on the claim that each geoflora has "maintained itself with only minor changes in composition for several epochs or periods of earth history." The extent to which Chaney had come to see each geoflora as a block of tightly defined vegetation was brought home during a visit I made to Berkeley in 1961. He had agreed to look at some unidentified material from my dissertation sites at Trout Creek and Succor Creek in southeastern Oregon. These were Miocene floras from the Columbia Plateau region that, along with numerous associated floras, were considered typical examples of the Arcto–Tertiary Geoflora. Chaney had arranged for Harry MacGinitie to be present to help with the identifications. [For a biography of MacGinitie, see Wolfe (1987)]. As I handed each specimen to Chaney, he gave it a cursory glance and passed it on to MacGinitie, who provided most of the identifications. The last specimen was a fossil leaf I suggested might be *Arctostaphylos*. Chaney did not even look at this one and passed it to MacGinitie with the comment, "It's not *Arctostaphylos*." I asked why, and he replied, "Because *Arctostaphylos* is not an Arcto–Tertiary element."

This represents a more rigid view of geofloras than held by many paleobotanists even at the time. MacGinitie, for example, who was a student of Chaney, "questioned whether any flora, as a unit, migrated during the Tertiary" and wrote that "The terms 'Arcto–Tertiary,' 'Madro–Tertiary,' and the like imply extremely useful concepts if we do not think of these terms as representing areas or centers from which mass migrations occurred. They picture to us in a general way the vegetation occupying an area . . ." (1982, p. 87).

Studies on the Quaternary history of vegetation have been particularly instructive on how different combinations of ecologically compatible species can emerge from successive climatic pertubations (Graham and Grimm, 1990; Webb, 1987). Davis (1976, 1981) calculated that during the Quaternary, forests seldom maintained a constant species composition for more than 2000–3000 years and that a given association may have originated at different times in different parts of its range. For example, the oak–chestnut association in the central Applachians "have included chestnut as a dominant for 5000 years or longer, while oak–chestnut forests in Connecticut have included chestnut for only 2000 years. Deciduous forests in Ohio were penetrated first by hickory and then by beech 4000 years later; in Connecticut beech arrived first, followed 3000 years later by hickory" (Davis, 1981). In the latest glacial maximum at 18 Kya, assemblages were present that have no modern analogs. In the central plains there were communities variously described as spruce woodland, spruce forest, prairie, spruce–oak woodland, and black ash tundra (Davis, 1989). As noted by Wright (1987), these were grouped into assemblages that are unknown today. The same applies to several faunal communities where species whose ranges do not overlap today apparently occurred together during glacial maxima (Graham, 1986; Graham and Grimm, 1990). The debate regarding the proper conceptual framework for envisioning Tertiary vegetational history is paralleled by similar discussions in the early 1900s about its Quaternary history. The Clementsian view was that plant communities moved mostly as intact units, while the Gleasonian view was that plants responded to environmental change as individuals.

If it is accepted that Quaternary climatic changes induced a dynamic response from biotas in terms of their composition and migratory history, it is instructive to look at the tempo of these changes as revealed by recent research in glacial geology. The traditional view of northern hemisphere glacial events was that there were four major glacial advances (Nebraskan, Kansan, Illinoian, Wisconsin), separated by four interglacials (Aftonian, Yarmouth, Sangamon, and the present interglacial or Holocene beginning ~11 Kya). Each phase was considered about equal in duration for the Quaternary (then viewed as encompassing a little over 1 m.y.), giving a relatively steady pace of ~175 Ky for each glacial and interglacial interval. The Quaternary has now been extended back to ~1.6 Ma, and the pace of Quaternary climatic change is much quicker than earlier conceived. There were 18–20 such cycles, with nine occurring within the past 800 Ky (Fig. 3.15; Davis, 1983; Johnson, 1982). If each unit of peak and valley in Fig. 3.15 represents a time of potential climate-induced change in the composition of North American plant associations, as was the case in the latest cycle, then the dynamic nature of these associations becomes apparent. Similar histories, on a slower and broader scale, must have extended back into the pre-Quaternary (Bennett, 1990); it seems unlikely that any vegetation type has ever migrated as a unit to the extent implied by the early versions of the geoflora concept.

The question arises as to whether the idea of geofloras should, or can, be abandoned. Wolfe (1975, 1977, 1994) is explicit on this point:

> "The conclusion is, I think, inescapable that the Arcto-Tertiary concept has never had a satisfactory stratigraphic foundation. . . . As knowledge of genetics and physiology increased, it should have been apparent, as Mason (1947) pointed out, that the Arcto–Tertiary concept was invalid. The discarding of this concept, which is indicated by the Alaskan and Siberian assemblages of fossil plants, is fundamental to an understanding of floristic and vegetational history." (1977)

Zhilin (1989) concurs, and Boulter (1984) notes that the theory has confused and delayed proper zonation of the central European Tertiary. A different view is expressed by Mai: "For all our palaeophytochronomical studies we confirm the old terms 'Arctotertiary' and 'Palaeotropical' and also the concept of the geofloras . . ." (1991).

It is likely that geoflora terminology will continue to be encountered in the literature. The concept is attractive because it adds a time dimension to extant plant formations. Also, some designation is convenient for North America Paleogene floras with prominent tropical elements, Middle and Late Cenozoic floras characterized more by temperate-deciduous elements, and Late Cenozoic floras of dry to arid aspect. Furthermore, just as the geoflora concept eventually suffered from overdefinition and rigidity, it is also possible, in marshalling support for its demise, to overstate the randomness of species and generic associations through time. The results from Quaternary paleoecology previously cited deal primarily with units of vegetation at the association level, and it is evident to many paleobotanists that the broader units of plant formations can be recognized over long periods of time. For example, Crane et al. acknowledge that the concept oversimplifies the complex history of plant communities, but add that

> "[T]he nearest living relatives of many extinct Paleocene plants are still associated in Recent mixed mesophytic forest and this suggests that the climatic and, perhaps, edaphic tolerances of some individual angiosperm lineages have either remained more or less constant, or have exhibited similar patterns of change over the last 60 million years." (1990, p. 59)

With regard to the fern *Onoclea sensibilis*, Rothwell and Stockey (1991) state that "the species has remained virtually unchanged in both structural features and ecological tolerances throughout the Cenozoic" (p. 123) and "thereby provide one of the most dramatic examples of structural and ecological stasis known for vascular plants" (p. 114). Similarly, Webb (1989) refers to intervals of stable faunal composition that last 10 m.y. or longer and suggests that many chronofaunas are roughly comparable to extant faunas living in similar situations. Boulter et al. (1993) note that, "The present-day ecological patterns recognisable in ecosystems more than twenty million years old suggest that most of the components [in Miocene plant assemblages from Bohemia] retained their ecological preferences throughout this time." The recent modeling experiments described in Chapter 2 (CO_2 Concentrations section) show

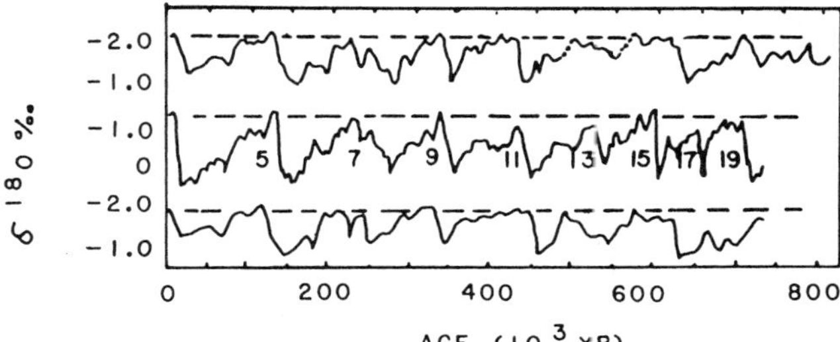

Figure 3.15. Oxygen isotope ratios in three deep-sea cores plotted against time. Low points on the curve represent times of glacial advance, and high points represent times of retreat. Arabic numerals designate interglacial stages. Nine glacial–interglacial cycles are recorded for the past 800,000 years. Reprinted from Davis [1983; based on Johnson (1982) and Shackleton and Opdyke (1976)] with the permission of Margaret Davis, the Missouri Botanical Garden, and the Geological Society of America.

that seasonal variation likely existed in parts of the high Arctic. This, along with the proposed limitation of tropical vegetation primarily to a coastal, transported community, lessens the necessity of viewing temperate and tropical elements as intermingled and having ecological requirements completely different from that of their modern descendents.

The geofloras terminology is entrenched in the literature and likely will continue to be used in the general sense described by MacGinitie (1962) and by those who were not involved in or concerned with its tumultuous past. Moreover, the continued use of a terminology or application of a concept that evolves over time is not uncommon. Huggett (1990) presents an interesting discussion of catastrophism and how its meaning has changed with increasing knowledge of past events in Earth history. The early proponents of catastrophism like Buckland and Cuvier and opponents like Hutton and Lyell would be amazed at how widely the idea is accepted today and how little it resembles the original version. Geofloras as a means of envisioning vegetational history is a heuristic concept; but if used, it should be in the general sense of MacGinitie (1962). Other concepts are being forged that emphasize the importance of Old World tropical lineages in the ancient high-latitude floras and the contribution of their descendents to the modern vegetation of the New World and that better reflect the dynamic and temporal nature of plant associations. One of these is especially relevant to the nature and history of high-latitude North American vegetation during the Paleocene.

BOREOTROPICS CONCEPT

Recent studies on Paleogene floras have revealed two characteristics of high-latitude paleocommunities. One is that even with some limited transport of Cenozoic exotic terranes along the west coast, there is still evidence for a tropical element associated with temperate deciduous vegetation in the Arctic Paleogene. The other is that the tropical component includes plants presently found in Old World vegetation. The interchange was facilitated by land connections and megathermal climates across the North Atlantic and Beringia, and the Old World connection is becoming more evident with continuing studies of Paleogene floras. For example, the southeast Asian *Sargentodoxa* (Sargentodoxaceae) has recently been added to the Early Miocene Brandon Lignite flora of Vermont (Tiffney, 1993, 1994), and fossil seeds of *Ensete* (Musaceae; Africa, Asia) are now known from the Middle Eocene nut bed locality of the Clarno Formation of north-central Oregon (Manchester and Kress, 1993). Wolfe (1975) proposed the name boreotropical flora to characterize this mix of high latitude deciduous and evergreen vegetation, unique by the distribution of extant congeners. It includes some elements that today are primarily Old World tropical in relationship, associated with early progenitors of temperate deciduous vegetation (see also Tiffney, 1985).

In addition to the increasing number of reports of Old World tropical elements in high-latitude Paleogene floras, the plausibility of the boreotropical concept can be assessed by testing some of the expected consequences. Lavin and Luckow (1993) suggest that because South America was isolated from North America during this time, "a prediction of this hypothesis posits that a taxon with a present-day center of diversity in tropical North America [Mexico, Central America, the Antilles], and with an early Tertiary fossil record from any region there, has a high probability of having sister-group relatives in the Paleotropics and derived relatives in South America." They looked at two taxa of legumes (the *Dichrostachys* group and Tribe *Robinieae* and allies) via area cladograms, and the results were consistent with the predictions. Furthermore, Wendt (1993) found that ~25% of the tree species in the Mexican lowland rain forests may be derived from Old World tropical progenitors that arrived from the north; it had previously been assumed that virtually all northern Latin American tropical tree species had come from South America. The tropical family Bombacaceae may have arrived from the north because the oldest fossil records are from the lower Maastrichtian of New Jersey (Wolfe, 1975), and it is not known from South America until the Paleocene. In contrast, Hammel and Zamora (1993) found no evidence for a boreotropical origin of the newly described *Ruptiliocarpon* (Lepidobotryaceae) based on the criteria of Lavin and Luckow (1993). This is apparently one of the majority of tropical genera that interchanged between North and South America through the Isthmus of Panama. More analyses are needed to better quantify the relative roles of different biogeographic pathways into North America.

Such data are providing new ways of envisioning the history of North American high-latitude vegetation. With reference to the modern deciduous forest formation, it is now clear that this community is a complex mixture of elements derived from various sources. Some certainly came from mesic Cretaceous ancestors and evolved in the high northern latitudes. Others are clearly derived from tropical groups that have evolved temperate components (e.g., *Diospyros virginiana*, the persimmon, a New World temperate representative of the primarily tropical family Ebenaceae). The contribution of Old World tropical elements to the composition of New World temperate deciduous and tropical forests is only now becoming fully recognized and constitutes support for the concept of a boreotropical flora.

Much of the information from the fossil record is used to reconstruct paleocommunities. For the task of establishing the composition, structure, and arrangement of an ancient plant community, the procedures are complex and beset with potential sources of error and some of the conclusions and theoretical considerations remain unsettled. If the data are to be used for the even more formidable task of estimating paleoenvironments and accurately tracing environmental–biotic interactions through time, additional independent information is essential. Paleotemperature, sea level, time, and faunal history, along with sound conceptual frameworks, illustrate some of the important contexts necessary, and currently being developed, for accurately interpreting the history of North American vegetation.

References

Ager, D. V. 1981. Major marine cycles in the Mesozoic. J. Geol. Soc. Lond. 138: 159–166.

Andrews, H. N., Jr. 1980. The Fossil Hunters. Cornell University Press, Ithaca, NY.

Axelrod, D. I. 1958. Evolution of the Madro–Tertiary Geoflora. Botan. Rev. 24: 433–509.

———. 1981. Altitudes of Tertiary forests estimated from paleotemperature. Proc. of the Symposium on Qinghai-Xizang (Tibet) Plateau (Beijing, China). Geological and Ecological Studies of Qinghai-Xizang Plateau. Geology, Geological History and Origin of Qinghai-Xizang Plateau. Science Press, Beijing and Gordon & Breach Science Publishers, New York. Vol. 1, pp. 131–137.

——— and H. P. Bailey. 1969. Paleotemperature analysis of Tertiary floras. Palaeogeogr., Palaeoclimatol., Palaeoecol. 6: 163–195.

Barnosky, A. D. 1989. The late Pleistocene event as a paradigm for widespread mammal extinction. In: S. K. Donovan (ed.), Mass Extinctions, Processes and Evidence. Columbia University Press, New York. pp. 235–254.

Barron, E. J., J. L. Sloan II, and C. G. A. Harrison. 1980. Potential significance of land–sea distribution and surface albedo variations as a climatic forcing factor; 180 M.Y. to the present. Palaeogeogr., Palaeoclimatol., Palaeoecol. 30: 17–40.

Barron, E. J. and W. M. Washington. 1984. The role of geographic variables in explaining paleoclimates: results from Cretaceous climate model sensitivity studies. J. Geophys. Res. 89: 1267–1279.

Bennett, K. D. 1990. Milankovitch cycles and their effects on species in ecological and evolutionary time. Paleobiology 16: 11–21.

Berger, W. H. 1982. Climatic steps in ocean history—lessons from the Pleistocene. In: Geophysics Study Committee, Climate in Earth History. National Academy Press, Washington, D.C. pp. 43–54.

Berggren, W. A., D. V. Kent, M.-P. Aubry, and J. Hardenbol (eds.). 1995. Geochronology, Time Scales and Global Stratigraphic Correlations. SEPM (Society for Sedimentary Geology), Tulsa, OK. Special Publ. 54.

Berggren, W. A., D. V. Kent, J. D. Obradovich, and C. C. Swisher III. 1992. Toward a revised Paleogene chronology. In: D. R. Prothero and W. A. Berggren (eds.), Eocene–Oligocene Climatic and Biotic Evolution. Princeton University Press, Princeton, NJ. pp. 29–45.

Berner, R. A. 1994. 3GEOCARB II: a revised model of atmospheric CO_2 over Phanerozoic time. Am. J. Sci. 294: 56–91.

Berner, R. A., A. C. Lasaga, and R. M. Garrels. 1983. The carbonate–silicate geochemical cycle and its effect on atmospheric carbon dioxide over the past 100 million years. Am. J. Sci. 283: 641–683.

Birkenmajer, J. 1990. Geochronology and climatostratigraphy of Tertiary glacial and interglacial succession on King George Island, South Shetland Islands (West Antarctica). In: H. Miller (ed.), Zentralblatt für Geologie und Palaontologie Teil I, Workshop on Antarctic Geochronology. E. Schweizerbart'sche Verlagsbuchhandlung, Stuttgart. pp. 141–151.

Boulter, M. C. 1984. Palaeobotanical evidence for land–surface temperature in the European Palaeogene. In: P. J. Brenchley (ed.), Fossils and Climate. Wiley, New York. pp. 35–47.

———, R. Hubbard, and Z. Kvaček. 1993. A comparison of intuitive and objective interpretations of Miocene plant assemblages from north Bohemia. Palaeogeogr., Palaeoclimatol., Palaeoecol. 101: 81–96.

Brown, J. H. and E. J. Heske. 1990. Control of a desert–grassland transition by a keystone rodent guild. Science 250: 1705–1707.

Chaney, R. W. 1920. The flora of the Eagle Creek Formation. Contrib. Walker Mus. 2: 115–181.

———. 1959. Miocene floras of the Columbia Plateau. Part I. Composition and Interpretation. Carnegie Inst. Wash. Contrib. Paleontol. 617: 1–134.

———. 1967. Miocene forests of the Pacific Basin; their ancestors and their descendants. Jubilee Publication Commemorating Professor Sasa's 60th Birthday. Hokkaido University, Sapporo. pp. 209–239.

Chappell, J. and N. J. Shackleton. 1986. Oxygen isotopes and sea level. Nature 324: 137–140.

Coates, A. G. et al. (+ 7 authors). 1992. Closure of the Isthmus of Panama: the near-shore marine record of Costa Rica and western Panama. Geol. Soc. Am. Bull. 104: 814–828.

Crane, P. R., S. R. Manchester, and D. L. Dilcher. 1990. A preliminary survey of fossil leaves and well-preserved reproductive structures from the Sentinel Butte Formation (Paleocene) near Almont, North Dakota. Fieldiana (Geol. n.s.) 20: 1–63.

Crowley, T. J., J. G. Mengel, and D. A. Short. 1987. Gondwanaland's seasonal cycle. Nature 329: 803–807.

Davis, M. B. 1976. Pleistocene biogeography of temperate deciduous forests. Geosci. Man 13: 13–26.

———. 1981. Quaternary history and the stability of forest communities. In: D. C. West, H. H. Shugart, and D. B. Botkine (eds.), Forest Succession: Concepts and Application. Springer-Verlag, Berlin. pp. 132–153.

———. 1983. Quaternary history of deciduous forests of eastern North America and Europe. Ann. Missouri Bot. Garden. 70: 550–563.

———. 1989. Insights from paleoecology on global change. Ecol. Soc. Am. Bull. 70: 222–228.

Dawson, M. R., R. M. West, W. Langston, Jr., and J. H. Hutchison. 1976. Paleogene terrestrial vertebrates: northernmost occurrence, Ellesmere Island, Canada. Science 192: 781–782.

Dorf, E. 1964. The use of fossil plants in palaeoclimatic interpretations. In: A. E. M. Nairn (ed.), Problems in Palaeoclimatology. Interscience Publishers, London. pp. 13–31.

Eglinton, G. et al. (+ 5 authors). 1992. Molecular record of secular sea surface temperature changes on 100-year timescales for glacial terminations I, II, and IV. Nature 356: 423–426.

Elias, S. A. 1991. Insects and climatic change. BioScience 41: 552–559.

Emiliani, C. 1955. Pleistocene temperatures. J. Geol. 63: 538–578.

Engler, A. 1879/1882. Versuch einer Entwicklungsgeschichte der Pflanzenwelt seit der Tertiärperiode I/II. W. Engelmann, Leipzig. p. 203.

Epstein, S., R. Buchsbaum, H. A. Lowenstam, and H. C. Urey. 1951. Carbonate–water isotopic temperature scale. Geol. Soc. Am. Bull. 62: 417–426.

Fairbanks, R. G. 1989. A 17,000-year glacio-eustatic sea level record: influence of glacial melting rates on the Younger Dryas event and deep-ocean circulation. Nature 342: 637–642.

FAUNMAP Working Group. 1996. Spatial response of mammals to Late Quaternary environmental fluctuations. Science 272: 1601–1606.

Geophysics Study Committee. 1990. Sea-Level Change. National Academy Press, Washington, D.C.

Graham, A. 1965. The Sucker Creek and Trout Creek Miocene floras of southeastern Oregon. Kent State Univ. Res. Bull. IX: 1–147.

———. 1988. Studies in neotropical paleobotany. VI. The lower Miocene communities of Panama—the Culebra Formation. Ann. Missouri Bot. Garden. 75: 1467–1479.

———. 1992. Utilization of the isthmian land bridge during the Cenozoic—paleobotanical evidence for timing, and the selective influence of altitudes and climate. Rev. Palaeobot. Palynol. 72: 119–128.

Graham, R. W. 1986. Response of mammalian communities to environmental changes during the late Quaternary. In: J. Diamond and T. J. Case (eds.), Community Ecology. Harper & Row, New York. pp. 300–313.

——— and E. C. Grimm. 1990. Effects of global climate change on the patterns of terrestrial biological communities. Trends Ecol. Evol. 5: 289–292.

Gray, J. and D. I. Axelrod. 1971. Memorial to Ralph Works Chaney 1890–1971. Geological Society of America, Boulder, CO.

Guilderson, T. P., R. G. Fairbanks, and J. L. Rubenstone. 1994. Tropical temperature variations since 20,000 years ago: modulating interhemispheric climate change. Science 263: 663–665.

Gurnis, M. 1990. Ridge spreading, subduction, and sea level fluctuations. Science 250: 970–972.

———. 1993. Phanerozoic marine inundation of continents driven by dynamic topography above subducting slabs. Nature 364: 589–593.

Hammel, B. E. and N. A. Zamora. 1993. *Ruptiliocarpon* (Lepidobotryaceae): a new arborescent genus and tropical American link to Africa, with a reconsideration of the family. Novon 3: 408–417.

Haq, B. U., J. Hardenbol, and P. R. Vail. 1987. Chronology of fluctuating sea levels since the Triassic. Science 235: 1156–1166 (See also, Technical Comments. 1988. Science, 241: 596–602).

Harrison, C. G. A. 1990. Long-term eustasy and epeirogeny in continents. In: Geophysics Study Committee (eds.), Sea-Level Change. National Academy Press, Washington, D.C. pp. 141–158.

Hays, J. D. and W. C. Pitman III. 1973. Lithospheric plate motion, sea-level changes and climatic and ecological consequences. Nature 246: 18–22.

Heer, O. 1869. Flora fossilis Alaskana. Kgl. Svenska Vetensk. Akad. Handl. 8.

Hollick, A. 1936. The Tertiary floras of Alaska. U.S. Geological Survey, Denver, CO, Prof. Paper 182, pp. 1–185.

Hsü, K. J. et al. (+ 9 authors). 1977. History of the Mediterranean salinity crisis. Nature 267: 399–403.

Hubbard, R. and M. C. Boulter. 1983. Reconstruction of Palaeogene climate from palynological evidence. Nature 301: 147–150.

Huggett, R. J. 1990. Catastrophism: Systems of Earth History. Edward Arnold, London.

Humphrey, S. R. 1974. Zoogeography of the nine-banded armadillo (*Dasypus novemcinctus*) in the United States. BioScience 24: 457–462.

Hutchison, J. H. 1982. Turtle, crocodilian, and champsosaur diversity changes in the Cenozoic of the north-central region of western United States. Palaeogeogr., Palaeoclimatol., Palaeoecol. 37: 149–164.

Janis, C. M. 1984. The use of fossil ungulate communities as indicators of climate and environment. In: P. J. Brenchley (ed.), Fossils and Climate. Wiley, New York. pp. 85–104.

Johnson, R. G. 1982. Brunhes–Matuyama magnetic reversal dated at 790,000 year B.P. by marine-astronomical correlations. Quatern. Res. 17: 135–147.

Kennett, J. P. and N. J. Shackleton. 1976. Oxygen isotopic evidence for the development of the psychrosphere 38 Myr ago. Nature 260: 513–515.

Kerr, R. A. 1984. Carbon dioxide and the control of the ice ages. Science 223: 1053–1054.

Koch, B. E. 1963. Fossil plants from the lower Paleocene of the Agatdalen (Angmartussut) area, central Nugssauq Peninsula, northwest Greenland. Medd. Gronland 172.

Koch, P. L., J. C. Zachos, and P. D. Gingerich. 1992. Correlation between isotope records in marine and continental carbon reservoirs near the Palaeocene/Eocene boundary. Nature 358: 319–322.

Kurtén, B. and E. Anderson. 1980. Pleistocene Mammals of North America. Columbia University Press, New York.

Lavin, M. and M. Luckow. 1993. Origins and relationships

of tropical North America in the context of the boreotropics hypothesis. Am. J. Bot. 80: 1–14.

Leg 119 Shipboard Scientific Party. 1988. Early glaciation of Antarctica. Nature 333: 303–304.

MacGinitie, H. D. 1962. The Kilgore flora, a late Miocene flora from northern Nebraska. Univ. Calif. Publ. Geol. Sci. 35: 67–158.

———. 1969. The Eocene Green River flora of northwestern Colorado and northeastern Utah. Univ. Calif. Publ. Geol. Sci. 83: 1–202.

Mai, D. H. 1991. Palaeofloristic changes in Europe and the confirmation of the Arctotertiary–Palaeotropical geofloral concept. Rev. Palaeobot. Palynol. 68: 29–36.

Manchester, S. R. and W. J. Kress. 1993. Fossil bananas (Musaceae): *Ensete oregonense* sp. nov. from the Eocene of western North America and its phytogeographic significance. Amer. J. Bot. 80: 1264–1272.

Martin, P. S. and H. E. Wright, Jr. (eds.). 1967. Pleistocene Extinctions. Yale University Press, New Haven, CT.

Markwick, P. J. 1994. "Equability," continentality, and Tertiary "climate": the crocodilian perspective. Geology 22: 613–616.

Mason, H. L. 1947. Evolution of certain floristic associations in western North America. Ecol. Monogr. 17: 201–210.

Matthews, R. K. and C. Frohlich. 1991. Orbital forcing of low-frequency glacioeustasy. J. Geophys. Res. 96: 6797–6803.

McKenna, M. C. 1975. Fossil mammals and early Eocene North Atlantic land continuity. Ann. Missouri Bot. Garden. 62: 335–353.

———. 1980. Eocene paleolatitude, climate, and mammals of Ellesmere Island. Palaeogeogr., Palaeoclimatol., Palaeoecol. 30: 349–362.

Miller, K. G., R. G. Fairbanks, and G. S. Mountain. 1987. Tertiary oxygen isotope synthesis, sea level history, and continental margin erosion. Paleoceanography 2: 1–19.

Miller, K. G., G. S. Mountain, the Leg 150 Shipboard Party, and Members of the New Jersey Coastal Plain Drilling Project. 1996. Drilling and dating New Jersey Oligocene–Miocene sequences: ice-volume, global sea level, and Exxon records. Science 271: 1092–1095.

Moore, T. C., Jr., T. S. Loutit, and S. M. Greenlee. 1987. Estimating short-term changes in eustatic sea level. Paleoceanography 2: 625–637.

Moore, T. C., Jr., N. G. Pisias, and L. D. Keigwin, Jr. 1982. Cenozoic variability of oxygen isotopes in benthic formaminifera. In: Geophysics Study Committee (eds.), Climate in Earth History. National Academy Press, Washington, D.C. pp. 172–182.

Opdyke, N. D. 1990. Magnetic stratigraphy of Cenozoic terrestrial sediments and mammalian dispersal. J. Geol. 98: 621–637.

Pitman W. C., III. 1978. Relationship between eustacy and stratigraphic sequences of passive margins. Geol. Soc. Am. Bull. 89: 1389–1403.

Poore, R. Z. and R. K. Matthews. 1984. Oxygen isotope ranking of Late Eocene and Oligocene planktonic foraminifera: Implications for Oligocene sea-surface temperatures and global ice-volume. Mar. Micropaleontol. 9: 111–134.

Prentice, K. C. and I. Y. Fung. 1990. The sensitivity of terrestrial carbon storage to climate change. Nature 346: 48–51.

Prentice, M. L. and R. K. Matthews. 1988. Cenozoic ice-volume history: development of a composite oxygen isotope record. Geology 16: 963–966.

———. 1991. Tertiary ice sheet dynamics: the snow-gun hypothesis. J. Geophys. Res. 96: 6811–6827.

Prothero, D. R. 1994. The late Eocene–Oligocene extinctions. Annu. Rev. Earth Planet. Sci. 22: 145–165.

——— and C. C. Swisher III. 1992. Magnetostratigraphy and geochronology of the terrestrial Eocene–Oligocene transition in North America. In: D. R. Prothero and W. A. Berggren (eds.), Eocene–Oligocene Climatic and Biotic Evolution. Princeton University Press, Princeton, NJ. pp. 46–73.

Roemmich, D. 1992. Ocean warming and sea level rise along the southwest U.S. Coast. Science 257: 373–375.

Rothwell, G. W. and R. A. Stockey. 1991. *Onoclea sensibilis* in the Paleocene of North America, a dramatic example of structural and ecological stasis. Rev. Palaeobot. Palynol. 70: 113–124.

Savage, D. E. and D. E. Russell. 1983. Mammalian Paleofaunas of the World. Addison-Wesley, London.

Savin, S. M. 1977. The history of the Earth's surface temperature during the past 100 million years. Annu. Rev. Earth Planet. Sci. 5: 319–355.

———. 1982. Stable isotopes in climatic reconstructions. In: Geophysics Study Committee (eds.), Climate in Earth History. National Academy Press, Washington, D.C. pp. 164–171.

——— and R. G. Douglas. 1985. Sea level, climate, and the Central American land bridge. In: F. G. Stehli and S. D. Webb (eds.), The Great American Biotic Interchange. Plenum, New York. pp. 303–324.

Savin, S. M. and F. Woodruff. 1990. Isotopic evidence for temperature and productivity in the Tertiary oceans. In: W. C. Burnett and S. R. Riggs (eds.), Phosphate Deposits of the World, Neogene to Modern Phosphorites. Cambridge University Press, Cambridge, U.K. Vol. 3, pp. 241–259.

Schlanger, S. O., H. C. Jenkyns, and I. Premoli-Silva. 1981. Volcanism and vertical tectonics in the Pacific Basin related to global Cretaceous transgressions. Earth Planet. Sci. Lett. 52: 435–449.

Schoell, M., S. Schouten, J. S. Sinninghe Damsté, J. W. de Leeuw, and R. E. Summons. 1994. A molecular organic carbon isotope record of Miocene climate changes. Science 263: 1122–1125.

Schrag, D. P., G. Hampt, and D. W. Murray. 1996. Pore fluid constraints on the temperature and oxygen isotopic composition of the glacial ocean. Science 272: 1930–1932.

Shackleton, N. J. 1974. Attainment of isotopic equilibrium between ocean water and the benthonic foraminiferal genus *Uvigerina*: isotopic changes in the ocean during the last glacial. Colloq. Int. C.N.R.S. 219: 203–210.

———. 1987. Oxygen isotopes, ice volume and sea level. Quatern. Sci. Rev. 6: 183–190.

——— and A. Boersma. 1981. The climate of the Eocene Ocean. J. Geol. Soc. Lond. 138: 153–157.

Shackleton, N. J. and J.-C. Duplessy. 1986. Temperature changes in ocean deep waters during the late Pleistocene. 2nd International Conference on Paleoceanography, Woods Hole, MA. Abstract.

Shackleton, N. J. and J. P. Kennett. 1975a. Paleotemperature history of the Cenozoic and the initiation of Antarctic glaciation: oxygen and carbon isotope analyses in DSDP sites 277, 279 and 281. Init. Rep. Deep Sea Drill. Proj., 29: 743–755.

———. 1975b. Late Cenozoic oxygen and carbon isotopic changes at DSDP site 284: implications for glacial history of the northern hemisphere and Antarctica. Initial Rep. Deep Sea Drilling Project, 29: 801–807.

Shackleton, N. J. and N. D. Opdyke. 1976. Oxygen-isotope and paleomagnetic stratigraphy of equatorial Pacific core V28-239 Late Pliocene to latest Pleistocene. In: R. M. Cline and J. D. Hays (eds.), Investigation of Late Quaternary Paleoceanography and Paleoclimatology. Geological Society of America, Boulder, CO. Mem. 145, pp. 449–464.

———. 1977. Oxygen isotope and palaeomagnetic evidence for early northern hemisphere glaciation. Nature 270: 216–219.

Sheehan, P. M., D. E. Fastovsky, R. G. Hoffmann, C. B. Berghaus, and D. L. Gabriel. 1991. Sudden extinction of the dinosaurs: latest Cretaceous, upper Great Plains, U.S.A. Science 254: 835–839.

Shotwell, J. A. 1968. Miocene mammals of southeast Oregon. Univ. Oregon Mus. Nat. Hist. Bull. 14.

Southam, J. R. and W. W. Hay. 1977. Time scales and dynamic models of deep-sea sedimentation. J. Geophys. Res. 82: 3825–3842.

———. 1981. Global sedimentary mass balance and sea level changes. In: C. Emiliani (ed.), The Ocean Lithosphere. Wiley, New York. pp. 1617–1684.

Stehli, F. G. and S. D. Webb (eds.). 1985. The Great American Biotic Interchange. Plenum, New York.

Taggart, R. E. and A. T. Cross. 1990. Plant successions and interpretations in Miocene volcanic deposits, Pacific northwest. In: M. G. Lockley and A. Rice (eds.), Volcanism and Fossil Biotas. Geological Society of America, Boulder, CO. Special Paper 244, pp. 57–68.

———, A. T. Cross, and L. Satchell. 1982. Effects of periodic volcanism on Miocene vegetation distribution in eastern Oregon and western Idaho. Proc. 3rd North Am. Paleontol. Conv. 2: 535–540.

Tanai, T. and K. Huzioka. 1967. Climatic implications of Tertiary floras in Japan. Symposium 25, 11th Pacific Science Congress, University of Tokyo, August–September, 1966, Tertiary Correlations and Climatic Changes in the Pacific. Tokyo, Japan. pp. 89–94.

Thompson, S. L. and E. J. Barron. 1981. Comparison of Cretaceous and present Earth albedos: implications for the causes of paleoclimates. J. Geol. 89: 143–167.

Tiffney, B. H. 1985. The Eocene North Atlantic land bridge: its importance in Tertiary and modern phytogeography of the northern hemisphere. J. Arnold Arb. 66: 243–273.

———. 1993. Fruits and seeds of the Tertiary Brandon lignite. VII. *Sargentodoxa* (Sargentodoxaceae). Am. J. Bot. 80: 517–523.

———. 1994. Re-evaluation of the age of the Brandon lignite (Vermont, USA) based on plant megafossils. Rev. Palaeobot. Palynol. 82: 299–315.

United States Department of Agriculture Yearbook. 1941. Climate and Man. U.S. Gov. Printing Office, Washington, D.C.

Urey, H. C. 1947. The thermodynamic properties of isotopic substances. J. Chem. Soc. Lond., 1947: 562–581.

———, H. A. Lowenstam, S. Epstein, and C. R. McKinney. 1951. Measurement of paleotemperatures and temperatures of the Upper Cretaceous of England, Denmark, and the southeastern United States. Geol. Soc. Am. Bull. 62: 399–416.

Vail, P. R. and J. Hardenbol. 1979. Sea-level changes during the Tertiary. Oceanus 22: 71–79.

Vail, P. R., R. M. Mitchum, Jr., and S. Thompson III. 1977. Seismic stratigraphy and global changes of sea level, part 4: global cycles of relative changes of sea level. In: C. E. Payton (ed.), Seismic Stratigraphy—Applications to Hydrocarbon Exploration. American Association of Petroleum Geologists, Tulsa, OK. Mem. 26, pp. 83–97.

Warrick, R. A., E. M. Barrow, and T. M. L. Wigley. 1993. Climate and Sea Level Change: Observations, Projections, and Implications. Cambridge University Press, Cambridge, U.K.

Watts, A. B. 1982. Tectonic subsidence, flexure and global changes of sea level. Nature 297: 469–474.

Webb, S. D. 1969. Extinction–origination equilibria in Late Cenozoic land mammals of North America. Evolution 23: 688–702.

———. 1977. A history of savanna vertebrates in the New World. Part I: North America. Annu. Rev. Ecol. Syst. 8: 355–380.

———. 1985. Main pathways of mammalian diversification in North America. In: F. G. Stehli and S. D. Webb (eds.), The Great American Biotic Interchange. Plenum, New York. pp. 201–217.

———. 1989. The fourth dimension in North American terrestrial mammal communities. In: D. W. Morris, Z. Abramsky, B. J. Fox, and M. R. Willig (eds.), Patterns in the Structure of Mammalian Communities. Museum of Texas Tech University, Lubbock, TX. Special Publ. 28, pp. 181–203.

———, R. C. Hulbert, Jr., and W. D. Lambert. 1995. Climatic implications of large-herbivore distributions in the Miocene of North America. In: E. S. Vrba, G. H. Denton, T. C. Partridge, and L. H. Burckle (eds.). 1995. Paleoclimate and Evolution, with Emphasis on Human Origins. Yale University Press, New Haven, CT. pp. 91–108.

Webb T., III. 1987. The appearance and disappearance of major vegetational assemblages: long-term vegetational dynamics in eastern North America. Vegetatio 69: 177–187.

Wendt, T. 1993. Composition, floristic affinities, and origins of the canopy tree flora of the Mexican Atlantic slope rain forests. In: T. P. Ramamoorthy, R. Bye, A. Lot, and J. Fa (eds.), Biological Diversity of Mexico: Origins and Distribution. Oxford University Press, Oxford, U.K. pp. 595–680.

West, R. M. and M. R. Dawson. 1978. Vertebrate paleontology and the Cenozoic history of the North Atlantic region. Polarforschung 48: 1103–1109.

Westgate, J. W. and C. T. Gee. 1990. Paleoecology of a middle Eocene mangrove biota (vertebrates, plants, and invertebrates) from southwest Texas. Palaeogeogr., Palaeoclimatol., Palaeoecol. 78: 163–177.

Wing, S. L., T. M. Bown, and J. D. Obradovich. 1991. Early Eocene biotic and climatic change in interior western North America. Geology 19: 1189–1192.

Wolfe, J. A. 1971. Tertiary climatic fluctuations and methods of analysis of Tertiary floras. Palaeogeogr., Palaeoclimatol., Palaeoecol. 9: 27–57.

———. 1975. Some aspects of plant geography of the northern hemisphere during the Late Cretaceous and Tertiary. Ann. Missouri Bot. Garden. 62: 264–279.

———. 1977. Paleogene floras from the Gulf of Alaska region. U.S. Geological Survey, Denver, CO, Professional Paper 997, pp. 1–108.

———. 1978. A paleobotanical interpretation of Tertiary climates in the northern hemisphere. Am. Sci. 66: 694–703.

———. 1987. Memorial to Harry D. MacGinitie (1896–1987). Ann. Missouri Bot. Garden. 74: 684–688.

———. 1994. Alaskan Paleogene climates as inferred from the CLAMP database. In: M. C. Boulter and H. C. Fisher (eds.), Cenozoic Plants and Climates of the Arctic. NATO ASI Series 1: Global Environmental Change. Springer-Verlag, Berlin. Vol. 27, pp. 223–238.

——— and D. M. Hopkins. 1967. Climatic changes recorded by Tertiary land floras in northwestern North America. In: K. Hatai (ed.), Tertiary Correlations and Climatic Changes in the Pacific. Sendai, Sasaki, Japan. pp. 67–76.

Wolfe, J. A., and R. Z. Poore. 1982. Tertiary marine and nonmarine climatic trends. In: Geophysics Study Committee (eds.), Climate in Earth History. National Academy Press, Washington, D.C. pp. 154–158.

Woodburne, M. O. (ed.). 1987. Cenozoic Mammals of North America: Geochronology and Biostratigraphy. University of California Press, Berkeley, CA.

Wright, H. E., Jr. 1987. Synthesis; the land south of the ice sheets. In: W. F. Ruddiman and H. E. Wright, Jr. (eds.), North America and Adjacent Oceans during the Last Deglaciation. DNAG (Decade of North American Geology) Geological Society of America, Boulder, CO. Vol. K-3, pp. 479–488.

Ye, Q., R. K. Matthews, W. E. Galloway, C. Frohlich, and S. Gan. 1993. High-frequency glacioeustatic cyclicity in the early Miocene and its influence on coastal and shelf depositional systems, NW Gulf of Mexico Basin. In: J. M. Armentrout, R. Bloch, H. C. Olson, and B. F. Perkins (eds.), Rates of Geologic Processes: Tectonics, Sedimentation, Eustasy and Climate: Implications for Hydrocarbon Exploration. 14th Annual Research Conference, Gulf Coast States Section, Society of Economic Paleontology and Mineralogy Foundation, Tulsa, OK. pp. 287–298.

Yemane, K. 1993. Contribution of Late Permian palaeogeography in maintaining a temperate climate in Gondwana. Nature 361: 51–54.

Zhilin, S. G. 1989. History of the development of the temperate forest flora in Kazakhstan, U.S.S.R. from the Oligocene to the early Miocene. Bot. Rev. 55: 205–330.

General Readings

Anderson, D. M. and R. S. Webb. 1994. Ice-age tropics revisited. Nature 367: 23–24.

Broecker, W. S. 1996. Glacial climate in the tropics. Science 272: 1902–1904.

Kerr, R. A. 1980. Changing global sea levels as a geologic index. Science 209: 483–486.

———. 1984. Vail's sea-level curves aren't going away. Science 226: 677–678.

———. 1987. Refining and defending the Vail sea level curve. Science 235: 1141–1142.

———. 1994. In tropical coral, signs of an ice age chill. Science 263: 173.

———. 1996. Ancient sea-level swings confirmed. Science 272: 1097–1098.

Lyle, M. 1992. Molecules record sea change. Nature 356: 385–386.

Warrick, R. A. 1993. Sea-level rise. EarthQuest, Office for Interdisciplinary Earth Studies, University Corporation for Atmospheric Research, Boulder, CO.

Notes

1. The standard selected was the ratio in a fossil belemnite from the Cretaceous Pee Dee Formation of North Carolina. The standard had a high isotopic Content and values are expressed as percent negative compared to the standard.

2. A linear regression is the conventional mathematical method for data derivation from values for constants used in expressions of simple (linear) relationships that involve two variables (in this case, fluctuating temperatures and changing isotope concentrations).

FOUR

Methods, Principles, Strengths, and Limitations

Methods of paleovegetation analysis can be grouped into two broad categories. Those that use plant microfossils for reconstructing terrestrial vegetation, past environments, migrations, and evolutionary histories constitute a part of paleopalynology that includes the study of pollen, spores, other acid-resistant microscopic structures, and phytoliths (distinctive, microscopic silicate particles produced by vascular plants). Those that use plant megafossils such as leaves, cuticles, cones, flowers, fruits, and seeds constitute paleobotany. Two important subdisciplines of paleobotany are dendrochronology (fossil woods) and analysis of packrat middens. The latter are sequences of nesting materials, constructed by packrats of the genus *Neotoma*, preserved in arid environments of the American southwest. The study of fossil fruits and seeds is a specialized field within paleobotany, and it is also used in studies on Quaternary vegetational history in the preparation of seed diagrams accompanying pollen and spore profiles from bog and lake sequences.

PALEOPALYNOLOGY (PLANT MICROFOSSILS)

Pollen and Spores

In 1916 Swedish geologist Lennart von Post demonstrated that pollen grains and spores were abundantly preserved in Quaternary peat deposits and could be used to trace recent forest history and climatic change (Davis and Faegri, 1967). The term palynology was subsequently introduced by Hyde and Williams in 1944 to include all studies concerned with pollen and spores. Paleopalynology has come to denote the study of acid-resistant microfossils generally, while pollen analysis designates those investigations dealing specifically with the Quaternary.

In the early 1950s researchers in the petroleum industry began to routinely apply paleopalynology to problems of stratigraphic correlation and the reconstruction of depositional environments in Tertiary and older strata (Hoffmeister, 1959). This added a practical dimension to a mostly academic pursuit and fostered interest in applied palynology and its use as a paleoecological research tool. This important development is reflected in the increased number of publications after about 1955 (Fig. 4.1; Jansonius and McGregor, 1996). As the history of other innovations might predict, there was a period of exuberant claims, isolated specialization, and exaggerated charges of deficiency in the method; but for palynology this seemingly inevitable period was mercifully brief. The different terminology, principles, and techniques involved in megafossil paleobotany and paleopalynology still result in specialization, but this limitation is frequently overcome by coordinated or collaborative projects, and an increasing number of practitioners work in both disciplines. Palynology is now one of several important sources of data available for decifering the complex interactions between environmental change and vegetational history.

The technique is based on several principles or generalizations. One is that essentially all plants produce either pollen grains or spores. This means that with few exceptions (e.g., Lauraceae, some aquatics), all taxonomic or ecologic groups of plants are potentially represented in the fossil record.

The second basis for the technique is that these pollen grains and spores are released to the atmosphere and fall as the pollen and spore rain. During transport, mixing occurs in the atmosphere and the settling particles provide a regional picture of the surrounding vegetation. In the early days of palynology it was voguish to point out that pollen and spores could be "blown in from anywhere" and that microfossil assemblages were an ecologically meaningless mixture of elements from various communities and distant

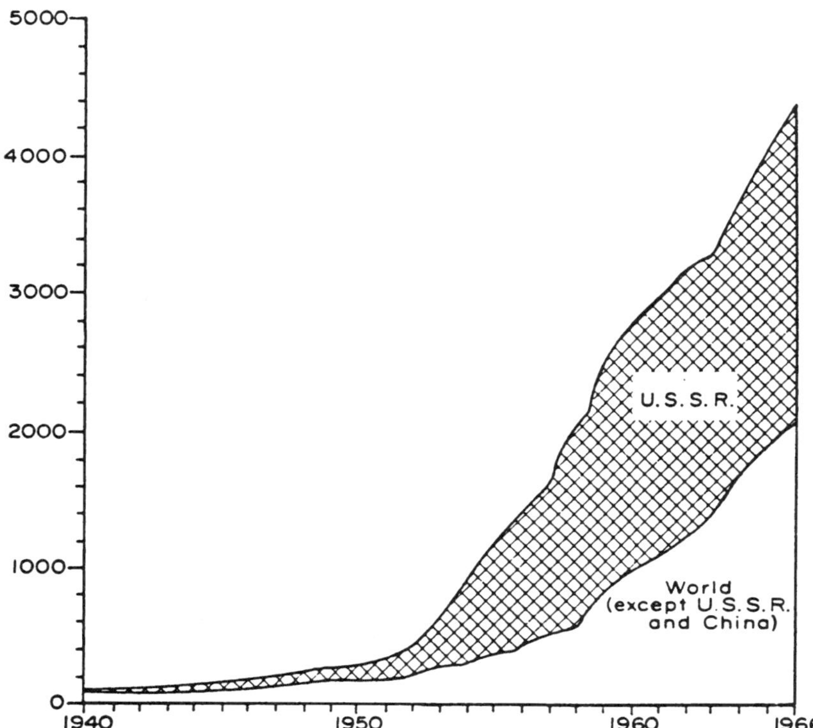

Figure 4.1. Number of papers on pre-Pleistocene palynology published between 1940 and 1966. Note the increase in the early 1950s correlated with the beginnings of extensive use of palynology in the petroleum industry. Reprinted from Kremp and Methvin (1968) with the permission of the Oklahoma Geological Survey.

regions. This is not the case; in fact, most pollen and spores fall within a few miles or less of the parent plant (see review in Farley, 1990). The exceptions, such as pine, are well known to palynologists (Kapp, 1961; Ritchie and Lichti-Federovich, 1967) and long-distance transport can be compensated for in pollen and spore assemblages.

The local to moderate aerial distribution of most pollen grains and spores can reveal the principal elements of upland associations growing some distance from the depositional basin and elements of understory vegetation. These plants are commonly absent or less well represented in the megafossil record. The pollen of several temperate genera has been recovered from Eocene floras in the Mississippi Embayment region, which on the basis of the megafossils, preserving mostly strand plants, was previously interpreted as coastal lowland tropical rain forest. The inland temperate component recorded by the pollen flora reveals a more complex landscape, a greater variety of habitats, and a warm temperate to subtropical climate.

Another positive result of the aerial distribution of pollen and spores is their incorporation into a variety of accumulating sediments. Like a synchronous ash fall, these microscopic particles settle into near-shore marine habitats and onto lake, bog, swamp, and floodplain surfaces. Thus, they may be recovered from sediments representing marine, brackish water, freshwater, and a variety of terrestrial habitats. Atmospheric mixing, rather than a deficiency, is one of the strengths of paleopalynology in reconstructing paleocommunities and past environments.

The third principle of paleopalynology is that pollen grains and spores are capable of being fossilized. Their durability is a result of the chemical composition of the wall. The chemistry is still not perfectly understood, but in pollen the exine (pollen wall) is probably composed of oxidative polymers of carotenoids and/or carotenoid esters called sporopollenin; the approximate empirical formula of which is $C_{90}H_{142}O_{27}$ (Brooks et al., 1971; Shaw, 1971). This substance is resistant to degradation under acid conditions but is destroyed by bases, extreme heat and pressure, mechanical corrosion, crystallization, enzymatic dissolution by fungi and bacteria, and by oxidation. As a result, well-preserved pollen and spores are not normally found in alkaline, igneous, metamorphic, or coarse-grained rocks because of sorting and oxidation; in sediments that accumulated very slowly or were agitated (aerated) for long periods before settling; or in deposits that accumulated under arid conditions or were exposed to postdepositional drying. The most favorable depositional environments are acidic, stagnant, poorly oxygenated, and/or with a rapid accumulation of fine-grained particles. Sediments deposited under these conditions are lithified into near-shore marine and deltaic shales; lacustrine volcanic shales, mudstones, claystones, and siltstones; and terrestrial (swamp) peat, lignite, and coal.

The variety of sediment types in which pollen and spores occur is the principal basis for their value in stratigraphic correlation. Most plants and animals are restricted to one kind of habitat (e.g., marine, freshwater, or terrestrial) and their fossils are found only in one or a few corresponding types of sediment. This makes correlation difficult between rocks of different lithologies on the basis of fossil content. The problem in geologic parlance is called transgressing the marine–nonmarine boundary. Pollen grains and spores can be used to correlate a wide variety of strata representing many kinds of depositional environments.

Because of their durability, acid-resistant plant microfossils are preserved in strata ranging in age from Precambrian to the present, and no measurable segment of time after ~3.8 billion years is unrepresented in the plant microfossil record. Also, palynomorphs can often be recovered from strata that lack plant megafossils, thus expanding the potential data base for studies on vegetational history.

The fourth principle is that techniques are available for the recovery of pollen and spores from the sediments. These techniques involve processing the samples through a series of acids (HCl, HF, HNO_3) that dissolve or oxidize various mineral (carbonate, silicate) and nonsporopollenin components (e.g., lignins), leaving a concentrated residue of acid-resistant microfossils (Gray, 1965; Traverse, 1988).[1]

The fifth principle or generalization is that pollen and spores vary in morphology such that plants can be recognized on the basis of their pollen and spore characters. The taxonomic level to which they can be recognized differs among various plant groups, but it is frequently to genus. There are some notable exceptions. In a few genera, such as *Cuphea* (Fig. 4.2) in the Lythraceae and *Bauhinia* in the caesalpinioid legumes (Fig. 4.3), many pollen types can be recognized to the species level. In many more instances, however, identifications can be made only to the family level [Cyperaceae (sedges), Poaceae–Gramineae (grasses)] or to groups of families (Chenopodiaceae, Amaranthaceae; Blechnaceae, Polypodiaceae, Pteridaceae). In families such as the Compositae–Asteraceae, Leguminosae–Fabaceae, Orchidaceae, Palmae–Areaceae, and others there is considerable pollen variation, but this is countered by the large size of the families, and only in certain instances can generic identifications be made. The occurrence of plant microfossils at localities that also preserve megafossils provides leads for the identification of unknowns in both groups.

Many fossil pollen grains and spores resemble those of modern genera back through about the Middle Eocene. In older deposits an increasing number of specimens are reminiscent of certain extant genera and families, but they cannot be assigned with certainty to modern taxa. The generic affinities of some earlier gymnosperm pollen and fern spores can be recognized, but with increasing age the likelihood that they are comparable biologically and ecologically to extant genera becomes more remote. The modern analog method for reconstructing paleoenvironments, based on fossil pollen and spore assemblages, is applicable to deposits from the present back through about the Middle Eocene.

The sixth principle is that plants are limited in their ecological tolerance. An important corollary to this fact is that therefore paleoecological conditions can be reconstructed on the basis of fossil pollen and spore assemblages.

An area where paleopalynology has proven especially useful is in providing a record of evergreen, herbaceous, annual, and suffrutescent plants (those that die back each year to a perrenial root stock). In most instances the leaves of these plants wither on the stem, or they are not shed in sufficient quantities or widely enough distributed to be extensively represented in the megafossil record. Fruits, seeds, and flowers are occasionally recovered and are of great value in biogeography and for tracing the evolutionary history of individual lineages. However, there are few such assemblages and the fossils are not usually found in sufficient quantities to make paleocommunity and environmental reconstructions. In contrast, many of these plants produce large amounts of pollen and spores, which provide a record of herbaceous and understory vegetation that otherwise would be meager to nonexistant. Such a record is important for providing an accurate inventory of plants from fossil floras and a true picture of plant diversity for a region, detecting the disturbance of vegetation associated with the early introduction of agriculture, tracing the time and direction of movement of ancient cultures into a new region, and determining the time and place of origin of important cultivars. The presence of pollen in the anthers of ancient flowers can also aid in evaluating taxonomic affinities (e.g., Crepet et al., 1992).

These principles impart to paleopalynology its particular set of strengths, as well as certain limitations. There are no morphological features of pollen or spores consistently associated with particular habitats or climate and therefore no "pollen or spore physiognomy" to aid in paleoenvironmental reconstructions. As a result, paleopalynological studies used for tracing vegetational history are limited to the modern analog method. As noted previously, pollen and spores can often be identified to the genus level, rarely to species, and in many instances only to a group of genera or higher taxonomic units. Thus, within the modern analog context, paleocommunity and paleoenvironmental reconstructions are further limited to inventories consisting mostly of generic identifications.

There are other aspects of the paleopalynological method that limit its precision in estimating the qualitative and quantitative characteristics of paleocommunities. Certain pollen and spore types lack sporopollenin or have thin exines that are destroyed in the fossilization and recovery processes, and they are frequently absent or underrepresented in fossil assemblages. Examples include the relatively delicate pollen of *Populus*, some Juncaceae (rushes), and Cyperaceae. Even when present, these grains are often folded and crumpled to the extent that identification and accurate counts are difficult. In some plant groups the wall is so delicate or the chemical composition is such that the grains are seldom preserved. The pollen of virtually all members of the order Zingiberales, with the exception of a few genera in the Costaceae (e.g., *Tapeinochilos*; Stone et al., 1979, 1981), is destroyed during fossilization or in the processing procedure. Of special significance in this regard

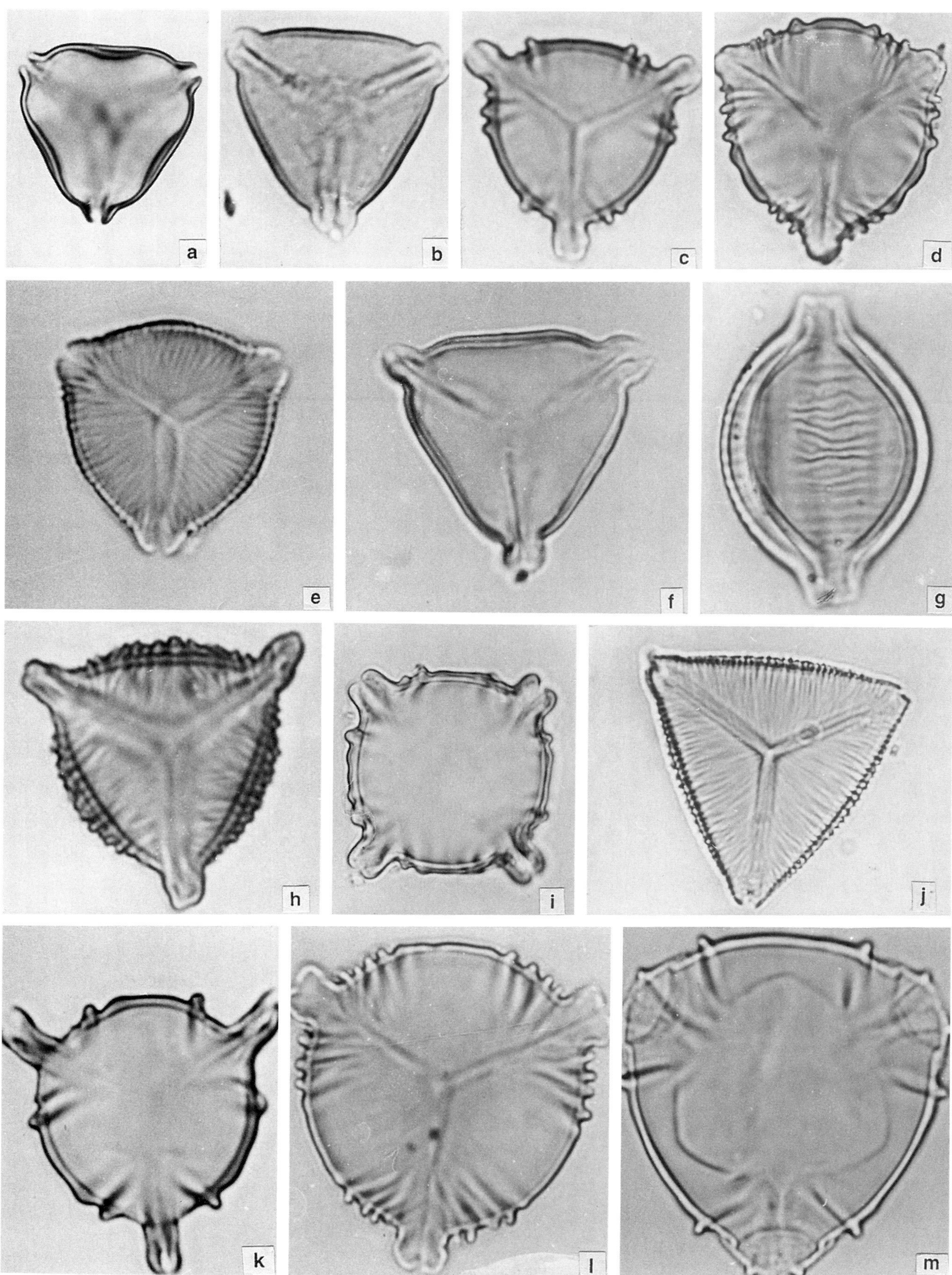

Figure 4.2. Light photomicrographs illustrating diversity in pollen morphology at the species level in *Cuphea* (family Lythraceae). (A) *C. patula* (25 µm), (B) *C. urbaniana* (28 µm), (C) *C. purpurescens* (28 µm), (D) *C. retroscabra* (32 µm), (E) *C. koehneana* (28 µm), (F) *C. reitzii* (32 µm), (G) *C. parsonsia* (25 x 32 µm), (H) *C. pseudosilene* (38 µm), (I) *C. ferrisiae* (32 µm), (J) *C. bustamanta* (36 µm), (K) *C. campestris* (32 µm), (L) *C. viscosa* (42 µm), (M) *C. glossostoma* (46 µm).

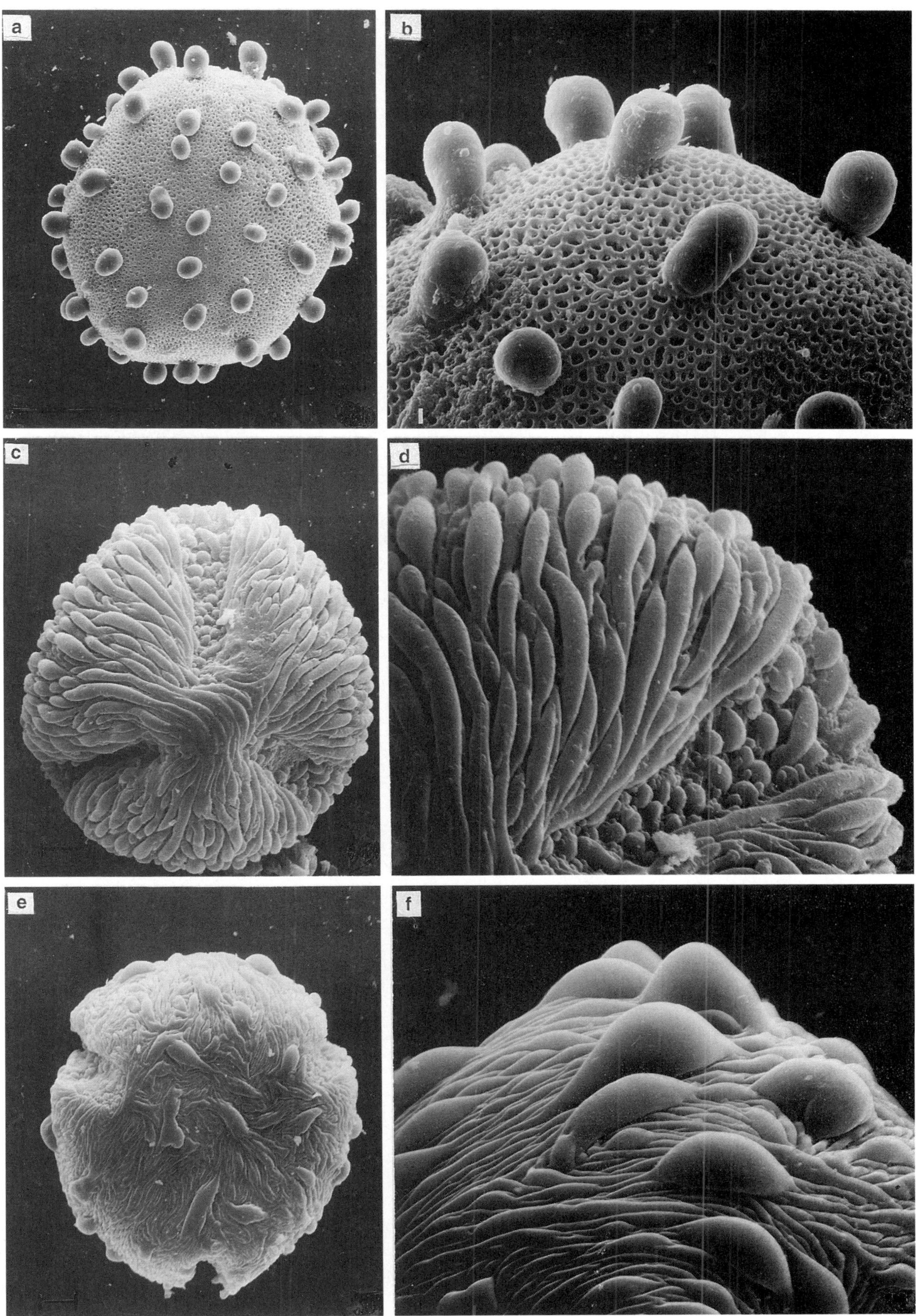

Figure 4.3. Scanning electron microscope photomicrographs illustrating diversity in pollen morphology at the species level in *Bauhinia* (family Leguminosae). (A,B) *B. aculata* (original magnifications x851, x2500), (C,D) *B. divaricata* (original magnifications x1750, x3000), (E,F) *B. dipetala* (original magnifications x1140, x3170).

is pollen of the family Lauraceae. It is a relatively minor component of the present temperate flora [*Lindera, Litsea, Nectandra, Sassafras, Persea, Umbellaria* (California laurel)], but the family is prominent in extant tropical vegetation and in the Early to Middle Cenozoic floras of North America. During the Late Cretaceous and Early Tertiary, when tropical communities extended into north temperate regions, or when exotic terranes were transported northward (e.g., the Miocene Carmel flora of California; D. I. Axelrod, personal communication, 1996), members of the Lauraceae were present, as documented by the megafossil record (Dilcher, 1963; Herendeen, 1991), but they are not found in plant microfossil assemblages. In contrast, thick-walled spores and pollen are often overrepresented in fossil assemblages. Many fern spores and the heavy-walled pollen of the Malpighiaceae and other groups are occasionally found to the near exclusion of thin-walled types. These examples collectively constitute differential preservation and this is one of the limitations of the paleopalynological method.

Another facet of the technique that limits quantitative reconstructions is differential production. Plants vary widely in the amount of pollen and spores produced, which complicates the estimation of the relative abundance of various components of the paleocommunity from microfossil assemblages. Differential production can be compensated for to some extent in Quaternary deposits by knowing the approximate amounts of pollen produced by modern plants. Using figures cited in Faegri et al. (1989), some estimates are

Phoenix dactylifera (date palm): 89,000 grains/anther
Cannabis (hemp): 70,000 grains/anther
Betula: 10,000 grains/anther
Acer: 1000 grains/anther
Linum catharticum (flax): 100 grains/anther
Malva: 64 grains/anther
10-year-old branch of *Pinus sylvestris* (scotch pine): 350 million grains
10-year-old branch of *Picea, Betula, Quercus*: 100 million
10-year-old branch of *Fagus sylvatica*: 28 million
Spruce forests of southern Sweden: 75,000 tons/year
Pinus sylvestris: 10–80 kg/ha/year (=~30,000–280,000 grains/cm^2/season)
Rybinsk Reservoir (USSR) vegetation: 6 kg/ha/year

Another factor in assessing the relative abundance of elements in a paleocommunity from pollen and spore evidence is the spatial arrangement of the species. A few individuals growing marginal to the basin of deposition may leave a numerical record similar to that from a larger number of individuals growing farther away. In Pleistocene and Holocene deposits accurate estimates of relative abundance and spatial arrangement can be obtained from modern pollen rain studies from different vegetation types, a knowledge of the ecology and composition of comparable modern associations, and by numerical methods of analysis such as pollen influx and absolute pollen frequencies (Chapter 8). For older sediments only general approximations of the composition, abundance, and arrangement of associations within plant formations can be gained from plant microfossil assemblages. The pollen of anemophilous (wind-pollinated) species is numerically overrepresented in pollen assemblages, while entomophilous (insect-pollinated) species are underrepresented. For example, all of the common pollen types recovered in one study of the latest Eocene to earliest Oligocene Florissant flora of Colorado were wind pollinated (Fig. 4.4).

In addition to limitations, there are other aspects of the method that constitute potential sources of error. These have been addressed extensively in the literature and are summarized by Faegri et al. (1989) and Traverse (1988). Contamination is the most important and it can occur during and after deposition and during the collection and processing of samples.

Redeposition is the process whereby sediments are eroded by wind or water, carried into a new depositional basin (allochthonous material), and deposited along with newly accumulating sediments (autochthonous material; Frederiksen, 1989). The exotic material usually represents a different age and often a different kind of environment. Some depositional conditions are particularly susceptible to contamination through redeposition. Deltas receive sediments transported by rivers that often erode a variety of substrates, and the resulting shales can contain a mixture of palynomorphs of widely different ages.

There are several ways to detect redeposited specimens. One is by noting the quality of preservation. Specimens that have been redeposited are frequently corroded or fragmented. The color of the microfossils usually darkens with age, and the presence of noticeably darker or lighter grains than is typical for the overall assemblage suggests possible contamination. The petroleum industry uses color charts to estimate the thermal maturity of organic matter (e.g., Pearson, 1984). At a certain critical temperature, referred to as the "window of petroleum formation," droplets of oil begin to form from kerogen. The older and most thermally altered specimens are darker, while younger palynomorphs are progressively lighter in color. The color charts, designed to estimate thermal maturity, also give a rough estimate of the relative age of different components in the sample. Another clue to redeposited pollen and spores is the morphology of the specimens. If palynomorphs of pre-Middle Eocene age have been redeposited into younger strata, experienced palynologists familiar with the section can often detect the contaminants through a subtle combination of morphological features atypical of younger specimens. These three characteristics of preservation quality, color, and morphology provide clues to possible contamination. A knowlege of the depositional environment for different rock types, the stratigraphic range of various palynomorphs, and a familiarity with the fossil content of adjacent strata can also aid in assessing the likelihood of redeposition. Cracks, fissures, and large sinks allow sediments

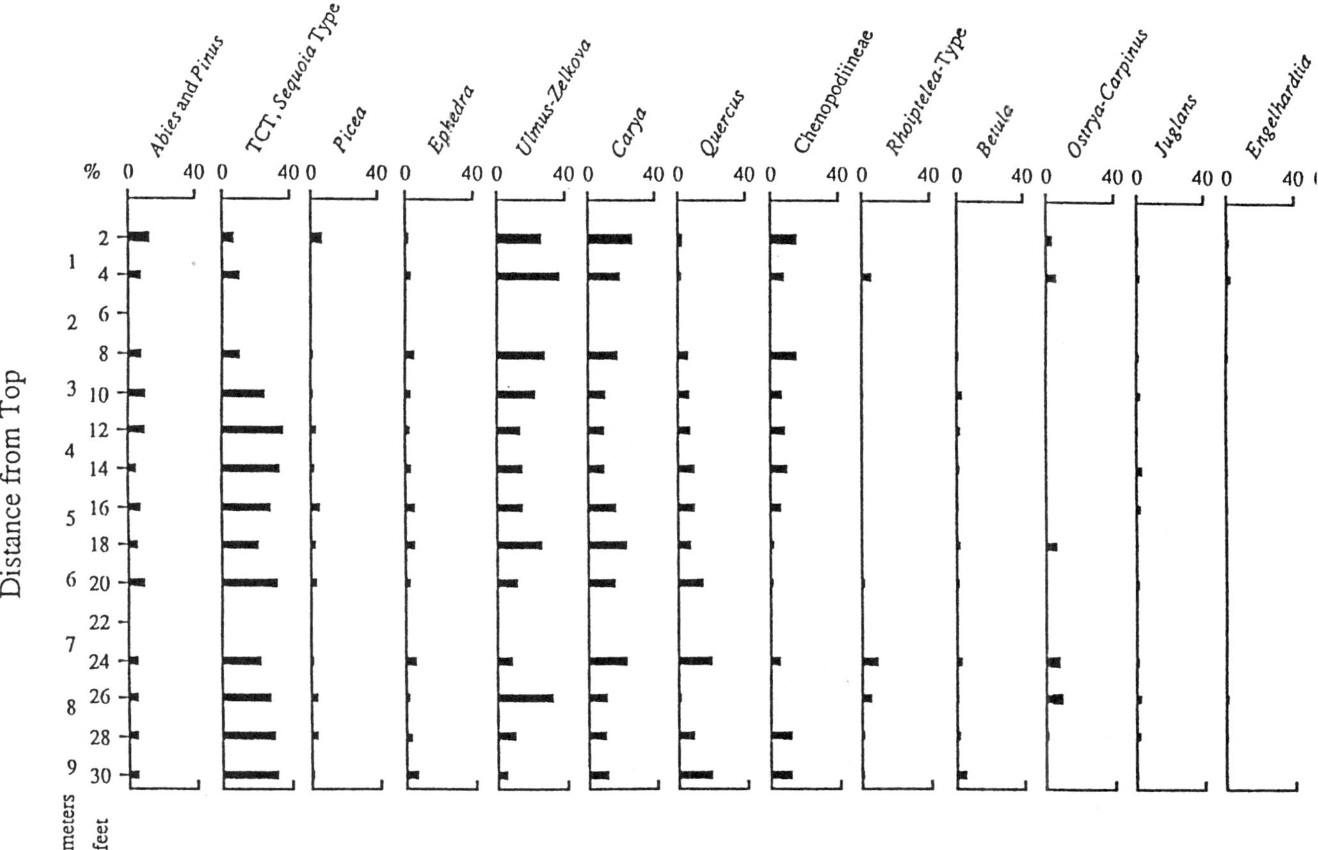

Figure 4.4. Chart of common pollen and spore types from the Late Eocene Florissant Beds National Monument, Colorado. All common pollen taxa are from wind-pollinated woody plants. Reprinted from Leopold et al. (1992) with the permission of Princeton University Press.

of one age to infiltrate older strata. The presence of these structures is usually revealed by abrupt horizontal changes in lithology.

A more subtle source of potential contamination, until it was brought to the attention of palynologists, was from drilling mud thinners (Traverse et al., 1961). At one of the early meetings of the American Association of Stratigraphic Palynologists I was asked to present a summary of work on the Tertiary history of vegetation in northern Latin America. One facet of this presentation dealt with the progressively later appearance of northern temperate pollen types into Mexico, Central America, and northern South America (*Alnus, Ilex, Myrica, Quercus,* and others). At the end of the presentation a palynologist working for one of the oil companies informed the audience that the purported trends were figments of the imagination and that these temperate types were widespread during the entire Cenozoic throughout the tropics of the western hemisphere. That has proved to not be the case. One possible explanation for the confusion was down-hole contamination of cuttings and side-wall contamination of cores from drilling mud thinners. This source of error was not universally recognized in the early days of palynology. During drilling operations a mud or flux is circulated into the drill hole as a lubricant to prevent clay particles from flocculating and making the mud too thick to circulate and to seal the walls of the hole. The thinning agent is manufactured from *Rhizophora* (mangrove) bark, quebracho (*Schinopsis lorentzii*) wood extract, and oxidized coal from several sources, including the Tongue River Formation of the Paleocene Fort Union Group of North Dakota. Traverse et al. (1961) recovered an estimated 600,000–4,000,000 fossil pollen and spores per gram from these muds. This constitutes a potential source of contamination in samples derived from commercial drilling operations and is so recognized by palynologists in the petroleum industry.

Phytoliths

By adjusting the processing procedure to exclude HF, a wide variety of plant and animal microfossils composed of silicate minerals often can be extracted from sedimentary rocks. Plant material of silicate composition representing terrestrial vegetation includes the phytoliths (Piperno, 1988), which are becoming more widely recognized as useful in paleoenvironmental studies, especially for the Quaternary and particularly for sediments associated with archeological sites (Rovner, 1971). Phytoliths are micro-

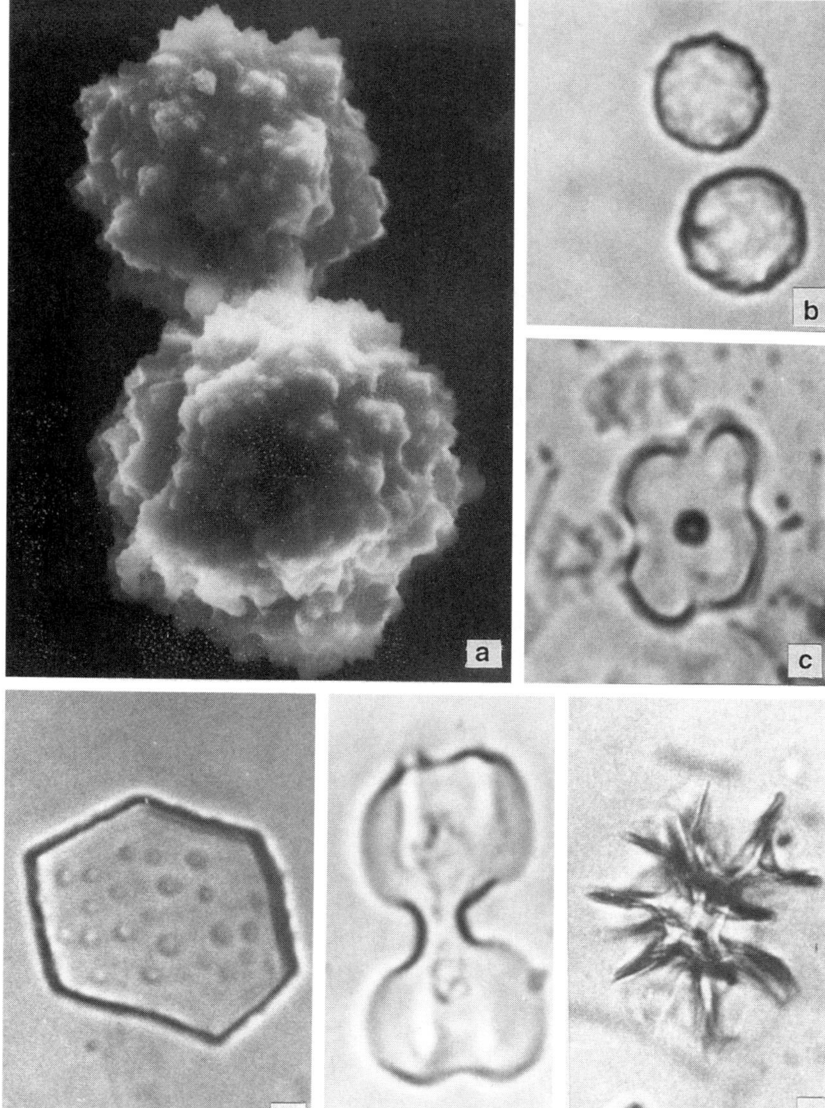

Figure 4.5. Phytoliths. (A) Spherical nodular phytolith from *Maranta arundinaceae* (9–18 µm), (B) spherical smooth phytolith from seeds of *Hirtella triandra* (6–13 µm), (C) variant 1 cross-shaped phytolith from *Zea Mays*, (D) seed phytolith from *Bursera simaruba*, (E) variant 5/6 dumbell phytolith from *Cenchrus echinatus*, (F) irregular pointed phytolith from *Odontocarya tamoides* (36–51 x 36–46 µm). Photographs courtesy of Dolores Piperno (see Piperno, 1988).

scopic particles of hydrated silica produced in the cells of many vascular plants. They vary in morphology among different plant groups (Fig. 4.5), are released upon decay of the parent plant into soils, and eventually become part of the sedimentary record. Phytoliths are common in the cells of grasses and other plants and have been used to study shifts in the North American deciduous forest–grassland ecotone during the Holocene; to identify savanna grasses in lake sediments from East Africa; to confirm cold paramo (alpine tundra) in pollen diagrams from the highlands of Colombia; and to document wet–dry cycles in the Holocene of Central America from aeolian particles (including phytoliths) recovered from DSDP and ODP cores. [See Piperno (1988) for references.) Phytoliths have been found in fossil plants of Miocene age in North America (Smiley and Huggins, 1981; Thomasson, 1983, 1984; Thomasson et al., 1986), but in kinds and abundance they are best represented in sediments of Quaternary age.

The study of fossil pollen, spores, and phytoliths constitutes one method for tracing the history of North American vegetation. The technique has inherent strengths and limitations, and a balanced appreciation of both is necessary to most effectively use the technique as a paleoecological research method.

PALEOBOTANY (PLANT MEGAFOSSILS)

The factors that determine the representation of megafossils in the geologic record are different from those that determine the representation of microfossils. In the Late Miocene Succor Creek flora of southeastern Oregon, approximately 26% of the taxa were found only as megafossils or as microfossils. Chaney (1959) presents quantitative data from a study of plant remains from 19 species accumulating in 42 pools along Redwood Creek in Muir Woods

National Monument, California. The most important factor that controlled entrance into the fossilization process was distance from the depositional basin. The four species comprising 91.26% of the leaves and fruits grew an average of 5.25 ft from the pools. The next group of four species, making up 8.34% of the specimens, grew at an average distance of 25.25 ft. Even the 11 fewest species were all within 50 ft.

The second most important factor was durability of the organ entering the pools. The leaves of *Rhododendron* and *Corylus rostrata* var. *californica* (hazel), which have a delicate spongy mesophyll, were abundant in December soon after the primary leaf fall, but nearly all had disintegrated by March. Both are rare in most Tertiary floras in the American west even when their associates are common to abundant. The leaves of *Acer negundo* (box elder) are also delicate and were not an important component of the litter even under the parent tree. The leaves of *Alnus rubra*, *Acer macrophyllum*, and *Lithocarpus densiflora* are more durable, were abundant in the pools, and are well represented in the fossil record.

A third factor was the weight-to-area ratio of the organ. In essence, lighter specimens that can travel farther were usually better represented than heavier ones. Also important were the height of the plant and its size (which determines the capacity for producing leaves and reproductive structures). Organs from large trees made up 94.25% of the specimens in the pools, small trees and shrubs 5.68%, and herbs 0.04%. Leaves of deciduous plants were better represented than those of evergreens.

The Late Cretaceous and Tertiary megafossil record for North America includes an abundance of leaf material of midsize to tall deciduous trees and large shrubs from mesic environments that have light, broad, durable leaves, and grow near the basins of deposition. The representation of these plants is sufficient to provide a basis for estimating the relative abundance and spatial arrangement of elements in these habitats and to allow environmental reconstructions with the modern analog and foliar physiognomy methods. Less material is available for small understory shrubs, herbs, and evergreens; plants with heavy leaves and reproductive structures; those growing removed from the depositional basin; and those growing in dry to arid habitats where conditions are not suitable for preservation. The fewer number and smaller size of these floras and the underrepresentation of arid elements in larger floras produces a more restricted data base for estimating their presence, abundance, arrangement, and environmental implication in a statistically meaningful sense. This limitation is compensated for, to some extent, by the large number of floras available in western North America, the midcontinent region, and the Mississippi Embayment. Fruits and seeds have a diagnostic morphology (Corner, 1976) and are occasionally abundant at some localities, like the Middle Eocene Clarno flora of Oregon, and seeds are particularly important in Quaternary peat deposits. Consideration of plant microfossils in conjunction with megafossils can also minimize some of the biases in the paleobotanical record.

Modern Analog Method

Two approaches are available to analyze fossil material for the purpose of reconstructing paleoenvironments. One is the modern analog method or nearest living relative (NLR) analogy. This approach involves identifying the fossils by comparing them to modern species having the most similar morphology. Using leaves as an example, this comparison is based on venation, size, shape, margins, petioles, accessory structures (bracts and stipules; extrafloral nectaries, Pemberton, 1992), and where possible, variations in these features through study of populations of the fossil material. The morphology and variation may be compared with that exhibited by cleared leaves (e.g., Todzia and Keating, 1991) and X-ray radiograms (Wing, 1992) of the presumed modern analog. Microscopic study of the leaf surface reveals details of the venation pattern and if cuticles are preserved, these may be examined with SEM to reveal additional morphological features such as stomata and trichome types and arrangement. A standardized set of definitions and terminology was developed by Hickey (1973, 1979; see also Dilcher, 1974; Hickey and Wolfe, 1975; Wolfe, 1989). Surveys of leaf venation patterns are under way to identify features characteristic of the lineage, rather than just ones that are the most obvious (e.g., Klucking, 1986; Roth, 1992). In the older literature identifications were often made on the basis of limited, fragmentary, poorly preserved specimens; and the level of accuracy has proven to be less than spectacular. The fossil leaf *Carya bendirei* has been described as four different species of *Salix*; two species of *Carya*, *Rhus*, and *Prunus*; and as *Magnolia*, *Cassia*, *Celastrus*, *Aesculus*, *Hicoria*, *Juglans*, *Ptelea*, *Quercus*, and *Arbutus*. The old procedure of "leaf matching" has been replaced by a critical examination of populations of well-preserved specimens utilizing more sophisticated techniques, with a corresponding increase in the accuracy of the identifications.

Once the specimens have been identified, the altitudinal and meterological conditions within the range of the modern species are recorded and the procedure continued for all elements of the fossil flora. Before paleoenvironments can be reconstructed, however, the assemblage must meet several criteria. First, the flora must be moderately large in terms of the number of species present. This is defined in relation to the size of other floras in the region; and for North America, where extensive Late Cretaceous and Cenozoic floras exist, moderately large would mean a minimum of about 30 species. Second, there must be a consistency in climatic requirements among several, but not necessarily all, members of the assemblage. In essence, this means that paleoenvironments are estimated on the basis of groups of species and not on individual "key" components. Third, the ecological heterogeneity of the flora must be accommodated through a reconstructed paleophysiography consistent with geological evidence from the region.

To properly evaluate the ecological import of the procedure, it is important to note that at this point there have been no definitive paleoclimatic reconstructions. Rather, a preliminary model of climate and physiography has been established for the flora. These preliminary reconstructions must be assessed within the context of results from independent lines of evidence. Other floras in the region, fossil faunas, and global paleotemperature and sea-level curves provide estimates of contemporaneous conditions and long-term trends within which the preliminary paleoenvironmental model can be evaluated. For North American floras at least some ancillary data are usually available. If such data are inconsistent with the botanical evidence (e.g., inconsistency in estimates derived from fossil mammalian faunas and the floras), attempts are made to provide a plausible explanation. However, for the Late Cretaceous and Cenozoic paleontological record of North America such situations are rare, and a hallmark of most current investigations is an overall consistency between results derived from the several lines of inquiry.

The modern analog method assumes that morphological similarity of isolated plant parts is indicative of taxonomic and ecological similarities to the extent that when the entire assemblage of medium to large sized fossil floras with at least moderate diversity are involved, enough of the elements will be comparable to make generalized reconstructions possible. Within limits, the assumption is valid, but there are obvious shortcomings inherent in the method. Many plant fossils resemble modern genera back to about the Middle Eocene, beyond which an increasing number represent extinct taxa. Therefore, the time interval for which the modern analog method can be used is limited mostly to the Middle Eocene to the present. There is also the possibility of new forms evolving that have different ecological parameters but retain the same or similar morphological features as the parents among those parts likely to be present in the fossil record (Chapter 3). Another complication is the virtual certainty of some change in the range and composition of plant formations and associations after each period of climatic reshuffling. As mentioned previously, where an organism does occur is not always the same as where it can occur, simply because it may not have had sufficient time or opportunity to expand to its maximum geographical and ecological range. Vegetational histories and paleoenvironmental reconstructions based exclusively on the morphology, distribution, and ecological parameters of presumed modern analogs tend to obscure the uniqueness of past vegetation types and dampen the curve of vegetation dynamics and past environmental change.

Foliar Physiognomy

The second approach to the reconstruction of paleoclimates is through the method of foliar physiognomy. It is a technique that can be used independently and, after the Middle Eocene, in conjunction with the modern analog method for studying leaf floras.

Physiognomy is the external morphology of a structure, organism, or community. With reference to leaves, and specifically to those features most relevant to fossil material, this morphology includes size, shape, texture, presence or absence of special structures (e.g., drip tips), and characteristics of the leaf margin. Leaf size is described by the Raunkiaer (1934) system as modified by Webb (1959) and Wolfe (1993).[2] It includes several size classes defined as follows:

Leptophyll: maximum size 0.25 cm^2
Nanophyll: 2.25 cm^2
Microphyll: 20.25 cm^2
Notophyll: 45 cm^2
Mesophyll: 182.25 cm^2
Macrophyll: 1640.2 cm^2
Megaphyll: no maximum size

Leaf shape includes linear, lanceolate (Fig. 4.6A), elliptical (Fig. 4.6B), ovate, rotund, spatulate, and cuneiform (wedge shaped; for a more complete listing of leaf forms and terminology, see Hickey, 1973, 1979; Radford, 1986; Stearn, 1983; Todzia and Keating, 1991; see illustrations in Wolfe, 1993). Leaf texture is often difficult to determine from fossil material and is commonly designated as thin, medium, or coriaceous (sclerophyllous). In general, large thick leaves are common among evergreens and are most abundant in megathermal and mesothermal temperature regimes; thinner ones in the mesophyll to notophyll size class are usually from deciduous plants and occur in mesothermal to microthermal climates. Drip tips (extended acuminate; Fig. 4.6B) are a prolongation of the midvein and associated laminar material beyond the terminus of the leaf and aid in draining excess water from the leaf surface. In tropical climates this is important in keeping epiphytic algae, fungi, bryophytes, small ferns, gametophytes, and other structures from developing to the point of causing leaf damage and interfering with photosynthesis. The most common types of leaf margin are entire (Fig. 4.6A), lobate, crenate (Fig. 4.6B), dentate, serrate (Fig. 4.7), and erosus (irregular).

It has long been observed that some of these features are associated in a general way with climate (Bailey and Sinnott, 1915, 1916; see also Givnish, 1979, 1986; Givnish and Vermeij, 1976). For example, large leaves are more prevalent among plants growing in the tropics and in temperate zones in the shaded understory and along stream banks. Smaller leaves are more frequent in open, climatically or physiologically drier habitats (e.g., tundra, hot and cold deserts). There is less obvious correlation between leaf shape and climate. Drip tips are a common but not a universal feature of evergreen leaves in tropical understory vegetation and are not found in plants of cold, mesic, or arid habitats.

A leaf-size index has been developed to compare leaf

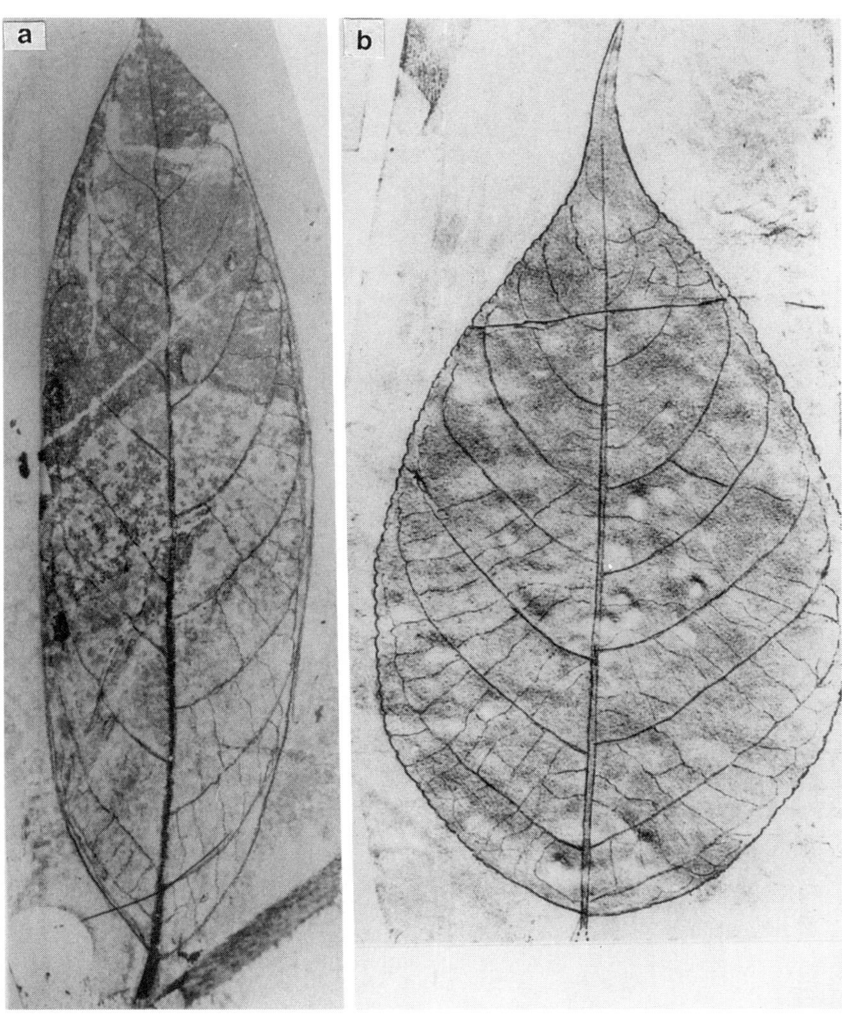

Figure 4.6. Fossil leaves illustrating size, entire/nonentire margins, and drip tip (or extended acuminate; see text for discussion). (A) *Eugenia americana*, Green River flora (late Middle Eocene), Colorado and Utah (MacGinitie, 1969, pl. 23, fig. 2, x1). (B) *Celastrus typica*, Florissant flora (latest Eocene), Colorado (MacGinitie, 1953, pl. 44, fig. 5, x1). Reprinted with the permission of the Carnegie Institution of Washington.

size of fossil and modern vegetation and between fossil floras (Wolfe and Upchurch, 1987):

$$\frac{4 \times MC + 3 \times ME + 2 \times NO + 1 \times MI - 100,}{2}$$

where MC is the percent of macrophyllous (or larger) species, ME is the percent of mesophyllous, NO is the percent of notophyllous, and MI is the percent of microphyllous. A low leaf-size index (e.g., 12–20; microphyllous) suggests cold or subhumid (dry) climates and full sunlight; a high index (e.g., 60–75; megaphyllous) reflects warm, moist, aseasonal climates and/or reduced (shaded) light levels.

Terms and values have also been assigned to temperature regimes (e.g., Wolfe, 1987):

Microthermal: MAT < 13°C
Mesothermal: MAT = 13–20°C
Megathermal: MAT > 20°C

The closest association between foliar physiognomy and climate is in the type of leaf margin (LMA). For purposes of estimating paleoclimates, the leaf margin categories are simplified to entire and nonentire. Consider the original tabulation of Bailey and Sinnott (1915) comparing leaf margins from temperate (east-central North America) and tropical (Amazon lowland) habitats:

Mesophytic-Cold-Temperate	% Entire	% Nonentire
Trees	10	90
Shrubs	14	86
Woody	13	87
Moist-Lowland-Tropical		
Trees	90	10
Shrubs	87	13
Woody	88	12

These observations have been further developed by Dilcher (1973), Dolph (1978, 1984), Dolph and Dilcher (1979), especially Wolfe (1971, 1977, 1978, 1993), and Wolfe and Hopkins (1967). Table 4.1 shows the percentage of entire-margined leaves from selected vegetation types

Figure 4.7. Fossil leaves illustrating comparatively small size and serrate margins. (A) *Betula vera*, Trout Creek flora (Late Miocene, Oregon; Graham, 1965, pl. 9, fig. 2, x1). (B) *Ulmus speciosa*, Kilgore flora (Late Miocene), Nebraska. Reprinted from MacGinitie (1962, pl. 10, fig. 4, x1) with the permission of the University of California Press.

Table 4.1. Percentages of species that have entire margins in some modern floras.

Flora	Percent	Vegetation
Brazil, lowland	88	Tropical rain forest
Malaya	86	Tropical rain forest
Philippine Islands, 200 m	82	Tropical rain forest
Ceylon, lowland	81	Tropical rain forest
Manila	81	Tropical rain forest
East Indies	79	Tropical rain forest
Philippine Islands, 450 m	76	Tropical rain forest
West Indies	76	Tropical rain forest
Hawaii, lowland	75	Paratropical rain forest
Ceylon, upland	74	Paratropical rain forest
Philippine Islands, 700 m	72	Paratropical rain forest
Hong Kong	72	Paratropical rain forest
Hainan, lowland	70	Paratropical rain forest
Philippine Islands, 1100 m	69	Montane rain forest
Taiwan, 0–500 m	61	Paratropical rain forest
Ceylon, upland	60	Paratropical rain forest
Hawaii, upland	57	Paratropical rain forest
Hainan, upland	55	Subtropical forest
Taiwan, 500–2000 m	45	Subtropical forest
Mixed mesophytic forest, China	30	Warm temperate forest
Northern China plain	22	Deciduous oak forest
Manchuria	10	Mixed no. hardwood forest

Adapted from Dilcher (1973) and Wolfe (1977).

extending from tropical rain forests to mixed northern hardwood forests. The range is from 88% in a tropical rain forest from the lowlands of Brazil to 10% in a northern hardwood forest in Manchuria. The closest correlation is between entire-margined leaves and MAT.

Wolfe and Upchurch (1987) believe that MAT can be reconstructed with an accuracy of ±5% if the assemblage has 30 or more species and within ±10% if 20–29 species are present. As they and others have recognized, however, the translation of quantitative measurements of modern leaf features into paleoclimates is complicated. Leaf margin types correlate with moisture and temperature, and it is difficult to sort out their individual effects on foliar physiognomy (Dilcher, 1973). Modifications of leaf texture, size, and margin are also a reflection of various defense strategies against herbivory (Brown and Lawton, 1991). The overprinting effect of genetic control may retard the response of leaf morphology to changes in climate to varying degrees among different lineages. For example, members of the mostly temperate family Tiliaceae in the tropics have serrate leaf margins and cordate bases, while species of the tropical Annonaceae and Lauraceae in temperate regions maintain entire-margined leaves (Wolfe, 1993). Assemblages from disturbed habitats (e.g., channel and lacustrine deposits) often contain a disproportionally high percentage of species with nonentire-margined leaves (Burnham, 1994). If the fossil flora represents a sere (a stage or phase) in a successional sequence leading toward climax vegetation from a series of disturbances (volcanism, flooding, etc.), changes in leaf margin percentages can be incorrectly ascribed to climatic trends. Similarly, taphonomic (depositional) factors[3] can select for different leaf sizes (Wolfe and Upchurch, 1987). For example, in tracing regional vegetational histories in the American west, it is important to recognize that many Early Eocene floras are preserved in fluvial (stream) or paludal (swamp) sediments (e.g., the Wilcox), while extensive Late Eocene floras (Florissant, Green River) are from lacustrine (lake) beds. Among other differences, streamside vegetation tends to have larger leaf size due to shading and high humidity than vegetation growing on exposed slopes.

After a major climatic change there is a lag time between the change and the introduction of species compatible with the altered conditions, the establishment of climax vegetation, and especially in the appearance of evolutionarily modified leaf types from in situ or adjacent progenitors. This affects the modern region and vegetation type selected as the standard for a particular leaf margin percentage–climate relationship and, hence, paleoclimatic reconstructions based on leaf physiognomy. Wolfe (1981) believes the deciduous forest formation of eastern North America is too depauperate for calibrating leaf margin percentages to climate due to the elimination of tropical forms after Late Eocene times and to increasing isolation from western European floras after about the Middle Eocene. Also, some larger leaf types potentially present in a deciduous forest are absent because of Arctic cold fronts that began moving across eastern North America in the Late Neogene. Instead, the mesic vegetation of eastern Asia was selected as the standard for relating leaf margin percentages to climates. By that standard, an increase of 3% entire-margined species equals an increase of 1°C in MAT, and an entire leaf margin percentage of about 60% equals the 20°C isotherm (Upchurch and Wolfe, 1987; Wolfe, 1979; Wolfe and Upchurch, 1987). Dolph (1990; see also Dolph, 1979; Dolph and Dilcher, 1979) believes that cold fronts and monsoon fluctuations have affected the recent vegetation of Asia. For Late Cretaceous analyses Wolfe and Upchurch (1987) selected a southern hemisphere scale because of the rarity of deciduous angiosperms and the prominence of conifers in both floras. Using the southern hemisphere scale, an increase of 4% of entire-margined species equals an increase of ~1°C in MAT, and an entire leaf margin percentage of ~68–70% equals the 20°C isotherm. There is now a question as to whether the modern southern hemisphere forests are a suitable analog for calibrating northern hemisphere Cretaceous leaf margin percentages.

The use of formulas and the assignment of specific climatic values to leaf size and leaf margin percentages may imply that a high degree of precision is possible by this method. In fact, a trend was developing to assign increasingly exact climatic parameters to percentages of entire-margined leaves. The lack of precise correspondence between the leaf composition in modern accumulating sediments compared to the surrounding vegetation (e.g., in lianas from an Indiana lake; Dolph, 1984) and other data suggest that single-variate leaf physiognomy alone provides a useful, albeit general, estimate of climate. Boyd (1990, 1994) believes that two different vegetation types can usually be inferred from most leaf margin ratios, "indicating the use of leaf-margin data for determining paleoclimates is not as precise as has been claimed" (1990, p. 194). Greenwood (1991, 1992) has found the strongest and most consistent climatic signal in leaf assemblages emerges only when taphonomic factors are adequately considered and when these signals are enhanced by multivariate statistical analysis of a large variety of community types analyzed from several floras. As noted by Wolfe (1993) much of the poor correlation between leaf physiognomy and climate can be attributed to single character analyses (e.g., leaf margin) that have not been subjected to statistical tests. When multiple characters are scored and analyzed by multivariate analysis (Kovach, 1989; particularly correspondence analysis), considerably better correlation is achieved.

Clearly, a synergistic approach is essential in paleoecology (Wing, 1987). Nonetheless, a ninefold difference in entire leaf margin percentages between the mesic and tropical climates as listed in Table 4.1 clearly documented foliar physiognomy as potentially an important methodology for assigning quantitative values to Late Cretaceous and Cenozoic climates and climatic change. In particular, it can provide a way of estimating paleoclimates that is independent of taxonomic identification or the assumption that the ecological requirements of species have remained constant over millions of years. It can further allow estimates of paleoclimate to be made from assemblages older than Middle Eocene where application of the modern analog method becomes increasing limited. Required was more sophisticated statistical analyses to achieve closer correspondence between leaf margin percentages and climate.

The statistical program CLAMP (Wolfe, 1990, 1993, 1994) better quantifies the relationship between leaf characters and climate. The program utilizes 29 character states of modern leaves from 25 to 30 species from 106 sites in North America, the Caribbean region, and Japan; Canonical Correspondence Analysis relates the samples to one another, to the character states, and to selected climatic parameters with a moderate to high degree of accuracy (Wolfe et al., 1996). A similar tabulation is made of leaf characters in fossil floras and correspondence analysis, together with multiple regression tests, provide an estimate of the climate (mainly temperature) for the fossil assemblages with an error range of ~2°C. In its present stage of development, Basinger et al. (1994) believe the program underestimates temperature and precipitation. In general, values of MAT from CLAMP are lower than those from LMA (Wolfe, 1994). Nonetheless, the development of LMA and CLAMP are innovative and important contributions to paleoecology, and their continuing refinement and broader application will add greater quantitative consistency to estimates of past climates. As noted by Wolfe and Hopkins, "[L]eaf-margin analyses of large, diversified fossil floras used in conjunction with consideration of their floristic composition, can yield reliable indications of climatic change" (1967, p. 68).

An application of fossil plant data that utilizes LMA and CLAMP is in estimating paleoelevations. This has traditionally been done by compiling the present-day altitudinal range of presumed modern analogs and proposing a paleophysiography sufficient to accommodate the upland elements. Considerable latitude is possible in such estimates, depending primarily on what MAT value and annual temperature range is assumed for the assemblage. Other variables include slope and exposure, precipitation (annual amounts and seasonal distribution), and, for progressively older floras, changes in ecological tolerances and the possibility that not all analogs are presently growing throughout their potential altitudinal range. A new approach is to estimate the MAT of a coastal or lowland fossil flora and a coeval upland fossil flora from the same physiographic region, then use an assumed lapse rate (e.g., 5°C/km) to calculate the paleoelevation of the upland assemblage (Chapter 6). These results depend on the accuracy of the MAT and lapse rate estimates. LMA and CLAMP are being used to provide better approximation of the MATs.

Stomatal Analysis

A recent innovation based on stomatal features has provided another paleobotanical method for estimating paleoclimates. Researchers have found that among modern plants the number of stomata decreases with increases in CO_2 concentration (Beerling, 1994; Beerling and Chaloner, 1994). This has been demonstrated from herbarium material of species collected over a 200-year period. Human-induced increases in CO_2 were found to correlate with a 40% reduction in stomata density in European forest trees, and the trend was confirmed by laboratory experiments using varying CO_2 concentrations. Specimens of *Quercus pseudocastanea*, the fossil representative of the modern European *Quercus petraea* (durmast oak) were then studied from the Late Miocene through the Pliocene (Van Der Burgh et al., 1993). Stomatal indices (stomata density/stomata density + epidermis cell density) varied between 9.5 and 11.5 in the older material to 16.2 in the Pliocene specimens. The increase was also evident in fossil leaves of *Fagus attenuata* (10.0 ± 0.8 to 14.1 ± 0.5), and both trends are associated with a decline in MAT of 2–3°C. Further study of herbarium material of *Q. petraea* representing a 120-year period allowed correlation of the stomatal index with known atmospheric CO_2 concentrations and gave an estimate of 280–370 ppm in the Late Neogene compared to 338 ppm in 1980. A similar relationship for the last glacial–interglacial cycle was found in leaves of *Pinus flexilis* preserved in packrat middens from the Great Basin (Van de Water et al., 1994). As the technique is refined and better quantified (Kürschner, 1997), one more biosensor will be available for estimating ancient CO_2 concentration and paleotemperature by calculating the stomatal index from fossil leaf material.

Dendrochronology

Dendrochronology is the study of the width, density, and isotopic composition of tree rings; it is used for reconstructing past climates (Cook and Kairiukstis, 1989; Filion et al., 1986; Fritts, 1966; Haugen, 1967; Hughes et al., 1982) and detecting and dating events that leave signals in the patterns of secondary wood development (Wiles et al., 1996). These events include fires, epidemics, anthropological activities (Baillie, 1982), hydrologic cycles, faulting (Page, 1970), ENSO events (Lough and Fritts, 1990; Swetnam and Betancourt, 1990), and landslides and flooding associated with seismic events (Atwater and Moore, 1992; Jacoby et al., 1988, 1992, 1995; Sheppard and Jacoby, 1989) and volcanic eruptions (LaMarche and Hirschboeck, 1984; Lough and Fritts, 1987; Oswalt, 1957).

Growth rings are formed by the vascular cambium that is activated in the stems of temperate trees and shrubs in the early spring and that continues to produce secondary wood (xylem) until late fall. Spring or early wood is formed during the favorable growing conditions of moisture, temperature, nutrient levels, and increasing day length. The resulting tracheids and vessel elements are comparatively large in diameter and form a band that is light colored in appearance. Summer or late wood develops under progressively less favorable conditions, the cells are typically smaller, and the band has a darker appearance. Growth rings are evident in most conifers and are enhanced in many woody dicotyledons by larger diameter vessels concentrated at the beginning of each ring (ring-porus condition). A series of rings with broad spring wood indicates that good growing conditions extended well into the summer months, while narrowing bands of spring wood across the section reflect deteriorating conditions of rainfall and/or temperature. However, the width of a growth ring is a function of several factors, including age, broad genetic control, a genetically influenced "memory" of past patterns, the local hydrologic cycle, and climate, particularly moisture and temperature. A considerable part of recent research in dendrochronology has been to develop sampling procedures, models, and statistical techniques designed to disentangle the various factors determining the width of growth rings.

The particular component of climate that can be measured through tree-ring analysis is, in part, a function of the habitat in which the tree grows. In subpolar and alpine regions temperature is the limiting factor, and growth rings in trees selected from this environment will primarily reflect changes in temperature. In the arid southwest where water is the limiting factor, the width of the spring wood can either be a function of the local hydrologic cycle if the trees are selected from floodplains or of precipitation if the trees are selected from exposed sites with shallow soils away from the fluctuating water levels of the floodplain.

Sampling procedures involve taking two cores per tree from a minimum of 10 trees per site. This allows patterns to be compared among trees and corrections made for missing and false or multiple rings. This process is called crossdating, and may encompass stands hundreds of kilometers apart. Another necessary procedure is standardization. As trees age, the width of the rings becomes narrower. It is necessary to apply a factor to correct for this and other external influences on growth, such as closing or opening of the canopy from wind damage, pathogens, and other factors. This is accomplished by calculating the mean width of rings in young and older portions of the stem, representing periods of comparable growing conditions, among several trees at a site. The exponential is then applied to the measurements of each ring width. A mean site chronology is established by combining the individual standardized chronologies.

Calibration of the series is necessary to allow association of a given width with the values of a particular climatic parameter (viz., annual rainfall). This is done by comparing each corrected ring width with known meterological data. The next step is verification, whereby meterological conditions are estimated from ring widths in other sections of the tree and then compared with actual weather information for that period.

With a series of tree-ring chronologies with standardized ring-width indices established from which precipitation and temperature values can be read, it is possible to estimate climatic changes by utilizing ancient, slow-growing trees common to the region. The oldest records are from the arid southwest, based on the early works of astronomer A. E. Douglas (1909), his student Edmund Schulman (1956), and H. C. Fritts (1976, 1991) at the Laboratory of Tree-Ring Research at the University of Arizona. By using living *Pinus aristata* (bristlecone pine, 4600 years old; Figs. 4.8, 4.9) and other trees, augmented by timbers and charcoal at archeological sites, and subfossils from caves, playa lakes, and bogs, it has been possible to establish a continuous chronology back 8800 years. Other laboratories at the Universities of Arkansas, Nebraska, and Washington, and at the Lamont-Doherty Earth Observatory (Columbia University), Oak Ridge National Laboratory, and the U.S. Geological Survey at Reston, Virginia, have established shorter chronologies for different parts of North America. The oldest living trees in eastern North America are 1700-year-old *Taxodium distichum* (bald cypress) growing along the Black River in North Carolina (Stahle et al., 1988). An International Tree-Ring Data Bank is based at the University of Arizona as part of a developing international tree-ring network (Stockton et al., 1985).

Measurements of ring widths utilizing appropriate statistical analyses is one procedure for estimating past climates. A second area of promising research is tree-ring isotope chemistry (Stuiver and Burk, 1985). This technique is based on the original observations by Urey (1947) that fluctuations in the isotopes of organic material may reflect the ambient temperature prevailing at the time the material was formed. The development and application of this observation to cellulose across the secondary xylem is called tree thermography (Libby, 1987). The determination of carbon ratios involves oxidizing the carbon in the sample to CO_2 and analyzing the gases with a mass spectrometer. The relative amounts of ^{13}C and ^{12}C in the wood is a function of their abundance in the atmosphere that in turn varies with temperature: increasing temperatures result in larger amounts of the heavier ^{13}C and a larger $^{13}C/^{12}C$ ratio. Current research involves determining a $^{13}C/^{12}C$ isotopic temperature coefficient whereby paleotemperatures can be estimated from a given ratio.

Isotopes of hydrogen (deuterium to hydrogen; Feng and Epstein, 1994) and oxygen (^{18}O to ^{16}O) from cellulose are also being used to estimate paleotemperatures. The relative amounts of these isotopes in the atmosphere varies according to ocean-surface temperatures. In warm oceans more of the heavier isotopes are distilled, while in a cooling ocean less are evaporated. These variations are preserved in the chemical composition of cellulose forming growth rings. To initially determine whether isotope ratios reflected ambient temperature, it was necessary to analyze the cellulose from trees growing near sites with long meteorological records. The longest mercury thermometer records are from Germany and date from 1690. The oldest tree near the meterological recording station was an oak that provided rings from 1700 to 1950. The tree-ring isotope patterns matched the actual recorded changes in temperature; both recorded such climatic events as the warm period of 1710–1740, a cold period from 1740 to 1800, and a slow warming until the present with a peak in 1930 (Libby, 1987).

One of the strengths of dendrochronology is that each climatic trend and event is encoded within a specific set of annually formed growth rings and it can be precisely dated. Also, the data are comparatively easy to obtain from any temperate forested region. The paleoclimatic information is not dependent on taxonomic identification or on the as-

Figure 4.8. *Pinus aristata* (bristlecone pine), Arizona. Photograph provided by Charles W. Stockton.

sumption of ecological consistency for a species over long periods of geologic time (Creber and Chaloner, 1985). The principal limitation is the relatively short time span over which the technique can be applied. Also, some aspects of tree-ring analysis cannot be applied to trees growing in tropical environments because of the lack or poor development of growth rings (Jacoby, 1989). However, just the presence or absence of rings in petrified wood can provide useful ancillary information for assessing past climates based on other approaches (Creber and Chaloner, 1984). Generally, the absence of distinct growth rings (Fig. 4.10) among several woods at a locality means nonseasonal tropical climates prevailed, while their presence (Fig. 4.11) suggests more temperate or seasonal climates. In the fossil record of North America the former are common in the Late Cretaceous and Paleocene in the midlatitudes, and the latter mostly characterize the Eocene and later times. In this connection, it is important to note, however, that the presence of growth rings in fossilized wood does not automatically exclude the existence of tropical to subtropical paleoenvironments. Tomlinson and Craighead (1972) examined the wood of 87 species growing in Florida south of latitude 26° N (viz., south of Ft. Lauderdale). Well-developed growth rings were present in 51 species, while 36 species produced no or poorly developed rings. It is also difficult sometimes to determine whether growth patterns in the secondary

Figure 4.9. Cross section of stem of *Pinus aristata* (bristlecone pine), illustrating growth rings. Photograph provided by Charles W. Stockton.

xylem are due to precipitation, temperature, or genetic memory. Some ecological implications of wood features are summarized in Fig. 4.12, and other aspects of wood anatomy and environment are discussed by Carlquist (1975, 1977, 1988) and Wheeler and Baas (1991, 1993).

Recent refinements in standardizing tree-ring chronologies, the development and application of statistical methods of analysis, and the emerging techniques of isotope chemistry have established dendrochronology as an increasingly precise and useful technique for the study of vegetational history and paleoenvironments.

Packrat Middens

> Part way up we came to a high cliff and in its face were niches . . . and in some of them we found balls of a glistening substance looking like pieces of variegated candy . . . it was evidently food of some sort, and we found it sweet but sickish, and those who were hungry, making a good meal of it, were a little troubled with nausea afterwards. From the diary of a lost prospector in the Gold Rush of 1849. (Jaroff, 1992)

Another method for studying the recent history of vegetation is through analysis of packrat (or woodrat) middens. These rodents belong to the genus *Neotoma* (Fig. 4.13); there are ~20 species living from sea level to 3350-m elevation across North America and ranging from the latitudes of British Columbia (~50° N) to Guatemala (~16° N). They forage for food and construct houses or dens from nearby vegetation, usually within a radius of less than 50 m from the site. The houses are frequently built in cliff recesses, rock shelters (Fig. 4.14), and caves. The middens also serve as urinal perches that become hardened over time (Fig. 4.15) and, in arid protected environments, they are preserved as consolidated layers of plant debris that reflect the local surrounding vegetation throughout the time the middens were being constructed. Because preservation is favored by dryness, the analysis of packrat middens as a method of paleovegetation analysis is confined to the arid southwest, even though the range of the organism is considerably greater. The principal species in the Chihuahuan Desert is the white-throated packrat (*N. albigula*; Fig. 4.13). In the Mohave Desert it is the desert packrat (*N. lepida*), and on the Colorado Plateau and elsewhere in the Great Basin it is the bushy-tailed packrat (*N. cinerea*).

The oldest known middens are more than 50 Ky from central Nevada (Spaulding, 1985) and 51 Ky from western Arizona (^{14}C finite date; K. L. Cole, unpublished data), but most date from the late glacial through the Holocene. Sites from the Chihuahuan Desert (Van Devender, 1990) have yielded about 480 plant taxa, most of which have been identified to species on the basis of megafossil remains (Fig. 4.16). The species found in the middens represent mostly escarpment vegetation because the middens that survive are those constructed in protected cliff-side cave sites and, as noted, the packrats range only ~50 m from the middens. The high organic content of the structures readily allows radiocarbon dating and about 1000 dates are now available from middens in the southwest. The relative abundance of identified megafossils from each level constitutes a spectrum, which provides a general picture of the surrounding vegetation for one point in time. The spectra can be arranged in a vertical sequence through the midden to constitute a profile for each taxon, and the profiles col-

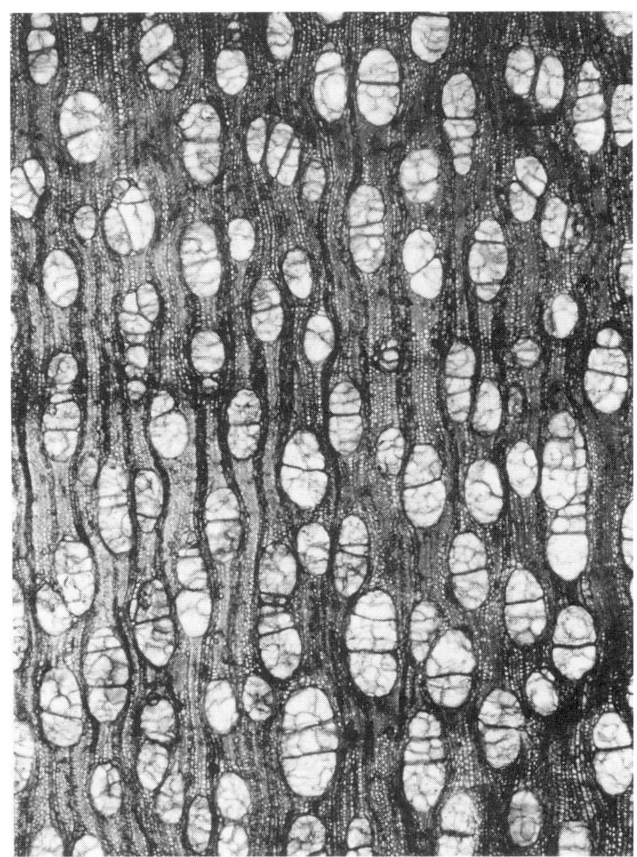

Figure 4.10. Fossil wood of *Paraphyllanthoxylon abbottii*, Paleocene, Big Bend National Park, Texas. Growth rings are indistinct to absent. Photograph courtesy of Elisabeth Wheeler (see Wheeler, 1991).

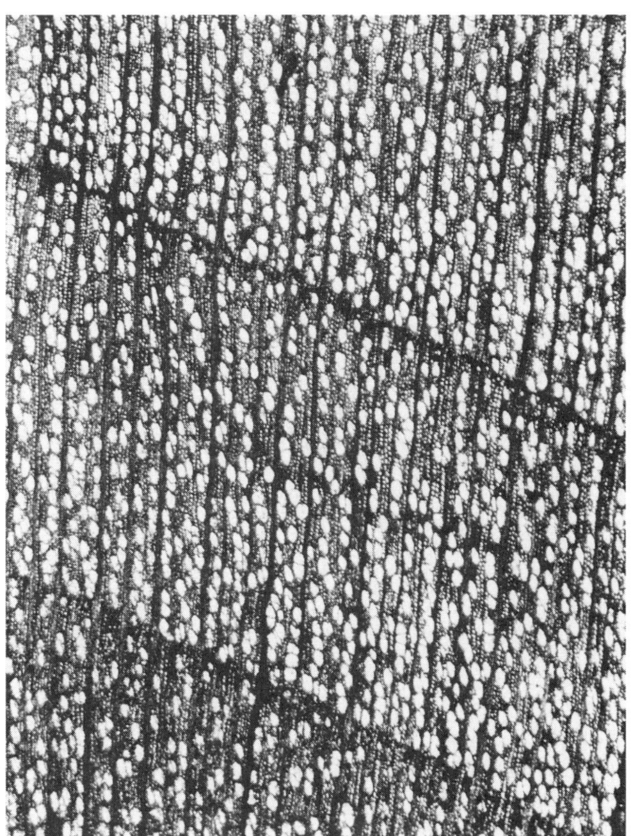

Figure 4.11. Fossil wood of *Magnolia longiradiata*, middle Eocene Clarno Formation, Oregon. Growth rings are distinct. Photograph courtesy of Elisabeth Wheeler (see Scott and Wheeler, 1982).

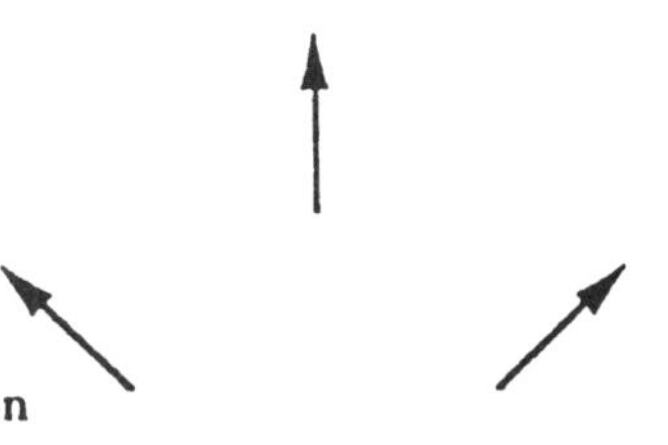

Figure 4.12. Trends in vessel element specialization reflecting different habitats. Reprinted from Wheeler and Baas (1991; adapted from Dickison, 1989) with the permission of the International Association of Wood Anatomists.

Figure 4.13. White-throated pack rat (*Neotoma albigula*), a mostly desert-inhabiting species. Photograph courtesy of Terry Vaughan (see Vaughan, 1990) and reprinted with the permission of the University of Arizona Press.

lectively make a diagram that reflects changes in the vegetation over time (Fig. 4.17). This proxy data can be used to infer changes in climate. The disappearance of the Anasazi tribe from Chaco Canyon in New Mexico ~1200 years B.P. correlates with depletion of pine debris in the middens. The inference is that the overuse of the woodland for construction and fuel may have contributed to erosion and loss of the land for farming (Betancourt and Van Devender, 1981). In addition, body size can be estimated from fecal pellets, and the size decreases during warm periods following Bergmann's Rule (Smith et al., 1995). Fluctuations in body size of *N. cinerea* from the Great Basin and Colorado Plateau closely track temperature changes simulated by the CCM at the NCAR.

A recent innovation is analysis of pollen and spores preserved in the acid environments provided by the middens. The plant microfossils mostly represent anemophilous pollen and spores blown into the caves by wind. Pollen and spore concentrations in middens have been reported as ranging from 17,000 to 197,000 grains/g by Thompson (1985), 4870 to 629,508 grains/g by Davis and Anderson (1987), and 23,003 to 533,500 grains/g by Anderson and Van Devender (1991). Other sources of microfossils are flowers, and pollen and spores adhering to plant surfaces brought into the middens as food or nesting material. Minor amounts may be brought in on the pelts and removed by preening, while others are preserved in fecal pellets. Pollen and spore concentrations in the latter range from about 12,350 to 408,350 grains/g and include zoophilous species. However, they may reflect dietary preferences more than a random sampling of the surrounding vegetation.

The procedure for isolating pollen and spores from the middens involves dissolving the binding urine matrix in water and treating the residue with 10% KOH and hot water washes that dissolve cuticular and cellulosic debris and clear the darkened specimens. Hydrofluoric acid removes silicates, and acetolysis (treatment with a mixture of

Figure 4.14. Midden of bushy-tailed pack rat (*Neotoma cinerea*), composed mostly of saltbrush (*Atriplex canescens*), Navajo County, Arizona. Photograph courtesy of Terry Vaughan (see Vaughan, 1990; photograph in original publication is upside down) and reprinted with the permission of the University of Arizona Press.

Figure 4.15. Indurated pack rat midden, Eleana Range, Nevada, 17.1–10.6 ka. Photograph courtesy of W. Geoffrey Spaulding (see Spaulding et al., 1990) and reprinted with the permission of the University of Arizona Press.

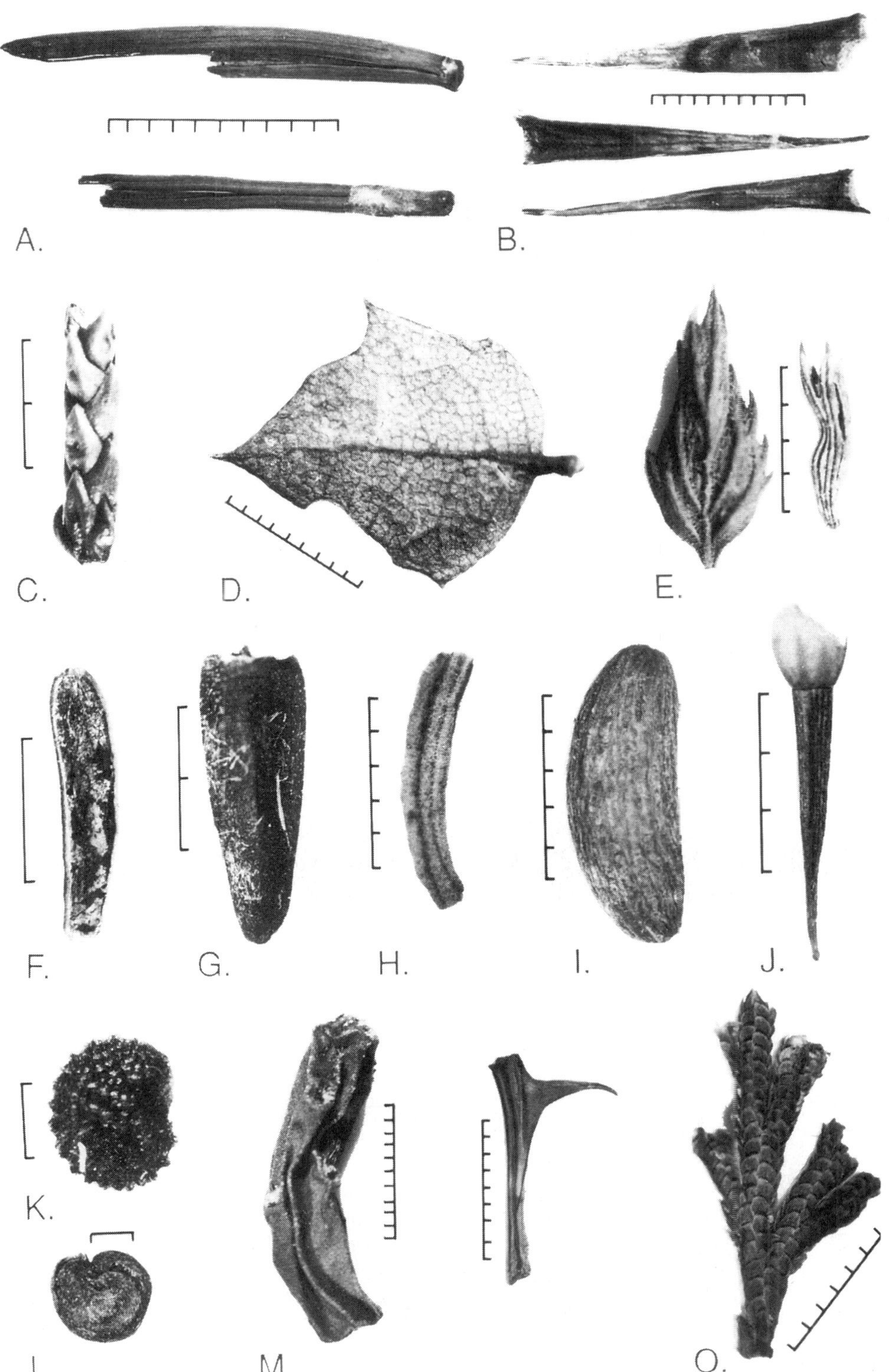

Figure 4.16. Plant megafossils from pack rat midden, Maravillas Canyon, Texas. Scale division in millimeters. (A) Papershell piñon needles, 21 ka; (B) Thompson/beaked yucca (*Yucca thompsoniana/rostrata*) leaf tips, 21 ka; (C) juniper twig, 21 ka; (D) shrub oak leaf, 20.6 ka; (E) mock pennyroyal (*Hedeoma plicatum*) leaf and fruit, 20.6 ka; (F) rock daisy (*Perityle* sp.) achene, 20.6 ka; (G) goldeneye (*Viguiera* cf. *dentata*), 16.2 ka; (H) winter fat (*Ceratoides lanata*) leaf, 16.2 ka; (I) desert olive (*Forestiera* sp.) seed, 16.6 ka; (J) gray daisy (*Bahia absinthifolia*) achene, 16.6 ka; (K) false nightshade (*Chamaesaracha* sp.) seed, 6 ka; (L) blind prickly pear (*Opuntia* sp.) seed, 5.1 ka; (M) prickly pear cactus (*Opuntia leptocaulis*) stem, 5.1 ka; (N) roemer catclaw (*Mimosa* sp.) twig, 5.1 ka; (O) resurrection plant (*Selaginella*) stem, 5.1 ka. Reprinted from Van Devender (1990) with the permission of the University of Arizona Press.

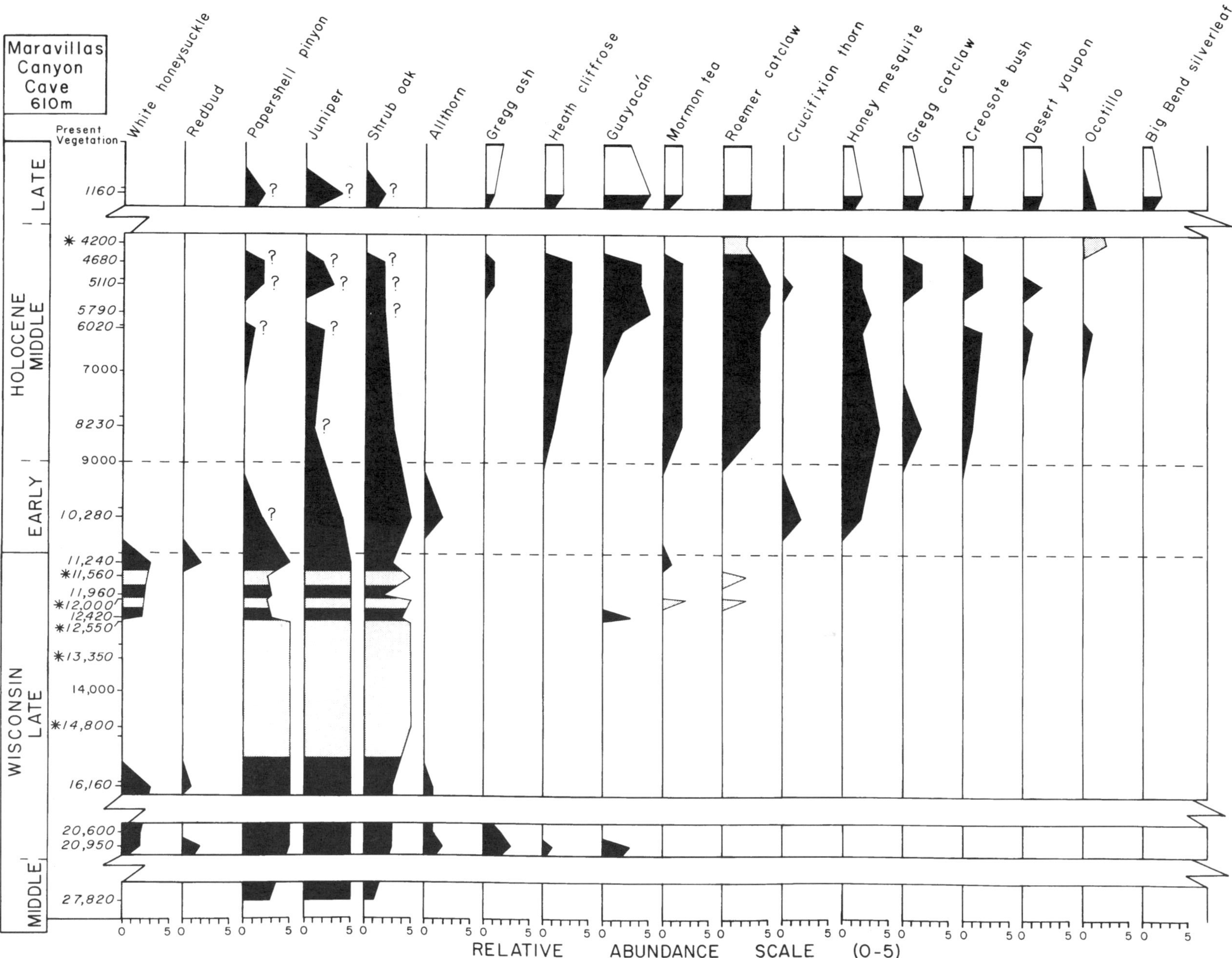

Figure 4.17. Megafossil profile from pack rat midden, Maravillas Canyon, Texas. Reprinted from Van Devender (1990) with the permission of the University of Arizona Press.

9 parts of acetic anhydride to 1 part of concentrated sulfuric acid) oxidizes other organic debris. The residue is mixed with a mounting medium, such as glycerine jelly or silicone oil, and mounted on slides for examination.

In addition to spores, pollen, and megafossils, the plant debris comprising middens also preserves changes in $^{13}C/^{12}C$ ratios in atmospheric CO_2. It has been found, for example, that the pattern of isotopically lighter atmospheric CO_2 during the last ice age, compared to interglacial periods, is preserved in the midden material (Marino et al., 1992).

One of the principal contributions of the study of packrat middens is the provision of a record of past vegetation from regions where other types of paleovegetation analysis cannot be made. For example, Anderson and Van Devender (1991) present a pollen record from the Waterman Mountains of southern Arizona. This is a hot, dry environment where plant mega- and microfossils preserved in middens are the only extensive source of data on the late-glacial and Holocene history of the vegetation. Another advantage is that the megafossils often can be identified to species. Midden material also provides abundant organic debris suitable for ^{14}C dating and establishing high-resolution chronologies.

A limitation of the technique is that it is applicable to a limited geographic region (the arid southwest) and to a relatively brief segment of time (mostly since 35 Kya). However, a 35 Ky record gives a detailed picture of vegetational change across a major glacial–interglacial boundary. A source of difficulty in interpretation is that the longer sequences are reconstructed by amalgamating results from several middens preserved in caves from regions that often differ slightly in topography, lithology, and vegetation. Behavioral differences among the species could result in some preferential selection of material for food and nest construction; and where composite sequences are built up from individual middens, this selectivity might be reflected in the analysis. There is often a lack of precise stratigraphic control between sites, and assemblages of different ages are occasionally mixed at a site. However, development of tandem accelerator mass spectrometry (TAMS), whereby particles as small as individual seeds can be dated, makes this less of a problem (Van Devender et al., 1985). In this technique ^{14}C atoms are counted directly, rather than having to count the rare ß particles emitted.

There are also the usual cadre of strengths and weaknesses inherent in the microfossil versus megafossil approaches. As noted previously, pollen grains and spores can mostly be identified to genus, or in some cases only to family, but they generally provide a more regional representation of the past vegetation. The megafossils can often be identified to genus and even to species, but they often provide a record biased toward the local canyon or cliffside vegetation. These habitats can include a disproportionately large number of relict species from previous climatic cycles and contain more mesic trees and shrubs than the surrounding treeless desert. Although the relative merits can be debated (Hall, 1985; Spaulding, 1990; Van Devender, 1988), new and valuable information is being derived from the study of megafossils and microfossils preserved in packrat middens (Anderson and Van Devender, 1991; Hall, 1986; Thompson, 1985).

The utilization of plant microfossils (pollen, spores, phytoliths), megafossils (stomatal analysis, dendrochronology, cellulose isotope chemistry), modern analogs, and leaf physiognomy (including newly developing computer programs and appropriate statistical tests) provides an impressive arsenal of botanical techniques available for tracing the development of vegetation through time. A balance is now emerging between the older preoccupation with "what is it?" and new opportunities to assess "what does it mean?". As greater use is made of other diverse and independent sources of context information (Chapters 2, 3), the descriptive and causal history of North American environments and biotas, from the Cretaceous to the present, can both be outlined in rather impressive detail. This history is recounted in the following chapters.

References

Anderson, R. S. and T. R. Van Devender. 1991. Comparison of pollen and macrofossils in packrat (*Neotoma*) middens: a chronological sequence from the Waterman Mountains of southern Arizona, U.S.A. Rev. Palaeobot. Palynol. 68: 1–28.

Atwater, B. F. and A. L. Moore. 1992. A tsunami about 1000 years ago in Puget Sound, Washington. Science 258: 1614–1617.

Bailey, I. W. and E. W. Sinnott. 1915. A botanical index of Cretaceous and Tertiary climates. Science 41: 831–834.

———. 1916. The climatic distribution of certain types of angiosperm leaves. Am. J. Bot. 3: 24–39.

Baillie, M. G. L. 1982. Tree-Ring Dating and Archeology. Croom Helm, London.

Basinger, J. F., D. R. Greenwood, and T. Sweda. 1994. Early Tertiary vegetation of Arctic Canada and its relevance to paleoclimatic interpretation. In: M. C. Boulter and H. C. Fisher (eds.), Cenozoic Plants and Climates of the Arctic. NATO ASI Series 1: Global Environmental Change. Springer-Verlag, Berlin. Vol. 27, pp. 175–198.

Beerling, D. J. 1994. Palaeo-ecophysiological studies on Cretaceous and Tertiary fossil floras. In: M. C. Boulter and H. C. Fisher (eds.), Cenozoic Plants and Climates of the Arctic. NATO ASI Series 1: Global Environmental Change. Springer-Verlag, Berlin. Vol. 27, pp. 23–33.

——— and W. G. Chaloner. 1994. Atmospheric CO_2 changes since the last glacial maximum: evidence from the stomatal density record of fossil leaves. Rev. Palaeobot. Palynol. 81: 11–17.

Betancourt, J. L. and T. R. Van Devender. 1981. Holocene vegetation in Chaco Canyon, New Mexico. Science 214: 656–658.

Boyd, A. 1990. The Thyra Ø flora: toward an understanding of the climate and vegetation during the Early Tertiary in the high Arctic. Rev. Palaeobot. Palynol. 62: 189–203.

———. 1994. Some limitations in using leaf physiognomic data as a precise method for determining paleoclimates with an example from the Late Cretaceous

Pautût flora of west Greenland. Palaeogeogr., Palaeoclimatol., Palaeoecol. 112: 261–278.

Brooks, J., P. R. Grant, M. D. Muir, P. van Gijzel, and G. Shaw (eds.). 1971. Sporopollenin. Academic Press, New York.

Brown, V. K. and J. H. Lawton. 1991. Herbivory and the evolution of leaf size and shape. Phil. Trans. R. Soc. Lond. B 333: 265–272.

Burnham, R. J. 1994. Paleoecological and floristic heterogeneity in the plant-fossil record—an analysis based on the Eocene of Washington. U.S. Geol. Surv. Bull. 2085B: 1–36.

Carlquist, S. 1975. Ecological Strategies of Xylem Evolution. University of California Press, Berkeley, CA.

———. 1977. Ecological factors in wood evolution: a floristic approach. Am. J. Bot. 64: 887–896.

———. 1988. Comparative Wood Anatomy. Springer-Verlag, Berlin.

Chaney, R. W. 1959. Miocene floras of the Columbia Plateau. Part I. Composition and interpretation. Carnegie Inst. Wash. Contrib. Paleontol. 617: 1–134.

Cook, E. R. and L. A. Kairiukstis (eds.). 1989. Methods of Dendrochronogy. Kluwer Academic Publishers, Boston.

Corner, E. J. H. 1976. The Seeds of Dicotyledons. Cambridge University Press, Cambridge, U.K.

Creber, G. T. and W. G. Chaloner. 1984. Climatic indications from growth rings in fossil woods. In: P. J. Brenchley (ed.), Fossils and Climate. Wiley, New York. pp. 49–74.

———. 1985. Tree growth in the Mesozoic and Early Tertiary and the reconstruction of palaeoclimates. Palaeogeogr., Palaeoclimatol., Palaeoecol. 52: 35–60.

Crepet, W. L., K. C. Nixon, E. M. Friis, and J. V. Freundenstein. 1992. Oldest fossil flowers of hamamelidaceous affinity, from the Late Cretaceous of New Jersey. Proc. Natl. Acad. Sci. USA 89: 8986–8989.

Davis, M. B. and K. Faegri. 1967. Translation: Forest tree pollen in south Swedish peat bog deposits, by Lennart von Post. Pollen et Spores 9: 375–401.

Davis, O. K. and R. S. Anderson. 1987. Pollen in packrat (*Neotoma*) middens: pollen transport and the relationship of pollen to vegetation. Palynology 11: 185–198.

Dickison, W. C. 1989. Steps toward the natural system of the dicotyledons: vegetative anatomy. Aliso 12: 555–566.

Dilcher, D. L. 1963. Cuticular analysis of Eocene leaves of *Ocotea obtusifolia*. Am. J. Bot. 50: 1–8.

———. 1973. A paleoclimatic interpretation of the Eocene floras of southeastern North America. In: A. Graham (ed.), Vegetation and Vegetational History of Northern Latin America. Elsevier Science Publishers, Amsterdam. pp. 39–59.

———. 1974. Approaches to the identification of angiosperm leaf remains. Bot. Rev. 40: 1–157.

Dolph, G. E. 1978. Variation in leaf size and margin type with respect to climate. Cour. Forsch. Inst. Senckenberg 30: 153–158.

———. 1979. Variation in leaf margin with respect to climate in Costa Rica. Bull. Torrey Bot. Club 106: 104–109.

———. 1984. Leaf form of the woody plants of Indiana as related to environment. In: N. S. Margaris, M. Arianoustou-Farragitaki, and W. C. Oechel (eds.), Being Alive on Land, Tasks for Vegetation Science. Dr. W. Junk, Publisher, The Hague. Vol. 13, pp. 51–61.

———. 1990. A critique of the theoretical basis of leaf margin analysis. Proc. Indiana Acad. Sci. 99: 1–10.

——— and D. L. Dilcher. 1979. Foliar physiognomy as an aid in determining paleoclimate. Palaeontograph. Abt. B, 170: 151–172.

Douglas, A. E. 1909. Weather cycles in the growth of big trees. Monthly Weather Review 37: 225–237.

Efremov, I. A. 1940. Taphonomy: a new branch of paleontology. Pan Am. Geol. 74: 81–93.

Faegri, K., P. E. Kaland, and K. Krzywinski. 1989. Textbook of Pollen Analysis, 4th edition. Wiley, New York.

Farley, M. B. 1990. Vegetation distribution across the Early Eocene depositional landscape from palynological analysis. Palaeogeogr., Palaeoclimatol., Palaeoecol. 79: 11–27.

Feng, X. and S. Epstein. 1994. Climatic implications of an 8000-year hydrogen isotope time series from bristlecone pine trees. Science 265: 1079–1081.

Filion, L., S. Payette, L. Gauthier, and Y. Boutin. 1986. Light rings in subarctic conifers as a dendrochronological tool. Quatern. Res. 26: 272–279.

Frederiksen, N. O. 1989. Late Cretaceous and Tertiary floras, vegetation, and paleoclimates of New England. Rhodora 91: 25–48.

Fritts, H. C. 1966. Growth-rings of trees: their correlation with climate. Science 154: 973–979.

———. 1976. Tree Rings and Climate. Academic Press, New York.

———. 1991. Reconstructing Large-Scale Climatic Patterns from Tree-Ring Data: A Diagnostic Analysis. University of Arizona Press, Tucson, AZ.

Gastaldo, R. A. 1988. A conspectus of phytotaphonomy. In: W. A. DiMichele and S. L. Wing (eds.), Methods and Applications of Plant Paleoecology. Paleontological Society, Knoxville, TN, Special Publ. 3. pp. 14–28.

Givnish, T. J. 1979. On the adaptive significance of leaf form. In: O. T. Solbrig, S. Jain, G. B. Johnson, and P. H. Raven (eds.), Topics in Plant Population Biology. Columbia University Press, New York. pp. 375–407.

———. 1986. On the Ecology of Plant Form and Function. Cambridge University Press, Cambridge, U.K.

——— and G. J. Vermeij. 1976. Sizes and shapes of liane leaves. Am. Naturalist 110: 743–770.

Graham, A. 1965. The Sucker Creek and Trout Creek Miocene Floras of Southeastern Oregon. Kent State University, Kent, OH. Research Ser. IX.

Gray, J. (Coordinator). 1965. Techniques in palynology. In: B. Kummel and D. Raup (eds.), Handbook of Paleontological Techniques. W. H. Freeman and Co., San Francisco. pp. 471–706.

Greenwood, D. R. 1991. The taphonomy of plant macrofossils. In: S. K. Donovan (ed.), The Processes of Fossilization. Belhaven Press, London. pp. 141–169.

———. 1992. Taphonomic constraints on foliar physiognomic interpretations of Late Cretaceous and Tertiary palaeoclimates. Rev. Palaeobot. Palynol. 71: 149–190.

Hall, S. A. 1985. Quaternary pollen analysis and vegetational history of the southwest. In: V. M. Bryant, Jr. and R. G. Holloway (eds.), Pollen Records of Late-Quaternary North American Sediments. American Association of Stratigraphic Palynologists Foundation, Dallas, TX. pp. 95–123.

———. 1986. Plant macrofossils from wood rat middens: vegetation or flora? American Quaternary Association 9th Biennial Meeting, Champaign–Urbana, June 2–4, 1986. Program and Abstracts, p. 136.

Haugen, R. K. 1967. Tree ring indices: a circumpolar comparison. Science 158: 773–775.

Herendeen, P. S. 1991. Lauraceous wood from the mid-Cretaceous Potomac Group of eastern North America: *Paraphyllanthoxylon marylandense* sp. nov. Rev. Palaeobot. Palynol. 69: 277–290.

Hickey, L. J. 1973. Classification of the architecture of dicotyledonous leaves. Am. J. Bot. 60: 17–33.

———. 1979. A revised classification of the architecture of dicotyledonous leaves. In: C. R. Metcalfe and L. Chalk (eds.), Anatomy of the Dicotyledons. Clarendon Press, Oxford, U.K. Vol. I, pp. 25–39.

——— and J. A. Wolfe. 1975. The bases of angiosperm phylogeny: vegetative morphology. Ann. Missouri Bot. Garden 62: 538–589.

Hoffmeister, W. S. 1959. Palynology's first 10 years as an aid to finding oil. Oil Gas J., August 17.

Hughes, M. K., P. M. Kelley, J. R. Pilcher, and V. C. Lamarche, Jr. (eds.). 1982. Climate from Tree Rings. Cambridge University Press, Cambridge, U.K.

Hyde, H. A. and D. W. Williams. 1944. Right word. Pollen Anal. Circ. 8: 6.

Jacoby, G. C. 1989. Overview of tree-ring analysis in tropical regions. Internatl. Assoc. Wood Anat. Bull. 10: 99–108.

———, G. Carver, and W. S. Wagner. 1995. Trees and herbs killed by an earthquake ~300 yr ago at Humboldt Bay, California. Geology 23: 77–80.

Jacoby, G. C., P. R. Sheppard, and K. E. Sieh. 1988. Irregular recurrence of large earthquakes along the San Andreas fault: evidence from trees. Science 241: 196–199.

Jacoby, G. C., P. L. Williams, and B. M. Buckley. 1992. Tree-ring correlation between prehistoric landslides and abrupt tectonic events in Seattle, Washington. Science 258: 1621–1623.

Jansonius, J. and D. C. McGregor (eds.). 1996. Palynology: Principles and Applications. American Association of Stratigraphic Palynologists Foundation, Dallas, TX.

Jaroff, L. 1992. Nature's time capsules. Time Magazine April 6: 61.

Kapp, R. O. 1961. Modern pollen rain studies in the central prairie of North America. Bull. Ecol. Soc. Am. 42: 110–111.

Klucking, E. P. 1986. Leaf Venation Patterns. Annonaceae. J. Cramer, Stuttgart. Vol. 1.

Kovach, W. L. 1989. Comparisons of multivariate analytical techniques for use in pre-Quaternary plant paleoecology. Rev. Palaeobot. Palynol. 60: 255–282.

Kremp, G. O. W. 1983. Fourth supplement, annotated references, Precambrian to Tertiary palynological literature. Paleo Data Banks 20: 1–68.

——— and J. G. Methvin. 1968. Taxonomic crisis in pre-Pleistocene palynology. Oklahoma Geol. Notes 28: 146–153.

Kürschner, W. M. 1997. The anatomical diversity of recent and fossil leaves of durmast oak (*Quercus petraea* Lieblein/*Q. pseudocastanea* Goeppert)—implications for their use as biosensors of palaeoatmospheric CO_2 levels. Rev. Palaeobot. Palynol. 96: 1–30.

LaMarche, V. C., Jr. and K. K. Hirschboeck. 1984. Frost rings in trees as records of major volcanic eruptions. Nature 307: 121–126.

Leopold, E. B., G. Liu, and S. Clay-Poole. 1992. Low-biomass vegetation in the Oligocene? In: D. R. Prothero and W. A. Berggren (eds.), Eocene–Oligocene Climatic and Biotic Evolution. Princeton University Press, Princeton, NJ. pp. 399–420.

Libby, L. M. 1987. Evaluation of historic climate and prediction of near-future climate from stable-isotope variations in tree rings. In: M. R. Rampino, J. E. Sanders, W. S. Newman, and L. K. Königsson (eds.), Climate-History, Periodicity, and Predictability. Van Nostrand Reinhold Co., New York. pp. 81–89.

Lough, J. M. and H. C. Fritts. 1987. An assessment of the possible effects of volcanic eruptions on North American climate using tree-ring data, 1602–1900 A.D. Climatic Change 10: 219–239.

———. 1990. Historical aspects of El Niño/Southern Oscillation-information from tree rings. In: P. W. Glynn (ed.), Global Ecological Consequences of the 1982–83 El Niño–Southern Oscillation. Elsevier Oceanography Series. Elsevier Science Publishers, Amsterdam. Vol. 52.

MacGinitie, H. D. 1953. Fossil plants of the Florissant Beds, Colorado. Carnegie Inst. Wash. Contrib. Paleontol. 599.

———. 1962. The Kilgore flora, a Late Miocene flora from northern Nebraska. Univ. Calif. Publ. Geol. Sci. 35: 67–158.

———. 1969. The Eocene Green River flora of northwestern Colorado and northeastern Utah. Univ. Calif. Publ. Geol. Sci. 83: 1–203.

Marino, B. D., M. B. McElroy, R. J. Salawitch, and W. G. Spaulding. 1992. Glacial-to-interglacial variations in the carbon isotopic composition of atmospheric CO_2. Nature 357: 461–466.

Oswalt, W. H. 1957. Volcanic activity and Alaskan spruce growth in A.D. 1783. Science 126: 928–929.

Page, R. 1970. Dating episodes of faulting from tree rings: effects of the 1958 rupture of the Fairweather fault on tree growth. Geol. Soc. Am. Bull. 81: 3085–3094.

Pearson, D. L. 1984. Pollen/Spore Color "Standard." Phillips Petroleum Co., Bartlesville, OK, Exploration Projects Section. Version 2, Privately circulated.

Pemberton, R. W. 1992. Fossil extrafloral nectaries, evidence for the ant-guard antiherbivore defense in an Oligocene *Populus*. Am. J. Bot. 79: 1242–1246.

Piperno, D. R. 1988. Phytolith Analysis. Academic Press, New York.

Radford, A. E. 1986. Fundamentals of Plant Systematics. Harper & Row, New York.

Raunkiaer, C. 1934. The Life-Forms of Plants and Statistical Plant Geography. Oxford University Press, Oxford, U.K.

Ritchie, J. C. and S. Lichti-Federovich. 1967. Pollen dispersal phenomena in Arctic–subarctic Canada. Rev. Palaeobot. Palynol. 3: 255–266.

Roth, I. 1992. Leaf Structure: Coastal Vegetation and Mangroves of Venezuela. Gebrüder Borntraeger, Berlin.

Rovner, I. 1971. Potential of opal phytoliths for use in paleoecological reconstruction. Quatern. Res. 1: 343–359.

Schulman, E. 1956. Dendroclimatic changes in semiarid America. University of Arizona Press, Tucson, AZ.

Scott, R. A. and E. A. Wheeler. 1982. Fossil woods from the Eocene Clarno Formation of Oregon. Internatl. Assoc. Wood Anat. Bull. 3: 135–154.

Shaw, G. 1971. The chemistry of sporopollenin. In: J. Brooks, P. R. Grant, M. Muir, P. van Gijzel, and G. Shaw (eds.), Sporopollenin. Academic Press, New York. pp. 305–350.

Sheppard, P. R. and G. C. Jacoby. 1989. Application of tree-ring analysis to paleoseismology: two case studies. Geology 17: 226–229.

Smiley, C. J. and L. M. Huggins. 1981. *Pseudofagus idahoensis*, N. Gen. et sp. (Fagaceae) from the Miocene Clarkia flora of Idaho. Am. J. Bot. 68: 741–761.

Smith, F. A., J. L. Betancourt, and J. H. Brown. 1995. Evolution of body size in the woodrat over the past 25,000 years of climate change. Science 270: 2012–2014.

Spaulding, W. G. 1985. Vegetation and Climates of the Last 45,000 Years in the Vicinity of the Nevada Test Site, South-Central Nevada. U.S. Geological Survey, Reston, VA, Professional Paper 1329, pp. 1–83.

———. 1990. Comparison of pollen and macrofossil based reconstructions of late Quaternary vegetation in western North America. Rev. Palaeobot. Palynol. 64: 359–366.

———, J. L. Betancourt, L. K. Croft, and K. L. Cole. 1990. Packrat middens: their composition and methods of analysis. In: J. L. Betancourt, T. R. Van Devender, and P. S. Martin (eds.). Packrat Middens. University of Arizona Press, Tucson, AZ. pp. 59–84.

Stahle, D. W., M. K. Cleaveland, and J. G. Hehr. 1988. North Carolina climate changes reconstructed from tree rings: A.D. 372 to 1985. Science 240: 1517–1519.

Staplin, F. L. 1962. Organic remains in meterorites—a review of the problem. J. Alberta Soc. Petrol. Geol. 10: 575–580.

Stearn, W. T. 1983. Botanical Latin. David and Charles Publishers, London.

Stockton, C. W., W. R. Boggess, and D. M. Meko. 1985. Climate and tree rings. In: A. D. Hecht (ed.), Paleoclimate Analysis and Modeling. Wiley-Interscience, New York. pp. 71–151.

Stone, D. E., S. C. Sellers, and W. J. Kress. 1979. Ontogeny of exineless pollen in *Heliconia*, a banana relative. Ann. Missouri Bot. Garden 66: 701–730.

———. 1981. Ontogenetic and evolutionary implications of a neotenous exine in *Tapeinochilos* (Zingiberales: Costaceae) pollen. Am. J. Bot. 68: 49–63.

Stuiver, M. and R. L. Burk. 1985. Appendix: Paleoclimatic studies using isotopes in trees. Appendix to Stockton et al., Climate and tree rings. In: A. D. Hecht (ed.), Paleoclimate Analysis and Modeling. Wiley-Interscience, New York. pp. 151–161.

Swetnam, T. W. and J. L. Betancourt. 1990. Fire–southern oscillation relations in the southwestern United States. Science 249: 1017–1020.

Thomasson, J. R. 1983. *Carex graceii* sp. N., *Cyperocarpus eliasii* sp. N., *Cyperocarpus terrestris* sp. N., and *Cyperocarpus pulcherrima* Sp. N. (Cyperaceae) from the Miocene of Nebraska. Am. J. Bot. 70: 435–449.

———. 1984. Miocene grass (Gramineae: Arundinoideae) leaves showing external micromorphological and internal anatomical features. Bot. Gazette 145: 204–209.

———, M. E. Nelson, and R. J. Zakrzewski. 1986. A fossil grass (Gramineae: Chloridoideae) from the Miocene with Kranz anatomy. Science 233: 876–878.

Thompson, R. S. 1985. Palynology and *Neotoma* middens. Am. Assoc. Stratigr. Palynol. Contrib. Ser. 16: 89–112.

Todzia, C. A. and R. C. Keating. 1991. Leaf architecture of the Chloranthaceae. Ann. Missouri Bot. Garden 78: 476–496.

Tomlinson, P. B. and F. C. Craighead, Sr. 1972. Growth-ring studies on the native trees of sub-tropical Florida. In: A. K. M. Ghouse and M. Yunus (eds.), Research Trends in Plant Anatomy, K. A. Chowdhury Commemoration Volume. Tata McGraw Hill, New Delhi. pp. 39–51.

Traverse, A. 1988. Paleopalynology. Unwin Hyman, Boston.

———, K. H. Clisby, and F. Foreman. 1961. Pollen in drilling-mud "thinners," a source of palynological contamination. Micropaleontology 7: 375–377.

Upchurch, G. R., Jr. and J. A. Wolfe. 1987. Mid-Cretaceous to Early Tertiary vegetation and climate: evidence from fossil leaves and woods. In: E. M. Friis, W. G. Chaloner, and P. R. Crane (eds.), The Origins of Angiosperms and Their Biological Consequences. Cambridge University Press, Cambridge, U.K. pp. 75–105.

Urey, H. C. 1947. The thermodynamic properties of isotopic substances. J. Chem. Soc. Lond. 562–581.

Van Der Burgh, J., H. Visscher, D. L. Dilcher, and W. M. Kürschner. 1993. Paleoatmospheric signatures in Neogene fossil leaves. Science 260: 1788–1790.

Van der Water, P. K., S. W. Leavitt, and J. L. Betancourt. 1994. Trends in stomatal density and $^{13}C/^{12}C$ ratios of *Pinus flexilis* needles during last glacial–interglacial cycle. Science 264: 239–243.

Van Devender, T. R. 1988. Pollen in packrat (*Neotoma*) middens: pollen transport and the relationship of pollen to vegetation. Palynology 12: 221–229.

———. 1990. Late Quaternary vegetation and climate of the Chihuhuan Desert, United States and Mexico. In: J. L. Betancourt, T. R. Van Devender, and P. S. Martin (eds.), Packrat Middens, the Last 40,000 Years of Biotic Change. University of Arizona Press, Tucson, AZ. pp. 104–133.

——— et al. (+ 7 authors). 1985. Fossil packrat middens and the tandem accelerator mass spectrometer. Nature 317: 610–613.

Vaughan, T. A. 1990. Ecology of living packrats. In: J. L. Betancourt, T. R. Van Devender, and P. S. Martin (eds.), Packrat Middens, the Last 40,000 Years of Biotic Change. University of Arizona Press, Tucson, AZ. pp. 14–27.

von Post, L. 1916. Om skogsträdpollen i sydsvenska torfmosslagerföljder. Geol. Fören. Stockholm Förhdl. 38: 384–390. [See Davis and Faegri (1967) for English translation.]

Webb, L. J. 1959. A physiognomic classification of Australian rain forests. J. Ecol. 47: 551–570.

Wheeler, E. A. 1991. Paleocene dicotyledonous trees from Big Bend National Park, Texas: variability in wood types common in the Late Cretaceous and Early Tertiary, and ecological inferences. Am. J. Bot. 78: 658–671.

——— and P. Baas. 1991. A survey of the fossil record for dicotyledonous wood and its significance for evolutionary and ecological wood anatomy. Internatl. Assoc. Wood Anat. Bull. 12: 275–332.

———. 1993. The potentials and limitations of dicotyledonous wood anatomy for climatic reconstructions. Paleobiology 19: 487–498.

Wiles, G. C., P. E. Calkin, and G. C. Jacoby. 1996. Tree-ring analysis and Quaternary geology: principles and recent applications. Geomorphology 16: 259–272.

Wing, S. L. 1987. Eocene and Oligocene floras and vegetation of the Rocky Mountains. Ann. Missouri Bot. Garden 74: 748–784.

———. 1992. High-resolution leaf x-radiography in systematics and paleobotany. Am. J. Bot. 79: 1320–1324.

Wolfe, J. A. 1971. Tertiary climatic fluctuations and methods of analysis of Tertiary floras. Palaeogeogr., Palaeoclimatol., Palaeoecol. 9: 27–57.

———. 1977. Paleogene Floras from the Gulf of Alaska Region. U.S. Geological Survey, Reston, VA. Professional Paper 997, pp. 1–108.

———. 1978. A paleobotanical interpretation of Tertiary climates in the northern hemisphere. Am. Sci. 66: 694–703.

———. 1979. Temperature parameters of humid to mesic forests of eastern Asia and relation of forests of other regions of the northern hemisphere and Australasia. U.S. Geological Survey, Reston, VA. Professional Paper 1106, pp. 1–37.

———. 1981. Paleoclimatic significance of the Oligocene and Neogene floras of the northwestern United States. In: K. J. Niklas (ed.), Paleobotany, Paleoecology, and Evolution. Praeger Publishers, New York. Vol. 2, pp. 79–101.

———. 1987. An overview of the origins of the modern vegetation and flora of the northern Rocky Mountains. Ann. Missouri Bot. Garden 74: 785–803.

———. 1989. Leaf-architectural analysis of the Hamamelididae. In: P. R. Crane and S. Blackmore (eds.), Evolution, Systematics, and Fossil History of the Hamamelidae, Vol. 1: Introduction and "Lower" Hamamelidae. Clarendon Press, Oxford, U.K. Systematics Association Special Vol. No. 40A, pp. 75–104.

———. 1990. Estimates of Pliocene Precipitation and Temperature Based on Multivariate Analysis of Leaf Physiognomy. U.S. Geological Survey, Reston, VA. Open file report 90-64, pp. 39–42.

———. 1993. A method of obtaining climatic parameters from leaf assemblages. U.S. Geol. Surv. Bull. 2040: 1–71.

———. 1994. Alaskan Palaeogene climates as inferred from the CLAMP database. In: M. C. Boulter and H. C. Fisher (eds.), Cenozoic Plants and Climates of the Arctic. NATO ASI Series 1: Global Environmental Change. Springer-Verlag, Berlin. Vol. 27, pp. 223–237.

———, A. B. Herman, and R. A. Spicer. 1996. Inferring climatic parameters for Cretaceous and Tertiary leaf assemblages using CLAMP (climate-leaf analysis multivariate program). International Organization of Palaeobotany Fifth Quadrennial Conference, Santa Barbara, CA, June/July, 1996. p. 113.

Wolfe, J. A. and D. M. Hopkins. 1967. Climatic changes recorded by Tertiary land floras in northwestern North America. In: K. Hatai (ed.), Tertiary Correlations and Climatic Changes in the Pacific. Sendai, Sasaki, Japan. pp. 67–76.

Wolfe, J. A. and G. R. Upchurch, Jr. 1987. North American non-marine climates and vegetation during the Late Cretaceous. Palaeogeogr., Palaeoclimatol., Palaeoecol. 61: 33–77.

General Readings

Chasan, R. 1996. What fossil woods say about climate. BioScience 46: 803–804.

Keir, R. 1992. Packing away carbon isotopes. Nature 357: 445–446.

Kern, R. A. and W. H. Schlesinger. 1992. Carbon stores in vegetation. Nature 357: 447–448.

Lewin, R. 1985. Plant communities resist climatic change. Science 228: 165–166 (discusses vegetational change in the Grand Canyon region over the past 24,000 years via analysis of packrat middens).

Niklas, K. J. 1987. Aerodynamics of wind pollination. Sci. Am. 257: 90–95.

Visscher, H. 1997. LPP part two: paleophysiology. Am. Assoc. Stratigr. Palynol. Newsletter 30: 9–10.

Yamaguchi, D. K., B. F. Atwater, D. E. Bunker, B. E. Benson, and M. S. Reid. 1997. Tree-ring dating the 1700 Cascadia earthquake. Nature 389: 922–923.

Notes

1. Claims of having recovered fossil organic microstructures from unanticipated sources (viz., metamorphic rocks; meteorites; see references in Kremp, 1983; Staplin, 1962) provide the lore and legend that is an entertaining part of any field. In 1970 I received a rock sample from Professor Wilhelm Klaus of the University of Vienna who asked to have it processed for pollen and spores. It was clearly a highly indurated metamorphic rock that proved virtually impossible to mechanically or chemically disintegrate and, as expected, it was barren of palynomorphs. Professor Klaus was an eminent paleobotanist and certainly aware that pollen and spores are not preserved in highly metamorphosed rocks. The explanation came when he informed me later that palynologists working in one of the former Eastern Block countries claimed to have developed a processing technique whereby plant microfossils could be obtained from such rocks. This claim reached the Director at the Geologische Bundesanstalt who inquired why scientists in other countries could do this while his staff could not. Professor Klaus divided a sample into sections, sent one to each of several laboratories throughout the world, and requested a report be sent to the Director. All the samples were barren and the matter was settled.

The novel technique was based on the assertion that palynomorphs were often present in metamorphic rocks but were so delicate that they were destroyed by standard processing procedures. The suggested alternative was to place the rock in a beaker of water, set it on a window sill or, even better, outdoors, and let the ultraviolet rays of the sun gently decompose the sample. The procedure worked best when the beaker was uncovered. The palynomorphs were remarkably well preserved and quite modern in aspect. Years later on a congress cruise on the River Ob in Siberian Russia, I talked with Professor Elena Zaklinskaya of the USSR, who knew the details of the procedure. She indicated that, as expected, the technicians were well aware of the deficiencies and had used this obvious ruse to placate overzealous supervisors then anxious for any claim of technological superiority over the West.

2. Wolfe's (1993) classification of leaf-size categories is leptophyll 1, 2; microphyll 1, 2, 3; and mesophyll.

3. Taphonomy is the study of the transition of organic remains from the living organism to fossil assemblages, that is, events involved in the fossilization process (Efremov, 1940; cited from Greenwood, 1991; see also Gastaldo, 1988).

FIVE

Late Cretaceous through Early Eocene North American Vegetational History

70–50 Ma

CONTEXT SUMMARY

At the end of the Cretaceous the Appalachian Mountains had undergone 180 m.y. of erosion since their principal uplift in the Middle Pennsylvanian through the Late Permian (300–250 Ma), but they were higher and more rugged than at present and provided a somewhat more diverse vertically zoned array of habitats (Figs. 5.1, 5.2). In contrast, the Rocky Mountains were only ~1 km above sea level at 65 Ma; the Coast Ranges, Sierra Nevada, and Cascade Mountains would not attain substantial heights until late in the Tertiary.

Computer models in the NM mode, which simulate conditions in western North America in the Late Cretaceous, show a nearly continuous westerly jet stream with relatively small amplitude between the troughs (low-pressure systems) and ridges (high-pressure systems). The present north–south seasonal meandering of the jet stream was also less. Thus, in the models precipitation and temperatures were more uniform throughout the year and there was less regional differentiation in climate.

CO_2 concentrations during the Cretaceous are estimated to have been 4–8 times to 10–12 times higher than at present. With a 2–5°C warming for each doubling of CO_2, this provides part of the explanation for the higher MAT documented for the Late Cretaceous and Early Tertiary.

High CO_2 concentrations near the end of the Cretaceous may have been a holdover from earlier intense volcanism in the South Pacific that began to subside at ~100 Ma. The term epeirogeny refers to vertical motions of the Earth's crust, and these movements affect the ocean floor, as well as the continents. There was a 50% increase in the production of ocean crust in the Middle Cretaceous compared to earlier times, as represented by the early Aptian Ontong Java Plateau, the Earth's largest oceanic plateau, now submerged over 2 km off the Solomon Islands. A sense of the magnitude of this structure can be gained by comparing its volume with that of the surface-exposed Deccan Traps of India (66 Ma). The latter are ~1 km thick and have a volume of 1.5×10^6 km^3. The Ontong Java Plateau is ~36 km thick and has a volume of 50×10^6 km^3. It is estimated to have formed in just a few million years and was only part of the midocean ridge, oceanic plateau, and flood basalt activity of the Middle to Late Cretaceous. The speculated origin of the structure is that a part of the lower mantle broke away and moved upward as a superplume beneath the Pacific. The waning of volcanism after ~100 Ma, and the associated reduced levels of CO_2, contributed to the decline of Late Cretaceous temperatures (Fig. 3.1). Although the oxygen isotope paleotemperature curve reveals that temperatures declined through the Late Cretaceous, values for benthic ocean waters were still 10–12°C higher at the K–T boundary than the present 1–2°C.

The term equable (warm, low seasonality) is often used to describe Cretaceous climates, and there was only sluggish equator to pole atmosphere and ocean circulation (Barron, 1983). Axelrod (1992a) notes the need for a better definition of the term, and recent modeling results show it may imply an overly simplistic view of Cretaceous and Early Tertiary environments. GCM simulations demonstrate that in the highlands at some far-northern interior sites, continental configuration and size facilitated a moderate ocean circulation and seasonal temperature regime, possibly reaching near freezing (Barron and Washington 1982a,b; Barron et al., 1980, 1981; Crowley et al., 1986). Rea et al. (1985) interpret eolian dust records to indicate an atmospheric circulation as vigorous as in the past few million years. Thus, the Hadley Cell was operating in the Late Cretaceous even though the continents were arranged differently, and there were extensive epicontinental seas (Horrell, 1991; Zeigler et al., 1984). However, as noted by Sloan and Barron (1990), models that predict Cretaceous and

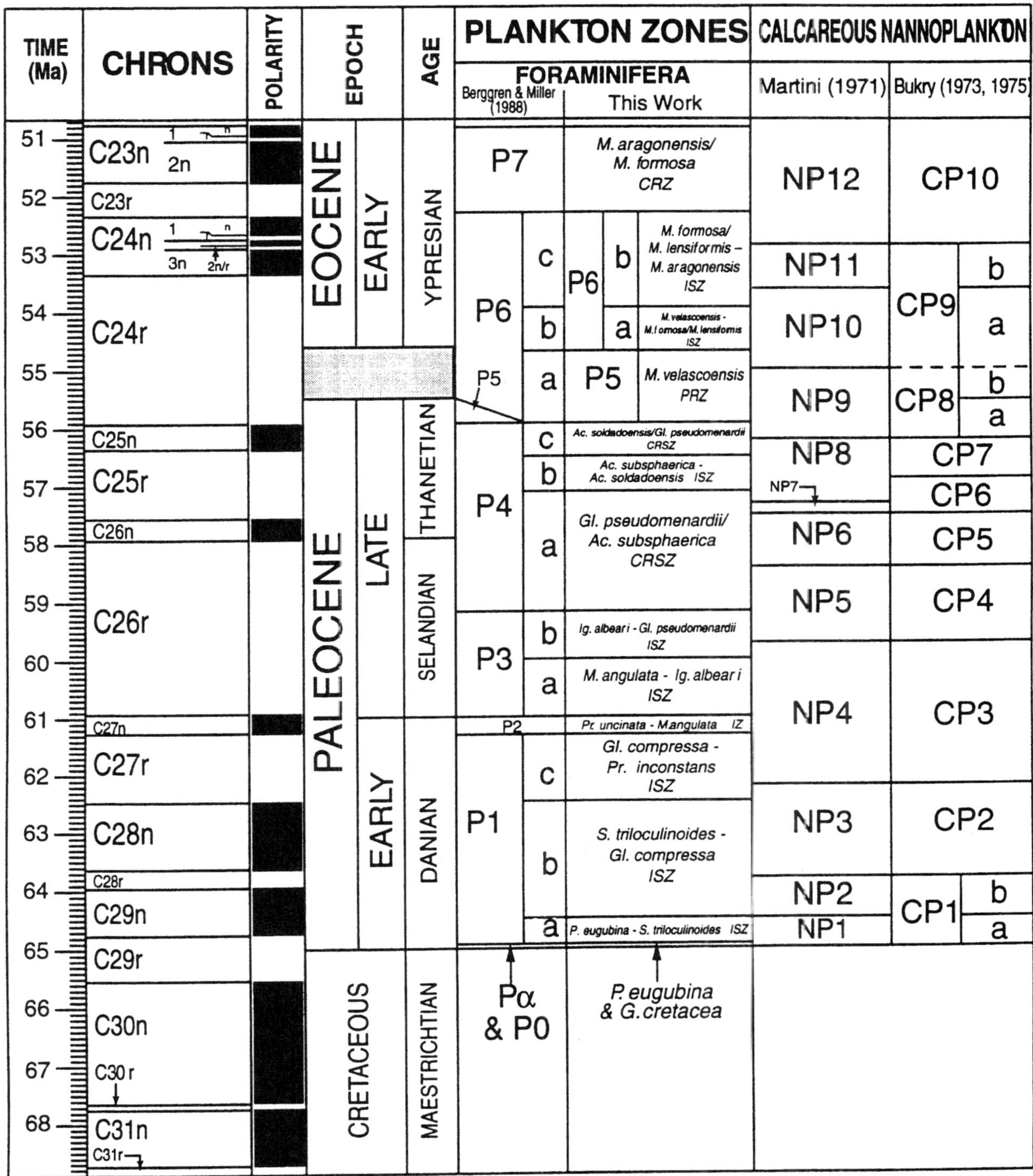

Figure 5.1. Paleocene time scale. Reprinted from Berggren et al. (1995) with the permission of the Society for Sedimentary Geology.

Early Tertiary temperatures remained substantially below freezing for long periods of time over extensive areas, even at the high interior northern latitudes, are not consistent with a wealth of paleontological data (Herman and Spicer, 1996). It is true that many of these data come from warmer coastal sites, as opposed to inland regions, but existing information is difficult to reconcile with frequent and prolonged periods of freezing. It has been suggested that paleontologists reevaluate their data and interpretations of the Cretaceous as essentially ice free (Barron et al., 1981). It is also possible that the models may be depicting temperatures that are too cold because coarse grid constraints do not fully incorporate local to regional physiographic features (e.g., the paleo-Brooks Range). The Cretaceous plant microfossil record from the North Slope of Alaska (Frederiksen, 1989a; Frederiksen et al., 1988) includes over 100 different kinds of pollen and spores. This means that forests and understory plants covered considerable parts of the continental interior and that regional vegetation was more extensive and diverse than suggested by some comparatively meager Arctic megafossil assemblages (Parrish and Spicer, 1988). Associated faunal evidence (e.g., dinosaur remains from the North Slope) further document the mesic nature of the climate. Computer simulations show that forest cover theoretically can warm high-latitude climates by 2.2°C, reducing discrepancies between the observed and modeled latest Cretaceus climates (Otto-Bliesner and Upchurch, 1997). The possibility, however, of seasonal, cool-temperate environments in the interior of the far north during the Cenomanian (Sellwood et al., 1994) and Maastrichtian, and even local glacial conditions above 1000 m at 85° N in the newly uplifted Brooks Range, is not precluded by the paleontological data. Seasonality in temperature, combined with a moderately diverse topography, allows a greater array of habitats to accommodate the more diverse vegetation now recognized from pollen evidence than does an oversimplified generalization that climates were "equable" across the entire continent.

In the Late Cretaceous through at least Middle Eocene times sea level was ~300 m higher that at present, showing a decline near the end of the Early Paleocene (60 Ma; Fig. 3.6). The highest levels of this interval would cover about 20% of the present Earth's surface. The high ocean level was due primarily to Mesozoic plate reorganization and to the absence of substantial glaciers. The formation of the mid-Atlantic Ridge beginning in the Jurassic, and the uplift of the Ontung Java Plateau in the Pacific, resulted in a decrease in ocean basin volume of about 2×10^{16} m^3. These events displaced water upon the continental margin and contributed to the flooding of the low-lying interior region of midcontinent North America. Global land–sea relationships at 100 Ma are shown in Fig. 5.3. At various times during the Cretaceous (e.g., in the Cenomanian), the waters of the present Gulf of Mexico and the Arctic Ocean connected to form a mostly shallow epicontinental sea that stretched some 5000 km through the continent and partitioned North America into an eastern and western portion. Among the effects of this sea were milder climates, muted seasonal variation in temperature, and the provincialization of the North American flora into eastern and western provinces.

Reduced land surfaces and the extensive midcontinent sea affected the Earth's albedo through greater absorption

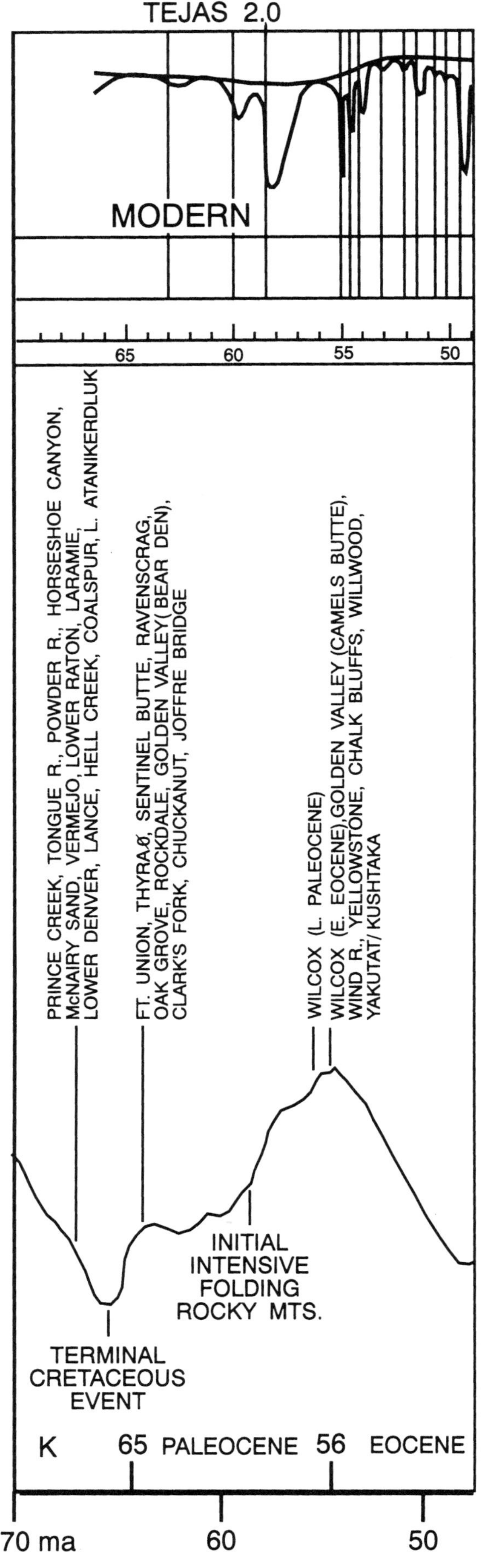

Figure 5.2. Principal fossil floras mentioned in the text plotted with general reference to the paleotemperature record (lower curve) and sea-level record (upper curve) for the Late Cretaceous through the Early Eocene. The Chalks Bluff flora is now dated at ~48 Ma.

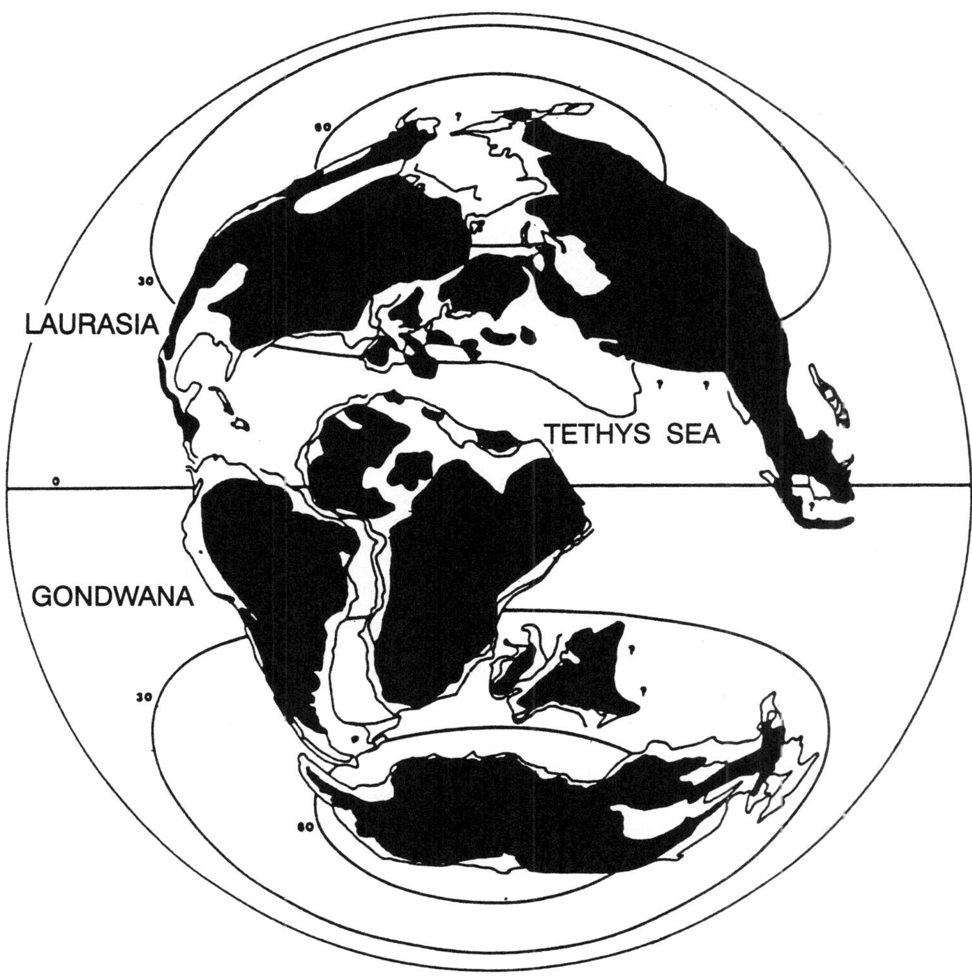

Figure 5.3. Generalized diagram of continental positions and land–sea relationships (maximum depth of 100–200 m) at 100 Ma. Modified from Crowley and North (1991) and based on Barron et al. (1980). Reprinted with permission of Elsevier Science-NL.

of solar heat. A sea level higher by 300 m would produce a change in albedo of 0.03, which would translate into a rise in surface MAT of ~3°C. More detailed calculations of the Cretaceous albedo in the region of North America are speculative because the important variable of the extent and thickness of cloud cover cannot be modeled accurately (Tselioudis et al., 1993), and there are no paleoclimatic indicators for cloud cover per se. If there was no cloud cover, present ambient temperatures would be ~20°C higher because water vapor clouds (as opposed to CO_2 clouds) are more effective at reflecting solar radiation than trapping heat through a greenhouse mechanism.

During the Mesozoic, North America was joined to Europe and Asia across the present North Atlantic and Pacific Oceans. The Asian connection is evident, for example, from the presence of *Triceratops* in eastern Asia and western North America. By the Cretaceous the continents had started to separate, and at ~66 Ma they had the configuration shown in Fig. 5.4. The North Atlantic connection had only begun to fragment in the Maastrichtian as revealed by the initial mixing of Arctic and Atlantic waters through the Davis Strait. Thus, in terms of plant and animal migrations there were no major physical barriers across the North Atlantic. Climates likely included coastal subtropical zones and warm-temperate to temperate ones locally in the northern regions. Similar climates prevailed through Beringia.

Late Cretaceous and Paleogene faunas are known from the margin of the epicontinental sea at localities from Texas to the North Slope of Alaska. They are particularly abundant and diverse from southern Kansas to southwestern Saskatchewan. The first report of Mesozoic mammals, in the form of abundant teeth, comes from the Lance Formation of Wyoming. Other faunas are known from the Hell Creek Formation of eastern Montana and adjacent regions and from the Scollard Formation of Alberta, Canada. The latter includes the Bug Creek fauna, which is the richest deposit of Mesozoic mammals in the world. These faunas contain Allotheria (multituberculates), Metatheria (marsupials), Eutheria (placentals), turtles, crocodilians, and champsosaurs (Hutchison, 1982). The presence of nonhibernating, nonburrowing species with limited migration capabilities further implies warm climates without pro-

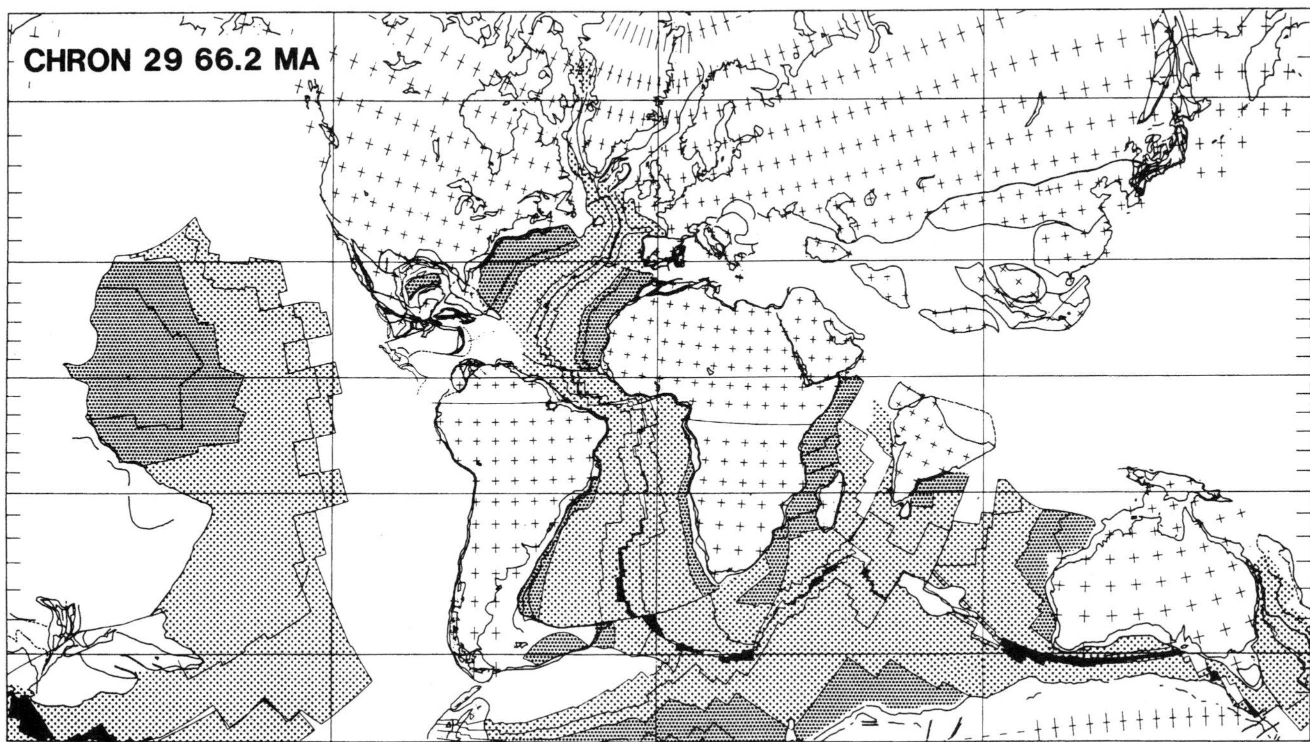

Figure 5.4. Plate reconstruction at chron 29 (Paleocene, 66.2 Ma). Reprinted from Scotese et al., (1989, in Scotese and Sager, 1989) with the permission of Elsevier Science-NL.

longed periods below freezing. Coral reefs were present 5°–15° N of their present distribution (Habicht, 1979).

At the end of the Cretaceous an asteroid collided with the Earth, leaving the Chicxulub crater now buried under the Yucatan Peninsula. About 44% of the genera and 70% of the species comprising the marine plankton and many large terrestrial animals, notably the dinosaurs, became extinct at this time. In contrast, many other terrestrial groups crossed the K–T boundary without extinction or their diversity had already begun to decline during the Late Campanian and Maastrichtian (Frederiksen, 1989a). Paleocene faunas still contain many elements found in the Late Cretaceous Bug Creek assemblage (Webb, 1989).

An asteroid impact is such an important and spectacular occurrence that it is tempting to interpret all kinds of data as consistent with this highly publicized event. As the duration of its effects are scaled down and its consequences found to be selective, less tailored claims likely will be forthcoming regarding the relationship between the event and vegetational history. [For a balanced summary of the spectrum of possible effects of the impact, see Crowley and North (1991).]

The principal biological consequence of the terminal Cretaceous event appears to have been extinction among the groups noted previously. Its most intense climatic consequences were brief (4–6 months), and the oxygen isotope record does not detect any major ocean temperature change bracketing the event. The impact may have temporarily increased CO_2 levels by 2–3 times (Hsü and McKenzie, 1985) within an estimated concentration of 10–12 times that of modern values. The trend from evergreen to deciduous vegetation resulting from a 4°C decline in temperature between the Middle to Late Cretaceous (Fig. 3.1) may have been strengthened by an impact winter and reduced light of 4–6 months' duration (Wolfe, 1987). The effects are most evident in the midlatitudes of western North America where the most complete fossiliferous transitional sections are available. The collision of an asteroid with the earth at the end of the Cretaceous is convincingly established. Present efforts are directed at assessing the relative roles of gradual environmental change and an abrupt catastrophy on climate and biotas.

In the Paleocene through the Early Eocene, the Rocky Mountains underwent their initial intensive folding to about one-half or possibly locally to near their present average elevation. In the northern Rocky Mountains there was considerable volcanic activity in the Early to Late Eocene. Two effects of the development of these uplands (to ~2500 m or more) were to divide the previously continuous lowlands of western North America into eastern and western zones and to reduce the flow of moist Pacific winds into the eastern Rocky Mountains and Plains area (Wing, 1987).

North America remained connected to Europe and Asia via the North Atlantic and Bering land bridges during the Paleocene and Early Eocene. Even though Greenland was

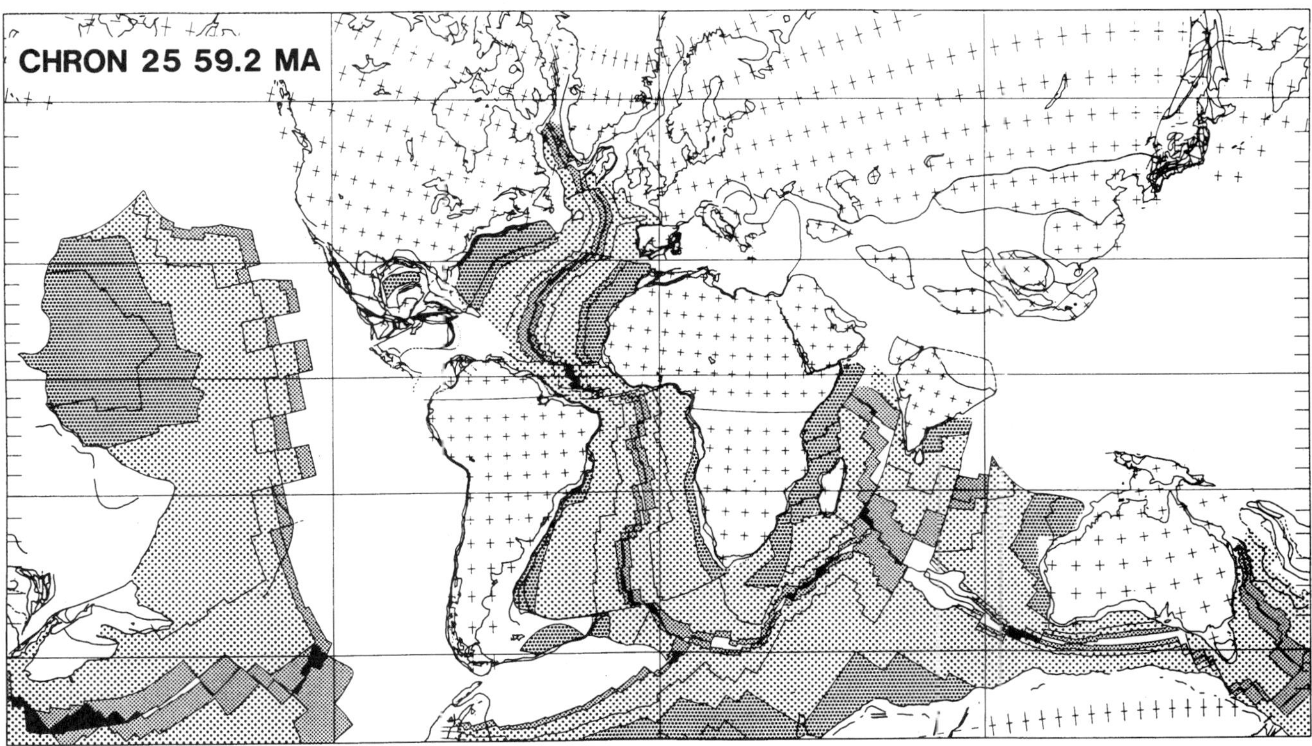

Figure 5.5. Plate reconstruction at chron 25 (Eocene, 59.2 Ma). Reprinted from Scotese et al. (1989; in Scotese and Sager, 1989) with the permission of Elsevier Science-NL.

mostly separated from Europe (Fig. 5.5), the distance was not great and migrations were facilitated by the mild climate. There were significant drops in sea level at 60, 49.5, and 40 Ma, representing periods of relative quiescence in ocean ridge activity and, at the latter dates, possibly early glaciations on Antarctica. The epicontinental sea was retreating, and the High Plains were beginning to appear due to continued uplift of the Rocky Mountains that tilted the Plains area by ~1 km over a distance of ~1400 km. Drainage was also affected by continental rebound as the sea disappeared and sediments began to erode. By the end of the Early Eocene the inland sea had mostly disappeared, and Gulf and Atlantic waters occupied the areas shown in Fig. 5.6.

Lower Tertiary faunas of southwest Texas include sharks, skates and rays, tarpon, and crocodylidae; those from Ellesmere Island included alligatorlike crocodilians, turtles, and perissodactyl mammals. Collectively, the North American Paleocene through Early Eocene faunas reflect widely distributed tropical to warm-temperate elements with strong affinities to Asia and Europe. As noted previously (Chapter 3), the Early Eocene represents the strongest mammalian similarity that ever existed between North America and Europe. There was little or no interchange as yet with the faunas of South America.

Ocean temperatures increased and reached a maximum for Cenozoic time in the Early Eocene (56 Ma; Fig. 3.1). The values are estimated at between 11 and 15°C warmer than at present for benthic waters and 21°C for surface waters in the high latitudes. There is no evidence for substantial continental glaciers during the interval between the Late Cretaceous and the end of the Early Eocene.

This varied and independent evidence provides a valuable set of parameters for interpreting the Late Cretaceous through Early Eocene vegetation of North America. It would be expected that the communities had broad geographic affinities, facilitated by the Thulean and DeGeer routes to eastern Laurasia and the Beringian land connection to eastern Asia, although North America itself was divided into eastern and western provinces by the Cretaceous midcontinental sea. The vegetation grew under MATs that were significantly warmer than at present. Under these conditions communities should be distributed along low thermal gradients from south to north with broad transitions and less differentiation into climatically or altitudinally zoned communities, except in the Appalachian Mountains that provided a moderate diversity of altitudinally zoned habitats. As a rough approximation of the south to north temperature cline, modeling results estimate a MAT of 23°C at 30° N latitude and 16°C at 50° N latitude (Thompson and Barron, 1981). Isotopic evidence yields a cline from 27 to 16°C, while present values at the same latitudes are from 20 to 6°C. Lloyd (1982) estimates the Cretaceous 18°C winter isotherm occurred at 40–50° latitude from the equator, compared to its present position of ~30°. Paleobotanical data suggest a +5°–10°C difference

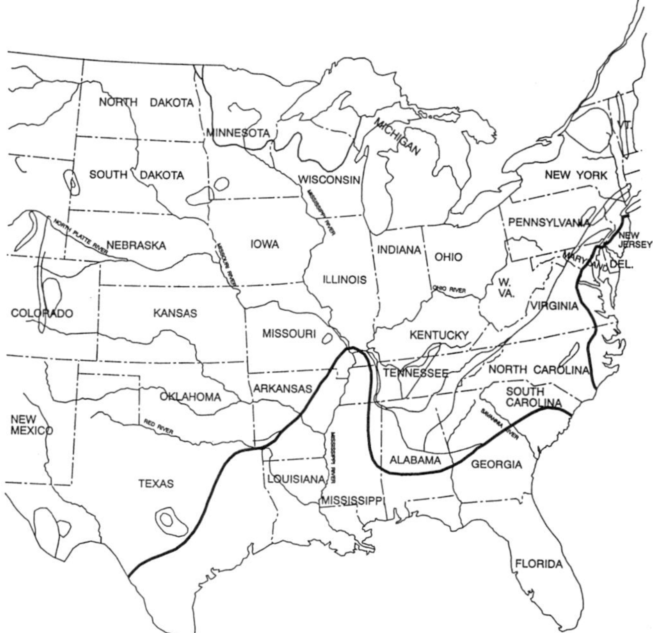

Figure 5.6. Approximate position of the Gulf–Atlantic coastline at the end of the Early Eocene. Based on the Geologic Map of North America compiled by Marshall Kay.

at paleolatitude (pl) 30° N (Upchurch and Wolfe, 1987) and +30°C at pl 80° N (Spicer and Parrish, 1986a), compared to modern surface air temperatures.

VEGETATION CLASSIFICATION

Paleobotanical information necessary for tracing the history of North American vegetation is not uniformly available either stratigraphically or geographically. As a consequence, this history must at times be extrapolated from existing data elsewhere or inferred from the various contexts previously discussed. This is particularly true for the Maastrichtian and Paleogene where a larger number of floras are known from the Western Interior basins than from eastern North America.

Interpretation of the plant data is facilitated when biological affinities can be established for the fossils to at least the level of family or genus. Artifical or form names must be used for fossils beyond the range of extant taxa or where the modern analog is not known. In many paleopalynological studies the primary goals are age determination, biostratigraphy, and correlation. For these geological applications, the biological affinities of the specimens are of secondary importance (Leffingwell, 1971; Norton and Hall, 1967, 1969; Oltz, 1969; Snead, 1969; Srivastava, 1970; Standley, 1965; Sweet, 1978; Tschudy, 1970). Palynological studies as part of petroleum industry research are proprietary, and the published material is often limited to taxonomy and general discussions of stratigraphy and depositional environments. Thus, even though paleobotanical information may be available for a particular age or locality, it is not always in the form necessary for tracing vegetational or phylogenetic histories; this is especially true for Late Cretaceous and Early Tertiary floras.

Another factor influencing the data base is the new more inclusive approach to the study of many Late Cretaceous and Tertiary floras. At the time of E. W. Berry and other early workers, it was conventional to consider fossil floras mostly as separate units with the goal of expeditiously providing a statement of the composition and paleoenvironment for each assemblage (viz., paleofloristics). One consequence of this approach was that variation within the same taxon from distant sites was not immediately evident, which contributed to a proliferation of names and a confused nomenclature. Also, modern methodologies such as electron microscopy, X-ray radiograms, and cleared leaf and pollen–spore reference collections were mostly not available. Identifications were frequently based on marginally preserved specimens, and many Cretaceous and Paleogene plants were erroneously assigned modern family or generic names on the basis of superficial resemblances in morphology. Knowledge of the distribution and ecological characteristics of the modern analogs, particularly for tropical plants, the dynamics of vegetation change (e.g., the fluidity of community composition over time), plate tectonics (for biogeographic studies essentially a post-1970 data base), and comparative material from extant plants were all meager by modern standards. The holdings at the United States National Herbarium, used by E. W. Berry for his study of southeastern United States and tropical American Tertiary plants, numbered ~1,223,400 specimens in 1925, compared to the present ~4,500,000 accessions. For these and other reasons the species lists for many of the older megafossil floras are suspect.

A recent trend is to focus on lineages preserved across

several floras, ages, and geographic regions. Although the initial aim may be the study of an individual flora, when exceptional material is encountered it is often used as a basis for investigating the worldwide fossil record of the lineage with the goal of providing a statement of its origin, phylogenetic relationships, and biogeographic history and to provide a more sound nomenclature (viz., paleomonographic studies). An example is the current project by David Dilcher and coworkers on the Middle Eocene Claiborne and older associated ("Wilcox") floras of the southeastern United States. The floras are being revised, but paleomonographic works are published as important material is discovered. Among these are treatments of the Poaceae (Gramineae; Crepet and Feldman, 1991), Gentianaceae (Crepet and Daghlian, 1981), Juglandaceae (Manchester, 1987), and Malpighiaceae (Taylor and Crepet, 1987). These paleomonographs are often done in collaboration with paleobotanists in several parts of the world and involve taxonomic specialists in the extant group (e.g., *Quercus*; Crepet, 1989; Crepet and Nixon, 1989a,b; Nixon, 1989; Nixon and Crepet, 1989). Similarly, Wolfe and Tanai (1987) have integrated information from fossil and extant *Acer* in a proposed new phylogeny and subgeneric classification, and Patrick Herendeen is providing similar information for the Leguminosale (see References, Chapter 6). A perusal of recent papers by Ruth Stockey and her coworkers on the Princeton flora of Alberta, Canada (Chapter 6), is witness to the varied techniques, time, and meticulous care now devoted to the identification of fossil plant remains. Such studies are of considerable value in providing a better understanding of the lineages and a reliable species list for the floras. In the meantime it is important to recognize that many North American Late Cretaceous and Tertiary floras are presently a mixture of forms that have been studied recently and intensely, together with those that have not been revised for 75 years or more.

In describing the past vegetation and environments of North America it is useful to follow a standardized terminology as far as possible. Tropical, subtropical, temperate, and cold temperate give a general but unquantified impression of climates; a comparable array of terms is applied to moisture regimes and vegetation types. Greater consistency in the way these terms are used is desirable and essential for statistical treatment of the data. On the other hand, it is futile to attempt overly precise definitions of entities and events that are by nature fluid and transitional.

Systems of vegetation classification have importance beyond the naming of communities. If versions of essentially modern vegetation types are viewed as extending back into the Paleogene, names such as tropical rain forest, subtropical forest, deciduous forest, mixed conifer forest, and various combinations will be used. This terminology is compatible with the concept of geofloras moving over the landscape in response to climatic change. If the paleobotanical record is interpreted as reflecting more fluid and dynamic assemblages, names will be preferred that imply the temporal nature of paleocommunities. The Late Cretaceous vegetation in western North America between latitudes 40°–50° N and 60°–65° N can be considered an ecotonal community between a subtropical evergreen and a deciduous forest, or it can be called a notophyllous broad-leaved evergreen forest. Concepts and semantics are often interwoven, and the system of classification can be used to imply a general similarity or a fundamental difference between paleovegetation and the modern flora. From the standpoint of vegetational history it is convenient to project modern vegetation formations as far back into the past as is justifiable, as do Axelrod et al. (1991), because it keeps a focus on the formation, its origin, and its development through time. It is also true that the structure and composition of vegetation have changed over time, and at some point the nomenclature should reflect these differences. The transition is variably around the Early Eocene, as revealed by the increasingly modern aspect of many megafossils and palynomorphs at the generic level in Middle Eocene and later times. A special nomenclature for the older communities has been developed by Wolfe, and it is used here to convey their ancient and unique character.

Tropical forest—MAT ~25°C at the transition from tropical to paratropical rain forest (Asia; Wolfe, 1979, p. 7), subhumid, estimated rain fall < 1650 mm, little seasonality, growth rings absent to poorly developed; an extinct, broad-leaved evergreen, single-tiered, open canopy vegetation, leaves mostly entire-margined, thick textured, few drip tips, leaf size indices below ~30.

Tropical rain forest—mean temperature of the coldest month does not fall below ~18°C, MAT generally above ~25°C, no pronounced or extended dry season; broad-leaved, evergreen, multistratal; drip tips, lianas, and buttressing common; leaves sclerophyllous, mostly mesophylls (megaphylls in substratum), entire-margined leaves average over ~75%.

Paratropical rain forest (= near-tropical; subtropical of many authors)—may experience some frost (–1°C to –3°C), MAT ranges from 20°C to 25°C, precipitation may be seasonal but there is no extended dry season; floristically allied to the tropical rain forest, predominantly broad-leaved evergreen with some deciduous plants, woody lianas diverse and abundant, buttressing present; entire-margined leaves ~57–75%.

A distinction frequently made between tropical and paratropical forests is that the former is typically 3-tiered while the latter has two canopy layers. This was based on the classic work of Richards (1952), but a 3-tiered canopy is no longer viewed as a defining feature of the tropical rain forest. The two forest types are similar, particularly as represented in the fossil record.

Subtropical forest—frosts are present but not severe, MAT is between 13 [and] 18°C, mean of the coldest month is between 0 [and] to 18°C, rainfall is more seasonal; sclerophylls are abundant, there are few lianas and no buttressing, predominantly a broad-leaved evergreen forest with some conifers and broad-leaved deciduous elements; entire-margined leaves ~39–55% (here mostly included in paratropical rain forest or notophyllous broad-leaved evergreen forest)

Notophyllous broad-leaved evergreen forest (ecotonal; oak–laural forest of eastern Asia)—mean of coldest month ~1°C, MAT ~13°C; some broad-leaved deciduous trees present, conifers not common, woody climbers abundant, buttressing virtually absent, sclerophyllous, no drip tips, small mesophyll (notophyll) size class; entire-margined leaves 40–60%.

Warm temperate forest—temperatures fall below 0°C during several months, there is a pronounced seasonality, mean annual temperature range is between 11 [and] 13°C, mean of coldest month is between −3 [and] −2°C; a broad-leaved deciduous forest which may have a significant percentage of conifers, broad-leaved evergreens are present but not dominant; entire-margined leaves ~30–38%.

Polar broad-leaved deciduous forest—MAT ~7–8°C to 15°C, distinct growth rings present; an extinct mesothermal to microthermal moist forest type related to the modern Deciduous Forest Formation; leaves large, thin-textured. (Wolfe, 1977, pp. 24–26; 1979; 1981a, p. 82)

Late Cretaceous Vegetation: Maastrichtian, 70–65 Ma

At the end of the Cretaceous (Fig. 5.7, Table 5.1) the North American continent was divided by an epicontinental sea (Jorgensen, 1993) into two floristic provinces (Batten, 1984; Frederiksen, 1987; Herngreen and Chlonova, 1981; Srivastava, 1981, 1994a; Fig. 5.8). In the east and extending into western Europe, the Normapolles group is characteristic of Middle to Upper Cenomanian (Cretaceous) to Early Oligocene deposits. Normapolles typically have three colpi or pores that are structurally complex and frequently protruding (Traverse, 1988; Fig. 5.9). They represent a complex of early Hamamelidae (Betulaceae, Juglandaceae, the related family Rhoipteleaceae, or an extinct family of the Juglandales; Friis, 1983). Near the end of the Cretaceous and into the Early Tertiary they became more diverse and widely distributed. In the Cenomanian worldwide there were about six genera, while by the Maastrichtian there were more than 50. Of the latter, ~26 have been reported from the Late Cretaceous of the Atlantic coastal plain and ~19 from the Mississippi Embayment (Tschudy, 1975, 1980, 1981). Recently Areces-Mallea (1990) reported the Normapolles *Basopollis* from the Paleogene of Cuba. The most widespread North American forms are *Atlantopollis*, *Complexiopollis*, and *Trudopollis* (Fig. 5.9). Many Normapolles have smooth exines suggesting wind pollination. This mechanism is most typical of plants in open habitats with at least moderate and/or seasonal dryness.

In the west and extending into Asia the microfossil floras include a unique group of pollen types known as the Triprojectacites. These are characterized by three prominent equatorial and up to two polar projections (Fig. 5.10). They range in age from the Turonian (the oldest records are from Siberia) through the Paleocene and reach their greatest diversity in the Campanian. Farabee (1990, 1991, 1993) recognizes 14 morphotypes (approximate genera) among 175 described species, and one genus (*Aquilapollenites*) is used to designate the western North American–eastern Asian province. As a group, the Triprojectacites are polyphyletic. The spinulose forms (e.g., *Aquilapollenites*) have some morphological similarities to the pollen of certain extant Loranthaceae and Santalaceae (Jarzen, 1977). In contrast to the Normapolles group, the Triprojectacites disappeared during the Early Tertiary. The southern terminus of both provinces in North America is marked by tropical megathermal vegetation. In the far north the Triprojectacites are found within the primarily Normapolles province (Greenland, the North Sea), suggesting that some parent plants may have been circumpolar.

The Late Cretaceous and earliest Tertiary plant fossils in North America were derived primarily from vegetation growing along the coastal plains and slopes of the Appalachian Mountains, Rocky Mountains, and the Alaska and Brooks ranges. Their distribution essentially outlines the margin of the epicontinental sea and adjacent oceans (Figs. 5.7, 5.8). In the southeastern United States the McNairy Sand Formation (or Member of Berry's 1925 Ripley Formation) of Tennessee (Tables 5.1, 5.2) is Maastrichtian in age; although it contains both leaves and wood, the material has not been studied taxonomically since Berry's (1925) original publication.[1] Thus, it would be unwise to attempt vegetational or paleoenvironmental reconstructions on the basis of the fossils he assigned or compared to modern genera *Acer*, *Aralia*, *Artocarpus*, *Bumelia*, *Cedrela*,

Figure 5.7, facing page. Landform map of the conterminous United States with overlay of principal Late Cretaceous through Early Eocene floras and basins mentioned in the text (Thelin and Pike, 1991). (See also Figs. 2.13–2.14.) (1) U.S. Geological Survey's Oak Grove core (Paleocene), (2) Wilcox flora of western Tennessee (early Eocene), (3) McNairy Sand Formation (Maastrichtian), (4) Rockdale Formation (microfossils; Paleocene), (5) Sentinel Butte flora (Paleocene), (6–8) Golden Valley Formation (Bear Den and Camels Butte members; Late Paleocene to Early Eocene), (9) Raton Basin (flora; Maastrichtian), (10) Denver Basin (flora; Maastrichtian), (11) Wind River flora (late Early Eocene), (12) Willwood flora (Early Eocene), (13) Yellowstone floras (Middle Eocene), (14) Clark's Fork flora (Paleocene–Eocene), (15) Fort Union floras (and eastward into the Dakotas; Paleocene), (16) Chalk Bluffs flora (Early Eocene), (17) Chuckanut flora (Paleocene?). a, Vermejo Basin; b, Laramie Basin; c, Lance Basin

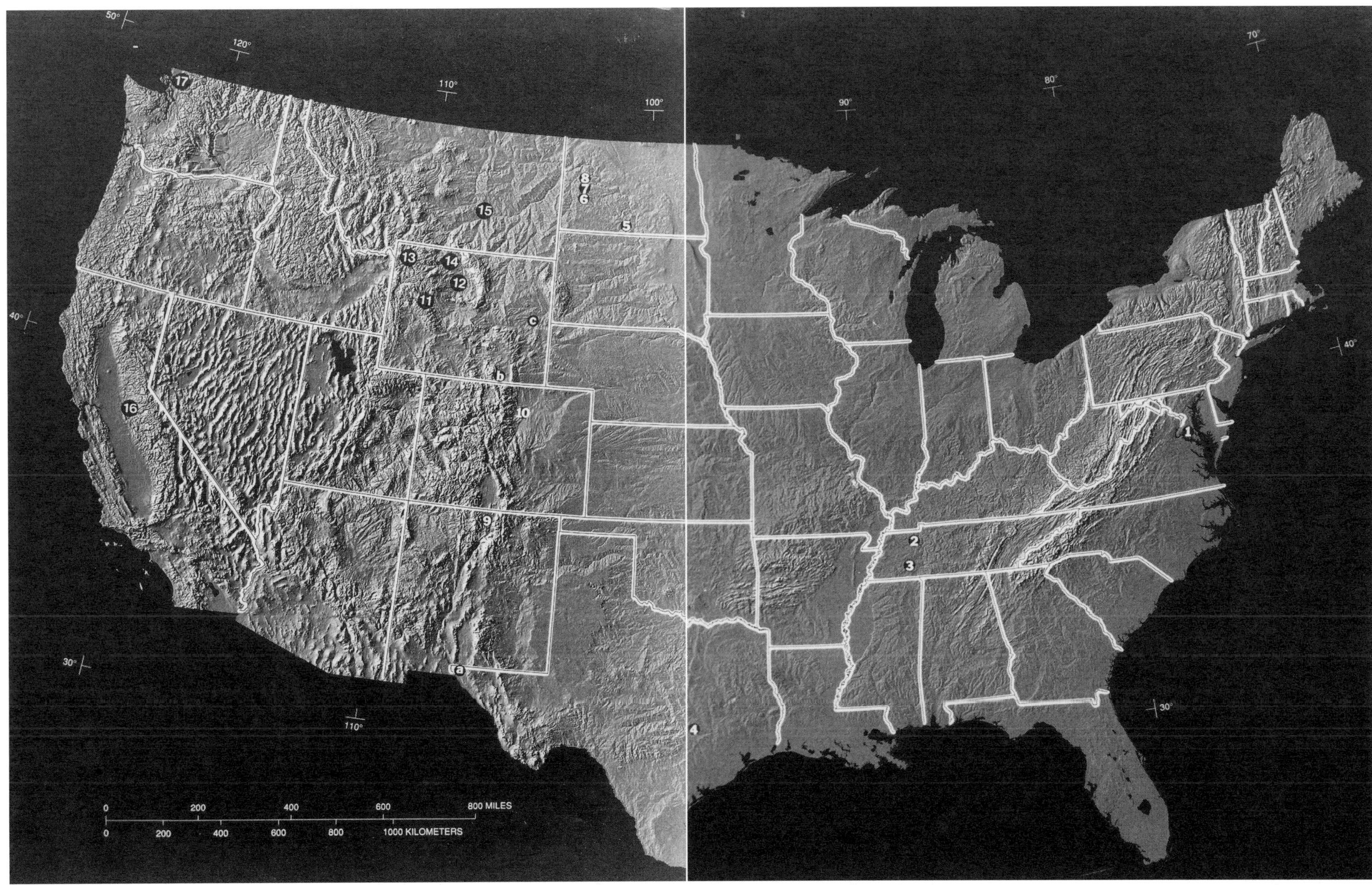

50°
120°
110°
100°
90°
80°
70°
40°
40°
30°
30°
110°
17
15
13
14
12
11
8
7
6
5
c
b
16
10
9
a
1
2
3
4
0 200 400 600 800 MILES
0 200 400 600 800 1000 KILOMETERS

Table 5.1. Geographic distribution of some Maastrichtian through early Eocene floras of North America.

	Maastrichtian	
Southeast	McNairy Sand	Wolfe and Upchurch (1987a), Berry (1925)
West Greenland	(?)Lower Atanekerdluk	Heer (1883)
Western interior		
Texas, New Mexico	Raton, Vermejo, Dawson, Arkose	Knowlton (1917, 1930)
Colorado	Denver, Laramie	Knowlton (1922), McGookey et al. (1972)
Wyoming, Montana	Hell Creek, Lance (Colgate), Medicine Bow	Dorf (1942), Nichols et al. (1990)
Alberta	Coalspur (Scollard)	Bell (1949)
	Lower Edmonton, St. Mary River	Bell (1949)
	Upper Edmonton, Whitemud	Berry (1935)
	Horseshoe Canyon	Serbet and Stockey (1991)
Pacific Coast	Patterson	Page (1979, 1980, 1981)
Alaska	Tongue River	Frederiksen (1989a)
	Prince Creek	Frederiksen et al. (1988), Parrish and Spicer (1988)
	Paleocene	
West Greenland	Upper Atanikerdluk	Heer (1883), Koch (1963, 1964), Pedersen (1976)
North Greenland	Thyra Ø	Boyd (1990)
Western interior		
North Dakota	Sentinel Butte	Crane et al. (1990)
Montana, Wyoming, North Dakota	Ft. Union	Brown (1962)
Alberta	Genesee	Chandrasekharam (1974)
Saskatchewan	Middle Ravenscrag,	Berry (1935)
	Upper Ravenscrag	Jarzen (1982a,b)
	Ravenscrag	McIver and Basinger (1993)
	Paskapoo	Bell (1949)
Alaska	Chickaloon	Wolfe (1966, 1972)
	Early Eocene	
Southeast	Wilcox (part)	See text
Western interior		
Wyoming	Wind River	Leopold and MacGinitie (1972)
	Yellowstone	Wheeler et al. (1977, 1978), Wing (1987)
Pacific coast		
California	Chalk Bluffs	MacGinitie (1941)
Alaska	Kushtaka	Wolfe (1977, 1985)

Adapted from Crane (1987, appendix) and Wolfe and Upchurch (1987a, table IV).

Celtis, *Chrysophyllum*, *Cinnamomum*, *Dalbergia*, *Eugenia*, *Euphorbiophyllum* (Fig. 5.11), *Fagus*, *Ficus*, *Grewiopsis*, *Juglans*, *Laurus*, *Liriodendron*, *Myrica*, *Nectandra*, *Potamogeton*, *Rhamnus*, *Salix*, and *Zizyphus*. Rather, the methodology developed by Jack Wolfe, described in the previous section, provides a more reliable basis for estimating climates and the type of vegetation. The Perry Place and Cooper Pit localities within the McNairy Sand Formation contain 53 and 50 species, respectively. Values for entire-margined leaves are 69 and 62% (and exceed 70% in other associated floras), most are thick textured, there are few species with drip tips, and the leaf-size indices are 28 and 29. These figures suggest a tropical forest–woodland (entire-margined leaves 57–75%; Fig. 5.12) dominated by broad-leaved evergreens and a megathermal temperature regime (MAT > 20°C) reaching 25°C during warm intervals. Little seasonality is inferred from the general absence or poor development of growth rings in the fossil woods, although anatomical features alone provide conflicting interpretations of Cretaceous climates (Wheeler and Baas, 1991). Low to moderate precipitation is indicated by the comparatively low leaf-size index. Extant megathermal vegetation with abundant rainfall and no pronounced dry season has a leaf-size index of ~75. Maastrichtian woods from southern Illinois have no growth rings (Wheeler et al., 1987), and wood anatomy suggests many eastern North American specimens were large trees typical of tropical angiosperms. Some trunks are up to a meter in diameter.

The type of vegetation suggested by these indices is not a typical multistratal, closed-canopy tropical rain forest, and it has no exact counterpart in modern New World communities. It is thus designated as a tropical forest (T in Fig. 5.8). A subhumid, open canopy where most members receive full sunlight is needed to explain the low leaf-size

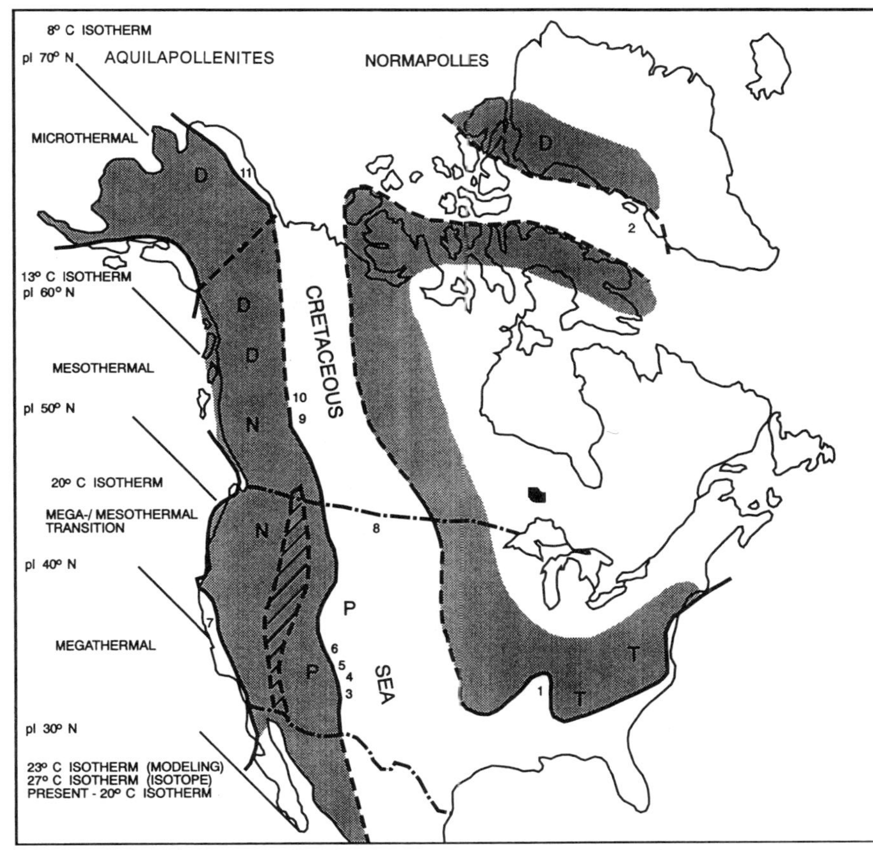

Figure 5.8. Generalized paleogeographic diagram of North America during the Late Cretaceous with approximate positions of paleolatitudes, principal formations–fossil floras mentioned in the text, and temperature zones. (1) McNairy; (2) Lower Atanekerdluk; (3) Vermejo; (4) Raton; (5) Denver; (6) Laramine; (7) Patterson; (8) Hell Creek–Lance (Colgate; Johnson, 1996); (9) Scollard, Horseshoe Canyon; (10) Coalspur; (11) Prince Creek (Colville River site). D, polar broadleaved deciduous forest; N, notophyllous broadleaved evergreen forest; P, paratropical forest; T, tropical forest. Area of continental inundation represents maximum extent during Late Cretaceous times.

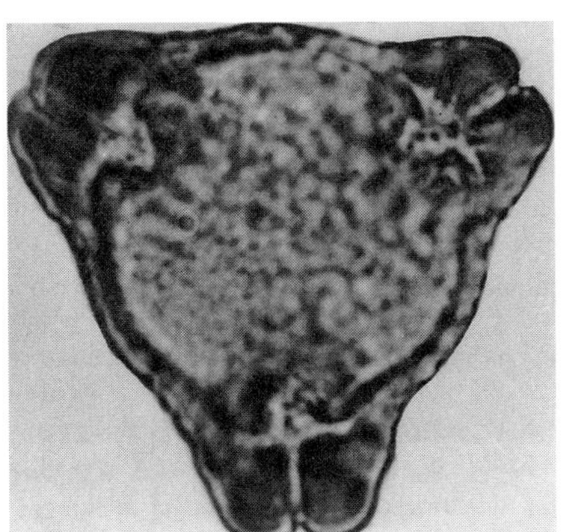

Figure 5.9. *Trudopollis variabilis*, a normapolles pollen from the Campanian Coffee Sand Formation, Tennessee (Tschudy, 1975 pl. 16, fig. 15).

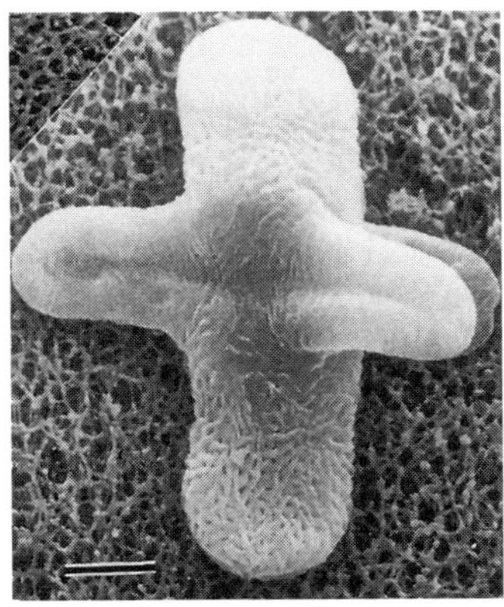

Figure 5.10. *Integricorpus rigidus*, a triprojectate pollen from the Campanian, Montana–Canada. Reprinted from Farabee (1990, fig. 16) with the permission of Elsevier Science-NL.

Table 5.2. Paleolatitudes and physiognomic data on North American Maastrichtian floras.

Assemblage	Formation	pl°	No. Sp.	Entire (%)	Leaf Index
Perry Place	McNairy Sand	37	53	69	28
Cooper Pit	McNairy Sand	37	50	62	29
L. Atanikerdluk[a]	Unnamed	55	56	77	12
Coalspur	Brazeau	66	6	—	84
Hell Creek	Hell Creek	56	21	62	53
Lance	Lance	52	40	58	38
Medicine Bow	Medicine Bow	50	42	67	55
Littleton	Denver	48	50	71	57
Laramie	Laramie	48	55	71	52
Lower Raton	Raton	46	43	72	34
Vermejo	Vermejo	46	63	71	34

pl, Paleolatitude. Adapted from Wolfe and Upchurch (1987a, tables V, VIII, IX).
[a]Lower Paleocene fide Koch (1964).

Figure 5.11. *Euphorbiophyllum cretaceum* (Euphorbiaceae) from the Late Cretaceous Ripley Formation, western Tennessee, illustrating the generalized nature and poor preservation of many specimens assigned to modern taxa in the older literature (Berry, 1925, pl. XII, fig. 6).

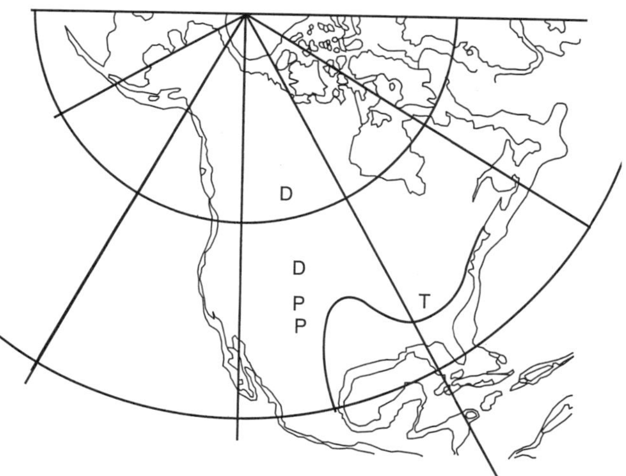

Figure 5.12. Generalized paleovegetation diagram for North America during the Early Paleocene. T, tropical rain forest; P, paratropical rain forest; D, broad-leaved deciduous forest. Reprinted from Upchurch and Wolfe (1987) with the permission of Cambridge University Press.

index; but regions supporting such extant megathermal vegetation typically have a pronounced dry season, the trees are shorter, and they grade into a semideciduous forest. Wolfe and Upchurch (1987a) provide the following estimates of precipitation for this unusual dry but not distinctly seasonal open-canopy woodland of southeastern United States. The estimates are based on climatic conditions and vegetation in New Caledonia and the vegetation of some porous sandy soils in the wet tropics, which have an abundance of conifers and many evergreen angiosperms: an annual rainfall of less than 1650 mm, a difference of less than 100 mm between the wet and drier season, and the driest month receiving at least 30 mm of rainfall. Recall from Chapter 4 that the suitability of a modern southern hemisphere analog for these Cretaceous floras is unsettled. The Ripley flora has recently been compared physiognomically to extant Sinaloan (Mexico) deciduous forest (Wolfe, in press). The distribution of coals and evaporites (Frakes, 1979) reflects rainfall values that were low in many regions during the Cretaceous and increased significantly in the Paleocene. Other sedimentological patterns are not so clear (e.g., Hallam, 1984), and interpretations from this line of evidence give a mixed picture of rainfall regimes (Parrish, 1987). Crepet (1981) and Batten (1984) also believe that the Late Cretaceous climate in the southeastern United States was dry; but based in part on the abundance of the presumably wind-pollinated Normapolles group, they also believe it was more seasonal (consistent with recent CLAMP analysis; J. A. Wolfe, personal communication, 1996). The early radiation of angiosperms from west Gondwana occurred from similar open areas in equable uplands with some seasonal drought (Axelrod, 1952, 1970; Stebbins, 1974). Wing and Tiffney (1987) cite grazing by dinosaurs as a factor in keeping the vegetation a low, seral forest, which they believe is indicated by small diaspore size. Wolfe and Upchurch (1987a) believe this small size is due to a dry climate. These differences in interpretation emphasize the approximate nature of the reconstructions. However, it is generally agreed that conditions in the southeastern United States during the Maastrichtian were warm and dry with at least moderate seasonality in rainfall.

To the far northeast there is a poorly known flora possibly of Maastrichtian age from western Greenland. It is from an unnamed formation and the depositional setting is not known, although it may have been along a floodplain. The Lower Atanikerdluk flora (Tables 5.1, 5.2) consists of 56 species; 77% have entire-margined leaves, requiring a megathermal climate, but with a leaf-size index of 12. This would indicate a tropical forest by Wolfe and Upchurch's (1987a) definition, but the leaf-size index is anomalously low for that community. Leaf fossils have been referred to the form genera *Proteoides*, *Dermatophyllites*, and *Chondrophyllum*, and to the extant genera *Magnolia*, *Andromeda*, and *Diospyros*. All of these also occur in Late Cretaceous floras in the southeastern United States and reflect the continuity of vegetation growing along low thermal gradients into the high northern latitudes. Coastal environments were certainly mild; but climatic reconstructions based on the flora are speculative, and nothing is known about Maastrichtian vegetation from the interior of Greenland.

Woods of Maastrichtian age are known from Big Bend National Park (*Javelinoxylon*, malvalean affinities; Wheeler et al., 1994). Sedimentological evidence and isotopic analysis of carbonate nodules from the paleosols suggest a warm dry climate with a MAT exceeding 15°C (Ferguson et al., 1991; Lehman, 1990).

On the western side of the epicontinetal sea, in the *Aquilapollenites* province, remnants of the Late Cretaceous and Early Tertiary vegetation are preserved in a series of basins extending from the southwestern United States to the North Slope of Alaska (Figs. 5.7, 5.8, Tables 5.1, 5.2). Leaf assemblages from the Vermejo, Lower Raton (Upchurch, 1995), Laramie, and Lower Denver formations were situated at about pl 40°–46° N in the Maastrichtian (presently ~37°–43° N). The number of species, percentage of entire-margined leaves, and leaf-size indices are given in Table 5.2. The MAT is estimated at 21–22°C (low megathermal) to 18°C (Nichols et al., 1989) for the Powder River Basin of Wyoming. Moisture conditions were generally subhumid (estimated ~1200 mm, Powder River Basin), but climates varied locally and/or over short spans of time. For example, leaf-size indices are higher in the Denver flora than in the Raton flora (800–1000 mm). The Denver assemblage also includes several species with moderate drip tips and others represent lianas (Menispermaceae), suggesting higher and less seasonal temperature and rainfall at this locality. Along the Pacific coast, woods of Maastrichtian age are known from near Patterson in central California (pl 45° N; Page 1979, 1980, 1981). Nine of the taxa exhibit no growth rings while in one they are well developed. This limited data from the coast suggests little temperature seasonality, although some slight seasonal differences in rainfall may have occurred. In general, rainfall values appear to be somewhat higher for western North America than in the east, and winds off the Pacific Ocean may have been the source. This paratropical megathermal vegetation (Fig. 5.8) probably extended somewhat further north, perhaps to pl 60°–70° N along the coast where January mean temperatures were likely above freezing.

It has long been recognized that a transition zone was present between ~pl 40° N and 50° N (Dorf, 1942), representing the broad boundary between megathermal and mesothermal temperatures (the 20°C isotherm is placed within this zone; Fig. 5.8). The floras on either side are different in composition, and those to the north have better representation of deciduous dicotyledons among the still dominant evergreens. In general, woods at and below ~pl 45° N have no or poorly developed growth rings, while those above ~pl 55° N mostly represent smaller trees and large shrubs that have better developed growth rings partially resulting from greater but slightly seasonal rainfall.

Between pl 50° N and 58° N (viz., in the northern Rocky Mountains) the vegetation was a notophyllous broad-leaved evergreen forest growing under a mesothermal and somewhat more moist but still subhumid climate than to the south and southeast (Fig. 5.8). Lianas and leaves with drip tips are rare and leaf size is relatively small, suggesting only limited closed-canopy forests. Some Araucariaceae and evergreen Taxodiaceae were present, broad-leaved deciduous vegetation grew mostly along streams, and Cycadophyte and Ginkgophyte holdovers from the Middle Cretaceous are better represented than in floras of similar age to the south and southeast.

Leaves from Maastrichtian floras located between pl 58° N and 66° N in the Western Interior have been described by Bell [1949; Coalspur (Scollard), Lower Edmonton, St. Mary River formations] and Berry (1935; Upper Edmonton, Whitemud formations). The assemblages include heterosporus ferns (*Hydropteris pinnata*; Rothwell and Stockey, 1994), aquatic angiosperms (*Triapago angulata*; Stockey and Rothwell, 1997), broad-leaved deciduous angiosperms, a variety of deciduous gymnosperms (*Glyptostrobus*, *Metasequoia*), and a few broad-leaved evergreens (palms). The percentage of lobed-leaved species is low, the leaves are generally of thin texture and moderately large in size, and diversity is low compared to modern extant microthermal broad-leaved deciduous vegetation. These are features of an extinct vegetation type described as a polar broad-leaved deciduous forest (Wolfe, 1985; Fig. 5.8) that grew under mesothermal to microthermal, moist climates. The MAT is estimated at ~15°C (Wolfe and Upchurch, 1987a). The climate was a balance between the Middle Cretaceous thermal maximum, declining temperatures of the Maastrichtian, and some northward movement of the craton that included Alaska in a more polar position.

Overall, fossil woods from the Maastrichtian of the Western Interior are poorly known, and the inadquate number of specimens studied gives a mixed climatic signal. Two conifer woods are known from the Vermejo Formation; in one growth rings are present and in the other they are poorly developed to absent. Coniferous wood from the Hell Creek Formation of eastern Montana has growth rings.[2] Ramanujan and Stewart (1969) describe four coniferous woods from the Maastrichtian of central Alberta [pl 65° N; Coalspur (Scollard)], all with well-developed growth rings. These data are consistent with evidence from Lance–Hell Creek mammals (the "*Triceratops* community" of Van Valen and Sloan, 1977) that suggests conditions warmer than those of the Bug Creek ("*Protungulatum* community") assemblage from Alberta, Canada.

The boundary between mesothermal and microthermal temperatures (MAT of 13°C) is placed at pl 65°–75° N. In the far northwest, floras of Late Cretaceous to Paleocene age are known from the North Slope of Alaska at pl 75°–85° N (the present Arctic Circle is at 66° 33' N). These are the highest paleolatitude Maastrichtian floras known for North America. The MAT is estimated at 10°C for the Cenomanian, 13°C in the Coniacian, 2–8°C (average 5°C, similar to the present-day MAT at Anchorage) in the Maastrichtian (~8°C by extrapolation of the latitudinal temperature gradient in Wolfe and Upchurch, 1987a), and 6–7°C in the Paleocene (Parrish and Spicer, 1988; Spicer and Parrish, 1986a, 1990). The Coniacian MAT for the North Slope of Alaska is now placed at ~12.5°C and the cold month mean at 5.7°C. For air temperature to be maintained near 5.7°C during the dark winter months, SSTs are estimated at between 6 and 8°C. The mechanism proposed for sustaining these warm polar temperatures is a significant poleward transport of heat (Herman and Spicer, 1996, 1997). If the deciduous Taxodiaceae had about the same ecological requirements as their modern descendants, the earlier MAT estimate of 5°C for the Maastrichtian would be minimum to slightly low because their vegetative and reproductive buds would have been damaged by frosts that would have occurred in the coldest month within a MAT of 5°C. At the present ecotone between mixed deciduous angiosperm and evergreen conifer forests at Prince Rupert, British Columbia, the MAT is 7.6°C and the coldest month average is 1.8°C (Horrell, 1991). Conditions were generally cooler in the high latitudes of the west than in the east. The precipitation trend was from driest in the southern United States to wettest in the Arctic; a sharper increase in precipitation at pl 46°–48° N was reflected by larger leaf sizes.

Pollen of *Cicatricosisporites*, *Sequoiapollenites*, *Taxodiapollenites*, *Aquilapollenites*, and *Cranwellia* has been reported from Late Cretaceous sediments in the Yukon–Tanana region of Alaska (~64° N; Foster and Igarashi, 1990). Frederiksen (1989a; see also Frederiksen et al., 1988) described Campanian to Maastrichtian palynofloras from the Tongue River Formation along the Arctic Slope of Alaska. The megafossil record has been interpreted as a relatively low-diversity angiosperm understory and a low-diversity ground cover of ferns and *Equisetum* (Parrish and Spicer, 1988). The assemblage upon which this reconstruction was based, however, consisted of only 10 taxa: two conifers, three angiosperms, one sphenophyte, two ferns, and two seed plants of unknown affinities (plus subsequently six taxa of coniferous wood). The microfossil record includes 110 types from 133 pollen-bearing samples through a stratigraphic section 1400 m thick. The value of the microfossil assemblage is that it includes a broader sampling of the regional vegetation and provides a more extensive record of the regional forest (Betulaceae–Myricaceae, Ulmaceae) and probable understory plants that are rare or absent as megafossils. The combined observations again support a synergistic approach for interpreting vegetational and paleoenvironmental history. The greater diversity revealed by the microfossils also argues for a MAT near the higher range of the 2–8°C estimate. Clemens and Allison (1985) reported remains of large dinosaurs from the Arctic Slope, which probably could not have existed there if the coldest month temperature was below ~7–8°C. However, this gets into the debate as to whether these dinosaurs were

Table 5.3. Dicotyledons from Late Cretaceous/Maastrichtian Floras of North America.

Taxon	Affinities	Oldest Occurrence	Reference
"*Myrtophyllum*" *torreyi*	Magnoliidae Magnoliales	Late K	Brown (1962) Upchurch and Dilcher (1990)
Fagopsis	Hamamelidae, Fagales Aff. Fagaceae	Late Campanian/early Maastrichtian	Wolfe and Upchurch (1986)
"*Carya*" *antiquorum*	Hamamelidae Juglandales Juglandaceae	Maastrichtian	Hickey (1980)
Aff. *Averrhoites*	Rosidae related to *Sapindopsis*?	Maastrichtian	Wolfe and Upchurch (1987a)
"*Acer*" *arcticum* complex	Rosidae Sapindales, aff. Aceraceae	Late Campanian or Early Maastrichtian	Wolfe and Upchurch (1986) Wolfe and Tanai (1987)
Theaceae	Dilleniidae Theales Theaceae	Campanian to Maastrichtian	B. H. Tiffney (unpublished data) J. A. Wolfe and G. R. Upchurch, Jr. (unpublished data)
"*Cissites*" *panduratus*, "*Ficus*" *leei*	Dilleniidae Euphorbiales Euphorbiaceae	Maastrichtian	Wolfe and Upchurch (1987a)
Parabombacaceaeoxylon	Dilleniidae Malvales no extant family	Maastrichtian	Wheeler et al. (1987)
Tiliaceae	Dilleniidae Malvales Tiliaceae	Maastrichtian	Wolfe and Upchurch (1987a)

Adapted from Upchurch and Wolfe (1993; see also Wolfe and Upchurch, 1986). For a listing of Maastrichtian plants from North America and other areas, including palynomorphs, see Horrell (1991).

warm or cold blooded (Barrick and Showers, 1994) and if they could have migrated seasonally (Axelrod, 1984). If they did not, it is unclear how they spent their time in the dark winter months (Crowley and North, 1991; but see Clemens and Nelms, 1993). Older Late Cenomanian woods show distinct growth rings with very narrow summer wood, suggesting rapid onset of winter conditions (Spicer and Parrish, 1986b). All the Late Cretaceous terrestrial vegetation of the North Slope of Alaska presently known was deciduous, died back to underground organs, or survived the winter as seeds (Spicer and Parrish, 1990), which suggests a seasonal temperature–light regime within a cool-temperate climate.

In the introductory section of this chapter it was suggested on the basis of context information that North American Late Cretaceous vegetation was distributed over lower thermal gradients than at present. It is now possible to quantify this gradient to some extent. Figure 5.8 shows an estimated MAT of 23°C at pl 30° N and 8°C at pl 75° N. Thus, the temperature values changed by ~15°C over ~45° of latitude for a lapse rate of ~0.3°C/1° latitude. This compares to the steeper modern gradient of ~0.5°C/1° of latitude (0.8–1.0° in eastern and central North America). Other evidence also suggests the Late Mesozoic thermal gradient was about one-half that of the present (Berggren, 1982). As noted previously, the Late Cretaceous boundary between megathermal and mesothermal temperatures (MAT 20°C) is placed between pl 40° N and 50° N, and between mesothermal and microthermal (MAT 13°C) it is placed between pl 65° N and 75° N. This compares with the present positions along the western coast of 30° N and 40° N, respectively (Upchurch, 1989).

The plants comprising North American vegetation during the Maastrichtian (Table 5.3) are difficult to relate to modern taxa, and the plant communities are difficult to reconstruct in detail because familiar modern analogs mostly do not exist. The angiosperms certainly diversified in the Late Cretaceous as shown by a doubling of extant families represented in the pollen record compared to the Campanian. All subclasses of the dicotyledons except the Asteridae are present by the end of the Maastrichtian (Crane, 1987, fig. 5.2d). Megathermal vegetation below pl 40°–50° N (Fig. 5.8) in both the east and west included a diverse gymnosperm element of Taxodiaceae [evergreens similar to *Sequoia*; deciduous forms assignable to *Parataxodium* (extinct) and *Metasequoia*] and possibly the Cupressaceae (remains similar to the extant *Callitris* and *Neocallitropis*, which presently has one species in New Caledonia). These coniferous trees probably formed an emergent, open-canopy woodland. Monocots were represented by palms, probably in a rosette growth form, which first appeared in the Santonian of megathermal regions and in the Campanian of mesothermal regions, and by the Zingiberales (first appearing in the Maastrichtian; *Spirematospermum*, Friis, 1987; *Zingiberopsis*, Hickey and Peterson, 1978). The dicots included members of the Magnoliidae, Hamamelidae [Plantanaceae, Fagaceae (*Fagopsis*), Juglandaceae], Rosidae (Aceraceae as represented by "*Acer*" *arcticum*, the inferred sister group of extant maples), Ranunculidae, and Dilleniidae. These formed a low, open understory beneath the conifers through the Early Maastrichtian (Upchurch and Wolfe, 1993). Vines were few, reflecting the subhumid conditions. The older Cycadophytes and Ginkgophytes were virtually absent from the southern region. By the Late Maastrichtian, angiosperms were dominant in megathermal climates and may have been the principal canopy-forming group in late successional to climax communities throughout most of North America. However, Tiffney (1984) and Crane (1987) believe the angiosperms were mostly small plants during most of the Cretaceous and that gymnosperms still played a prominent role. In the northern transition zone at ~pl 40°–50° N, megathermal vegetation extended into regions of higher precipitation and probably formed a more closed-canopy forest, as suggested by the larger leaf size (~50 vs. 28–40). In assemblages where facies analyses have been made, angiosperms were dominant in many floodplain, swamp, and lacustrine habitats by the beginning of the Late Maastrichtian.

Between pl 40°–50° N and 60°–65° N the vegetation was a notophyllous broad-leaved evergreen forest. In the *Aquilapollenites* province Araucariaceae and pollen of *Tricolpites reticulatus*, considered similar to *Gunnera* (Jarzen, 1980; Jarzen and Dettmann, 1989), have been reported. Permineralized cones with pollen have been recovered from the Maastrichtian Horseshoe Canyon Formation near Drumheller, Alberta, Canada (Serbet and Stockey, 1991). The cones are described as *Drumhellera kurmanniae* of the *Taxodium–Metasequoia–Sequoia–Sequoiadendron* complex that was abundant in swampy, wetland habitats. Other remains include Cycadophytes, *Ginkgo*, *Equisetum* (horsetail), lycopod megaspores, and cercidiphyllaceous fruits and leaves. Broad-leaved deciduous plants were common in successional or disturbed habitats, especially along streams. The presence of a few vines and some species with drip tips and the somewhat larger size of the leaves suggest slightly more humid conditions, and woods reflect some seasonality in temperature or precipitation. In several respects the notophyllous broad-leaved evergreen forest is similar to an ecotonal paratropical (subtropical) community near the limits of its northern distribution and one transitional between subtropical forests to the south and deciduous forests to the north.

North of pl 60°–65° N to as far as 80°–82° N, there was a polar broad-leaved deciduous forest. Deciduous gymnosperms were common; angiosperms with large, thin-textured leaves formed an understory vegetation indicative of a microthermal climate and polar light regimes. The presence of fusain (charred organic material) suggest that fires may have contributed to an open-canopy structure (Spicer and Parrish, 1990; Spicer et al., 1994). The emerging picture of Late Cretaceous climate at these high latitudes is one that was somewhat cooler and more winter seasonal than previously suggested (Smiley, 1967). This coolness and its associated increase in wetness also better accounts for the presence of extensive Late Cretaceous and Paleocene peat and coal deposits on the North Slope. At the same time it limits the role of frequent fires as an explanation for the openness of the vegetation.

Plants of the polar broad-leaved deciduous forest were dominated by angiosperms (since earlier Cretaceous times) and included the Taxodiaceae as the most abundant and widespread gymnosperm. As in the region immediately to the south, *Ginkgo* was present in the southern part of the polar deciduous forest, along with trochodendroids, "*Viburnum*"-type leaves, platanoids (Herman, 1994), and other Hamamelidae. Megafossils from the Kogosukruk Tongue Member of the Prince Creek Formation along the Colville River, North Slope of Alaska (Campanian and Maastrichtian), also reveal a vegetation that included deciduous coniferous trees. The microfossils add members of the Betulaceae–Myricaceae and Ulmaceae complexes that diversified during the Campanian, Maastrichtian, and Paleocene, even though overall diversity in the far north declined toward the end of the Maastrichtian. The decline was due primarily to fewer pollen types of entomophilous species (Frederiksen, 1991a), suggesting that temperatures were near the minimum for pollinators.

Two problems that complicated reconstruction of far-northern biotas earlier were recently resolved. A significant alteration was proposed in the above-described south (tropical) to north (temperate) pattern of North American Late Cretaceous and Paleogene communities based on paleomagnetic data from Campanian to Early Eocene deposits in Arctic Canada (Axel Heiberg, Ellesmere islands, Bay Fiord area; Hickey et al., 1983). Fossils associated with the Eureka Sound Formation were interpreted as documenting latest Cretaceous megathermal plants in the Arctic that preceded their midlatitude occurrences by ~18 m.y., while Eocene vertebrate fossils supposedly appeared in the Arctic 2–4 m.y. earlier than in the south. If so, the original dispersal center for many tropical species would have to be placed in the Cretaceous Arctic because of their earliest occurrence there. Studies have now shown that the parent rocks containing the fossils had been altered by secondary magnetization (magnetic overprinting), the stratigraphic ranges used for certain palynomorphs were incomplete, and some microfossils had been redeposited (Kent et al., 1984; Norris and Miall, 1984; Opdyke, 1990). These plant-bearing sections within the Eureka Sound Formation are now regarded as Eocene in age and are compatible with the generally accepted latitudinal distribution pattern of North American paleocommunities.

A second problem was envisioning a forested vegetation growing under the low light intensities and long periods of darkness in the polar environment. At the high northern latitudes above the Arctic Circle the sun is at to just below the horizon for about 6 months of the year and at to just above for the other 6 months (to a maximum of 43° during the Arctic summer). The present vegetation is mostly tundra. A higher CO_2 concentration in the Cretaceous and Paleogene has been proposed as one mechanism for increased warmth and, as a consequence, for more available moisture. The dim light, however, remained an enigma and required some innovative thinking. An early suggestion was that the tilt of the Earth's polar axis (presently at 23.5°) may have been significantly less in the past (5°–15°) and that it varied in concert with vegetational changes read from the fossil record: "These are precisely the changes that are inferred from paleobotanical data for the Oligocene and Neogene and would indicate that a significant decrease in the earth's inclination has occurred during the last 30 million years" (Wolfe, 1978, p. 701; 1980). However, a significantly reduced obliquity has no known basis in celestial mechanics because a plausible mechanism for the change is unknown, and modeling experiments actually produced a substantial cooling (by ~10°C at an obliquity of 15°; Barron, 1984). Creber and Chaloner (1985) cite evidence that the proposed analogy to obliquity variations on Mars does not apply to Earth because of its different shape and the stablizing influence of the moon. Other difficulties have been reviewed by Axelrod (1984) and Crowley and North (1991). The idea persists, however, and recently Muller and MacDonald (1997) suggested that changes in inclination periodically bring the earth into the cosmic dust cloud which then alters climate. The study was proposed as an alternative to or an amplification of the 100 Ky Milankovitch eccentricity cycles of the Quaternary, but the mechanism could apply to more ancient times. The theory is still not widely accepted, but no alternative explanation was immediately apparent to explain the existence of forests under polar light regimes. An obstacle has always been the underlying assumption that even with warmer temperatures and more available moisture, present light levels are too low to support trees. That assumption was never tested, and the review by Creber and Chaloner (1985) indicates that it may not be true. Accumulating evidence from plant physiology suggests that the primary limitation to large plants in the Arctic is low temperature, which reduces the rate of photosynthesis (see references in Creber and Chaloner, 1985). Measurements of light levels show that the June solar input on Cornwallis Island (27.8 MJ/m^2/day; 75.15° N) is about the same as for Washington, D.C. (31.5 MJ/m^2/day; 38.54° N) because during the summer season at the poles there is sunlight over the full 24 h. The calculations indicate that trees most suitable for intercepting maximum amounts of limited light would be conical in shape and widely spaced. For a tree 17 m tall with a basal radius of 1.6 m, the crown would have a light-intercepting surface of ~25 m^2. This surface would receive over 80,000 MJ of radiation during the illuminated part of the year under ideal conditions of no shading and limited cloud cover, which is more than sufficient to support tree growth. Measurements of tree density in Eocene floras of Ellesmere Island (Iceberg Bay Formation) actually suggest spacing about the same as in mesic forests of the midlatitudes (Francis, 1988). Nonetheless, the consensus is that a deciduous vegetation of trees and shrubs could have existed in the dim Arctic light under present inclination values, assuming higher temperatures and available moisture.

One of the major changes in the modernization of the North American vegetation has been the replacement of this Late Cretaceous and earliest Tertiary polar broad-leaved deciduous forest and associated deciduous conifers by the boreal coniferous forest of evergreen conifers (*Abies*, *Picea*) in later Tertiary and modern times. As will be seen later, this transition is evident at midaltitudes (700–900 m) in the Middle Eocene of the Pacific Northwest (Thunder Mountain flora of Idaho; Republic flora of Washington). It intensified in the Late Eocene, consolidated during the Oligocene, and was accomplished by the Mio–Pliocene.

THE K–T BOUNDARY: 65 MA

The bolide collision with the Earth at the end of the Cretaceous produced the terrestrial biological consequences discussed earlier (notably selective extinctions). Beyond this clear effect, it is difficult to separate changes in terrestrial vegetation at the K–T boundary due to the impact from those resulting from orogenic and climatic change. Further complications are that complete sections through the transition are not widely available, the uniformly drastic and worldwide effects implied by the original scenario have been scaled down, the effects differed at various sites and latitudes, and broad-leaved evergreen megathermal vegetation suffered more than microthermal deciduous communities. The present data allow a range of interpretations regarding the immediate effect of the impact, as well as its long-term consequences on climate and biotic history.

The marine ^{18}O record is equivocal at the K–T boundary. Isotopic studies show gradually cooling temperatures across the boundary (Crowley and North, 1991; Savin, 1977). Thus, it is possible to read this record as a trend of lowering temperatures that, along with the worldwide marine regressions and lowering CO_2 levels, reached a threshold that was primary in determining the vegetational history (Hickey, 1980, 1981). Wolfe and Upchurch note that

> [D]uring the Paleocene, a marked increase in diversity of deciduous dicotyledons occurred in mesothermal and adjacent megathermal vegetation. This anomalous deciduousness resulted in vegetation that was not physiognomically synchronized with temperature. Thus, leaf-margin analyses of particularly mesothermal, and to some extent megathermal, Paleocene vegetation yield temperature estimates that are far too low, which have resulted in invalid suggestions of a major temperature decline from the late Maastrichtian into the Paleocene. (1987a)

[See also Wolfe and Upchurch (1986).] If the impact is considered primary, then the ^{18}O record can be interpreted as reflecting a temperature lowering insufficient and too gradual to produce the changes observed.

A complex series of fluctuations have been proposed that parallel many of the intuitive or anticipated biotic effects of a bolide impact (Wolfe, 1990). Application of the CLAMP program to leaf assemblages from western North America on each side of the boundary has given results interpreted as evidence for profound changes in climate, including first a cooling (impact winter) of 1–2 months then an increase in MAT of 10°C (greenhouse effect; Wolfe, 1990), but there is no evidence for freezing in the form of frost heaving in paleosols in eastern Montana (Retallack, 1994). A 400% increase in rainfall was also proposed. This reconstruction is consistent with an earlier suggestion that because there was no known crater, the bolide may have fallen in the deep ocean, thus increasing moisture and seeding the atmosphere with microscopic debris. A crater of the proper age has now been found on the Yucatan Peninsula, Mexico, and this complicates the proposed explanation as primary for such a dramatic increase in rainfall and coordinated cooling–warming phases. However, the possible occurrence of hypercanes (runaway hurricanes) caused by various events, including an asteroid impact, are theoretically capable of putting large amounts of water and aerosols into the atmosphere (Emanuel et al., 1995), and more tropical vegetation did develop in the Early Paleocene. A distant impact (Chicxulub crater) and immediately thereafter a closer one (Manson crater; Koeberland and Anderson, 1996) has been read from the paleobotanical record, and the time of the event has been placed in early June (Wolfe, 1991; but see Hickey and McWeeney, 1992; Nichols, 1992). The Manson crater is now dated at ~73.8 Ma. As noted by Crowley and North, the ^{18}O record is ambiguous in its support of the greenhouse model, and there is even less confirming evidence that there were subsequent long-term trends in climate directly attributable to a bolide impact.

> Although there may have been a very severe environmental disruption at the K–T, and perhaps a CO_2 aftermath lasting tens of thousands of years, there does not seem to be any strong evidence that the impact affected the long-term evolution of the climate of the earth. The "mean background climate state" of the latest Maastrichtian (~67 Ma, i.e., just before the impact) does not seem greatly different from the "postaftermath" Early Paleocene (~64 Ma). (1991, p. 178)

The greatest disturbance in vegetation is evident in floras from the Western Interior of North America where complete K–T boundary sequences are preserved. The boundary in the Raton Basin is marked by a 2-cm thick clay layer with anomalously high amounts of iridium and shocked quartz. This was the first site where a terrestrial iridium anomaly was reported in conjunction with a palynologically defined K–T boundary (Orth et al., 1981). Vegetation in the Raton Basin included plants with small leaf size (index ~34), thick cuticles, and few drip tips and lianas. The MAT is estimated at ~21–22°C. Above the clay layer, fern spores appear in abundance (the "fern spike"; Fleming and Nichols, 1990; Nichols et al., 1986; Tschudy et al., 1984), followed by an angiosperm recolonization phase (Wolfe and Upchurch, 1987b) where the leaf-size index of the angiosperms doubles (70) and many of the leaves have drip tips and thinner cuticles. This reflects greater rainfall, estimated at 1500 mm or more, and a paratropical rain forest (Fig. 5.12) that persisted through the Early Paleocene. Mass kills are also interpreted from a section at Teapot Dome (J. A. Wolfe, personal communication, 1996), and K–T boundary extinctions are prominent between the Upper Hell Creek Formation (late Maastrichtian) and the Ludlow Member of the Fort Union Formation (Paleocene; Johnson, 1992). A pattern consistent with some effect of a bolide impact may also be evident south of pl 48° N in the southeastern United States between the Maastrichtian McNairy Sand flora and the lower Paleocene Naborton assemblage of Louisiana with the development of a tropical rain forest resulting from increased precipitation (Fig. 5.12; Wolfe, 1978). In the Denver Basin a paratropical rain forest appears with more deciduous dicotyledons, possibly reflecting an initial brief lowering of temperature. In the central Alberta Coalspur Formation there is a decrease in diversity (higher extinction levels) between the Maastrichtian (abundant aquiloid pollen) and the Early Paleocene (abundant gymnosperm pollen and no fern spike; Jerzykiewicz and Sweet, 1986).

Changes interpreted as resulting from the impact are also present in the mesothermal vegetation above pl 50° N. In the Maastrichtian, predominantly evergreen vegetation occurred up to pl 58° N, and a polar deciduous forest was further north. By the Early Paleocene the vegetation between 50° N and 58° N was almost entirely a broad-leaved deciduous forest (Fig. 5.12; trochodendroids, hamamelidaleans, Tiliaceae). Evergreen angiosperms were essentially eliminated from North American mesothermal climates, along with araucarian holdovers. As noted, Wolfe and Upchurch (1986, 1987a; see also Wolfe, 1987) attribute much of this change to a drop in temperature to near freezing as a result of the bolide impact. Others emphasize already declining temperatures (Hickey, 1981, 1984) and increasing seasonality in the Late Cretaceous (Batten, 1984; Frederiksen, 1989a; Srivastava, 1994b; Sweet et al., 1990).

In other regions to the north, in Japan, and in the southern hemisphere K–T sections are not complete and the effect of the bolide impact appears minor (Askin, 1988; Johnson, 1993; Keller, 1993; Keller et al., 1993). Asian floras reflect slightly cooling temperatures at the boundary, while floras in the Hell Creek Formation of Montana and Wyoming suggest a warming trend (Johnson and Hickey, 1990). Members of the northern floras were already adapted to cool and seasonally dark conditions, and in the southern hemisphere the effects of the impact apparently did not subject the floras to freezing temperatures. In these regions the K–T vegetation shows gradual changes more attributable to climate control. At Bug Creek in eastern Montana there is no striking difference in pedotypes across the K–T boundary. Annual rainfall for the Upper Cretaceous Hell Creek Formation is estimated at 900–1200 mm, while for the Lower Paleocene Tullock Formation it is also estimated at 900–1200 mm (Retallack, 1994).

Another aspect of the complex explanation for biotic changes at the boundary is that the deciduous habit of northern plants probably rendered them better capable of surviving and recovering from the impact winter than broad-leaved evergreen plants whose history is recorded in relatively great detail in the Raton Basin sequence. It has been suggested that Cretaceous dinosaurs from the North Slope of Alaska (pl 70°–85° N) were also adapted to seasonally cold temperatures, as well as to periods of low light levels, and that the bolide impact was not a major factor in their extinction there (Brouwers et al., 1987; see also Responses, 1988).

The methodology used to detect extinctions can influence interpretations toward either a gradual or a more sudden change in diversity. The pollen record, based on identifications mostly at a level equivalent to biological genera, reveals extinctions of 20–30% within the region extending from northern New Mexico to central Alberta. The megafossil record, where identifications are mostly to a level equivalent to biological species, show extinctions as high as 79% in megathermal vegetation (Johnson et al., 1989), declining to 25% in the polar broad-leaved deciduous forest from central Alberta (Upchurch, 1989; Wolfe and Upchurch, 1986). Another factor likely accounting for the differences in the two records is that pollen of the Lauraceae does not preserve as fossils, while leaves are prominently represented in megafossil floras of the southern Rocky Mountains region and are one of the components showing a major decline in diversity at the boundary. In contrast, some rare and declining species may be better and longer represented as pollen, implying a more gradual rate of extinction. The value of a multifaceted approach is particularly worthwhile in elucidating any such complex problem, as noted by Johnson and Hickey with reference to zonation of K–T boundary sites in the northern Rocky Mountains and Great Plains: "The fact that the megafloral zones are not reflected by palynostratigraphy argues for using an integrated approach to biostratigraphy that combines the high stratigraphic resolution of palynomorphs with the high taxonomic resolution of megafossils" (1990, p. 433; see also Lidgard and Crane, 1990). As a general assessment, the overall evidence seems consistent with at least local to regional biotic changes more rapid than can be attributed solely to climatic and orogenic trends. This argues for a bolide impact as one component of vegetational change at the K–T boundary in the midlatitudes of North America (Nichols and Fleming, 1990).

A somewhat similar view, but attributing even less long-term and widespread effect to the bolide, is expressed by Beeson, who worked on the Cretaceous Corsicana and Tertiary Kincaid K–T boundary formations in Falls County, Texas. "The concept of a single catastrophic event as the operative mechanism for these types of changes is probably simplistic. It is more reasonable to expect that events such as bolide impacts only represent a relatively small contribution to the more significant intrinsic terrestrial forces acting on biotic change" (1992, p. 18). Hoffman believes that "The contribution of the environmental consequences of impact and volcanism to the Cretaceous-Tertiary mass extinction can hardly be disentangled" and that

> It thus appears that the Cretaceous–Tertiary mass extinction cannot be plausibly interpreted as caused solely by a bolide impact and its environmental consequences, and it is more feasible to invoke a coincidence of at least two different factors which were causally unrelated to each other (since the impact occurred at the end of a period of intense volcanism and extinctions). (1989, p. 9)

Keller et al. (1993) believe the destructive effects were negligible in the high latitudes. My own assessment of the vegetation change at the K–T boundary is that it was, in part, a consequence of climatic trends already in motion, which the impact accentuated especially in the midnorthern latitudes of the western hemisphere through alteration of regional environments.[3]

An indirect effect of discussions of biotic change at the K–T boundary has been a reassessment of the precision of leaf physiognomy in reconstructing past climates. It is becoming recognized that older univariate statistics (viz., a given percent of leaf sizes or entire margins equals a relatively specific rainfall and temperature value) is too simplistic (Wolfe, 1990) and that a number of other factors also affect leaf form (higher levels of CO_2, defense against herbivory, etc.). If the 10°C increase in temperature proposed for just after the K–T boundary is accepted, this must nullify, to some extent, the use of modern southern hemisphere floras as a standard for interpreting Late Cretaceous and Paleogene leaf margin percentages because these calculations suggest values in the cool megathermal range (for the Denver and Raton Basins). Also, if the increase is correct, this alters the previously assumed relationship between leaf margin percentages and temperature because these percentages provide a MAT reading of ~10°C at the K–T boundary as opposed to 23°C derived from CLAMP.

ORIGIN AND SPREAD OF DECIDUOUSNESS

Deciduous angiosperms are known from at least the Middle Cretaceous, while events at the K–T boundary have been cited as a factor in the spread of deciduousness among the angiosperms. Three factors have already been discussed that contributed to its origin and development: gradual cooling from mid-Cretaceous highs that reached a threshhold at the K–T boundary, seasonal instability of streamside habitats, and abrupt lowering of temperatures resulting from the bolide impact. A fourth factor involves the migratory history of the angiosperms after their origin in the early Cretaceous. The group spread from their presumed region of origin in the tropical uplands of West Gondwana and radiated into areas that were just fragmenting into the various crustal plates. To the north the angiosperms encountered the seasonally dry climates of the late Mesozoic. According to this view, deciduousness originated on the northern margin of broad-leaved evergreen vegetation (Axelrod et al., 1991). This preadapted the angiosperms to the prolonged winter-dark conditions (a fifth factor) and the cooler conditions at the high northern latitudes, which they had reached by the Middle- to Late Cretaceous, and to the cooling climates of post Late Eocene times that faciliated their subsequent southern expansion.

With temperature, light, instability of habitats, and drought already available as long-term factors in the evolution of deciduousness, it is difficult to ascribe a central role to the bolide impact with maximum effects of mostly only a few months' to a few years' duration. Leaf fall is a complex, genetically controlled physiological and anatomical process that involves the synthesis and functioning of growth hormones whose effects are coordinated with seasonal changes, formation of abscission layers, dormancy, and rejuvenation. That the evolution of such processes would be significantly affected by so brief, albeit dramatic and intense, an event is difficult to envision. Axelrod et al. conclude that "The origin and abundance of the deciduous habit in the Northern Hemisphere, is linked with the restriction of seaways, widespread volcanism, and development of more continental, cooler climates from the Middle Paleocene onward, rather than the postulated impact of a bolide at the close of the Cretaceous . . ." (1991, p. 413).

Regardless of the probable multifaceted, integrated causes for alteration in climate, the Cretaceous–Paleocene interval encompasses some important changes in the North American vegetation. The epicontinental sea was regressing during the Late Maastrichtian (Kauffman, 1977), initiating removal of the barrier between the eastern and western floristic provinces. Triprojectacites begin to appear more in eastern North America and Europe, and Normapolles are found in increasing numbers in western North America and Asia. The Late Cretaceous was a time of warm, subhumid (dry), nearly aseasonal climate over the lower third of the continent with an estimated annual rainfall of ~1100 mm (drier in the southeast).

In the high northern latitudes, events at the K–T boundary contributed to the diversification of deciduous taxa during the Paleocene (e.g., Betulaceae, Fagaceae, Hamamelidaceae, Juglandaceae; Ulmoideae; Manchester, 1987; Nichols and Ott, 1978). As with the spread of deciduousness, it is difficult to determine the relative importance of the bolide impact and lower threshold temperatures in explaining changes in vegetation evident in the fossil record. Major trends in biotic history are the result of endogenic (tectonic–orogenic–volcanic–evolutionary–migratory) and exogenic (solar-induced global climate) trends periodically influenced by catastrophic events. As with any complex system, a multiplicity of factors are usually involved, and the only explanations to be rejected are those that force a "mine or thine" alternative.

Paleocene Vegetation: 65–56.5 Ma

Many of the difficulties encountered in reconstructing Late Cretaceous vegetation and paleoenvironments also apply to the Paleocene (Figs. 5.1, 5.2, 5.5, 5.12, Table 5.1). Although there are many floras of this age in the Rocky Mountains region and several along the Atlantic and Gulf coastal plain, overall the floras are few in number in relation to the geographic extent (the North American continent north of Mexico) and the time interval involved (~10 Ma). This precludes making detailed reconstructions for any one time interval or providing high-resolution histories from closely spaced floras within individual biotic–physiographic provinces. Many floras are in the process of being studied or revised, their age limits use of the modern analog method, some include only microfossils (Elsik, 1968a,b, 1970), and among these the primary goal is often stratigraphy and correlation (Frederiksen, 1991b; Pocknall, 1987). For these reasons the history is general and interpretations are continually being refined.

Beyond the complex changes at the Maastrichtian–Danian boundary (Fig. 5.1), two climatic trends mark the Paleocene. After an initial lowering in the earliest Paleocene, MAT rose steeply (Fig. 3.1) and there was an increase in precipitation, especially south of pl 48° N. These changes are evident in the widespread formation of peat, which was subsequently altered to the Tertiary lignites and coals that now characterize many parts of the midcontinent region. Recall from Chapter 2 (Orogeny section) that the warm, moist winds that presently bring precipitation into the southeastern United States developed in the model simulations with the Paleocene–Eocene uplift of the Rocky Mountains. The increasing warmth and moisture is clearly reflected in the floras. Overall diversity increases, there is rapid diversification of juglandaceous pollen, new morphotypes appear [*Carya*, *Symplocos*, thick-walled *Alangium*, Restionaceae, *Proxapertites* (an extinct relative of the Old World coastal tropical palm *Nipa*)], and Normapolles decline (Frederiksen, 1989b). Leaf size in Paleocene floras is almost always larger than in Late Cretaceous assemblages, and half or more of the angiosperm components in the southern United States floras have drip tips (Upchurch, 1989). Forests covered the landscape, and there is little evidence of extensive open country (Axelrod, 1992b). Microthermal vegetation was restricted to the higher northern latitudes, and in the far north deciduousness included response to low winter light.

Increased precipitation had its greatest effect on the already warm but dry biota in the southern United States, while rising temperatures had a significant effect on the already moist but cool polar communities. In the southern region an important consequence of greater moisture was the initial appearance there of a community recognizable in physiognomy and climatic parameters as a closed-canopy, multistratal, tropical rain forest (Fig. 5.12). At Naborton and Mansfield, Louisiana, the estimated MAT for the Early Paleocene is ~27°C.

Several consequences may be expected from the appearance and spread of a tropical rain forest (Wolfe and Upchurch, 1987a). Favorable niches would exist for the development of tall trees in response to high precipitation, luxuriant growth, and dense spacing; buttressing; liana habit; epiphytic habit; understory habit; and seeds capable of shade germination. Fossil representatives of families that include tall canopy-forming trees in extant megathermal vegetation (Anacardiaceae, Lauraceae, Rutaceae; pollen of the Bombacaceae–Sterculiaceae–Tiliaceae complex) are rare or absent in Late Cretaceous vegetation, become increasingly prominent in the Paleocene, and are abundant and diverse by the Early Eocene. Families that include numerous lianas (Icacinaceae, Menispermaceae, Vitaceae) show a similar pattern of diversification. Diaspore size increases from the Late Cretaceous to the Paleocene, probably in response to the larger food reserve needed to compensate for limited seeding photosynthesis in shaded habitats (Tiffney, 1984). Other adaptations, such as buttressing, epiphytism, and understory habit, are difficult to detect from the fossil record. The appearance of this novel tropical rain forest formation was a major step in the modernization of the Earth's vegetation, and its history for North America north of Mexico begins in the Paleocene.

Associated with the tropical rain forest were present-day temperate plants that may have lived in the Appalachian highlands. The U.S. Geological Survey's upper Paleocene Oak Grove core in northern Virginia (Frederiksen, 1991a) contains pollen similar to *Alnus* and *Betula*, already present in older Cretaceous deposits, small forms of *Carya*, the *Alfaroa–Engelhardia–Oreomunnea* complex (presently growing in upland temperate habitats in tropical latitudes), *Platycarya* (an eastern Asian temperate tree of the Juglandaceae; essentially terminal Paleocene), *Ilex* (widespread in warm-temperate to subtropical regions), and the Ulmaceae. A similar type of vegetation is reflected by the pollen flora of the Paleocene Rockdale lignite from Milam County, Texas (Elsik, 1968a,b, 1970) that includes spores referable to the bryophytes (*Sphagnum*), lycopods, a diverse fern component, palms, Taxodiaceae, winged gymnosperm pollen, and numerous angiosperm pollen types whose biological affinities to family and genus are unknown. The overall plant diversity trend in the southeastern United States was from a low level at the beginning of the Paleocene, possibly due partly to the aftermath of the asteroid impact, to higher diversity in the Middle Paleocene. The increase was likely a result of new and less competitive habitats and increased rainfall (Frederiksen, 1994). Fossil mammalian faunas of the Middle to Late Paleocene show more rapid radiation and include Pantodonta, Taeniodonta, and Dinocerata, representing browsers probably of open woodland. The faunal evidence frequently suggests more open vegetation than does the paleobotanical record (see Chapter 3).

To the far northeast, plant megafossils from the early Paleocene Atanikerdluk flora of Greenland continue to reflect a polar broad-leaved deciduous forest of trochodendroids, platanoids, and other Hamamelidae, mixed with deciduous coniferous elements such as *Metasequoia* (Koch, 1963).[4] Some megathermal elements occupied coastal regions and expanded in the later Paleocene and Early Eocene. The Late Paleocene–Early Eocene Thyra Ø flora (Ø= island) from northeastern Greenland (Prinsesse Dagmar Ø; pl 77°–79° N) consists of 30–31 megafossil species, including 22–23 angiosperm leaf types. Identifications include *Equisetum*, *Dennstaedtia*, *Cephalotaxus*, Cupressaceae, *Elatocladus*, *Fokienia*, *Ginkgo*, *Metasequoia*, Betulaceae, *Cercidiphyllum*, cf. *Corylus*, *Platanus*, and several unidentified leaves. Large entire-margined specimens were assigned to *Musophyllum* (now *Musopsis groenlandicum*; Boyd, 1990, 1992) and are interpreted as an evergreen monocotyledon of the Heliconiaceae–Musaceae–Strelitziaceae complex. They are suggested as indicating a nearly frost-free, warm-temperate to subtropical climate along the coast with a MAT of 15–20°C (Boyd, 1990). This is warmer than other estimates of 10–15°C for the region north of 65°–70° N based on leaf physiognomy (Wolfe, 1985, 1987) and 12–15°C (Basinger et al. 1994) and 7.9–9.3 ± 2°C for Paleocene and Eocene MAT based on multiple regression analyses (Spicer and Parrish 1990). Boyd (1990) believes that leaf physiognomy yields an anomalously low temperature estimate, at least for the Late Paleocene of northern Greenland, and Axelrod et al. (1991) suggest seasonal darkness as a factor in the high percentages of lobed leaves.

In the west, fossil wood of *Paraphyllanthoxylon abbottii* from the Paleocene Black Peaks Formation of Big Bend National Park, Texas (Torrejonian–Tiffanian NALMA or provincial ages), lacks distinct growth rings (Wheeler, 1991). In the midcontinent region, the Fort Union Formation (or Group) of Wyoming, Montana, and North Dakota is an extensive series of strata ranging in age from Early through Middle Paleocene. Fossils described by Brown (1962) from the Fort Union Formation are mostly older than those from the Sentinel Butte Formation (Crane et al., 1990) and the Golden Valley Formation (Bear Den and Camels Butte members; Hickey, 1977). The four floras constitute a series from the Early Paleocene to the Early Eocene. Brown (1962) recognized 170 species of plant megafossils, including diverse ferns, *Ginkgo* (Fig. 5.13A), Taxodiaceae (*Glyptostrobus, Metasequoia*; Fig. 5.13B), araucarias, palms, and numerous notophyllous angiosperm leaves with serrate margins and larger entire-margined types. Although many generic identifications are wrong, there are specimens that resemble *Sabal* (Fig. 5.13C), Juglandaceae (*Carya, Juglans, Pterocarya*; *Cruciptera*, Manchester, 1991; *Polyptera*, Man-

Figure 5.13. (a) *Ginkgo adiantodes* from the Paleocene Fort Union Formation, eastern northern Rocky Mountains–Great Plains region (Brown, 1962, pl. 10, fig. 5). (b) *Metasequoia occidentalis* from the Paleocene Fort Union Formation, eastern northern Rocky Mountains–Great Plains region (Brown, 1962, pl. 12, fig. 5). (c) *Sabal imperialis* from the Paleocene Fort Union Formation, eastern northern Rocky Mountains–Great Plains region (Brown, 1962, pl. 14, fig. 2). (d) *Platanus raynoldsi* from the Paleocene Fort Union Formation, eastern northern Rocky Mountains–Great Plains region (Brown, 1962, pl. 31, fig. 6). Plant microfossils from the Paleocene–Lower Eocene Fort Union/Wasatch formations, Powder River Basin, Wyoming–Montana. (e) *Caryapollenites veripites.* (f) *Momipites wyomingensis.* (g) *Ulmipollenites krempii.* (h) *Alnipollenites verus.* (i) *Intratriporopollenites* sp. Reprinted from Pocknall (1987) with the permission of the American Association of Stratigraphic Palynologists Foundation.

chester and Dilcher, 1967—the oldest unequivocal Juglandaceae), Betulaceae (*Betula, Corylus; Palaeocarpinus,* Manchester, 1996), Cornaceae (*Cornus*), Fagaceae (*Castanea, Quercus*), platanoids (Fig. 5.13D), Ulmaceae (*Celtis, Ulmus, Zelkova*), Moraceae (?; *Ficus*), Lauraceae (*Cinnamomum, Lindera, Persea, Sassafras*), *Cercidiphyllum, Magnolia,* Vitaceae, *Viburnum*(?), and others. *Ceratophyllum* (*C. furcatispinum*) occurs in the Fort Union Group in Montana (Herendeen et al., 1990). Pollen assemblages (Pocknall, 1987) include microfossils resembling Juglandaceae (*Carya* = *Caryapollenites*, Fig. 5.13E; *Platycarya*, see also Wing and Hickey, 1984; the *Alfaroa–Engelhardia–Oreomunnea* complex = *Momipites*, Fig. 5.13F), Ulmaceae (*Ulmus–Zelkova* = *Ulmipollenites*, Fig. 5.13G), *Alnus* (=*Al-*

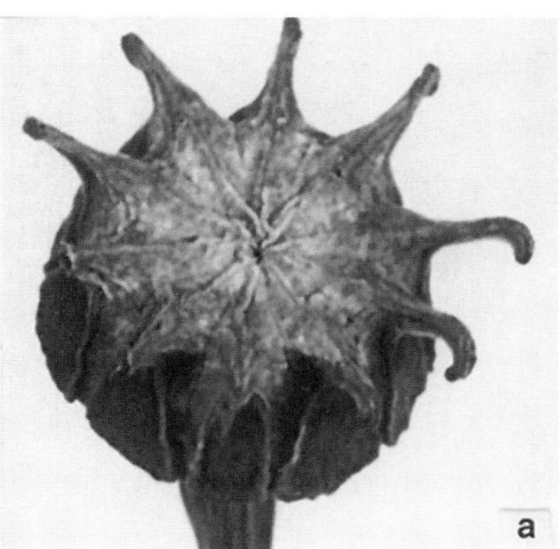

Figure 5.14. Fruits of *Trochodendron–Nordenskioldia*. (A) Modern fruit of *Trochodendron aralioides* (original magnification x7). (B) Fossil fruit of *Nordenskioldia borealis* (original magnification x5) from the Paleocene Tongue River Member, Fort Union Formation, near Melville, Montana (Crane, 1989, fig. 5c,e). (C) Fossil fruit of *Nordenskioldia borealis* from the Paleocene Sentinel Butte Member, Ft. Union Formation, near Almont, North Dakota (Crane, 1989, fig. 7d; Crane et al., 1990, fig. 6c). Reprinted with permission of Springer-Verlag (Fig. 5.14A–C) and the University of Chicago Press (Fig. 5.14C).

nipollenites, Fig. 5.13H), and the Bombacaceae–Sterculiaceae–Tiliaceae complex (*Intratriporopollenites*, Fig. 5.13I). In general, localities south of about central Wyoming include *Cinnamomum*, *Menispermum*, and palms; those north of the Colorado–Wyoming border include *Ginkgo*, *Acer*, *Betula*, *Corylus*, *Viburnum*, and other temperate genera. Physiographic relief was moderate, but recall from Chapter 2 that nearly horizontal subduction of the Farallon Plate extended as far east as the Black Hills and produced uplands in the region. Also, in a temporal sense the Fort Union vegetation was growing during the transition between cooler Late Cretaceous–earliest Paleocene temperatures and the onset of warmer conditions. In a geographic sense the region was near the ecotone between the northern microthermal polar broad-leaved deciduous forest and the megathermal dry tropical forest–evergreen tropical rain forest expanding from the south (Fig. 5.12). Wolfe (1985) interprets the flora of the Tongue River member of the Fort Union Group (Powder River Basin) as a paratropical rain forest, while Wing (1981) believes the slightly younger early Willwood flora of the Big Horn Basin represents a notophyllous broad-leaved evergreen forest.

The Sentinel Butte flora near Almont in central North Dakota is Late Paleocene in age. It presently comprises 30–35 species, including *Ginkgo adiantoides* (Ginkgoaceae), cf. *Parataxodium* (Taxodiaceae), cf. *Canticocculus* (Menispermaceae), *Cornus* (Cornaceae), *Cyclocarya brownii* (Juglandaceae; Manchester, 1987; Manchester and Dilcher, 1982), *Meliosma rostellata* (Meliosmaceae–Sabiaceae), *Nordenskioldia borealis* (an extinct genus of the Trochodendraceae; Crane et al., 1990; Fig. 5.14A–C), *Nyssidium–Joffrea* complex (Cercidiphyllaceae; Crane, 1989; Fig. 5.15A, B), *Paleomyrtinaea princetonensis* (Myrtaceae; Pigg et al., 1993), *Paleocarpinus* (Betulaceae), *Palaeophytocrene* (Icacinaceae), Platanaceae, *Psidium* (Myrtaceae), *Spirematospermum* (Zingiberaceae; Manchester and Kress, 1993), and others whose familial affinities are uncertain (e.g., *Porosia verrucosa*; araceous?).

The Golden Valley Formation in the Williston Basin of western North Dakota provides a sampling of the easternmost Paleocene and Eocene floras of the Rocky Mountain region (Hickey, 1977). It consists of the Bear Den Member with 41 plant species of latest Paleocene age (56 Ma) and the Camels Butte Member with 37 plant species of Early Eocene age (53–55 Ma). The most abundant fossils identified from the former unit are *Metasequoia* (Taxodiaceae), *Corylus* (probably *Paleocarpinus*; Betulaceae), "*Meliosma*" (probably a platanoid), "*Cocculus*" (Menispermaceae; now *Nordenskioldia*, Manchester et al., 1991), *Cercidiphyllum*, and *Chaetoptelea* (Ulmaceae); they represent a lowland forest. Important associates include *Platanus* and extinct members of the Menispermaceae (*Menispermites*), Rosidae (*Averrhoites*), Hamamelidaceae ("*Viburnum*" *cupanoides*), and a laurel close to the modern genus *Persea* (L. J. Hickey, personal communication, 1996). Owing to the development of an extensive shallow lake in the earliest Eocene, swampy and aquatic vegetation, including *Equisetum*, *Glyptostrobus* (Taxodiaceae), *Sparganium* (Sparganiaceae), and *Porosia* (a floating aquatic) become more important toward the top of the Bear Den Member. The Bear Den vegetation shows closest affinity with the immediately older Paleocene floras from the region. It is a notophyllous broad-leaved evergreen forest mixed with some deciduous species.

The woody flora of the slightly younger, Early Eocene, Camels Butte Member is dominated by *Platycarya americana*, "*Meliosma*" (Platanaceae), and *Ternstroemites* (Theaceae), associated with *Averrhoites*, *Dombeya* (Sterculiaceae),

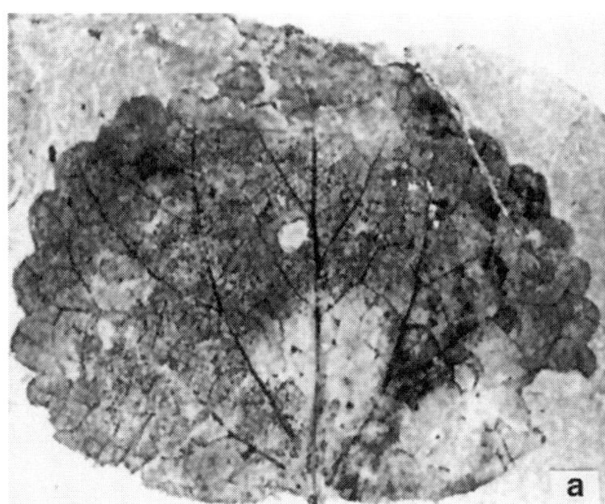

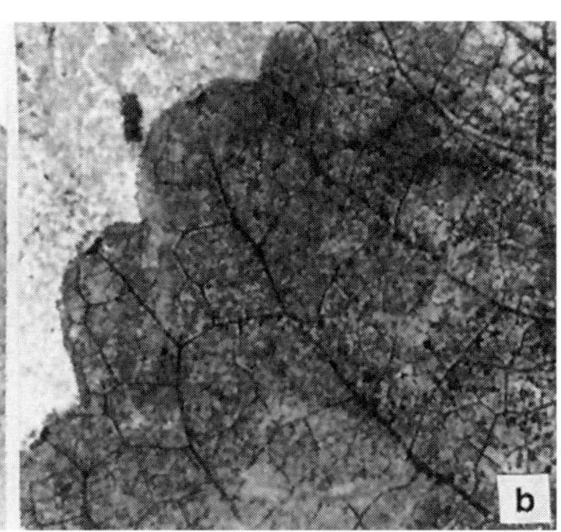

Figure 5.15. Fossil leaf of Cercidiphyllaceae (*Trocho dendroidesarctica–Nyssidium–Joffrea* complex) from the Paleocene Tongue River Member, Fort Union Formation, near Melville, Montana (Crane, 1989, fig. 8b,c). (A) original magnification ×1. (B) original magnification ×3. Reprinted with permission of Springer-Verlag.

Metasequoia, and other extinct Theaceae (*Paraternstroemia*). The floating fern *Salvinia preauriculata* is the most conspicuous element of the swamp and aquatic flora of the Camels Butte Member; all of the paludal and lacustrine Paleocene forms except *Osmunda* persist, and these are joined by *Isoetites* (Isoetaceae), *Lygodium* (Schizeaceae), *Nelumbo*, and *Zingiberopsis* (Zingiberaceae). The Camels Butte flora shares most of its species with other Eocene floras and, in general, shows little similarity to the immediately older Paleocene floras. [For a complete listing of plants identified from the Golden Valley Formation see Hickey (1977) with the revisions and additions noted above.]

The Willwood flora from the Bighorn Basin of northwestern Wyoming is of latest Paleocene through Early Eocene age (52.8 ± 0.3 Ma; earliest lost Cabinian/earliest (Wasatchian NALMA or provincial age) and shows a progressive increase in entire-margined leaves (41–52%). This reflects a change from a deciduous forest in the Late Paleocene toward a mixed evergreen and deciduous broad-leaved forest with seasonal rainfall, but no widespread aridity, and nearly frost-free conditions (Wing, 1980, 1981, 1984, 1987; Wing and Bown, 1985; Wing et al., 1991; Table 5.4; Fig. 5.12). The MAT is estimated at a maximum near 18°C and a cold-month average above 0°C (probably above 5°C) for the late Early Eocene. Evidence from leaf physiognomy suggests a cold-month mean temperature of 1–8°C. Computer models for the continental interior at ~50 Ma hindcast harsh winters, but the widespread and consistent paleobotanical and faunal evidence indicate a need for revision in these sensitivity tests. The Lower Eocene part of the Willwood Formation contains an extensive fossil mammalian fauna presently under study (crocodile relatives, land turtles, tree-dwelling mammals; Bown et al., 1994) and, together with the contemporaneous flora, an extensive data base is becoming available for reconstructing the biota and paleoenvironments.

The Clark's Fork Basin floras range through the Puercan (Early Paleocene), Tiffanian (latest Paleocene), and Clarkforkian (latest Eocene) NALMA or provincial ages. The reported composition is similar to that of other Paleocene floras in the region (Hickey, 1980; S. L. Wing, personal communication, 1996): *Ginkgo*, *Metasequoia*, *Glyptostrobus*, *Cercidiphyllum*, *Corylus*, Juglandaceae, *Meliosma* (=Platanaceae), *Platanus*, and others. The Tiffanian MAT at Clark's Fork is estimated at 10°C and the Clarkfordian MAT at Clark's Fork at 13.5°C (Hickey, 1980).

The stratigraphic relationship between floras in the isolated basins of the Great Plains is complex (Fig. 5.16). A summary of paleoclimatic estimates for the Sentinel Butte, Bear Den, Camels Butte, Clarks Fork, and other assemblages in the upper Great Plains has been provided by Crane et al. (1990) and Wing et al. (1991). The Sentinel Butte flora contains few if any broad-leaved evergreens, only one conifer (cf. *Parataxodium*), and is dominated by deciduous plants indicating a temperature-seasonal climate. The most similar modern analog is the mixed mesophytic forest of Asia and vegetation along the south part of the Atlantic coastal plain of eastern North America. Entire-margined leaves constitute 45% of the flora, which is higher than in a mixed mesophytic forest and lower than in a tropical or paratropical forest. The closest matching figures from data plotted by Dolph and Dilcher (1979) are from the piedmont of North and South Carolina, where the MAT is 15°C, the coldest month average is 4.5°C, and the warmest month average is 24°C. For Bear Den time the MAT is estimated from 13°C at 56 Ma to 15–16° C at ~55 Ma and at 17.8°C for the Camels Butte flora (Hickey, 1977). The MAT for the Powder River Basin in northern Wyoming is estimated at 16–18°C between 54 and 50 Ma (Wing et al., 1991) and 18–21°C for the Raton Basin further south at the New Mexico–Colorado boundary (Wolfe, 1990; Wolfe and Upchurch, 1987b).

If these regional MAT estimates are placed in sequence (cf. Fig. 5.16), the values are 10°C (Tiffanian), 13.5°C (Clarkfordian), 15°C (Sentinel Butte), 13°C (Bear Den, 56 Ma), 15–16° C (Bear Den, 55 Ma), and 17.8°C (Camels Butte, 55–53 Ma). In general, they show recovery from Early Paleocene lows and the trend toward the Cenozoic thermal maximum reached in the Early Eocene. However, the trend was not an uninterrupted progression from cooler to warmer conditions. Wing (1996) found that high-resolution

Table 5.4. Plant megafossils from Early Eocene part of Willwood Formation, Bighorn Basin, northwestern Wyoming.

Pond	Swamp	Levee/Crevasse Splay
Zingiberopsis	*Zingiberopsis*	*Zingiberopsis*
Menispermites	*Menispermites*	*Menispermites*
Amesoneuron (palm leaf fragment)	*Amesoneuron*	*Amesoneuron*
Platanus		*Platanus*
Salvinia		*Salvinia*
Betulaceous leaves		Betulaceous leaves
Cercidiphyllum		*Cercidiphyllum*
Populus		*Populus*
	?Typhaceae	?Typhaceae
	Equisetum	*Equisetum*
	Allantoidopsis	*Allantoidopsis*
	Alnus	*Alnus*
	Averrhoites	*Averrhoites*
	Glyptostrobus	*Glyptostrobus*
Leguminosites	*Nemaleison*	*Meliosma*
Salix	*Lygodium*	?*Ampelopsis*
Spirodela	*Cnemidaria*	aff *Carya*
"*Sparganium*" (Cyperaceae)	*Phoebe*	*Woodwardia*
?Rutaceae	*Platycarya*	?Tiliaceae
?Berberidaceae	*Dombeya*	*Cornus*
?*Dalbergia*	*Thelypteris*	?*Eugenia*
Potamogeton	"*Sequoia*"	*Fagopsis*
Leguminosae	?Magnoliaceae	*Ginkgo*
Menispermaceae	"*Apocynophyllum*"	*Metasequoia*
Azolla		*Ulmus*
		(*Chaetoptelea*)
	Placement unknown	
		Penosphyllum
Persites		
	Polyptera	
?*Schoepfia*		

Adapted from Bown et al. (1994), Farley (1990), and Wing (1981, 1984).

sampling, combined with critical taxonomic studies of fossil floras in the northern Rocky Mountains region, reveals a cool excursion of 1–1.5 m.y. at the Paleocene–Eocene transition. This important work brings paleobotanical data into closer agreement with emerging theoretical considerations about the nature of climatic change. It is becoming increasingly evident that trends in climate are characterized by reversals and abrupt threshold changes. These are frequently obscured in older floras by the uneveness of the fossil record and broad sampling intervals. Another interesting new result is that fossil faunas in the region reflect the same cool interval (K. D. Rose, personal communication, 1996), further demonstrating the value of synergistic approaches to paleoenvironmental studies and the importance of context information.

Faunal evidence also favors progressively warmer MAT (Hutchison, 1982). Koch et al. (1992; see also Rea et al., 1990) have studied the isotopic carbon ratios in tooth enamel of *Phenacodus* and paleosol carbonates from the Fort Union (65–55 Ma) and Willwood formations (55–52 Ma) in the Bighorn Basin of northwestern Wyoming, which spans the Paleocene–Eocene boundary. A decrease in ^{13}C occurs across the boundary and probably reflects increasing temperatures. Paleocene vegetation was mesothermal–megathermal broad-leaved deciduous forest giving way to more megathermal evergreen tropical vegetation later in the Paleocene and Early Eocene (cf. Fig. 3.1), as indicated by such frost-sensitive indicators as the tree fern *Cnemidaria*, palms, Zingiberaceae, and cycads.

Pabst (1968) described the sphenophytes, ferns, and gymnosperms from the Paleogene (Eocene?) Chuckanut Formation of northwestern Washington (latitude ~48° N), which are generally similar to Paleocene floras previously described (*Equisetum*, numerous ferns, *Glyptostrobus*, *Metasequoia*, *Taxodium*).

North of about pl 50° the earliest Paleocene assemblages have leaf-margin percentages indicative of a microthermal climate with possible January temperatures near freezing. A flora from the Paleocene Ravenscrag Formation in southwestern Saskatchewan, Canada grew on a broad alluvial plain with meandering streams, ponds, and swamps. The vegetation included a broad-leaved deciduous forest (*Platanites canadensis*; *Trochodendroides speciosa*, Fig. 5.17A; *Amelanchites similis*; *Cornophyllum newberryi*; *Dicotylophyllum anomalum*; *Ginkgo*), needle-leaved deciduous forest (*Fokienia ravenscragensis*, Fig. 5.17B; *Glyptostrobus*

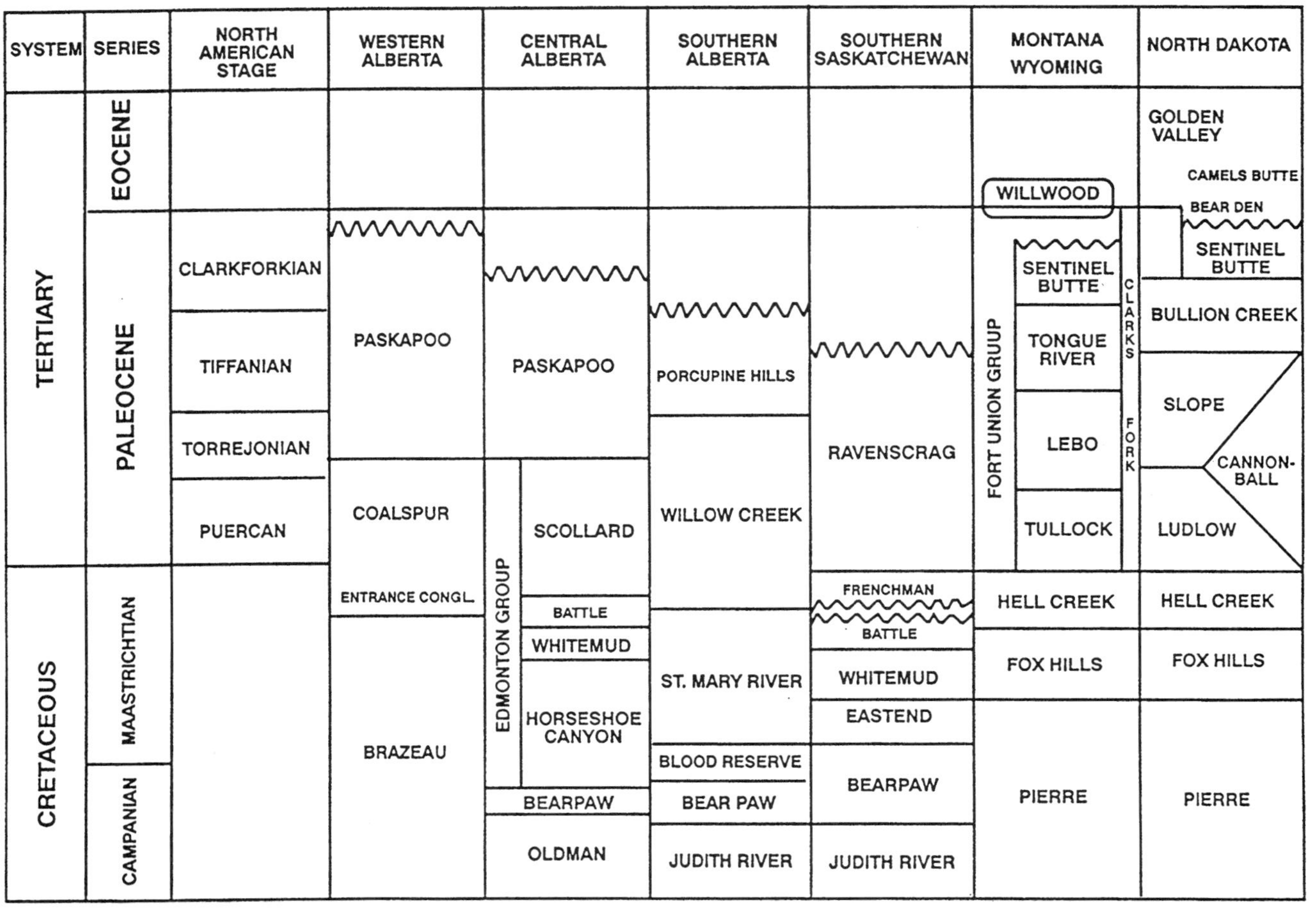

SYSTEM	SERIES	NORTH AMERICAN STAGE	WESTERN ALBERTA	CENTRAL ALBERTA	SOUTHERN ALBERTA	SOUTHERN SASKATCHEWAN	MONTANA WYOMING	NORTH DAKOTA
TERTIARY	EOCENE						WILLWOOD	GOLDEN VALLEY; CAMELS BUTTE; BEAR DEN
	PALEOCENE	CLARKFORKIAN	PASKAPOO	PASKAPOO			FORT UNION GRUUP: SENTINEL BUTTE; CLARKS FORK	SENTINEL BUTTE
		TIFFANIAN	PASKAPOO	PASKAPOO	PORCUPINE HILLS	RAVENSCRAG	TONGUE RIVER	BULLION CREEK; SLOPE
		TORREJONIAN	COALSPUR	EDMONTON GROUP: SCOLLARD	WILLOW CREEK	RAVENSCRAG	LEBO	SLOPE; CANNON-BALL
		PUERCAN	COALSPUR	SCOLLARD	WILLOW CREEK	RAVENSCRAG	TULLOCK	LUDLOW
CRETACEOUS	MAASTRICHTIAN		ENTRANCE CONGL.; BRAZEAU	BATTLE; WHITEMUD; HORSESHOE CANYON	ST. MARY RIVER	FRENCHMAN; BATTLE; WHITEMUD; EASTEND	HELL CREEK; FOX HILLS	HELL CREEK; FOX HILLS
	CAMPANIAN		BRAZEAU	BEARPAW; OLDMAN	BLOOD RESERVE; BEAR PAW; JUDITH RIVER	BEARPAW; JUDITH RIVER	PIERRE	PIERRE

Figure 5.16. Stratigraphic relationship between Upper Cretaceous and Lower Tertiary formations of the midcontinent region. Modified from McIver and Basinger (1993, fig. 4) and based on references cited therein. Reprinted with the permission of the Canadian Society of Petroleum Geologists.

nordenskioldii; *Metasequoia occidentalis*), with a few needle-leaved evergreens and possibly a rare broad-leaved evergreen (*Myexiophyllum americanum*). The arrangement of the principal species are shown in Fig. 5.18, and the MAT is estimated at 16°C. The Joffre Bridge flora (latitude ~65° N) near Red Deer in Alberta, Canada, is from the Paskapoo Formation of Late Paleocene (Late Tiffanian) age (Bell, 1949). It is presently under study but known to contain *Botrychium* (Rothwell and Stockey, 1989), *Onoclea sensibilis* (Rothwell and Stockey, 1991), the *Cercidiphyllum*-like *Joffrea speirsii* (Crane and Stockey, 1985), *Limnobiophyllum* (Lemnaceae; Stockey et al., 1997), *Paleocarpinus joffrensis* (Betulaceae, aff. *Corylus*; Sun and Stockey, 1992), platanoid remains [various parts described as *Platanus* (or *Macginitiea*; leaves), *Macginicarpa* (pistillate inflorescences), and *Platananthus* (staminate inflorescences; Pigg and Stockey, 1991)], and abundant *Azolla* (Hoffman and Stockey, 1992). Associated remains include *Equisetum*, *Osmunda*, *Woodwardia*, *Ginkgo*, *Glyptostrobus*, *Metasequoia*, *Acer*-like fruits, *Chaetopetala*, *Porosia*, and *Viburnum*-like leaves; deciduous Pinaceae (*Pseudolarix*) are present in other related floras.

A Late Paleocene palynoflora from the Eureka Sound Group, Northwest Territories, includes *Sphagnum*, *Lycopodium*, *Osmunda*, *Schizaea*, Cupressaceae–Taxodiaceae, *Picea*, *Pinus*, *Sparganium*, cf. *Acer*, *Alnus*, *Betula*, *Corylus*, Ericaceae, Juglandaceae (*Momipites–Caryopollenites*, *Pterocarya*), *Liquidambar*, *Pachysandra*, cf. *Quercus*, *Tilia*, and *Ulmus* (McIntyre, 1989). Absent from the region are palms and broad-leaved evergreens and entire-margined species referred to *Cinnamomum*, *Ficus*, *Magnolia*, and *Persea* that were so abundant to the south. In the Bonnet Plume Formation of the Yukon Territory, Rouse and Srivastava (1972) note a floristic change across the K–T boundary, indicating cooler Paleocene conditions. Angiosperms (*Alnus*, *Betula*, *Corylus*, *Fraxinus*, *Myrica*, *Ulmus*, trochodendroids, platanoids and other hamamelids) and conifers (*Cedrus*, *Glyptostrobus*, *Metasequoia*, *Picea*, *Pinus*, *Taxodium*) were apparently codominant at the high northern latitudes. These shifted further north and to higher elevations as temperatures began to rise in the Paleocene. The Late Paleocene (53.3 ± 1.5 to 55.8 ± 1.7 Ma; Triplehorn et al., 1984) Chickaloon flora of Alaska is a broad-leaved evergreen and coniferous forest of Taxodiaceae (*Glyptostrobus*, *Metasequoia*), trochodendroids (*Nordenskioldia*, *Joffrea*), and several members of uncertain affinities

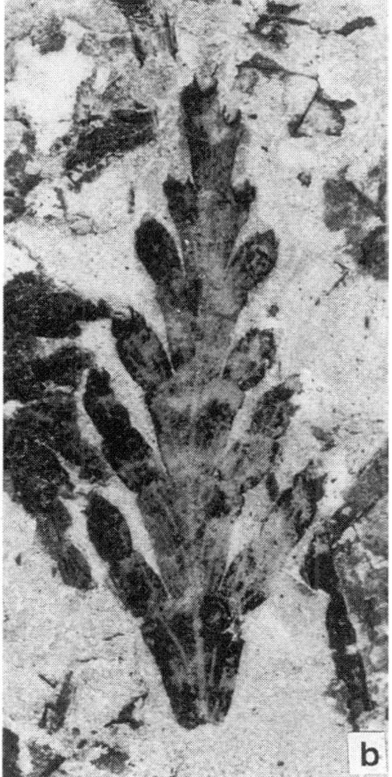

Figure 5.17. (A) Leaf of *Trochodendroides speciosa* from the Paleocene Ravenscrag Formation, southwestern Saskatchewan, Canada. Reprinted from McIver and Basinger (1993, pl. 21, fig. 1) with the permission of the Canadian Society of Petroleum Geolgists. (B) Leafy branch of *Fokienia ravenscragensis* from the Paleocene Ravenscrag Formation, southwestern Saskatchewan, Canada. Reprinted from McIver and Basinger (1993, pl. 13, fig. 2) with the permission of the Canadian Society of Petroleum Geologists.

(*"Carya" antiquora, Averrhoites affinis, Macginitiea, "Grewiopsis"* sp., *"Meliosma" longifolia, Fagopsis groenlandica*; Wolfe, 1994). Broad-leaved evergreen vegetation eventually extended to ~60° N in the Early Eocene. As noted previously, Wolfe (1985, 1987) estimates a MAT of 10–15°C (CLAMP 12.3°C) poleward from ~65° N to 70° N, while Boyd (1990, 1992) places the range at a seemingly warm 15–20°C in northernmost Greenland to account for *Musopsis* (*Musophyllum*).

The number of extant angiosperm families doubles from that of the Maastrichtian based on fossil pollen evidence. Prominent families in the Paleocene are the Betulaceae (only *Alnus* and possibly *Betula* in the Late Cretaceous), Fagaceae, Hamamelidaceae, Ulmoideae, and Juglandaceae (Manchester, 1987), which diversified during this period. Lianas probably assignable to the Tiliaceae (*Grewia* type) also appear. Conifers at the lower latitudes are generally rare.

Evidence for herbaceous vegetation during the period from the Maastrichtian through the Paleocene is meager in both the micro- and megafossil record. Spores resembling the Blechnaceae–Polypodiaceae–Pteridaceae complex are represented and megaspores of the lycopods, Gleicheniaceae, and Schizaeaceae are present. Crane (1987) suggests that lycopods, ferns, and angiosperm shrubs played a prominent role in understory and colonizing vegetation during the Late Cretaceous and Paleocene. [See Spicer et al. (1985) for a discussion of *Pityrogramma calomelanos* as the primary colonizer on Volcán Chichonal, Chiapas, Mexico).] As a generalization, the Paleocene vegetation was relatively uniform across the Holarctic region and typically included *Ginkgo, Metasequoia, Glyptostrobus, MacGinitiea* and other members of the *Platanus* complex, early representatives of the *Carya* and *Ampelopsis* lineages, and various members of the *Cercidiphyllum* complex (Wing, 1987). In a generalized south to north direction, Paleocene vegetation consisted of broad-leaved tropical rain forest (Gulf Coast), warm-temperate broad-leaved evergreen forest with few deciduous hardwoods (Raton, Denver floras), mixed deciduous and evergreen broad-leaved forest north of Colorado–Wyoming, and mixed deciduous–coniferous forests with a few montane conifers further north (Axelrod et al., 1991). Later in the Eocene many of the present-day Asian genera were eliminated from the eastern Rocky Mountains and central Plains region because of seasonal dryness. These floras will show greater affinities with the modern vegetation of Mexico and Central America in a complex pattern that may have begun in the latest Paleocene and Early Eocene.

Early Eocene Vegetation: 56.5–50 Ma

During the early Eocene temperatures continued to rise, reached their highest values in all of Cenozoic time at 50–55 Ma, and persisted at these values for 2–5 Ma (Koch et al., 1992) (Figs. 5.1, 5.2, 5.7, 5.19). The temperature gradient for the continental interior is estimated at ~0.4°C/1° latitude (Greenwood and Wing, 1995) and was probably less along the coasts. Rainfall also continued at high levels. An increase in volcanic activity in the Late Paleocene at ~56 Ma identifies CO_2 as a factor in the warming trend (Vogt, 1979; see also Berner et al., 1983; Owen and Rea, 1985; Sloan and Rea, 1995). Another warm period at 17–15 Ma (Fig. 3.1), also corresponding to a peak in volcanic activity, further supports an interaction between volcanism, CO_2, and temperature. There is a marked decrease in the ocean carbon reservoir (increasing temperatures). The Norwegian–Greenland Sea was opening at this time and

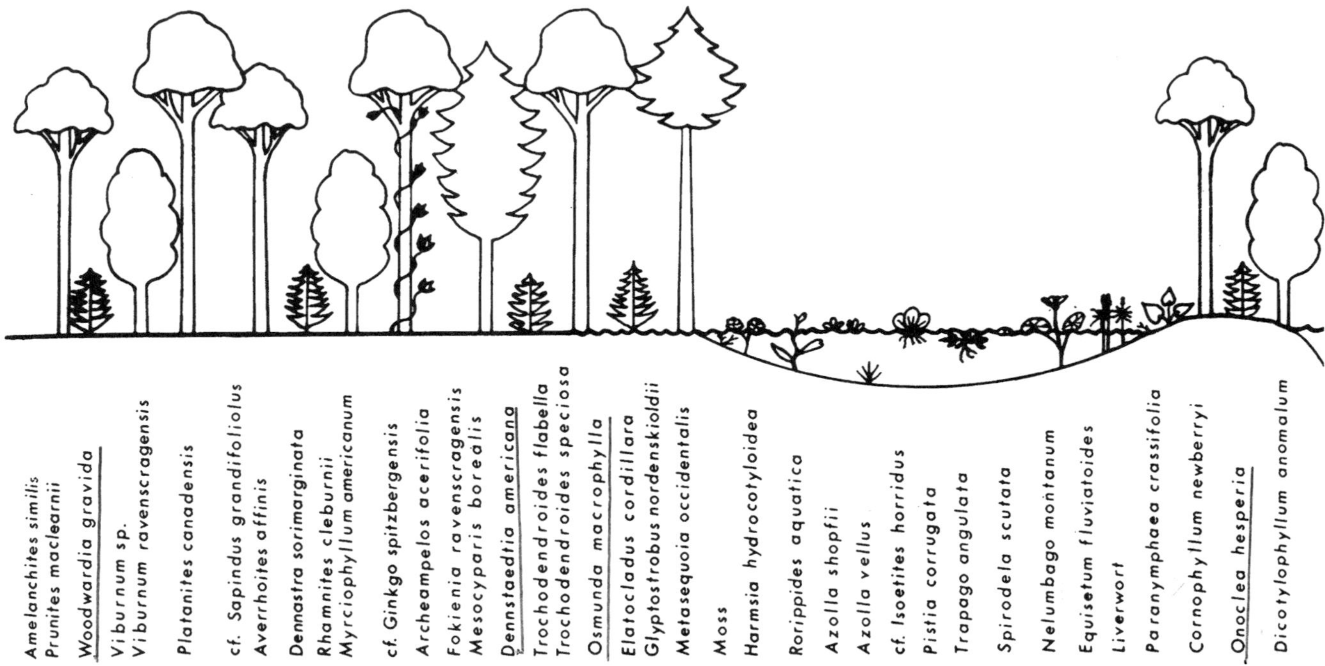

Figure 5.18. Profile diagram of the Ravenscrag Butte Paleocene forest community. Reprinted from McIver and Basinger (1993, fig. 5) with the permission of the Canadian Society of Petroleum Geologists.

the Thulean volcanic episode was in full force. An abrupt decrease in the size of eolian particles reflects moist conditions and more dense vegetation cover (Rea et al., 1985). Megathermal tropical rain forest vegetation attained its greatest latitudinal extent, reaching to near 45°–50° N (further north along coasts), and lateritic paleosols are found at the same latitudes. Paratropical rain forest extended to 60°–65° N, and fossils of a meso- to microthermal notophyllous broad-leaved evergreen forest occur on an exotic terrane at 70°–75° N along the northeast Pacific (Wolfe, 1972, 1977, 1985), or (Axelrod et al., 1991) a polar broad-leaved deciduous forest occupied the continental interior. Several taxa are found in North America for the first time during this interval. The first appearnace of *Platycarya* approximately marks the Paleocene–Eocene transition, and megafossils of the ferns *Cnemidaria magna* and *Lygodium kaulfussi* indicate an Eocene age for the interior basins (Bown et al., 1994). Leaves of *Quercus* are found in the Early Eocene of the Sierra Nevada (MacGinitie, 1941). The Paleocene–Eocene transition was also marked by extinctions in both plant and mammalian groups, and overall there was a decrease in diversity (Wing, 1997; Wing et al., 1995).

The Eocene history of vegetation in the southeastern United States is difficult to trace because the floras are currently being revised on a paleomonographic basis and their composition is not fully known. The floras were initially studied by Berry (1916, 1922, 1924, 1930, 1941), but Dilcher (1971, 1973) has found that ~60% of the identifications are incorrect. The fossil reported as *Taxodium* is a *Podocarpus*; the family Proteaceae, represented by four genera fide Berry, is not present; eight of 10 species of *Sapindus* are one form and it is not *Sapindus*; and one form described as *Aralia* is *Dendropanax*. Also, the dating of the fossil-bearing strata has had a complex history. Berry thought that his specimens came from the Wilcox Formation, dated as latest Paleocene to Early Eocene, and represented a coastal strand depositional environment. The most extensive fossil-bearing deposits are from the Bell City pottery clay pit near Bell City in western Kentucky and from the Puryear clay pit near Puryear, Henry County, Tennessee. The fossils at these sites are now assigned to the Claiborne Formation of Middle Eocene age, dated primarily on the basis of palynostratigraphy, and were deposited in ox-bow lakes and channel fills of meandering rivers several miles inland from the coast. Furthermore, until recently other plant-bearing deposits from western Tennessee (e.g., the Buchanan clay pit) were thought to be in the Paleocene part of the Wilcox. These deposits are now dated as Early Eocene Wilcox (N. O. Frederiksen, personal communication, 1992), which affects by some 5 m.y. reports of the early to earliest fossil record of various taxa (e.g., grasses; Crepet and Feldman, 1991; see also Crepet and Taylor, 1985, 1986). The relationship between stages of the Paleocene through the Oligocene and Gulf Coast provincial stages are shown in the right column of Fig. 5.20.

Floras of the Lower Eocene Wilcox Formation, as presently defined, have not been extensively studied. Reports from western Tennessee include the grass spikelets and inflorescence fragments previously mentioned; the legume *Protomimosoidea buchananensis* (Fig. 5.21), inter-

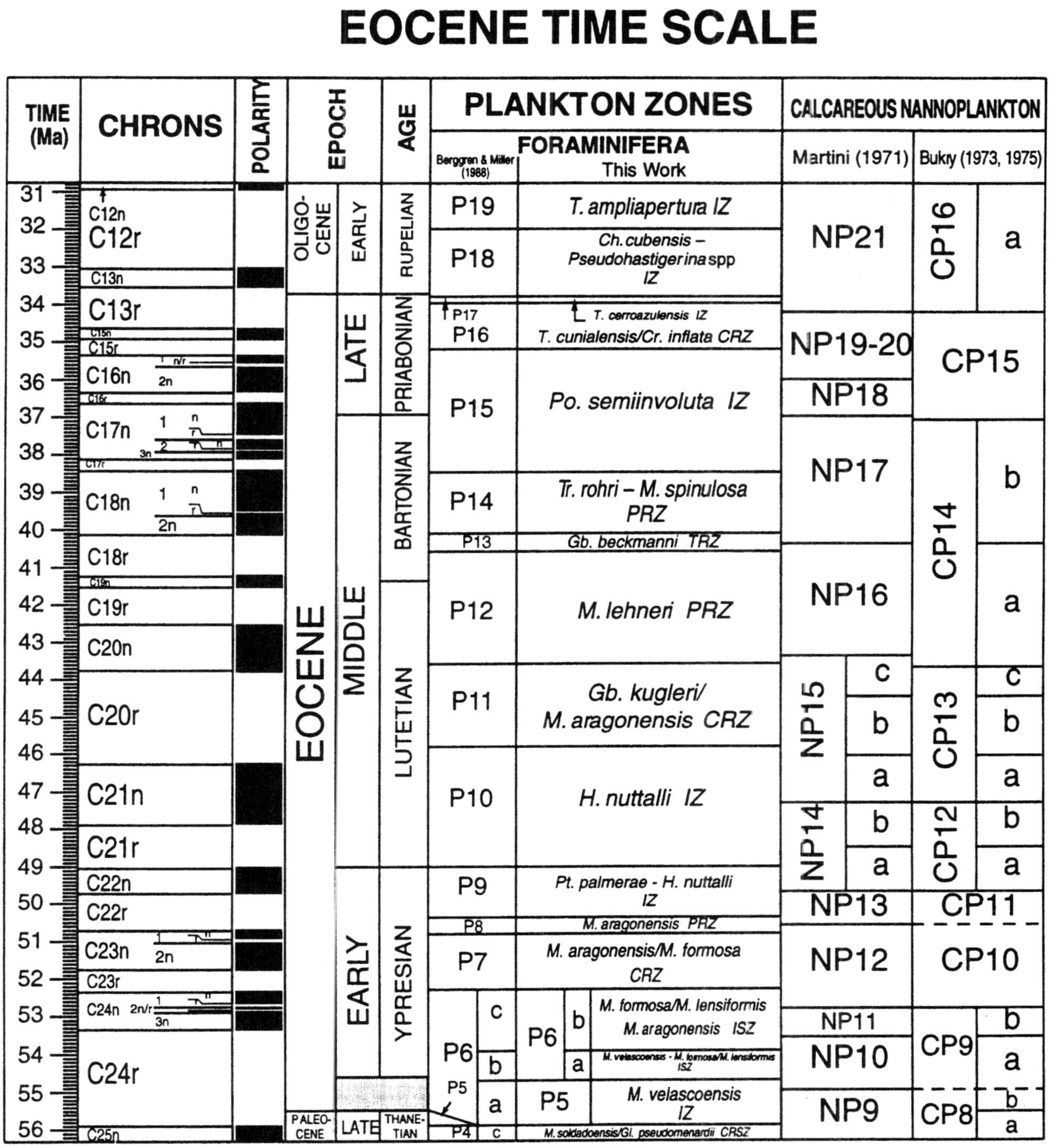

EOCENE TIME SCALE

Figure 5.19. Eocene time scale. Reprinted from Berggren et al. (1995) with the permission of the Society for Sedimentary Geology.

preted as combining features of the subfamily Mimosoideae and the *Dimorphandra* group of the subfamily Caesalpinioideae (Crepet and Taylor, 1985, 1986); *Barnebyanthus buchananensis*, an extinct genus of the papilionoid tribe Sophoreae (Crepet and Herendeen, 1992); *Pistillipollenites macgregorii* (Gentianaceae fide Crepet and Daghlian, 1981, but see Stockey and Manchester, 1988); and various unidentified mesophyllous, entire-margined leaves suggesting a megathermal climate and tropical rain forest vegetation. The extinct *Dryophyllum* occurs in many Paleogene floras in the southeast and is possibly the complex from which other Fagaceae were derived (Jones and Dilcher, 1988). It ranges from about the Early Eocene into the Late Eocene.

Palynofloras from the southeastern United States record a decline in plant diversity at the Paleocene–Eocene boundary (upper NP9–lower NP10), followed by a sharp recovery and a significant turnover in vegetation during the Early Eocene. Numerous Paleocene and earliest Eocene angiosperm pollen types disappear and are replaced by new

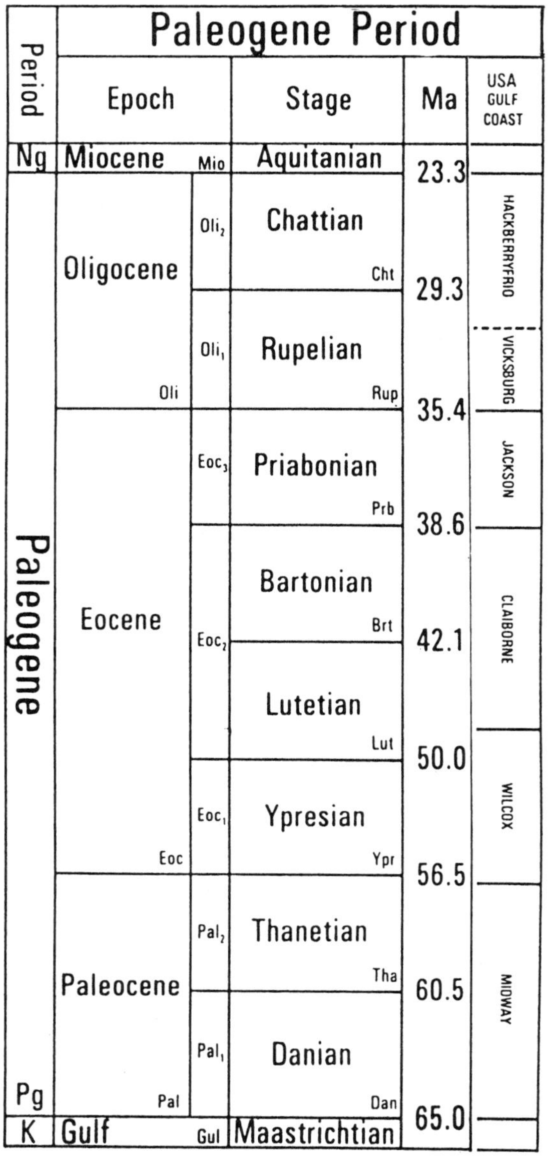

Period	Epoch		Stage	Ma	USA Gulf Coast
Paleogene Period					
Ng	Miocene (Mio)		Aquitanian	23.3	
Paleogene	Oligocene (Oli)	Oli_2	Chattian (Cht)	29.3	HACKBERRYFRIO
		Oli_1	Rupelian (Rup)	35.4	VICKSBURG
	Eocene (Eoc)	Eoc_3	Priabonian (Prb)	38.6	JACKSON
		Eoc_2	Bartonian (Brt)	42.1	CLAIBORNE
			Lutetian (Lut)	50.0	
		Eoc_1	Ypresian (Ypr)	56.5	WILCOX
	Paleocene (Pal)	Pal_2	Thanetian (Tha)	60.5	MIDWAY
Pg		Pal_1	Danian (Dan)	65.0	
K	Gulf (Gul)		Maastrichtian		

Figure 5.20. Age and stage equivalency of Paleogene strata in the Gulf Coast region. Note that the Eocene–Oligocene boundary is now placed at 34 Ma (Chapter 6). Modified from Axelrod (1992b) and based on Harland et al. (1990). Reprinted with the permission of Cambridge University Press.

Figure 5.21. *Protomimosoidea buchananensis* from the Lower Eocene Wilcox Formation, Buchanan, Tennessee. Reprinted from Crepet and Taylor, 1986, fig. 4) with the permission of the Botanical Society of America.

forms (Frederiksen, 1988). This does not seem attributable solely to climate, emigration, extinction, or evolutionary change. Rather, it involved immigration of new elements over the North Atlantic land bridge. According to Frederiksen, "It now appears that the early Eocene event is the most distinct floral immigration event documented for the Eocene of North America." (1988, p. 44). *Platycarya* (Juglandaceae) and *Eucommia* (Eucommiaceae) migrated from Europe into North America (but see Call and Dilcher, 1997). They mostly became extinct in the United States and Canada in the Late Eocene–Early Oligocene, persisted in Central Mexico (Puebla) into the late Cenozoic (Magallón-Puebla and Cevallos-Ferriz, 1994), and are now native only to eastern Asia. This parallels the strong North American–European affinity seen in Lower Eocene faunas and documents that the North Atlantic region posed no serious physical or climatic barrier to migration. Eocene faunas from Ellesmere Island (Dawson et al., 1976; West and Dawson, 1978) contain alligator and arboreal prosimian primates, including an extinct dermopteran distantly related to the modern flying foxes (lemurs) of the Old World tropics (Estes and Hutchison, 1980). The locality is presently at latitude 78° N and the paleolatitude was near 79° N (McKenna, 1980). Thus, very little of the change in environment and biota can be attributed to plate movement. The climates at these highest northern latitudes supported both megathermal (southern coastal Thulean route) and more temperate (northern inland DeGeer route) vegetation, allowing an array of plants and animals to move into the southern United States. Early Eocene warmth is widely documented in the northern hemisphere, as indicated by the London Clay flora of England. The MAT there is estimated at 25°C (77°F) compared to the present 10°C (50°F). The coldest month mean temperature for the Eocene southern Arctic was 10°C, and limited frost was possible (Tiffney, 1994).

In the central Rocky Mountains region the Paleocene and Early Eocene was a time of intense deformation. Most of the highlands and major basins were present by the Early Eocene, reaching one-half or more of their present average elevation. New estimates based on paleobotanical evidence suggest that, at least locally, elevations may have

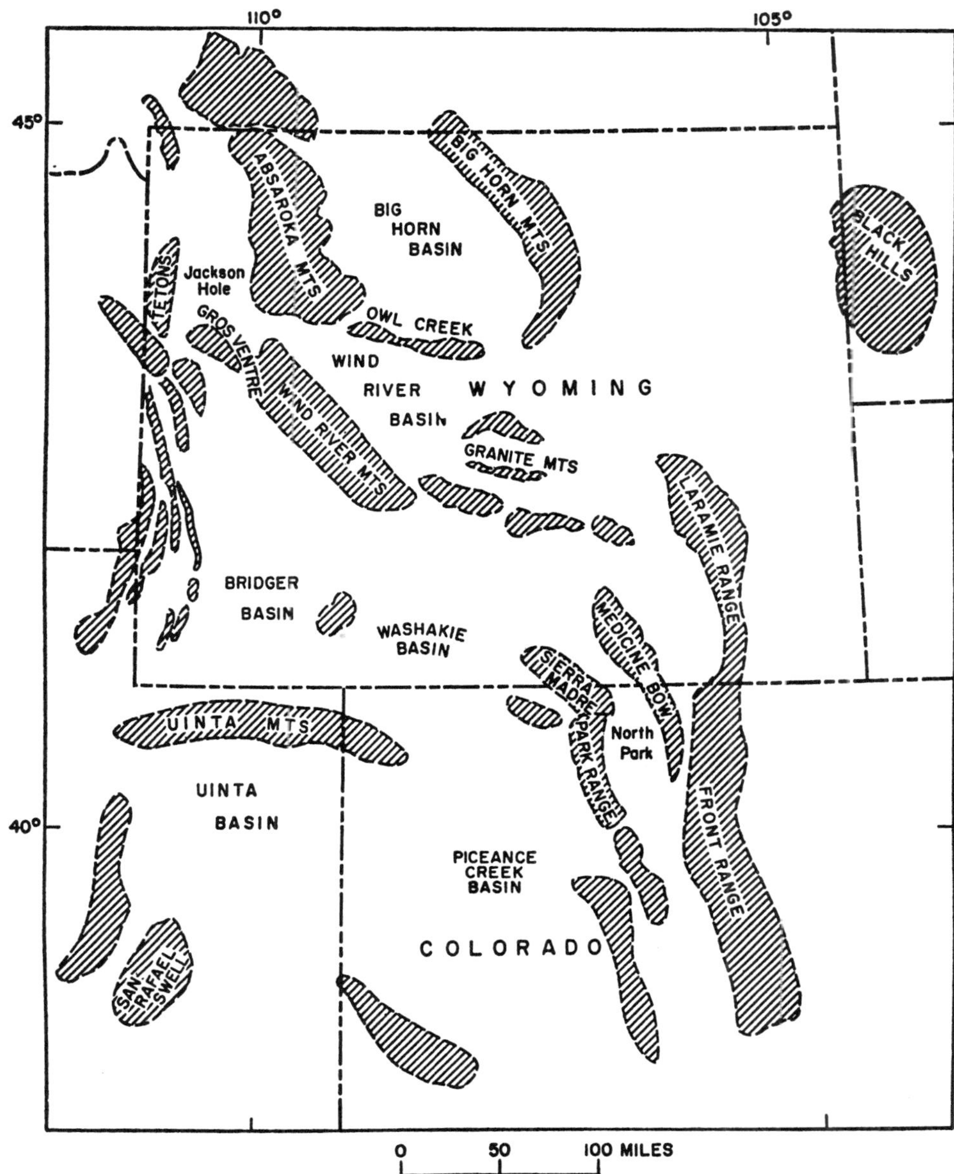

Fig. 5.22 Major mountain ranges and basins in the central Rocky Mountains region during the Paleocene and Early Eocene (Leopold and MacGinitie, 1972).

been near modern values (Chapter 6, Paleoelevation Analysis). Numerous lakes formed by the blockage of drainage systems occupied the basins (Fig. 5. 22), creating favorable conditions for the preservation of extensive floras and faunas. During the latest Eocene–earliest Oligocene these large basin lakes disappeared (Franczyk et al., 1992).

In general, the Early Eocene floras of the region represent mesothermal sclerophyllous broad-leaved evergreen (tropical to paratropical rain forest) and mixed mesophytic elements with distinct Asian affinities: *Aleurites* (Euphorbiaceae), *Dipteronia* (Aceraceae), *Euptelea* (Eupteleaceae), and *Platycarya* (Juglandaceae; Leopold and MacGinitie, 1972). The flora of the late Early Eocene Wind River Formation (~50–51 Ma) has not been studied in detail, but Leopold and MacGinitie (1972) provide a preliminary list that includes southeast Asian genera (*Actinidia*, *Alangium*, *Cinnamomum*, *Euptelea*, *Eustigma*, *Firmiana*, *Limacia*, *Litsea*, *Machilus*, *Maesa*, *Millettia*, *Neolitsea*, *Paliurus*, *Platycarya*, *Pterocarya*), a few genera now present in tropical northern Latin America (*Cedrela*, *Luehea*, *Mabea*, *Oreopanax*, *Serjania*, *Thouinia*, *Ungnadia*), and a number occurring in both regions. The plants suggest humid subtropical conditions. A few microthermal temperate elements were present (*Acer*, *Alnus*, *Populus*), but these are not numerous or diverse. In contrast, floras of similar to slightly younger age west of the Rocky Mountains are often dominated in both diversity and abundance by microthermal taxa (Copper Basin, Republic, Thunder Mountain; Chapter 6), probably due to the higher elevation (Wing, 1987) and unimpeded moist winds from the Pacific.

The Wind River Formation also preserves one of the most diverse Early Eocene vertebrate faunas in North America. More than 100 species including mammals (65 species), reptiles (22 lizards, three snakes, two turtles); crocodylians (two), birds (two), amphibians (four), and fishes (three) have been identified. [Recent parts in the published series are Dawson et al. (1990), Rose et al. (1991), and Stucky and Krishtalka (1990).] The fossils are from the Lost Cabin Member of the Wind River formation in the latest Wasatchian NALMA or provincial age (50.5 Ma). The fauna represents communities living along flood plains, small ponds, and well-drained swamps. The lithology of the sediments shows evidence of some periodic but not extensive drying (e.g., filled mud cracks). Species diversity and composition suggest a humid, multistoried tropical to subtropical gallery forest with only slight seasonal differences in rainfall (Stucky et al., 1990). This is consistent with the more limited paleobotanical data. Fossil plants from slightly younger deposits are known from Yellowstone National Park (50–51 Ma, early Bridgerian NALMA), including the famous silicified forests at Specimen Ridge and Amethyst Mountain (Dorf, 1960; Knowlton, 1899; Wheeler et al., 1977, 1978); but these have not been studied in detail. [For a review of the Yellowstone floras see Wing (1987).] Computer modeling of Early Eocene climates in the midcontinent region predicted a cold-month mean temperature 3–10°C below freezing, or about the same as at present. Increasing the temperatures via a greater CO_2 concentration for areas such as Wyoming and the Dakotas overheats the globe generally in the model predictions. In contrast, a wealth of paleontological evidence from plants to crocodiles indicates winter temperatures above 5°C (1–8°C) in Wyoming and Montana (Wing and Greenwood, 1993). Efforts are underway to reconcile the differences through simplified models that more efficiently carry heat from the equator and incorporate the effect of large lakes in the region (Chapter 6, Middle through Late Eocene Vegetation, Green River flora). This is an example of the important role of paleontological verification data in improving the accuracy of modeling results.

The Middle Eocene Chalk Bluffs flora (Ione Formation, 49–52 Ma, Wolfe 1981b; 50–52 Ma, Wing and Greenwood, 1993; ~48 Ma, J. A. Wolfe, personal communication, 1996) occurs along the west slope of the Sierra Nevada in California. It is tropical in aspect, but it has not been revised since MacGinitie's (1941) original study. Generally, the lowland vegetation in the Pacific Northwest was megathermal rain forest.

To the far north, palynofloras are known from the Early Eocene part of the Eureka Sound Group of Cornwall and Amund Ringnes islands (Sverdrup Group), Northwest Territories. The fossiliferous deposits were initially dated as Paleocene–Eocene, but they are now regarded as Early Eocene (Kalgutkar and McIntyre, 1991; McIntyre and Ricketts, 1989). Identifications include *Sphagnum*, Cupressaceae–Taxodiaceae, *Picea*, *Pinus*, *Alnus*, *Betula*, *Carya*, *Corylus*, *Engelhardia* (*Momipites*), Ericaceae, *Ilex*, *Pachysandra*, *Pterocarya*, *Tilia* (rare), and *Ulmus*. Late Paleocene–Early Eocene palynofloras are also known from the Eureka Sound Group (Iceberg Bay Formation) of Ellesmere Island (Kalkreuth et al., 1993). These add *Glyptostrobus*, *Metasequoia*, *Sequoia*, *Taxodium*, *Acer*, *Cercidiphyllum*, *Diervilla*, *Fraxinus*, *Liquidambar*, *Nyssa*, *Quercus*, and *Viburnum*. Megafossils from the Eocene Buchanan Lake Formation, Axel Heiberg Island in the Canadian high Arctic, are from swamp-coal, fluvio-lacustrine shale, and channel-sand lithofaces (Greenwood and Basinger, 1994). The vegetation was a mosaic of *Metasequoia* and *Glyptostrobus* (taxodiaceous swamp), a mixed coniferous community (*Pseudolarix*, LePage and Basinger, 1995a; ?*Chamaecyparis*; *Larix*, LePage and Basinger 1991a,b, 1995b; *Pinus*), *Alnus*–fern (*Osmunda*) bogs, and various broad-leaved deciduous gymnosperms (*Ginkgo*) and angiosperms (cf. *Betula*, aff. *Quercus*, cf. *Zelkova*, Juglandaceae, platanoids, *Cercidiphyllum* complex).

To the northwest, late Early Eocene floras (originally dated as Middle Eocene) are preserved within the Yakutat block of Alaska (62° N; e.g., Kushtaka flora; Wolfe, 1977, 1985). The coastal localities include floras with broad-leaved evergreens, 65% have entire margins, drip tips are common, and lianas are well represented. Recall that this is the time of maximum warmth in the Cenozoic and the period when megathermal tropical vegetation attained its northernmost extent. The floras include many plants of tropical Asian affinity identified as *Alangium*, *Alnus*, *Anamirta*, *Barringtonia* (a tropical Pacific beach plant), *Cananga*, *Clerodendrum*, *Kandelia* (a mangrove), Lauraceae (five species), *Luvunga*, *Limacia*, *Melanorrhoea*, *Meliosma*, Menispermaceae (five paleotropical genera), *Myristica*, the dipterocarp *Parashorea*, *Phytocrene*, *Platycarya*, *Pyrenancantha*, *Sageretia*, *Saurauia*, *Stemonurus*, and *Tetracentron*. Paleoclimatic conditions are difficult to assess. At the warmest end of the range the fossils along coastal areas were earlier interpreted as a megathermal paratropical rain forest growing under a MAT of ~21–25°C. The new estimate from CLAMP is 19.4°C (Wolfe, 1994), supporting a paratropical rain forest–microthermal mixed broad-leaved evergreen and coniferous forest. The coolest range would be 10–20°C (average 15°C). On the basis of LMA, the slightly older Kupreanof Island flora could represent a notophyllous broad-leaved evergreen forest or a paratropical rain forest.

The difficulty in interpreting these floras is partly because the Yakutat block is one of several suspect terranes or microplates separated from the Wrangell Mountains by the Denali Fault (Fig. 2.13) and transported from the south. The extent of the transport is uncertain. Some estimates place the terrane ~20° latitude further south during the Eocene (Bruns, 1983; Keller et al., 1984). Axelrod (1984, see also Axelrod et al., 1991) believes Paleogene climates in Alaska were temperate, and that strata containing tropical elements from the coastal side of the Denali Fault were carried up to 1500 km from the south. McKenna notes that

> Southwestern Alaska has been the site of tectonic unrest throughout the Cenozoic as various terranes have arrived from the south. Essentially all of the terranes south of the Denali Fault were formed at sites many degrees south of their present position and have been mashed against the North American Plate as the Kula Plate and part of the Pacific Plate moved north to destruction under North America and eastern Asia [Fig. 2.16]. Thus, environmental information derived from Paleogene and Mesozoic floras south of the Denali fault system tells us not about ancient Alaska but about somewhere else. (1983, p. 469)

Certainly an interpretation of extensive and widespread tropical floras in the Arctic based on these enigmatic coastal Paleogene communities (pl 65°–70° N) is not consistent with general global symmetry of vegetation at the time or with the more meager tropical element found in other high northern latitude floras (e.g., *Musophyllum* in the Thyra Ø flora of Greenland). The adjacent Eocene floras of northern Japan and Kamschatka are more temperate.

However, other evidence suggests the Yakutat block may have moved only 5° northward from about the latitude of British Columbia (Plafker, 1984; Wolfe, 1985; Wolfe and McCoy, 1984). J. A. Wolfe (personal communication, 1996) notes that to have moved 20°, the Yukutat block somehow would have had to jump a triple plate junction. The current uncertainty is whether movement was from the vicinity of British Columbia or from as far south as Oregon or beyond (D. Stone, University of Alaska, personal communication, 1994). The question is still open (Tiffney, 1985; see also Kerr, 1995). However, there is no paleomagnetic evidence to suggest extensive transport during the Paleogene from the western Pacific, so coastal areas can be interpreted as supporting tropical vegetation with an Old World component as least as far north as 45° (Oregon) or 57° (British Columbia). Northward of 70° N and probably further south in the continental interior the vegetation was a polar broad-leaved deciduous forest.

These floras indicate that during the Paleogene Beringia was occupied by a polar broad-leaved deciduous forest with a narrow southern fringe of evergreen megathermal plants (Tiffney, 1985), although the latter were transported at least some distance from the south. This tropical flora with a prominent Old World component is one basis for the boreotropical flora hypothesis, in contrast to the Arcto–Tertiary geoflora concept (Chapter 3).

VEGETATION SUMMARY

At the end of the Cretaceous the vegetation of North America consisted of four principal plant formations (Table 5.5). A tropical forest occupied southeastern United States and some version extended northward toward Greenland; there is little data on Maastrichtian vegetation from the intervening region. Genera common to the southern and northern floras include the form genera *Chondrophyllum*, *Dermatophyllites*, and *Proteoides*, and the extant genera *Andromeda*, *Diospyros*, and *Magnolia*. It was an open-canopy forest with relatively tall trees with poorly developed growth rings, and it grew under subhumid conditions. The emergent stratum consisted mostly of early representatives of the Taxodiaceae (evergreen forms similar to *Sequoia*; deciduous ones similar to *Metasequoia*, and the extinct *Parataxodium*) and the Cupressaceae (similar to *Callitris* and *Neocallitropis*). Understory vegetation included rosette palms, Zingiberales (*Spirematospermum*, *Zingiberopsis*), Magnoliidae (Chloranthaceae), Plantaceae, Fagaceae (*Fagopsis*), Juglandaceae, Aceraceae, Ranunculidae, and Dilleniidae (Theaceae, Euphorbiales, Malvales).

A paratropical forest grew in the southern and middle regions of the western United States. It was a broad-leaved evergreen community of subhumid habitats but more moist than in the southeast. In composition it was similar to the tropical forest.

The notophyllous broad-leaved evergreen forest occupied areas north of the paratropical forest, and it is recognized by leaves generally smaller than those in the evergreen community to the south. Members included *Equisetum*, lycopods, Cycadophyte and Ginkgophyte holdovers from earlier Cretaceous vegetation, the Araucariaceae, evergreen Taxodiaceae, *Nyssidium–Joffrea* (related to *Cercidiphyllum*), *Nordenskioldia* (related to *Trochodendron–Tetracentron*), pollen similar to *Gunnera*, and *Drumhellera* of the *Taxodium–Metasequoia–Sequoia–Sequoiadendron* complex. Broad-leaved deciduous species grew along streams and in other disturbed habitats.

A polar broad-leaved deciduous forest occupied the higher latitudes in North America during the Maastrichtian. In the lowlands and swamps, forests of deciduous Taxodiaceae were dominant; in better drained areas there was *Ginkgo* (along the southern transition margin), trochodendroids, *Viburnum*-type plants, platanoids and other Hamamelididae, Betulaceae–Myricaceae, and Ulmaceae.

By the end of the Cretaceous declining temperatures from mid-Cretaceous highs and drainage of the epicontinental sea (greater seasonality, less poleward transfer of heat to the interior) and polar light regimes were favoring diversification and expansion of deciduous microthermal vegetation in the high latitudes, in disturbed habitats, and in the few moderately high-altitude sites that existed at the time. The bolide impact may have intensified this trend regionally.

Although difficult to quantify, four refinements are emerging in the interpretation of Maastrichtian and Paleogene floras of the high northern latitudes. There is recognition that tropical vegetation from coastal sites did not extend widely over inland areas. This more limited emphasis on the tropical elements has reduced the degree of implied random ecological heterogeneity of Paleogene vegetation. It also indicates that climates and physical conditions were more diverse and less uniformly equable than earlier thought, allowing the arrangement of plants into

Table 5.5. Summary of North American vegetation types and estimated MATs: Late Cretaceous through Early Eocene. Names for plant communities follow terminology discussed under Vegetation.

Region	Vegetation	
	Late Cretaceous	MAT
Southeast (30°–?° N)	Tropical forest	20°C (25° warm intervals)
Northeast (65°–75° N)	?	?10°C
West (40°–46° N)	Paratropical	21–22°C
Northwest (50°–58° N)	Notophyllous broad-leaved evergreen	<20°C
North (58°–66° N)	Polar broad-leaved deciduous	15°C
North (65°–75° N)	Polar broad-leaved deciduous	8°C
	K–T Boundary Change	
West–northwest	Notophyllous broad-leaved evergreen to polar broad-leaved deciduous	20°C 15°C
	Paleocene	
Southeast	Tropical rain forest	27°C
Northeast	Notophyllous broad-leaved evergreen to polar broad-leaved deciduous; coastal megathermal communities	15–20°C
West (High Plains)		
Ft. Union, Sentinel Butte,	Ecotonal and trending (N–S and Early to Late Paleocene)	15°C
Bear Den to	from polar broad-leaved deciduous forest to	13–15/16°C
Camels Butte	paratropical rain forest–notophyllous broad-leaved evergreen forest	17.8°C
Powder River Basin	Notophyllous broad-leaved evergreen forest	16–18°C
Northwest	Trending Early to Late Paleocene	
Joffre Bridge (Alberta) to Donnet Plume (Yukon Territory)	Polar broad-leaved deciduous to notophyllous broad-leaved evergreen	10–15°C (from ~8°C in Late Cretaceous)
	Early Eocene	
Southeast	Tropical to subhumid paratropical rain forest	>27°C
Northeast	?Notophyllous broad-leaved evergreen (coastal) to restricted polar broad-leaved deciduous (inland, uplands)	Avg. ?18°C
Plains, eastern Rocky Mts.	Subhumid paratropical rain forest	?25°C
West to northwest	Tropical to paratropical rain forest (coastal);	?25°C
	notophyllous broad-leaved evergreen to polar broad-leaved deciduous (70°–75°, N interior and highlands); or more extensive	?12–15° C
	paratropical rain forest–broad-leaved evergreen (70°–75° N) forest (Kushtaka flora, suspect terrane?)	?21–25°C

habitats and communities more comparable to modern associations. Also, evidence from plant microfossils reveals a greater diversity in forest and understory vegetation than originally suggested by megafossil evidence alone. In terms of lineages, an emerging generalization is that many Paleocene through Middle Eocene angiosperms, assigned to modern taxa in the older literature, represent extinct or intermediate forms; from the Middle Eocene onward an increasing number can be placed in modern genera based on details of leaf architecture and the study of associated multiple organs.

In the earliest Paleocene there was an increase in rainfall, which had its most notable effect on the subhumid vegetation of the southeastern United States. With greater moisture, the tropical forest was replaced by a more closed-canopy, multistratal tropical rain forest. Subsequent climatic changes have left remnants or reintroductions in southern peninsula Florida and along the coasts of the southeastern United States (e.g., *Rhizophora*) and as elements of tropical origin that evolved temperate descendents (e.g., *Diospyros*, the only temperate representative of the primarily tropical family Ebenaceae). The Appalachian highland provided habitats for more temperate assemblages (*Alnus*, *Betula*, *Carya*, and others). Toward the northeast the vegetation was a notophyllous broad-leaved evergreen forest, especially along the coast of the North Atlantic land bridge, grading into a polar broad-leaved deciduous forest further inland and in the uplands.

The moist lowlands of the Plains area during the Paleocene and Early Eocene were occupied by a paratropical rain forest–notophyllous broad-leaved evergreen forest, grading northward into a notophyllous broad-leaved deciduous forest.

The warming trend continued into the Early Eocene and culminated with the highest MAT in all of Cenozoic time. The tropical rain forest persisted in the southeastern United States but was trending toward a subhumid paratropical rain forest near the Middle Eocene. Few paleobotanical data are available for the region to the northeast, but fossil faunas and some evidence from the Eureka Sound Formation suggest that a paratropical rain forest–notophyllous broad-leaved evergreen forest likely grew in the area. The polar broad-leaved deciduous forest, if present, was restricted to uplands in the higher latitudes. Paratropical rain forest extended into the Plains area and along the eastern slopes of the Rocky Mountains. On the western side of the mountains, midlatitude coastal areas were occupied by a tropical to paratropical rain forest. Interpretation of the vegetation toward the northwest depends on the resolution of the Alaskan terrane problem. The longitudinal arrangement of vegetation was from notophyllous broad-leaved evergreen (coastal) to polar broad-leaved deciduous forest (interior) at 70°–75° N if there was extensive transport of terranes. If there was less transport, the cline was from paratropical rain forest to notophyllous broad-leaved evergreen forest with more restricted deciduous forest.

Thus, at the end of the Early Eocene the principal plant communities of North America were tropical rain forest, paratropical rain forest, notophyllous broad-leaved evergreen forest, and a restricted polar broad-leaved deciduous forest. Tropical, subtropical, and warm-temperate vegetation had reached the greatest geographic extent it would achieve; subsequent events favored expansion of microthermal deciduous communities. As yet there is no extensive tundra formation, coniferous forest formation (boreal or montane associations: Appalachian or western montane coniferous forests), grassland formation, shrubland/chaparral–woodland–savanna formation, or desert formation.

References

Andrews, H. N., Jr. 1980. The Fossil Hunters, in Search of Ancient Plants. Cornell University Press, Ithaca, NY.

Areces-Mallea, A. E. 1990. *Basopollis krutzchi* Kedves: primera determinación de un normapolles en el Paleógeno de Cuba. Cienc. Tierra Espacio 17: 27–32.

Askin, R. A. 1988. The palynological record across the Cretaceous/Tertiary transition on Seymour Island, Antarctica. In: R. M. Feldmann and M. O. Woodburne (eds.), Geology and Paleontology of Seymour Island, Antarctic Peninsula. Geol. Soc. Amer. Mem. 169: 155–162.

Axelrod, D. I. 1952. A theory of angiosperm evolution. Evolution 6: 29–60.

———. 1970. Mesozoic paleogeography and early angiosperm history. Bot. Rev. 36: 277–319.

———. 1984. An interpretation of Cretaceous and Tertiary biota in polar regions. Palaeogeogr., Palaeoclimatol., Palaeoecol. 45: 105–147.

———. 1992a. What is an equable climate? Palaeogeogr., Palaeoclimatol., Palaeoecol. 91: 1–12.

———. 1992b. Climatic pulses, a major factor in legume evolution. In: P. S. Herendeen and D. L. Dilcher (eds.), Advances in Legume Systematics: Part 4. The Fossil Record. Royal Botanic Gardens, Kew, U.K. pp. 259–279.

———, M. T. K. Arroyo, and P. H. Raven. 1991. Historical development of temperate vegetation in the Americas. Rev. Chil. Hist. Nat. 64: 413–446.

Barrick, R. E. and W. J. Showers. 1994. Thermophysiology of *Tyrannosaurus rex*: evidence from oxygen isotopes. Science 265: 222–224.

Barron, E. J. 1983. A warm, equable Cretaceous: the nature of the problem. Earth-Sci. Rev. 19: 305–338.

———. 1984. Climatic implications of the variable obliquity explanation of Cretaceous–Paleogene high-latitude floras. Geology 12: 595–598.

———, J. L. Sloan II, and C. G. A. Harrison. 1980. Potential significance of land–sea distribution and surface albedo variations as a climatic forcing factor; 180 M.Y. to the present Palaeogeogr., Palaeoclimatol., Palaeoecol. 30: 17–40.

Barron, E. J., S. L. Thompson, and S. H. Schneider. 1981. An ice-free Cretaceous? Results from climate model simulations. Science 212: 501–508.

Barron, E. J., and W. M. Washington. 1982a. Cretaceous climate: a comparison of atmospheric simulations with the geologic record. Palaeogeogr., Palaeoclimatol., Palaeoecol. 40: 103–133.

———. 1982b. Atmospheric circulation during warm geologic periods: is the equator-to-pole surface-temperature gradient the controlling factor? Geology 10: 633–636.

Basinger, J. F., D. R. Greenwood, and T. Sweda. 1994. Early Tertiary vegetation of Arctic Canada and its relevance to paleoclimatic interpretation. In: M. C. Boulter and H. C. Fisher (eds.), Cenozoic Plants and Climates of the Arctic. NATO ASI Series 1: Global Environmental Change. Springer-Verlag, Berlin. Vol. 27, pp. 175–198.

Batten, D. J. 1984. Palynology, climate and the development of Late Cretaceous floral provinces in the northern hemisphere: a review. In: P. Brenchley (ed.), Fossils and Climate. Wiley, New York. pp. 127–164.

Beeson, D. C. 1992. High resolution palynostratigraphy

across a marine Cretaceous–Tertiary boundary interval, Falls County, Texas. Am. Assoc. Stratig. Palynol. Newslett. 25: 17–18.

Bell, W. A. 1949. Uppermost Cretaceous and Paleocene floras of western Alberta. Geol. Surv. Can. Bull. 13: 1–231.

Berggren, W. A. 1982. Role of ocean gateways in climatic change. In: Geophysics Study Committee (eds.), Climate in Earth History. National Academy Press, Washington, D.C. pp. 118–125.

———, D. V. Kent, M.-P. Aubry, and J. Hardenbol (eds.). 1995. Geochronology, Time Scales and Global Stratigraphic Correlation. Society of Sedimentary Geology, Tulsa, OK. Special Publ. 54.

Berner, R. A., A. C. Lasaga, and R. M. Garrels. 1983. The carbonate-silicate geochemical cycle and its effect on atmospheric carbon dioxide over the past 100 million years. Am. J. Sci. 283: 641–683.

Berry, E. W. 1916. The Lower Eocene Floras of Southeastern North America. U.S. Geological Survey, Reston, VA. Professional Paper 91, pp. 1–481.

———. 1922. Additions to the flora of the Wilcox Group. U.S. Geological Survey, Reston, VA. Professional Paper 131-A, pp. 1–21.

———. 1924. The Middle and Upper Eocene Floras of Southeastern North America. U.S. Geological Survey, Reston, VA. Professional Paper 92, pp. 1–206.

———. 1925. The Flora of the Ripley Formation. U.S. Geological Survey, Reston, VA. Professional Paper 136, pp. 1–94.

———. 1930. Revision of the Lower Eocene Wilcox Flora of the Southeastern States. U.S. Geological Survey, Reston, VA. Professional Paper 156, pp. 1–196.

———. 1935. A preliminary contribution to the floras of the Whitemud and Ravenscrag formations. Geol. Surv. Can. Mem. 182: 1–107.

———. 1941. Additions to the Wilcox Flora from Kentucky and Texas. U.S. Geological Survey, Reston, VA. Professional Paper 193-E, pp. 83–99.

Bown, T. M., K. D. Rose, E. L. Simons, and S. L. Wing. 1994. Distribution and Stratigraphic Correlation of Upper Paleocene and Lower Eocene Fossil Mammal and Plant Localities of the Fort Union, Willwood, and Tatman Formations, Southern Big Horn Basin, Wyoming. U.S. Geological Survey, Reston, VA. Professional Paper 1540, pp. 1–103.

Boyd, A. 1990. The Thyra Ø flora: toward an understanding of the climate and vegetation during the Early Tertiary in the high Arctic. Rev. Palaeobot. Palynol. 62: 189–203.

———. 1992. *Musopsis* N. Gen.: a banana-like leaf genus from the Early Tertiary of eastern north Greenland. Am. J. Bot. 79: 1359–1367.

Brouwers, E. M. (+ 5 authors). 1987. Dinosaurs on the North Slope, Alaska: high latitude, latest Cretaceous environments. Science 237: 1608–1610.

Brown, R. W. 1962. Paleocene flora of the Rocky Mountains and Great Plains. U.S. Geological Survey, Reston, VA. Professional Paper 375, pp. 1–119.

Bruns, T. R. 1983. Model for the origin of the Yakutat block, an accreting terrane in the northern Gulf of Alaska. Geology 11: 718–721.

Call, V. B. and D. L. Dilcher, 1997. The fossil record of *Eucommia* (Eucommiaceae) in North America. Am. J. Bot. 84: 798–814.

Chandrasekharam, A. 1974. Megafossil flora from the Genesee locality, Alberta, Canada. Palaeontograph., Abt. B, 147: 1–41.

Clemens, W. A. and C. W. Allison. 1985. Late Cretaceous terrestrial vertebrate fauna, North Slope, Alaska. Geol. Soc. Am. Abstr. Programs 17: 548.

——— and L. G. Nelms. 1993. Paleoecological implications of Alaskan terrestrial vertebrate fauna in latest Cretaceous time at high paleolatitudes. Geology 21: 503–506.

Crane, P. R. 1987. Vegetational consequences of the angiosperm diversification. In: E. M. Friis, W. G. Chaloner, and P. R. Crane (eds.), The Origins of Angiosperms and Their Biological Consequences. Cambridge University Press, Cambridge, U.K. pp. 107–144.

———. 1989. Paleobotanical evidence on the early radiation of nonmagnoliid dicotyledons. Plant Syst. Evol. 162: 165–191.

———, S. R. Manchester, and D. L. Dilcher. 1990. A preliminary survey of fossil leaves and well-preserved reproductive structures from the Sentinel Butte Formation (Paleocene) near Almont, North Dakota. Fieldiana (Geology) 20: 1–63.

Crane, P. R. and R. A. Stockey. 1985. Growth and reproductive biology of *Joffrea speirsii* gen. et sp. nov., a *Cercidiphyllum*-like plant from the Late Paleocene of Alberta, Canada. Can. J. Bot. 63: 340–364.

Creber, G. T. and W. G. Chaloner. 1985. Tree growth in the Mesozoic and Early Tertiary and the reconstruction of palaeoclimates. Palaeogeogr., Palaeoclimatol., Palaeoecol. 52: 35–60.

Crepet, W. L. 1981. The status of certain families of the Amentiferae during the Middle Eocene and some hypotheses regarding the evolution of wind pollination in dicotyledonous angiosperms. In: K. J. Niklas (ed.), Paleobotany, Paleoecology, and Evolution. Praeger, New York. Vol. 2, pp. 103–128.

———. 1989. History and implications of the early North American fossil record of Fagaceae. In: P. R. Crane and S. Blackmore (eds.), Evolution, Systematics, and Fossil History of the Hamamelidae, Vol. 2, "Higher" Hamamelidae. Clarendon Press, Oxford, U.K. Special Vol. 40B, pp. 45–66.

——— and C. P. Daghlian. 1981. Lower Eocene and Paleocene Gentianaceae: floral and palynological evidence. Science 214: 75–77.

Crepet, W. L. and G. D. Feldman. 1991. The earliest remains of grasses in the fossil record. Am. J. Bot. 78: 1010–1014.

Crepet, W. L. and P. S. Herendeen. 1992. Papilionoid flowers from the Early Eocene of southeastern North America. In: P. S. Herendeen and D. L. Dilcher (eds.), Advances in Legume Systematics: Part 4. The Fossil Record. Royal Botanic Gardens, Kew, U.K. pp. 43–55.

Crepet, W. L. and K. C. Nixon. 1989a. Earliest megafossil evidence of Fagaceae: phylogenetic and biogeographic implications. Am. J. Bot. 76: 842–855.

———. 1989b. Extinct transitional Fagaceae from the Oligocene and their phylogenetic implications. Am. J. Bot. 76: 1493–1505.

Crepet, W. L. and D. W. Taylor. 1985. The diversification of the Leguminosae: first fossil evidence of the Mimosoideae and Papilionoideae. Science 228: 1087–1089.

———. 1986. Primitive mimosoid flowers from the Paleocene–Eocene and their systematic and evolutionary implications. Am. J. Bot. 73: 548–563.

Crowley, T. J. and G. R. North. 1991. Paleoclimatology. Oxford Monographs in Geology and Geophysics. Oxford Univ. Press, Oxford, U.K. No. 16, Chap. 9.

Crowley, T. J., D. A. Short, J. G. Mengel, and G. R. North. 1986. Role of seasonality in the evolution of climate during the past 100 million years. Science 231: 579–584.

Dawson, M. R., L. Krishtalka, and R. K. Stucky. 1990. Revision of the Wind River faunas, Early Eocene of central Wyoming. Part 9. The oldest known Hystricomorphous rodent (Mammalia:Rodentia). Ann. Carnegie Mus. 59: 135–147.

Dawson, M. R., R. M. West, W. Langston, Jr., and J. H. Hutchison. 1976. Paleogene terrestrial vertebrates: northernmost occurrence, Ellesmere Island, Canada. Science 192: 781–782.

Dilcher, D. L. 1971. A revision of the Eocene flora of southeastern North America. Palaeobotanist 20: 7–18.

———. 1973. A paleoclimatic interpretation of the Eocene floras of southeastern North Amerca. In: A. Graham (ed.), Vegetation and Vegetational History of Northern Latin America. Elsevier Science Publishers, Amsterdam. pp. 39–59.

Dolph, G. E. and D. L. Dilcher. 1979. Foliar physiognomy as an aid in determining paleoclimate. Palaeontograph., Abt. B, 170: 151–172.

Dorf, E. 1942. Upper Cretaceous floras of the Rocky Mountain region. Carnegie Inst. Wash. Contrib. Paleontol. 508.

———. 1960. Tertiary fossil forests of Yellowstone National Park, Wyoming. Billings Geological Society of America Guidebook, Geological Society of America, Denver, CO, 11th Annual Field Conference, pp. 253–260.

Elsik, W. C. 1968a. Palynology of a Paleocene Rockdale lignite, Milam County, Texas. I. Morphology and Taxonomy. Pollen Spores 10: 263–314.

———. 1968b. Palynology of a Paleocene Rockdale lignite, Milan County, Texas. II. Morphology and Taxonomy (end). Pollen Spores 10: 599–664.

———. 1970. Palynology of a Paleocene Rockdale lignite, Milam County, Texas. III. Errata and taxonomic revisions. Pollen Spores 12: 99–101.

Emanuel, K. A., K. Speer, R. Rotunno, R. Srivastava, and M. Molina. 1995. Hypercanes: a possible link in global extinction scenarios. J. Geophys. Res. 100D: 13755–13765.

Estes, R. and J. H. Hutchison. 1980. Eocene lower vertebrates from Ellesmere Island, Canadian Arctic Archipelago. Palaeogeogr., Palaeoclimatol., Palaeoecol. 30: 325–347.

Farabee, M. J. 1990. Triprojectate fossil pollen genera. Rev. Palaeobot. Palynol. 65: 341–347.

———. 1991. Botanical affinities of some Triprojectacites fossil pollen. Am. J. Bot. 78: 1172–1181.

———. 1993. Morphology of triprojectate fossil pollen: form and distribution in space and time. Bot. Rev. 59: 211–249.

Farley, M. B. 1990. Vegetation distribution across the early Eocene depositional landscape from palynological analysis. Palaeogeogr., Palaeoclimatol., Palaeoecol. 79: 11–27.

Ferguson, K. M., T. M. Lehman, and R. T. Gregory. 1991. C- and O-isotopes of pedogenic soil nodules from two sections spanning the K/T transition in west Texas. Geol. Soc. Am., Abstr. Program 23: 302.

Fleming, R. F. and D. J. Nichols. 1990. The fern-spore abundance anomaly at the Cretaceous–Tertiary boundary: a regional bioevent in western North America. In: E. G. Kauffman and O. H. Walliser (eds.), Extinction Events in Earth History. Springer-Verlag, Berlin. pp. 347–350.

Foster, H. L. and Y. Igarashi. 1990. Fossil pollen from nonmarine sedimentary rocks of the eastern Yukon–Tanana region, east-central Alaska. U.S. Geol. Surv. Bull. 1946: 11–20.

Frakes, L. A. 1979. Climates Throughout Geologic Time. Elsevier Science Publishers, Amsterdam.

Francis, J. E. 1988. A 50 million-year old forest from Strathcona Fiord, Ellesmere Island, Arctic Canada: evidence for a warm polar climate. Arctic 41: 314–318.

Franczyk, K. J., T. D. Fouch, R. C. Johnson, C. M. Molenaar, and W. A. Cobban. 1992. Cretaceous and Tertiary paleogeographic reconstructions for the Uinta–Piceance Basin study area, Colorado and Utah. U.S. Geol. Surv. Bull. 1787-Q: 1–37.

Frederiksen, N. O. 1987. Tectonic and paleogeographic setting of a new latest Cretaceous floristic province in North America. Palaios 2: 533–542.

———. 1988. Sporomorph Biostratigraphy, Floral Changes, and Paleoclimatology, Eocene and Earliest Oligocene of the Eastern Gulf Coast. U.S. Geological Survey, Reston, VA. Professional Paper 1448.

———. 1989a. Changes in floral diversities, floral turnover rates, and climates in Campanian and Maastrichtian time, North Slope of Alaska. Cretaceous Res. 10: 249–266.

———. 1989b. Late Cretaceous and Tertiary floras, vegetation, and paleoclimates of New England. Rhodora 91: 25–48.

———. 1991a. Rates of floral turnover and diversity change in the fossil record. Palaeobotanist 39: 127–139.

———. 1991b. Pulses of Middle Eocene to earliest Oligocene climate deterioration in southern California and the Gulf Coast. Palaios 6: 564–571.

———. 1994. Paleocene floral diversities and turnover events in eastern North America and their relation to diversity models. Rev. Palaeobot. Palynol. 82: 225–238.

———, T. A. Ager, and L. E. Edwards. 1988. Palynology of Maastrichtian and Paleocene rocks, lower Colville River region, North Slope of Alaska. Can. J. Earth Sci. 25: 512–527.

Friis, E. M. 1983. Upper Cretaceous (Senonian) floral structures of juglandalean affinity containing normapolles pollen. Rev. Palaeobot. Palynol. 39: 161–188.

———. 1987. *Spriematospermum chandlerae* sp. nov., an extinct species of Zingiberaceae from the North American Cretaceous. Tert. Res. 9: 7–12.

Greenwood, D. R. and J. F. Basinger. 1994. The paleoecology of high-latitude Eocene swamp forests from Axel Heiberg Island, Canadian high Arctic. Rev. Palaeobot. Palynol. 81: 83–97.

Greenwood, D. R. and S. L. Wing. 1995. Eocene continental climates and latitudinal temperature gradients. Geology 23: 1044–1048.

Habicht, J. K. A. 1979. Paleoclimate, paleomagnetism, and

continental drift. In: American Association Petroleum Geology (eds.), Studies in Geology 9. American Association of Petroleum Geology, Tulsa, OK.

Hallam, A. 1984. Continental humid and arid zones during the Jurassic and Cretaceous. Palaeogeogr., Palaeoclimatol., Palaeoecol. 47: 195–223.

Harland, W. B. (+ 5 authors). 1990. A Geologic Time Scale 1989. Cambridge University Press, Cambridge, U.K.

Heer, O. 1883. Den Zweiten Theil der fossilen Flora Grönlands. Flora Fossilis Arctica. J. Wurster, Zurich. Vol. 7.

Herendeen, P. S., D. H. Les, and D. L. Dilcher. 1990. Fossil *Ceratophyllum* (Ceratophyllaceae) from the Tertiary of North America. Am. J. Bot. 77: 7–16.

Herman, A. B. 1994. Late Cretaceous Arctic platanoids and high latitude climate. In: M. C. Boulter and H. C. Fisher (eds.), Cenozoic Plants and Climates of the Arctic. NATO ASI Series 1: Global Environmental Change. Springer-Verlag, Berlin. Vol. 27, pp. 151–159.

——— and R. A. Spicer. 1996. Palaeobotanical evidence for a warm Cretaceous Arctic Ocean. Nature 380: 330–333.

———. 1997. New quantitative palaeoclimate data for the Late Cretaceous Arctic: evidence for a warm polar ocean. Palaeogeogr., Palaeoclimatol., Palaeoecol. 128: 227–251.

Herngreen, G. F. W. and A. F. Chlonova. 1981. Cretaceous microfloral provinces. Pollen Spores 23: 441–555.

Hickey, L. J. 1977. Stratigraphy and paleobotany of the Golden Valley Formation (Early Tertiary) of western North Dakota. Geol. Soc. Am. Mem. 150: 1–183.

———. 1980. Paleocene stratigraphy and flora of the Clark's Fork Basin. In: P. D. Gingerich (ed.), Early Cenozoic Paleontology and Stratigraphy of the Bighorn Basin, Wyoming. Univ. Mich. Papers Paleontol. 24: 33–49.

———. 1981. Land plant evidence compatible with gradual, not catastrophic, change at the end of the Cretaceous. Nature 292: 529–531.

———. 1984. Changes in the angiosperm flora across the Cretaceous–Tertiary boundary. In: W. A. Berggren and J. A. Van Couvering (eds.), Catastrophes and Earth History, the New Uniformitarianism. Princeton University Press, Princeton, NJ. pp. 279–337.

Hickey, L. J. and L. J. McWeeney. 1992. Plants at the K/T boundary. Nature (Sci. Corresp.) 356: 295–296.

Hickey, L. J. and R. K. Peterson. 1978. *Zingiberopsis*, a fossil genus of the ginger family from Late Cretaceous to Early Eocene sediments of Western Interior North America. Can. J. Bot. 56: 1136–1152.

Hickey, L. J., R. M. West, M. R. Dawson, and D. K. Choi. 1983. Arctic terrestrial biota: paleomagnetic evidence of age disparity with mid-northern latitudes during the Late Cretaceous and Early Tertiary. Science 221: 1153–1156.

Hoffman, A. 1989. Changing palaeontological views on mass extinction phenomena. In: S. K. Donovan (ed.), Mass Extinctions, Processes and Evidence. Columbia University Press, New York. pp. 1–18.

Hoffman, G. L. and R. A. Stockey. 1992. Vegetative and fertile remains of *Azolla* from the Paleocene Joffre Bridge locality, Alberta. Am. J. Bot. 79(Suppl.): 102–103.

Horrell, M. A. 1991. Phytogeography and paleoclimatic interpretation of the Maestrichtian. Palaeogeogr., Palaeoclimatol., Palaeoecol. 86: 87–138.

Hsü, K. J. and J. A. McKenzie. 1985. A "Strangelove" ocean in the earliest Tertiary. In: E. T. Sundquist and W. S. Broecker (eds.), The Carbon Cycle and Atmospheric CO2: Natural Variations Archean to Present. Geophysical Monograph 32. American Geophysical Union, Washington, D.C. pp. 487–492.

Hutchison, J. H. 1982. Turtle, crocodilian, and champsosaur diversity changes in the Cenozoic of the north-central region of western United States. Palaeogeogr., Palaeoclimatol., Palaeoecol. 37: 149–164.

Jarzen, D. M. 1977. *Aquilapollenites* and some Santalalean genera. Grana 16: 29–39.

———. 1980. The occurrence of *Gunnera* pollen in the fossil record. Biotropica 12: 117–123.

———. 1982a. Angiosperm pollen from the Ravenscrag Formation (Paleocene), southern Saskatchewan, Canada. Pollen Spores 14: 119–155.

———. 1982b. Palynology of Dinosaur Provincial Park (Campanian) Alberta. Syllogeus 38: 1–69.

——— and M. E. Dettmann. 1989. Taxonomic revision of *Tricolpites reticulatus* Cookson ex Couper, 1953 with notes on the biogeography of *Gunnera* L. Pollen Spores 31: 97–112.

Jerzykiewicz, T. and A. R. Sweet. 1986. The Cretaceous–Tertiary boundary in the central Alberta foothills. I. Stratigraphy. Can. J. Earth Sci. 23: 1356–1374.

Johnson, K. R. 1992. Leaf-Fossil evidence for extensive floral extinction at the Cretaceous–Tertiary boundary, North Dakota, USA. Cretaceous Res. 13: 91–117.

———. 1993. Extinctions at the Antipodes. Nature 366: 511–512.

———. 1996. Description of seven common leaf species from the Hell Creek Formation (Upper Cretaceous: Upper Maastrichtian), North Dakota, South Dakota, and Montana. Proc. Denver Museum Natural History, Ser. 3, No. 12: 1–47.

Johnson, K. R. and L. J. Hickey. 1990. Megafloral Change Across the Cretaceous/Tertiary Boundary in the Northern Great Plains and Rocky Mountains, U.S.A. Geological Society of America, Boulder, CO. Special Paper 247, pp. 433–444.

Johnson, K. R., D. J. Nichols, M. Attrep, Jr., and C. J. Orth. 1989. High-resolution leaf-fossil record spanning the Cretaceous/Tertiary boundary. Nature 340: 708–711.

Jones, J. H. and D. L. Dilcher. 1988. A study of the "*Dryophyllum*" leaf forms from the Paleogene of southeastern North America. Palaeontograph., Abt. B, 208: 53–80.

Kalgutkar, R. M. and D. J. McIntyre. 1991. Helicosporous fungi and Early Eocene pollen, Eureka Sound Group, Axel Heiberg Island, Northwest Territories. Can. J. Earth Sci. 28: 364–371.

Kalkreuth, W. D., D. J. McIntyre, and R. J. H. Richardson. 1993. The geology, petrology and palynology of Tertiary coals from the Eureka Sound Group at Strathcona Fiord and Bache Peninsula, Ellesmere Island, Arctic Canada. Int. J. Coal Geol. 24: 75–111.

Kauffman, E. G. 1977. Geological and biological overview: western interior Cretaceous basin. Mt. Geol. 14: 75–100.

Kay, M. Geologic Map of North America. Wiley, New York. [Distributed by the Geological Society of America. no scale.]

Keller, G. 1993. The Cretaceous–Tertiary boundary transition in the Antarctic Ocean and its global implications. Mar. Micropaleontol. 21: 1–45.

———, E. Barrera, B. Schmitz, and E. Mattson. 1993. Gradual mass extinction, species survivorship, and long-term environmental changes across the Cretaceous–Tertiary boundary in high latitudes. Geol. Soc. Am. Bull. 105: 979–997.

Keller, G., R. Von Huene, K. McDougall, and T. R. Bruns. 1984. Paleoclimatic evidence for Cenozoic migration of Alaska terranes. Tectonics 3: 473–495.

Kent, D. V., M. C. McKenna, N. D. Opdyke, J. J. Flynn, and B. J. MacFadden. 1984. Arctic biostratigraphic heterochroneity. Sci. (Lett.): 224: 173–174.

Kerr, R. A. 1992. Did an asteroid leave its mark in Montana bones? Science 256: 1395–1396.

Kerr, R. A. 1995. How far did the west wander? Science 268: 635–637.

Knowlton, F. H. 1899. Fossil flora of the Yellowstone National Park. U.S. Geol. Surv. Monogr. 32: 651–882.

———. 1917. Fossil floras of the Vermejo and Raton formations of Colorado and New Mexico. U.S. Geological Survey, Reston, VA. Professional Paper 101, pp. 223–455.

———. 1922. The Laramie flora of the Denver Basin, with a review of the Laramie problem. U.S. Geological Survey, Reston, VA. Professional Paper 130, pp. 1–175.

———. 1930. The flora of the Denver and associated formations of Colorado. U.S. Geological Survey, Reston, VA. Professional Paper 155, pp. 1–142.

Koch, B. E. 1963. Fossil plants from the lower Paleocene of the Agatdalen (Angmartussut) area, central Nûgssuaq Peninsula, northwest Greenland. Meddelelser Grønland 172: 1–120.

———. 1964. Review of fossil floras and nonmarine deposits of west Greenland. Geol. Soc. Am. Bull. 75: 535–548.

Koch, P. L., J. C. Zachos, and P. D. Gingerich. 1992. Correlation between isotope records in marine and continental carbon reservoirs near the Palaeocene/Eocene boundary. Nature 358: 319–322.

Koeberl, C. and R. R. Anderson (eds.). 1996. The Manson impact structure, Iowa: anatomy of an impact crater. Geol. Soc. Am. Sp. Paper 302: 1–482.

Leffingwell, H. A. 1971 (1970). Palynology of the Lance (Late Cretaceous) and Fort Union (Paleocene) formations of the type Lance area, Wyoming. In: R. M. Kosanke and A. T. Cross (eds.), Symposium on Palynology of the Late Cretaceous and Early Tertiary. Geological Society of America, Boulder, CO. Special Paper 127, pp. 1–64.

Lehman, T. M. 1990. Paleosols and Cretaceous/Tertiary transition in the Big Bend region of Texas. Geology 18: 362–364.

Leopold, E. B. and H. D. MacGinitie. 1972. Development and affinities of Tertiary floras in the Rocky Mountains. In: A. Graham (ed.), Floristics and Paleofloristics of Asia and Eastern North America. Elsevier Science Publishers, Amsterdam. pp. 147–200.

LePage, B. A. and J. F. Basinger. 1991a. A new species of *Larix* (Pinaceae) from the Early Tertiary of Axel Heiberg Island, Arctic Canada. Rev. Palaeobot. Palynol. 70: 89–111.

———. 1991b. Early Tertiary *Larix* from the Buchanan Lake Formation, Canadian Arctic Archipelago and a consideration of the phytogeography of the genus. In: R. L. Christie and N. J. McMillan (eds.), Tertiary Fossil Forests of the Geodetic Hills, Axel Heiberg Island, Arctic Archipelago. Geological Survey of Canada Bull. 403: 67–82.

———. 1995a. Evolutionary history of the genus *Pseudolarix* Gordon (Pinaceae). Internat. J. Plant Sci. 156: 910–950.

———. 1995b. The evolutionary history of the genus *Larix* (Pinaceae). U.S. Dept. Agriculture, Forest Service, Gen. Tech. Rept. GTR-INT-319: 19–29.

Lidgard, S. and P. R. Crane. 1990. Angiosperm diversification and Cretaceous floristic trends: a comparison of palynofloras and leaf macrofloras. Paleobiology 16: 77–93.

Lloyd, C. R. 1982. The mid-Cretaceous Earth: paleogeography; ocean circulation and temperature; atmospheric circulation. J. Geol. 90: 393–414.

Lowenstam, H. A. and S. Epstein. 1954. Paleotemperatures of the post-Aptian Cretaceous as determined by the oxygen isotope method. J. Geol. 62: 207–248.

MacGinitie, H. D. 1941. A Middle Eocene flora from the central Sierra Nevada. Carnegie Inst. Wash. Contrib. Paleontol. 534.

Magallón-Puebla, S. and S. Cevallos-Ferriz. 1994. *Eucommia constans* n. sp. Fruits from upper Cenozoic strata of Puebla, Mexico: morpholgical and anatomical comparison with *Eucommia ulmoides* Oliver. Internat. J. Plant Sci. 155: 80–95.

Manchester, S. R. 1987. The fossil history of the Juglandaceae. Monogr. Syst. Bot. Missouri Bot. Garden 21: 1–137.

———. 1991. *Cruciptera*, a new juglandaceous winged fruit from the Eocene and Oligocene of western North America. Syst. Bot. 16: 715–725.

Manchester, S. R. and Z.-D. Chen. 1996. *Palaeocarpinus aspinosa* sp. Nov. (Betulaceae) from the Paleocene of Wyoming, U.S.A. Internat. J. Plant Sci. 157: 644–655.

———, P. R. Crane, and D. L. Dilcher. 1991. *Nordenskioldia* and *Trochodendron* (Trochodendraceae) from the Miocene of northwestern North America. Bot. Gaz. 152: 357–368.

Manchester, S. R. and D. L. Dilcher. 1982. Pterocaryoid fruits (Juglandaceae) in the Paleogene of North America and their evolutionary and biogeographic significance. Am. J. Bot. 69: 275–286.

Manchester, S. R. and D. L. Dilcher. 1997. Reproductive and vegetative morphology of *Polyptera* (Juglandaceae) from the Paleocene of Wyoming and Montana. Am. J. Bot. 84: 649–663.

Manchester, S. R. and W. J. Kress. 1993. Fossil bananas (Musaceae): *Ensete oregonense* sp. nov. from the Eocene of western North America and its phytogeographic significance. Am. J. Bot. 80: 1264–1272.

McGookey, D. P. et al. (+ 6 authors). 1972. Cretaceous system. In: W. W. Mallory (ed.), Geologic Atlas of the Rocky Mountain Region. Rocky Mountain Association of Geology, Denver, CO. pp. 190–228.

McIntyre, D. J. 1989. Paleocene palynoflora from northern Somerset Island, District of Franklin, N.W.T. Current Res., Geol. Surv. Can. 89–1G: 191–197.

——— and B. D. Ricketts. 1989. New palynological data from Cornwall Arch, Cornwall and Amund Ringnes islands, District of Franklin, N.W.T. Current Res., Geol. Surv. Can. 89–1G: 199–202.

McIver, E. E. and J. F. Basinger. 1993. Flora of the Ravenscrag Formation (Paleocene), southwestern Saskatchewan, Canada. Palaeontograph. Can. 10: 1–167.

McKenna, M. C. 1980. Eocene paleolatitude, climate, and mammals of Ellesmere Island. Palaeogeogr., Palaeoclimatol., Palaeoecol. 30: 349–362.

———. 1983. Holarctic landmass rearrangement, cosmic events, and Cenozoic terrestrial organisms. Ann. Missouri Bot. Garden 70: 459–489.

Muller, R. A. and G. J. MacDonald. 1997. Glacial cycles and astronomical forcing. Science 277: 215–218.

Nichols, D. J. 1992. Plants at the K/T boundary. Nature (Sci. Corresp.): 356: 295.

——— and R. F. Fleming. 1990. Plant microfossil record of the terminal Cretaceous event in the western United States and Canada. In: V. L. Sharpton and P. D. Ward (eds.), Global Catastrophes in Earth History. Geological Society of America, Boulder, CO. Special Paper 247, pp. 445–455.

——— and N. O. Frederiksen. 1990. Palynological evidence of effects of the terminal Cretaceous event on terrestrial floras in western North America. In: E. G. Kauffman and O. H. Walliser (eds.), Extinction Events in Earth History. Springer-Verlag, Berlin. pp. 351–364.

Nichols, D. J., D. M. Jarzen, C. J. Orth, and P. Q. Oliver. 1986. Palynological and iridium anomalies at the Cretaceous–Tertiary boundary, south-central Saskatchewan. Science 231: 714–717.

Nichols, D. J. and H. L. Ott. 1978. Biostratigraphy and evolution of the *Momipites–Caryapollenites* lineage in the Early Tertiary in the Wind River Basin, Wyoming. Palynology 2: 93–112.

Nichols, D. J., J. A. Wolfe, and D. T. Pocknall. 1989. Latest Cretaceous and Early Tertiary history of vegetation in the Powder River Basin, Montana and Wyoming. 28th Int. Geol. Congr. Guidebook T132: 28–33.

Nixon, K. C. 1989. Origins of Fagaceae. In: P. R. Crane and S. Blackmore (eds.), Evolution, Systematics, and Fossil History of the Hamamelidae, Vol. 2, "Higher" Hamamelidae. Clarendon Press, Oxford, U.K. Special Vol. 40B, pp. 23–43.

——— and W. L. Crepet. 1989. *Trigonobalanus* (Fagaceae): taxonomic status and phylogenetic relationships. Am. J. Bot. 76: 828–841.

Norris, G. and A. D. Miall. 1984. Arctic biostratigraphic heterochroneity. Sci. (Lett.): 224: 174–175.

Norton, N. J. and J. W. Hall. 1967. Guide sporomorphae in the Upper Cretaceous–Lower Tertiary of eastern Montana (U.S.A.). Rev. Palaeobot. Palynol. 2: 99–110.

———. 1969. Palynology of the Upper Cretaceous and Lower Tertiary in the type locality of the Hell Creek Formation, Montana, U.S.A. Palaeontograph., Abt. B, 128: 1–64.

Oltz, D. F., Jr. 1969. Numerical analyses of palynological data from Cretaceous and Early Tertiary sediments in east central Montana. Palaeontograph., Abt. B, 128: 90–166.

Opdyke, N. D. 1990. Magnetic stratigraphy of Cenozoic terrestrial sediments and mammalian dispersal. J. Geol. 98: 621–637.

Orth, C. J., J. S. Gilmore, J. D. Knight, C. L. Pillmore, and R. H. Tschudy. 1981. An iridium abundance anomaly at the palynological Cretaceous–Tertiary boundary in northern New Mexico. Science 214: 1341–1343.

Otto-Bliesner, B. L. and G. R. Upchurch, Jr. 1997. Vegetation-induced warming of high-latitude regions during the Late Cretaceous period. Nature 385: 804–807.

Owen, R. M. and D. K. Rea. 1985. Sea-floor hydrothermal activity links climate to tectonics: the Eocene carbon dioxide greenhouse. Science 227: 166–169.

Pabst, M. B. 1968. The flora of the Chuckanut Formation of northwestern Washington, the Equisetales, Filicales, and Coniferales. Univ. Calif. Publ. Geol. Sci. 76: 1–85.

Page, V. M. 1979. Dicotyledonous wood from the Upper Cretaceous of central California. J. Arnold Arb. 60: 323–349.

———. 1980. Dicotyledonous wood from the Upper Cretaceous of central California, II. J. Arnold Arb. 61: 723–748.

———. 1981. Dicotyledonous wood from the Upper Cretaceous of central California. III. Conclusions. J. Arnold Arb. 62: 437–455.

Parrish, J. T. 1987. Global palaeogeography and palaeoclimate of the Late Cretaceous and Early Tertiary. In: E. M. Friis, W. G. Chaloner, and P. R. Crane (eds.), The Origins of Angiosperms and Their Biological Consequences. Cambridge University Press, Cambridge, U.K. pp. 51–73.

——— and R. A. Spicer. 1988. Late Cretaceous terrestrial vegetation: a near-polar temperature curve. Geology 16: 22–25.

Pedersen, K. R. 1976. Fossil floras of Greenland. In: A. Escher and W. Swatt (eds.), Geology of Greenland. Geological Survey of Greenland, Copenhagen. pp. 519–535.

Pigg, K. B. and R. A. Stockey. 1991. Platanaceous plants from the Paleocene of Alberta, Canada. Rev. Palaeobot. Palynol. 70: 125–146.

——— and S. L. Maxwell. 1993. *Paleomyrtinaea*, a new genus of permineralized myrtaceous fruits and seeds from the Eocene of British Columbia and Paleocene of North Dakota. Can. J. Bot. 71: 1–9.

Plafker, G. 1984. Comments and replies on "Model for the origin of the Yakutat block, an accreting terrane in the northern Gulf of Alaska." Geology 12: 563.

Pocknall, D. T. 1987. Palynomorph biozones for the Fort Union and Wasatch formations (Upper Paleocene–Lower Eocene), Powder River Basin, Wyoming and Montana, U.S.A. Palynology 11: 23–35.

Ramanujan, C. G. K. and W. N. Stewart. 1969. Fossil woods of Taxodiaceae from the Edmonton Formation (Upper Cretaceous) of Alberta. Can. J. Bot. 47: 115–124.

Rea, D. K., M. Leinen, and T. R. Janecek. 1985. Geologic approach to the long-term history of atmospheric circulation. Science 277: 721–725.

Rea, D. K., J. C. Zachos, R. M. Owen, and P. D. Gingerich. 1990. Global change at the Paleocene–Eocene boundary: climatic and evolutionary consequences of tectonic events. Palaeogeogr., Palaeoclimatol., Palaeoecol. 79: 117–128.

Responses. 1988. Science 239: 10–11.

Retallack, G. J. 1994. A peodtype approach to latest Cretaceous and earliest Tertiary paleosols in eastern Montana. Geol. Soc. Am. Bull. 106: 1377–1397.

Richards, P. W. 1952. The Tropical Rain Forest, an Ecological Study. Cambridge University Press, Cambridge, U.K.

Rose, K. D., L. Krishtalka, and R. K. Stucky. 1991. Revision of the Wind River faunas, Early Eocene of central

Wyoming. Part 11. Palaeanodonta (Mammalia). Ann. Carnegie Mus. 60: 63–82.

Rothwell, G. W. and R. A. Stockey. 1989. Fossil Ophioglossales in the Paleocene of western North America. Am. J. Bot. 76: 637–644.

———. 1991. *Onoclea sensibilis* in the Paleocene of North America, a dramatic example of structural and ecological stasis. Rev. Palaeobot. Palynol. 70: 113–124.

Rothwell, G. W. and R. A. Stockey. 1994. The role of *Hydropteris pinnata* Gen. et Sp. Nov. in reconstructing the cladistics of heterosporous ferns. Am. J. Bot. 81: 479–492.

Rouse, G. E. and S. K. Srivastava. 1972. Palynological zonation of Cretaceous and Early Tertiary rocks of the Bonnet Plume Formation, northeastern Yukon, Canada. Can. J. Earth Sci. 9: 1163–1179.

Savin, S. M. 1977. The history of the Earth's surface temperature during the past 100 million years. Annu. Rev. Earth Planet. Sci. 5: 319–355.

Scotese, C. R., L. M. Gahagan, and R. L. Larson. 1989. Plate tectonic reconstructions of the Cretaceous and Cenozoic ocean basins. In: C. R. Scotese and W. W. Sager (eds.), Mesozoic and Cenozoic Plate Reconstructions. Elsevier Science Publishers, Amsterdam. pp. 27–48.

Scotese, C. R. and W. W. Sager (eds.). 1989. Mesozoic and Cenozoic Plate Reconstructions. Elsevier Science Publishers, Amsterdam.

Sellwood, B. W., G. D. Price, and P. J. Valdes. 1994. Cooler estimates of Cretaceous temperatures. Nature 370: 453–455.

Serbet, R. and R. A. Stockey. 1991. Taxodiaceous pollen cones from the Upper Cretaceous (Horseshoe Canyon Formation) of Drumheller, Alberta, Canada. Rev. Palaeobot. Palynol. 70: 67–76.

Sheehan, P. M. and D. E. Fastovsky. 1992. Major extinctions of land-dwelling vertebrates at the Cretaceous–Tertiary boundary, eastern Montana. Geology 20: 556–560.

Sloan, L. C. and E. J. Barron. 1990. "Equable" climates during Earth history? Geology 18: 489–492.

Sloan, L. C. and D. K. Rea. 1995. Atmospheric carbon dioxide and early Eocene climate: a general circulation modeling sensitivity study. Palaeogeogr., Palaeoclimatol., Palaeoecol. 119: 275–292.

Smiley, C. J. 1967. Paleoclimatic interpretations of some Mesozoic floral sequences. Am. Assoc. Petrol. Geol. Bull. 51: 849–863.

Snead, R. G. 1969. Microfloral diagnosis of the Cretaceous–Tertiary boundary, central Alberta. Res. Council Alberta Bull. 25: 1–148.

Spicer, R. A., R. J. Burnham, P. Grant, and H. Glicken. 1985. *Pityrogramma calomelanos*, the primary, post-eruption colonizer of Volcán Chichonal, Chiapas, Mexico. Am. Fern J. 75: 1–5.

Spicer, R. A., K. S. Davies, and A. B. Herman. 1994. Circum-Arctic plant fossils and the Cretaceous–Tertiary transition. In: M. C. Boulter and H. C. Fisher (eds.), Cenozoic Plants and Climates of the Arctic. NATO ASI Series 1: Global Environmental Change. Springer-Verlag, Berlin. Vol. 27, pp. 161–174.

Spicer, R. A. and J. T. Parrish. 1986a. Paleobotanical evidence for cool north polar climates in Middle Cretaceous (Albian–Cenomanian) time. Geology 14: 703–706.

———. 1986b. Fossil woods from northern Alaska and climate near the Middle Cretaceous North Pole. Geol. Soc. Am. Abstr. Program 18: 759.

———. 1990. Late Cretaceous–Early Tertiary palaeoclimates of northern high latitudes: a quantitative view. J. Geol. Soc. Lond. 147: 329–341.

Srivastava, S. K. 1970. Pollen biostratigraphy and paleoecology of the Edmonton Formation (Maestrichtian), Alberta, Canada. Palaeogeogr., Palaeoclimatol., Palaeoecol. 7: 221–276.

———. 1981. Evolution of Upper Cretaceous phytogeoprovinces and their pollen flora. Rev. Palaeobot. Palynol. 35: 155–173.

———. 1994a. Evolution of Cretaceous phytogeoprovinces, continents and climates. Rev. Palaeobot. Palynol. 82: 197–224.

———. 1994b. Palynology of the Cretaceous–Tertiary boundary in the Scollard Formation of Alberta, Canada, and global KTB events. Rev. Palaeobot. Palynol. 83: 137–158.

Standley, E. A. 1965. Upper Cretaceous and Paleocene plant microfossils and Paleocene dinoflagellates and hystrichosphaerids from northwestern South Dakota. Bull. Am. Paleontol. 49: 1–384.

Stebbins, G. L., Jr. 1974. Flowering Plants: Evolution above the Species Level. Belknap Press, Harvard University. Cambridge, MA.

Stockey, R. A., G. L. Hoffman, and G. W. Rothwell. 1997. The fossil monocot *Limnobiophyllum scutatum*: resolving the phylogeny of Lemnaceae. Am. J. Bot. 84: 355–368.

Stockey, R. A. and S. R. Manchester. 1988. A fossil flower with in situ *Pistillipollenites* from the Eocene of British Columbia. Can. J. Bot. 66: 313–318.

Stockey, R. A. and G. W. Rothwell. 1997. The aquatic angiosperm *Trapago angulata* from the Upper Cretaceous (Maastrichtian) St. Mary River Formation of southern Alberta. Internat. J. Plant Sci. 158: 83–94.

Stucky, R. K. and L. Krishtalka. 1990. Revision of the Wind River faunas, Early Eocene of central Wyoming. Part 10. *Bunophorus* (Mammalia, Artiodactyla). Ann. Carnegie Mus. 59: 149–171.

——— and A. D. Redline. 1990. Geology, vertebrate fauna, and paleoecology of the Busk Spring quarries (Early Eocene, Wind River Formation), Wyoming. In: T. M. Bown and K. D. Rose (eds.), Dawn of the Age of Mammals in the Northern Part of the Rocky Mountain Interior, North America. Geological Society of America, Boulder, CO. Special Paper 243, pp. 169–186.

Sun, F. and R. A. Stockey. 1992. A new species of *Palaeocarpinus* (Betulaceae) based on infructescences, fruits, and associated staminate inflorescences and leaves from the Paleocene of Alberta, Canada. Int. J. Plant Sci. 153: 136–146.

Sweet, A. R. 1978. Palynology of the Ravenscrag and Frenchman formations. In: J. A. Irvin, S. H. Whitaker, and P. L. Broughton (eds.), Coal Resources of Southern Saskatchewan: A Model for Evaluation Methodology. Geological Survey of Canada, Ottawa, Economic Geological Report. Vol. 30, pp. 29–38.

———, D. R. Braman, and J. F. Lerbekmo. 1990. Palynofloral response to K/T boundary events; a transitory interruption within a dynamic system. In: V. L. Sharpton and P. D. Ward (eds.), Global Catastrophes in Earth History. Geological Society of America, Boulder, CO. Special Paper 247, pp. 457–469.

Taylor, D. W. and W. L. Crepet. 1987. Fossil floral evidence of Malpighiaceae and an early plant–pollinator relationship. Am. J. Bot. 74: 274–286.

Thelin, G. P. and R. J. Pike. 1991. Landforms of the Conterminous United States—A Digital Shaded-Relief Portrayal. U.S. Geological Survey Miscellaneous Investigation Series Map I–2206. U.S. Geological Survey, Reston, VA.

Thompson, S. L. and E. J. Barron. 1981. Comparison of Cretaceous and present Earth albedos: implications for the causes of paleoclimates. J. Geol. 89: 143–167.

Tiffney, B. H. 1984. Seed size, dispersal syndromes, and the rise of the angiosperms: evidence and hypothesis. Ann. Missouri Bot. Garden 71: 551–576.

———. 1985. The Eocene North Atlantic land bridge: its importance in Tertiary and modern phytogeography of the northern hemisphere. J. Arnold Arb. 66: 243–273.

———. 1994. An estimate of the early Tertiary paleoclimate of the southern Arctic. In: M. C. Boulter and H. C. Fisher (eds.), Cenozoic Plants and Climates of the Arctic. NATO ASI Series 1: Global Environmental Change. Springer-Verlag, Berlin. Vol. 27, pp. 267–295.

Traverse, A. 1988. Paleopalynology. Unwin Hyman, Boston.

Triplehorn, D. M., D. L. Turner, and C. W. Naeser. 1984. Radiometric age of the Chickaloon Formation of south-central Alaska: location of the Paleocene–Eocene boundary. Geol. Soc. Am. Bull. 95: 740–742.

Tschudy, R. H. 1970 (1971). Palynology of the Cretaceous–Tertiary boundary in the northern Rocky Mountain and Mississippi Embayment regions. In: R. M. Kosanke and A. T. Cross (eds.), Symposium on Palynology of the Late Cretaceous and Early Tertiary. Geological Society of America, Boulder, CO. Special Paper 127, pp. 65–112.

———. 1975. Normapolles pollen from the Mississippi Embayment. U.S. Geological Survey, Reston, VA. Professional Paper 865, pp. 1–42.

———. 1980. Normapolles pollen from *Aquipollenites* province, western United States. N. Mex. Bur. Mines Min. Resources Circ. 170: 1–14.

———. 1981. Geographic distribution and dispersal of Normapolles genera in North America. Rev. Palaeobot. Palynol. 35: 283–314.

———, C. L. Pillmore, C. J. Orth, J. S. Gilmore, and J. D. Knight. 1984. Disruption of the terrestrial plant ecosystem at the Cretaceous–Tertiary boundary, Western Interior. Science 225: 1030–1032.

Tselioudis, G., A. A. Lacis, D. Rind, and W. B. Rossow. 1993. Potential effects of cloud optical thickness on climate warming. Nature 366: 670–672.

Upchurch, G. R., Jr. 1989. Terrestrial environmental changes and extinction patterns at the Cretaceous–Tertiary boundary, North America. In: S. K. Donovan (ed.), Mass Extinctions: Processes and Evidence. Columbia University Press, New York. pp. 195–216.

———. 1995. Dispersed angiosperm cuticles: their history, preparation, and application to the rise of angiosperms in Cretaceous and Paleocene coals, southern Western Interior of North America. Int. J. Coal Geol. 28: 161–227.

——— and D. L. Dilcher. 1990. Cenomanian angiosperm leaf megafossils, Dakota Formation, Rose Creek locality, Jefferson County, southeastern Nebraska. U.S. Geol. Surv. Bull. 1915: 1–55.

Upchurch, G. R., Jr. and J. A. Wolfe. 1987. Mid-Cretaceous to Early Tertiary vegetation and climate: evidence from fossil leaves and woods. In: E. M. Friis, W. G. Chaloner, and P. R. Crane (eds.), The Origins of Angiosperms and Their Biological Consequences. Cambridge University Press, Cambridge, U.K. pp. 75–105.

———. 1993. Cretaceous vegetation of the Western Interior and adjacent regions of North America. In: W. G. E. Caldwell and E. G. Kauffman (eds.), Evolution of the Western Interior Basin. Geological Association of Can., St. Johns, Newfoundland, Special Paper 39, pp. 243–281.

Van Valen, L. M. and R. E. Sloan. 1977. Ecology and extinction of the dinosaurs. Evol. Theory 2: 37–64.

Vogt, P. R. 1979. Global magmatic episodes: new evidence and implications for the steady-state mid-oceanic ridge. Geology 7: 93–98.

Webb, S. D. 1989. The fourth dimension in North American terrestrial mammal communities. In: D. W. Morris, Z. Abramsky, B. J. Fox, and M. R. Willig (eds.), Patterns in the Structure of Mammalian Communities. Museum of Texas Tech University, Lubbock, TX. Special Paper 28, pp. 181–203.

West, R. M. and M. R. Dawson. 1978. Vertebrate paleontology and the Cenozoic history of the North Atlantic region. Polarforschung 48: 103–119.

Wheeler, E. A. 1991. Paleocene dicotyledonous trees from Big Bend National Park, Texas: variability in wood types common in the Late Cretaceous and Early Tertiary, and ecological inferences. Am. J. Bot. 78: 658–671.

——— and P. Baas. 1991. A survey of the fossil record for dicotyledonous wood and its significance for evolutionary and ecological wood anatomy. Int. Assoc. Wood Anat. Bull. 12: 275–332.

Wheeler, E. A., M. Lee, and L. C. Matten. 1987. Dicotyledonous woods from the Upper Cretaceous of southern Illinois. Bot. J. Linn. Soc. 95: 77–100.

Wheeler, E. A., T. M. Lehman, and P. E. Gasson. 1994. *Javelinoxylon*, an Upper Cretaceous dicotyledonous tree from Big Bend National Park, Texas, with presumed malvalean affinities. Am. J. Bot. 81: 703–710.

Wheeler, E. A., R. A. Scott, and E. S. Barghoorn. 1977. Fossil dicotyledonous woods from Yellowstone National Park. J. Arnold Arb. 58: 280–302.

———. 1978. Fossil dicotyledonous woods from Yellowstone National Park, II. J. Arnold Arb. 59: 1–26.

Wing, S. L. 1980. Fossil floras and plant-bearing beds of the central Bighorn Basin. In: P. D. Gingerich (ed.), Early Cenozoic Paleontology and Stratigraphy of the Bighorn Basin. University of Michigan Museum Paleontol., Ann Arbor, MI, 119–125.

———. 1981. A Study of Paleoecology and Paleobotany in the Willwood Formation (Early Eocene, Wyoming). Ph.D. dissertation, Yale University, New Haven, CT.

———. 1984. Relation of paleovegetation to geometry and cyclicity of some fluvial carbonaceous deposits. J. Sediment. Petrol. 54: 52–66.

———. 1987. Eocene and Oligocene floras and vegetation of the Rocky Mountains. Ann. Missouri Bot. Garden 74: 748–784.

———. 1996. Paleocene–Eocene floral and climate change in the northern Rocky Mountains, USA. International Organization of Palaeobotany Fifth Quadrennial Conference, Santa Barbara, CA, June–July, 1996. p. 113.

Wing, S. L. 1997. Global warming and plant species richness: a case study of the Paleocene/Eocene boundary. In: M. L. Reaka-Kudla, D. E. Wilson, and E. O. Wilson (eds.), Biodiversity II. Understanding and Protecting our Natural Resources. Joseph Henry Press, Washington, D.C. pp. 163–185.

Wing, S. L., J. Alroy, and L. J. Hickey. 1995. Plant and mammal diversity in the Paleocene to early Eocene of the Bighorn Basin. Palaeogeogr., Palaeoclimatol., Palaeoecol. 115: 117–155.

——— and T. M. Bown. 1985. Fine scale reconstruction of late Paleocene–Early Eocene paleogeography in the Bighorn Basin of northern Wyoming. In: R. M. Flores and S. S. Kaplan (eds.), Cenozoic Paleogeography of the West-Central United States. Rocky Mountain Section, Society of Economic Paleontology Mineral., Golden, CO. pp. 93–106.

——— and J. D. Obradovich. 1991. Early Eocene biotic and climatic change in interior western North America. Geology 19: 1189–1192.

Wing, S. L. and D. R. Greenwood. 1993. Fossils and fossil climate: the case for equable continental interiors. Phil. Trans. R. Soc. Lond. B 341: 243–252.

Wing, S. L. and L. J. Hickey. 1984. The *Platycarya* perplex and the evolution of the Juglandaceae. Am. J. Bot. 71: 388–411.

Wing, S. L. and B. H. Tiffney. 1987. Interactions of angiosperms and herbivorous terapods through time. In: E. M. Friis, W. G. Chaloner, and P. R. Crane (eds.), The Origins of the Angiosperms and Their Biological Consequences. Cambridge University Press, Cambridge, U.K. pp. 203–224.

Wolfe, J. A. 1966. Tertiary plants from the Cook Inlet Region, Alaska. U.S. Geological Survey, Reston, VA. Professional Paper 398-B, pp. 1–32.

———. 1972. An interpretation of Alaskan Tertiary floras. In: A. Graham (ed.), Floristics and Paleofloristics of Asia and Eastern North America. Elsevier Science Publishers, Amsterdam. pp. 201–233.

———. 1977. Paleogene floras from the Gulf of Alaska Region. U.S. Geological Survey, Reston, VA. Professional Paper 997, pp. 1–108.

———. 1978. A paleobotanical interpretation of Tertiary climates in the Northern Hemisphere. Am. Sci. 66: 694–703.

———. 1979. Temperature Parameters of Humid to Mesic Forests of Eastern Asia and Relation to Forests of Other Regions of the Northern Hemisphere and Australasia. U.S. Geological Survey, Reston, VA. Professional Paper 1106, pp. 1–37.

———. 1980. Tertiary climates and floristic relationships at high latitudes in the northern hemisphere. Palaeogeogr., Palaeoclimatol., Palaeoecol. 30: 313–323.

———. 1981a. Paleoclimatic significance of the Oligocene and Neogene floras of the northwestern United States. In: K. J. Niklas (ed.), Paleobotany, Paleoecology, and Evolution. Praeger, New York. Vol. 2, pp. 79–101.

———. 1981b. A chronologic framework for Cenozoic megafossil floras of northwestern North America and its relation to marine geochronology. Geological Society of America, Boulder, CO. Special Paper 184, pp. 39–47.

———. 1985. Distribution of major vegetational types during the Tertiary. In: E. T. Sundquist and W. S. Broecker (eds.), The Carbon Cycle and Atmospheric CO2: Natural Variations Archean to Present. Geophysical Monograph 32. American Geophysical Union, Washington, D.C. pp. 357–375.

———. 1987. Late Cretaceous–Cenozoic history of deciduousness and the terminal Cretaceous event. Paleobiology 13: 215–226.

———. 1990. Palaeobotanical evidence for a marked temperature increase following the Cretaceous/Tertiary boundary. Nature 343: 153–156.

———. 1991. Palaeobotanical evidence for a June "impact winter" at the Cretaceous/Tertiary boundary. Nature 352: 420–423.

———. 1994. Alaskan Palaeogene climates as inferred from the CLAMP database. In: M. C. Boulter and H. C. Fisher (eds.), Cenozoic Plants and Climates of the Arctic. NATO ASI Series 1: Global Environmental Change. Springer-Verlag, Berlin. Vol. 27, pp. 223–237.

———. In press. Relations of environmental change to angiosperm evolution during the Late Cretaceous and Tertiary. In: I. Motomi and H. Nishida (eds.), Prospect for Systematic Botany in the 21st Century. Springer-Verlag, Berlin.

——— and S. McCoy. 1984. Comments and Replies on "Model for the origin of the Yakutat block, an accreting terrane in the northern Gulf of Alaska." Geology 12: 564–565.

Wolfe, J. A. and T. Tanai. 1987. Systematics, phylogeny, and distribution of *Acer* (maples) in the Cenozoic of western North America. J. Fac. Sci. Hokkaido Univ. 22: 1–246.

Wolfe, J. A. and G. R. Upchurch, Jr. 1986. Vegetation, climatic and floral changes at the Cretaceous–Tertiary boundary. Nature 324: 148–152.

———. 1987a. North American nonmarine climates and vegetation during the Late Cretaceous. Palaeogeogr., Palaeoclimatol., Palaeoecol. 61: 33–77.

———. 1987b. Leaf assemblages across the Cretaceous–Tertiary boundary in the Raton Basin, New Mexico and Colorado. Proc. Natl. Acad. Sci. USA 84: 5096–5100.

Ziegler, A. M., M. L. Hulver, A. L. Lottes, and W. F. Schmachtenberg. 1984. Uniformitarianism and palaeoclimates: inferences from the distribution of carbonate rocks. In: P. Brenchley (ed.), Fossils and Climate. Wiley, New York. pp. 3–25.

General Readings: K–T Boundary

Archibald, J. D. 1981. The earliest known Palaeocene mammal fauna and its implications for the Cretaceous–Tertiary transition. Nature 291: 650–652. (An example of an earlier essay arguing for gradual, and presumably climatically controlled, transition and extinctions at the K–T boundary.)

Lerbekmo, J. F., A. R. Sweet, and R. M. St. Louis. 1987. The relationship between the iridium anomaly and palynological floral events at three Cretaceous–Tertiary boundary localities in western Canada. Geol. Soc. Am. Bull. 99: 325–330.

McIver, E. E., A. R. Sweet, and J. F. Basinger. 1991. Sixty-five-million-year-old flowers bearing pollen of the extinct triprojectate complex—a Cretaceous–Tertiary boundary survivor. Rev. Palaeobot. Palynol. 70: 77–88.

Pillmore, C. L., R. H. Tschudy, C. J. Orth, J. S. Gilmore, and J. D. Knight. 1984. Geologic framework of nonmarine Cretaceous–Tertiary boundary sites, Raton Basin, New Mexico and Colorado. Science 223: 1180–1183. (See also three adjacent essays in this issue on the same subject.)

Sweet, A. R. 1986. The Cretaceous–Tertiary boundary in the central Alberta foothills: II. Miospore and pollen taxonomy. Can. J. Earth Sci. 23: 1375–1388. [For part I see Jerzykiewicz and Sweet (1986).]

General Readings

Barron, E. J. 1994. Chill over the Cretaceous. Nature 370: 415.

Chaloner, W. G. and A. Hallam (eds.). 1989. Evolution and Extinction. Cambridge University Press, Cambridge, U.K.

Dorf, E. 1964. The petrified forests of Yellowstone Park. Sci. Am. 210: 106–114.

Kerr, R. A. 1991. Did a burst of volcanism overheat ancient Earth? Science 251: 746–747.

———. 1992a. Did an asteroid leave its mark in Montana bones? Science 256: 1395–1396.

———. 1992b. When climate twitches, evolution takes great leaps. Science 257: 1622–1624.

———. 1993a. Fossils tell of mild winters in an ancient hothouse. Science 261: 82.

———. 1993b. A bigger death knell for the dinosaurs? Science 261: 1518–1519.

———. 1997. Upstart ice age theory gets attentative but chilly hearing. Science 277: 183–184.

McGowran, B. 1990. Fifty million years ago. Sci. Am. 78: 30–39.

Morell, V. 1993. How lethal was the K-T impact? Science 261: 1518–1519.

———. 1994. Warm-blooded dino debate blows hot and cold. Science 265: 188.

Roush, W. 1996. Probing flower's genetic past. Science 273: 1339–1340.

Notes

1. On a historical note, samples from the McNairy Sand (Ripley) Formation in Tennessee were part of the original material used in the development of the oxygen isotope technique for paleotemperature analysis (Lowenstam and Epstein, 1954). Their estimate of MAT, made over 40 years ago, was 20–29°C.

2. The Hell Creek Formation consists of sediments deposited ~2 Ma before the terminal Cretaceous extinctions at 65 Ma. The overlying Tullock Formation was deposited immediately afterward. For an interesting discussion of the paleontological analyses at this important site regarding the bolide impact, see Sheehan and Fastovsky (1992) and the reference to Kerr (1992a).

3. The paleontological literature on K–T boundary events is extensive and cannot be reviewed in detail here. For those interested in reading further on this subject, a supplementary list of essays with emphasis on paleobotany and paleopalynology is provided under General Readings: K–T Boundary in the References section.

4. Oswald Heer (1883) described the extensive fossil floras of the Arctic in a seven-volume work published between 1868 and 1883 [For an account of his life see Andrews (1980).]

SIX

Middle Eocene through Early Miocene North American Vegetational History

50–16.3 Ma

CONTEXT SUMMARY

During the Middle Eocene through the Early Miocene, erosion of the Appalachian Mountains exceeded uplift and there was a net reduction in elevation (Figs. 5.19, 6.1). In the Rocky Mountains uplift continued through the Middle Eocene (end of the Laramide orogeny), waned in the Middle Tertiary, and then increased beginning at about 10 Ma. Earlier reconstructions placed paleoelevations in the Rocky Mountains during the Middle Eocene through the Early Miocene at approximately half the present relief. The maximum elevation in the Front Ranges during the latest Eocene was estimated at ~2500 m (~8000 ft; MacGinitie, 1953). Recent approximations are for nearly modern elevations in several areas by the Eocene–Oligocene. Extensive Eocene volcanism deposited ash and blocked drainage systems, augmenting uplift and facilitating the preservation of extensive fossil floras and faunas. In the far west the beginning of Tertiary volcanism in the Sierra Nevada is dated at ~33 Ma near the Eocene–Oligocene boundary.

A drying trend becomes evident in the Middle Eocene and reduced moisture, along with the waning of volcanic activity in the Oligocene, restricted conditions favorable to fossilization. The number of Oligocene floras in the northern Rocky Mountains is considerably fewer than in younger deposits to the west.

In the absence of extensive plate reorganization and orogeny, CO_2 concentration decreased, which contributed to a temperature decline that continued through the Cenozoic and intensified in the Late Tertiary. Recall from Chapter 2 (sections on orogeny and volcanism) that uplift plays a role in determining long-term climate by creating rainshadows, altering atmospheric circulation patterns, and increasing the erosion of silicate rocks that causes a drawdown of CO_2. This allows heat to escape from the troposphere and results in lower temperatures. Marine benthic temperatures were ~10°C in the early Late Eocene and ~2°C near the Eocene–Oligocene boundary, assuming an essentially ice-free Earth during that time, and increased to ~5–6°C near the end of the Early Miocene (Fig. 3.1). Temperatures over land in the midnorthern latitudes are estimated to have dropped by ~12°C between the Late Eocene and Early Oligocene (Wolfe, 1992a). The annual range in temperature also widened from an estimated 3–5°C in the Pacific Northwest in the Middle Eocene to 25°C in the Oligocene. From the Middle Eocene through the Late Eocene the decline correlates with decreasing CO_2 concentration; after that time it was additionally influenced in the high latitudes by full opening of the North Atlantic seaway, which allowed an intermingling of Arctic and Atlantic Ocean waters with the formation of NADW.

In the southern hemisphere there was also a change in ocean circulation in the Oligocene caused by partial opening of the Drake Passage (55–50 Ma) and the further separation of Australia and Tasmania from Antarctica (30–25 Ma). This led to the development of a circum-Antarctic current that thermally isolated Antarctica from warmer equatorial waters and contributed to continental glaciation. Glacial deposits as old as the Middle Eocene have been reported from King George Island off the Antarctic Peninsula (Birkenmajer, 1987). The oldest unequivocal evidence for a substantial continental ice sheet on Antarctica is the Early Oligocene (chron C13N; 33.5 Ma) in the form of ice-rafted detritus on the East Antarctic margin of the Weddell Sea (ODP Leg 113) and on the southern and central parts of the Kerguelen Plateau (ODP Leg 119, 120). This correlates with a 1% increase in ^{18}O (Berggren and Prothero, 1992). Among the results were an enhancement of cold deep waters, steeper pole to equator temperature gradients, and strengthened high-pressure systems with concomitant increased and seasonal dryness. During the Oligocene to the end of the Early Miocene temperatures

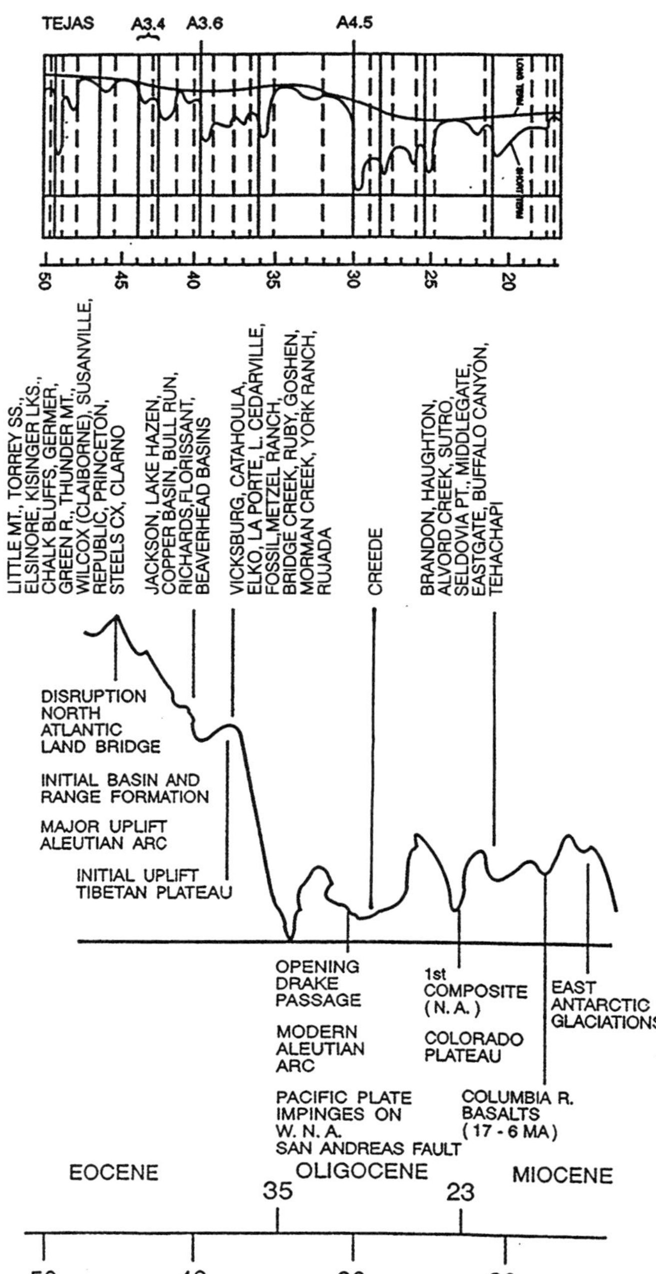

Figure 6.1. Principal fossil floras mentioned in the text plotted with general reference to the paleotemperature record (lower curve) and sea-level record (upper curve) for the Middle Eocene through the Early Miocene.

fluctuated near the line between glacial and ice-free conditions (Fig. 3.1; Zachos et al., 1997). The Miocene thermal optimum occurred between ~21 and 15 Ma, corresponding to planktic foraminifera zones N5–N10. In the Middle Miocene (15–4 Ma) another temperature drop ushered in extensive continental glaciations on Antarctica and early lowland glaciation in the Arctic.

Sea level fell during the interval between the Middle Eocene and Early Miocene, and the change shows a pattern consistent with temperatures fluctuating along the boundary between intermittent glacial and ice-free conditions. A core from near Mays Landing on the coastal plain of New Jersey preserves erosion surfaces near the Lower Eocene–Middle Eocene boundary, near the Middle Eocene–Upper Eocene boundary, within the Middle Eocene, and in the Middle Oligocene (Owens et al., 1988). The most significant sea-level event was in the Oligocene at 31–29 Ma. Other lowerings occurred at 21, 16.5, and 15.5 Ma (Fig. 3.6). These changes correlate with hiatuses in other continental shelf, slope, and epicontinental settings and are therefore suggestive of eustatic changes in sea level (Miller et al., 1990). For example, the fall of about 30 m at 31–29 Ma correlates approximately with the evidence from ice-rafted detritus of increasing ice on Antarctica.

The epicontinental sea continued to retreat from mid-continent North America and by the end of the Early

Miocene had reached a point about halfway between the Middle Eocene position shown in Fig. 5.6 and the present Gulf Coast (viz., along a line running midway between Austin and Houston, Texas, to just south of Jackson, Mississippi, and around the southern terminus of the Appalachian Mountains). The depositional setting of a gradually retreating shoreline bordered by swamps and lakes facilitated the preservation of extensive fossil floras of coastal and lowland vegetation, and pollen was also being blown in from the adjacent Appalachian highlands. The retreat resulted in an increased albedo, an increase in the erosion of silicate and other rocks, and greater continentality in the interior regions.

Precipitation decreased and became more seasonal. Paleosols (fossil soil horizons) from the Badlands of South Dakota show a drying trend from an estimated annual precipitation of 1000 mm in the early Late Eocene (38 Ma), to 500–900 mm in the Early Oligocene (32 Ma), to 450–500 mm at 30.5 Ma, and to 250–450 mm in the Late Oligocene–Early Miocene (29.5 Ma; Retallack, 1983a,b, 1990).

Some Ir anomalies have been reported from the Late Eocene (~35 to ~40 Ma; Hut et al., 1987) and from the Late Miocene (~10 Ma; Asaro et al., 1988); but the concentrations are much lower than at the K–T boundary and they do not correspond to extinction events in the marine realm, which are presently viewed as mostly endogenic rather than exogenic in origin. Microtektites are associated only with a minor extinction event at Chron C15R (~36.5 Ma). Impact structures of Late Eocene age (35.5 Ma) have been discovered in the Chesapeake Bay region and in Siberia (Chapter 2), but the biological consequences of the event have not been assessed (Koeberl et al., 1996).

Summarizing events during the Paleogene is complicated by the fact that the termination date for the Eocene has been the most unsettled of all the Tertiary epochs. Estimates for the Eocene–Oligocene boundary range from 38 Ma to as late as 32 Ma (Wolfe, 1981a), but results from a recently developed single crystal laser fusion (SCLF) $^{40}Ar/^{39}Ar$ radiometric dating technique are focusing on an age of ~34 Ma (Berggren et al., 1995; 33.5 Ma, Prothero, 1994):

> The Eocene/Oligocene boundary was recently stratotypified at the 19 m level in the Massignano section of the northern Apennines. It is denoted by the last occurrence (LO) of the planktonic foraminiferal genus *Hantkenina* which occurs at a stratigraphic level in Magnetozone C13R, estimated to lie at a level 0.14 of the interpolated distance between the base of C13N and the top of C15N, with an estimated numerical age of 34.0 Ma. (Prothero and Berggren, 1992, p. 3)

New dates for other epoch–stage boundaries and fossil biotas are also being published, and erroneous dates are being corrected. The movement of epoch boundaries can affect the age name used for assemblages. For example, if a fossil flora is correctly dated at 35 Ma, and the Eocene–Oligocene boundary is placed at 36 Ma, the flora belongs to the Oligocene. If the boundary is placed at 34 Ma, the flora then belongs to the Eocene. Similar situations arise if the boundary is established but new and different dates are obtained for the flora. Both kinds of age "changes" are partly semantic, and a flora can be found under different epoch names in the literature. In the following chapters new dates (some published and others unpublished, in press, or estimated and/or extrapolated) are indicated in brackets along with previously published dates.

In discussing the climates of the Early Eocene, it was noted that the trend toward maximum warmth was accomplished in abrupt, steep, short-term increments. Accumulating evidence suggests that this mode is characteristic of long-term trends generally, and that climate jumps from one stable state or threshold to another in short pulses. The temperature decline from the Middle Eocene through the Early Miocene fell in the same stepwise fashion. This period is one of the most complex in North American vegetational history and also one of the more challenging to outline because of previous uncertainty as to when the decline began and a resulting confused timetable.

The phrase "Oligocene deterioration" was initially used to characterize a temperature decline and associated vegetation changes evident in fossil floras from Alaska, the Pacific Northwest, and the Gulf Coast (Wolfe, 1971). Subsequently, these changes were labeled the "Terminal Eocene Event" (Wolfe, 1978), implying that there was a single event and that it occurred at the end of the Eocene. The term became entrenched in the literature, in part, because of the title of a book, *Terminal Eocene Events* (Pomerol and Premoli-Silva, 1986), resulting from a UNESCO sponsored project, "Geological Events at the Eocene–Oligocene Boundary" as part of the International Geological Correlation Program. Ironically, among the conclusions in the book were "1) the Eocene–Oligocene boundary considered as a major evolutionary break of the Cenozoic is characterized by a gradual, and not abrupt, environmental change which started in the Middle Eocene and continued into the earliest Oligocene for some 300,000 years"; that is, it began in the Middle Eocene rather than at the end of the Eocene and extended over a span of about 10 Ma, and "2) There is not a single terminal Eocene event, but a series of events concentrated close to the Eocene–Oligocene boundary as testified by an acceleration of the rate of overturn among most of the taxonomic groups, by an intensification of the climatic deterioration and by a major reorganization of the oceanic water masses" (Pomerol and Premoli-Silva, 1986, p. vi; see also Pomerol, 1981).

Evidence presented in 1989 at a Penrose Conference and a Theme Session of the Geological Society of America (Prothero and Berggren, 1992) confirm that the most significant cooling event in the Paleogene began in the Middle Eocene (Bartonian–Priabonian boundary; see also Axelrod, 1992; Marty et al., 1988), was augmented by a slighter cooling at the Eocene–Oligocene boundary and by more pronounced cooling in the mid-Oligocene (32.5 Ma in the revised chronology used by Berggren and Prothero, 1992; Fig. 3.1), and occurred in a stepwise fashion. Eocene floras from southern England show a similar pattern (Collinson et al., 1981). Planktonic foraminifera from deep-sea sites record five extinction events between the Lutetian and the Late Rupelian, as summarized by Prothero (1989; new dates fide Prothero, personal communication, 1996):

The Lutetian-Bartonian (Middle Eocene) event occurred at the boundary between planktonic foraminifera zones P12 and P13, at the top of magnetic Chron C18R (~41 Ma). This was a minor event near the beginning of the cooling trend of the Middle Eocene and corresponds to a drop in sea level termed Tejas A3.4 by Haq et al. (1987; Fig. 6.1).

The Bartonian–Priabonian (Middle–Late Eocene) event occurred at the boundary at the P14–P15 boundary at Chron C17R (~37 Ma). This was a major event involving the extinction of spinose tropical foraminifera and migration toward the equator of midlatitude species. Among the calcareous nanoplankton,[1] nearly 50% of the species disappeared. A cooling event is also evident in the oxygen isotope record, and a major drop in sea level occurred at the top of Tejas A3.6 (Fig. 6.1).

The Late Priabonian event (Late Eocene) occurred at the P15–P16 boundary at Chron C15R (~35 Ma). This was a minor event that involved continued cooling, lowering of sea level, and the only extraterrestrial impacts.

The Terminal Eocene Event is now mostly considered as Early Oligocene, P17–P18 boundary at mid-Chron C13R (~33.5 Ma). Few extinctions are evident among the planktonic foraminifera and nanoplankton, but more occurred in the ostracodes and benthic foraminifera. Oxygen isotopes record a cooling of ~2°C and a slight drop in sea level.

The mid-Oligocene was at the Late Rupelian at the top of P19 and mid-Chron C11R ~30.5 Ma; this signified Oligocene deterioration (Wolfe, 1971) and was later called the Terminal Eocene Event (Wolfe, 1978; see also Prothero, 1989). This event essentially eliminated most holdovers from the warmer Late Eocene and transitional Early Oligocene as marked by the extinction of many cool-water foraminifera and nanoplankton, oxygen isotope records, and a fall in sea level. From ~32 through 30 Ma (Tejas A4.5) the largest sea-level drop in Tertiary history is recorded, and it likely was the result of the beginning of full continental glaciation on Antarctica.

Thus, there are five extinction events recorded in the marine realm (41, 37, 35, 33.5, 30.5 Ma) spread over about 10 m.y., which precludes any single catastrophic event as the causal mechanism for biotic changes during this period. As would be expected, there was a panoply of causes, the principal ones being continued decline in CO_2 concentration as plate reorganization and erosion rates waned and changes in ocean circulation.

It is not possible to reconstruct the terrestrial plant record with the same resolution as in the marine realm, where near-continuous records are frequently available from DSDP and ODP cores. The plant megafossil record is particularly discontinuous, but plant microfossils from wells approximate the sequences from marine cores in some instances. The terrestrial plant record is summarized by Frederiksen:

> [I]t now appears that, at least in the Gulf Coast, the "terminal Eocene" event took place entirely at the beginning of the Oligocene and that little if any climatic change can be detected in this region at the end of the Eocene. (1988, p. 46)

The timing of faunal history for this period is unsettled because current revisions in the isotopic time scale affect the NALMAs for the Paleogene (older by ~2 m.y.), and these are currently being recalibrated to the Global Polarity Time Scale (GPTS; Figs. 3.12, 3.13, 5.19):

> The Uintan land mammal "age" (long thought to be [L]ate Eocene) is now placed in the Middle Eocene, and the Uintan/Duchesnean boundary appears to correspond to the [M]iddle/[L]ate Eocene (Bartonian/Priabonian) boundary [the Uintan and Duchesnean are now both [M]iddle Eocene fide D. R. Prothero, personal communication, 1996]. The Chadronian, long thought to be Early Oligocene, now appears to be [L]ate Eocene, and the Chadronian/Orellan boundary (at about 34 Ma) appears to correspond to the Eocene–Oligocene boundary. The Orellan and Whitneyan, once considered [M]iddle and Late Oligocene, are now placed in the [E]arly Oligocene. Since these discoveries are so recent, even faunal summaries like those of Prothero (1985, 1989) and Stucky (1990) are now out of date." (Prothero and Berggren, 1992, p. 16; see also Prothero, 1995)

The events between the Middle Eocene and Early Oligocene culminated in a rapid change in European faunas known as the *Grande Coupure* (the great break). Sixty percent of the mammals of Europe became extinct at about 32.5 Ma (NP22/23; Early Oligocene rather than the previous assignment to the Eocene–Oligocene boundary). In North America the older protrogomorph rodents were replaced by modern families (heteromyids, geomyids, castorids, sciurids, cylindrodontids, cricetids). The first appearances of larger mammals in North America include Canidae, Mustelidae, Tapiridae, Rhinocerotidae, Anthracotheriidae, and Tayassuidae spaced through the Duchesnean and Chadronian. This was due in large part to invasions across Beringia from the west as latest Eocene–Early Oligocene glaciers in Antarctica, reaching up to 70% of their modern volume, drew down sea levels in a surge that lasted only a few hundred thousand years. Similarities between Late Eocene and Oligocene Asian and North American assemblages include, in particular, groups adapted to open woodland and savanna habitats. Examples are *Mytonolagus* (a rabbit); lophodont rodents—such as eomyids, cricetids, and zapodids among the small herbivores; and piglike entelodonts, crescent-toothed artiodactyls (Camelidae, Hypertragulidae, Leptomerycidae), and perissodactyls (tapiroids, rhinocerotids, chalicotheres) among the larger herbivores (Webb, 1985). The relationship is evident in assemblages from the Badlands of South Dakota, the Big Bend region of Texas, and southern California in the small-mammal forest faunas of the mid-Eocene to large-mammal woodland and savanna faunas of the Eocene–Oligocene. The latter includes herding ungulates (*Mesohippus*, *Leptomeryx*, *Poebrotherium*, *Merycoidodon*) and hypsodont small herbivores (*Palaeolagus*, *Ischyromys*). The environmental implication is an expansion of woodland and savanna habitats resulting from cooling and drying climates. As might be expected from the Oligocene section of Fig. 3.1, after the shifts between the Middle Eocene and Early Oligocene, faunal history during the rest of the Oligocene was less eventful. The Late Eocene–earliest Oligocene marks the end of extensive land-mammal interchange with Europe across the North Atlantic land bridge.

During the Middle Eocene (Early Uintan) a number of archaic forms disappeared from North America, further reducing prior similarities with European faunas. Lost were the Condylarthra, Taeniodonta, Tillodontia, Dinocerata, and northern Notoungulata; and prosimian primates declined (Webb, 1989). About 60% of Early Oligocene genera were new to North America.

In the Middle Oligocene open country faunas included *Palaeolagus* (a rabbit) and the ruminants *Leptomeryx*, *Hypertragulus*, and *Poebrotherium*; stream borders were inhabited by *Protoceras* (a relative of the camelids), *Elomeryx* (anthracothere), *Agnotocastor* (a beaver), and *Subhyracodon* (a rhino). *Mesohippus* (horse) and *Merycoidodon* (oreodont) were wide-ranging.

During the Early Miocene (Arikareean NALMA) immigrations from Asia, which had been occurring intermittently, increased significantly. They include *Cynelos* (amphicyonid), *Menoceras* (rhinocerotid), *Moropus* (chalicothere), and the smaller *Plesiosminthus* (zapodid) and *Pseudotheridomys* (eomyid). New introductions from Asia reached a peak in the Hemingfordian with 16 new genera. These included ruminants (pronghorn antilocaprids, giraffe-horned dromomerycids, hornless moschids, representing forest to savanna and browsing to grazing mammals); brachypotherine rhinocerotid ungulates (*Teleoceras*); and especially carnivores (cats, *Pseudaelurus*; bears, *Ursavus*; *Hemicyon*; and various species of *Cynelos*, *Cephalogale*, *Leptarctus*, *Potamotherium*, *Amphictis*, and *Mionictis*). The principle rodents were beaver (*Anchitheriomys*), eomyid (*Eomys*), and petauristine flying squirrel (*Blackia*; Webb, 1985). The faunal record documents a fully functioning Bering land bridge because the animal traffic included such transport modes as ambulatory, cursorial, amphibious, and volant. The rich and varied fauna utilizing the bridge is paralleled by equally diverse plant assemblages. The faunal evidence suggests that widespread shrub savanna habitats continued until about 20 Ma, when Miocene grassland savanna began to increase.

MIDDLE THROUGH LATE EOCENE VEGETATION: 50–35.4 (34) MA

As noted in Chapter 5, the climate of the Late Cretaceous through the Early Eocene was warm, equable, and maritime. Coastal assemblages show that megathermal conditions (MAT > 20°C) extended to about 48° N (Figs. 6.1, 6.2, Table 6.1). Areas of microthermal temperature were highly restricted and at times possibly limited only to mountains in the polar latitudes (Wolfe, 1987). After the thermal highs of the Early Eocene, in the Middle Eocene a complex series of changes began that varied in intensity in different parts of the continent but trended toward cooler (especially lower minimum winter temperatures) and more seasonal winter dry climates. The Middle Eocene represents essentially a transition from the hothouse conditions of the Early Tertiary to the icehouse conditions of the later Cenozoic. This is an important time in the modernization of North American plant communities from ancient Cretaceous and Early Tertiary predecessors, and several modern plant formations and associations appear during this interval. The vegetation classification given in Table 1.1 for modern communities at the formation level can be used for Middle Eocene and later assemblages.

In discussing vegetational history for the southeastern United States, provincial stage names are shown in the far-right column of Fig. 5.20. Middle Eocene through Early Miocene megafossil floras are known from the Gulf Coast region, but these have not been revised in recent times; only certain taxa have been treated through paleomonographic surveys. In his initial study Berry (1916) listed 341 species for the "Wilcox" flora, but at the western Kentucky–Tennessee sites (now assigned to the Middle Eocene Claiborne Formation) only a few have been studied recently (Table 6.2). The prominence of the Leguminosae is partly a reflection of the climatic changes that were taking place in the southeastern United States. The megathermal tropical rain forest of the Paleocene and Early Eocene was being replaced by a semideciduous tropical dry forest (Fig. 6.3) in response to reduced and more seasonally distributed rainfall. A corresponding increase in legumes is evident by the number of specimens having cross-striated pulvini characteristic of the group. The family is presently an important component of the tropical dry forest of northern Latin America. Mangroves grew along the coasts, as represented by pollen of the Old World tropical mangrove palm *Nipa* in the Laredo (Claiborne) Formation of south Texas (Gee, 1990; Westgate and Gee, 1990) and *Spinizonocolpites prominanthus* on the eastern Gulf Coast thoughout most of the Eocene (Frederiksen, 1988, pl. 17).

In contrast to the discontinuous and mostly unrevised megafossil record, pollen and spore assemblages have been studied extensively in this petroleum-rich area; and much of the vegetational history is presently based on plant microfossils (Frederiksen, 1981, 1988, 1991; Gray, 1960; Jones, 1961; Nichols, 1970). Claiborne palynomorphs include aff.

Table 6.1. Geographic distribution of Middle Eocene through Early Miocene floras of North America.

	Eocene	
Southeast	Claiborne	Berry (1916) Frederiksen (1981, 1988) (see text)
	Jackson	Frederiksen (1980a, 1981, 1991)
Mid-Atlantic	Mays Landing	Owens et al. (1988)
Northeast	Lake Hazen (Ellesmere Isl.)	Christe (1964), Christe and Rouse (1976)
West		
Wyoming	Little Mt., Tipperary, Kisinger Lakes	Leopold and MacGinitie (1972) MacGinitie (1974)
Colorado–Utah	Green River	MacGinitie (1969)
	Florissant	MacGinitie (1953)
Montana: latest Eocene to earliest Oligocene	Beaverhead Basins	Becker (1969)
	Ruby	Becker (1961)
	Metzel Ranch	Becker (1972)
	Morman Creek	Becker (1960)
	York Ranch	Becker (1973)
Idaho	Thunder Mt.	Axelrod (1990)
	Germer	Axelrod et al. (1991)
Nevada	Elko	Axelrod (1992)
	Copper Basin	Axelrod (1966b)
	Bull Run	Axelrod (1966b, c)
California	Elsinore	Frederiksen (1991)
	Susanville	Wolfe (1978, 1985)
	La Porte	Potbury (1935), Wolfe (1992b)
	Lower Cedarville	
	Chalks Bluff	MacGinitie (1941)
Oregon	Clarno	Manchester (1990), Scott (1954)
	Goshen	Chaney and Sanborn (1933)
	Comstock	Sanborn (1935)
Washington	Republic	Wolfe and Wehr (1987)
British Columbia	Princeton	Cevallos-Ferriz et al. (1991); Table 6.9 and see text
	Oligocene	
Southeast		
Texas	Catahoula	Berry (1916); see text
Mississippi	Vicksburg	Frederiksen (1981)
West		
Colorado	Creede	Axelrod (1987), Wolfe and Schorn (1989, 1990); Table 6.12
Oregon	Bridge Creek	Chaney (1924, 1927)
	Fossil	Manchester and Meyer (1987); Table 6.13
	Rujada	Lakhanpal (1958)
Northwest		
Mackenzie Delta	Richards	Norris (1982)

table continued

Figure 6.2, facing page. Landform map of the conterminous United States with overlay of principal Middle to Late Eocene (1–18), Oligocene (19–23), and Early Miocene (24–27) floras mentioned in the text. (1) Mays Landing; (2) Wilcox (Claiborne); (3) Laredo (Claiborne); (4) Little Mountain; (5) Kisinger Lakes; (6) (and vicinity) Green River; (7) Thunder Mountain, Challis Volcanics; (8) Florissant; (9) Copper Basin; (10) Bull Run; (11) Elsinore; (12) Susanville; (13) La Porte; (14) Comstock, Goshen; (15) Lower Cedarville; (16) Clarno, Bridge Creek, Fossil; (17) Republic; (18) Princeton; (19) Catahoula; (20) Brandon; (21) Creede; (22) Beaverhead Basins, Metzel Ranch, Mormon Creek, Ruby, York Ranch; (23) Rujada; (24) Buffalo Canyon; (25) Alvord Creek; (26) Tehachapi; (27) Sutro (Thelin and Pike, 1991).

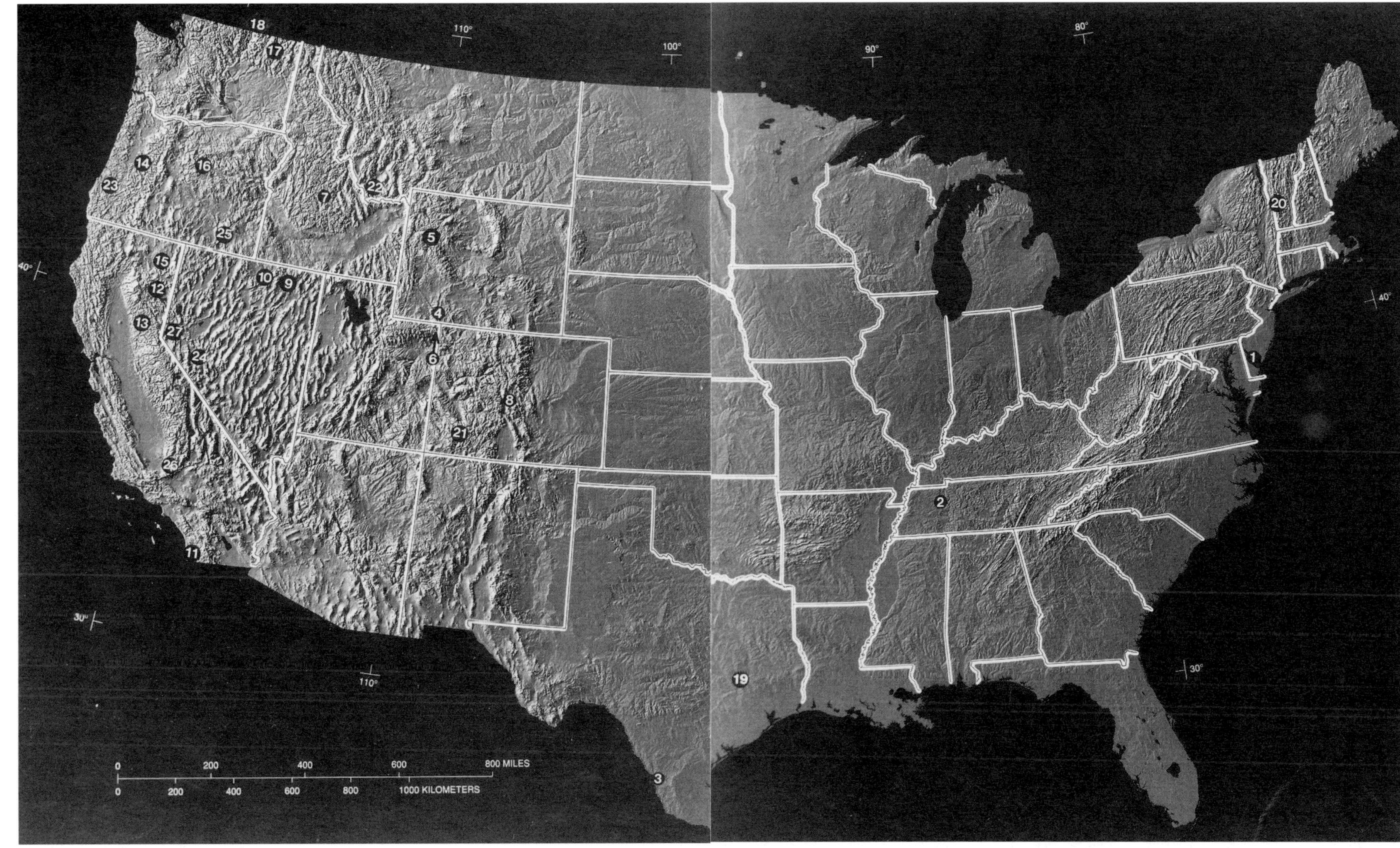

110°
100°
90°
80°
40°
30°
0 200 400 600 800 MILES
0 200 400 600 800 1000 KILOMETERS

Table 6.1. continued

	Early Miocene	
Northeast		
Vermont	Brandon	Tiffney (1994), Traverse (1994)
Devon Island	Haughton	Hickey et al. (1988)
		Whitlock and Dawson (1990)
Banks Island	Mary Sachs	Matthews and Ovenden (1990)
West		
Nevada	Buffalo Canyon	Axelrod (1991)
	Sutro	Axelrod (1991)
Oregon	Alvord Creek	Axelrod (1944)
California	Tehachapi	Axelrod (1939)
Northwest		
Alaska	Seldovia Point	Wolfe and Tanai (1980)

Table 6.2. Megafossils from Middle Eocene Claiborne Formation, western Kentucky–Tennessee.

Podocarpaceae	
Podocarpus (as *Taxodium*, Berry, 1916)	Dilcher (1969)
Palmae	
Sabal	Dilcher (1968), Daghlian and Dilcher (1972)
Araceae	
Philodendron	Dilcher and Daghlian (1977)
Araliaceae	
Dendropanax (as *Aralia*, Berry, 1916)	Dilcher and Dolph (1970)
Ceratophyllaceae	
Ceratophyllum	Herendeen et al. (1990)
Euphorbiaceae	
Crepetocarpon	Dilcher and Manchester (1988)
Hippomaneoidea	Crepet and Daghlian (1982)
Fagaceae	
Castaneoidea	Crepet and Daghlian (1980)
Juglandaceae	
Eokachyra	Crepet et al. (1975)
Eoengelhardia	Crepet et al. (1980)
Oreoroa	Dilcher and Manchester (1986)
Paraoreomunnea	Dilcher et al. (1976)
Lauraceae	
Ocotea	Dilcher (1963)
Leguminosae	
Caesalpinioideae	
Crudia	Herendeen and Dilcher (1990a)
Caesalpinia	Herendeen and Dilcher (1991)
Mimosoideae	
Eomimosoidea	Crepet and Dilcher (1977)
	Daghlian et al. (1980)
Duckeophyllum	Herendeen and Dilcher (1990b)
Eliasofructus	Herendeen and Dilcher (1990b)
Parvileguminophyllum	Herendeen and Dilcher (1990b)
Papilionoideae	
Diplotropis	Herendeen and Dilcher (1990c)
Nyssaceae	
Nyssa	Dilcher and McQuade (1967)
Oleaceae	
Fraxinus	Call and Dilcher (1992)
Ulmaceae	
Eoceltis	Zavada and Crepet (1981)

Acrostichum, *Anemia* (cf. *Mohria*), Gramineae, *Nipa* and other palms, *Ephedra* (presumably growing in sandy xeric coastal habitats), *Pinus*, *Alangium*, *Alnus*, probably Araceae (*Proxapertites*), Bombacaceae, *Betula*, *Carya*, *Castanea*, *Celtis*, *Eucommia*, *Fagus*, *Cyrilla–Cliftonia*, *Ilex*, *Quercus*?, *Corylus*, *Nyssa*, *Juglans*, *Platycarya*, *Pterocarya* and other Juglandaceae (*Momipites*), Olacaceae, Sapindaceae? (*Nuxpollenites*), Onagraceae, *Ostrya–Carpinus*, *Platanus*, Sapotaceae, *Symplocos*, *Tilia*, and *Ulmus*. The temperate plants were mostly part of the upland vegetation of the southern Appalachian Mountains. The Fagaceae (probably Castaneoideae, Quercoideae, and the extinct *Dryophyllum*) are abundant throughout the Middle Eocene, while *Quercus* pollen especially characterizes the latest Claiborne–Vicksburg vegetation and was likely abundant on the Gulf Coastal Plain (Frederiksen, 1981). Like many taxa in North American Tertiary floras, *Eucommia* presently grows in eastern Asia (one species, *E. ulmoides*, in China), but it was widespread during the Eocene and Early Oligocene. Megafossils are known from the Beaverhead Basins, Claiborne, Clarno, Florissant, Green River, John Day, Klondike Mountain (Republic), and other floras (Call and Dilcher, see also Magallón-Puebla and Cevallos-Ferriz, 1994). The stratigraphic range is from late Early Eocene to Early Oligocene in North America (north of Mexico). Microfossils range from the Late Paleocene (Powder River Basin, Wyoming) to the Oligocene.

A summary of the principal palynomorphs from the succeeding Jacksonian provincial stage is given in Table 6.3. The nature of the vegetation is difficult to reconstruct from palynological evidence because a number of extinct or unknown forms are present, many are of a generalized morphology and can be referred only provisionally to clusters of genera or families, and tropical pollen is comparatively poorly known (Frederiksen, 1981). By the Late Eocene a significant number of fossil botanical taxa are similar morphologically to modern forms and can be sorted into ecologicaly consistent and meaningful assemblages. This is in contrast to the impression from some lit-

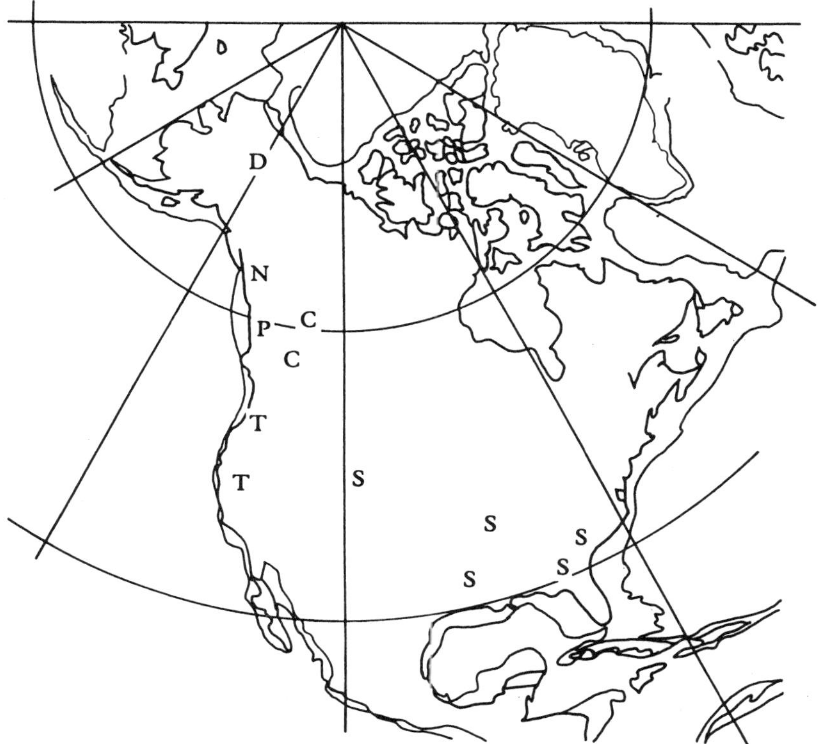

Figure 6.3. Generalized paleovegetation diagram for North America during the late Middle Eocene. C, mixed coniferous forest; D, polar broad-leaved deciduous forest; N, notophyllous broad-leaved forest; P, paratropical rain forest; S, tropical semideciduous forest; T, tropical rain forest. Reprinted from Upchurch and Wolfe (1987) with the permission of Cambridge University Press.

erature on Tertiary and Quaternary paleoecology that ecological requirements and generic composition of all but the most modern of floras have so shifted that the modern analog method has virtually no value. This has not been my experience and I do not believe it is true. The Middle Eocene, however, is a transition period in the modernization of floras and lineages. Consequently, there is a range in the confidence level of biological identifications expressed in the nomenclature from relatively certain (e.g., *Nipa*), to probable (cf. *Carya*), to possible (*Clethra*?), to merely designating a comparable morphotype to convey information on pollen and spore structure (*Eugenia*/*Myrtus*?, *Fagus*-type). If a specimen is not exactly comparable to a modern analog, it may later be found to represent a completely unrelated taxon. These annotations are important in evaluating the taxonomic and biogeographic significance of the prefaced names.

In the absence of *Avicennia*, *Laguncularia*, *Rhizophora*, and other New World mangroves, coastal brackish-water habitats were occupied by *Nipa*, which was possibly associated with *Acrostichum* that presently occurs with mangroves. *Nipa* was widespread in the northern hemisphere during the Eocene (Tralau, 1964), but it became extinct there and in the Caribbean region and Europe, near the Eocene–Oligocene boundary, and occurs today only in tropical southeast Asia. On sandy flats, but removed from the tidal influence, *Pinus* and an understory of *Sabal*–*Serenoa*-type palms and *Ephedra* were present (early sand pine scrub association). *Nyssa*, members of the Taxodiaceae complex, and probably *Carya* grew in and marginal to freshwater swamps (flood-plain forest association). Most of the inland, well-drained, low to midelevation landscape apparently was covered by a mosaic of paratropical and warm-temperate genera. Paratropical representatives include *Anemia*–*Mohria*, *Lygodium*, various Palmae, and possibly some Anacardiaceae, Bignoniaceae, Bombacaceae, Loranthaceae, Rutaceae, Sapindaceae, and Sapotaceae that cannot be identified to genus. Jacksonian-age pollen of genera that presently range into warm-temperate zones includes *Podocarpus*, *Acer*?, *Alnus*, various Fagaceae, *Ilex*, *Platanus*, *Pterocarya*, and *Ulmus*-type. At the end of the Eocene near the Jackson (Upper Eocene)–Vicksburg (Lower Oligocene) boundary (34–35 Ma, calcareous nanofossil zone NP21), pollen of *Dryophyllum*–*Quercus* (oak) increases greatly as a result of a cooler and drier climate compared to Wilcox and Claiborne time (Frederiksen, 1981). Many of the plants probably ranged into cool-temperate upland habitats where *Picea* and *Tsuga* were present. This warm-temperate to subtropical forest is reminiscent of the drier phase of present-day vegetation along the lower eastern slopes of the Sierra Madre Oriental in Mexico between 1000 and 2000 m elevation where *Podocarpus*, *Acer*, *Alfaroa*, *Alnus*, *Carya*, *Celtis*, *Ilex*, *Juglans*, *Myrica*, *Platanus*, *Tilia*, *Ulmus*, and others are associated with tree ferns, *Clethra*, *Ficus* and other Moraceae, various Myrtaceae, Leguminosae, and other tropical trees and shrubs. Mangroves fringe the coast there, and cool-temperate plants grow in the adjacent highlands [presently *Abies* and species of *Pinus* (e.g., *P. montezumae*, *P. rudis*) and *Picea* until recent times]. Huatusco is located in this zone at an

Table 6.3. Palynomorphs from Jacksonian provincial stage (Upper middle and Upper Eocene, ~40–37 Ma) of Gulf Coast United States. After Frederiksen, 1980a, 1981.

Palynomorphs	Affinities
Bryophytes	
Sphagnum	Sphagnaceae
Pteridophytes	
Lycopodium	Lycopodiaceae
Selaginella	Selaginellaceae
Acrostichum?	Pteridaceae
Cicatricosisporites dorogensis	*Anemia–Mohria*, Schizaeaceae
Laevigatosporites haardtii	Various Filicales (ferns)
Lygodiumsporites adriennis	Schizaeaceae, *Lygodium*
Verrucatosporites alienus	Various Filicales (ferns)
Gymnosperms	
Cedrus	Pinaceae
Cryptomeria–Metasequoia–Sequoia	Taxodiaceae
Ephedra	Ephedraceae
Picea	Pinaceae
Pinus tenuextima	Pinaceae
Podocarpus	Podocarpaceae
Taxodiaceae–Cupressaceae–Taxaceae	
Tsuga	Pinaceae
Monocotyledon Angiosperms	
Aglaoreidia pristina	Unknown
Arecipites pseudotranquillus	Palmae
Liliacidites tritus	*Pseudophoenix*?, Palmae
Milfordia sp.	Prob. Restionaceae
Monocolpopollenites tranquillus	Palmae
Nipa sp.	Palmae
Proxapertites sp.	Unknown, Araceae or Annonaceae?
Pseudophoenix? sp.	Palmae
Sabal	Palmae
Serenoa?	Palmae
Dicotyledon Angiosperms	
Acer?	Aceraceae
Ailanthipites berryi	Unknown
Alangium?	Alangiaceae
Albertipollenites? *araneosus*	Bignoniaceae
Alnus	Betulaceae
Anacolosa–Cathedra–Ptychopetalum	Olacaceae
Bumelia?	Sapotaceae
Caprifoliipites sp.	*Viburnum*?, Caprifoliaceae
Carya	Juglandaceae
Celtis	Ulmaceae
Cupanieidites orthoteichus	*Loranthus*?, Loranthaceae
	Cupania?, Sapindaceae
Cupuliferoidaepollenites liblarensis	*Dryophyllum*?, Fagaceae
C. certus	*Cassia*?, Leguminosae
C. brevisulcatus	*Chrysophyllum*?, Sapotaceae
Cyrillaceaepollenites megaexactus	*Cyrilla–Cliftonia*
Diospyros?	Ebenaceae
Fraxinoipollenites variabilis	Unknown
F. sp.	Unknown
Intratriporopollenites sp.	Tiliaceae–Bombacaceae

continued

Table 6.3. continued

Juglans	Juglandaceae
Ludwigia	Onagraceae
Manilkara?	Sapotaceae
Momipites coryloides	Juglandaceae, *Engelhardia* group
M. microfoveolatus	Juglandaceae, *Engelhardia* group
Myrica propria	Myricaceae
Myriophyllym	Haloragidaceae
Myrtaceidites sp.	*Eugenia–Myrtus*?, Myrtaceae
Nuxpollenites	Sapindaceae?
Nyssa	Nyssaceae
Parthenocissus?	Vitaceae
Platanus	Platanaceae
Pollenites modicus	Unknown
P. pseudocingulum granulatus	Fagaceae?
Pseudolaesopollis ventosa	*Costaea*?, Cyrillaceae
Pterocarya	Juglandaceae
Quercoidites microhenricii	Fagaceae
Rhoipites	*Rhus*? (Anacardiaceae), Cornaceae, Nyssaceae
Salixipolllenites sp.	*Fraxinus*?, Oleaceae
Siltaria sp. (<25μm)	Fagaceae
Striopollenites terasmaei	Unknown
Symplocos	Symplocaceae
Tetracolporopollenites sp.	Sapotaceae
Tilia	Tiliaceae
Tricolporopollenites illiacus	*Ilex*, Aquifoliaceae
Ulmus–Planera–Zelkova type	Ulmaceae
Verrutricolporites sp.	Unknown
Zanthoxylum?	Rutaceae

Adapted from Frederiksen (1980a, 1981).

elevation of 1344 m. There the MAT is 15.7°C, the mean monthly range is 7°C [between 15°C (December–January) and 22°C (May)], and the annual rainfall is 1745 mm. The MAT extrapolated downward 1 km to 344 m is 18.7°C using a lapse rate of 3°C/km (Wolfe, 1992b) or 21.6°C at a lapse rate of 5.9°C/km (Meyer, 1992). Sedimentological evidence also indicates that from the Late Eocene to the Miocene the rainfall in Texas was gradually decreasing and becoming more seasonal (winter dry; Folk, 1955).

The vegetation of the Mississippi Embayment region, or any other area of North America, was not static even at the time intervals of stages. Frederiksen (1991) has shown that the general deterioration of terrestrial climate between the Middle Eocene (Lutetian, Bartonian) to earliest Oligocene (Rupelian) encompassed four events of rapid vegetation change. One was a turnover at the beginning of the Luetian (early Middle Eocene) as reflected by peaks in pollen of first occurrences (*Platanus*, *Cyrilla–Cliftonia*, large grains of *Carya*) and last occurrences (mostly Paleocene species; *Jarzenipollis trina*, *Pistillipollenites macgregorii*, *Momipites tenuipolus* group). The second event, occurring between the latest Lutetian and early Bartonian, has been termed the Middle Eocene Diversity Decline (MEDD) by Frederiksen (1991; Cook Mountain Formation, Texas–Louisiana; Lisbon Formation, Alabama; P13–P14, NP16). Some new taxa appeared, including the first grasses, and *Pterocarya* and *Ephedra* migrated in from the west. Overall, however, the elimination of many Paleocene and Early Eocene arborescent taxa in the Palmae, Juglandaceae, Ulmaceae, and probable Eucommiaceae resulted in a net decrease in diversity as a result of reduced and more seasonal rainfall. The third event was an increase in pollen similar to modern *Quercus* in the upper Priabonian (Late Eocene), reflecting the temperature decline and redistribution of rainfall. The continuation and intensification of this trend eventually produced the fourth event, which was a further reduction in diversity as more tropical elements were eliminated from the vegetation of the southeastern United States (the Terminal Eocene Event; now Early Oligocene; NP21). Evidence of these stepwise changes in climate, and the paralleled response by the biota, is preserved in the palynological and megafossil records (Wolfe, 1978; Wolfe and Poore, 1982), but they are out of phase and occurred later in the megafossil analyses. The differences between the two records are further discussed by Frederiksen (1988), but the reason for the disparity is not known. There were also local differences in vegetation and other changes that occurred over comparatively short periods of time. For example, swamp vegetation in the Claibornian is dominated by Fagaceae, while in the Jacksonian the prominent pollen types in swamp deposits are Juglandaceae (*Momipites*).

The purpose of this excursion into detailed changes in

climate and communities is to avoid the misconception that vegetational history followed an even course as might be suggested by the generalized benthic paleotemperature curve (Fig. 3.1). A similar impression might be created by the emphasis in this text on major trends that lead to the modernization of North American vegetation at the formation and association levels. Significant alterations in the environment were also occurring on a much finer time scale, including the Milankovitch variations, which were periodically overprinted or enhanced by such factors as continentality, variations in CO_2 concentration, alterations and new threshold levels in ocean circulation, but which are less apparent at the resolution of tens of thousands to millions of years. As noted earlier in Chapter 5, Wing (1996) has shown that in the northern Rocky Mountains a previously unrecognized cool interval of 1–1.5 m.y. duration occurred at the Paleocene–Eocene boundary. This was within an overall trend toward the Early Eocene thermal maximum. When a sufficient number of floras are carefully studied from a region to give high resolution, trends can often be seen that are obscured by general collecting. New data are emerging that suggest abrupt climatic changes superimposed on long-term trends are the rule. The pace probably did increase in the Late Cenozoic, but the frequency and intensity of the changes are also obscured in older sediments.

It is now recognized that these fluctuations often correlate with rapid peaks in evolution (Rea et al., 1990). A sharp decrease in ^{18}O in seafloor sediments off Antarctica indicates that during the seemingly gradual warming at 55 Ma, deep-sea temperatures rose from 10 to 18°C in 2 Ky (Kennett and Stott, 1991). This not only caused an extinction of 40% of the benthic foraminifera, but the event correlates with a rapid change from archaic terrestrial mammals to modern forms in the Bighorn Basin of Wyoming and Montana in the Early Eocene. The correlation between these two geographically distant events is preserved in carbon isotope patterns. The temperature rise caused a release of CO_2 from the ocean surface to the atmosphere, which was incorporated into the tissue of land plants via photosynthesis and eventually made its way into the teeth and bone of grazing and browsing land animals. When carbon isotopes were measured from the rapidly evolving Bighorn Basin herbivores, the temperature spike was found at the same time as the ^{18}O detected increase in the latest Paleocene–Early Eocene temperature gradient from Antarctica. Koch et al. (1992) report that the $^{12}C/^{13}C$ ratio in enamel apatite and paleosol carbonate decreased by 5.9 and 4.5%, respectively (increasing temperatures) in less than 50 Ky and persisted for less than 100 Ky. An appreciation of the nature of biotic history warrants an occasional reminder that stasis was no more a characteristic of past times than it is of the present, that what appears as a gradual trend on a broad scale is actually punctuated by rapid shifts in climate, and that these shifts are now being recognized as forcing mechanisms for plant and animal evolution (Bennett, 1990).

The same general sequence of Late Eocene to Early Oligocene vegetational change described for the Gulf Coast is also evident in the Mays Landing core from southeastern New Jersey. By the Late Oligocene common palynomorphs there included *Pinus*, *Podocarpus*, *Tsuga*, Taxodiaceae–Cupressaceae type, *Alnus*, *Betula*, *Carya*, *Fagus*, *Ilex*, *Liquidambar*, *Momipites* (*Alfaroa–Engelhardia–Oreomunnea*), *Nyssa*, *Quercus* (Owens et al., 1988), and other dominants of the modern deciduous forest formation.

The climatic sequence suggested by the plant microfossil record (Frederiksen, 1991; Owens et al., 1988) is similar but not identical to that from the marine realm (Prothero, 1989; Table 6.4). Further studies may eventually bring the patterns into closer accord, allowing for lag time and differences in the nature and degree of forcing mechanisms between marine and terrestrial environments. In the meantime, it is clear that temperatures during the Middle Eocene to Middle Oligocene downshifted in stages, in part due to decreases in CO_2 concentration and the initiation of glaciation on Antarctica, which cooled marine waters (less evaporation) and the atmosphere (less available moisture), and altered ocean and atmosphere circulation.

As a generalization, the Late Eocene and later vegetation of the southeastern United States records the gradual elimination of many tropical species; the reorganization, expansion, and diversification of temperate species in response to cooling; a widening between cold–warm monthly averages; and reduced and greater seasonality in rainfall. The Middle Eocene through the earliest Oligocene is the transition phase between tropical and temperate conditions in the midnorthern latitudes, and the Jacksonian and related microfossil assemblages reflect this period of change in climate and vegetation.

Quantitative estimates of the Middle Eocene climate for the southeastern United States, dating back to the time of Berry in the early 1900s, range widely and include imprecisely defined and overlapping terms noted earlier. More recently, the vegetation has been characterized as a tropical dry forest with a MAT fluctuating between 20 and 28°C (Wolfe, 1978; 22 and 30°C, Wolfe and Poore, 1982). Dilcher (1973) suggests a climate like that of southern peninsular Florida or coastal Louisiana (MAT 24°C, 21°C, respectively), with occasional frosts in the highlands, as does Axelrod (1966a) and Frederiksen (1980b; see previous Huatusco data extrapolated to sea level; 18.7–21.6°C). The tropical dry forest was eliminated in the southeastern United States after the Eocene by cooling temperatures and as changes in atmospheric circulation, induced by continued uplift of the Rocky Mountains, began bringing greater precipitation and warm moist winds from the south.

To the far northeast on Ellesmere Island (latitude 81° 30' N), sediments correlated with the Eocene portion of the Late Cretaceous to Eocene Eureka Sound Group have yielded a rich mammalian fauna, along with large salamanders, anguid and varamid lizards, a boid snake, turtles, tortoises, and alligator (Estes and Hutchison, 1980;

Table 6.4. Relationship between marine, Mississippi Embayment, and Mays Landing events, documenting stepwise nature of Middle Eocene to Middle Oligocene climatic change and vegetation response.

Marine[a]	Mississippi Embayment[b]	Mays Landing[c]
Middle Eocene Lutetian–Bartonian ~41 Ma	1st pollen event Early Middle Eocene 2nd pollen event Middle Eocene (MEDD)	Early Eocene–Middle Eocene boundary Middle Eocene
Middle Eocene–Late Eocene boundary Bartonian–Priabonian ~37 Ma		Middle Eocene–Late Eocene boundary
Late Eocene Late Priabonian ~35 Ma	3rd pollen event Late Eocene	
Eocene–Oligocene boundary 33.5 Ma	4th pollen event Early Oligocene	
Middle Oligocene Late Rupelian ~30.5 Ma		Middle Oligocene 31–29 Ma

[a]Data from Prothero (1989).
[b]Data from Frederiksen (1991).
[c]Data from Owens et al. (1988).

McKenna, 1980). A pollen flora from Lake Hazen, northern Ellesmere Island (MacGregor, in Christe, 1964; Christe and Rouse, 1976) includes *Cedrus*, *Picea*, *Pinus*, *Tsuga*, *Acer*, *Betula*, *Carya*, *Corylus–Carpinus*, *Castanea*, *Fagus*, *Pterocarya*, and *Quercus*. On Axel Heiberg Island, the Buchanan Lake Formation preserves fossil floras from a floodplain environment. The palynoflora adds *Larix*, *Metasequoia*, *Alnus*, *Diervilla*, *Engelhardia*, *Fraxinus*, *Juglans*, *Lonicera*, *Tilia*, and *Viburnum* to the Ellesmere assemblage (McIntyre, 1991). The vegetation was a temperate mixed hardwood–conifer forest similar to the modern lake states forest and Appalachian highlands to the south. An adjacent megafossil flora preserves holdovers from the warm-temperate broad-leaved deciduous forest that had occupied these latitudes in the Early Paleocene (Fig. 5.12), including *Larix*, *Pinus*, *Taxodium*, *Acer*, *Betula*, *Cercidiphyllum*, *Corylus*, *Platanus*, *Populus*, and *Ulmus*. Records of *Cruciptera*, *Paleocarya*, and *Hooleya* (extinct genera of the Juglandaceae) in the Tertiary of western North America, in the Middle Eocene pipe clays of southern England, and in Messel, Germany, document that the North Atlantic land bridge was fully functioning (Manchester et al., 1994). Cold-month temperatures were 0–4°C (generally above freezing), warm-month temperatures ~25°C, and MAT is estimated at 12–13°C (Axelrod, 1984; 12–10°C; Basinger et al., 1994). McKenna (1983) places coastal water temperature bordering Ellesmere Island in the Middle Eocene at 15°C.

A MAT of 24°C for the southeastern United States at 35° N and 13°C for Ellesmere Island at 81° N calculates to a Middle Eocene gradient of 0 28°C /1° latitude. This compares with 0.3°C /1° latitude in the Maastrichtian, indicating that by the Middle Eocene the MAT had declined episodically from Early Eocene maxima to about what they were at the end of the Cretaceous (Fig. 3.1). The trend continued through Middle and Late Cenozoic time to the present gradient of ~0.5°C /1° latitude.

Little is known of Late Eocene vegetation of the Plains region, although woods of *Robinia*, *Prunus*, and *Zelkova* are known from the Chadronian of Nebraska and indicate a seasonal climate (Wheeler and Landon, 1992). Paleosols, root diameters, and tree spacing in the Badlands of South Dakota suggest rainfall decreased from 1000 mm (38 Ma, Late Eocene) to 250–450 mm (29.5 Ma, Late Oligocene). The shift in vegetation inferred from this sequence is from moist forest (38 Ma), to dry forest (34 Ma), to dry woodland (33 Ma), to wooded grassland with gallery forests (32 Ma), to more open savanna with an increasing understory of grasses by 30 Ma. Land snails show a similar shift at the Eocene–Oligocene boundary from large-shelled subtropical forms (MAT of 16°C and annual precipitation of 450 mm) to the smaller warm-temperate ones presently occupying open woodlands with a pronounced dry season (Evanoff et al., 1992). In the High Plains and northern Rocky Mountains, Early Eocene megathermal tropical and paratropical vegetation, which had extended north to 60°–65° N, started to be replaced by microthermal deciduous communities.

In the central Rocky Mountains crustal deformation slowed, but volcanism deposited ash up to 1700 m thick in the basins. Elevations had reached a point where vegetation east and west of the continental divide was developing differently (Leopold and MacGinitie, 1972; Wing, 1987; Fig. 5.8, hatched area). To the east in the earliest part of the Middle Eocene, a seasonally dry, subtropical open woodland–savanna phase is well documented (34.8% in the playa lake Little Mountain flora of southeastern Wyoming, 51–50 Ma; Leopold and MacGinitie, 1972) and paleotropical elements constituted 13%, declining from 16.7% in the Wind River flora. Members included *Alchornea*, *Cardiospermum*, *Populus*, and microphyllous leguminosae. Slightly later in the early Middle Eocene, tropical and subtropical components reached a peak of ~50% in the Tipperary and Kisinger lakes flora of the Aycross Formation in the Wind River Basin of northeastern Wyoming (MacGinitie, 1974; ~49–50 Ma) due to an increase in neotropical (15.5%) and pantropical (22.0%) species. Non-entire-margined leaves reached just over 50%, which signifies a subtropical forest (wet summers, dry winters) in the leaf physiognomy system, and about 60% were deciduous. Members included *Acrostichum*, *Thelypteris*, *Chamaecyparis*, *Glyptostrobus*, *Apeiba*, *Canavalia*, and *Dendropanax*. In the middle Middle Eocene (Green River time, 45–48 Ma) tropical and subtropical species declined to 33.8%, and New World taxa constituted 23.4% of the flora.

The change was gradual and complex from an Early Eocene and older tropical–paratropical vegetation with primarily Asian affinities in the central Rocky Mountains and Plains to a seasonally dry subtropical Middle Eocene vegetation with Mexican–Central American affinities. Some seasonally dry subtropical elements from the present-day Latin American flora actually began to appear in the Paleocene–Eocene (Clarkforkian, Wasatchian). Examples include *Chaetoptela microphylla*, *Woodwardia gravida*, and *Populus* sect. *Abaso* (Wing, 1987). Some Asian elements were already locally extinct in the eastern Rocky Mountains, including *Metasequoia* and members of the *Cercidiphyllum* complex. Other records that are exceptions to the overall trend are the Asian *Platycarya* and members of the Flacourtiaceae, Icacinaceae, and Menispermaceae, which were present by the Wasatchian and continued through the Bridgerian. *Ailanthus* is recorded in the Chalk Bluffs and Green River floras (Early to Middle Eocene) and persists through the Late Eocene–Oligocene (Ruby and Beaverhead Basins floras of southwestern Montana; Becker, 1961, 1969) and into later Tertiary time. Nonetheless, the trend from primarily Old World to New World tropical–subtropical affinities is evident among the floras. Cooling temperatures that reduced migration through Beringia was a factor.

The Green River flora and the Thunder Mountain flora are of approximately the same age, but were deposited at different elevations on the east and west side of the Rocky Mountains, and give some insight into the effect of the developing continental divide and the zonation of western American vegetation during the Middle Eocene. The Green River flora (~45 Ma, Roehler, 1992a,b; 45–48 Ma, Wing and Greenwood, 1993) of northwestern Colorado and northeastern Utah is preserved in lacustrine (lake) sediments. The formation covers an area of ~65,000 km^2 (25,000 mi^2); and some of the lake sediments are varved, reflecting seasonal deposition. The sediments were laid down near the end of the Laramide orogeny at ~pl 35° N (5°–8° south of its present location; Roehler, 1993), placing it near the juncture of the Hadley and Rossby circulations where high pressures dominated and winds were primarily from the west. Surrounding mountains were 1–1.3 km higher than the basin, and rainshadows likely contributed to the dryness of the intermontane basin. Seventy-seven depositional cycles are present in the basin and they appear to conform to the 23 Ky precession and the 100 Ky eccentricity cycles. The Green River Formation is similar in age to the Claiborne Formation of the Mississippi Embayment region, indicating that the seasonally dry subtropical climate of the Gulf region extended to the central Rocky Mountains during the Middle Eocene.

The fossils were deposited in three principal basins occupied by great shallow lakes: Uinta Basin (Lake Uinta), Green River Basin (Lake Gosiute), and Fossil Basin (Fossil Lake; Figs. 6.4, 6.5). The Green River Formation is known for its exceptionally well-preserved fish fauna and for the greatest reserves of oil shale presently known (carbonate rocks rich in kerogen). Genera reported by MacGinitie (1969) are given in Table 6.5. The most abundant specimens are of *Lygodium* (abundant), palm leaves (abundant), *Typha* (common), *Musophyllum* (common), *Zingiberopsis* (common), *Canavalia* (common), and *Sapindus* (abundant). *Ceratophyllum* (Herendeen et al., 1990) and *Populus* (Fig. 6.6A; Manchester et al., 1986) have been confirmed, and *Caesalpinia* (Herendeen and Dilcher, 1991) and *Eucommia* (Call and Dilcher, 1997) have been added. Other members include *Acer* (Fig. 6.6C), *Engelhardia* (Fig. 6.6B), *Platanus* (Fig. 6.6D), *Pterocarya* (Fig. 6.6F), and *Zelkova* (Fig. 6.6E). Pollen (mostly Wodehouse, 1932, 1933) includes *Ephedra*, *Picea*, *Alnus*, *Betula*, *Carya*, *Nyssa*, *Platycarya*, and *Tilia*. The vegetation occupied four habitats: lake margins, swamps, and floodplains at 300 m elevation (*Platanus*, *Populus*, *Salix*); drier, well-drained, bordering slopes up to 700 m above the lake level (*Ephedra*, some *Pinus* species, *Cardiospermum*, *Celtis*, *Quercus*, *Sapindus*, *Vauquelinia*; savanna, oak–piñon pine woodland); uplands 1300 m above the lake level (mostly broad-leaved deciduous trees; *Acer*, *Carya*, *Engelhardia*, *Gymnocladus*, *Liquidambar*, *Liriodendron*); and a mixed hardwood–coniferous forest zone at 1000 m and extending possibly to 1700 m (*Picea*, other *Pinus* species, *Betula*). Rainfall is estimated to have been ~700 mm (28 in.; subhumid) and seasonal.

MacGinitie (1969) interpreted the slope vegetation during Green River time as a unique type of savanna woodland with widely spaced trees and shrubs but with little or

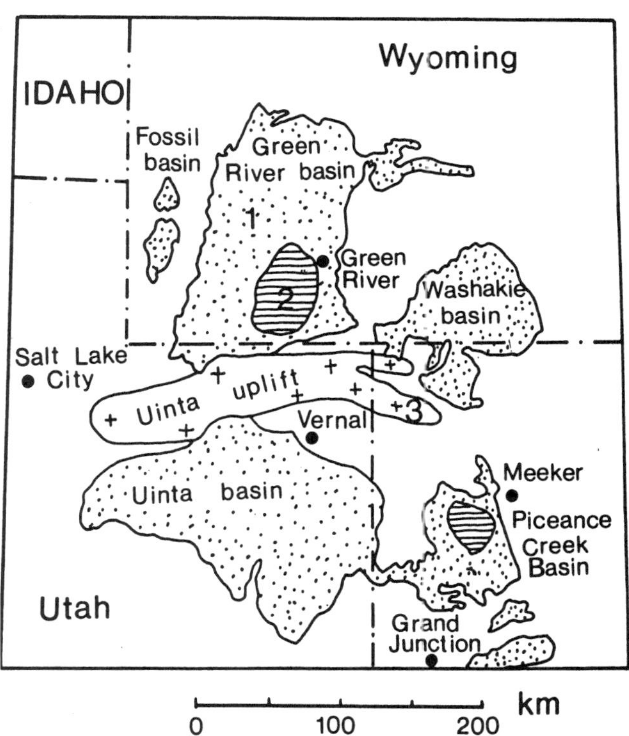

Figure 6.4. Sedimentary deposits of Eocene Lakes Uinta (Uinta and Piceance Creek Basins) and Gosiute (Green River and Washakie Basins) and Fossil Lake (Fossil Basin). (1) Green River Formation; (2) bedded salts; (3) basement. Reprinted from Eugster (1982; based on Eugster and Hardie, 1978) with the permission of the Geophysics Study Committee (1982) of the National Academy of Sciences. Courtesy the National Academy Press.

Figure 6.5. (A) Carr Creek Canyon from Wardell Ranch, Green River flora, northwestern Colorado. (B) Fossil plant locality southeast of Bonanza, Utah. Evacuation Creek Member, Green River flora. Repinted from MacGinitie (1969) with the permission of the University of California Press.

Table 6.5. Megafossils from late Middle Eocene Green River flora of northwestern Colorado and northeastern Utah.

Pteridophytes

Aspleniaceae
Asplenium delicatula
A. serraforme
Equisetaceae
Equisetum winchesteri
Isoetaceae
Isoetites horridus
Schizaeaceae
Lygodium kaulfussii
Salviniaceae
Azolla berryi
Pteridaceae
Acrostichum hesperium

Gymnosperms

Pinaceae
Pinus balli
P. florissanti
Taxodiaceae
Sequoia cf. *affinis*

Monocotyledon Angiosperms

Potamogetonaceae
Potamogeton rubus
Sparganiaceae
Sparganium antiquum
S. eocenicum
Typhaceae
Typha lesquereuxi

Dicotyledon Angiosperms

Aceraceae
Acer lesquereuxi
Dipteronia insignis
Anacardiaceae
Anacardites schinoloxus
Astronium truncatum
Rhus nigricans
Toxicodendron winchesteri
Apocynaceae
Apocynospermum coloradensis
Araliaceae
Araliophyllum quina
Oreopanax elongatum
Aristolochiaceae
Aristolochia mortua
Berberidaceae
Mahonia eocenica
Bombacaceae
Ochroma murata
Burseraceae
Bursera inaequalateralis
Celastraceae
Celastrus winchesteri
Euphorbiaceae
Aleurites glandulosa
Fagaceae
Quercus cuneatus
Q. petros
Hammamelicaceae
Distylium eocenica
Liquidambar callarche
Juglandaceae
Engelhardtia uintaensis
Pterocarya roanensis
Lauraceae
Beilschmiedia eocenica
Lindera allardi
Ocotea coloradensis
Persea coriacea
Leguminosae
Caesalpinites falcata
C. pecorae
Erythrina roanensis
Gymnocladus hesperia
Leguminosites lesquereuxiana
L. regularis
Mimosites coloradensis
Parvileguminophyllum coloradensis
Swartzia wardelli
Malvaceae
Hibiscus roanensis
Meliaceae
Cedrela trainii
Menispermaceae
Menispermites limaciodes
Myrtaceae
Eugenia americana
Oleaceae
Osmanthus praemissa
Platanaceae
Platanus wyomingensis
Proteaceae
Lomatia lineatulus
Rhamnaceae
Berchemiopsis paucidentata
Rosaceae
Prunus stewarti
Rosa hilliae
Vauquelinia comptonifolia
Rutaceae
Ptelea cassioides
Salicaceae
Populus cinnamomoides
P. wilmattae
Salix cockerelli
S. longiacuminata
Sapindaceae
Allophylus flexifolia
Athyana balli
Cardiospermum coloradensis
Koelreuteria viridifluminis
Sapindus dentoni
Thouinia eocenica
Simarubaceae
Ailanthus lesquereuxi
Sterculiaceae?
Sterculia coloradensis
Styracaceae
Styrax transversa
Symplocaceae
Symplocos exilis
Tiliaceae
Triumfetta ovata
Ulmaceae
Celtis mccoshii
Zelkova nervosa

Adapted from MacGinitie (1969). The specimen reported as *Fraxinus flexifolia* (Brown, 1940) is now considered to be a mimosoid leaflet *Parvileguminophyllum coloradensis* (Call and Dilcher, 1994).

Figure 6.6. (A) *Populus wilmattae*, Green River flora. (B) *Engelhardia uintaensis*, Green River flora. (C) *Acer lesquereuxi*, Green River flora. (D) *Platanus wyomingensis*, Green River flora. (F) *Pterocarya roanensis*, Green River flora. (E) *Zelkova nervosa*, Green River flora. Reprinted from MacGinitie (1969) with the permission of the University of California Press.

no grass. He called it Orizaban–subtropical, by analogy with seasonally dry woodland savanna on Mt. Orizaba, Mexico. Wolfe (1985) believes it was a version of the semi-deciduous tropical dry forest that characterized the southeastern United States during the Middle and Late Eocene. Independent evidence from sedimentology confirms the subhumid aspect of the climate (Eugster, 1982). Evaporation exceeded inflow most of the time, and at least 25 episodes are recorded where Lake Gosiute dried up completely and became a salt pan. In general, the Green River differs from the older tropical Wind River flora with distinct Asian affinities in being dry subtropical with greater Latin American affinities. Varves in the lake sediments indicate that long warm dry summers alternated with cool moist winters.

It has been suggested that the meager representation of grasses in the Eocene and Oligocene fossil record may not be solely due to their rarity in the vegetation. Thomasson (1985) believes the later Miocene record is too diverse to exclude their presence at least by the Oligocene and that their appearance marks the development of preservable silicified tissue. However, fossil pollen is also rare, and until some tangible evidence for a grass understory is found, the vegetation can best be described as a woodland–savanna or tropical dry (oak–piñon pine) forest or woodland.

Early computer models for Eocene climates of the continental interior yielded simulations that were in conflict with results from paleontology. The CCM and energy balance model hindcasted winters with temperatures of −30°C (Barron, 1983; Sloan and Barron, 1990), while a mass of biological and geological information clearly indicated minimum winter temperatures at or above freezing. However, the models did not incorporate the boundary condition of large lakes. When the Green River lakes were added to the simulations, with a CO_2 concentration twice that of the present, the difference between the model and ground data was largely resolved. In fact, the model responded as strongly to the presence of the lakes as to the level of CO_2 (Sloan, 1994).

A flora from the Germer Member of the Challis Volcanics (47–46 Ma) was deposited at an elevation of ~600 m and illustrates the difference in vegetation east and west of the continental divide. While the Green River flora to the east at lower elevations was a woodland–savanna or tropical dry oak–piñon pine woodland, the Germer flora represents a mixed deciduous hardwood–coniferous forest of *Aesculus*, *Alnus*, *Betula*, *Carya*, *Cercidiphyllum*, *Eucommia*, *Juglans*, and *Ostrya*, associated with frequent to abundant *Metasequoia*, *Pseudolarix*, and *Sequoia* and lesser amounts of *Abies*, *Chamaecyparis*, *Picea*, *Pinus*, and *Tsuga* (Axelrod et al., 1991). This ecotonal forest type was common at the transition between the deciduous forest and the western montane coniferous forest in the Middle Tertiary when broad-leaved deciduous angiosperms were more prominent. It now has been mostly eliminated as the deciduous forest in the west declined with the drying conditions associated with uplift of the coastal mountains. Clearly by the Middle Eocene the Rocky Mountains were directly influencing the course of vegetational history by atmospheric circulation and rainshadow mechanisms, while the coastal mountains were not yet interfering with moisture transport from the Pacific Ocean.

The Thunder Mountain flora of Middle Eocene age (~45 Ma; Axelrod, 1990) in central Idaho, 90 km northwest of the Germer flora, was deposited at an elevation of ~1550 m. It illustrates some of the elevational zonation evident in floras on the west side of the Rocky Mountains. The Road Florule is lowest in elevation and includes *Abies*, *Larix*, *Picea*, *Pinus*, *Tsuga*, *Sequoia*, *Chamaecyparis*, and *Thuja*, along with *Alnus*, *Amelanchier*, *Arctostaphylous*, *Castanea*, *Comptonia*, *Cornus*, *Mahonia*, *Malus*, *Populus*, *Paxistima*, *Pterocarya*, *Quercus*, *Rhamnus*, *Rhododendron*, *Salix*, and *Viburnum*. It represents the upper elevational zone of a mixed hardwood–coniferous forest. The Dewey Florule is a higher elevation assemblage that descended to lower elevations along streams and canyons through cold-air drainage. It includes *Abies*, *Larix*, *Picea*, and *Pinus*, but only a few deciduous hardwood species, and is a western montane coniferous forest. The MAT is estimated at 8.5°C, tree line at 2195 m, and precipitation between 1100 and 1150 mm (mostly summer rain and light winter snow).

The presence of *Abies* and *Picea* reveals that elements of the western montane coniferous forest were already established in western North America by the Middle Eocene. These began to coalesce into the modern formation as elevations increased and temperatures declined. The presence of *Ephedra*, *Celtis*, and *Ocotea* in the Green River flora further documents that plants adapted to subhumid habitats were also available for later assemblage into desert and other dry to arid vegetation with the rise of the Coast Ranges, Sierra Nevada, and Cascade Mountains, principally in the Pliocene. Vegetational history is a record of once-isolated preadapted elements or restricted early versions of a vegetation type present in edaphically or physiographically controlled sites, responding to environmental change by assemblage, diversification, and expansion into new kinds of communities. This conceptual view of the dynamics of community evolution is nicely presented by Axelrod (1992) with regard to the Eocene Elko flora (36 Ma) of northeastern Nevada and three Miocene floras (15–16 Ma) from western Nevada (Middlegate, Eastgate, and Buffalo Canyon; Wolfe et al. 1997, Table 1).

Slightly younger vegetation to the east of the central Rocky Mountains is preserved in lacustrine shales of the latest Eocene to earliest Oligocene Florissant Formation of central Colorado (Fig. 6.7, Table 6.6) dated at ~35 Ma (Epis and Chapin, 1974; adjusted to new decay constant). Western North America rotated southward during the Cenozoic, and the Florissant locality was ~2° north of its present location. The lake sediments represent an estimated 5000 years of deposition (McLeroy and Anderson, 1966). MacGinitie (1953) recognizes 114 species distributed among the princi-

Figure 6.7. Fossil plant locality at Florissant, Colorado. People from left to right: Richard A. Scott, L. Imogene Doher, Alan Graham, and Estella B. Leopold. Photograph by Chester A. Arnold.

Table 6.6. Megafossils from latest Eocene–Early Oligocene Florissant flora of Colorado.

Bryophyta	
Musci (mosses)	
Grimmiaceae	
Plagiopodopsis scudderi	
P. cockerelliae	
Pteridophytes	
Equisetaceae	Polypodiaceae
Equisetum florissantense	*Dryopteris guyottii*
Gymnosperms	
Gnetaceae	Pinaceae cont.
Ephedra miocenica	*P. hambachi*
Cupressaceae	*P. wheeleri*
Chamaecyparis linguaefolia	Taxaceae
Pinaceae	*Torreya geometrorum*
Abies longirostris	Taxodiaceae
Picea lahontense	*Sequoia affinis*
P. magna	
Pinus florissanti	
Monocotyledon Angiosperms	
Cyperaceae(?)	Liliaceae
Cyperacites lacustris	*Smilax labidurommae*
Gramineae	Typhaceae
Stipa florissanti	*Typha lesquereuxi*

continued

Table 6.6. continued

Dicotyledon Angiosperms

Aceraceae
- *Acer coloradense*
- *A. florissanti*
- *A. heterodentatum*
- *A. oregonianum*

Anacardiaceae
- *Astronium truncatum*
- *Cotinus fraterna*
- *Rhus lesquereuxi*
- *R. obscura*
- *R. stellariaefolia*
- *Schmaltzia vexans*

Araliaceae
- Araliaceous fruits
- *Oreopanax dissecta*

Aristolochiaceae
- *Aristolochia mortua*

Berberidaceae
- *Mahonia marginata*
- *M. obliqua*
- *M. subdenticulata*

Betulaceae
- *Paracarpinus fraterna*
- *Fagopsis longifolia*

Burseraceae
- *Bursera serrulata*

Caprifoliaceae
- *Sambucus newtoni*

Celastraceae
- *Celastrus typica*

Convolvulaceae
- *Convolvulites orichitus*

Euphorbiaceae
- *Euphorbia minuta*

Fagaceae
- *Castanea dolichophylla*
- *Quercus dumosoides*
- *Q. knowltoniana*
- *Q. lyratiformis*
- *Q. mohavensis*
- *Q. orbata*
- *Q. peritula*
- *Q. predayana*
- *Q. scottii*
- *Q. scudderi*

Juglandaceae
- *Carya libbeyi*

Lauraceae
- *Lindera coloradica*
- *Persea florissantia*
- *Sassafras hesperia*

Leguminosae
- *Caesalpinites acuminatus*
- *C. coloradicus*
- *Cercis parvifolia*
- *Conzattia coriacea*
- *Leguminosites lespedezoides*
- *Phaca wilmattae*
- *Phaseolites dedal*
- *Prosopis linearifolia*
- *Robinia lesquereuxi*
- *Vicia*

Meliaceae
- *Cedrela lancifolia*
- *Trichilia florissanti*

Moraceae
- *Morus symmetrica*

Myrtaceae
- *Eugenia arenaceaeformis*

Oleaceae
- *Osmanthus praemissa*

Platanaceae
- *Platanus florissanti*

Proteaceae
- *Lomatia lineata*

Rhamnaceae
- *Colubrina spireaefolia*
- *Zizyphus florissanti*
- *Zizyphus florissanti*

Rosaceae
- *Amelanchier scudderi*
- *Cercocarpus myricaefolius*
- *Crataegus copeana*
- *C. hendersoni*
- *C. nupta*
- *Malus florissantensis*
- *M. pseudocredneria*
- *Prunus gracilis*
- *Rosa hilliae*
- *Vauquelinia coloradensis*
- *V. liniara*

Rutaceae
- *Ptelea cassiodes* (see legend)

Salicaceae
- *Populus crassa*
- *Salix coloradica*
- *S. libbeyi*
- *S. ramaleyi*
- *S. taxifolioides*

Sapindaceae
- *Athyana haydenii*
- *Cardiospermum terminalis*
- *Dodonaea umbrina*
- *Koelreuteria alleni*
- *Sapindus coloradensis*
- *Thouinia straciata*

Saxifragaceae
- *Hydrangea fraxinifolia*
- *Philadelphus minutus*
- *Ribes errans*

Simarubaceae
- *Ailanthus americana*

Staphyleaceae
- *Staphylea acuminata*

Styracaceae
- *Halesia reticulata*

Thymelaeaceae
- *Daphne septentrionalis*

Tiliaceae
- *Tilia populifolia*

Ulmaceae
- *Celtis mccoshii*
- *Ulmus tenuinervis*
- *Zelkova drymeja*

Verbenaceae
- *Holmskioldia speirii* (=*Florissantia*, Malvales)
- *Petrea perplexans* (= *Asterocarpinus perplexans*, Betulaceae)

Vitaceae
- *Parthenocissus osbornii*
- *Vitis florissantella*

Adapted from MacGinitie (1953). *Ptelea cassiodes* (Rutaceae) = *Diplodipelta reniptera* (Manchester and Donoghue, 1995). For Aceraceae, see revision by Wolfe and Tanai (1987).

pal families Leguminosae (9 genera/10 species), Rosaceae (7/11), Sapindaceae (6/6), Anacardiaceae (4/6), Pinaceae (3/6), Ulmaceae (3/3), Rhamnaceae (3/3), Lauraceae (3/3), and Saxifragaceae (3/3). The most abundant specimens are identified as *Fagopsis longifolia* (Fig. 6.8; Betulaceae, now Fagaceae; Manchester and Crane, 1983; 30%), *Zelkova drymeja* (9.6%; =*Cedrelospermum Lineatum*; Manchester, 1989), *Chamaecyparis linguaefolia* (6.3%), *Typha lesquereuxi* (5.8%), *Populus crassa* (5%), *Rhus stellariaefolia* (2.8%), *Staphylea acuminata* (2.7%), *Athyana haydenii* (2.3%), *Sequoia affinis* (Fig. 6.9A; 2.2%), *Cercocarpus myricaefolius* (2.1%), *Quercus scottii* (1.5%), *Rhus obscura* (1.3%), *Pinus florissanti* (1.3%), *Sapindus coloradensis* (1.1%), *Carya libbeyi* (1.0%), and *Bursera serrulata* (1.0%). Others include *Ailanthus* (Fig. 6.9C), *Castanea* (Fig. 6.9D), *Eucommia* (Call and Dilcher, 1997), *Holmskioldia speirii* (Verbenaceae; now *Florissantia* of the Bombacaceae–Tiliaceae–Sterculiaceae complex; Manchester, 1992), *Salix* (Fig. 6.9E), and the moss *Plagiopodopsis* (Fig. 6.9B). The megafossil component of the Florissant flora has been up-

Figure 6.8. *Fagopsis longifolia*, Florissant flora. Reprinted from MacGinitie (1953) with the permission of the Carnegie Institution of Washington.

dated by Manchester (in press). The vegetation is interpreted as a rich mesic forest along stream and lake sides and a drier evergreen oak–pine woodland on higher ground, similar to the present dry savannalike community of western Texas to northeastern Mexico (oak–laurel forest of Asia or California woodland in western North America). One of the characteristic features of the Florissant flora is the small size of many leaves compared to their most similar modern species. The leaves are also mostly coriaceous and not entire margined. Annual rainfall was estimated at 508–635 mm (20–25 in.) with ample summer rain, a pronounced winter dry season, and a MAT of 18°C (65°F; i.e., mostly frost free). CLAMP yields a MAT of ~12.5°C (Wolfe, 1992b), and analysis of a new collection of 29 species gives 10.7 ± 1.5°C (K. M. Gregory, personal communication, 1995). The estimated precipitation trend between the Kisinger Lakes (~48 Ma), Green River (45 Ma), and Florissant (35 Ma) floras of Middle to Late Eocene age is shown in Fig. 6.10. An unresolved problem is the paleoelevation of the Florissant and other floras in western North America.

PALEOELEVATION ANALYSIS

Estimates of paleoelevation can be made on the basis of the modern analog method. However, new ways are being devised in an effort to more accurately determine and standardize the estimates (Axelrod, 1966a; Axelrod and Bailey, 1976; Gregory, 1996; Gregory and Chase, 1992; Gregory and McIntosh, 1996; Meyer, 1992; Wolfe, 1992b). Axelrod and Bailey (1976; Axelrod, 1966a) developed the following equation for estimating the paleoelevation of Late Eocene to Middle Oligocene floras of the Rio Grande Depression:

$$\text{elevation} = \frac{(\text{sea-level MAT - fossil flora MAT})}{\text{terrestrial lapse rate}} + \text{sea level,}$$

where sea-level MAT is the MAT of a sea-level fossil flora coeval with the inland fossil flora (°C), fossil flora MAT is the MAT of the inland fossil flora (°C), terrestrial lapse rate is the vertical cooling rate of the atmosphere above a land mass (°C/km), and sea level is the sea level at the time of deposition of the inland fossil flora relative to modern sea level (km). In this equation there is little accommodation for climatic change or compensation for latitude, MAT is based on modern analogs, and a lapse rate of 5.5°C/km is assumed valid for almost all situations. Their results suggest that paleoelevation in the Rio Grande Depression region increased by ~2000 m in the 6–8 Ma interval between the Eocene (38–40 Ma) and the Early Oliogcene (32 Ma).

Lapse rate is the change in temperature as a function of latitude or altitude. The altitudinal lapse rate has been calculated for various sites throughout the world and aver-

Figure 6.9. (A) *Sequoia affinis*, Florissant flora. (B) *Plagiopodopsis cockerelliae*, Florissant flora. (C) *Ailanthus americana*, Florissant flora. (D) *Castanea dolichophylla*, Florissant flora. (E) *Salix coloradica*, Florissant flora. Reprinted from MacGinitie (1953) with the permission of the Carnegie Institution of Washington.

aged into a worldwide mean of 5.0–6.0°C /km. However, the figure is higher for windward than leeward slopes and varies within a site over time, depending on the strength of incursions of frigid Arctic air that in turn are influenced by rising mountains and plateaus. Other factors include high base levels, fog, cold-air drainage, and local upwelling.

Meyer (1986) determined temperature differences between stations in the same vicinity as the fossil flora and applied a lapse rate of 5.9°C /km (more widely >4°C and <8°C). By this calculation Eocene elevations in the Rio Grande Depression region were already 3000–4700 m. Wolfe derived lapse rates

> [B]y calculating temperature differences between coastal and upland and (or) interior areas, which is the basis for the methodology proposed by Axelrod (1966[a]). I have, moreover, included all lower altitude, interior stations, most of which were excluded by Meyer (1986); such sites probably are present in topographic analogues of Tertiary depositional basins, which are basic to paleoaltitudinal estimates. (1992b)

A lapse rate of 2.0–3.0°C is suggested for western conterminous North America (Wolfe, 1992b). The CLAMP program can be applied to leaf physiognomic data to refine estimates of MAT for coastal and upland fossil floras. It should be noted, however, that the MATs for coastal fossil floras are estimates (CLAMP= ± 2°C), the MATs for upland floras are estimates, and the lapse rate is an approximation. Thus, a range of values is presently possible for the calculated paleoelevations.

The paleoelevation of the Florissant flora was estimated at ~900 m based on the modern analog method (MacGinitie, 1953). Using the newer method, and basing estimates

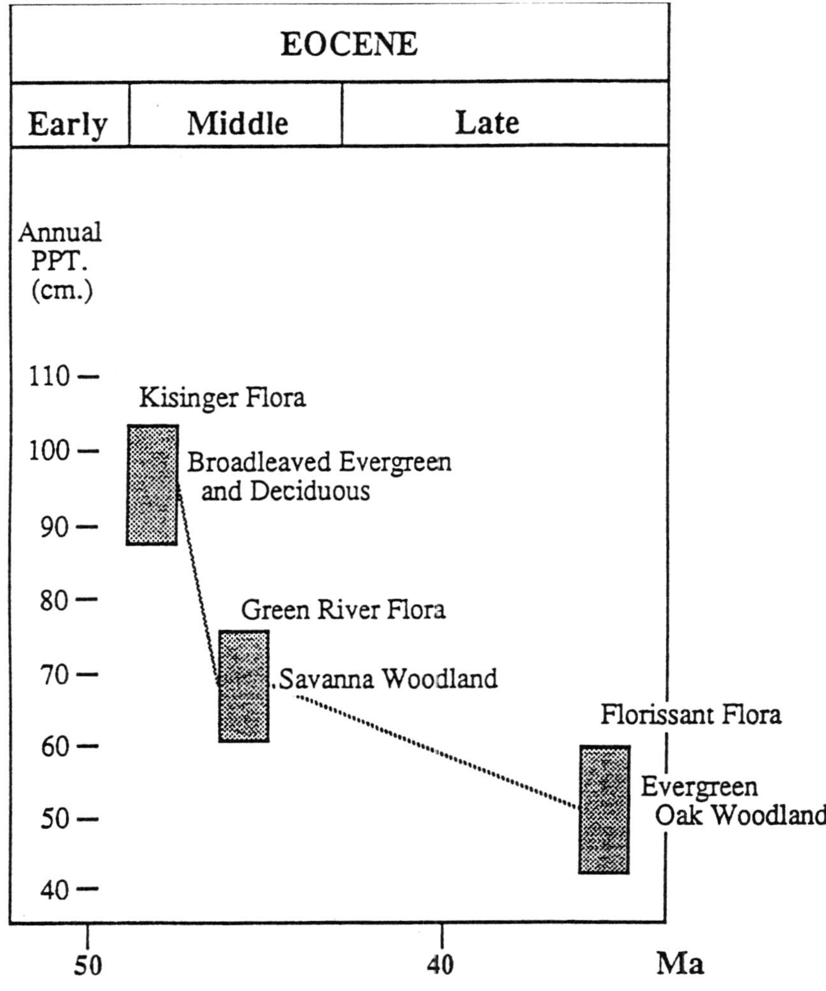

Figure 6.10. Trend in annual precipitation estimated from Middle and Late Eocene Rocky Mountains floras. Reprinted from Leopold et al. (1992) with the permission of Princeton University Press.

of MAT on foliar physiognomy, Wolfe (1992b) compares the MAT of the inland flora with that of sea-level flora(s) of the same age and applies a lapse rate of 3.0°C /km, considered most accurate when beginning with a high base level (Wolfe, 1994b). Meyer (1992) estimates paleoelevation by projecting the sea-level MAT inland to the MAT of the inland flora (e.g., Florissant) using the worldwide average lapse rate of ~5.9°C /km and incorporating other compensating factors. These various approaches give an estimated paleoelevation for the Florissant flora of 2300–3300 m (Gregory, 1994), 2450 m (Meyer, 1986, 1992), and 2700–2900 m (Wolfe, 1992b), or essentially modern elevations by the Eocene–Oligocene. (The present elevation is ~2500 m.)

In another study, Gregory and McIntosh (1996) used the Goshen flora (MAT = 19.5°C) as the sea-level assemblage in calculating the paleoelevation of the Pitch–Pinnacle flora. This flora is from the west flank of the Sawatch Range, Colorado, west of Florissant, and is dated at 32.9–29 Ma. Using foliar physiognomy as a basis, Early Oligocene paleoelevation is estimated at near-modern values. In the Basin and Range province paleoelevations were calculated for two Late Eocene floras in northeastern Nevada (Povey et al., 1994). Sea-level data was based on fossil floras in the Puget Group of Washington and MAT was calculated using CLAMP. The results also suggest paleoelevations similar to those of today.

Estimates of paleoelevation involving MATs and lapse rates (2–5.9°C/km) are presently imprecise. In addition to the approximations used in the calculations, another problem is that changes in either elevation or climate can produce similar signals from fossil floras; in much of Cenozoic North America both were occurring simultaneously. A possible way of avoiding errors introduced by uncertainties in MAT and lapse rates is to use the paleobotanical data to infer enthalpy (a measure of the energy content of a system per unit mass). This approach yields a paleoelevation for the Florissant flora of 2.9 ± 0.7 km (Forest et al., 1995), and paleoelevations in the Rio Grande Rift area of New Mexico are estimated at ~3000–4700 m. Gregory-Wodzicki (1997) estimates the paleoelevation of the Late Eocene House Ranch flora (31.4 ± 0.5 Ma) of western Utah at 3.6 ± 0.7 km (MAT-based) or 2.9 ± 1.5 km (enthalpy-based; presently 1.7 km). Analyses of 12 Middle Miocene floras from western Nevada indicate a paleoelevation of ~3 kms above sea level at 15–16 Ma. This is 1–1.5 km higher than at present, and the difference is attributed to collapse by ~13 Ma (Wolfe et al., 1997). By all these data, sections of the Rocky Mountains in central Colorado, southwestern Montana, New Mexico, and in the Basin and Range Province locally had reached at least modern elevations by the latest Eocene and would have been adequate to cast rainshadows over parts of the eastern slopes and High Plains and to influence atmospheric circulation. These data imply a reduced role for uplift in Late Cenozoic cooling (Chapter 2, Orogeny).

THE FLORA OF THE BEAVERHEAD BASINS in southwestern Montana was assigned to the Oligo–Miocene (Becker, 1969), but it is now placed in the latest Eocene to earliest Oligocene and is approximately coeval with the Florissant flora (Wing, 1987; S. L. Wing, personal communication, 1994). The change is based on Chadronian or older vertebrate faunas from the Renova Formation that includes the Beaverhead Basins flora. It comprises 110 genera and 160 species as described by Becker (1969). Prominent genera include *Quercus* (nine species); *Mahonia* (eight); *Acer* (six); *Salix* (four); and three species each of *Abies*, *Picea*, *Betula*, *Cassia*, *Cercocarpus*, *Fraxinus*, and *Zelkova*. Taxonomic revisions for *Ulmus* and *Zelkova* in Middle to Late Tertiary floras of western North America have been presented by Tanai and Wolfe (1977). The communities include aquatic and lakeshore vegetation (*Taxodium*, *Typha*), flood-plain and humid forest (*Chamaecyparis*, *Sequoia*, *Thuja*, *Cercidiphyllum*, *Salix*), deciduous and mixed deciduous hardwood–coniferous forest on the slopes (*Acer*, *Corylus*, *Ulmus*), a high montane forest (*Abies*, *Picea*), and a subhumid chaparral formation on drier exposed slopes (*Juniperus*, *Arctostaphylos*, *Berberis*, *Cercocarpus*, *Mahonia*, *Potentilla*). Plants presently of eastern North America, western North America, and Asia (*Ginkgo*, *Glyptostrobus*, *Metasequoia*, *Ailanthus*, *Eucommia*) are also represented. The paleoelevation of the lake basin is estimated at 500–700 m (1500–2000 ft), and highlands were up to 2000 m (6000 ft) in the vicinity. The presence of *Sequoia* and *Cedrela* suggests minimum temperatures were at to just above freezing, and annual precipitation is placed at 1000–1300 mm (40–50 in.).

One specimen from the Beaverhead Basins flora, *Viguiera cronquistii* (Compositae–Asteraceae), warrants special comment. This is an important identification because it would be the oldest megafossil record of the family in North America (the oldest verified pollen is Oligo–Miocene; Graham, 1996), and it has been used as evidence for positioning the Heliantheae as the primitive tribe in the Compositae. Also, modern species of *Viguiera* are mostly polyploids, a relatively advanced cytological feature, which would favor arguments that the family had a much earlier origin. A reexamination of the specimen revealed that although it is composite-like in aspect (only bracts of the supposed capitulum are preserved), the fossil "cannot be considered unequivocally to be the remains of a composite" (Crepet and Stuessy, 1978). Its affinities are presently unknown.

The Metzel Ranch flora in the Upper Ruby River Basin of southwestern Montana, just to the east of the Beaverhead Basins, was also considered Oligo–Miocene in age (Becker, 1972); but recent estimates place it, along with the Mormon Creek (Becker, 1960) and York Ranch (Becker, 1973) floras of Montana, in the Early Oligocene Orellan NALMA (32.2–30.8 Ma, Wing, 1987; Fig. 3.13). It is distinct in composition and paleoecological import from the Beaverhead Basins and Ruby floras. It differs primarily in the poorer representation of gymnosperms and in the absence of oaks. The prominence of Rosaceae (10.9%), Leguminosae (6.53%), and Rhamnaceae (6.53%), along with the presence of *Juniperus*, Gramineae, *Crataegus*, and *Mahonia*, suggest dryness and the absence of *Abies* and *Picea* indicates low elevations in the immediate vicinity. The Metzel Ranch flora is an example of the local drier conditions and the move toward better definition and expansion of a shrubland/chaparral–woodland–savanna to the west of the Rocky Mountains as the Pacific sierras just began to affect moisture regimes inland. The trend was intensified as more pronounced highs settled over the region.

Another Late Eocene to Early Oligocene flora in southwestern Montana is preserved in intermontane lacustrine shales of the Ruby Basin. The Ruby flora is slightly younger (Whitneyan NALMA; 30.8–29.2 Ma; Wing, 1987) and somewhat drier than the Mormon Creek, Metzel Ranch, and York Ranch floras. Becker (1961) recognized 61 genera and 82 species. Forty percent of the species also occur in the Florissant flora. The two floras reflect a similar vegetation, but the Ruby flora is slightly more mesic and upland in aspect. Genera present in the Ruby vegetation but absent from the Florissant include *Glyptostrobus*, *Metasequoia*, *Pseudotsuga*, *Alnus*, *Betula*, *Cornus*, *Fagus*, *Fraxinus*, *Myrica*, and *Nyssa*. Tropical warm-temperate and drier elements present in the Florissant and absent from the Ruby assemblage include *Ephedra*, *Bursera*, *Cedrela*, *Lindera*, *Persea*, *Prosopis*, and *Trichilia*. All of these Late Eocene to Early Oligocene floras from southwestern Montana preserve a mixed coniferous and broad-leaved deciduous forest and shrubland that grew under temperate to dry-temperate conditions and show no clear climatic trend during the interval (Wing, 1987; Wolfe, 1992a). The CLAMP estimate of MAT is 11–12°C with an annual range of 16°C.

The Late Eocene (~40 Ma) Copper Basin flora in northeastern Nevada reveals that the higher elevations and more moist conditions west of the Rockies reached by the Middle Eocene (Wind River, Green River vs. Thunder Mountain, Germer floras) persisted through the Late Eocene and later time. The Copper Basin flora (Axelrod, 1966b), in contrast to the Florissant, is a microthermal mixed deciduous hardwood–coniferous assemblage. The 42 species represent three communities adjacent to a small lake. The bordering vegetation consisted of *Acer*, *Alnus*, *Amelanchier*, *Crataegus*, *Lithocarpus*, *Mahonia*, *Prunus*, and *Salix*. The slope forest included *Cephalotaxus*, *Chamaecyparis*, *Pseudotsuga*, *Sequoia*, *Acer*, *Aesculus*, *Euonymus*, *Prunus*, *Rho-*

dodendron, *Sassafras*, and *Ulmus*. The montane conifers were *Abies*, *Picea*, and *Pinus*, associated with *Acer* and *Ribes*, and a subalpine community of *Picea*, *Pinus*, *Larix*, and *Tsuga*. Annual precipitation is estimated at 1270–1524 mm (50–60 in.), MAT at 11°C (51°F; cool temperate), and the elevation ~1200 m (3600 ft; modern elevation 7100 ft). The study is of interest because at the time, mixed deciduous hardwood–coniferous forests were unreported from the Eocene of the western United States. The Copper Basin and Republic floras (see later discussion) show that the community was developing by the Middle Eocene (viz., *Abies*, *Picea*). The Bull Run flora (Axelrod, 1966b,c) lies just to the west of Copper Basin. It is probably somewhat younger in age, but it also preserves a near pure western montane coniferous forest of *Abies*, *Picea*, *Pinus*, *Pseudotsuga*, *Tsuga*, *Chamaecyparis*, and *Thuja* at the highest elevations (1250–1500 m) and a few moist deciduous angiosperms with small leaf indices further downslope.

Until recently, the Middle Eocene vegetation throughout southern California had to be generalized by extrapolation from floras well outside the area. Tropical rain forest was present along the coast to the north (Wolfe, 1985, 1989), and it could have extended southward. However, modeling results by Parrish et al. (1982) simulate precipitation in the Lutetian (50–42.1 Ma) as somewhat lower in southern California and northern Mexico than in areas to the north. This is consistent with evidence from sedimentology, vertebrate paleontology, and a fossil pollen record from the San Diego area (Elsinore flora) that shows some winter drying, and possibly a cooling trend, beginning in the Middle and Late Eocene (Frederiksen, 1991). Black fusinite and dark brown semifusinite from the uplands are present in the samples, indicating forest fires (Cope, 1981). The climatic deterioration between the Early Lutetian and the Priabonian shows a similar stepwise pattern as recorded in the southeastern United States. The limited evidence suggests a winter-dry mixed deciduous hardwood–coniferous forest in the uplands and the beginning of a complex trend leading to the replacement of the Paleocene to earliest Lutetian tropical to paratropical lowland vegetation by more seasonally dry communities. The dry tropical forest proposed by Axelrod (1979) for the early Lutetian, based on some questionable pollen identifications (Ting, 1975; W. S. Ting, unpublished data; see also Frederiksen, 1991), probably had not yet developed. Analysis of paleosols in San Diego County and northwestern Mexico reveal an annual rainfall of 1250–1900 mm (Abbott, 1981; Peterson and Abbott, 1979), and a megafossil assemblage from the Lower Lutetian Torrey Sandstone of San Diego is interpreted as growing under a paratropical climate with an annual rainfall of 1200–1500 mm (Myers, 1990).

This drying trend is evident in the early Lutetian by a turnover peak (first pollen event) wherein the last appearances of *Milfordia* (Restionaceae), *Diporoconia* (Palmae?), *Lanagiopollis* (*Alangium*), and *Triatriopollenites* (probably *Myrica*) correspond to the first records of *Ruellia* (?; Acanthaceae), *Bursera* (Burseraceae), *Pachysandra–Sarcococca* (Buxaceae), *Lonicera* type (Caprifoliaceae), Onagraceae, other Restionaceae, *Chiranthodendron– Fremontodendron–Triumfetta* (Bombacaceae–Sterculiaceae–Tiliaceae), Cyrillaceae(?), and Juglandaceae. The drying may have intensified during the Middle and Late Lutetian as shown by caliche and salt crystal formation, which suggest an estimated annual rainfall of ~500–750 mm in the lowlands (500–1000 mm based on faunal evidence; Axelrod, 1979; Novacek and Lillegraven, 1979). The vegetation during the Middle Lutetian has been interpreted as a form of savanna, but with understory shrubs and herbs rather than grasses and more mesic trees growing as gallery forests. The plant microfossil evidence confirms the unusual nature of the "savanna" here, as in the Green River Formation, in that grasses were not prominent in the understory as in modern savannas. Pollen evidence further reveals a greater diversity of vegetation with more trees [gallery forests; Palmae, Bombacaceae–Sterculiaceae–Tiliaceae, Eucommiaceae, Fagaceae(?), Juglandaceae, Myrtaceae(?), Symplocaceae(?), Ulmaceae] than suggested by the sedimentological and faunal evidence from the interplains, which emphasizes greater aridity. Coarse detritus from the highlands indicates drier conditions there with a longer and more pronounced dry season.

By the Bartonian the drying trend, and perhaps cooling temperatures, was having a more pronounced effect on the vegetation of southern California as evidenced by the MEDD (second pollen event), which is also present in the southeastern United States. The last appearances during this interval include Acanthaceae, Aquifoliaceae, Bombacaceae, Euphorbiaceae, Palmae(?), Symplocaceae(?), Tiliaceae, and Ulmaceae(?). In contrast to the southeast, however, the MEDD was not followed by abundant introductions (e.g., there is no evidence of grasses in the Eocene of southern California), so the net decrease in pollen diversity is greater.

In the Lower Priabonian a third pollen event is expressed by a rapid increase in Quercoideae forms, as in the southeastern United States, which is paralleled by the beginning of redbed deposition in the Sespe Formation of Ventura County, California (Dickinson and Leventhal, 1988). The fourth event is a continued decline in diversity and corresponds to the Late Eocene–Early Oligocene cooling. The culmination of these trends produced a winter dry paratropical forest or a type of savanna with gallery forests along streams and in the lowlands and a more distinctly seasonally dry mixed coniferous forest in the uplands. This marks an early stage in the development of the Madrean (scrubland/chaparral–woodland–savanna) vegetation of the present arid southwestern United States and northwestern Mexico. The rainfall pattern was summer wet, so that a major change after the Eocene was the development of the summer dry Mediterranean climate that has characterized the region since Plio–Pleistocene times.

The late Middle Eocene Susanville flora of northern Cal-

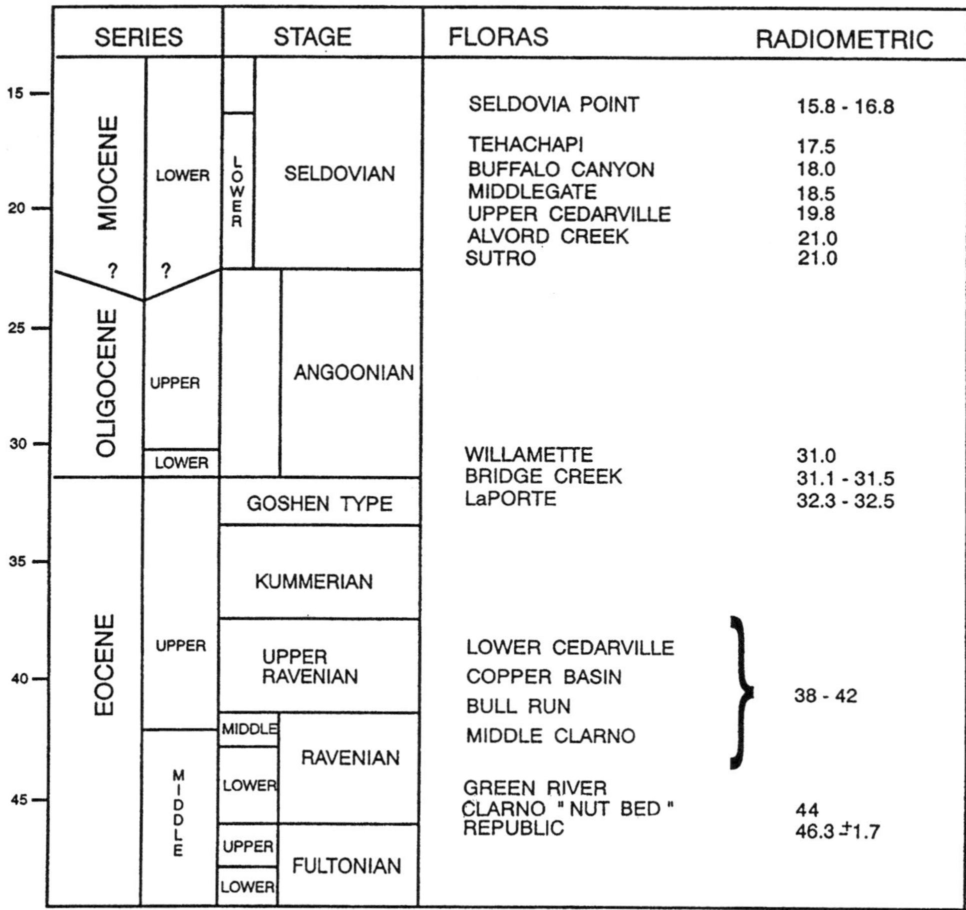

Figure 6.11. Floral stage names proposed for western North America (Wolfe, 1981a) and some radiometric dates cited for the floras. See text for revised radiometric dates.

ifornia (40° N latitude) includes notophyllous to megaphyllous broad-leaved evergreens (some with finely serrate margins) that are interpreted as a coastal tropical rain forest growing under a MAT of ~27°C (Wolfe, 1978, 1985). The various Platanaceae remains (*Macginitiea*, *Macginicarpa*, *Platananthus*) described from the Late Paleocene Joffre Bridge flora also occur in several floras from California (Chalk Bluffs, late Early Eocene), Oregon (Clarno Formation, West Branch Creek locality, Middle to Late Eocene), and Washington (Steels Crossing, Middle Eocene; Manchester, 1986).

The latest Eocene–Early Oligocene La Porte flora (Potbury, 1935; 32–33.2 Ma), south of Susanville in northwestern California, is similar in aspect, but it has not been revised since the original publication. The MAT was originally estimated at ~24°C, and the CLAMP MAT estimate is ~22.3°C (Wolfe, 1993). These values are near those of the Comstock and Goshen floras 8° latitude to the north, suggesting deposition at a slightly higher elevation (Wolfe, 1992b). The Lower Cedarville flora of far northeastern California is older but also Late Eocene in age. Wolfe (1981a) has proposed a series of botanically defined floral stages based on these Pacific Northwest communities (but see Axelrod, 1987). The stages for the period ~50–16.3 Ma are shown in Fig. 6.11, along with some floras discussed in the text and their original radiometric dates (see text for revised dates).

The Middle–Late Eocene Clarno flora is located in the John Day Basin of north-central Oregon. One locality dated at ~44 Ma consists of numerous fruits and seeds with some leaf and wood material and is known as the "nut beds" by local collectors. These stream and lake delta deposits have been studied most extensively among the Clarno localities (Manchester, 1990, 1994; Scott, 1954) and comprise 145 genera and 173 species (Table 6.7). Most are trees and lianas. Among the 69 species for which good estimates of growth form could be made, 43% are lianas from such families as the Icacinaceae, Menispermaceae, and Vitaceae. About 52% of the dicot leaves are entire margined, and broad-leaved evergreens are abundant. The flora is tropical to paratropical in composition with only a few temperate genera (e.g., *Ginkgo*, *Pinus*, *Taxus*, *Emmenopterys*, *Parthenocissus*). Twenty-four percent of the genera are found in Early to Middle Eocene floras of Europe, consistent with geological and other biological evidence for land connections across the North Atlantic and

Table 6.7. Composition of Middle Eocene Nut Beds flora, Clarno Formation, Oregon.

Gymnosperms

Pinaceae
Pinus sp.

Taxaceae
Diploporus torreyoides
Taxus masonii
Torreya clarnensis

Monocotyledon Angiosperms

Musaceae
Ensete oregonense

Palmae
Sabal bracknellensis
S. jenkinsii

Dicotyledon Angiosperms

Actinidiaceae
Actinidia oregonensis
Alangiaceae
Alangium eydei
A. rotundicarpum
Anacardiaceae
Pentoperculum minimus
Rhus rooseae
Annonaceae
Anonaspermum cf. *pulchrum*
A. bonesii
A. rotundum
Araliaceae
Paleopanax oregonensis
Betulaceae
Coryloides hancockii
Kardiasperma parvum
Burseraceae
Bursericarpum oregonense
B. sp.
Cornaceae
Cornus clarnensis
Langtonia bisculcata
Mastixia sp.
Mastixioidiocarpum oregonense
Mastixicarpum occidentale
Nyssa scottii
N. spatulata
N. sp.
Fagaceae
Castanopsis crepetii
Quercus paleocarpa
Flacourtiaceae
Saxifragispermum tetragonalis
Hamamelidaceae
Fortunearites endressii
Hydrangeaceae
Hydrangea knowltonii
Icacinaceae
Iodeae
Comicilabium atkinsii
Iodes chandlerae
I. multireticulata
Iodicarpa ampla
I. lenticularis
Phytocreneae
Palaeophytocrene hancockii
P. pseudopersica
Pyrenacantha occidentalis

Juglandaceae
Juglandeae
Cruciptera simsonii
Juglans clarnensis
Engelhardieae
cf. *Palaeocarya clarnensis*
Platycaryeae
Paleoplatycarya? hickeyi
Lauraceae
Laurocalyx wheelerae
Laurocarpum hancockii
L. nutbedensis
L. raisinoides
Lindera clarensis
Leguminosae
Leguminocarpon sp.
Lythraceae
Decodon sp.
Magnoliaceae
Magnolia muldoonae
M. paroblonga
M. tiffneyi
Menispermaceae
Coscineae
Anamirta leiocarpa
Menispermeae
Diploclisia auriformis
Eohypserpa scottii
Davisicarpum limacioides
Palaeosinomenium venablesii
Tinosporeae
Atriaecarpum clarnense
Calycocarpum crassicrustae
Chandlera lacunosa
Curvitinospora formanii
Odontocaryoidea nodulosa
Thanikaimonia geniculata
Tinomiscoidea occidentalis
Tinospora elongata
T. hardmanae
Platanaceae
Macginicarpa glabra
Platanus hirticarpa
Tanyoplatanus cranei
Rosaceae
Prunus weinsteinii
P. olsonii
Rubiaceae
Emmenopterys dilcheri

Sabiaceae
Meliosma beusekomii
M. bonesii
M. elongicarpa
M. cf. *jenkinsii*
M. leptocarpa
Sabia americana
Sapindaceae
Deviacer wolfei
Palaeoallophylus globosa
P. gordonii
Sapotaceae
Bumelia? globosa
B.? subangularis
Schisandraceae
Schisandra oregonensis
Staphyleaceae
Tapiscia occidentalis
Symplocaceae
Symplocos nooteboomii
Theaceae
Cleyera grotei
Ulmaceae
Celtidoideae
Amphananthe maii
Celtis burnhamae
C. sp.
Trema nucilecta
Ulmoideae
Cedrelospermum lineatum
Vitaceae
Ampelocissus auriforma
A. scotti
Ampelopsis rooseae
Parthenocissus angustisulcata
P. clarnensis
Vitis magnisperma
V. tiffneyi

Adapted from Manchester (1994); Incertae Sedae are not included. For *Mastixia–Mastixioidiocarpum–Mastixicarpum* (Cornaceae), see Tiffney and Haggard (1996).

the North Pacific during the Early Tertiary. The flora is primarily Asian in affinity, however, and includes *Actinidia, Alangium, Ampelocissus, Celtis*, Cornaceae (*Cornus, Langtonia, Mastixia, Nyssa*), *Decodon, Ensete* (Manchester and Kress, 1993; the only herbaceous angiosperm), *Eucommia Florissantia* (malvalean affinity; Manchester, 1992), *Hydrangea, Iodes*, Juglandaceae (*Cruciptera*; Manchester, 1991), *Magnolia, Meliosma, Musa*, Platanaceae, *Quercus, Sabia, Schisandra*, and Tiliaceae (*Pteleaecarpum*; Bůžek et al., 1989; Kvaček et al., 1991). The latter fossil occurs at only one probable Clarno locality and is more characteristic of the Bridge Creek flora. It is similar to the extant Chinese genus *Craigia* of the Tiliaceae. The presence of *Quercus* is noteworthy because this record, together with that from the Middle Eocene Chalks Bluff flora (MacGinitie, 1941; Wolfe, 1973), are among the earliest occurrences of the genus. The southeast Asian genus *Eneste* (Musaceae) presently grows where the MAT is between 20 and 28°C and the annual range is between 2 and 19°C (viz., frost free). The CLAMP MAT estimate is 16°C. Among the vertebrate fossils is one crocodilian. Because of the proximity of the flora to the paleocoast of Oregon ~100 km to the west and the absence of the Cascade Mountains and Coast Ranges, moisture was greater and more uniformly distributed than it was to the east of the Rocky Mountains where the vegetation was tropical but seasonally dry (Leopold and MacGinitie, 1972; Wing, 1987; each side of the hatched area in Fig. 5.8). Some seasonality in rainfall may be inferred from the woods (e.g., *Vitaceoxylon tiffneyi, V. carlquistii*; Vitaceae) that contain growth rings (Wheeler and LaPasha, 1994). However, in *Clarnoxylon blanchardii* (Juglandaceae; Manchester and Wheeler, 1993), although growth rings are present, the vessel elements lack helical thickenings and are thus more characteristic of tropical than temperate species (Van der Graaff and Baas, 1974).

The Late Eocene Goshen (~31 [39] Ma; Chaney and Sanborn, 1933) and Comstock (Sanborn, 1935) floras of west-central Oregon have not been revised since the original publications, but they generally represent moist forests similar to the modern upland vegetation of southeastern Mexico (*Allophylus, Annona, Astronium, Cupania, Drimys, Ficus, Hydrangea, Ilex, Inga, Nectandra, Ocotea, Platanus, Quercus, Sapium, Siparuna, Smilax*). The specimen described as *Viburnum palmatum* (Chaney and Sanborn, 1933) is now *Florissantia ashwillii* (Manchester, 1992). The CLAMP MAT estimate is 20°C (Wolfe, 1992a).

The early Middle Eocene Republic flora of northeastern Washington (Klondike Mountain Formation) was deposited at a paleoelevation of 700–900 m and is dated at 48–49 Ma (Wolfe and Wehr, 1987; see also Wehr, 1995; and a series of papers in *Washington Geology*). The depositional basin at Republic was associated with a raised structure called the Eocene Interior Arc (Myers, 1996). The flora further emphasizes differences between the landscape and vegetation of the eastern Rocky Mountains–Great Plains area and regions to the west (Fig. 5.8). In the east, a microthermal element was present but not dominant in the primarily tropical to subtropical seasonally dry climate and generally low elevations. To the west, microthermal elements are present and diverse due to the higher paleoelevation (Axelrod, 1966a). The vegetation was a microthermal mixed deciduous hardwood–coniferous forest (Table 6.8). The MAT is estimated at 12–13°C, and the mean annual range of temperature (MART) at 5–6°C.

The Republic flora is the oldest known assemblage with co-occurring members of a western montane coniferous forest of Pinaceae (*Abies, Picea, Pinus, Pseudolarix, Tsuga*) and Cupressaceae (*Chamaecyparis, Thuja*). The oldest known *Betula* related to *B. papyrifera–B. occidentalis* and the oldest *Tilia* are from the Middle Eocene Republic flora. Also present is "*Holmskioldia*" *speirii* (Verbenaceae), now considered as *Florissantia quilchensia* (Manchester, 1992). According to Wing (1987), the early diversification of many modern microthermal lineages likely took place during the Eocene in the volcanic highlands of the northwestern United States (see also Wolfe, 1986, 1987). The western Montane Coniferous Forest association further developed and continued to expand in Late Eocene and Neogene times. Although the number of floras is few, it is probable that the northern Rocky Mountains were occupied primarily by a coniferous forest for most of the Oligocene and later times, replacing the older megathermal evergreen vegetation. The Middle Eocene Republic flora records the beginnings of the transition.

An extensive flora of Middle Eocene age (~45–50 Ma) is preserved in the Allenby Formation near Princeton, southern British Columbia, Canada (~49° N). Results of the study are being published separately for each taxon by Ruth Stockey and coworkers and when completed a detailed picture will be available of the Middle Eocene vegetation of southwestern Canada (Pigg and Stockey, 1996). The members identified to date are listed in Table 6.9. Provisional identifications pending further study include *Abies, Acer, Aesculus, Castanea, Cercidiphyllum, Fagus*, and Ulmaceae (Crane and Stockey, 1987). [In addition to references listed in Table 6.9, see also Cevallos-Ferriz et al. (1991), Erwin (1987), and Erwin and Stockey (1991b).] Other fossils are of unknown family affinity (e.g., *Princetonia allenbyensis*, Magnoliopsida, Fig. 6.12; *Ethela sargantiana*, monocotyledon, Erwin and Stockey, 1992; *Eorhiza arnoldii*). The flora is presently interpreted as a shallow, shoreline, lacustrine ecosystem (marsh, swamp, lake margin) growing under moist, subtropical to warm-temperate conditions and it seems to represent a mixed deciduous hardwood (with deciduous coniferous elements)–evergreen coniferous forest that extended southward through Washington (Republic flora) and northeastern Nevada (Copper Basin flora). In the Middle Eocene generally across the northern region other plants include *Glyptostrobus, Larix, Picea, Taxodium, Alnus, Betula, Nyssa, Platanus, Populus, Pterocarya*, and *Quercus* (Axelrod, 1984).

In the late Middle Eocene, megathermal tropical and

Table 6.8. Composition of early Middle Eocene Republic flora, northeastern Washington.

Gymnosperms

- Cephalotaxaceae
  - *Cephalotaxus* type
- Cupressaceae
  - *Chamaecyparis*
  - *Thuja*
- Ginkgoaceae
  - *Ginkgo biloba*
- Pinaceae
  - *Abies milleri*
  - *Picea*
  - *Pinus*
  - *Pseudolarix americana*
  - *Tsuga*
- Taxaceae
  - *Amentotaxus* type
  - *Taxus* type
  - *Torreya*
- Taxodiaceae
  - *Cryptomeria* type
  - *Glyptostrobus* type
  - *Metasequoia occidentalis*
  - *Sequoia*

Monocotyledon Angiosperms

- Musaceae
  - *Ensete*

Dicotyledon Angiosperms

- Aceraceae
  - "*Acer*" *arcticum* (*Deviacer*, Sapindaceae)
  - *Acer* sp.
- Anacardiaceae
  - *Rhus malloryi*
- Betulaceae
  - *Alnus parvifolia*
  - *Betula leopoldae*
  - *Corylus*
  - *Paleocarpinus*
  - aff. *Corylus*
  - aff. *Carpinus*
- Bignoniaceae
- Burseraceae
  - *Barghoornia oblongifolia*
- Cercidiphyllaceae
  - *Cercidiphyllum obtritum*
- Cornaceae
  - *Cornus* sp.
- Davidiaceae
  - *Tsukada davidiifolia*
- Eucommiaceae
  - *Eucommia montana*
- Fagaceae
  - *Castaneophyllum*
  - *Fagopsis undulata*
- Grossulariaceae
  - *Ribes*
- Hamamelidaceae
  - *Corylopsis*
  - *Langeria magnifica*
  - *Liquidambar*
- Hydrangeaceae
  - *Hydrangea*
- Icacinaceae
  - *Palaeophytocrene*
- Iteaceae
  - *Itea* sp.
- Juglandaceae
  - *Carya*? sp.
  - *Carya*/*Juglans*
  - *Cruciptera*
  - *Pterocarya* sp.
- Lauraceae
  - *Phoebe* sp.
  - *Sassafras hesperia*
- Leguminosae
- Menispermaceae
  - *Calycocarpum*
- Myricaceae
  - *Comptonia columbiana*
- Myrtaceae
  - *Paleomyrtinaea*
- Nymphaeaceae
  - *Nuphar*
- Olacaceae
  - *Schoepfia republicensis*
- Platanaceae
  - *Macginicarpa*
  - *Macginitiea gracilis*
- Rosaceae
  - *Crataegus*
  - *Photinia pageae*
  - *Potentilla pageae*
  - *Prunus*
  - *Spiraea*
- Rubiaceae
  - cf. *Emmenopterys*
- Sabiaceae
  - *Meliosma*
  - *Sabia*
- Salicaceae
  - *Populus*
  - *Salix*
- Sapindaceae
  - *Bohlenia americana*
  - *Deviacer* ("*Acer*" *arcticum*)
  - *Koelreuteria arnoldi*
- Smilacaceae
  - *Smilax*
- Sterculiaceae
  - *Florissantia quilchenensis*
- Theaceae
  - *Ternstroemites*
- Tiliaceae
  - *Craigia*
  - *Tilia johnsoni*
- Trochodendroid group
  - *Cercidiphyllum*
  - *Joffrea*
  - *Nordenskioldia*
  - *Trochodendron*
- Ulmaceae
  - *Cedrelospermum*
  - *Ulmus*
- Verbenaceae?
  - "*Holmskioldia*" *speirii*
- Vitaceae
  - *Vitis*
- Incertae sedis
  - *Calycites ardtunesis*
  - *Republica hickeyi*
  - *Pteronepelys wehri*

Adapted from Gandolfo (1996), Schorn and Wehr (1996), Wehr and Hopkins (1994), Wehr and Manchester (1996), and Wolfe and Wehr (1987). *Cruciptera* (Juglandaceae) is added by Manchester (1991); *Eucommia montana* is added by Call and Dilcher (1997).

Table 6.9. Composition of Middle Eocene Princeton Chert flora (Allenby Formation) near Princeton, southern British Columbia, Canada.

Taxon	Reference
Pteridophytes	
Blechnaceae	
Blechnoid fern	
Dryopteridaceae	
Diplazium	Rothwell et al. (1994)
Onocleoid fern	
Osmundaceae	
cf. *Osmunda*	
Polypodiaceae	
Dennstaedtiopsis aerenchemata	Basinger and Rothwell (1977)
Gymnosperms	
Pinaceae	
Pinus andersonii	Stockey (1984)
P. arnoldii	Miller (1973); Stockey (1984)
P. princetonensis	Stockey (1984)
P. similkameenensis	Miller (1973), Stockey (1984)
P. sp.	Phipps et al. (1995)
Taxodiaceae	
Metasequoia milleri	Basinger (1981, 1984), Basinger and Rothwell (1977), Rothwell and Basinger (1979)
Monocotyledon Angiosperms	
Alismataceae	
Heleophyton helobiaeoides	Erwin and Stockey (1989)
Araceae	
Keratosperma allenbyensis	Cevallos-Ferriz and Stockey (1988a)
Arecaceae (Palmae)	
Uhlia allenbyensis	Erwin and Stockey (1991a, 1994)
Juncaceae–Cyperaceae	
Ethela sargantiana	Erwin and Stockey (1992)
Liliaceae	
Soleredera rhizomorpha	Erwin and Stockey (1991a)
Dicotyledon Angiosperms	
Cornaceae	
Mastixicarpum	Wehr (1995)
Grossulariaceae	
Ribes	Cevallos-Ferriz (1995)
Lauraceae	Sun and Stockey (1991)
Lythraceae	
Decodon allenbyensis	Cevallos-Ferriz and Stockey (1988b)
cf. *Lythrum*	Cevallos-Ferriz and Stockey (1988b)
Myrtaceae	
Paleomyrtinaea princetonensis	Pigg et al. (1993)
Magnoliaceae	
Liriodendroxylon princetonensis	Cevallos-Ferriz and Stockey (1990b)
Nymphaeaceae	
Allenbya collinsonae	Cevallos-Ferriz and Stockey (1989)
Rosaceae	
Paleorosa similkameenensis	Basinger, 1976, Cevallos-Ferriz et al. (1993)
Prunus allenbyensis	Cevallos-Ferriz and Stockey (1990c)
Prunus (types 1, 2, 3)	Cevallos-Ferriz and Stockey (1991)
Sapindaceae	
Wehrwolfea striata	Erwin and Stockey (1990)
Vitaceae	
Ampelocissus similkameenensis	Cevallos-Ferriz and Stockey (1990a)
Unknown Dicotyledonae	
Eorhiza arnoldii	Robison and Person (1973), Stockey and Pigg (1994)
Princetonia allenbyensis	Stockey (1987), Stockey and Pigg (1991)

Adapted from Pigg and Stockey (1996) and Cevallos-Ferriz et al. (1991). Taxa without references are R. A. Stockey, personal communication, 1996).

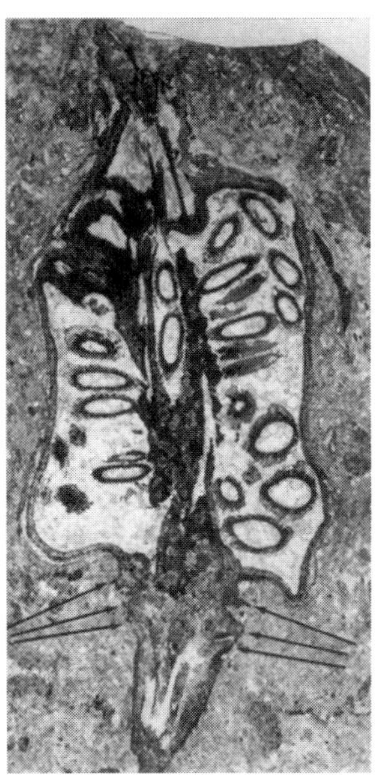

Figure 6.12. Longitudinal section through the fruit *Princetonia allenbyensis*, Princeton flora, British Columbia. Reprinted from Stockey and Pigg (1991) with the permission of Elsevier Science–NL.

paratropical forests were reduced in extent as temperatures (especially minimum winter temperatures) continued to decline. The Paleogene floras from the main part of Alaska (Wolfe, 1977, 1992a) record a change from a late Early Eocene Asian paratropical vegetation to a Middle Eocene cooler broad-leaved evergreen subtropical community (*Engelhardia* type, *Cedrela–Melia*, *Ilex*; CLAMP MAT estimate ~17.2–16.2°C, originally reported as ~18°C) to an Oligocene temperate forest (*Metasequoia*, *Alnus*; 4.5–7°C), for a decline of 8–10.5°C. The Eocene coast was fringed with evergreen Lauraceae, and the MAT is estimated at ~16°C. The Late Eocene Rex Creek flora of Alaska includes *Metasequoia*, *Acer*, *Alnus*, *Betula*, *Carya*, *Castanea*, *Cedrela*, *Fagus*, *Juglans*, *Liquidambar*, *Pachysandra*, *Platanus*, *Prunus*, *Pterocarya*, *Quercus*, and *Ulmus*. The CLAMP estimate is 12.3°C. These values are broadly consistent with a MAT of >13°C for the notophyllous broad-leaved evergreen forest (oak–laurel type) from the Late Eocene Puget Group flora of northeastern Washington (Burnham, 1994). There entire leaf margin species average 50%, and the average leaf-size index is 57. A summary of high-latitude floras and the inferred climates is given in Table 6.10.

The climatic patterns deduced from the proxy data of plant fossils just described are also consistent with recent independent context information from clay minerals (Robert and Chamley, 1991). The paleobotanical results suggest the following sequence:

Paleocene: after a brief initial low, temperatures began to rise; precipitation increased.

Early Eocene: temperatures increased to maximum values; continued high precipitation.

Middle Eocene: temperatures began to decline; rainfall decreased and became more seasonal.

In an analysis of 15 localities in the Atlantic and Pacific Oceans, values for kaolinite, a hydrous aluminum silicate indicative of humid conditions, are high at the K–T boundary and decrease in the Middle Eocene. The abundance of this and other clay minerals also covaries with ^{18}O values that were relatively high at the K–T boundary (cooling from mid-Cretaceous highs), lowest in the Middle Eocene (warmest climates; kaolinite increases), and higher at the end of the Middle Eocene with the onset of cooling (kaolinite decreases). Studies are also underway to model Eocene climates, but at this early stage some differences still exist between inferred paleoclimatic information and model results (Sloan and Barron, 1992; see previous discussion of the Green River flora). Events that may prove relevant to changes at the Eocene–Oligocene boundary are the near simultaneous impacts of asteroids at Chesapeake Bay and in Siberia (Chapter 2).

OLIGOCENE VEGETATION: 35.4–23.3 MA

There are few recently published studies on plant megafossil or microfossil assemblages of Oligocene age from the southeastern United States (Fig. 6.13). Megafossils are known from the Catahoula Formation of eastern Texas

Table 6.10. Alaskan Paleogene floras and inferred climates.

Assemblage	Age	MAT (°C)	CMMT (°C)	WMMT (°C)	MART (°C)	MGSP (cm)
Redoubt	E Oligo	9.0	−4.0	22.0	26.0	60?
Rex Creek	L Eo	12.3	0.3	24.3	24.0	40?
Katalla	L Eo	16.2	6.6	25.6	19.2	120
Carbon Mt.	M–L Eo	8.7	−0.8	18.2	19.0	?
Aniakchak	M Eo	13.7	3.3	24.1	20.8	>145
Charlotte Ridge	M Eo	17.2	8.6	25.8	17.2	>145
Kulthieth	E Eo	19.4	12.6	26.2	13.6	>145
Chickaloon	L Paleo	12.3	1.1	23.5	22.4	>145

MAT, mean annual temperature; CMMT, cold-month mean temperature; WMMT, warm-month mean temperature; MART, mean annual range of temperature; MGSP, mean growing season precipitation. Adapted from Wolfe (1994a, p. 230).

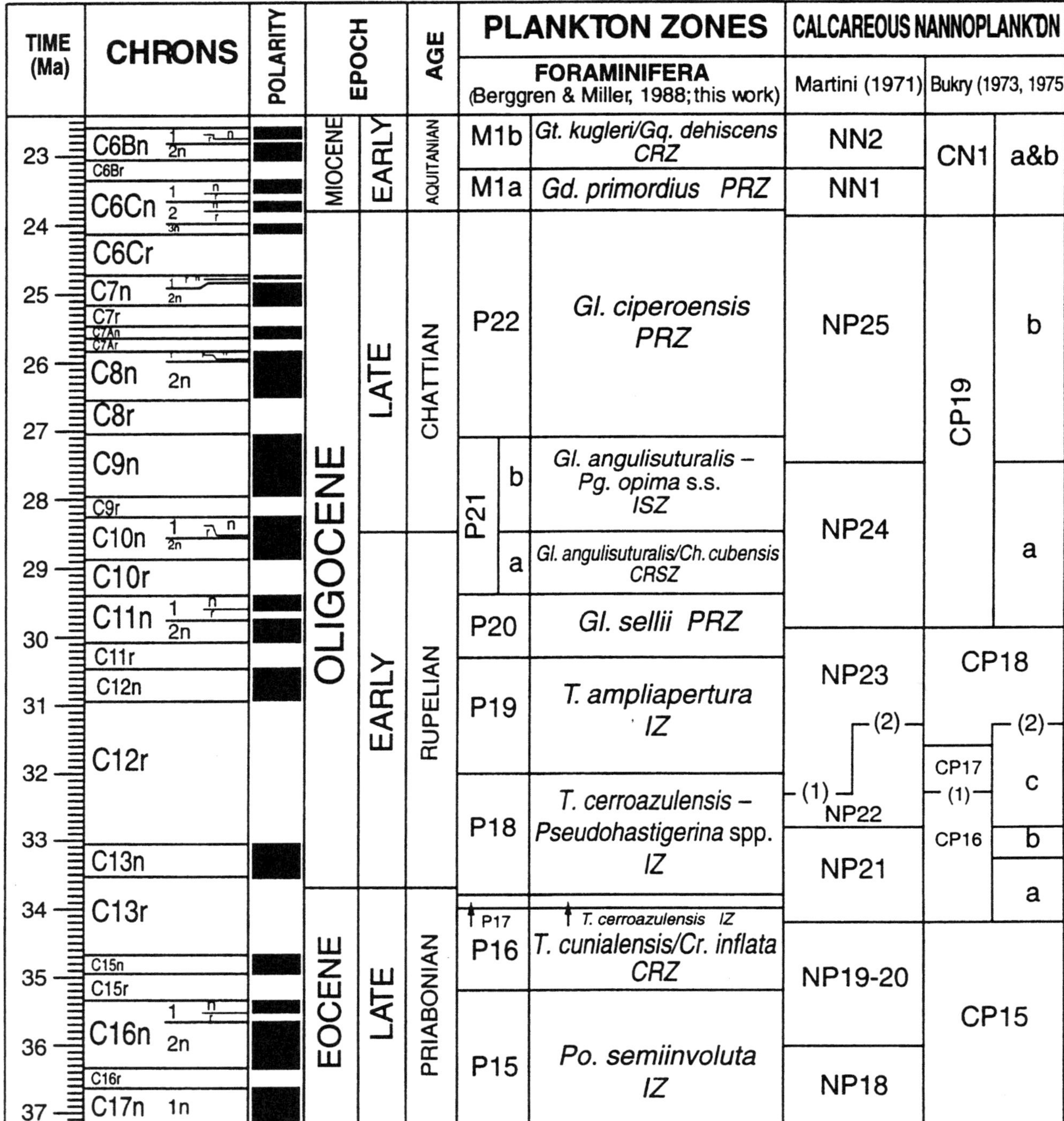

Figure 6.13. Oligocene time scale. Reprinted from Berggren et al. (1995) with the permission of the Society for Sedimentary Geology.

(Berry, 1916); however, the preservation is poor, most identifications have not been confirmed, and the age of the Catahoula Sandstone within the Oligocene is uncertain. Some recently collected material includes *Eomimosoidea plumosa* (Leguminosae; Daghlian et al., 1980), also known from the Middle Eocene Claiborne Formation of western Kentucky–Tennessee, and *Quercus* (Fagaceae; Daghlian and Crepet, 1983). The Vicksburgian pollen–spore flora (Table 6.11) records a continued decline in diversity as more tropical elements were eliminated from the region due to cooling temperatures and more seasonal rainfall (Frederiksen's, 1991 fourth vegetation event). The Early Oligocene overall was a time of cooling and increased seasonal dryness, corresponding to the first definitive evidence for substantial ice sheets on Antarctica.

Oligocene vegetation in the Plains region is not well known (*Celtis*, hackberry, is common), although Retallack (1990), using evidence from paleosols, believes the vegeta-

Table 6.11. Palynomorphs from Vicksburgian provincial stage (Early Oligocene) of Gulf Coast of United States.

Palynomorphs	Affinities
Gymnosperms	
Cupressacites hiatipites	Taxodiaceae–Cupressaceae–Taxaceae
Monocotyledon Angiosperms	
Aglaoreidia pristina	Unknown
Milfordia sp.	Unknown
Palmae	
Sparganium/*Typha* type	Sparganiaceae–Typhaceae
Dicotyledon Angiosperms	
Cupuliferoipollenites sp.	Fagaceae
Fraxinoipollenites sp.	Unknown
Momipites	Juglandaceae, *Engelhardia* group
Myrtaceidites sp.	Unknown
Quercus?	Fagaceae
Siltaria sp.	Fagaceae
Striopollenites terasmaei	Unknown

Adapted from Frederiksen (1981).

tion was scrubland growing under a regime of low rainfall that was seasonally distributed.

In the west one of the most throughly analyzed megafossil assemblages is the Late Oligocene Creede flora of southwestern Colorado about 180 km southwest of Florissant. Axelrod (1987) recognized 73 species (19 gymnosperms, 54 angiosperms), and *Pinus* (10 species) and *Ribes* (eight species) were the most diverse genera (Table 6.12). The depositional setting was a structural moat surrounding a caldera (crater). Two major communities were recognized: a mixed hardwood–coniferous forest and a piñon–juniper woodland on exposed drier and warmer slopes. (Modern analog species are *Pinus edulis* and *Juniperus osteosperma*.) Other elements record a riparian community and aquatic vegetation. The paleoelevation was placed near 1200–1400 m, and the walls of the caldera rose to a nearby volcanic plateau at 2100–2400 m. Annual precipitation was estimated at 460–635 mm (summer wet), the MAT at 11.5°C (1.9°C today), and the annual range of temperature at 16°C (25°C today). Notable features of this higher elevation mixed coniferous forest and drier piñon–juniper woodland scrub were the rarity of present-day mid- to lower elevation mesic Asian or eastern North American exotics that characterize other western Oligocene and Miocene floras and the small size of the leaves. This was attributed to reduced summer rain enhanced by the local physiographic setting. Affinities of the Creede flora were cited as primarily with the local modern vegetation and that of the southern Rocky Mountains of Arizona, New Mexico, and northwestern Mexico (the Madrean region). The area is an important center of diversification for *Pinus* and *Quercus*. With the development of cold winters and drier summers in the southern Rocky Mountains in Miocene and later times, many of the pines and oaks were eliminated there and are now found in the Sierra Madre Occidental of Mexico.

In a restudy of the Creede flora (dated at 27.2 Ma incorporating new decay constants), and using mostly the same collections, only 34 taxa were recognized (Wolfe and Schorn, 1989, 1990; Table 6.12). As would be expected from these taxonomic treatments, the paleoenvironmental reconstructions are also different. The latter study proposes four major communities: fir–spruce forest, fir–pine forest, pine–juniper forest or woodland, and mountain mahogany (*Cercocarpus*) chaparral, with several species having affinities with eastern North America and eastern Asia. The associations are considered distinct from similar modern communtities, and the Creede forests are described as having no exact modern analog. The MAT derived from the CLAMP program is ~4.5°C (4.2°C ; Wolfe, 1993), which is fully 7°C cooler than the estimate by Axelrod (1987).

Among other things, these two taxonomic treatments illustrate the latitude that still exists in the identification of paleobotanical material. The revisions provided by Wolfe and Schorn (1990) are followed here to characterize the vegetation during Creede time. One generalization provided by both studies is that the montane coniferous forest, which was evident early in the Princeton and Republic floras, and drier shrubland–woodland vegetation present to the east of the Rocky Mountains (e.g., in the Green River flora) continued to develop from precursors that were pres-

Table 6.12. Megafossils from Late Oligocene Creede flora, Colorado.

Axelrod (1987)	
Gymnosperms	
Cupressaceae	Pinaceae con't.
Juniperus creedensis	*Pinus crossii*
J. gracillensis	*P. engelmannoides*
Pinaceae	*P. florissantii*
Abies concoloroides	*P. macginitieii*
A. rigida	*P. ponderodoides*
Picea coloradensis	*P. riogrande*
P. lahontensis	*P. sanjuanensis*
P. sonomensis	*P. wasonii*
Pinus alvordensis	*Pseudotsuga glaucocoides*
P. coloradensis	*Tsuga pentranensis*
Monocotyledon Angiosperms	
Cyperaceae	
Cyperacites creedensis	
Dicotyledon Angiosperms	
Aceraceae	Ranunculaceae
Acer riogrande	*Ranunculus creedensis*
Betulaceae	Rhamnaceae
Betula smithiana	*Condalia mohavensis*
Berberidaceae	*Zizyphus florissantii*
Berberis coloradensis	Rosaceae
B. riogrande	*Cercocarpus holmesii*
Mahonia creedensis	*C. linearifolius*
M. obliqua	*Chamaebatiaria creedensis*
Bignoniaceae	*Crataegus creedensis*
Chilopsis coloradensis	*Fallugia lipmanii*
Caprifoliaceae	*Holodiscus harneyensis*
Symphoricarpos wassukana	*H. idahoensis*
Elaeagnaceae	*Peraphyllum septentrionale*
Shepherdia creedensis	*Physocarpus petiolaris*
Ericaceae *P*	*P. triloba*
Arbutus stewartii	*Prunus chaneyii*
Fagaceae	*P. creedensis*
Quercus creedensis	*Rosa hilliae*
Grossulariaceae	*Rubus riogrande*
Ribes birdseyii	*Sorbus potentilloides*
R. creedensis	Salicaceae
R. dissecta	*Populus cedrusensis*
R. lacustroides	*P. creedensis*
R. riogrande	*P. pliotremuloides*
R. stevenii	*Salix creedensis*
R. wasonii	*S. venosiuscula*
R. webbii	Sambucaceae
Hippuridaceae	*Sambucus longifolius*
Hippurus coloradensis	Sapindaceae
Leguminosae	*Sapindus coloradensis*
Cercis buchananensis	Saxifragaceae
Robinia californica	*Fendlera coloradensis*
Nymphaeaceae	*Jamesia caplanii*
Nuphar coloradensis	*Philadelphus creedensis*
Oleaceae	Vacciniaceae
Fraxinus creedensis	*Vaccinium creedensis*

Table 6.12. continued

Wolfe and Schorn (1990)	
Abies rigida	*Stockeya creedensis*
Picea magna	*Holodiscus stevenii*
Pinus crossii	*Crataegus creedensis*
P. sanjuanensis	*Sorbus potentilloides*
P. riogrande	*Potentilla creedensis*
P. sp. 1	*Cercocarpus hendricksonii*
P. sp. 2	*C. nanophyllus*
Juniperus creedensis	*Osmaronia*? *stewartiae*
Berberis coloradensis	*Prunus creedensis*
Mahonia aceroides	*P.* sp.
Dilleniaceae	Rosoideae
Populus larsenii	*Cercis* sp.
Salix sp.	Legume
Jamesia caplanii	*Catalpa coloradensis*
Ribes lacustroides	*Eleopoldia lipmanii*
R. obovatum	*Monocotylophyllum* sp
R. robinsonii	
Eleiosina praeconcinna	

Adapted from Axelrod (1987) and Wolfe and Schorn (1990).

ent in the Middle Eocene. Subtropical elements were essentially eliminated from the central Rocky Mountains in the Oligocene, and overall the forests became more modern in aspect during this time (Leopold and MacGinitie, 1972).

To the west in north-central Oregon Early Oligocene plants are known from the John Day Formation in the Crooked River and John Day basins in the same vicinity as the Late Eocene Clarno flora. The classic locality at Bridge Creek is Early Oligocene and has been radiometrically dated at 31.8 and 32.3 Ma (Evernden et al., 1964; corrected to new constant; more recently 32.24–33.62 Ma, Meyer and Manchester, 1996); therefore, it is younger than the Clarno (44–43 Ma). The Bridge Creek flora was studied by Chaney (1924, 1927), individual elements are part of wider revisions (Brown, 1939, 1946, 1959; Manchester, 1992, *Florissantia speirii*; Manchester and Crane, 1987; Tanai and Wolfe, 1977; Wolfe, 1977; Wolfe and Tanai, 1987), and there is a recent revision by Meyer and Manchester (1997; 91 genera, 110 species). Current studies are focused on a locality at Fossil, Oregon (32 Ma, SCLF technique), just to the northeast of Clarno, and about 35 mi. north of the principal Bridge Creek locality. Manchester and Meyer (1987) presently recognize 35 species from the lacustrine shales at fossil (Table 6.13), including *Metasequoia*, *Acer*, *Alnus*, *Paracarpinus* (leaves), *Asterocarpinus* (extinct, fruit; Manchester and Crane, 1987), and *Pteleaecarpum* (*Craigia*; see Clarno flora) as the most abundant. *Eucommia* occurs at other localities in the John Day Formation (Call and Dilcher, 1997). Of particular interest is *Cercidiphyllum* (katsura, eastern Asia), which was widespread in mid-Tertiary floras across the northern hemisphere, because the Early Oligocene occurrence in the Bridge Creek flora is among the oldest known; earlier similar forms are now assigned to related but extinct genera (Crane and Stockey, 1985). The vegetation is principally a temperate broad-leaved deciduous hardwood forest with many elements similar to those in the mixed northern hardwood forest of North America and eastern Asia. The MAT is estimated at 9–11°C with a mean annual range of ~23°C. Broad-leaved evergreens are rare. Compared to the Late Eocene Clarno flora in the same area (predominantly broad-leaved evergreen), the Early Oligocene Bridge Creek–Fossil floras (predominantly broad-leaved deciduous) reflect a trend toward a cooler, drier, more seasonal climate.

The Rujada flora of west-central Oregon is assigned to the Late Oligocene by Lakhanpal (1958) and to the Early Oligocene by Wolfe (1981b). It preserves remnants of the vegetation growing along the shores of Oregon adjacent to the rising Cascade Mountains. The name derives whimsically from the initials of two loggers (R. Upton and J. Anderson) and a forestry project undertaken by the Department of Agriculture. The flora preserves a familiar assemblage of mesic deciduous hardwood elements and higher altitude gymnosperms with affinities to the modern vegetation of eastern North America, western North America, and eastern Asia. The presence of a few warm-temperate to subtropical elements and broad-leaved evergreens (*Annona*, *Machilus*, *Nectandra*, *Ocotea*), along with some wider ranging plants also found in warm-temperate vegetation (*Alangium*, *Engelhardia*), reflect the maritime environments of the coast rather than the increasingly dry climates of the intermountain basins and the south-facing or leeward slopes to the west. The presence of broad-leaved deciduous, broad-leaved evergreen, and coniferous elements is similar to the mixed broad-leaved evergreen and coniferous forests of eastern Asia. The MAT is estimated at

Table 6.13. Megafossils from John Day Formation, Fossil, Oregon.

Pteridophytes	
Cf. *Polypodium*	
Gymnosperms	
Pinaceae	Taxodiaceae
Pinus sp.	*Metasequoia occidentalis*
Abies sp.	
Monocotyledon Angiosperms	
Monocotyledonous leaf	
Dicotyledon Angiosperms	
Aceraceae	Leguminosae
Acer ashwilli	*Cercis* sp.
A. cranei	*Cladrastis* sp.
A. manchesteri	Meliaceae
A. osmonti	*Cedrela merilli*
Berberidaceae	Moraceae
Mahonia simplex	*Morus*-like leaves
Betulaceae	Platanaceae
Alnus hollandiana	Platanaceous fruits
Asterocarpinus perplexans	*Platanus aspera*
Bombacaceae–Tiliaceae–Sterculiaceae	*P. condoni*
Florissantia speirii	Rosaceae
Cercidiphyllaceae	*Crataegus newberryi*
Cercidiphyllum crenatum	Cf. *Sorbus*
Fagaceae	Tiliaceae
Fagus pacifica	*Plafkeria obliquifolia*
Quercus consimilis	*Pteleaecarpum* (*Craigia*) *oregonense*
Hydrangeaceae	*Tilia circularis*
Hydrangea florissantia	Ulmaceae
Juglandaceae	*Tremophyllum hesperium*
Cruciptera	*Ulmus pseudo-americana*
Cf. *Engelhardia olsoni*	
Juglans sp.	
Pterocarya sp.	

Adapted from Bůžek et al. (1989), Kvaček et al. (1991), Manchester (1991, 1992), and Manchester and Meyer (1987).

12–13°C. The presence of *Abies* suggests that the Cascade Mountains, having reached an elevation to support this high-elevation conifer, were likely beginning to cast a rain-shadow locally across the adjacent interior basins (e.g., the drier Bridge Creek and Fossil floras). The Late Oligocene Yaquina flora of western Oregon (McClammer, 1978) is preserved in delta deposits and represents a humid to mesic coastal rain forest. The CLAMP estimate of MAT is 15.5°C (Wolfe, 1993), and some adjustment in interpretation is necessary between the floras at Rujada (age uncertain, MAT based on modern analogs) and Yaquina (MAT based on CLAMP).

To the far northwest, development of temperate deciduous and coniferous forests was favored by the Middle Eocene through Early Oligocene cooling. Palynomorphs from the lower part of the Richards Formation in the Mackenzie Delta region of northern Canada record both cool (*Sphagnum*, *Picea*, *Alnus*, *Betula*) and warmer (*Metasequoia*, *Pinus*, *Sequoia*, *Tsuga*, *Castanea*, *Pterocarya*, *Quercus*) elements in the Late Eocene (Norris, 1982), likely resulting from wind transport of pollen from higher altitude vegetation into the lowlands. In the upper part of the formation (Early and Middle Oligocene) a decrease in diversity and a change in composition of the assemblage is evident. Because the change is not associated with facies differences, it is likely attributable to climate. The common palynomorphs are *Sphagnum*, *Lycopodium*, *Osmunda*, *Picea*, *Pinus*, *Tsuga*, *Alnus*, *Betula*, *Carpinus*, and *Pterocarya*; the more thermophylic *Sequoia*, *Castanea*, *Quercus*, *Tilia*, and *Ulmus* disappear. [Compare these palynomorph assemblages with elements of the modern closed forest phase of the boreal coniferous forest formation and the deciduous forest formation of the southern Appalachians (Chapter 2).] These return later in the Oligocene, and the changes in vegetation probably reflect the same fluctuations previously discussed that occurred during the Middle Eocene through the Oligocene in the southeastern United States and southern California and that are paralleled in the marine-based paleotemperature record (Fig. 3.1). In Alaska the Salicaceae (*Populus, Salix*) diversified during the Eocene–Oligocene temperature decline to become the dominant streamside element and then spread southward to characterize midnorthern latitude gallery forests by the Early Miocene. This characteristic riverine assemblage first appears as a well-defined community in the mid- to Late Eocene of the high northern latitudes.

The reassignment in age of several North American floras has affected the data base available for interpreting Oligocene vegetation and terrestrial environments. The placement of the Brandon lignite flora from the Oligocene to the Early Miocene, and the Bridge Creek, Fossil, Florissant, Beaverhead Basin, Metzel Ranch, York Ranch, Mormon Creek, Ruby, and other floras from the Oligocene to the Late Eocene–Early Oligocene leaves few assemblages for the period between ~36–23 Ma. This emphasizes the importance of the Creede flora, the need to explore for other floras of Oligocene age, and the enhanced role of faunas and context information in paleoenvironmental reconstructions (Figs. 3.1, 3.13, 6.1).

EARLY MIOCENE VEGETATION: 23.3–16.3 MA

The climatic patterns of the Oligocene, induced in part by early continental glaciations on Antarctica, continued into the Early Miocene but with a slight warming trend that is evident until about 15 Ma (Figs. 3.1, 6.14). Recently studied Early Miocene floras are not widely available from the southeastern United States, but the existing information suggests continued warm-temperate to subtropical conditions.

To the northeast few Tertiary floras are known from this mostly Paleozoic and Mesozoic province. An exception is the Brandon Lignite in west-central Vermont that was discovered in 1848 during mining operations and was subsequently used to generate steam for the Brandon Iron and Car Wheel Company. The lignite has long been known for its fossils, which include pollen, spores, other kinds of microfossils (e.g., colonial algae), fruits, seeds, and wood. Palynomorphs from the Brandon Lignite identified as to biological affinities by Traverse (1955) are listed in Table 6.14 (see also Traverse, 1994). The most abundant taxa are *Carya, Castanea, Corylus*?, *Cyrilla, Engelhardia, Gordonia, Ilex, Manilkara, Morus, Nyssa, Planera, Quercus, Rhamnus, Siltaria* (extinct genus), *Ulmus*, and *Vitis*. Megafossils (mostly fruits and seeds) are under study and presently include *Alangium* (along with pollen, Fig. 6.15; Eyde et al., 1969), *Caldesia* (Alismataceae; Haggard and Tiffney, 1997; also in the Clarkia flora of Idaho), Cyperaceae (B. H. Tiffney, personal communication, 1996), *Caricoidea* (extinct, monocotyledonae), Illiciaceae (Tiffney and Barghoorn, 1979), Magnoliaceae (Tiffney, 1977), *Microdiptera* (Tiffney, 1981), Moraceae (Moroideae), *Nyssa* (Eyde and Barghoorn, 1963; see also Eyde, 1997), Rutaceae (*Euodia, Zanthoxylum*; Tiffney, 1980), Theaceae (cf. *Cleyera*), *Sargentodoxa* (Sargentodoxaceae, a deciduous vine presently restricted to southeast Asia; Tiffney, 1993), *Turpinia* (Staphyleaceae; Tiffney, 1979), and Vitaceae (Tiffney and Barghoorn, 1976). A complete listing of the megafossils identified from the Brandon flora is given in Tiffney (1994). The habitat was river swamps similar to those of the mesothermal Atlantic–Gulf Coastal Plain further to the south, as indicated by *Cyrilla, Gordonia, Persea*, and *Symplocos*, which presently do not extend northward into areas with significant frost. The climate was warm-temperate to subtropical, essentially frost free (MAT 17°C; presently 7.6°C) with ~1660 mm of annual precipitation (presently 950 mm; Tiffney, 1993). Present-day Asian genera such as *Glyptostrobus, Alangium*, and *Pterocarpus* were present, as in other mid-Tertiary floras across the northern hemisphere. In New England the vegetation had changed from the tropical forest of the Early Eocene to deciduous broad-leaved or mixed mesophytic forest by the Early Miocene.

A challenging problem has been to determine the age of the lignite that is underlain by the Lower Cambrian Cheshire Quartzite, overlain by Quaternary drift, and lacks radiometrically datable material and vertebrate remains. An innovative approach was developed by Barghoorn (1950, 1951, 1953) and Spackman (1949) based on Reid (1920). Percentages of exotic genera in Tertiary floras of known age were plotted against time, and the resulting curve was used to estimate the age of various floras. In the case of the Brandon Lignite this value was 72%, which corresponded to an age of Late Oligocene (Traverse, 1955, fig. 5). Other age estimates are Eocene–Oligocene to earliest Miocene (35.4–23.3 Ma; Tiffney, 1993), but the most recent assignment is to the Early Miocene (Tiffney, 1994; Tiffney and Traverse, 1994; Traverse, 1994).

Shortly after publication of the age–area curve, Axelrod (1957) published a critique to which Wolfe and Barghoorn (1960) responded. The arguments presented in support of these and other often diametrically opposed conclusions (e.g., Axelrod, 1952, 1961; Scott et al., 1960) reflect the debated nature of many aspects of vegetational history. In the meantime, considerable new information, innovative ideas, and revisions are generated and progress is made. In retrospect, the age–area curve gave an age estimate (Late Oligocene) that was near the most recent one (Early Miocene).

To the far north on Devon Island, Northwest Territories (75° 22' N, 89° 40' W), lacustrine sediments of the Haughton Formation from an impact crater 16 km wide at 22.4 ± 1.4 to 23.4 ± 1 Ma (Early Miocene, Aquitanian; Omar et al., 1987) contain plant micro- and megafossils and faunal remains of a mixed coniferous–hardwood assemblage dominated by *Larix, Picea, Pinus, Tsuga, Alnus, Betula, Corylus*, Ericales, and *Ulmus–Zelkova* (Table 6.15). The closest modern analog is the cool-temperate conifer–hardwood vegetation transitional between the boreal forest and the deciduous forest in northeastern maritime North America.

TIME (Ma)	CHRONS	EPOCH		AGE
	C5ADn	MIOCENE	MIDDLE	SER.
	C5ADr			
15	C5Bn (1n, 1r, 2n)			LANGHIAN
	C5Br			
16	C5Cn (1n, 1r, 2n, 2r, 3n)			
	C5Cr		EARLY	BURDIGALIAN
17	C5Dn			
18	C5Dr			
	C5En			
19	C5Er			
	C6n			
20	C6r			
21	C6An (1n, 1r, 2n)			AQUITANIAN
	C6Ar			
22	C6AAn			
	C6AAr (1r, 2n, 2r, 3n, 3r)			
23	C6Bn (1n, 1r, 2n)			
	C6Br			
24	C6Cn (1n, 1r, 2n, 2r, 3n)	OLIGO-CENE	LATE	CHATTIAN
	C6Cr			

PLANKTONIC FORAMINIFERA

(SUB)TROPICAL — Berggren (this work); Blow (1969)

Zone	Subzone	Berggren (this work)		Blow (1969)
M7		*Gt. peripheroacuta* Lin. Z		N10
M6		*O. sutur. – Gt. peripher.* IZ		N9
M5	b	*Pr. glomerosa – O. suturalis* ISZ	*Pr. sicana – O. suturalis* IZ	N8
	a	*Pr. sicana – Pr. glomerosa* ISZ		
M4	b	*Gd. bispherica* – PRSZ	*C. dissimilis – Pr. sicana* IZ	N7
	a	*Cat. dissimilis – Gt. birnageae* ISZ		
M3		*Globigerinatella insueta – Catapsydrax dissimilis* Conc. RZ		N6
M2		*Catapsydrax dissimilis* IZ		N5
M1	b	*Gt. kugleri – Gq. dehiscens* Conc. RSZ	*Gt. kugleri* TRZ	N4
	a	*Gd. primordius* ISZ		
P22		*G. ciperoensis* IZ		P22

TRANSITIONAL — Berggren and others (1983a); this work

Zone	Subzone	Berggren and others (1983a); this work	
Mt6		*O. suturalis/ Gt. peripheroronda* Conc. RZ	
Mt5	b	*Pr. golmerosa – O. suturalis* ISZ	*Pr. sicana – O. suturalis* IZ
	a	*Pr. sicana – Pr. glomerosa* ISZ	
Mt4		*Gt. miozea* PRZ	
Mt3		*Gt. praescitula - Gt. miozea* IZ	
Mt2		*Globorotalia incognita – Globorotalia semivera* PRZ	
Mt 1	b	*Gt. kugleri – Gq. dehiscens* Conc. RSZ	*Gl. kugleri* TRZ
	a	*Gd. primordius* ISZ	
P22		*G. ciperoensis* IZ	

(SUB)ANTARCTIC — Berggren (1992)

Zone	Berggren (1992)
AN4	*Gt. miozea* IZ
AN3	*Gt. praescitula* IZ
AN2	*Gt. incognita* PRZ
AN1	*Gl. brazieri* PRZ
AP16	*G. euapertura* IZ

CALCAREOUS NANNOPLANKTON

Martini (1971)	Bukry (1973, 1975)
NN5	CN4
NN4	CN3
NN3	CN2
NN2	CN1 b
NN1	CN1 a&b
NP25	CP19

Figure 6.14. Early Miocene time scale. Reprinted from Berggren et al. (1995) with the permission of the Society for Sedimentary Geology.

Table 6.14. Fossils from Early Miocene Brandon Lignite, Vermont.

Algae

Chlorophyta
Botryococcus

Dinoflagellates
Peridinium

Pteridophytes

Fern spores (unidentified)

Gymnosperms

Pinaceae
Pinus (haploxylon type)
Pinus (sylvestris type)

Taxodiaceae
Glyptostrobus

Monocotyledon Angiosperms

Gramineae

Dicotyledon Angiosperms

Alangiaceae
Alangium
Anacardiaceae
Rhus
Aquifoliaceae
Ilex
Betulaceae
Corylus?
Clethraceae
Clethra(?)
Ericaceae
Ericaceae (x)
Lyonia(?)
Oxydendrum(?)
Rhododendron
Vaccinium
Zenobia (x)
Fagaceae
Castanea
Fagus
Quercus
Siltaria (extinct genus)
Hamamelidaceae
Liquidambar
Juglandaceae
Carya
Engelhardia
Juglans
Pterocarya
Lauraceae
Cinnamomum-type (x)
Ocotea-type (x)
Persea (x)
Magnoliaceae
Illicium (x)
Magnolia (x)

Moraceae
Morus
Nyssaceae
Nyssa
Onagraceae
Jussiaea
Rhamnaceae
Rhamnus
Rosaceae
Rubus (x)
Rosaceae(?)
Rutaceae
Horniella (extinct genus)
Phellodendron (x)
Toddalieae
Santalaceae
Nestronia(?)
Sapotaceae
Manilkara
Mimusops
Symplocaceae
Symplocos
Theaceae
Gordonia
Tiliaceae
Tilia
Ulmaceae
Ulmus
Planera
Vitaceae
Vitis (Subg. *Muscadinia* group (x)

Adapted from Traverse (1955); x, megafossil only). See Traverse (1994) for list of palynomorphs assigned to form taxa and text for other megafossils described later.

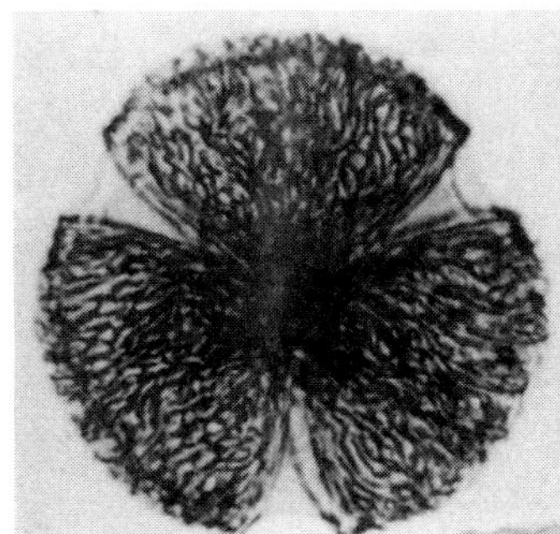

Figure 6.15. Pollen of *Alangium barghoornianum*, Brandon Lignite, Vermont. Adapted from Traverse (1955).

The MAT estimates range widely between 0 and 11°C (Hickey et al., 1988) and 15–20°C (Whitlock and Dawson, 1990). Eliminating the high and low extremes, the MAT figures are 11–15°C (compare, e.g., 17°C for the Brandon lignite). In the Canadian Maritime Provinces precipitation is ~1000 mm/year.

To the west on Banks Island, a flora of late Early to early Middle (Seldovian) Miocene age is preserved in the Mary Sachs Gravel. This informal unit was named to differentiate it from the Beaufort Formation (type locality Prince Patrick Island) that, through imprecise usage, had come to include unrelated deposits with a wide range of ages. The flora is one of the richest in the Neogene of Arctic North America (Matthews and Ovenden, 1990; Table 6.16); it also reflects mixed coniferous–hardwood forest having affinities with eastern Asia (*Glyptostrobus*, *Metasequoia*), eastern (including southeastern) North America (*Phyllanthus*), and extant local vegetation (*Abies*, *Thuja*, *Alnus*, various Ericaceae, *Menyanthes*). In addition to the taxa listed in Table 6.16, others from smaller floras in the region include *Chara–Nitella* type, cf. *Tubela*, *Nuphar*, *Nymphaea*, cf. *Vitis*, *Viola* sp., *Decodon* sp., *Hippuris* sp., *Vaccinium* sp., *Nymphoides*, and *Lycopus*. In some sections of the Beaufort Formation younger than 22 Ma (Banks Island) and ~3 Ma (Meighen Island), nonarboreal taxa are common (Gramineae, *Artemisia*, various herbs). In one section from Meighen Island an early and local forest–tundra vegetation was present (Matthews, 1987). Tree taxa are fewer in the Late Pliocene–Early Pleistocene (Matuyama age) floras of the Kap København Formation. These floras record the development of tundra from ~3 Ma to the present and are discussed in Chapter 7.

Little is known of the Early Miocene vegetation in the Plains area. A shrubland–savanna likely occupied the region, and a few grass fossils representing extinct genera begin to appear. These persist into the Middle and Late Miocene when they are augmented by additional grass genera (Stipidae and Poidae) and prairie herbs (e.g., Boraginaceae). A small flora from the Marsland Formation in northwest Nebraska (19–18 Ma; Hemingfordian NALMA) includes *Acer* cf. *negundo* (box elder), *Crataegus* (hawthorn), *Robinia* (locust), and *Ulmus* (elm). Annual rainfall is estimated at 900 mm (Axelrod, 1985a).

Table 6.15. Pollen and spores from Early Miocene Haughton Formation, Devon Island, Arctic Canada.

Bryophyta	Dicotyledon Angiosperms
Sphagnum type	*Acer*
Pteridophytes	*Alnus*
Botrychium type	*Betula*
Dryopteris type	*Carya*
Lycopodium type	*Castanea*
Osmunda type	Chenopodiineae
Pteridium type	*Corylus* type
Selaginella	Cruciferae
Gymnosperms	Ericales
Abies	Cf. *Fagus*
Cupressaceae	Cf. *Fraxinus*
Larix type	*Ilex*
Picea	*Juglans*
Pinus	*Liquidambar*
Pinus strobus type	*Ostrya–Carpinus*
Tsuga	*Populus*
Monocotyldeon Angiosperms	*Pterocarya*
Cyperaceae	*Quercus*
Potamogeton	*Salix*
Sparganium type	*Ulmus–Zelkova*

Adapted from Whitlock and Dawson (1990).

In the central Rocky Mountains tectonic activity had waned in the Middle Tertiary, but in the Miocene it resumed. Erosion from uplift combined with volcanism deposited extensive sandstones and tuff (water-lain volcanic ash). In Jackson Hole, Wyoming, up to 2300 m (7000 ft) of volcanics were deposited, and in the San Juan Mountains of southern Colorado an area of 25,000 km^2 was filled with 20,000 km^3 of ash and tuff. The climatic history for the northern Rocky Mountains is sketchy because few Neogene floras are available. The region was likely a western montane coniferous forest with an understory and streamside component of broad-leaved deciduous angiosperms; but as emphasized previously, apparent stasis is usually an artifact of poor resolution. Sedimentary patterns, paleosol mineralogy, and vertebrate assemblages suggest fluctuations in precipitation: wet to dry in the Late Eocene, dry to wet in the late Early Miocene, wet to dry in the early Middle Miocene, and dry to wet in the Late Miocene–Pliocene transition (Thompson et al., 1982). These fluctuations contributed to the dynamic nature of the vegetation and undoubtedly to evolutionary diversification.

In the western Great Basin, vegetation near the Early to Middle Miocene boundary is preserved in the Buffalo Canyon flora of western Nevada (15.6 Ma; Axelrod, 1991; Wolfe et al., 1997; Table 6.17). The largest families are Pinaceae (10 species), Rosaceae (11), Salicaceae (10), Aceraceae (5), and Betulaceae (5). The largest genera are *Salix* (7 species), *Acer* (5), *Betula* (4), *Populus* (4), *Ribes* (4), *Picea* (3), and *Prunus* (3). Most species presently occur in the western United States, and a few occur in the eastern United States and eastern Asia. It is a mixed conifer–deciduous hardwood forest with representatives of the adjacent

Table 6.16. Plant megafossils from Mary Sachs Gravel southern Banks Island, N.W.T.

Gymnosperms

Cupressaceae
- cf. *Thuja occidentalis*

Pinaceae
- cf. *Abies grandis*
- cf. *Larix omoloica*
- *Larix* sp.
- *Picea banksii*
- *P.* sp.
- *Pinus* five-needle type
- *P. itelmenorum*

Pinaceae cont.
- *P. palaeodensiflora*
- cf. *P. funebris*
- *Tsuga* sp.

Taxodiaceae
- *Glyptostrobus* sp.
- *Metasequoia disticha*
- *M.* sp.
- *Taxodium* sp.

Monocotyledon Angiosperms

Alismataceae
- cf. *Sagisma*

Araceae
- *Epipremnum crassum*

Cyperaceae
- *Carex* sp.

Potamogetonaceae
- cf. *Potamogeton richardsonii*

Sparganiaceae
- *Sparganium* sp.

Dicotyledon Angiosperms

Actinidiaceae
- *Actinidia* sp.

Aizoaceae
- cf. *Sesuvium*

Araliaceae
- *Aralia* sp.

Betulaceae
- *Alnus* (*Alnobetula*) sp.
- Cf. *Alnus incana*
- Cf. *Betula apoda*
- *Betula* dwarf shrub type
- *B.* arboreal type

Capparidaceae
- *Cleome* sp.
- *Polanisia* sp.

Caprifoliaceae
- *Diervilla* sp.
- *Sambucus* sp.
- *Weigela* sp.

Chenopodiaceae
- *Chenopodium* sp.

Crassulaceae
- *Sedum* type

Ericaceae
- *Andromeda polifolia*
- *Arctostaphylos alpina/rubra* type
- *Chamaedaphne* sp.

Euphorbiaceae
- *Phyllanthus* (*Phyllanthus*) sp.

Gentianaceae
- *Menyanthes* small form

Hypericaceae
- *Hypericum* sp.

Juglandaceae
- *Juglans eocineria*

Labiateae
- *Teucrium* sp.

Lythraceae
- *Microdiptera/Mneme* type

Magnoliaceae
- *Liriodendron* sp.

Moraceae
- *Morus* sp.

Myricaceae
- *Comptonia* sp.
- *Myrica* (Gale) sp.

Onagraceae
- *Ludwigia* sp.

Polygonaceae
- *Rumex* sp.

Ranunculaceae
- *Ranunculus* (*Batrachium*) sp.
- *R. hyperboreus*

Rhamnaceae
- Cf. *Paliurus* sp.

Rosaceae
- *Potentilla* sp.
- *Rubus* sp.

Saxifragaceae
- Genus?

Solanaceae
- *Solanum/Physalis* type

Verbenaceae
- *Verbena* sp.

Adapted from Matthews and Ovenden (1990).

Table 6.17. Plant megafossils from Early–Middle Miocene Buffalo Canyon flora, western Nevada.

Gymnosperms

Cupressaceae
Chamaecyparis cordillerae
Juniperus desatoyana
Pinaceae
Abies concoloroides
A. laticarpus
Larix churchillensis
Picea lahontense
P. magna
Pinaceae cont.
P. sonomensis
Pinus balfouroides
P. ponderosoides
Pseudotsuga sonomensis
Tsuga mertensioides

Monocotyledon Angiosperms

Typhaceae
Typha lesquereuxii

Dicotyledon Angiosperms

Aceraceae
Acer medianum
A. negundoides
A. oregonianum
A. trainii
A. tyrrellii
Berberidaceae
Mahonia macginitiei
M. reticulata
Betulaceae
Alnus latahensis
Betula ashleyii
B. desatoyana
B. idahoensis
B. thor
Caprifoliaceae
Symphoricarpos wassukana
Carpinaceae
Carpinus oregonensis
Ericaceae
Arbutus trainii
Fabaceae (Leguminosae)
Amorpha stenophylla
Robinia bisonensis
Fagaceae
Chrysolepis sonomensis
Quercus hannibalii
Q. wislizenoides
Grossulariaceae
Ribes barrowsae
R. bonhamii
R. stanfordianum
R. webbii
Hydrangeaceae
Hydrangea bendirei
Juglandaceae
Carya bendirei
Juglans desatoyana
Meliaceae
Cedrela trainii
Myrtaceae
Eugenia nevadensis
Nymphaeaceae
Nymphaeites nevadensis
Oleaceae
Fraxinus desatoyana
F. eastgatensis
Rosaceae
Amelanchier desatoyana
Cercocarpus ovatifolius
Chamaebatia nevadensis
Crataegus middlegatei
Heteromeles desatoyana
Lyonothamnus parvifolius
Prunus chaneyii
P. moragensis
P. treasheri
Rosa harneyana
Sorbus cassiana
Salicaceae
Populus cedrusensis
P. eotremuloides
P. payettensis
P. pliotremuloides
Salix churchillensis
S. desatoyana
S. laevigatoides
S. owyheeana
S. pelviga
S. storeyana
Ulmaceae
Ulmus speciosa
Zelkova brownii
Vacciniaceae
Vaccinium sophoroides

Adapted from Axelrod (1991).

broad-leaved evergreen sclerophyll vegetation from lower elevations on warmer and drier sites. The MAT is estimated at 10°C (CLAMP 8.7°C), mean annual range at 14°C, and annual precipitation at 900–1000 mm (35–40 inches) with ample summer rain. The paleoelevation of the lake basin was estimated at ~1280 m (4200 ft; presently 1900 m). However, many of the Early and Middle Miocene floras of the Basin and Range Province may have been considerably higher based on leaf physiognomy and mean annual enthalpy calculations (e.g., Buffalo Canyon, 3.2 km; Eastgate, 2.7 km; Middlegate, 2.8 km; Wolfe et al., 1997, table 1). Axelrod (1985b) attributes differences between the Middlegate and Eastgate floras (15.5 Ma) to slope exposure. The Middlegate flora is dominated by sclerophyllous trees and shrubs (shrubland–chaparral) living on drier south-facing slopes. The vegetation included *Arbutus prexalapensis*, *Cedrela trainii*, *Cercocarpus antiquus*, *C. pacifica*, *Lithocarpus nevadensis*, *Quercus hannibali*, *Q. shrevoides*, and *Q. simulata*. The MAT is estimated at 10.2°C (CLAMP). The Eastgate flora has a better representation of mixed coniferous–hardwood species from north-facing canyons (*Abies concoloroides*, *Larix cassiana*, *L. nevadensis*, *Amelanchier grayi*, *Aesculus preglabra*). The MAT is estimated at 10.3°C (CLAMP).

To the north the Alvord Creek flora (~21 [21–23] Ma, southeastern Oregon; Axelrod, 1944) is also an upland coniferous–hardwood forest. To the south the Tehachapi flora (17.5 [16.1] Ma; southeastern end of the Sierra Nevada; Axelrod, 1939) is an evergreen oak–cypress–piñon woodland and associated arid subtropical scrub growing at lower elevations in a warmer climate. To the west, the Sutro flora (21 Ma) near Silver City, Nevada, is a richer mesophytic exotic broad-leaved evergreen and conifer vegetation on the windward side of the low Sierra Nevada (Axelrod, 1991).

In the far northwest the Seldovia Point flora from the Kenai Group, Cook Inlet region of Alaska, is Late Seldovian in age (late Early to early Middle Miocene, ~15.8–16.8 [14.5–14.2] Ma; Fig. 6.11). As with the vegetation to the east across the high northern latitudes, it represents a mixed mesophytic hardwood forest of *Celtis*, *Fagus*, *Juglans*, *Liquidambar*, *Magnolia*, *Nyssa*, *Platanus*, *Pterocarya*, and *Zelkova* (megafossil evidence), with coniferous forests close by (microfossil evidence; Wolfe and Tanai, 1980). The MAT is estimated at 6–7°C (CLAMP 9°C) with a mean annual range of ~26°C. A summary of MAT and MART for western Washington and Oregon for the Late Eocene through the Miocene is given in Fig. 6.16. It shows the dramatic increases in MART across the Eocene–Oligocene boundary.

As shown on the global benthic paleotemperature curve (Fig. 3.1), the Earth's biotas experienced another downshift in temperature and other environmental changes in the Middle Miocene (15–14 Ma) that ushered in a new round of evolutionary activity, a reshuffling of distributions, and continued modernization of the North American vegetation.

VEGETATION SUMMARY

During the Middle Eocene the southeastern United States was occupied by a semideciduous tropical dry forest (Table 6.18). Mangrove vegetation of *Nipa* and associated *Acrostichum* fringed the coast. *Nipa* disappeared at the end of the Eocene and the brackish-water habitats eventually became occupied by elements of the modern neotropical vegetation: *Avicennia nitida*, *Laguncularia racemosa*, and *Rhizophora mangle* associated with *Conocarpus erecta*. On sandy sites *Pinus* and palmetto-like palms (*Sabal–Serenoa* type) were present and represent an early version of the modern sand pine scrub of the pine woods association. Warm-temperate deciduous elements of Fagaceae and Juglandaceae occurred with more tropical elements in the low to midaltitudes and graded into the cooler temperate vegetation in the Appalachian highlands (*Picea*, *Tsuga*). The drying and cooling trend in the Middle and Late Eocene (Fig. 6.10, 6.16) that supported dry deciduous forest later became too cool, and moisture increased in the southeast with redirected pressure systems derived, in part, from continued rise of the western mountains. This gradually eliminated the tropical dry forest, favoring the development and expansion of warm-temperate deciduous vegetation at lower altitudes, and the Appalachian montane coniferous forest in the highlands. Components of associations of the deciduous forest formation (beech–maple; oak–chestnut, oak–hickory; southern mixed hardwood, flood-plain forest, *Nyssa*, *Taxodium*, early maple(?)–basswood) in the eastern United States were present in the Claiborne–Jackson floras; their beginnings, along with the various edaphic and fire-controlled pine woodlands, can be traced to the Middle Eocene. *Picea* is present in the Late Eocene Jackson flora. The broad sequence of communities in the southeastern United States has been a Late Cretaceous community termed a tropical forest with no exact modern analog; a Paleocene through Early Eocene tropical rain forest; a Middle Eocene tropical dry forest; and with subsequent decline of the dry tropical components, the appearance, diversification, and expansion of various associations comprising the modern deciduous forest formation and the modern Appalachian coniferous forest formation.

Temperature gradients were becoming steeper but were still lower in the Middle Eocene through the Early Miocene than at present. Many elements of a warm-temperate broad-leaved deciduous forest, including present-day Asian exotics, extended through the New England area (Brandon Lignite flora) and continued as far north as Ellesmere Island (81° N latitude). *Acer*, *Carya*, *Corylus–Carpinus*, *Castanea*, and *Fagus*, along with cooler temperate mixed coniferous forest species of *Larix*, *Picea*, *Tsuga*, and *Betula* formed a version of the lake states–Canadian maritime forest.

The semideciduous tropical dry forest of the southeastern United States extended northwest around the remnant of the Cretaceous–Paleocene sea at low to midelevations, across the Plains, along the eastern side of the Rocky

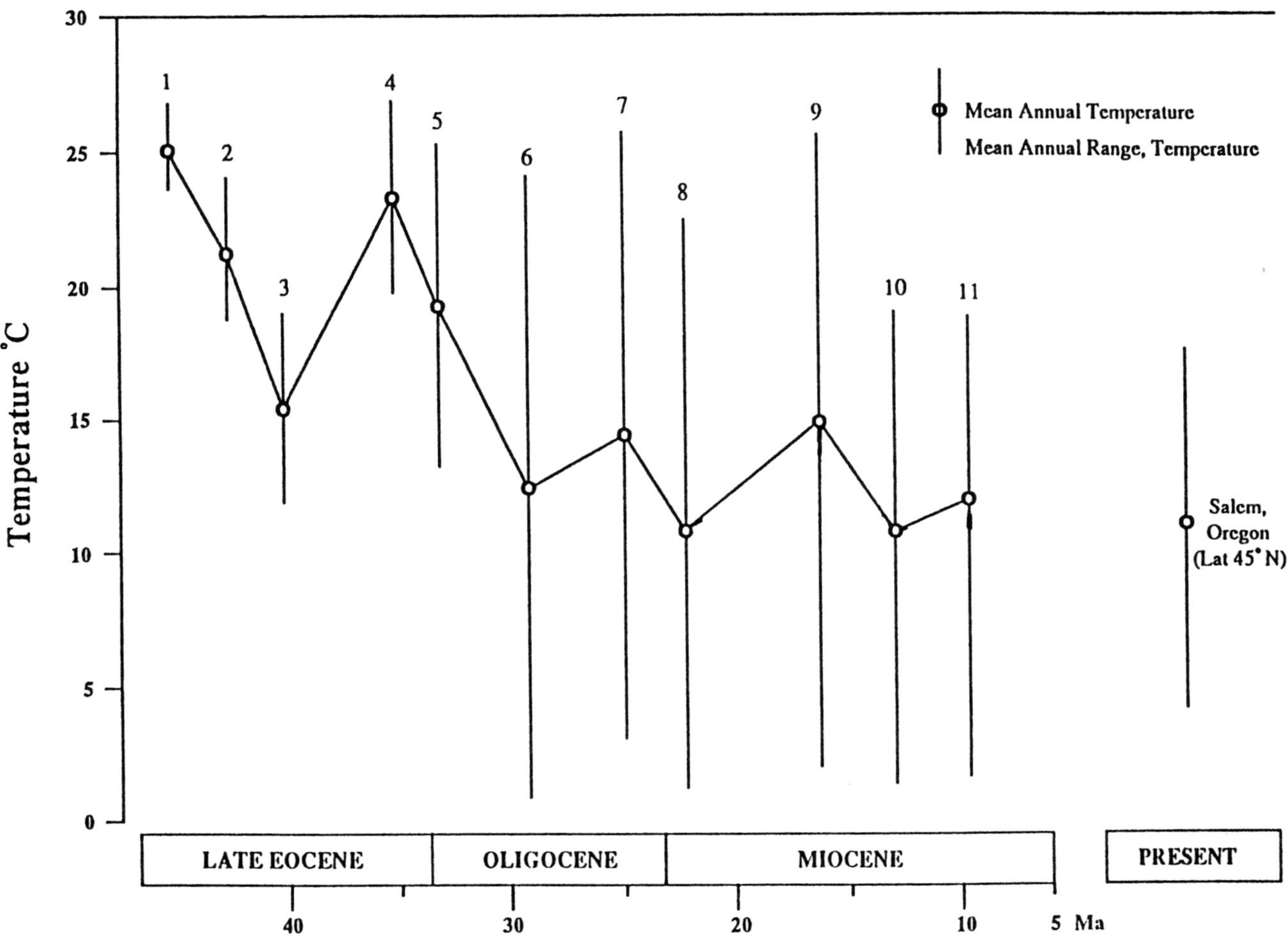

Figure 6.16. Estimated average annual temperatures (circles) and range of temperatures (lines) for western Oregon and western Washington based on Cenozoic floras. Localities (1–5) Puget Sound Group; (6–11) near Salem, Oregon; (6) Bridge Creek, Willamette; (7) Yaquina; (8) Eagle Creek; (9) Latah and other sites; (10) Faraday; (11) Troutdale. Reprinted from Leopold et al. (1992; based on Wolfe 1978, 1981b, 1992a) with the permission of Princeton University Press.

Mountains, and into southern California (San Diego microfossil floras). In the Plains and along the eastern Rocky Mountains the general sequence of vegetation was from an Early Eocene moist subtropical community with abundant paleotropical elements (Wind River flora), a drier subtropical phase in the Middle Eocene with transition to one with greater neotropical affinities (Little Mountain flora), a dry cooling period with a decrease in tropical and subtropical species (Green River flora), and a Late Eocene phase when the vegetation was subhumid temperate.

It is becoming clear from the evidence surveyed to this point that changes in climate proceed in stepwise fashion and each new level ushers in some modification of the existing vegetation. Such events allow the perpetuation of restricted, shifting communities growing in edaphically or slope-controlled environments (e.g., subhumid elements within an overall more mesic environment). Eventually a point is reached, however, where further changes cause not just a modification of existing vegetation, but cross a threshold wherein the region can no longer support the formation. Such changes may be accelerated, slowed, or temporarily redirected by interplay of the myriad factors shown at the end of Chapter 2. Nonetheless, if the overall environmental trend is directional, even though interrupted, a major change in regional vegetation will occur. This was the case in the the Middle Eocene, and especially in the Oligocene through the Early Miocene, when forested areas in the Plains region gave way to a community of smaller trees, shrubs, herbs, and an increasing number of grasses, with arborescent forms confined mostly to riverine habitats. During the Oligocene the canopy became more open and grasses increased to approach a savanna in the modern sense. The grass–herb component become increasingly prominent with the continued rise of the Rocky Mountains and with the next major temperature decline beginning in the Middle Miocene. An important factor in the development and maintenance of savanna is reduced and strongly seasonal rainfall, and the dry season occurs in

Table 6.18. Summary of North American vegetation types and estimated MATs, Middle Eocene through Early Miocene.

Region	Vegetation	MAT
	Middle Eocene	
Southeast (~30° N)		
Claiborne, Jackson		
Coastal	Tropical	
Lowland	Tropical dry, sand pine scrub, flood plain, deciduous [elements of mixed mesophytic, southern mixed hardwood, beech (Claiborne)–maple (*Acer*?, Jackson), maple (*Acer*?, Jackson)–basswood (*Tilia*, Claiborne), oak–hickory, oak–chestnut]	24°C (20– 28°C)
Upland	Deciduous	
Highland	Early Appalachian montane coniferous (*Picea*, *Tsuga*)	Occasional frost
Northeast		
Ellesmere Island (81° N)	Mixed hardwood–coniferous (lake states type)	12–13°C
West		
High Plains, eastern Rocky Mts.		
Little Mountain	Savanna, woodland	
Tipperary, Kisinger lakes	Subtropical	
Green River		
300 m	Savanna, woodland (oak, piñon), tropical dry forest	13°C
1300 m	Deciduous	
1700 m	Mixed hardwood–coniferous	
Volcanic highlands		
Germer (600 m)	Mixed hardwood–coniferous	
Thunder Mt. (1550 m)	Mixed hardwood–coniferous to western montane coniferous	8.5°C
S. California		
Elsinore		
Lowland	Paratropical, deciduous	
Upland	Mixed hardwood–coniferous	
N. California		
Susanville	Tropical rain forest	27°
Northwest		
Oregon		
Clarno	Tropical to paratropical	CLAMP, 16°C
Washington		
Republic	Mixed hardwood–coniferous, early western montane coniferous	12–13°C
British Columbia		
Princeton	Mixed hardwood–coniferous	
Alaska		
Lowland	Notophyllous broad-leaved evergreen	CLAMP, 15–16°C
Inland, upland	Deciduous, early mixed hardwood–coniferous	
	Late Eocene	
West		
High Plains, eastern Rocky Mts.		
Colorado		
Florissant		
1200 m	Savanna-woodland (oak, pine), tropical dry	18°C (CLAMP, 12.5°C)
Montana (Late Eocene to earliest Oligocene)		
Beaverhead Basins, Metzel Ranch, Mormon Creek, York, Ruby		
Lowland	Chaparral, woodland, deciduous	
Upland	Mixed hardwood–coniferous	
	Western montane coniferous	

continued

Table 6.18. continued

Region	Vegetation	MAT
Volcanic highlands		
Nevada		
Copper Basin, Bull Run (1200–1500 m)	Mixed hardwood–coniferous, western montane coniferous	11°C
N. California		
La Porte, Lower Cedarville	Tropical rain?	24°C (CLAMP 20°C)
Northwest		
Oregon		
Comstock, Goshen	Tropical rain	24°C (CLAMP 20°C)
Alaska		
Coastal	Subtropical, notophyllous broad-leaved evergreen?	16°C (CLAMP)
Inland, upland	Deciduous, mixed hardwood–coniferous	15°C (CLAMP)
	Oligocene	
Southeast		
Catahoula		
Vicksburg		
Coastal	Tropical	
Inland	Tropical toward deciduous (mixed mesophytic)	
Upland	Deciduous (mixed mesophytic)	
West		
High Plains, eastern	Shrubland, woodland, savanna	
Rocky Mts.		
Colorado		
Creede		
Lowland	Chaparral (mountain mahogany), woodland (piñon–juniper)	
Upland	Western montane coniferous (fir–pine, fir–spruce)	4.5°C
Northwest		
Oregon		
Bridge Creek, Fossil	Deciduous	10–12°C
Rujada		
Coastal	Notophyllous broad-leaved evergreen	
Inland	Deciduous	
Upland	Western montane coniferous	
Alaska	Deciduous	4.5–7°C
MacKenzie Delta		
Richards	Deciduous	
	Early Miocene	
Southeast	Warm-temperate to subtropical?	
Northeast		
Vermont		
Brandon	Deciduous (mixed mesophytic)	17°C
Devon Island		
Haughton	Mixed hardwood–coniferous (lake states type)	0–11°C
Banks Island		
Mary Sachs Gravel	Mixed hardwood–coniferous (lake states type)	8–12°C
West		
High Plains	Grassy savanna, woodland?	
Northern Rocky Mts.		
Lowland	Deciduous	
Upland	Western montane coniferous?	
Nevada		
Buffalo Canyon		
Lowland	Woodland, shrubland, notophyllous broad-leaved evergreen	10°C
Upland	Mixed hardwood–coniferous	
Sutro		
Lowland	Notophyllous broad-leaved evergreen	
Upland	Western montane coniferous	

Table 6.18. continued

Region	Vegetation	MAT
S. California		
Tehachapi		
Lowland	Shrubland	
Upland	Woodland (evergreen oak, cypress, piñon)	
Oregon		
Alvord Creek	Mixed hardwood–coniferous	
Northwest		
Alaska		
Seldovia Point	Deciduous, mixed hardwood–coniferous	6–7°C

Most plant formation and association names follow terminology used for modern communities discussed in Chapter 1 and listed in Table 1.1. Remnants of older communities are named according to the terminology discussed in Chapter 5 (Vegetation).

the cooler half of the year (Buffalo Canyon and related floras). Savanna vegetation has mostly disappeared from North America through continued development of the Grassland Formation and more recently by spread of adjacent woodlands. Fires that normally constrained woodland expansion are now mostly prevented. The greatest extent of savanna was in recent times as a transition community between the grassland and the oak–hickory association of the deciduous forest formation (Chapter 2).

By the Middle Eocene, scattered highlands in the Rocky Mountains supported *Picea*, *Pinus*, and *Betula*. In the moister volcanic highlands to the west *Abies*, *Larix*, *Picea*, *Thuja*, *Tsuga*, *Alnus*, *Castanea*, *Populus*, *Quercus*, and others were already established as a broad-leaved deciduous forest, a mixed hardwood–conifer forest with many present-day Asian representatives, and a western montane coniferous forest at the highest elevations (above 1200 m in Nevada–Idaho). The Republic flora preserves the oldest record of co-occurring coniferous forest elements (*Abies*, *Picea*, *Pinus*, *Pseudolarix*, *Tsuga*, *Chamaecyparis*, *Thuja*, and *Betula* of the *B. papyrifera–B. occidentalis* complex). Thus, the earliest presence of the Rocky Mountain Montane Coniferous Forest, like the less extensive and well-defined Appalachian counterpart, can be placed in the Middle Eocene. Many of these elements extended into the Arctic at progressively lower elevations down to ~300 m. The mixed hardwood–coniferous forest is an ecotonal community between the lower elevation deciduous forest and the higher elevation montane coniferous forest. It is more restricted in distribution today because of higher elevations, sharper climatic gradients, and greater community definition; however, in the Middle Eocene through the Early Miocene it was widespread as a transitional vegetation type. During the Middle Eocene and Early Oligocene microthermal angiosperm families such as the Aceraceae, Betulaceae, and Rosaceae underwent rapid diversification and throughout North America modernization of the vegetation to the generic level was well underway. In the central Rocky Mountains, subtropical elements disappeared during the Oligocene and the Cordilleran flora took on a more modern aspect.

To the west from about central California northward to just beyond the present Canadian border, more tropical vegetation with Old World affinities fringed the coast.

During the drying trend of the Middle Eocene, early preadapted components of a dry vegetation type appeared (e.g., *Ephedra*, *Celtis*, *Ocotea* in the Green River flora). These dry-habitat plants formed the shrubland/chaparral–woodland–savanna formation in the western United States during the Middle to Late Eocene, individual elements of which eventually diversified, coalesced, and expanded into the desert formation with the rise of the Coast Ranges, Cascade Mountains, and Sierra Nevada in the Neogene.

Thus, by the end of the Early Miocene the older tropical dry forest and much of the notophyllous broad-leaved evergreen vegetation had disappeared. The principal plant communities of North America were the remaining elements of a tropical community along the southern coasts, deciduous forest (sand pine scrub and other components of the pine woods association; oak–chestnut; oak–hickory; southern mixed hardwoods; flood-plain forest), elements of an Appalachian montane coniferous forest, lake states forest (to the far north), shrubland/chaparral–woodland–savanna, mixed hardwood–conifer forest, and western montane coniferous forest. There is no evidence as yet for extensive tundra, grassland, or desert; elements of the boreal forest were only just being assembled in the western volcanic highlands and in the high latitudes, preadapted for the next downshift in temperatures that would begin in the Middle Miocene.

References

Abbott, P. L. 1981. Cenozoic paleosols, San Diego, California. Catena 8: 223–237.

Asaro, F. et al. (+ 6 authors). 1988. Possible world-wide Middle Miocene iridium anomaly and its relationship to periodicity of impacts and extinctions. Global Catastrophes in Earth History: An Interdisciplinary Conference on Impacts, Volcanism, and Mass Mortality. Lunar Planetary Institute, Houston, Tx. pp. 6–7.

Axelrod, D. I. 1939. A Miocene flora from the western border of the Mohave Desert. Carnegie Inst. Wash. Contrib. Paleontol. 516: 1–128.

———. 1944. The Alvord Creek flora. Carnegie Inst. Wash. Contrib. Paleontol. 553: 225–262.

———. 1952. A theory of angiosperm evolution. Evolution 6: 29–60.

———. 1957. Age-curve analysis of angiosperm floras. J. Paleontol. 31: 273–280.

———. 1961. How old are the angiosperms? Am. J. Sci. 259: 447–459.

———. 1966a. A method for determining the altitudes of Tertiary floras. Palaeobotanist 14: 144–171.

———. 1966b. The Eocene Copper Basin flora of northeastern Nevada. Univ. Calif. Publ. Geol. Sci. 59: 1–125.

———. 1966c. Potassium-argon ages of some western Tertiary floras. Am. J. Sci. 264: 497–506.

———. 1979. Age and origin of Sonoran Desert vegetation. California Academy of Sciences, San Francisco. Occasional Papers 132.

———. 1984. An interpretation of Cretaceous and Tertiary biota in polar regions. Palaeogeogr., Palaeoclimatol., Palaeoecol. 45: 105–147.

———. 1985a. Rise of the grassland biome, central North America. Bot. Rev. 51: 163–201.

———. 1985b. Miocene floras from the Middlegate Basin, west-central Nevada. Univ. Calif. Publ. Geol. Sci. 129.

———. 1987. The Late Oligocene Creede flora, Colorado. Univ. Calif. Publ. Geol. Sci. 130: 1–235.

———. 1990. Environment of the Middle Eocene (45 Ma) Thunder Mountain flora, central Idaho. Natl. Geogr. Res. 6: 355–361.

———. 1991. The Early Miocene Buffalo Canyon flora of western Nevada. Univ. Calif. Publ. Geol. Sci. 135: 1–76.

———. 1992. Climatic pulses, a major factor in legume evolution. In: P. S. Herendeen and D. L. Dilcher (eds.), Advances in Legume Systematics: Part 4. The Fossil Record. Royal Botanic Gardens, Kew, U.K. pp. 259–279.

———, M. T. K. Arroyo, and P. H. Raven. 1991. Historical development of temperate vegetation in the Americas. Rev. Chil. Hist. Nat. 64: 413–446.

Axelrod, D. I. and H. P. Bailey. 1976. Tertiary vegetation, climate, and altitude of the Rio Grande depression, New Mexico–Colorado. Paleobiology 2: 235–254.

Barghoorn, E. S. 1950. The Brandon lignite. Vermont Bot. Bird Clubs Joint Bull. 18: 21–36.

———. 1951. Age and environment: a survey of North American Tertiary floras in relation to paleoecology. J. Paleontol. 25: 736–744.

———. 1953. Evidence of climatic change in the geologic record of plant life. In: H. Shapley (ed.), Climatic Change—Evidence, Causes, and Effects. Harvard University Press, Cambridge, MA. pp. 235–248.

Barron, E. J. 1983. A warm, equable Cretaceous: the nature of the problem. Earth-Sci. Rev. 19: 305–338.

Basinger, J. F. 1976. *Paleorosa similkameenensis*, gen. et sp. nov., permineralized flowers (Rosaceae) from the Eocene of British Columbia. Can. J. Bot. 54: 2293–2305.

———. 1981. The vegetative body of *Metasequoia milleri* from the Middle Eocene of southern British Columbia. Can. J. Bot. 59: 2379–2410.

———. 1984. Seed cones of *Metasequoia milleri* from the Middle Eocene of British Columbia. Can. J. Bot. 62: 281–289.

———, D. R. Greenwood, and T. Sweda. 1994. Early Tertiary vegetation of Arctic Canada and its relevance to paleoclimatic interpretations. In: M. C. Boulter and H. C. Fisher (eds.), Cenozoic Plants and Climates of the Arctic. NATO ASI Series 1:Global Environmental Change. Springer-Verlag, Berlin. Vol. 27, pp. 175–198.

——— and G. W. Rothwell. 1977. Anatomically preserved plants from the Middle Eocene (Allenby Formation) of British Columbia. Can. J. Bot. 55: 1984–1990.

Becker, H. F. 1960. The Tertiary Mormon Creek flora from the upper Ruby River Basin in southwestern Montana. Palaeontograph., Abt. B 107: 83–126.

———. 1961. Oligocene Plants from the Upper Ruby River Basin, Southwestern Montana. Geological Society of America, Boulder, CO. Mem. 82.

———. 1969. Fossil plants of the Tertiary Beaverhead Basins in southwestern Montana. Palaeontograph., Abt. B 127: 1–142.

———. 1972. The Metzel Ranch flora of the Upper Ruby River Basin, southwestern Montana. Palaeontograph., Abt. B 141: 1–61.

———. 1973. The York Ranch flora of the upper Ruby River Basin, southwestern Montana. Palaeontograph. Abt. B, 143: 1–93.

Bennett, K. D. 1990. Milankovitch cycles and their effects on species in ecological and evolutionary time. Paleobiology 16: 11–21.

Berggren, W. A., D. V. Kent, M.-P. Aubry, and J. Hardenbol (eds.). 1995. Geochronology, Time Scales and Global Stratigraphic Correlation. Society of Sedimentary Geology, Tulsa, OK. Special Publ. 54.

——— and D. R. Prothero. 1992. Eocene–Oligocene climatic and biotic evolution: an overview. In: D. R. Prothero and W. A. Berggren (eds.), Eocene–Oligocene Climatic and Biotic Evolution. Princeton University Press, Princeton, N.J. pp. 1–28.

Berry, E. W. 1916. The Lower Eocene Floras of Southeastern North America. U.S. Geological Survey, Reston, VA. Prof. Paper 91, pp. 1–481.

Birkenmajer, K. 1987. Tertiary glacial and interglacial deposits, South Shetland Islands, Antarctica: geochronology versus biostratigraphy (a progress report). Bull. Pol. Acad. Sci. Earth Sci. 36: 133–145.

Brown, R. W. 1939. Fossil leaves, fruits, and seeds of *Cercidiphyllum*. J. Paleontol. 13: 485–499.

———. 1940. New species and changes of name in some American fossil floras. J. Wash. Acad. Sci. 30: 344–356.

———. 1946. Alterations in some fossil and living floras. J. Wash. Acad. Sci. 36: 344–355.

———. 1959. A bat and some plants from the Upper Oligocene of Oregon. J. Paleontol. 33: 125–129.

Burnham, R. J. 1994. Paleoecological and floristic heterogeneity in the plant-fossil record—an analysis based on the Eocene of Washington. Bull. U.S. Geol. Surv. 2085–B: 1–36.

Bůžek, C., Z. Kvaček, and S. R. Manchester. 1989. Sapindaceous affinities of the *Pteleaecarpum* fruits from the Tertiary of Eurasia and North America. Bot. Gaz. 150: 477–489.

Call, V. B. and D. L. Dilcher. 1992. Investigations of angiosperms from the Eocene of southeastern North America: samaras of *Fraxinus wilcoxiana* Berry. Rev. Palaeobot. Palynol. 74: 249–266.

———. 1994. *Parvileguminophyllum coloradensis*, a new combination for *Mimosites coloradensis* Knowlton, Green River Formation of Utah and Colorado. Rev. Palaeobot. Palynol. 80: 305–310.

———. 1997. The fossil record of *Eucommia* (Eucommiaceae) in North America. Am. J. Bot. 84:798–814.

Call, V. B., D. M. Erwin, and R. A. Stockey. 1993. Further observations on *Paleorosa similkameenensis* (Rosaceae) from the Middle Eocene Princeton Chert of British Columbia, Canada. Rev. Palaeobot. Palynol. 78: 277–291.

Call, V. B. and R. A. Stockey. 1988a. Permineralized fruits and seeds from the Princeton Chert (Middle Eocene) of British Columbia: Araceae. Am. J. Bot. 75: 1099–1113.

———. 1988b. Permineralized fruits and seeds from the Princeton Chert (Middle Eocene) of British Columbia: Lythraceae. Can. J. Bot. 66: 303–312.

———. 1989. Permineralized fruits and seeds from the Princeton Chert (Middle Eocene) of British Columbia: Nymphaeaceae. Bot. Gaz. 150: 207–217.

———. 1990a. Permineralized fruits and seeds from the Princeton Chert (Middle Eocene) of British Columbia: Vitaceae. Can. J. Bot. 68: 288–295.

———. 1990b. Vegetative remains of the Magnoliaceae from the Princeton Chert (Middle Eocene) of British Columbia. Can. J. Bot. 68: 1327–1339.

———. 1990c. Vegetative remains of the Rosaceae from the Princeton Chert (Middle Eocene) of British Columbia. Int. Assoc. Wood Anat. Bull. 11: 261–280.

———. 1991. Fruits and seeds from the Princeton Chert (Middle Eocene) of British Columbia: Rosaceae (Prunoideae). Bot. Gaz. 152: 369–379.

——— and K. B. Pigg. 1991. The Princeton Chert: evidence for in situ aquatic plants. Rev. Palaeobot. Palynol. 70: 173–185.

Cevallos-Ferriz, S. R. 1995. Fruits of *Ribes* from the Princeton Chert, British Columbia, Canada. Am. J. Bot. 82: 84.

Chaney, R. W. 1924. A comparative study of the Bridge Creek flora and the modern redwood forest. Carnegie Inst. Wash. Contrib. Palaeontol. 349: 1–22.

———. 1927. Geology and palaeontology of the Crooked River Basin, with special reference to the Bridge Creek flora. Carnegie Inst. Wash. Contrib. Palaeontol. 346: 45–138.

——— and E. I. Sanborn. 1933. The Goshen flora of west central Oregon. Carnegie Inst. Wash. Contrib. Paleontol. 439.

Christe, R. L. 1964. Geological Reconnaissance of Northeastern Ellesmere Island, District of Franklin, Canada. Geological Survey of Canada, Ottawa, Memoir 331.

——— and G. E. Rouse. 1976. Eocene Beds at Lake Hazen, Northern Ellesmere Island. Geological Survey of Canada, Ottawa, Report on Activities Paper 76–3C, pp. 153–156.

Collinson, M. E., K. Fowler, and M. C. Boulter. 1981. Floristic changes indicate a cooling climate in the Eocene of southern England. Nature 291: 315–317.

Cope, M. J. 1981. Products of natural burning as a component of the dispersed organic matter of sedimentary rocks. In: J. Brooks (ed.), Organic Maturation Studies and Fossil Fuel Exploration. Academic Press, New York. pp. 89–109.

Crane, P. R. and R. A. Stockey. 1985. Growth and reproductive biology of *Joffrea speirsii* gen. et sp. nov., a *Cercidiphyllum*-like plant from the Late Paleocene of Alberta, Canada. Can. J. Bot. 63: 340–364.

———. 1987. *Betula* leaves and reproductive structures from the Middle Eocene of British Columbia, Canada. Can. J. Bot. 65: 2490–2500.

Crepet, W. L. and C. P. Daghlian. 1980. Castaneoid inflorescences from the Middle Eocene of Tennessee and the diagnostic value of pollen (at the subfamily level) in the Fagaceae. Am. J. Bot. 67: 739–757.

———. 1982. Euphorbioid inflorescences from the Middle Eocene Claiborne Formation. Am. J. Bot. 69: 258–266.

——— and M. Zavada. 1980. Investigations of angiosperms from the Eocene of North America: a new juglandaceous catkin. Rev. Palaeobot. Palynol. 30: 361–370.

Crepet, W. L. and D. L. Dilcher. 1977. Investigations of angiosperms from the Eocene of North America: a mimosoid inflorescence. Am. J. Bot. 64: 714–725.

Crepet, W. L., P. S. Herendeen, and F. W. Potter, Jr. 1975. Investigations of angiosperms from the Eocene of North America: a catkin with juglandaceous affinities. Am. J. Bot. 62: 813–823.

Crepet, W. L. and T. F. Stuessy. 1978. A reinvestigation of the fossil *Viguiera cronquistii* (Compositae). Brittonia 30: 483–491.

Daghlian, C. P. and W. L. Crepet. 1983. Oak catkins, leaves and fruits from the Oligocene Catahoula Formation and their evolutionary significance. Am. J. Bot. 70: 639–649.

Crepet, W. L. and T. Delevoryas. 1980. Investigations of Tertiary angiosperms: a new flora including *Eomimosoidea plumosa* from the Oligocene of eastern Texas. Am. J. Bot. 67: 309–320.

Crepet, W. L. and D. L. Dilcher. 1972. Middle Eocene sabaloid palms. Indiana Acad. Sci. Proc. 81: 94.

Dickinson, K. A. and J. S. Leventhal. 1988. The geology, carbonaceous materials, and origin of the epigenetic uranium deposits in the Tertiary Sespe Formation in Ventura County, California. Bull. U.S. Geol. Surv. 1771: 1–18.

Dilcher, D. L. 1963. Cuticular analysis of Eocene leaves of *Ocotea obtusifolia*. Am. J. Bot. 50: 1–8.

———. 1968. Revision of Eocene palms from southeastern North America based upon cuticular analysis. Am. J. Bot. 55: 725.

———. 1969. *Podocarpus* from the Eocene of North America. Science 164: 299–301.

———. 1973. A paleoclimatic interpretation of the Eocene floras of southeastern North America. In: A. Graham (ed.),Vegetation and Vegetational History of Northern Latin America. Elsevier Science Publishers, Amsterdam. pp. 39–59.

——— and C. P. Daghlian. 1977. Investigations of angiosperms from the Eocene of southeastern North America: *Philodendron* leaf remains. Am. J. Bot. 64: 526–534.

Dilcher, D. L. and G. E. Dolph. 1970. Fossil leaves of *Dendropanax* from Eocene sediments of southeastern North America. Am. J. Bot. 57: 153–160.

Dilcher, D. L. and S. R. Manchester. 1986. Investigations of angiosperms from the Eocene of North America: leaves of the Engelhardieae (Juglandaceae). Bot. Gaz. 147: 189–199.

———. 1988. Investigations of angiosperms from the Eocene of North America: a fruit belonging to the Euphorbiaceae. Tert. Res. 9: 45–58.
Dilcher, D. L. and J. F. McQuade. 1967. A morphological study of *Nyssa* endocarps from Eocene deposits in western Tennessee. Bull. Torrey Bot. Club 94: 35–40.
Dilcher, D. L., F. W. Potter, Jr., and W. L. Crepet. 1976. Investigations of angiosperms from the Eocene of North America: juglandaceous winged fruits. Am. J. Bot. 63: 532–544.
Epis, R. C. and C. E. Chapin. 1974. Stratigraphic nomenclature of the Thirtynine Mile volcanic field, central Colorado. Bulletin U.S. Geological Survey 1395–C: 1–22.
Erwin, D. M. 1987. Silicified palm remains from the Middle Eocene (Allenby Formation) of British Columbia. Am. J. Bot. 74: 681.
——— and R. A. Stockey. 1989. Permineralized monocotyledons from the Middle Eocene Princeton Chert (Allenby Formation) of British Columbia: Alismataceae. Can. J. Bot. 67: 2636–2645.
———. 1990. Sapindaceous flowers from the Middle Eocene Princeton Chert (Allenby Formation) of British Columbia, Canada. Can. J. Bot. 68: 2025–2034.
———. 1991a. *Soleredera rhizomorpha* gen. et sp. nov., a permineralized monocotyledon from the Middle Eocene Princeton Chert of British Columbia, Canada. Bot. Gaz. 152: 231–247.
———. 1991b. Silicified monocotyledons from the Middle Eocene Princeton Chert (Allenby Formation) of British Columbia, Canada. Rev. Palaeobot. Palynol. 70: 147–162.
———. 1992. Vegetative body of a permineralized monocotyledon from the Middle Eocene Princeton Chert of British Columbia. Cour. Forsch.-Inst. Senckenberg 147: 309–327.
———. 1994. Permineralized monocotyledons from the Middle Eocene Princeton Chert (Allenby Formation) of British Columbia, Canada: Arecaceae. Paleontograph., Abt. B 234: 19–40.
Estes, R. and J. H. Hutchison. 1980. Eocene lower vertebrates from Ellesmere Island, Canadian Arctic Archipelago. Palaeogeogr., Palaeoclimatol., Palaeoecol. 30: 325–347.
Eugster, H. P. 1982. Climatic significance of lake and evaporite deposits. In: Geophysics Study Committee (eds.), Climate in Earth History. National Academy Press, Washington, D.C. pp. 105–111.
——— and L. A. Hardie. 1978. Saline lakes. In: A. Lerman (ed.), Lakes—Chemistry, Geology, Physics. Springer-Verlag, Berlin. pp. 237–293.
Evanoff, E., D. R. Prothero, and R. H. Lander. 1992. Eocene–Oligocene climatic change in North America: the White River Formation near Douglas, east-central Wyoming. In: D. R. Prothero and W. A. Berggren (eds.), Eocene–Oligocene Climatic and Biotic Evolution. Princeton University Press, Princeton, N.J. pp. 116–130.
Evernden, J. F., D. E. Savage, G. H. Curtis, and G. T. James. 1964. Potassium-argon dates and the Cenozoic mammalian chronology of North America. Am. J. Sci. 262: 145–198.
Eyde, R. H. 1997. Fossil record and ecology of *Nyssa* (Cornaceae), Bot. Rev. 63: 97–123.
Eyde, R. H. and E. S. Barghoorn. 1963. Morphological and paleobotanical studies of the Nyssaceae, II the fossil record. J. Arnold Arb. 44: 328–370.
Eyde, R. H., A. Bartlett, and E. S. Barghoorn. 1969. Fossil record of *Alangium*. Bull. Torrey Bot. Club 96: 288–314.
Folk, R. L. 1955. Petrographic Stratigraphy of the Texas Gulf Coast Tertiary. Guidebook, Tertiary Field Trip. University of Texas Geological Society, Austin, TX. pp. 15–18.
Forest, C. E., P. Molnar, and K. A. Emanuel. 1995. Palaeoaltimetry from energy conservation principles. Nature 374: 347–350.
Frederiksen, N. O. 1980a. Sporomorphs from the Jackson Group (Upper Eocene) and Adjacent Strata of Mississippi and Western Alabama. U.S. Geological Survey, Reston, VA. Professional Paper 1084, pp. 1–75.
———. 1980b. Mid-Tertiary climate of southeastern United States: the sporomorph evidence. J. Paleontol. 54: 728–739.
———. 1981. Middle Eocene to Early Oligocene plant communities of the Gulf Coast. In: J. Gray, A. J. Boucot, and W. B. Berry (eds.), Communities of the Past. Hutchinson Ross Publ. Co., Stroudsburg, PA. pp. 493–549.
———. 1988. Sporomorph Biostratigraphy, Floral Changes, and Paleoclimatology, Eocene and Earliest Oligocene of the Eastern Gulf Coast. U.S. Geological Survey, Reston, VA. Professional Paper 1448, pp. 1–68.
———. 1991. Pulses of Middle Eocene to earliest Oligocene climatic deterioration in southern California and the Gulf Coast. Palaios 6: 564–571.
Gandolfo, M. 1996. The presence of Fagaceae (oak family) in sediments of the Klondike Mountain Formation (Middle Eocene), Republic, Washington. Wash. Geol. 24: 20–21.
Gee, C. T. 1990. On the fossil occurrences of the mangrove palm *Nypa*. Proceedings of the Symposium, Paleofloristic and Paleoclimatic Changes in the Cretaceous and Tertiary, Prague, 1989. pp. 315–319.
Geophysics Study Committee. 1982. Climate in Earth History. National Academy Press, Washington, D.C.
Graham, A. 1996. A contribution to the geologic history of the compositae. In: D. J. N. Hind and H. J. Beentje (eds.), Compositae. Royal Botanic Gardens, Kew. pp. 123–140.
Gray, J. 1960. Temperate pollen genera in the Eocene (Claiborne) flora, Alabama. Science 132: 808–810.
Gregory, K. M. 1994. Palaeoclimate and palaeoelevation of the 35 Ma Florissant flora, Front Range, Colorado. Palaeoclimates 1: 23–57.
———. 1996. Paleoclimate estimates biased by foliar physiognomic responses to increased atmospheric CO_2? Palaeogeogr., Palaeoclimatol., Palaeoecol. 124: 39–51.
——— and C. G. Chase. 1992. Tectonic significance of paleobotanically estimated climate and altitude of the Late Eocene erosion surface, Colorado. Geology 20: 581–585.
Gregory, K. M. and W. C. McIntosh. 1996. Paleoclimate and paleoelevation of the Oligocene Pitch-Pinnacle flora, Sawatch Range, Colorado. Geol. Soc. Amer. Bull. 108: 545–561.
Gregory-Wodzicki, K. M. 1997. The Late Eocene House Ranch Flora, Sevier Desert, Utah: Paleoclimate and Paleoelevation. Palaios 12: 552–567.
Haq, B. U., J. Hardenbol, and P. R. Vail. 1987. Geochronol-

ogy of fluctuating sea levels since the Triassic. Science 235: 1156–1166.

Herendeen, P. S. and D. L. Dilcher. 1990a. Reproductive and vegetative evidence for the occurrence of *Crudia* (Leguminosae, Caesalpinioideae) in the Eocene of southeastern North America. Bot. Gaz. 151: 402–413.

———. 1990b. Fossil mimosoid legumes from the Eocene and Oligocene of southeastern North America. Rev. Palaeobot. Palynol. 62: 339–361.

———. 1990c. *Diplotropis* (Leguminosae, Papilionoideae) from the Middle Eocene of southeastern North America. Syst. Bot. 15: 526–533.

———. 1991. *Caesalpinia* subgenus *Mezoneuron* (Leguminosae, Caesalpinioideae) from the Tertiary of North America. Am. J. Bot. 78: 1–12.

Herendeen, P. S., D. H. Les, and D. L. Dilcher. 1990. Fossil *Ceratophyllum* (Ceratophyllaceae) from the Tertiary of North America. Am. J. Bot. 77: 7–16.

Hickey, L. J., K. R. Johnson, and M. R. Dawson. 1988. The stratigraphy, sedimentology, and fossils of the Haughton Formation: a post-impact crater fill, Devon Island, N.W.T., Canada. Meteoritics 23: 221–231.

Hut, P. et al. (+ 7 authors). 1987. Comet showers as a cause of mass extinctions. Nature 329: 118–126.

Jones, E. L. 1961. Plant microfossils of the Laminated Sediments of the Lower Eocene Wilcox Group in South-Central Arkansas. Thesis, University of Oklahoma, Norman, OK.

Kennett, J. P. and L. D. Stott. 1991. Abrupt deep-sea warming, palaeoceanographic changes and benthic extinctions at the end of the Paleocene. Nature 353: 225–229.

Koch, P. L., J. C. Zachos, and P. D. Gingerich. 1992. Correlation between isotope records in marine and continental carbon reservoirs near the Palaeocene/Eocene boundary. Nature 358: 319–322.

Koeberl, C., C. Poag, W. Reimold, and D. Brandt. 1996. Impact origin of the Chesapeake Bay structure and the source of the North American tektites. Science 271: 1263–1266.

Kvaček, Z., C. Bůžek, and S. R. Manchester. 1991. Fossil fruits of *Pteleaecarpum* Weyland—Tiliaceous, not Sapindaceous. Bot. Gaz. 152: 522–523.

Lakhanpal, R. N. 1958. The Rujada flora of west central Oregon. Univ. Calif. Publ. Geol. Sci. 35: 1–66.

Leopold, E. B., G. Liu, and S. Clay-Poole. 1992. Low-biomass vegetation in the Oligocene? In: D. R. Prothero and W. A. Berggren (eds.), Eocene–Oligocene Climatic and Biotic Evolution. Princeton University Press, Princeton, NJ. pp. 399–420.

——— and H. D. MacGinitie. 1972. Development and affinities of Tertiary floras in the Rocky Mountains. In: A. Graham (ed.), Floristics and Paleofloristics of Asia and Eastern North America. Elsevier Science Publishers, Amsterdam.pp. 147–200.

MacGinitie, H. D. 1941. A Middle Eocene flora from the central Sierra Nevada. Carnegie Inst. Wash. Contrib. Paleontol. 534.

———. 1953. Fossil plants of the Florissant beds, Colorado. Carnegie Inst. Wash. Contrib. Paleontol. 599.

———. 1969. The Eocene Green River flora of northwestern Colorado and northeastern Utah. Univ. Calif. Publ. Geol. Sci. 83: 1–203.

———. 1974. An early Middle Eocene flora from the Yellowstone–Absaroka volcanic province, northwestern Wind River Basin, Wyoming. Univ. Calif. Publ. Geol. Sci. 108: 1–103.

Magallón-Puebla, S. and S. R. S. Cevallos-Ferriz. 1994. *Eucommia constans* N. Sp. fruits from upper Cenozoic strata of Puebla, Mexico: Morphological and anatomical comparison with *Eucommia ulmoides* Oliver. Internat. J. Plant Sci. 155: 80–95.

Manchester, S. R. 1986. Vegetative and reproductive morphology of an extinct plane tree (Platanaceae) from the Eocene of western North America. Bot. Gaz. 147: 200–226.

———. 1989. Attached reproductive and vegetative remains of the extinct American–European genus *Cedrelospermum* (Ulmaceae) from the Early Tertiary of Utah and Colorado. Am. J. Bot. 76: 256–276.

———. 1990. Eocene to Oligocene floristic changes recorded in the Clarno and John Day Formations, Oregon, USA. Proceedings of the Symposium, Paleofloristic and Paleoclimatic Changes in the Cretaceous and Tertiary, Prague, 1989. pp. 183–187.

———. 1991. *Cruciptera*, a new juglandaceous winged fruit from the Eocene and Oligocene of western North America. Syst. Bot. 16: 715–725.

———. 1992. Flowers, fruits, and pollen of *Florissantia*, an extinct malvalean genus from the Eocene and Oligocene of western North America. Am. J. Bot. 79: 996–1008.

———. 1994. Fruits and seeds of the Middle Eocene Nut Beds flora, Clarno Formation, Oregon. Palaeontograph. Am. 58: 1–205.

———. In press. Update on the Megafossil Flora of Florissant, Colorado, USA. Geological Society of America, Boulder, CO. Special Paper.

Manchester, S. R., M. E. Collinson, and K. Goth. 1994. Fruits of the Juglandaceae from the Eocene of Messel, Germany, and implications for Early Tertiary phytogeographic exchange between Europe and western North America. Int. J. Plant Sci. 155: 388–394.

Manchester, S. R. and P. R. Crane. 1983. Attached leaves, inflorescences, and fruits of *Fagopsis*, an extinct genus of fagaceous affinity from the Oligocene Florissant flora of Colorado, U.S.A. Am. J. Bot. 70: 1147–1164.

———. 1987. A new genus of Betulaceae from the Oligocene of western North America. Bot. Gaz. 148: 263–273.

Manchester, S. R., D. L. Dilcher, and W. D. Tidwell. 1986. Interconnected reproductive and vegetative remains of *Populus* (Salicaceae) from the Middle Eocene Green River Formation, northeastern Utah. Am. J. Bot. 73: 156–160.

Manchester, S. R. and M. J. Donoghue. 1995. Winged fruits of Linnaceae (Caprifoliaceae) in the Tertiary of western North America: *Diplodipelta* gen nov. Int. J. Plant Sci. 156: 709–722.

Manchester, S. R. and W. J. Kress. 1993. Fossil bananas (Musaceae): *Ensete oregonense* sp. nov. from the Eocene of western North America and its phytogeographic significance. Am. J. Bot. 80: 1264–1272.

Manchester, S. R. and H. W. Meyer. 1987. Oligocene fossil plants of the John Day Formation, Fossil, Oregon. Oregon Geol. 49: 115–127.

Manchester, S. R. and E. A. Wheeler. 1993. Extinct juglandaceous wood from the Eocene of Oregon and its implications for xylem evolution in the Juglandaceae. Int. Assoc. Wood Anat. Bull. 14: 103–11.

Marty, R., R. Dunbar, J. B. Martin, and P. Baker. 1988. Late Eocene diatomite from the Peruvian coastal desert, coastal upwelling in the eastern Pacific, and Pacific circulation before the terminal Eocene event. Geology 16: 818–822.

Matthews, J. V., Jr. 1987. Plant Macrofossils from the Neogene Beaufort Formation on Banks and Meighen islands, District of Franklin. Current Research, part A. Geological Survey of Canada, Ottawa, Paper 87–1A, pp. 73–87.

——— and L. E. Ovenden. 1990. Late Tertiary plant macrofossils from localities in Arctic/Subarctic North America: a review of the data. Arctic 43: 364–392.

McClammer, J. U. 1978. Paleobotany and Stratigraphy of the Yaquina Flora (Latest Oligocene–Earliest Miocene) of Western Oregon. Thesis, University of Maryland, College Park, MD.

McIntyre, D. J. 1991. Pollen and spore flora of an Eocene forest, eastern Axel Heiberg Island, N.W.T. Geol. Surv. Can. Bull. 403: 83–97.

McKenna, M. C. 1980. Eocene paleolatitude, climate, and mammals of Ellesmere Island. Palaeogeogr., Palaeoclimatol., Palaeoecol. 30: 349–362.

———. 1983. Holarctic landmass rearrangement, cosmic events, and Cenozoic terrestrial organisms. Ann. Missouri Bot. Garden 70: 459–489.

McLeroy, C. A. and R. Y. Anderson. 1966. Laminations of the Oligocene Florissant Lake deposits, Colorado. Geol. Soc. Am. Bull. 77: 605–618.

Meyer, H. W. 1986. An Evaluation of Methods for Estimating Paleoaltitudes Using Tertiary Floras from the Rio Grande Rift Vicinity, New Mexico and Colorado. Ph.D. dissertation, University California, Berkeley, CA.

———. 1992. Lapse rates and other variables applied to estimating paleoaltitudes from fossil floras. Palaeogeogr., Palaeoclimatol., Palaeoecol. 99: 71–99.

——— and S. R. Manchester. 1996. Paleobotanical significance of Eocene flowers, fruits, and seeds from Republic, Washington. Wash. Geol. 24: 25–27.

———. 1997. The Oligocene Bridge Creek Flora of the John Day Formation, Oregon. Univ. Calif. Publs. Geol. Sci. 141: 1–19.

Miller, C. N., Jr. 1973. Silicified cones and vegetative remains of *Pinus* from the Eocene of British Columbia. Contrib. Univ. Mich. Mus. Paleontol. 24: 101–118.

Miller, K. G. et al. (+ 6 authors). 1990. Eocene–Oligocene sea-level changes on the New Jersey coastal plain linked to the deep-sea record. Geol. Soc. Am. Bull. 102: 331–339.

Myers, J. A. 1990. A Bridgerian Age Flora from Del Mar, California. Master's thesis, San Diego State University, San Diego, CA.

———. 1996. Volcanic arcs and vegetation. Wash. Geol. 24: 37–39.

Nichols, D. J. 1970. Palynology in relation to depositional environments of lignite in the Wilcox Group (Early Tertiary) in Texas. Thesis, Pennsylvania State University, University Park, PA.

Norris, G. 1982. Spore-pollen evidence for Early Oligocene high-latitude cool climatic episode in northern Canada. Nature 297: 388–389.

Novacek, M. J. and J. A. Lillegraven. 1979. Terrestrial vertebrates from the later Eocene of San Diego County, California: a conspectus. In: P. L. Abbott (ed.), Eocene Depositional Systems, San Diego, California. Pacific Section, Society for Economic Paleontology and Mineralogy, Los Angeles. pp. 69–79.

Omar, G. et al. (+ 5 authors). 1987. Fission-track dating of Haughton astrobleme and included biota, Devon Island, Canada. Science 237: 1603–1605.

Owens, J. P. et al. (+ 5 authors). 1988. Stratigraphy of the Tertiary Sediments in a 945-Foot-Deep Corehole Near Mays Landing in the Southeastern New Jersey Coastal Plain. U.S. Geological Survey, Reston, VA. Professional Paper 1484, pp. 1–39.

Parrish, J. T., A. M. Ziegler, and C. R. Scotese. 1982. Rainfall patterns and the distribution of coals and evaporites in the Mesozoic and Cenozoic. Palaeogeogr., Palaeoclimatol., Palaeoecol. 40: 67–101.

Peterson, G. L. and P. L. Abbott. 1979. Mid-Eocene climatic change, southwestern California and northwestern Baja California. Palaeogeogr., Palaeoclimatol., Palaeoecol. 26: 73–87.

Phipps, C. J., J. M. Osborn, and R. A. Stockey. 1995. *Pinus* pollen cones from the Middle Eocene Princeton Chert (Allenby Formation) of British Columbia, Canada. Int. J. Plant Sci. 156: 117–124.

Pigg, K. B. and R. A. Stockey. 1996. The significance of the Princeton Chert permineralized flora to the Middle Eocene upland biota of the Okanogan Highlands. Wash. Geol. 24: 32–36.

——— and S. L. Maxwell. 1993. *Paleomyrtinaea*, a new genus of permineralized myrtaceous fruits and seeds from the Eocene of British Columbia and Paleocene of North Dakota. Can. J. Bot. 71: 1–9.

Pomerol, C. (ed.). 1981. Paleogene paleogeography and the geological events at the Eocene–Oligocene boundary. Palaeogeogr., Palaeoclimatol., Palaeoecol. 36: 155–362.

——— and I. Premoli-Silva (eds.). 1986. Terminal Eocene Events. Elsevier Science Publishers, Amsterdam.

Potbury, S. S. 1935. The La Porte flora of Plumas County, California. Carnegie Inst. Wash. Contrib. Paleontol. 465: 29–81.

Povey, D. A. R., R. A. Spicer, and P. C. England. 1994. Palaeobotanical investigation of Early Tertiary palaeoelevations in northeastern Nevada: initial results. Rev. Palaeobot. Palynol. 81: 1–10.

Prothero, D. R. 1985. North American mammalian diversity and Eocene–Oligocene extinctions. Paleobiology 11: 389–405.

———. 1989. Stepwise extinctions and climatic decline during the later Eocene and Oligocene. In: S. K. Donovan (ed.), Mass Extinctions: Processes and Evidence. Columbia University Press, New York. pp. 217–234.

———. 1994. The Late Eocene–Oligocene extinctions. Annu. Rev. Earth Planet. Sci. 22: 145–165.

———. 1995. Geochronology and magnetostratigraphy of Paleogene North American Land Mammal "Ages": an update. In: W. A. Berggren, D. V. Kent, M.-P. Aubry, and J. Hardenbol (eds.), Geochronology, Time Scales and Global Stratigraphic Correlation. Society for Sedimentary Geology, Tulsa, OK. Special Publ. 54, pp. 305–315.

——— and W. A. Berggren (eds.). 1992. Eocene–Oligocene Climatic and Biotic Evolution. Princeton University Press, Princeton, NJ.

Rea, D. K., J. C. Zachos, R. M. Owen, and P. D. Gingerich. 1990. Global change at the Paleocene–Eocene boundary: climatic and evolutionary consequences of tectonic events. Palaeogeogr., Palaeoclimatol., Palaeoecol. 79: 117–128.

Reid, E. M. 1920. A comparative review of Pliocene floras, based on the study of fossil seeds. Quarterly Geol. Soc. Lond. 76: 145–161.

Retallack, G. J. 1983a. Late Eocene and Oligocene Paleosols from Badlands National Park, South Dakota. Geological Society of America, Boulder, CO. Special Paper 193.

———. 1983b. A paleopedological approach to the interpretation of terrestrial sedimentary rocks: the mid-Tertiary fossil soils of Badlands National Park, South Dakota. Geol. Soc. Am. Bull. 94: 823–840.

———. 1990. Soils of the Past: An Introduction to Paleopedology. Unwin-Hyman, Boston.

Robert, C. and H. Chamley. 1991. Development of Early Eocene warm climates, as inferred from clay mineral variations in oceanic sediments. Palaeogeogr., Palaeoclimatol., Palaeoecol. 89: 315–331.

Robison, C. R. and C. P. Person. 1973. A silicified semiaquatic dicotyledon from the Eocene Allenby Formation of British Columbia. Can. J. Bot. 51: 1373–1377.

Roehler, H. W. 1992a. Introduction to Greater Green River Basin Geology, Physiography, and History of Investigations. U.S. Geological Survey, Reston, VA. Professional Paper 1506-A, pp. 1–14.

———. 1992b. Correlation, Composition, Areal Distribution, and Thickness of Eocene Stratigraphic Units, Greater Green River Basin, Wyoming, Utah, and Colorado. U.S. Geological Survey Professional Paper 1506–E, pp. 1–49.

———. 1993. Eocene Climates, Depositional Environments, and Geography, Greater Green River Basin, Wyoming, Utah, and Colorado. U.S. Geological Survey, Reston, VA. Professional Paper 1506–F, pp. 1–74.

Rothwell, G. W. and J. F. Basinger. 1979. *Metasequoia milleri* n. sp., anatomically preserved pollen cones from the Middle Eocene (Allenby Formation) of British Columbia. Can. J. Bot. 57: 958–970.

Rothwell, G. W., R. A. Stockey, and H. Nishida. 1994. Filicaleans of the Middle Eocene Princeton Chert: I. A dryopterid species. Am. J. Bot. 81: 101–102.

Sanborn, E. I. 1935. The Comstock flora of west central Oregon. Carnegie Inst. Wash. Contrib. Paleontol. 465: 1–28.

Schorn, H. E. and W. C. Wehr. 1996. The conifer flora from the Eocene uplands at Republic, Washington. Wash. Geol. 24: 22–24.

Scott, R. A. 1954. Fossil fruits and seeds from the Eocene Clarno Formation of Oregon. Palaeontograph., Abt. B 96: 66–97.

———, E. S. Barghoorn, and E. B. Leopold. 1960. How old are the angiosperms? Am. J. Sci. 258A: 284–299.

Sloan, L. C. 1994. Equable climates during the Early Eocene: significance of regional paleogeography for North American climate. Geology 22: 881–884.

——— and E. J. Barron. 1990. "Equable" climates during Earth history? Geology 18: 489–492.

———. 1992. A comparison of Eocene climate model results to quantified paleoclimatic interpretations. Palaeogeogr., Palaeoclimatol., Palaeoecol. 93: 183–202.

Spackman, W. 1949. The flora of the Brandon lignite: geological aspects and a comparison of the flora with its modern equivalents. Dissertation, Harvard University, Cambridge, MA.

Stockey, R. A. 1984. Middle Eocene *Pinus* remains from British Columbia. Bot. Gaz. 145: 262–274.

———. 1987. A permineralized flower from the Middle Eocene of British Columbia. Am. J. Bot. 74: 1878–1887.

——— and K. B. Pigg. 1991. Flowers and fruits of *Princetonia allenbyensis* (Magnoliopsida; family indet.) from the Middle Eocene Princeton Chert of British Columbia. Rev. Palaeobot. Palynol. 70: 163–172.

———. 1994. Vegetative growth of *Eorhiza arnoldii* Robison and Person from the Middle Eocene Princeton Chert locality of British Columbia. Int. J. Plant Sci. 155: 606–616.

Stucky, R. K. 1990. Evolution of land mammal diversity in North America during the Cenozoic. Curr. Mammal. 2: 375–432.

Sun, Z. and R. A. Stockey. 1991. Lauraceous inflorescences from the Middle Eocene Princeton Chert (Allenby Formation) of British Columbia, Canada. Am. J. Bot. 78: 125.

Tanai, T. and J. A. Wolfe. 1977. Revisions of *Ulmus* and *Zelkova* in the Middle and Late Tertiary of Western North America. U.S. Geological Survey, Reston, VA. Professional Paper 1026, pp. 1–14.

Thelin, G. P. and R. J. Pike. 1991. Landforms of the conterminous United States—a digital shaded-relief portrayal. U.S. Geological Survey, Reston, VA. Ser. Map I–2206.

Thomasson, J. R. 1985. Miocene fossil grasses: possible adaptation in reproductive bracts (lemma and palea). Ann. Missouri Bot. Garden 72: 843–851.

Thompson, G. R., R. W. Fields, and D. Alt. 1982. Land-based evidence for Tertiary climatic variations: northern Rockies. Geology 10: 413–417.

Tiffney, B. H. 1977. Fruits and seeds of the Brandon lignite: Magnoliaceae. Bot. J. Linn. Soc. 75: 299–323.

———. 1979. Fruits and seeds of the Brandon lignite III. *Turpinia* (Staphyleaceae). Brittonia 31: 39–51.

———. 1980. Fruits and seeds of the Brandon lignite, V. Rutaceae. J. Arnold Arb. 61: 1–40.

———. 1981. Fruits and seeds of the Brandon lignite, VI. *Microdiptera* (Lythraceae). J. Arnold Arb. 62: 487–516.

———. 1993. Fruits and seeds of the Tertiary Brandon lignite. VII. *Sargentodoxa* (Sargentodoxaceae). Am. J. Bot. 80: 517– 523.

———. 1994. Re-evaluation of the age of the Brandon lignite (Vermont, USA) based on plant megafossils. Rev. Palaeobot. Palynol. 82: 299–315.

———and E. S. Barghoorn. 1976. Fruits and seeds of the Brandon lignite. I. Vitaceae. Rev. Palaeobot. Palynol. 22: 169–191.

———. 1979. Flora of the Brandon lignite. IV. Illiciaceae. Am. J. Bot. 66: 321–329.

Tiffney, B. H. and K. K. Haggard. 1996. Fruits of Mastixioideae (Cornaceae) from the Paleogene of western North America. Rev. Palaeobot. Palynol. 92: 29–54.

Tiffney, B. H. and A. Traverse. 1994. The Brandon lignite (Vermont) is of Cenozoic, not Cretaceous, age! Northeast. Geol. 16: 215–220.

Ting, W. S. 1975. Eocene pollen assemblage from Del Mar, southern California. Geosci. Man 11: 159.

Tralau, H. 1964. The genus *Nypa* van Wurmb. Konigl. Svenska Vetensk-Akad. Handl. 10: 1–29.

Traverse, A. 1955. Pollen Analysis of the Brandon Lignite of Vermont. U.S. Dept. Interior, Washington, D.C. Bureau of Mines Rept. Invest. 5151.

———. 1994. Palynofloral geochronology of the Brandon lignite of Vermont, USA. Rev. Palaeobot. Palynol. 82: 265–297.

Upchurch, G. R., Jr. and J. A. Wolfe. 1987. Mid-Cretaceous to Early Tertiary vegetation and climate: evidence from fossil leaves and woods. In: E. M. Friis, W. G. Chaloner, and P. R. Crane (eds.), The Origins of Angiosperms and Their Biological Consequences. Cambridge University Press, Cambridge, U.K. pp. 75–105.

Van der Graaff, N. A. and P. Baas. 1974. Wood anatomical variation in relation to latitude and altitude. Blumea 22: 101–121.

Vrba, E. S., G. H. Denton, T. C. Partridge, and L. H. Burckle (eds.), 1995. Paleoclimate and Evolution, with Emphasis on Human Evolution. Yale University Press, New Haven, CT.

Washington Geology. 1996. 24.

Webb, S. D. 1985. Main pathways of mammalian diversification in North America. In: F. G. Stehli and S. D. Webb (eds.), The Great American Biotic Interchange. Plenum, New York. pp. 201–217.

———. 1989. The fourth dimension in North American terrestrial mammal communities. In: D. W. Morris, Z. Abramsky, B. J. Fox, and M. R. Willig (eds.), Patterns in the Structure of Mammalian Communities. Texas Tech University, Lubbock, TX. Special Publ. 28, pp. 181–203.

Wehr, W. C. 1995. Early Tertiary flowers, fruits, and seeds from Washington state and adjacent areas. Wash. Geol. 23: 3–16.

——— and D. Q. Hopkins. 1994. The Eocene orchards and gardens of Republic, Washington. Wash. Geol. 22: 27–34.

Wehr, W. C. and S. R. Manchester. 1996. Paleobotanical significance of Eocene flowers, fruits, and seeds from Republic, Washington. Wash. Geol. 24: 25–27.

Westgate, J. W. and C. T. Gee. 1990. Paleoecology of a middle Eocene mangrove biota (vertebrates, plants, and invertebrates) from southwest Texas. Palaeogeogr., Palaeoclimatol., Palaeoecol. 78: 163–177.

Wheeler, E. A. and J. Landon. 1992. Late Eocene (Chadronian) dicotyledonous woods from Nebraska: evolutionary and ecological significance. Rev. Palaeobot. Palynol. 74: 267–282.

——— and C. A. LaPasha. 1994. Woods of the Vitaceae—fossil and modern. Rev. Palaeobot. Palynol. 80: 175–207.

Whitlock, C. and M. R. Dawson. 1990. Pollen and vertebrates of the Early Neogene Haughton Formation, Devon Island, Arctic Canada. Arctic 43: 324–330.

Wing, S. L. 1987. Eocene and Oligocene floras and vegetation of the Rocky Mountains. Ann. Missouri Bot. Garden. 74: 748–784.

———. 1996. Paleocene–Eocene floral and climate change in the northern Rocky Mountains, USA. International Organization of Palaeobotanists Fifth Quadrennial Conference, Santa Barbara, CA, June–July, 1996. p. 113.

——— and D. R. Greenwood. 1993. Fossils and fossil climate: the case for equable continental interiors in the Eocene. Phil. Trans. R. Soc. Lond. (B) 341: 243–252.

Wodehouse, R. P. 1932. Tertiary pollen; I. Pollen of the living representatives of the Green River flora. Bull. Torrey Bot. Club 59: 313–340.

———. 1933. Tertiary pollen; II. Pollen of the Green River oil shales. Bull. Torrey Bot. Club 60: 479–524.

Wolfe, J. A. 1971. Tertiary climatic fluctuations and methods of analysis of Tertiary floras. Palaeogeogr., Palaeoclimatol., Palaeoecol. 9: 27–57.

———. 1973. Fossil forms of Amentiferae. Brittonia 25: 334–355.

———. 1977. Paleogene floras from the Gulf of Alaska Region. U.S. Geological Survey, Reston, VA. Professional Paper 997, pp. 1–108.

———. 1978. A paleobotanical interpretation of Tertiary climates in the Northern Hemisphere. Am. Sci. 66: 694–703.

———. 1981a. A Chronologic Framework for Cenozoic Megafossil Floras of Northwestern North America and Its Relation to Marine Geochronology. Geological Society of America, Boulder, CO. Special Paper 184, pp. 39–47.

———. 1981b. Paleoclimatic significance of the Oligocene and Neogene floras of the northwestern United States. In: K. J. Niklas (ed.), Paleobotany, Paleoecology, and Evolution. Praeger, New York. Vol. 2, pp. 79–101.

———. 1985. Distribution of major vegetational types during the Tertiary. In: E. T. Sundquist and W. S. Broecker (eds.), The Carbon Cycle and Atmospheric CO_2: Natural Variations Archean to Present. American Geophysical Union, Washington, D.C. Monograph 32, pp. 357–376.

———. 1986. Tertiary floras and paleoclimates of the Northern Hemisphere. Univ. Tennessee Dept. Geol. Sci. Studies Geol. 15: 182–196.

———. 1987. An overview of the origins of the modern vegetation and flora of the northern Rocky Mountains. Ann. Missouri Bot. Garden. 74: 785–803.

———. 1989. North American Eocene vegetation and its climatic implications. EOS (Trans. Am. Geophys. Union) 70: 375.

———. 1992a. Climatic, floristic, and vegetational changes near the Eocene/Oligocene boundary in North America. In: D. R. Prothero and W. A. Berggren (eds.), Eocene–Oligocene Climatic and Biotic Evolution. Princeton University Press, Princeton, N.J. pp. 421–436.

———. 1992b. An analysis of present-day terrestrial lapse rates in the western conterminous United States and their significance to paleoaltitudinal estimates. U.S. Geol. Surv. Bull. 1964: 1–35.

———. 1993. A method of obtaining climatic parameters from leaf assemblages. U.S. Geol. Surv. Bull. 2040: 1–71.

———. 1994a. Alaskan Paleogene climates as inferred from the CLAMP database. In: M. C. Boulter and H. C. Fisher (eds.), Cenozoic Plants and Climates of the Arctic. NATO ASI Series 1: Global Environmental Change. Springer-Verlag, Berlin. Vol. 27, pp. 223–234.

———. 1994b. Tertiary climatic changes at middle latitudes of western North America. Palaeogeogr., Palaeoclimatol., Palaeoecol. 108: 195–205.

——— and E. S. Barghoorn. 1960. Generic change in Tertiary floras in relation to age. Am. J. Sci. 258A: 388–399.

Wolfe, J. A. and R. Z. Poore. 1982. Tertiary marine and nonmarine climatic trends. In: Geophysics Study Committee (eds.), Climate in Earth History. National Academy Press, Washington, D.C. pp. 154–158.

Wolfe, J. A. and H. E. Schorn. 1989. Paleoecologic, paleoclimatic, and evolutionary significance of the Oligocene Creede flora, Colorado. Paleobiology 15: 180–198.

———. 1990. Taxonomic revision of the Spermatopsida of the Oligocene Creede flora, southern Colorado. U.S. Geol. Surv. Bull. 1923: 1–40.

Wolfe, J. A., H. E. Schorn, C. E. Forest, and P. Molnar. 1997. Paleobotanical evidence for high altitudes in Nevada during the Miocene. Science 276: 1672–1675.

Wolfe, J. A. and T. Tanai. 1980. The Miocene Seldovia Point flora from the Kenai Group, Alaska. U.S. Geological Survey, Reston, VA. Professional Paper 1105, pp. 1–52.

———. 1987. Systematics, phylogeny, and distribution of *Acer* (maples) in the Cenozoic of western North America. J. Fac. Sci., Hokkaido Univ. 22: 1–246.

Wolfe, J. A. and W. Wehr. 1987. Middle Eocene dicotyledonous plants from Republic, northeastern Washington. U.S. Geol. Surv. Bull. 1597: 1–25.

Zachos, J. C., B. P. Flower, and H. Paul. 1997. Orbitally paced climate oscillations across the Oligocene/Miocene boundary. Nature 388: 567–570.

Zavada, M. S. and W. L. Crepet. 1981. Investigations of angiosperms from the Middle Eocene of North America: flowers of the Celtidoideae. Am. J. Bot. 68: 924–933.

General Readings

Gregory, K. M. 1994. New prospects in old bubbles. Nature 372: 407–408.

Prothero, D. R. 1994. The Eocene–Oligocene Transition, Paradise Lost. Columbia University Press, New York.

Retallack, G. J., E. A. Bestland, and T. J. Fremd. 1996. Reconstructions of Eocene and Oligocene plants and animals of central Oregon. Oregon Geol. 58: 51–69.

Valdes, P. 1994. Damping seasonal variations. Nature 372: 221.

Note

1. Calcareous nanoplankton is the floating aquatic community comprising small organisms (~5µm) with calcitic walls. The Coccolithoporidae (unicellular green algae) are the principal component and leave remains (nanofossils) in the form of wall platelets called coccoliths and discoasters.

SEVEN

Middle Miocene through Pliocene North American Vegetational History

16.3–1.6 Ma

CONTEXT SUMMARY

During the Middle Miocene through the Pliocene the Appalachian Mountains underwent continued erosion and approached modern elevations (Figs. 7.1–7.3). The Rocky Mountains had undergone uplift to half or more of their present elevation during the Late Cretaceous to Middle Eocene Laramide Revolution; after a lull during the Middle Eocene through the Early Miocene, there was increased tectonic activity beginning ~12 Ma and especially between 7 and 4 Ma. Locally some highlands may have approached or attained modern elevations. The increasingly high mountains and plateaus of Asia and North America deflected the major air streams southward, bringing colder polar air into the middle latitudes of North America. An extensive Antarctic ice sheet further cooled ocean waters and contributed to the spread of seasonally dry climates. The elimination of most of the Asian exotics from the North American flora dates to the Late Miocene–Pliocene as a result of a decline in summer rainfall. The Sierra Nevada attained about two-thirds of their present elevation within the past 10 Ma. They were appreciably elevated at ~5 Ma, stood at ~2100 m at 3 Ma, and have risen ~950 m since 3 Ma (Huber, 1981). The California Coast Ranges and Cascade Mountains attained significant heights by 3 Ma, and there was a rapid rise of the Alaska Range at ~6 Ma.

Temperatures increased between ~18 and 16 Ma (Fig. 3.1). In the absence of major plate reorganization and intense volcanic activity and with increased erosion from continued replacement of the dense evergreen forest by deciduous forest and shrubland (increasing albedo), atmospheric CO_2 concentration decreased and a sharp lowering of temperature occurred in the Middle Miocene between 15 and 10 Ma (Figs. 3.1, 7.3). Eolian dust deposits increased in the Late Cenozoic, suggesting greater aridity (Rea et al., 1985). This is supported by kaolinite records from North Atlantic deep sea sediments (Chamley, 1979). At ~4.8–4.9 Ma global cooling and a marine regression of ~40–50 m combined to isolate the Mediterranean Basin from the ocean and to concentrate large volumes of salt as water evaporated. The biota was destroyed, giving rise to the term Messinian salinity crisis. The evaporation involved ~40 times the present water volume of the Mediterranean. The large amounts of salts stored in the basin reduced ocean salinity by ~6%. This allowed northern hemisphere seawater to freeze at a higher temperature, creating more extensive sea ice and providing positive feedback to the trend of lowering temperatures.

The cooling trend continued into the Early Pliocene, then warmer than present marine waters were evident by Planktic Foraminiferal Zone N19 (Middle Pliocene, ~4 Ma; Cronin and Dowsett, 1991); that was the last time the Earth was significantly warmer than at present (by ~3.5°C; see collection of papers on Pliocene climates in Marine Geology, Vol. 27, 1996). This period culminating in mid-Pliocene warmth, especially evident at the mid- and high latitudes, correlates with an increase in CO_2 concentration to near modern levels (~350 ppm; Crowley, 1991) and also involved greater ocean heat transport (Raymo et al., 1996). At ~3.1 Ma the sea level was an estimated 35 ± 18 m higher along the Atlantic Coast than it is now (Dowsett and Cronin, 1990), and transgressive events were the largest of the Late Neogene (Krantz, 1991). The shoreline extended further inland than at any time since the Eocene (Cronin, 1991). The formation of the Bering Strait at ~3.4 Ma correlates with this rise. There was only seasonal ice in the Arctic Basin, likely occurring during the months of winter darkness. Estimates of winter temperatures north of 70° N (e.g., Baffin Island, Canadian Archipelago) are 20–22°C warmer than at present, and summer temperatures were 6–8°C warmer (Zubakov and Borzenkova, 1990). One effect was wetter (pluvial) climates in the Middle Pliocene

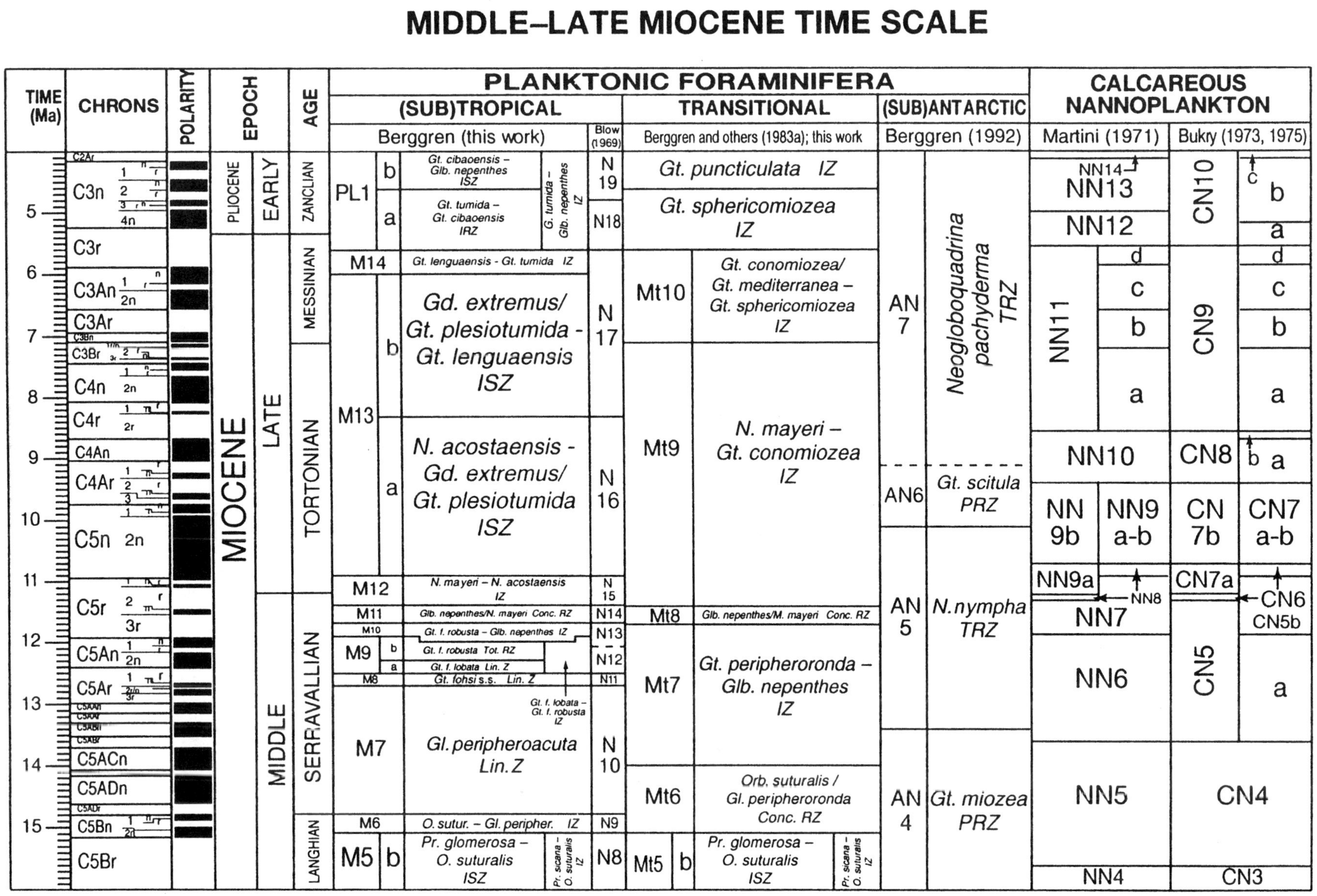

MIDDLE–LATE MIOCENE TIME SCALE

Column	Units (top to bottom)
TIME (Ma)	5; 6; 7; 8; 9; 10; 11; 12; 13; 14; 15
CHRONS	C2Ar; C3n (1, 2, 3, 4n); C3r; C3An (1, 2n); C3Ar; C3Bn; C3Br (2, 3r); C4n (1, 2n); C4r (1, 2r); C4An; C4Ar (1, 2, 3); C5n (1, 2n); C5r (1, 2, 3r); C5An (1, 2n); C5Ar (1, 2n, 3r); C5AAn; C5AAr; C5ABn; C5ABr; C5ACn; C5ADn; C5ADr; C5Bn (1, 2n); C5Br
POLARITY	
EPOCH	PLIOCENE – EARLY; MIOCENE – LATE; MIOCENE – MIDDLE
AGE	ZANCLIAN; MESSINIAN; TORTONIAN; SERRAVALLIAN; LANGHIAN
PLANKTONIC FORAMINIFERA – (SUB)TROPICAL – Berggren (this work)	PL1 b: *Gt. cibaoensis – Glb. nepenthes* ISZ; PL1 a: *Gt. tumida – Gt. cibaoensis* IRZ; *G. tumida – Glb. nepenthes* IZ; M14: *Gt. lenguaensis - Gt. tumida* IZ; M13 b: *Gd. extremus/ Gt. plesiotumida - Gt. lenguaensis* ISZ; M13 a: *N. acostaensis - Gd. extremus/ Gt. plesiotumida* ISZ; M12: *N. mayeri – N. acostaensis* IZ; M11: *Glb. nepenthes/N. mayeri* Conc. RZ; M10: *Gt. f. robusta – Glb. nepenthes* IZ; M9 b: *Gt. f. robusta* Tot. RZ; M9 a: *Gt. f. lobata* Lin. Z; M8: *Gt. fohsi s.s.* Lin. Z; *Gt. f. lobata – Gt. f. robusta* IZ; M7: *Gl. peripheroacuta* Lin. Z; M6: *O. sutur. – Gl. peripher.* IZ; M5 b: *Pr. glomerosa – O. suturalis* ISZ; *Pr. sicana – O. suturalis* IZ
Blow (1969)	N19; N18; N17; N16; N15; N14; N13; N12; N11; N10; N9; N8
TRANSITIONAL – Berggren and others (1983a); this work	*Gt. puncticulata* IZ; *Gt. sphericomiozea* IZ; Mt10: *Gt. conomiozea/ Gt. mediterranea – Gt. sphericomiozea* IZ; Mt9: *N. mayeri – Gt. conomiozea* IZ; Mt8: *Glb. nepenthes/M. mayeri* Conc. RZ; Mt7: *Gt. peripheroronda – Glb. nepenthes* IZ; Mt6: *Orb. suturalis / Gl. peripheroronda* Conc. RZ; Mt5 b: *Pr. glomerosa – O. suturalis* ISZ; *Pr. sicana – O. suturalis* IZ
(SUB)ANTARCTIC – Berggren (1992)	AN 7: *Neogloboquadrina pachyderma* TRZ; AN6: *Gt. scitula* PRZ; AN 5: *N.nympha* TRZ; AN 4: *Gt. miozea* PRZ
CALCAREOUS NANNOPLANKTON – Martini (1971)	NN14; NN13; NN12; NN11 (d, c, b, a); NN10; NN 9b; NN9 a-b; NN9a; NN8; NN7; NN6; NN5; NN4
CALCAREOUS NANNOPLANKTON – Bukry (1973, 1975)	CN10 (c, b, a); CN9 (d, c, b, a); CN8 (b, a); CN 7b; CN7 a-b; CN7a; CN6; CN5 (CN5b, a); CN4; CN3

Figure 7.1. Middle–Late Miocene time scale. Reprinted from Berggren et al. (1995) with the permission of the Society for Sedimentary Geology.

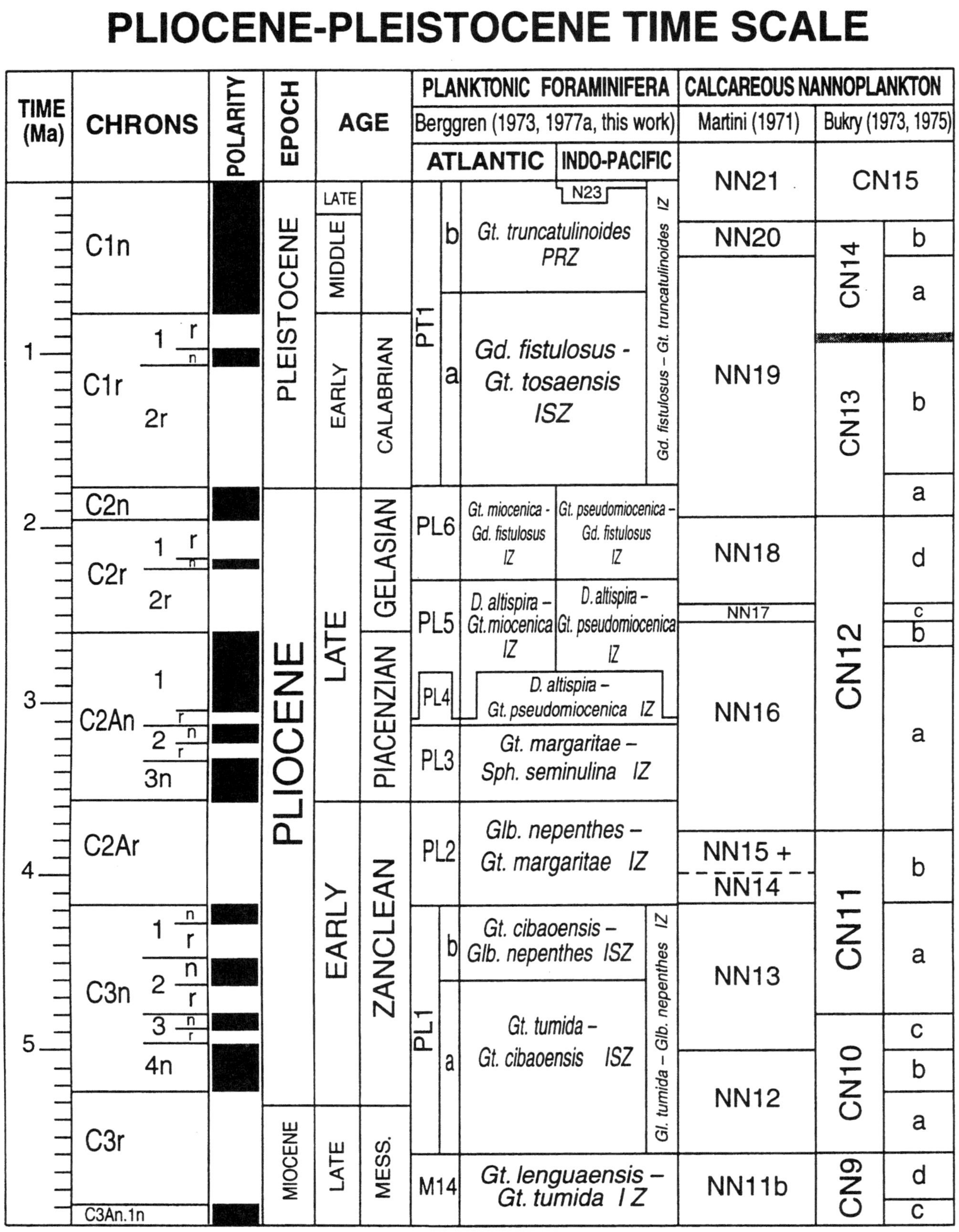

PLIOCENE-PLEISTOCENE TIME SCALE

TIME (Ma)	CHRONS	POLARITY	EPOCH	AGE		PLANKTONIC FORAMINIFERA Berggren (1973, 1977a, this work) ATLANTIC	INDO-PACIFIC	CALCAREOUS NANNOPLANKTON Martini (1971)	Bukry (1973, 1975)
	C1n		PLEISTOCENE	LATE		PT1 b *Gt. truncatulinoides* PRZ	N23	NN21	CN15
				MIDDLE			*Gd. fistulosus – Gt. truncatulinoides* IZ	NN20	CN14 b
								NN19	CN14 a
1	C1r 1 r, n; 2r			EARLY	CALABRIAN	PT1 a *Gd. fistulosus - Gt. tosaensis* ISZ			CN13 b
	C2n		PLIOCENE	LATE	GELASIAN	PL6 *Gt. miocenica - Gd. fistulosus* IZ	*Gt. pseudomiocenica - Gd. fistulosus* IZ		CN13 a
2	C2r 1 r, n; 2r							NN18	CN12 d
						PL5 *D. altispira - Gt. miocenica* IZ	*D. altispira - Gt. pseudomiocenica* IZ	NN17	CN12 c
	C2An 1				PIACENZIAN			NN16	CN12 b
3	C2An 2 r, n, r					PL4 *D. altispira – Gt. pseudomiocenica* IZ			CN12 a
	C2An 3n					PL3 *Gt. margaritae – Sph. seminulina* IZ			
4	C2Ar			EARLY	ZANCLEAN	PL2 *Glb. nepenthes – Gt. margaritae* IZ		NN15 + NN14	CN11 b
	C3n 1 n, r; 2 n, r					PL1 b *Gt. cibaoensis – Glb. nepenthes* ISZ	*Gl. tumida – Glb. nepenthes* IZ	NN13	CN11 a
5	C3n 3 n, r; 4n					PL1 a *Gt. tumida – Gt. cibaoensis* ISZ			CN10 c
								NN12	CN10 b
	C3r		MIOCENE	LATE	MESS.				CN10 a
	C3An.1n					M14 *Gt. lenguaensis – Gt. tumida* IZ		NN11b	CN9 d, c

Figure 7.2. Pliocene–Pleistocene time scale. Reprinted from Berggren et al. (1995) with the permission of the Society for Sedimentary Geology.

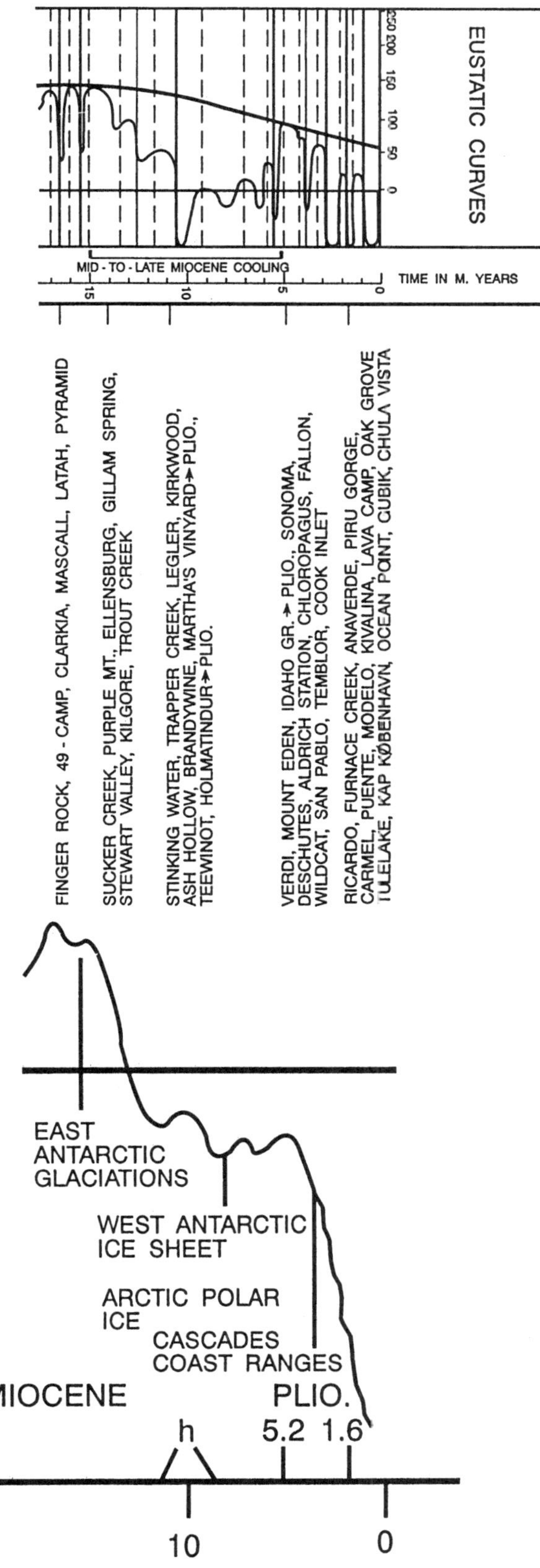

Figure 7.3. Principal fossil floras mentioned in the text plotted with general reference to the paleotemperature record (lower curve) and sea level (upper curve) for the Late Cretaceous through the Early Eocene. h, see Fig. 3.1.

compared to widespread Mio–Pliocene aridity. There is debate as to the extent that this warming affected the East Antarctic ice sheet (Barrett et al., 1992; Marchant et al., 1993).

Although local glaciers dating to the Late Cretaceous were possibly present intermittently at high elevations in the paleo-Brooks Range of Alaska, ice cover in the Arctic Ocean began in the Late Miocene (6–5 Ma) and intensified in the Late Pliocene when ice volume reached about one-half to two-thirds that of the Quaternary glaciations (Crowley and North, 1991; Curry and Miller, 1989; 3.0–2.1 Ma, Repenning and Brouwers, 1992). Some ice was present on Greenland at ~7 Ma (Larsen et al., 1994), and the Greenland ice sheet may have formed by ~3.2 Ma (Leg 105 Shipboard Scientific Party, 1986). There was cooling between ~3.0 and 2.1 Ma, and the eastern North American Laurentide ice sheet began to form ~2.6 Ma, correlating with the abrupt appearance of ice-rafted debris in Arctic Ocean sediments that was probably from mountain glaciers. The first record of permafrost, a frigid Arctic Ocean, and the largest Pliocene glaciation is at 2.4 Ma; between 2.4 and 2.1 Ma perennial ice may have developed at the poles. There are tillites older than 2.1 m.y. in the Puget Sound region of Washington (reversed-magnetized tills of the lower Matuyama Chron), older than 1.75 m.y. in the Cascade Range, and ~2 m.y. in the Yellowstone region of Wyoming. Glaciation intensified between 2.2 and 2.1 Ma when ice briefly extended down the Mississippi River Valley as far south as Iowa (type C till, Afton, Iowa). This cold period ended abruptly at ~2.1 Ma and fluctuating warmer conditions existed between ~1.7 and 1.2 Ma. At ~1.1 Ma cold periods became more severe, the Arctic Ocean had periods of permanent ice cover, and by 850 ka ice was again extending down the Mississippi River Valley initiating the multiple glaciations in the conterminous United States.

The growing ice sheets increased albedo, which served as a positive feedback to cooling temperatures. The Milankovitch cycles are increasingly evident in the younger, widespread, and better preserved sediments; climates downshifted in the now familiar stepwise fashion (Dowsett and Loubere, 1992). As noted by Krantz, "During the [L]ate Pliocene, global climate settled into a phase of low-amplitude glacial–interglacial fluctuations with a 41-ka period forced by the obliquity cycle. This pattern persisted until the [M]iddle Pleistocene transition to a dominant 100-ka cycle beginning at ~0.7 Ma." (1991, p. 169).

By the end of the Pliocene the Gulf of Mexico coastline had retreated to within ~120 km (80 m) inland of its present position and extended from just west of Houston, Texas, along a line from Baton Rouge, Louisiana, to Mobile, Alabama, and along the west coast of Florida. Much of the Florida Peninsula below ~27° N latitude was still inundated. The Atlantic Ocean extended inland ~45 km (30 m) up to Savannah, Georgia, then more narrowly inland through the mid-Atlantic States and New England. The

Hadley and Rossby atmospheric circulation patterns and the seasonal meandering of the polar fronts were near their modern positions and subsequently varied with the Milankovitch cycles.

Ungulate faunas were prominent in North America, doubling every 3–5 m.y. from the Hemingfordian to the Clarendonian NALMA. This suggests a change from woodland savanna toward grassland savanna in parts of the continent. A major change in the composition of North American mammal communities took place in the Mio–Pliocene when two-thirds of the new appearances were introductions from Asia, including some microtid rodents, a flying squirrel, a relative of the lesser panda, a brown bear, elk, hyaenid, and rhinocerotid. Also making their first appearance near the Gauss–Matuyama magnetopolarity boundary (~2.8 Ma) were new immigrants from South America (sloth, armadillo, porcupine). By the end of the Pliocene there was a decline in the rich savanna ungulate fauna, which was gradually replaced by smaller herbivores.

VEGETATION

The southeastern United States and the adjacent continental shelf includes fossiliferous sediments in a petroleum-rich region that has been drilled extensively for oil and gas (Fig. 7.4, Table 7.1). Yet little is known about the vegetational history or terrestrial paleoenvironments of the Gulf Coast region after Vicksburgian (Early Oligocene) time. It is probable that exotic elements continued to disappear, tropical elements decreased, warm-temperate deciduous forest expanded at low to middle elevations, and the Appalachian montane coniferous forest developed further in the highlands. The deciduous forest reached its maximum southern extent during and after the Middle Miocene cooling, and elements were established in eastern Mexico by the Pliocene (Graham, 1976a,b). Pollen of the xerophyte *Ephedra* is present in eastern coastal plain deposits through the Middle Miocene (Frederiksen, 1984). If it was actually growing there, it likely persisted in areas of physiologically arid, deep sandy soil, even during times of increasing precipitation in the southeast during the Middle Tertiary, but was eliminated with the additional impact of lowering temperatures. Modern pollen rain studies show that *Ephedra* pollen can be transported long distances.

In the Pliocene, changes in the vegetation of southwestern peninsular Florida are evident from pollen and spore assemblages in the Tamiami Formation (Pinecrest Beds, 3.5–2 Ma; Table 7.2). Others can be inferred from coastal ostracode and benthic foraminifera faunas (Willard et al., 1993). In the interval between ~4.5 and 3.5 Ma pollen of *Pinus*, *Betula*, *Carya*, *Fagus*, *Fraxinus*, *Juglans*, *Liquidambar*, and *Ulmus* indicate conditions cooler than at present. At about 3.5 Ma climates warmed to the mid-Pliocene maximum, slightly above to modern values, as shown by an increase in *Pinus* and by the decline or disappearance of deciduous hardwoods. The Pliocene warmth was due to some increase in atmospheric CO_2 and to an increase in oceanic heat transport from the tropics. An appreciable increase in CO_2 would probably warm tropical and subtropical regions generally and allow the introduction of tropical elements into subtropical Florida. This is not shown in the pollen and spore assemblages, and according to Raymo and Rau (1992, 1993) the increase in CO_2 in the mid-Pliocene was not appreciable. If the mechanism was oceanic heat transport, it is likely coastal sites would warm but not the interior to the extent of supporting tropical vegetation. There is evidence for this mechanism in that the Isthmus of Panama closed at about this time (3.5 Ma; Coates et al., 1992; Graham, 1992), which strengthened the flow of the Gulf Stream.

The reading of paleoclimates from coastal assemblages is complex, however, because an increase in the pace of ocean currents produces another effect that cools local climates. Upwelling brings cold bottom waters to the surface and can create a local cool interval during an otherwise warming cycle. This was the case in southeastern coastal Mexico where upwelling contributed to the presence of a cool-temperate element in the Middle Pliocene Paraje Solo flora (Graham, 1976a,b). Similarly, in Florida a catastrophic death assemblage of seabirds (extinct cormorant and others) is preserved in the Pinecrest beds where red tides associated with increased upwellings probably caused toxic poisoning (Emslie and Morgan, 1994). The implication is that tropical holdovers from earlier times were periodically eliminated and occasionally reintroduced by fluctuations in climate, sea level, ocean circulation, and edaphic changes associated with continued emergence of peninsular Florida. The trend, however, was toward pine, grass, rosette palms, and scrub vegetation that could cope with physiologically dry deep sandy soil; the coast was fringed by tropical mangrove communities.

Along the Atlantic coastal plain, sediments of Neogene age are limited, but some data are available from drilling operations that penetrated the Middle Miocene Kirkwood Formation at Mays Landing, New Jersey (Owens et al., 1988). Palynomorphs include abundant *Pinus*, *Carya*, and

Figure 7.4, facing page. Land form map of the conterminous United States with overlay of principal Middle Miocene through Pliocene floras mentioned in the text: (1) Mays Landing, (2) Brandywine, (3) Martha's Vineyard, (4) Ash Hollow, (5) Kilgore, (6) Teewinot, (7) Pyramid, (8) Mount Eden, (9) Chula Vista, (10) Sonoma, (11) Trapper Creek, (12) Mascall, (13) Clarkia, (14) Succor Creek, (15) Trout Creek, (16) Stinking Water, (17) Ellensburg, (18) Oak Grove, (19) Tulelake, (20) Deschutes, (21) Fingerrock, (22) 49-Camp, and (23) Latah (Thelin and Pike, 1991).

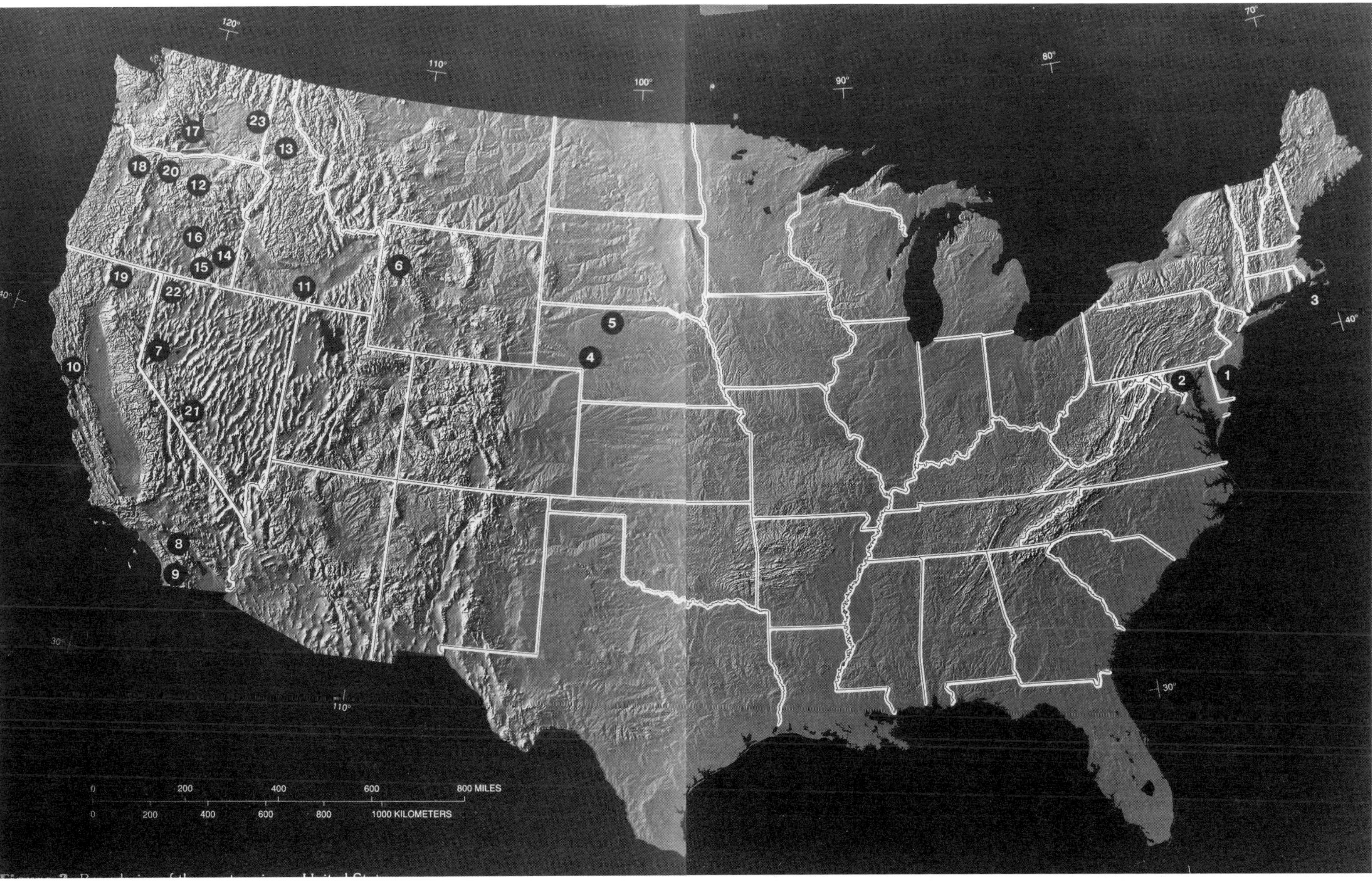

120°
110°
100°
90°
80°
70°
40°
40°
30°
110°
30°
1
2
3
4
5
6
7
8
9
10
11
12
13
14
15
16
17
18
19
20
21
22
23
0 200 400 600 800 MILES
0 200 400 600 800 1000 KILOMETERS

Table 7.1. Geographic distribution of Middle Miocene through Pliocene floras of North America.

Region	Flora	Reference
	Middle to Late Miocene (Including Mio–Pliocene)	
Mid-Atlantic	Kirkwood	Owens et al. (1988)
	Legler	Greller and Rachele (1983), Rachele (1976)
	Brandywine	Groot (1991), McCartan et al. (1990)
Northeast	Holmatindur	Mudie and Helgason (1983)
Western Interior		
Nebraska	Ash Hollow	Thomasson et al. (1990)
	Kilgore	MacGinitie (1962)
West		
Wyoming	Tweeinot	Barnowsky (1984)
Nevada	Pyramid	Axelrod (1992a)
	Verdi	Axelrod (1958)
	Fingerrock	Axelrod (1992b)
	Stewart Valley	Axelrod (1992b)
	Purple Mountain	Axelrod (1992b)
	49-Camp	Axelrod (1992b)
	Gillam Spring	Axelrod (1992b)
Idaho	Trapper Creek	Axelrod (1964)
	Clarkia	Smiley and Rember (1981, 1985)
		Smiley et al. (1975)
		Manchester et al. (1991)
California	Mount Eden	Axelrod (1937, 1950c)
Oregon	Succor Creek	Cross and Taggart (1982), Fields (1990, 1992),Graham (1963, 1965), Taggart et al. (1982)
	Trout Creek	Graham (1963, 1965)
		MacGinitie (1933)
	Stinking Water	Chaney (1959), Chaney and Axelrod (1959)
	Deschutes	Chaney (1938)
	Mascall	Chaney (1959), Chaney and Axelrod (1959)
Washington	Ellensburg	Smiley (1963)
	Latah	Knowlton (1926)
Northwest		
Alaska	Cook Inlet	Wolfe (1966, 1972), Wolfe and Hopkins (1967), Wolfe and Leopold (1967), Wolfe et al. (1966)
	Pliocene	
Southeast		
Florida	Pinecrest Beds	Willard (1993), Willard et al. (1993)
Northeast	Martha's Vineyard	Frederiksen (1984, 1989)
	Kap København	Ovenden (1993)
West		
Nevada	Aldrich Station	Axelrod (1956)
	Chloropagus	
	Fallon	
California	Ricardo	Webber (1933)
	Furnace Creek	Axelrod (1940)
	Anaverde	Axelrod (1950a)
	Piru Gorge	Axelrod (1950b)
	Chula Vista	Axelrod and Deméré (1984)
	Carmel	Axelrod and Cota (1993)
	Puente	
	Modelo	
	San Pablo	Condit (1938)
	Sonoma	Axelrod (1944, 1950d)
	Wildcat	Dorf (1930)
	Temblor	Renney (1969)
Oregon	Tulelake	Adam et al. (1989, 1990a,b)
	Oak Grove	Wolfe (1990)
Northwest		
Alaska	Lava Camp	Hopkins et al. (1971)
	Kivalina	Hopkins et al. (1971)
	Gubick	Repenning and Brouwers (1992)
	Ocean Point	Nelson and Carter (1985, 1992)

Table 7.2. Palynomorphs from Pliocene Pinecrest Beds (Tamiami Formation), near Sarasota, Florida.

Arboreal Pollen	Nonarboreal Pollen	Spores
Juniperus	*Ephedra*	*Isoetes*
Pinus	Cyperaceae	*Lycopodium*
TCT	Poaceae	*Equisetum*
Acer	*Ambrosia*	*Osmunda*
Alnus	*Artemisia*	*Polypodium*
Betula	*Aster*	Monolete spores
Bumelia	Asteraceae indet.	Trilete spores
Carya	Chenopodiaceae	Trilete, zonate spores
Castanea	Ericaceae	
Celtis	*Ilex*	
Cercis	*Iva*	
Corylus	Labiateae	
Diospyros	Leguminosae	
Fagus	Liguliflorae	
Fraxinus	*Polygonum*	
Juglans	Ranunculaceae	
Liquidambar	Tubuliflorae	
Magnolia	Umbelliferae	
Myrica		
Nyssa		
Ostrya–Carpinus		
Platanus		
Quercus		
Salix		
Ulmus		

Adapted from Willard (1993).

Quercus, with lesser amounts of TCT (pollen of the families Taxodiaceae–Cupressaceae–Taxaceae, which cannot be distinguished), *Alnus*, *Betula*, *Fagus*, *Ilex*, *Liquidambar*, *Momipites* (*Alfaroa–Engelhardia–Oreomunnea*), *Nyssa*, Sapotaceae, *Tilia*, *Ulmus*, and exotics such as *Podocarpus* and *Pterocarya*. Near the top of the section small amounts of Cyperaceae, Gramineae, chenoam (pollen of the families Chenopodiaceae–Amaranthaceae, which cannot be distinguished), and Compositae–Asteraceae pollen appear, suggesting local open habitats.

The overlying Cohansey Sand Formation is slightly younger, but it is also Middle Miocene in age (estimated at ~11 Ma; Greller and Rachele, 1983; Rachele, 1976). Pollen from the Legler Lignite lens is similar to that from the Kirkwood Formation and includes *Tsuga*, as well as the exotics *Podocarpus*, *Alangium*, *Momipites*, *Pterocarya*, *Cyrilla*, *Gordonia*, and *Cyathea*.

The Late Miocene Brandywine flora of southern Maryland also reflects a warm-temperate climate (Groot, 1991; McCartan et al., 1990; Table 7.3), as do the molluscan faunas (Ward and Powars, 1989; Ward and Strickland, 1985). There is one extinct genus (*Mneme*, Lythraceae), two no longer occur in Maryland (*Sophora*, west of the Appalachians; *Larix*, south only to New Jersey), and four are exotic to North America (the Asian *Alangium*, *Pterocarya*, *Trapa*, and *Zelkova*). Several thermophilic taxa present in the Legler flora are absent from the Brandywine assemblage (*Cyathea*, *Podocarpus*, *Cyrilla*, *Gordonia*, *Momipites*), suggesting cooler and more seasonal climates. The palynoflora of the Middle to Late Miocene Bryn Mawr Formation of Maryland also suggests climates warmer than at present, but cooling from earlier times (*Pinus*, *Taxodium*, *Carya*, *Ilex*, *Quercus*), and some remaining exotics (*Sciadopitys*, *Alangium*, *Engelhardia*; Pazzaglia et al., 1997).

In New England, plant microfossils of Middle Miocene and Pliocene age are known from Plymouth County and Martha's Vineyard in eastern Massachusetts (Frederiksen, 1984, 1989). The Middle Miocene vegetation was a rich, warm-temperate mixed mesophytic forest with abundant conifers, while the Pliocene forest was less rich and more cool temperate (Table 7.4). Pollen of most exotics had disappeared from eastern North America by the Pliocene, giving a more modern aspect to the vegetation. The absence of *Abies*, *Picea*, and *Betula* in the Early Miocene Brandon Lignite flora of Vermont attests to the Middle to Late Tertiary cooling in northeastern North America since Brandon time.

Evidence for climatic deterioration is further evident in high latitude floras. The vegetation sequence in the Holmatindur Tuff floras of eastern Iceland (65° N; 10.3–9.5 Ma; Mudie and Helgason, 1983) is from a temperate hardwood forest (*Larix*, *Picea*, *Pinus*, *Sequoia*, *Taxodium*, *Carpinus–Ostrya*, *Carya*, *Fagus*, *Juglans*, *Nyssa*, *Quercus*, *Ulmus*, and others), to a cool-temperate deciduous hardwood–conifer forest (increase in *Picea* and *Pinus*, decrease in *Fagus*), to a cold-temperate forest and subarctic *Alnus* woodland (abundant *Picea*, *Betula* sect. *Nanae*, *Alnus viridis*; Mudie and Helgason, 1983). In the complex strata of Banks Island (73° N) involving the Beaufort Formation and the Mary Sachs Gravel (Chapter 6; Fyles et al., 1994; Matthews, 1987; Matthews and Ovenden, 1990), the sequence is from mixed deciduous–conifer forest (*Picea*, *Pinus* dominant, *Juglans eocinerea*; Hills et al., 1974) to a decrease then disappearance of broad-leaved deciduous elements and an increase in *Larix* and *Picea*. About 80 species of Late Tertiary mosses are known from the Arctic islands (Banks, Ellesmere, Meighen, Prince Patrick, Queen Elizabeth Islands, Fig. 2.14; Beaufort Formation, Prince Patrick Island; Kap København Formation, northern Greenland, Bennike and Böcher, 1990; Mary Sachs Gravel, Banks Island; Ovenden, 1993). They indicate a mosaic of diverse conifer forest, stunted conifer woodland, and depauperate forest–tundra vegetation. The most tundralike assemblages are the youngest and are found in Plio–Pleistocene and Pleistocene deposits (~1.7 Ma; Repenning and Brouwers, 1992). The Kap København flora (82° 30' N; ~2 Ma) suggests that the Arctic tree line was 2500 km north of its present position in northeastern North America (Funder et al., 1985), although it probably fluctuated with the temperature changes described by Repenning and Brouwers (1992) and summarized by Crowley and North (1991). The assemblage consists of remains similar to modern *Larix occidentalis*, *Picea mariana*, *Taxus*, *Thuja occidentalis*, *Betula*, and *Cornus stolonifera*. The mosses present, such as *Schis-*

Table 7.3. Composition of Late Miocene Brandywine palynoflora, southern Maryland.

Isoetes	*Cornus*	*Ptelea*
Selaginella	*Euonymus*	*Pterocarya*
Ceratophyllum	*Fagus*	*Quercus*
Larix	*Gleditsia*	Sp. 1
Pinus	*Hamamelis?*	Sp. 2
Taxodium	*Ilex*	Sp. 3
Potamogeton	*Juglans*	*Salix*
Smilax	*Liquidambar*	*Sambucus?*
Typha	*Liriodendron*	*Sophora*
Acer	*Mneme*	*Toxicodendron*
Alangium	*Nuphar*	*Trapa*
Alnus	*Nyssa*	*Ulmus*
Ambrosia	*Parthenocissus?*	*Vitis*
Artemisia	*Platanus*	*Zelkova* type
Betula	*Populus*	
Carpinus–Ostrya	Sp. 1	
Carya	Sp. 2	
Celastrus	*Prunus*	

Adapted from McCarten et al. (1990).

tidium maritimum, are more typical of boreal forest than open tundra. These were beginning to disappear from northern Greenland by ~2 Ma, and tundra was expanding in Siberia (Wolfe, 1985). The flora bordered on an Arctic Ocean without perennial sea-ice cover, as indicated by sediments well sorted by wave action.

Fossil faunas of Tertiary age are extensively preserved in midcontinent and western North America, and NALMAs serve as the conventional temporal framework for describing the biotic history. In the Plains area the trend toward lower temperatures and decreasing seasonal (winter dry) precipitation continued, especially after ~16 Ma. A mosaic of open deciduous forest and grassland was present by the Middle Miocene (Barstovian, ~13 Ma) and was widespread by the Late Miocene (Clarendonian–Hemphillian, ~8 Ma). The broad vegetation sequence was from deciduous forest with prairie elements, to dry grassy woodland (savanna) with riparian forests, to more extensive grassland at the Miocene–Pliocene transition, but not yet pure prairie (Axelrod, 1985; Leopold and Denton, 1987; Wolfe, 1985). Although Elias (1942) proposed that the present Plains region was grassland by the late Miocene, more recent studies suggest it was savanna. At ~10 Ma a major volcanic eruption centered in southwestern Idaho deposited 3 m of ash as far east as Nebraska and preserved the extensive Ash Bed fauna. It includes camel, five species of horses, zebra, and other warm dry-land browsing and grazing mammals. At Big Springs the fauna by ~2 Ma includes lemming, shrew, and mastodon.

A number of fossil fruit and seed floras are known from northern Texas to South Dakota, ranging in age from Hemingfordian through Hemphillian (~18–4 Ma; Chaney and Elias, 1938; Elias, 1932, 1935, 1942). Grasses and prairie herbs became more abundant, widespread, and diverse through the sequences. In northern Texas identifications of Clarendonian plants include *Juniperus*, *Arctostaphylos*, *Celtis*, *Gymnocladus*, *Salix*, *Sapindus*, and *Ulmus*; the associated fauna consisted of browsing and grazing horses, oreodonts, and camels. In the panhandle of western Oklahoma the fauna included tortoise, rhinoceros, peccary, and an oreodont (Hesse, 1936). The estimated regional precipitation was ~510 mm (20 in), similar to the present, and represents the drier part of the Tertiary (Mio–Pliocene transition, 6–5 Ma; Axelrod, 1985). The area was sparsely wooded with scattered open grasslands. Many of the older grass identifications have been confirmed and augmented by Thomasson (1978a,b, 1979, 1980a–c, 1982, 1983, 1984, 1985) using SEM and phytoliths. The Late Miocene Ash Hollow Formation of Nebraska (6–7 Ma; Thomasson et al., 1990) includes the sedges *Carex*, *Cyperocarpus*, and *Eleofimbris*; the grasses *Archaeoleersia*, *Berriochloa*, *Graminophyllum*, *Nassella*, and *Panicum*; the borages *Biorbia* and *Cryptantha*; *Celtis* (Ulmaceae), *Chenopodium* (Chenopodiaceae), and Compositae–Asteraceae. Many of these floras are small and preservation is often poor. The associated faunas provide evidence of age, general vegetation type, and environmental change (Clarendonian chronofaunal decline; Webb, 1983) that is especially valuable for characterizing the Neogene of the Plains region. It was noted in Chapter 2 (CO_2 concentration section) that at ~6 Ma there was a change in ^{13}C in the plant debris from paleosols and from the tooth enamel of ungulates that suggests a shift from C3 plants (many trees and shrubs) to C4 plants (grasses; Cerling et al., 1997). This, in turn, suggests a continued decline in CO_2 concentration, more pronounced seasonality in rainfall, and lower temperatures.

Table 7.4. Miocene and Pliocene palynomorphs from eastern Massachusetts. X = Taxa also present in the Pliocene.

Miocene	Pliocene
Sphagnum	X
Lycopodium	X
Selaginella	
Gleicheniidites or *Neogenisporis* (reworked?)	X
Osmunda	
Ephedripites hungaricus	
Abies	
Larix	
Picea	X (?)
Pinus	X
Podocarpus	
Sciadopitys	
TCT	
Tsuga	
Cyperaceae	
Gramineae	X
Sparganium–Typha	
Alnus	X
Betula	X
Boehlensipollis hohlii (reworked?)	
Carya	
Castanea	
Chenopodiaceae–Amaranthaceae	
Compositae	
Short spine	X
Long spine	
Comptonia	
Cornaceae	
Corylus	
Cyrilla–Cliftonia	
Ericaceae	X
Fagus	
Gordonia	
Ilex	
Juglans type	
Liquidambar	
Ludwigia	
Malvaceae	
Momipites spackmanianus complex (reworked?)	
Morus	
Myrica	X (?)
Myriophyllum	
Nyssa	
Platanus	
Pseudolaesopollis sp. (reworked?)	
Pterocarya type	
Quercus	
Salix	
Sapotaceae	
Symplocos	
Tilia	
Ulmus–Zelkova	
Umbelliferae	
	Botrychium
	Palmae
	Carya (Paleogene type)
	Caryophyllaceae(?)

Adapted from Frederiksen (1984).

A more extensive assemblage is the Late Miocene Kilgore flora preserved in lacustrine sediments of the Valentine Formation of northern Nebraska (Barstovian, 13–14 Ma; MacGinitie, 1962; Table 7.5). The most abundant components are megafossils of *Platanus*, *Meliosma* (Fig. 7.5A), *Ulmus*, *Cedrela*, *Acer* cf. *negundo*, and *Robinia*; and microfossils of *Pinus*, *Carya*, *Ulmus* (Fig. 7.5B), *Sarcobatus*(?; greasewood), *Liquidambar*, *Alnus* (Fig. 7.5C), and *Artemisia* (Fig. 7.5D). It is a streamside assemblage of broad-leaved deciduous forest bordered by a woodland of *Pinus*, *Artemisia*, *Diospyros*, *Quercus*, and *Sarcobatus*, and patches of grassland (Axelrod, 1985). The climate was dry (estimated precipitation 760–890 mm) and temperatures were declining, but winter temperatures were still higher than at present as indicated by crocodilians and large nonburrowing tortoises in associated fossil faunas (Hutchison, 1982) and by the Mexican exotics *Cedrela*, *Cordia*, and *Meliosma*.

Along the eastern slopes of the southern Rocky Mountains, the woodland–grassland vegetation of the Plains graded into piñon–juniper woodland; at higher elevations

Table 7.5. Composition of Late Miocene Kilgore flora of northern Nebraska.

Megafossils	
Chamaecyparis linguaefolia	*Platanus vitifolia*
Monocotyledonous leaf	*Populus crassa*
Acer minor	*P. gallowayi*
Carya libbeyi	*P. washoensis*
Cedrela trainii	*Prunus acuminata*
Celtis kansana	*Pterocarya oregoniana*
Cladrastis prelutea	*Quercus argentum*
Cocculus rotunda	*Q. parvula*
Cordia prealba	*Q. preturbinella*
Crataegus nupta	*Q. remingtoni*
Diospyros miotexana	*Ribes infrequens*
Fraxinus coulteri	*Robinia lesquereuxi*
Mahonia marginata	*Ulmus speciosa*
Meliosma predentata	*Vitis pannosa*
Nyssa copeana	
Microfossils	
Algal spores	*Cedrela*
Fungal spores	*Celtis*
Lycopodium	*Compositae*
Polypod	*Ilex*
Picea	*Juglans*
Pinus	*Liquidambar*
Cf. *Sequoia*	*Meliosma*
Cyperaceae	Onagraceae
Gramineae	*Pterocarya*
Acer	*Quercus*
Alnus	*Salix*
Ambrosia	Cf. *Sarcobatus*
Artemisia	*Tilia*
Cf. *Betula*	Cf. *Ulmus*
Carya	

Adapted from MacGinitie (1962).

Figure 7.5. (A) *Meliosma predentata*, Kilgore flora. (B) Pollen of *Ulmus*, Kilgore flora. (C) Pollen of *Alnus*, Kilgore flora. (D) Pollen of *Artemisia*, Kilgore flora (A–D). Reprinted from MacGinitie with the permission of the University of California Press. (E) *Pseudotsuga longifolia*, Trapper Creek. Reprinted from Axelrod (1964) with the permission of the University of California Press.

it graded through *Quercus*, *Artemisia*, and various herbs into a western montane coniferous forest. The Teewinot Lake flora at Jackson Hole, Wyoming (late Clarendonian, ~8 Ma; Barnowsky, 1984), shows a shrubland to near desert with *Ephedra* and *Sarcobatus*, through which riparian vegetation extended (*Carya*, *Pterocarya*, *Salix*, *Ulmus–Zelkova*), and a western montane coniferous forest of *Abies*, *Picea*, *Pinus*, and *Tsuga* at higher elevations. Asian exotics had nearly disappeared from the eastern slopes of the Rocky Mountains by Clarendonian time, but they persisted later west of the Rocky Mountains. In the western Plains and adjacent foothills, as in the central Plains, grasses were present but not widespread or abundant in the Middle and Late Miocene.

The central and northern Rocky Mountains became a greater barrier between vegetation to the east and west during Middle Miocene through Pliocene time. Only six species from the Kilgore flora occur in floras of similar age of the Columbia Plateau (Leopold and Denton, 1987). Fossil grasses from the Plains have affinities mostly to the east and south into Mexico, while those west of the Rocky Mountains have affinities to the north.

To the south and west of the Rocky Mountains in the southwestern United States the Miocene vegetation was mostly a piñon pine–juniper woodland and shrubland with evergreen oak, grading upward into depauperate western montane coniferous forest. There were exceptions to the dry vegetation, which were created by local topographic lows across the Sierra Nevada. The Pyramid flora in western Nevada (15.6 Ma; Axelrod, 1992a) is to the west of the Buffalo Canyon, Middlegate, and Eastgate floras discussed earlier (Chapter 6). It is dominated by deciduous hardwoods (*Acer*, *Alnus*, *Betula*, *Cladrastis*, *Fraxinus*, *Ostrya*, *Platanus*, *Populus*), including members of a floodplain association (*Taxodium*, *Quercus simulata*). The estimated MAT is ~13.5°C (CLAMP 7.5° C), annual mean range 15°C, annual precipitation 900 mm (35–40 in.), and the paleoelevation ~550–600 m (1800–2000 ft; presently 1.4 km). However, new calculations using leaf physiognomy and mean annual enthalpy suggest higher paleoelevations from many Middle Miocene floras in the Basin and Range Province (e.g., 2.8 km for the Pyramid flora; Wolfe et al., 1997). Recall from Chapter 1 (Pacific coastal and Rocky Mountains coniferous forests) that there is a modern analog for the effect of topographic lows across the western mountains on vegetation. A similar opening along which the Columbia River flows extends today through the Cascade Mountains, allowing heavy rains and cold temperatures to penetrate onto the western slopes of the Rocky Mountains near the Columbia Plateau. These forests share species that elsewhere are used to define different montane coniferous forest associations. The Pyramid flora illustrates one aspect of the importance of local topography on the composition of western North American paleovegetation. Younger Mio–Pliocene floras in the region record the partly orographic decrease in rainfall and the development of increasingly arid vegetation from the rise of the Sierra Nevada and coastal mountains (Aldrich Station, 12.4–13.3 Ma; Chloropagus, 14.3 Ma; Fallon, 15.4 Ma; Axelrod, 1956; Wolfe et al. 1997; see also Axelrod, 1958, 1980, for index maps of the many Neogene floras in Nevada and California). The four florules of the Purple Mountain flora of western Nevada (~14.8 Ma) grew under an estimated rainfall of 760–890 mm (30–35 in.; oldest), decreasing to 625–760 mm (25–30 in.; Axelrod, 1995). The Verdi flora to the west of Middlegate in western Nevada consists of flood–plain vegetation, with oak woodland, chaparral, and savanna on the slopes, and pine and fir in the bordering hills. Rainfall in the lowlands is estimated at 450–500 mm (18–20 in.; Axelrod, 1958).

Several floras of Pliocene age are known from the arid region of interior southern California (Furnace Creek: Axelrod, 1940; Anaverde: Axelrod, 1950a; Piru Gorge: Axelrod, 1950b; Ricardo: Webber, 1933). During parts of the Pliocene the Central Valley of California was a marine embayment (Stanton and Dodd, 1970), and the Gulf of California extended further north than at present. One of the most extensive assemblages is at Mount Eden (~5.3 Ma) in the San Jacinto Mountains southeast of Los Angeles (Axelrod, 1937, 1950c; Table 7.6). It was then ~30 km east of the marine embayment in the Salton Sink Basin (Axelrod and Cota, 1993). The species comprise six communities that reflect semiarid conditions and low to moderate elevations in the interior. The near-desert community included members typical of the more mesic parts of the present Sonoran Desert. The shrubland elements are mostly holdovers from the subtropical vegetation that already characterized interior southern California during the Miocene (e.g., Tehachapi flora). Elements available to occupy these habitats were, in turn, present as local, edaphic and slope-controlled, preadapted types since Middle Eocene Green River time. In the Pliocene these were coalescing and expanding in response to drying climates. Rainfall is estimated at 300–450 mm (12–18 in.), but with some summer rain, in contrast to the summer-dry conditions that developed later in the Pliocene and Pleistocene. It is likely that at least some elements of the present dry vegetation of southern coastal California (Diegan sage; Axelrod, 1978) spread from the arid interior. The California grassland becomes recognizable as a distinct association in the Pleistocene.

Table 7.6. Composition and paleocommunities of Mount Eden flora.

Desert	
Baccharis beaumontii	*Eysenhardtia pliocenica*[a]
Cercidium edenensis[a]	*Ficus edenensis*[a]
Chilopsis sp.	*Forestiera buchananensis*
Condalia coriacea	*Prunus (Emplectocladus) preandersonii*
Dodonaea californica[a]	*P. (Emplectocladus) prefremontii*
Ephedra sp.	
Shrubland–Chaparral	
Arctostaphylos preglauca	*Eysenhardtia pliocenica*[a]
A. prepungens	*Ficus edenensis*[a]
Ceanothus edenensis	*Fraxinus edenensis*
C. precuneatus	*Prunus (Laurocerasus) mohavensis*
C. prespinosus	*Quercus pliopalmeri*
Cercidium edenensis[a]	*Rhamnus moragensis*
Cercocarpus cuneatus	*Rhus (Malosma) prelaurina*
Dodonaea californica[a]	*R. (Schmaltzia) moragensis*
Woodland	
Pinus pieperi	*Quercus convexa*[a]
Arbutus prexalapensis	*Q. dayana*[a]
Eysenhardtia pliocenica[a]	*Q. douglasoides*
Juglans beaumontii	*Q. lakevillensis*[a]
J. nevadensis	*Salix edenensis*[a]
Persea coalingensis	*S. truckeana*[a]
Platanus paucidentata[a]	*S. wildcatensis*[a]
Populus sonorensis	*Sapindus oklahomensis*
Western Montane Coniferous Forest	
Cupressus preforbesii	*Amorpha oblongifolia*
Pinus hazeni	*Philadelphus nevadensis*
P. pretuberculata	*Populus pliotremuloides*
Pseudotsuga premacrocarpa	*Quercus dayana*[a]
Amorpha oblongifolia	

Adapted from Axelrod (1950c).
[a] These species occur in more than one unit.

The Mediterranean (winter-wet) climate that presently characterizes southern coastal California is evident from the N–S precipitation cline. Annual rainfall at Mt. Baker in northern Washington is 2818 mm (110.96 in.), at Eureka (northern California) it is 954 mm (37.58), at San Francisco (~central California) it is 514 mm (20.23), and at San Diego (southern California) it is 257 mm (10.11). In San Diego the winter months receive 46.7 mm (1.84 in.) of precipitation in December, 50 mm (1.97) in January, 56 mm (2.22) in February, and 40 mm (1.59) in March, or 75% of the annual precipitation. In Tucson, Arizona, to the west and at about the same latitude (32° N), the annual precipitation is similar (283 mm; 11.16 in.); but the winter months receive (26 mm; 1.02 in), (21 mm; 0.81 in), (23 mm; 0.89 in), and (19 mm; 0.76 in.) or 34% of the annual precipitation.

The explanation for Mediterranean climates and sclerophyllous vegetation in southern California involves a complex of factors. One component are summer storms off the northeast Pacific Ocean that move progressively southward and do not reach southern California until the winter season. This pathway is a result of the coastal mountains that deflect east-trending storm tracks to the south. This places the origin of Mediterranean climates late in the Tertiary (Late Pliocene and Pleistocene) by which time the coastal mountains had reached substantial heights. Another component is the presence of the southern Rocky Mountains, the Sierra Nevada, and the high pressure arm of the Hadley regime that prevent rain-generating summer lows off the Caribbean Sea and Gulf of Mexico from penetrating to the west coast. Elements of dry sclerophyllous vegetation, already present in western North America in the Eocene (e.g., Green River flora), coalesced after the Middle to Late Miocene drying trend and were available to occupy the Mediterranean-type habitats of southern California by the Pliocene.

The broad temperature trend during the Miocene along the southern California coast is reflected in the ^{13}C content of hopane and sterane compounds preserved in the Monterey Formation at Naples Beach (Schoell et al., 1994; Chapter 3). The decreasing ^{13}C values parallel increasing amounts of ^{18}O and show a prominent cooling at ~15 Ma. Coastal waters changed from well mixed to a highly stratified photic zone (upper 100 m) as at present. The decrease in temperature in the zone between the Early and Late Miocene is estimated at 3.1–4.3°C. Near the end of the Miocene the lower photic-zone temperature was near the present 10°C. One effect of these cooling coastal waters, comparable to that of the cold Humboldt Current observed today along western South America (Chapter 2), was reduced precipitation along the borderlands. The ^{13}C-based cooling trend reversed briefly at 7–8 Ma, consistent with the warming shown on the ^{18}O-derived benthic temperature curve (Chapter 3).

On the California coast near the Mexican border a small flora is preserved in the San Diego Formation at Chula Vista (~3 Ma; Axelrod and Deméré, 1984). It includes *Pinus diegensis* (cf. the modern *P. radiata*), *P. jeffreyoides* (*P. jeffreyi*), *P. pieperi* (*P. sabiniana*, digger pine), *Palmae*, *Persea coalingensis* (avocado), *Populus alexanderi* (*P. trichocarpa*), *Salix wildcatensis* (*S. lasiolepis*), *Platanus paucidentata* (*P. racemosa*), and *Quercus lakevillensis* (*Q. agrifolia*, California live oak). It is a pine–oak woodland with palms, avocado (*Persea*), cottonwood, willow, and sycamore along the streams. Environments were more mesic than in the interior, as shown by other floras in the region (Carmel: an exotic terrane flora, Puente, Modelo; Axelrod and Cota, 1993). The MAT is estimated at 16°C (presently 12–13°C), and the annual precipitation in the lowlands was 500–580 mm (20–23 in.) compared to the present 220 mm (9 in.). The flora grew during the mid-Pliocene warm period, and a MAT 3–4°C warmer than at present brought more summer rain. The summer-dry winter-wet Mediterranean climate (Arroyo et al., 1995) that presently characterizes the area developed in the Late Pliocene and Pleistocene.

In the San Francisco Bay area several floras span the interval between the Mio–Pliocene and the Late Pliocene (e.g., Sonoma: Axelrod, 1944, 1950d; San Pablo: Condit, 1938). These are west of the Sierra Nevada and received moisture from the Pacific Ocean. The Sonoma flora includes several florules between Napa and Santa Rosa about 50 km to the northwest. The Napa florule is slightly more interior and represents a pine (cf. *P. ponderosa*)–Douglas fir (*Pseudotsuga*) forest near sea level and an oak woodland (*Quercus dayana*, *Q. wislizenoides*)–chaparral (*Arctostaphylos fergusoni*, *Cercocarpus cuneatus*, *C. linearifolius*, *Photinia sonomensis*) on the slopes, together with several elements that now border the redwood forest on drier sites (*Amorpha condoni*, *Holodiscus harneyana*). At Santa Rosa the more coastal vegetation included better representation of the modern redwood forest (*Sequoia affinis*, *Alnus rubroides*, *Castanopsis sonomensis*, *Ceanothus edensis*, *Lithocarpus klamathensis*, *Mahonia simplex*, *Salix boisiensis*, *Umbellularia salicifolia*) and a higher elevation coastal coniferous forest of *Abies sonomensis*, *Picea sonomensis*, *Tsuga sonomensis*, and associated *Rhododendron* sp. and *Vaccinium sonomensis*. Rainfall is estimated at 1000 mm (40 in., 280 mm more than at present; Axelrod, 1944) with greater summer rain and a ~3°C warmer MAT. Wolfe (1990) estimates rainfall at 100 mm higher and a 2–4°C warmer MAT. Other fossil floras in the region showing similar vegetation include the Wildcat (Dorf, 1930), San Pablo (Condit, 1938), and Temblor (Renney, 1969).

To the north around the Columbia Plateau in the Middle to Late Miocene a rich broad-leaved deciduous forest grew at low to midelevations (below ~600 m; *Metasequoia*, *Taxodium*, *Ailanthus*, *Betula*, *Cedrela*, *Diospyros*, *Fagus*, *Hamamelis*, *Ilex*, *Magnolia*, *Persea*, *Phoebe*, *Pterocarya*, *Quercus*, *Sassafras*, *Ulmus*, *Zelkova*). This graded through a mixed conifer–hardwood forest (*Abies*, *Chamaecyparis*, *Picea*, *Pinus*, *Acer*, *Betula*, *Quercus*, *Ulmus*, *Zelkova*) into a western montane coniferous forest at the highest elevations. The Tertiary floras of the Columbia Plateau are among the richest and most extensive in the world. The forests grew under warm-temperate summer-wet conditions and are prominent until ~4.5 Ma (Hemphillian), after which progressively arid summer-dry conditions developed to the lee of the rising Sierra Nevada and Cascade Mountains. The vegetation was part of a broad zone of deciduous forest and mixed conifer–hardwood forest that extended into Alaska (Seldovia Point flora) and across the midlatitudes of the northern hemisphere into Europe and Asia. Throughout the sequences, pollen of herbs and xeric elements are present but not yet abundant.

The Trapper Creek flora of south central Idaho, earlier estimated to be 15–16 m.y. in age (Axelrod, 1964), is now placed at 10.5–12 m.y. (Clarendonian; Armstrong et al., 1975; Fields, 1983). It is a summer-wet deciduous forest with some broad-leaved evergreens and a montane conifer–

Table 7.7. Suggested taxonomic and geographic affinities of common members of late Middle Miocene Stinking Water flora, Harney County, Oregon.

Fossil	Similar Modern Species		
	East America	East Asia	West America
Glyptostrobus oregonensis		*G. pensilis*	
Alnus harneyana (Fig. 7.10C)		*A. hirsuta*	
Alnus relatus	*A. maritima*	*A. japonica*	
Platanus dissecta	*P. occidentalis*		*P. racemosa*
Populus lindgreni (Fig. 7.10B)	*P. heterophylla*	*P. rotundifolia*	
Quercus dayana (Fig. 7.10A)	*Q.virginiana*		
Q. hannibali			*Q. chrysolepis*
Q. prelobata (Fig. 7.10D)			*Q. lobata*
Q. pseudolyrata	*Q. borealis*		*Q. kelloggii*
Q. simulata		*Q. stenophylla, Q.myrsinaefolia*	
Ulmus speciosa	*U. fulva*		

Adapted from Chaney (1959).

hardwood forest. Components of the deciduous forest included swamp forest (*Taxodium, Nyssa*), lake border vegetation (*Acer, Populus, Salix*), valley forest (*Alnus, Amelanchier, Carya, Cornus, Fraxinus, Parthenocissus, Prunus, Pterocarya, Sophora, Ulmus*; broad-leaved evergreens *Berberis, Ilex, Quercus*), and mountain slope vegetation (*Pinus*, cf. *P. ponderosa, P. monticola*; *Calocedrus, Cedrus, Garrya, Rhus*). Some frost-sensitive plants (e.g., *Liquidambar*) began disappearing in the Middle Miocene, although *Taxodium* swamps persisted in Idaho until ~12 Ma and in eastern Washington until ~8 Ma (Leopold and Denton, 1987). Conifers included *Abies, Keteleeria, Picea, Pseudotsuga* (Fig. 7.5E), *Sequoiadendron*, and *Tsuga*. The flora suggests higher and cooler conditions than do other floras in the region (Succor Creek, Stinking Water, see below), and the dominant vegetation was coniferous forest; the broad-leaved community is interpreted as mostly recovery vegetation following disruption by at least five episodes of volcanic activity (Taggart and Cross, 1990).

The Mascall flora is from along the East Fork of the John Day River in northeastern Oregon, and it is Early Barstovian in age (~16 Ma). The most abundant megafossils (>1%) in order are *Taxodium dubium, Quercus pseudolyrata, Carya bendirei, Quercus dayana, Platanus dissecta, Quercus merriami, Acer bolanderi, Metasequoia occidentalis, Ginkgo adiantoides, Ulmus speciosa, Acer minor* (*A. medianum*), *Cedrela trainii* (*C. pteraformis*), *Ulmus paucidentata, Betula thor*, and *Acer scottiae*. The flora reflects a swamp cypress and deciduous forest in the lowlands and on the slopes and a hardwood–conifer forest in the uplands (Chaney, 1959; Chaney and Axelrod, 1959). Affinities are with the present vegetation of eastern North America, western North America, and eastern Asia. (For a complete listing of a Columbia Plateau flora, see the more recently studied Succor Creek flora described below; examples of taxonomic and geographic affinities of some Columbia Plateau fossils with modern species are given in Table 7.7.)

To the north in Idaho, the Clarkia flora (~16 Ma), about 55 mi northeast of Moscow, is preserved in lacustrine deposits resulting from damming of the drainage system by lava flows to form the proto-St. Maries River and Clarkia Lake. Nearby basalts and numerous ash layers (2 mm to 50 cm thick in the section) indicate frequent volcanic activity similar to that at Trout Creek (Figs. 3.10, 3.11); the effects were probably similar to those described below for the Succor Creek flora. The systematics of the Clarkia flora have not been presented in detail and identifications are at the generic level. Three communities are represented (Smiley and Rember, 1981, 1985; Smiley et al., 1975): a bottomland swamp and riparian forest (*Taxodium, Chamaecyparis, Populus, Salix, Nyssa*), a mesic slope forest (*Metasequoia, Sequoia, Thuja, Acer, Aesculus, Alnus, Betula, Carya, Castanea, Cercidiphyllum, Cornus, Corylus, Diospyros, Fagus, Fraxinus, Hamamelis, Juglans, Liquidambar, Liriodendron, Ostrya, Paulownia, Pterocarya, Quercus, Sassafras, Ulmus, Zelkova*), and a drier slope forest (*Pinus, Amorpha, Amelanchier, Celtis, Crataegus, Quercus, Rhus, Rosa*). A western montane coniferous forest was also present in the vicinity as shown by the presence of pollen and megafossils of *Abies* and pollen of *Picea* and *Tsuga*. The extinct genera *Nordenskioldia* (infructescences, fruits; widespread in the Paleogene of the Northern Hemisphere) and *Zizyphoides* (leaves) have been reported from the Clarkia flora (Manchester et al., 1991). The flora also contains fossil infructescences and fruits of *Trochodendron* (Manchester et al., 1991). The principal vegetation association was a mixed mesophytic forest of the deciduous forest formation.

Ancient DNA

Leaf compressions of unoxidized cuticular and mesophyll tissue in the Clarkia flora contain original pigmentation. This imparts to the assemblage a feature rare in fossil floras: the potential preservation of ancient DNA and other organic compounds (Golenberg, 1991; Golenberg et al.,

Figure 7.6. (A, right) Succor Creek locality, southeastern Oregon. (B) *Glyptostrobus oregonensis*,vegetative branch. (C) *Glyptostrobus oregonensis*, cone. (D) *Fraxinus coulteri*. (E) *Sassafras columbiana*. (F) *Hydrangea*-like calyx. (G) *Ulmus speciosa*. (H) *Acer chaneyi* (*A. benderi* in Graham, 1965; see Wolfe and Tanai, 1987). (I) Pollen of *Picea*. (J) Pollen of *Abies*. (K) Pollen of *Ambrosia*. (L) Pollen of Chenopodiaceae–Amaranthaceae. (M, N) Pollen of *Pinus*. (O) Pollen of *Carya*. (P) Pollen of *Juglans*. Adapted from Graham (1965).

1990; Soltis and Soltis, 1993; Soltis et al., 1992). A partial sequence [759 base pairs (bp)] was obtained from the chloroplast DNA *rbc*L gene from tissues that were part of an extinct species of *Magnolia* (*M. latahensis*). The gene encodes for ribulose 1,5-bisphosphate carboxylase/oxygenase, which is an enzyme responsible for fixing atmospheric carbon. Comparisons with *rbc*L sequences of modern *Magnolia* species revealed affinities with that genus. In addition, there were differences between the modern and extinct species of 17 base substitutions of which 12 were transitions and 13 were silent third-position substitutions (Soltis and Soltis, 1993). Similar analyses were made of fossil *Taxodium* from the Clarkia flora and compared with modern *Taxodium*, *Metasequoia*, and *Pseudotsuga* (Soltis et al., 1992). These suggested the ancient DNA was from *Taxodium*. Amplified products (780 and 1380 bp) were larger than achieved with ancient animal DNA, which were interpreted as a consequence of less degradation under the favorable preservation conditions at Clarkia.

DNA sequences have also been obtained from fossil *Juniperus* and *Symphoricarpus* (snowberry, Caprifoliaceae) from packrat middens older than 45,000 yrs (Rogers and Bendich, 1985) and from Oligo-Miocene bee and termite remains preserved in Dominican amber that are ~22 Ma in age (Cano et al., 1992; DeSalle et al., 1992). However, a new study based on the extent of racemization of aspartic acid, alanine, and leucine suggests that chloroplast DNA from the Clarkia material has undergone extensive microbial degradation (Poinar et al., 1996), and another study (Austin et al., 1997) casts doubt on other reports of ancient DNA from insects preserved in amber.

The preservation quality of the Clarkia specimens has allowed another innovation in the study of fossil leaves. Measurements of stomatal guard cells have been used to estimate chromosome number and ploidy level of the fossils (Platanaceae, Magnoliaceae, Lauraceae; Masterson, 1994).

The Succor Creek flora (Fig. 7.6) of Barstovian age (14–15 [14.8–15.5] Ma; Downing and Swisher, 1993) in southeastern Oregon (Malheur County) and adjacent Idaho (Owyhee County) was deposited at ~1000 m (Cross and Taggart, 1982) with greater highlands to the south. The flora has been studied extensively (Brooks, 1935; Cross and Taggart, 1982; Fields, 1992; Graham, 1963, 1965; Smith, 1938, 1939; Taggart et al., 1982) and presently consists of ~175 megafossil and ~100 microfossil taxa for a total of ~275 named taxa (Table 7.8). However, when parallel names for different organs of the same taxon are taken into account, a more realistic number is ~160 species (P. F. Fields, personal communication, 1996). It is one of the largest fossil floras known for North America. About 25–30% of the species have entire-margined leaves (temperate climate), although the number of specimens required for leaf physiogonomic analysis (30 dicot leaves) and their percentage varies among the different localities.

The principal communities were warm-temperate forest with evergreen elements in the lowlands (*Cedrela*, *Mahonia*, *Oreopanax*, *Quercus*) and progressively cool-temperate deciduous forest, mixed conifer–hardwood forest, and western montane coniferous forest at higher elevations. Lake margin, marsh, and riparian vegetation bordered the depositional basin (Fig. 7.7). The effects of frequent volcanic activity are evident in plant microfossil assemblages from closely spaced samples. At the Whiskey Creek section

Figure 7.6. continued

of the Succor Creek Formation, the interval between samples 13 and 14 reveals a significant disruptive event. The forest abruptly declines and is replaced by a herbaceous vegetation of Gramineae, Amaranthaceae–Chenopodiaceae, Compositae–Asteraceae, and Malvaceae. This is interpreted as a postdisruption sere (successional stage) following destruction of the forest by volcanic ash fall. The succession progressed intermittently through a pine parkland back to deciduous forest. Mats of conifer needles associated with charcoal may reflect direct blast effects, outgassing, or heat. The volcanism that created conditions favorable to fossilization (Chapter 2, Figs. 2.21, 2.22) also periodically set succession back in local sites to bare ground, comparable to that witnessed in the May 18, 1980 eruption of Mount St. Helens. Recognition of these events is important in distinguishing between cyclic vegetational changes due to climate and those resulting from succession and repeated regional catastrophies.

In an early demonstration that organic compounds such as flavonoids do preserve for millions of years in plant fossils, Niklas and Giannasi (1977; Niklas, 1981) recovered kaempferol, dihydrokaempferol, an n-alkane chain length range of 10–32 carbons, hydroxy acids, steranes, triterpenoids, and methyl pheophorbide a from *Zelkova oregoniana* leaves from Succor Creek (Fig. 7.8).

The Trout Creek flora in Harney County, southeastern Oregon, is preserved in diatomite with numerous lenses of volcanic ash (Figs. 3.10, 3.11; Graham, 1963, 1965; MacGinitie, 1933). It is slightly younger (13.8 Ma) than the Succor Creek flora, as reflected by the absence of the Mexican

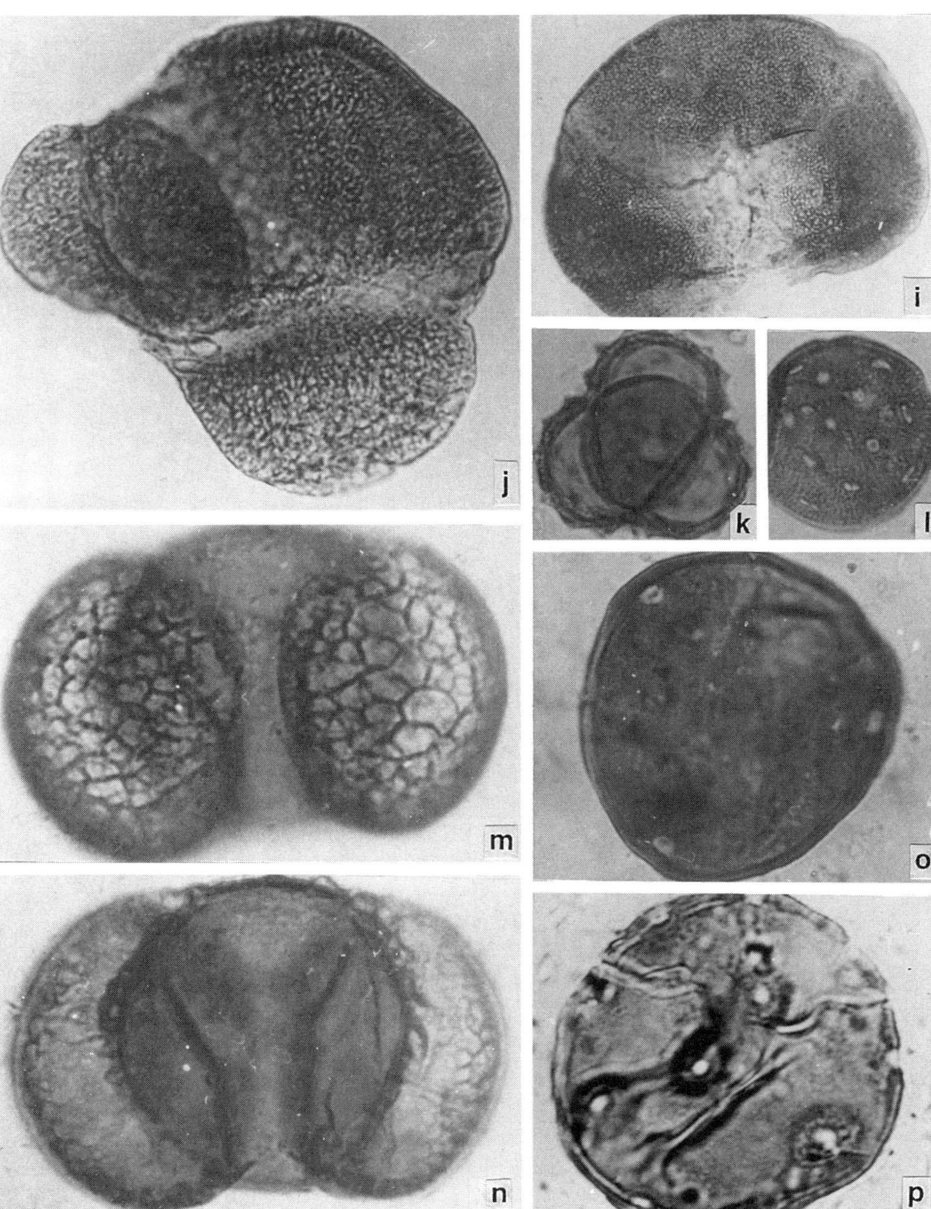

Figure 7.6. continued

exotics *Cedrela* and *Oreopanax* and the Asian *Ginkgo*, *Glyptostrobus*, and *Pterocarya*. It also grew under cooler conditions, because in addition to the absence of the warm temperate to subtropical *Cedrela* and *Oreopanax*, conifers are more diverse and abundant at Trout Creek than those of the lower paleoelevations in northern Succor Creek localities. However, they do compare well with conifers from the upper paleoelevations from southern Succor Creek sites.

Conifers are rare in the Stinking Water flora (11 [12.5–13.4] Ma) just to the north in Harney County. This is a lowland flora with abundant oaks, together with *Glyptostrobus oregonensis*, *Alnus harneyana*, *A. relatus*, *Platanus dissecta*, *Populus lindgreni*, and *Ulmus speciosa*. Swamp habitats are evident by the abundant fruits of *Trapa americana* and rootstocks of *Nymphaeites nevadensis* (*Nymphaea*, fide P. F. Fields, personal communication, 1996) in some facies. The suggested affinities of these fossils with living species (fide Chaney, 1959) are listed in Table 7.7. The distribution of the modern species illustrates the broad geographic occurrence of temperate plants across the midnorthern latitudes during the Middle Tertiary and the relictual nature of many present-day taxa.

The Ellensburg flora of south-central Washington is ~14 m.y. in age (Axelrod, 1992b; Evernden and James, 1964; Smiley, 1963; [10.5–13.5, depending on the horizon]), and preserves a piedmont vegetation from the eastern flank of the Cascade Range. The flora consists of a series of florules that record an early lowland swamp cypress forest (*Taxodium dubium*), a midaltitude mixed deciduous hardwood forest (*Acer columbianum*, *Betula thor*, *Fagus washoensis*, *Fraxinus coulteri*, *Platanus dissecta*, *Quercus bretzi*, *Sassafras hesperia*, *Ulmus newberryi*), and a younger post-

Table 7.8. Composition of Late Miocene Succor Creek flora of southeastern Oregon.

Megafossils		Microfossils	
Equisetum octangulatum	*Gymnocladus dayana*	*Alternaria*	*Carpinus*
E. sp.	*Hiraea knowltoni*	Eumycophyta	*Carpinus-Ostrya*
Davallia solidites	*Hydrangea bendirei*	Chytridaceae	*Carya* (Fig. 7.6O)
Osmunda claytonities	*H.*-like calyx (Fig. 7.6F)	Fungi Imperfecti	*Castanea*
Polypodium sp.	*Ilex fulva*	Fungal remains (aff. uncertain)	*Celtis*
Woodwardia deflexipinna	*Juglans* sp.	*Micrhystridium*	Chenopodiaceae–Amaranthaceae (Fig. 7.6L)
Abies sp.	Legume (fruit, leaf)	*Psophosphaera*	Compositae
Cephalotaxus californica	Legume (thorny branches)	*Sigmapollis*	Low spine
Cupressus-Juniperus	*Liquidambar* sp.	Acritarchs undifferentiated	High spine
Ginkgo adiantoides	*Lithocarpus klamathensis*	*Botryococcus*	*Cornus*
Glyptostrobus oregonensis (Fig. 7.6B, C)	*Magnolia ovulata*	*Ovoidites*	*Corylus*
Keteleeria sp.	*Mahonia macginitiei*	Pediastrum	Eleagnaceae
Librocedrus masoni	*M. malheurensis*	Algal remains (aff. uncertain)	*Epilobium*(?)
Metasequoia occidentalis	*M. reticulata*	Monolete spores	Ericaceae
Picea lahontense	*M. simplex*	Trilete spores	*Fagus*
P. magna	*M. trainii*	*Lycopodium*	*Fraxinus*
Pinus harneyana	*M.* sp.	*Equisetum*	*Ilex*
P. sp. (leaf)	*Nymphaea* sp.	*Davallia*	*Juglans* (Fig. 7.6P)
P. sp. (cones)	*Nyssa copeana*	*Osmunda*	Leguminosae (polyad)
Pseudotsuga longifolia	*N. hesperia*	*Polypodium*	*Liquidambar*
Sequoia sp.	*Oreopanax precoccinea*	*Woodwardia*	*Lithocarpus*
Taxodium sp.	*Ostrya oregoniana*	Unknown fern spore	*Mahonia*
Thuja dimorpha	*Persea pseudocarolinensis*	*Ephedra*	Malvaceae
Tsuga sonomensis	*Photinia* sp.	*Abies* (Fig. 7.6J)	*Nymphaea*
Carex sp.	*Platanus bendirei*	*Cedrus*	Nymphaeaceae
Cyperites sp. (Gramineae)	*P. youngii*	Cupressaceae	*Nyssa*
Smilax sp.	*Populus eotremuloides*	*Keteleeria*	Onagraceae
Typha lesquereuxi	*P. lindgreni*	*Picea* (Fig. 7.6I)	*Ostrya*
Acer busamarum busamarum	*P. payettensis*	*Pinus* (Fig. 7.6M,N)	*Pachysandra*
A. b. fingerrockense	*P. pliotremuloides*	*Podocarpus*	*Platanus*
A. chaneyi (Fig. 7.6H)	*P. voyana*	*Pseudotsuga–Larix*	*Populus*
A. latahense	*P. washoensis*	Taxaceae	*Pterocarya*
A. medianum	*Ptelea miocenica*	Taxodiaceae	*Quercus*
A. negundoides	*Pterocarya mixta*	TCT	Rosaceae
A. schorni	*Prunus* sp.	*Tsuga*	*Salix*
A. scottiae	*Pyrus mckenziei*	Gramineae	*Sarcobatus*
A. septilobatum	*Quercus dayana*	*Potamogeton*	*Shepherdia*
A. tyrellense	*Q. eoprinus*	*Typha*	*Tilia*
Ailanthus indiana	*Q. hannibali*	*Acer*	*Ulmus*
Alnus hollandiana	*Q. prelobata*	*Alnus*	Umbelliferae
A. relatus	*Q. pseudolyrata*	*Ambrosia* (Fig. 7.6K)	*Zelkova*(?)
A. sp.	*Q. simulata*	*Artemisia*	
Amelanchier couleeana	*Q.* subsp. *erythrobalanus*	*Betula*	
Anoda suckerensis	*Q.* sp.	Caprifoliaceae	
Arbutus idahoensis	*Rhus* sp.		
A. trainii	*Ribes* sp.		
Betula thor	*Salix hesperia*		
B. sp. (leaf, wood)	*S. succorensis*		
B. sp. (seed)	*S.* sp.		
Carpinus sp.	*Sassafras columbiana* (Fig. 7.6E)		
Carya sp.	*Symphoricarpos salmonensis*		
Castanea spokanensis	*Tilia aspera*		
Castanopsis sp.	*Ulmus knowltoni*		
Cedrela pteraformis	*U. newberryi*		
Celtis sp.	*U. owyhensis*		
Cornus ovalis	*U. paucidentata*		
Crataegus gracilens	*U. speciosa* (Fig. 7.6G)		
C. sp.	*Umbellularia* sp.		
Diospyros oregoniana	*Vaccinium sonomensis*		
Evodia sp.	*Zelkova browni*		
Fagus washoensis			
Fraxinus coulteri (Fig. 7.6D)			

Adapted from Fields (1990). For *Amelanchier* see Schorn and Gooch (1994); for *Ptelea* see Call and Dilcher (1995). A revised list based on Fields (1996) is in preparation (P. F. Fields, personal communication, 1996).

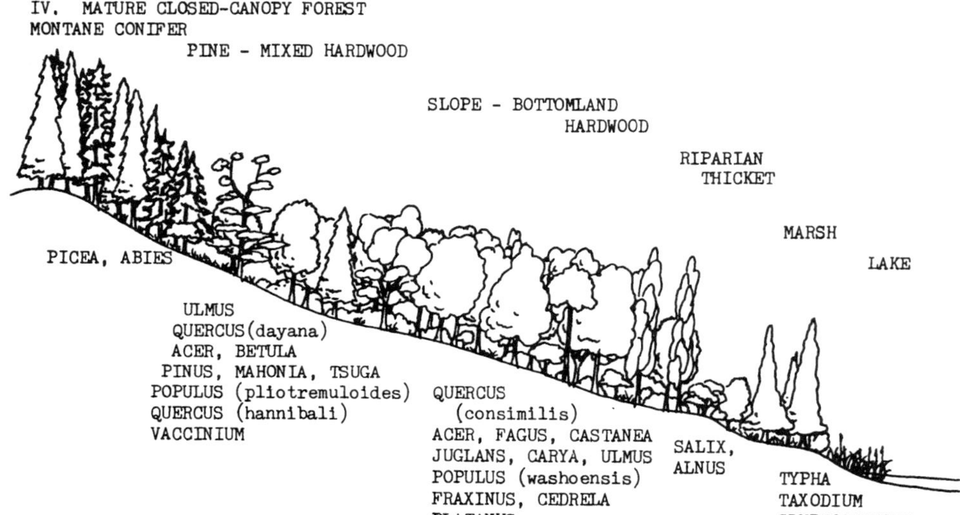

Figure 7.7. Generalized reconstruction of climax paleovegetation, Succor Creek, southeastern Oregon. Reprinted from Cross and Taggart (1982) with the permission of Aureal Cross, Ralph Taggart, and the Missouri Botanical Garden.

Ellensburg semiarid oak forest (*Quercus prelobata*; small-leaved *Ceanothus precuneatus*, *Ilex opacoides*, *Ulmus moorei*, *U. paucidentata*). The climate changed from a humid summer-wet to a semiarid summer-dry regime. The Ellensburg and associated floras document that through ~10–8 Ma the vegetation was not yet affected by decreasing moisture from the rise of the Cascade Mountains. The Latah flora at Spokane, Washington–Coeur d'Alene, Idaho (15.8 Ma), has not been revised since Knowlton's (1926) study, but it also represents a mixed mesophytic forest.

Pollen from a series of strata in the Idaho Group of the Snake River Plain record the decline of the deciduous forest from the Late Miocene through the Pliocene (Leopold and Wright, 1985). The older assemblages include *Carya*, *Juglans*, *Quercus* (megafossils), *Pterocarya*, and *Ulmus*, along with diverse woods of deciduous hardwoods. *Nyssa* disappeared from western North America after the Late Miocene (Eyde and Barghoorn, 1963; Wen and Stuessy, 1993). Near the end of the sequence (~3–2 Ma) the flora is impoverished pine and other conifers, and hardwoods are rare. In still younger Plio–Pleistocene floras, components of a Great Basin desert–shrubland with grasses (to 60%; *Sar-*

Figure 7.8. Pigmented leaf of ~14–15 m.y. old *Zelkova oregoneniana*, Succor Creek. Photograph courtesy of Karl J. Niklas.

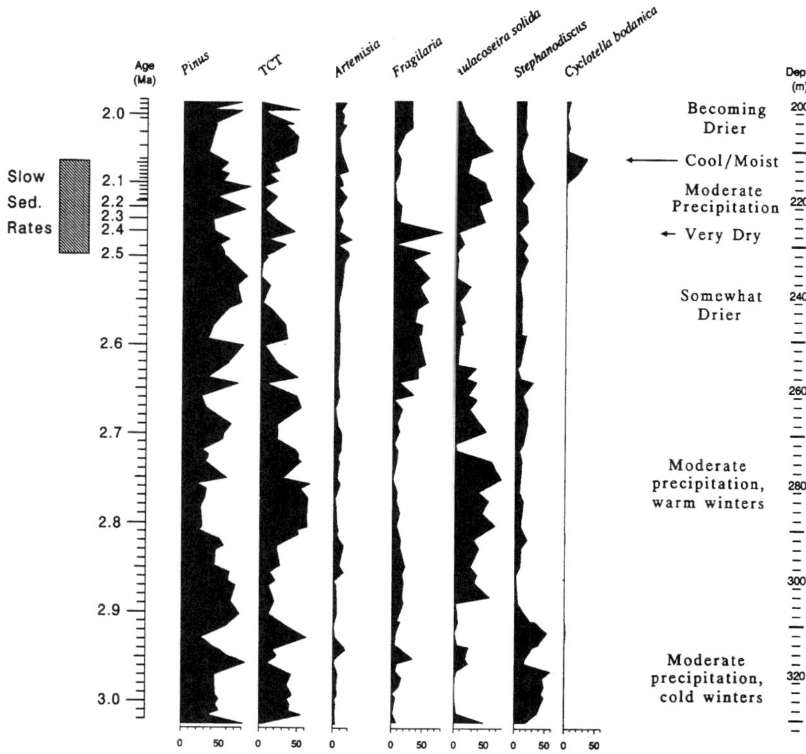

Figure 7.9. Pollen percentage diagram illustrating paleoenvironmental fluctuations at Tulelake, California from 3 to 2 Ma. Reprinted from Thompson (1991; based on Adam et al., 1990b) with the permission of Quaternary Science Reviews, Pergamon Press, and Elsevier Science Ltd.

cobatus, Artemisia) increase, and there is further decline of the deciduous forest (trace amounts of *Carya* and *Ulmus*). Fish faunas and oxygen isotopic composition of fish otoliths from the Snake River Plain record a change from warm moist with cool summers and mild winters (Late Miocene) to cooler summers (Smith and Patterson, 1994). In the Great Salt Lake Basin of northern Utah, pollen from an Amoco Production core revealed the vegetation of the eastern Great Basin for the last ~4 Ma (Davis, in press). Cold-desert shrubs of the Chenopodiaceae (*Amaranthus* or *Sarcobatus*) were present on the slopes, and Cyperaceae (sedges) and *Typha* (cattail) were along the streams and lake margin. Only a small amount of pollen from a few holdovers of the deciduous forest was recovered (*Juglans, Ostrya–Carpinus, Shepherdia, Ulmus*). The Early Pliocene climate is estimated at 3–4°C warmer and 50–100 mm drier than at present. Estimates for the Oak Grove Fork flora of northwestern Oregon (~3 Ma) are for temperatures 2–4°C higher and precipitation 1000 mm lower (Wolfe, 1990). Pollen records reveal abundant conifers and more mesic plants (*Abies, Cedrus, Alnus*) than do samples of similar age inland, suggesting greater continentality toward the interior.

In northern California near the Oregon border a core from Tulelake, just to the east of the southern Cascade Range, includes a Pliocene section (3–2 Ma; Adam et al., 1989, 1990a,b; Fig. 7.9). The lower portion of the core reflects a mixed conifer forest similar to that on the lower western slopes of the Sierra Nevada 500 km to the south and suggests temperatures 5°C warmer than at present and significant summer drought.

The Deschutes flora of northern Oregon (~4–5 Ma; Chaney, 1938) is a low-diversity, riparian vegetation of deciduous hardwoods bordered by shrubland [*Acer negundoides, Populus washoensis, P. alexanderi* (*P.* aff. *trichocarpa*), *Prunus irvingii* (*P.* aff. *emarginata*), *Quercus prelobata, Salix* sp. (*S.* cf. *florissantii*), *Salix* aff. *caudata, Ulmus affinis*]. The flora reflects a decline in summer precipitation from earlier Miocene times. The Palouse Prairie of Idaho, eastern Washington, and eastern Oregon appears in the Late Pliocene ~3 Ma, fully 10 Ma after the Plains grassland. Its history is associated with the rise of the Cascade Mountains and the development of different rainfall regimes east and west of the Rocky Mountains. To the east, summer rains are brought into the Plains from tropical lows moving northward across the region in June and July. To the west, the Pacific summer air is dry from the rainshadow effect of the Cascade Mountains and precipitation comes from winter lows. In general, the Columbia Plateau floras indicate that components of grassland and near-desert vegetation became important during the Pliocene (Leopold and Wright, 1985) but were not widespread until the Plio–Pleistocene. This is in contrast to an earlier development suggested by grazing mammals (see Chapter 3, Faunas). In the Pacific Northwest winter and summer average temperatures in the Neogene differed by ~20°C (Wolfe, 1978).

The effects of the temperature and rainfall decline with greater seasonality at ~15 Ma on western North American vegetation is recorded in numerous tectonic basins where sediments of Tertiary age preserve an extensive sequence of floras (Axelrod, 1992b). These floras can be arranged in pairs, one just before and one just after the temperature decline at ~15 Ma, and along a line from south (Nevada) to north (Washington). The Fingerrock flora (16.4 Ma; 15.5 Ma Wolfe et al., 1997) of southwestern Nevada has numerous deciduous hardwoods that include exotics from eastern North America and eastern Asia (48%; species of *Acer*, *Alnus*, *Betula*, *Carya*, *Diospyros*, *Eugenia*, *Gymnocladus*, *Malus*, *Photinia*, *Populus*, *Quercus*, *Robinia*, *Sophora*, *Ulmus*, and *Zelkova*), along with *Abies*, *Picea*, *Pinus*, and *Tsuga* from the highlands and *Arbutus*, *Cercocarpus*, *Garrya*, *Lithocarpus*, and *Quercus* from a drier broad-leaved evergreen sclerophyll woodland. The Stewart Valley flora from the same region is ~14.5 m.y. in age and contains none of the eastern exotic hardwoods. Rather, it is a Madrean vegetation typical of present-day Arizona, southern California, and especially northern Mexico where many dry-habitat species of *Pinus* and *Quercus* likely had their origin. The fossil flora consists of *Arbutus*, *Garrya*, *Juglans*, *Lyonothamnus*, *Populus*, *Quercus*, *Rhus*, and *Sapindus*, with *Abies*, *Chamaecyparis*, *Pinus*, *Amelanchier*, *Arctostaphylos*, *Holodiscus*, *Philadelphus*, *Prunus*, and *Ribes* on the cooler slopes and *Salix* along river margins. A similar pattern is seen in assemblages at Pyramid (15.6 Ma; 44% exotic species) and Purple Mountain (14.7 Ma; 14% exotics) in west-central Nevada. Further north in northwestern Nevada the 49-Camp flora (~16 Ma) has 55% exotics among species of *Acer*, *Ailanthus*, *Carya*, *Cedrela*, *Cercidiphyllum*, *Cocculus*, *Fagus*, *Nyssa*, *Ptelea*, *Oreopanax*, *Populus*, *Quercus*, *Sassafras*, *Tilia*, and *Ulmus*, with an upland flora of *Abies*, *Chamaecyparis*, *Pinus*, *Acer*, *Fraxinus*, *Populus*, *Prunus*, and *Rosa*, and a meager broad-leaved, evergreen sclerophyll vegetation of *Arbutus* and *Quercus*. The Gillam Spring flora has 26% exotics. A decrease in summer precipitation, estimated at 35–40% below the pre–15 Ma annual total, is inferred from the Gillam Spring assemblage (Axelrod and Schorn, 1994). In central Oregon floras at Mascall (~16 Ma; 70% exotic species) and Stinking Water (11 Ma; 35%: Fig. 7.10) show a similar trend, as do the Latah (15.8 Ma; 75% exotics) and Ellensburg (14 Ma; 44% exotics) floras of Washington.

These data indicate that the assemblages older than ~15 Ma are richer in mesic, Asian exotics and that the percentages increase to the north (Mascall and Latah floras are richest) where precipitation was higher. Floras in the same region that are younger than ~15 m.y. have fewer exotics that were rapidly eliminated beginning in the south as summer rainfall decreased. Modern rainfall records and the distribution of woody taxa in the western Great Plains suggest that when precipitation falls below 75–90 mm (3.0–3.5 in.) during each of the summer months, deciduous hardwoods rapidly decrease (Axelrod, 1992b). This is probably near the threshhold level reached in the arid southwest after ~15 Ma. The cause for decreasing rainfall in the southwest is complex and not all factors have been integrated into a fully satisfactory model. The $^{18}O/^{16}O$ marine record shows a sharp decrease in benthic temperatures at ~15 Ma associated with the spread of the East Antarctic ice sheet. This allowed cold waters to spread into the world's ocean basins, which caused an upwelling of nutrient-rich cold waters along some coastal regions. That such an upwelling did occur is documented by widespread blooms of diatoms preserved as extensive deposits of diatomite at about the same time in California and other parts of the world. Winds blowing across these cooler waters lost moisture and provided less rainfall to the interior. The cooling trend was augmented at ~8–7 Ma by the spread of the West Antarctic ice sheet (intensification of high-pressure systems under the descending arm of the Hadley convection cell), early glaciation in the northern Coast Ranges and the Arctic Basin, and continued rise of the western mountains. Axelrod (1992b) identifies the Mio–Pliocene as the driest part of the Tertiary, and woodland–shrubland/chaparral spread more widely in the west and exotics persisted only as relicts along the mild coast. Thompson (1991, 1996) provides a summary of general climatic conditions inferred from megafossil and microfossil floras and associated faunas from the western United States: 4.8–2.4 Ma, mostly wetter and warmer than at present; 2.4–2.0 Ma, colder and drier; 2.0–1.8 Ma, return to warmer and more moist.

One consequence of the Late Tertiary increase in the height of the Rocky Mountains and other western mountain systems was the blockage of warm air into the western Arctic region of North America. Prior to the Alaska Range attaining significant heights at ~4–5 Ma, southern Alaska was relatively flat. The subsequent uplift contributed to the Neogene cooling in the high latitudes. The Seldovia Point flora of Alaska (late Early to early Middle Miocene; Chapter 6) is a deciduous forest, and a coniferous forest is in the vicinity (Wolfe and Tanai, 1980). Although the data are meager, the aggregation of *Abies*, *Larix*, *Picea*, *Tsuga*, and others appears to have taken place there beginning in the Late Miocene and Pliocene. Because several of these elements, along with *Betula*, are present in the Middle Eocene Republic flora of northeastern Washington, the coniferous forest probably spread north and south and into lower elevations from upland centers in the northern Rocky Mountains (Axelrod et al., 1991). In floras of Mio–Pliocene age from the Alaska Range, *Abies*, *Larix*, *Picea*, *Pinus*, *Tsuga*, *Populus*, and *Salix* were present; in the lowlands of the Cook Inlet region of Alaska, some boreal conifers were present by the Pliocene (*Picea*, *Tsuga*; Wolfe, 1966, 1972; Wolfe and Hopkins, 1967; Wolfe and Leopold, 1967; Wolfe et al., 1966). However, the Cook Inlet flora also includes some holdovers from older Miocene vegetation, such as *Glyptostrobus*, and possibly *Pterocarya*, *Tilia*, and *Ulmus* (extinct in Alaska in the Late Miocene). Wolfe et al. (1966) propose three paleobotanical stages for the Seldovia Point Kenai Group of the Cook Inlet region: Seldovian (deciduous forest), Homerian (mixed conifer and hardwood forest), and Clamgulchian (conifer forest with occasional hardwoods). The floras are similar to the modern northern mixed

Figure 7.10. (A) *Quercus dayana*, Stinking Water. (B) *Populus lindgreni*, Stinking Water. (C) *Alnus harneyana*, Stinking Water. (D) *Quercus prelobata*, Stinking Water. Reprinted from Chaney and Axelrod (1959) with the permission of the Carnegie Institution of Washington.

conifer–hardwood forest of New England and the lake states, with some added Asian elements, and grew under an estimated MAT of ~9°C decreasing to ~3°C toward the top of the sequence. [See papers in Thompson, 1994; fossil spruce and larches from near Circle, east-central Alaska, were studied by Miller and Ping (1994); and larches from other parts of the Arctic are under study by Schorn (1994).] At Fort Yukon the present MAT is −6.4°C and annual precipitation is 168 mm. A fossil palynoflora from the Porcupine River region of east-central Alaska (67° 20' N) was dated at 15.2 Ma (Seldovian) by the $^{40}Ar/^{39}Ar$ method (White and Ager, 1994). The vegetation is deciduous forest and includes TCT, *Abies*, *Larix–Pseudotsuga*, *Picea*, *Pinus*, *Sciadopitys* (presently Asian), *Tsuga*, cf. *Juncus*, *Acer*(?), *Alnus*, *Betula*, *Cornus*, cf. *Corylus*, cf. *Carpinus*, *Carya*, *Castanea*-type, *Cercidiphyllum*, *Fagus*, cf. *Galium*, *Ilex*, *Juglans*, *Liquidambar*, *Ludwigia*, *Nymphaea*, *Nyssa*, *Pterocarya* (Asian), *Quercus*, *Rhus* types, *Salix*, *Tilia*-type, and *Ulmus*-type. Another pollen–spore sequence from the Nenana coal field of Alaska (63.5° N) correlates with the Upper Seldovian stage (late–early to early Middle Miocene) and is also mixed northern hardwood forest with eastern United States–Asian affinities and thermophilous taxa (e.g., *Fagus*, *Quercus*). The Middle Miocene MAT is estimated at ~9°C, falling to ~5°C in the Late Miocene (Leopold and Liu, 1994). An early Middle Pliocene pollen–spore assemblage at Circle has increased boreal conifers and suggests a MAT of ~0.6–2.6°C, or ~7–9°C warmer than the present MAT of −6.4°C (Ager et al., 1994). Paleosol evidence is consistent with deciduous forest vegetation and a temperate climate (Smith et al., 1994). By the Late Miocene thermophilous genera such as *Carya*, *Fagus*, *Liquidambar*, and *Nyssa* are rare or absent. The Pliocene Lava Camp flora on the Seward Peninsula, about 75 km south of the Arctic Circle, includes *Picea glauca*, *P. mariana*, *P. sitchensis*, *Pinus monticola*, *Tsuga heterophylla*, *Carex*, *Cyperus*, *Alnus*, *Betula*, *Corylus*, *Epilobium*, Ericaceae, *Oenothera*, *Salix*, *Symphoricarpos*, and *Vaccinium* (Hopkins et al., 1971). Fossil woods in the early Middle Miocene from the Porcupine River region are Taxodiaceae and pines with growth rings greater than 1 cm wide, while in the Late Pliocene the woods are *Picea* and *Larix* with growth rings less than 1 mm wide (Wheeler and Arnette, 1994). Axelrod et al. (1991) note that this is not a taiga community and compare it to the modern vegetation along the northern border of the coast coniferous forest. The associated insect fauna is not a tundra or boreal forest assemblage but is similar to that occurring in the Pacific coast coniferous forest of southern British Columbia–

Table 7.9. Pollen and spore counts of Colvillian age, North Slope of Alaska.

	Site 1 Sample			Site 2 Sample	
	A	B	C	A	B
Abies	—	+	—	+	+
Tsuga mertensiana	—	+	+	—	—
Other *Tsuga*	—	+	+	—	+
Pinus	2.3	8.8	8.5	10.2	1.2
Picea	25.1	30.5	18.7	33.8	13.9
Alnus	5.1	6.6	5.7	7.5	16.3
Betula	36.0	27.4	33.5	27.5	32.2
Ericales	5.1	6.0	8.2	6.9	6.6
Salix	1.6	1.6	2.2	—	+
Populus	1.3	2.5	+	1.3	+
Other woody taxa	Co	Ca,La	Co	—	Co
Gramineae	7.7	4.4	8.9	4.6	14.8
Cyperaceae	10.3	6.6	10.4	3.9	9.9
Other herbs	3.2	1.9	1.6	2.6	+
Unknown	+	1.9	+	+	1.2
Indeterminate	9.6	7.2	8.5	13.8	9.3
Sphagnum	10.0	14.5	19.6	17.4	14.5
Lycopodium selago	—	+	+	+	+
Other *Lycopodium*	6.8	4.7	2.2	7.8	4.5
Selaginella selaginoides	+	—	+	1.3	+
Other *Selaginella*	+	—	—	+	—
Cf. *Osmunda*	2.6	1.9	+	—	+
Other trilete spores	19.0	11.9	23.4	8.8	11.1
Monolete fern spores	3.9	2.2	7.3	5.6	1.2
Pediastrum colonies	+	—	+	—	2.7

All values are expressed as a percent of the total pollen excluding indeterminate grains. (—) Not present; (+) present but <1%; Co, *Corylus*; Ca, *Carya*; La, *Larix*. Adapted from Nelson and Carter (1992).

northern Washington. This forest extended to near the Arctic Circle in the Pliocene, as shown by the Kivalina flora 100 km north of Lava Camp (Hopkins et al., 1971). A boreal forest of modern aspect was not yet assembled or widespread. Rather, vegetation in the Arctic began changing from a mixed mesophytic hardwood forest to a coniferous forest at ~15 Ma followed by a more prominent herb flora at ~7 Ma (White et al., 1997). By the end of the Tertiary average July temperatures in the Cook Inlet region dropped by ~7°C from the Early Miocene thermal high (Wolfe and Leopold, 1967; Wolfe and Tanai, 1980). It was not until ~4 Ma that the lowland boreal coniferous forest became established.

On the North Slope of Alaska on the Arctic Ocean borderland (~70° N), a series of coastal plain sediments in the Gubik Formation were deposited during the Late Pliocene and Pleistocene. Several transgressions are evident; the late Pliocene ones are called the Colvillian (oldest), Bigbendian, and Fishcreekian. The time interval from the Colvillian through the Bigbendian is from ~3 to ~2.4 Ma when the latest Bigbendian was being deposited near the beginning of the Matuyama Reversed polarity chron. The end of the Fishcreekian is placed at ~2 Ma. Ostracodes from the Colvillian and Bigbendian indicate that waters were still warmer than at present and comparable to those today along the coasts of northern Nova Scotia and southern Labrador (45°–50° N; viz., not frozen to the shore any time during the year; Repenning and Brouwers, 1992). In the Fishcreekian the ostracode fauna suggests cooling to ~5°C, similar to temperatures now prevailing in the southern Bering Sea (60°–65° N).

Pollen from the Ocean Point section on the Colville River (Table 7.9; Fig. 7.11) also indicates a climate warmer than at present (Nelson and Carter, 1985, 1992). *Tsuga* and *Pinus* were present in the Colvillian but decreased or disappeared in the Bigbendian, leaving a spruce–birch parkland with some *Pinus*, *Tsuga*, *Larix*, *Alnus*, and *Populus*. There was no extensive tundra or permafrost as there is today, but open areas did include grasses and tundra elements (forest–tundra mosaic). In the late Fishcreekian (~2 Ma) the boreal forest still extended further north than at present, as suggested by occasional pollen of *Larix*, *Picea*, and *Pinus*; but the local vegetation was changing through a shrub tundra to herb-dominated tundra, and permafrost was present.

A summary of the age of these western North American floras between 16 and 2 Ma, along with proposed floral stages (Wolfe, 1981), is given in Fig. 7.12.

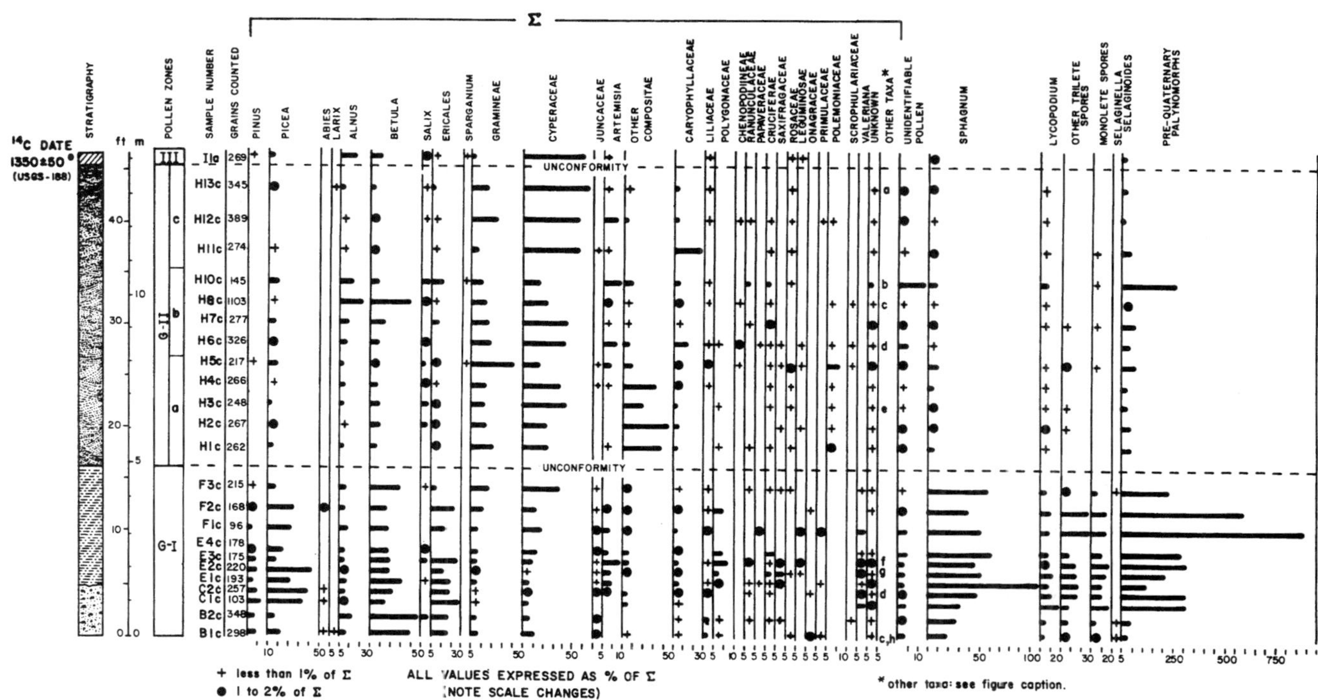

Figure 7.11. Pollen percentage diagram from the Gubik Formation at the Ocean Point site. Note changes in scale. *Other taxa: a, *Populus*; b, *Menyanthes*; c, *Potamogeton*; d, *Plantago*; e, *Listera* type (Orchidaceae); f, *Drosera*; g, *Myriophyllum*; and h, Rubiaceae. Reprinted from Nelson and Carter (1985) with the permission of Quaternary Research.

Age	SERIES		STAGE		FLORAS	RADIOMETRIC
1.6	PLIOCENE	UPPER			OCEAN POINT	3.0 - 2.0
					GUBIK	3.0 - 2.0
					LAVA CAMP	-
		LOWER	UPPER	CLAMGULCHIAN	TULELAKE	3 - 2
					CHULA VISTA	3.0
					DESCHUTES	4 - 5
5	MIOCENE	UPPER	L		MOUNT EDEN	5.3
				HOMERIAN	SONOMA	-
					TWEEINOT LAKE	8.0
					ALDRICH STATION	10.7 - 11.2
					TRAPPER CREEK	10.5 - 12.0
					STINKING WATER	11
12		MIDDLE	UPPER	SELDOVIAN	TROUT CREEK	13
					ELLENSBURG	14
					SUCKER CREEK	14 - 15
					MASCALL	15.4
					PYRAMID	15.6
16					CLARKIA	16

Figure 7.12. Summary of the relative ages, floral stages, and radiometric dates for the principal western North American floras between ~16 and 2 m.y. in age discussed in the text.

VEGETATION SUMMARY

Although published information on vegetational history is surprisingly meager for the Middle Miocene through the Pliocene of the southeastern United States, it is likely that modernization of the vegetation accelerated with the Middle to Late Miocene cooling event (Table 7.10). The principal effects were the disappearance of most Asian and neotropical exotics by the Pliocene, modernization and expansion of the mixed mesophytic forest, the temporal association of different forest genera to constitute various other deciduous forest associations, and development of the Appalachian coniferous forest. In Chapter 6 it was noted that by Claiborne (Middle Eocene)–Jackson (Late Eocene) time, elements were available for assemblage into the mixed mesophytic, southern mixed hardwood, oak–hickory, oak–chestnut, sand pine scrub (*Pinus*, *Sabal*, other rosette palms), flood-plain (*Taxodium*, *Nyssa*), and mangrove (*Acrostichum*, *Nipa*) associations. *Tilia* is reported from the Claiborne Formation, but *Acer*(?) first occurs in the Jackson Formation. Definite *Acer* pollen is known from the Brandywine flora, so an initial maple–basswood association in the region could date from the Late Miocene. *Nipa* disappeared by the end of the Eocene, but it is not known when *Rhizophora* first appeared in the southeast. In the intervening period, brackish habitats were probably occupied by *Acrostichum* and various palms. *Picea* is known from the Jackson Formation, *Tsuga* occurs later in the Middle Miocene Legler Lignite, and *Abies* is found in Pliocene deposits but to the north at Martha's Vineyard. Thus, the limited fir–spruce–hemlock Appalachian coniferous forest in the south is likely of Plio–Pleistocene origin. It may have formed earlier to the north, but *Abies* and *Picea* are not known from the Early Miocene Brandon Lignite of Vermont. Today *Abies balsamea* (balsam fir) extends south on isolated peaks to Virginia and West Virginia, and *A. fraseri* (Fraser fir) reaches only as far south as North Carolina and Tennessee at elevations of 1400–2100 m (4500–6900 ft).

To the north, the deciduous forest intermingled with an increasing number of cold-temperate conifers to form a version of the lake states forest. This community was more broadly distributed than at present; however, with later development of the boreal coniferous forest, it gradually became limited and better defined as the transition vegetation between the deciduous forest and boreal coniferous forest in the Great Lakes region. Further to the north, fossil floras record a mixed temperate hardwood forest (Middle Miocene), a cool-temperate coniferous-hardwood forest (Mio–Pliocene), and a cold-temperate conifer forest and *Alnus* woodland (late Pliocene). In the Arctic islands the sequence was from conifer forest, to stunted conifer woodland, to depauperate forest–tundra vegetation (Plio–Pleistocene, ~1.7 Ma).

Boreal coniferous forest elements such as *Abies*, *Picea*, *Thuja*, *Tsuga*, and *Betula* are first recorded together in the Middle Eocene Republic flora of northwestern Washington; these volcanic highlands were likely a center of dispersal for the formation. The forest was moving into lower elevations by ~15 Ma, and by the Mio–Pliocene a version was occupying the lowlands of the Cook Inlet region of Alaska; but it still included some Asian exotics and deciduous hardwoods (*Glyptostrobus*, *Pterocarya*, *Tilia*, *Ulmus*). These disappeared in the Pliocene, and the Lava Camp and Kivalina floras preserve a near-taiga community most similar to the modern coast coniferous forest of northern Washington and southern British Columbia. The boreal coniferous forest of essentially modern aspect appears later in the Pliocene at ~4 Ma, and by 2 Ma it had reached the Colville River region on the north slope of Alaska. With the continuing trend toward colder climates it began disappearing along its northern limits at this time and was locally mixed with shrub tundra and herb-dominated tundra growing on permafrost. Widespread tundra of modern aspect did not develop until near the onset of extensive Arctic continental glaciations at ~850 ka.

In the continental interior, the trend toward colder and winter-dry climates had produced a mosaic of open deciduous forest–woodland with patches of grassland by ~13 Ma (Kilgore flora). Between 13 and 6–5 Ma grassland expanded at the expense of forest that was confined mostly to riparian habitats. Near-prairie vegetation with *Juniperus* and *Celtis* occupied much of the region. Grasslands of modern aspect appeared with the next principal temperature decline beginning ~1.6 Ma and later in the Quaternary.

The woodland–grassland of the Plains graded eastward into a piñon–juniper woodland toward the eastern slopes of the central and northern Rocky Mountains. The highlands supported a western montane coniferous forest. The woodland extended around the southern end of the Rocky Mountains into the southwest where it was associated with evergreen oak to form Madrean woodland–chaparral. By the Pliocene the southwest was arid and the vegetation resembled the more mesic parts of the Sonoran Desert. Along the moister southern California coast the vegetation was a pine–oak woodland that included palms and *Persea* (Chula Vista flora). To the north in the San Francisco Bay area (Sonoma flora) the lowland vegetation near the coast was a Pacific coast coniferous forest that included *Sequoia*, while further inland a pine–douglas fir forest grew in moister habitats with an oak woodland–chaparral on the slopes.

The principal Miocene communities of the Columbia Plateau region during the Miocene (Succor Creek flora) were a mixed mesophytic association of the deciduous forest generally below ~600 m that included Asian and eastern North American exotics, a transition conifer–hardwood forest, and an upland western montane coniferous forest. The conifer–hardwood forest extended northward into Alaska (Seldovia Point flora) and downward in elevation during the Middle Tertiary. The decline of the mixed mesophytic association in western North America is recorded in the Idaho Group floras with the rise of the

Table 7.10. Summary of North American vegetation types and estimated MATs, Middle Miocene through Pliocene.

Region	Vegetation	MAT
	Middle to Late Miocene	
Southeast	Warm-temperate deciduous	
Mid-Atlantic	Warm-temperate deciduous	
New Jersey	(mixed mesophytic, beech–maple,	
Kirkwood	maple–basswood, lake states,	
Legler (11 Ma)	oak–chestnut, oak–hickory,	
Maryland	southern mixed hardwood, sand	
Brandywine	pine scrub, floodplain); Appalachian montane coniferous (north), early version (south)	
Northeast		
Massachusetts		
Martha's Vineyard	Warm-temperate deciduous	
Iceland		
Holmatindur (10.3–9.5 Ma)	Sequence: deciduous to subarctic *Alnus* woodland	
Western interior		
Nebraska		
Kilgore (13–14 Ma)	Woodland–grassland, riparian deciduous	
Ash Hollow (6–7 Ma)	Wooded grassland	
West		
Idaho		
Trapper Creek (10.5–12 Ma)	Deciduous, western montane coniferous	
Idaho Group (Miocene part)	Deciduous, pine woodland	
Wyoming		
Teewinot (8 Ma)	Shrubland to near desert, riparian, western montane coniferous	
Nevada		
Fingerrock (16.4) Ma), Pyramid (15.6 Ma)	Deciduous	13.5°C
Verdi	Oak woodland, chaparral, savanna	
Stewart Valley (14 Ma)	Pine-oak woodland (Madrean)	
California		
Mount Eden (5.3 Ma)	Near desert	
Northwest		
Oregon		
Succor Creek (14–15 Ma)	Deciduous, swamp cypress (floodplain), riparian, mixed conifer–hardwood, western montane coniferous	
Washington		
Ellensburg	Swamp cypress (floodplain), deciduous, oak woodland (drier sites)	
Alaska		
Seldovia, Cook Inlet, Lava Camp, Kivalina	Deciduous, mixed conifer–hardwood, Pacific coast coniferous, early boreal coniferous	9–3°C
	Pliocene	
Northeast		
Massachusetts		
Martha's Vineyard	Cold-temperate deciduous	
Greenland		
Kap København (82° 30' N; 2 Ma)	Boreal forest, forest tundra	
West		
Idaho		
Idaho Group (Pliocene part)	Desert–shrubland with grasses	
California		
Chula Vista (3 Ma)	Pine–oak woodland, riparian warm temperate	16°C
Sonoma		
Napa	Western montane coniferous, oak–woodland–chaparral	
Santa Rosa	Pacific coast coniferous (*Sequoia*)	
Tulelake (3–2 Ma)	Western montane coniferous	
Alaska		
Colville	Boreal coniferous, forest–tundra	

Plant formation and association names follow terminology used for modern communities discussed in Chapter 1 and listed in Table 1.1.

coastal mountains. Impoverished pine forest, then Great Basin desert–shrubland (*Sarcobatus*, *Artemisia*) with grasses appear at ~3–2 Ma.

Important events in the modernization of the North American vegetation during the period from ~16.3–1.6 Ma were:

1. the disappearance of Asian and neotropical exotics by the Pliocene (~5 Ma);
2. the continued but temporal grouping of various genera into associations that define the present version of the deciduous forest formation (e.g., maple–basswood);
3. the persistence of tropical elements along the Florida coast;
4. the further association of *Abies*, *Picea*, *Tsuga*, and others to constitute the Appalachian montane forest (progressively later toward the south);
5. the appearance of a near-modern boreal coniferous forest at ~4 Ma;
6. the gradual restriction of the transition vegetation between the boreal coniferous forest and the deciduous forest into the present lake states forest;
7. the development of near-grasslands in the continental interior by ~2 Ma;
8. the formation of near-modern deserts during the Pliocene (~5 Ma) with the continued rise of the Sierra Nevada, initial appearance of the high Cascade Mountains and the Coast Ranges, and redirection of atmospheric circulation;
9. the appearance of the high-altitude coastal coniferous forest in the midlatitudes by ~4 Ma;
10. the consequent disruption of the earlier near-continuous deciduous forest formation across the temperate latitudes of the northern hemisphere;
11. the appearance of the California grasslands and Palouse Prairie at ~3 Ma; and
12. the appearance of near-modern tundra and sustained permafrost at ~2 Ma.

Beginning at ~1.6 Ma, these Late Tertiary communities were subjected to the increasingly turbulent and fast-paced climatic events of the oncoming ice age.

References

Adam, D. P. et al. (+ 5 authors). 1989. Tulelake, California: the last 3 million years. Palaeogeogr., Palaeoclimatol., Palaeoecol. 72: 89–103.

J. P. Bradbury, and H. J. Rieck. 1990a. Pliocene environmental changes at Tulelake, Siskiyou County, California, from 3 to 2 Ma. In: L. B. Gosnell and R. Z. Poore (eds.), Pliocene Climates: Scenario for Global Warming. U.S. Geological Survey, Reston, VA. Open-File Report 90–64, pp. 3–5.

——— and A. M. Sarna-Wojcicki. 1990b. Environmental changes in the Tule Lake Basin, Siskiyou and Modoc Counties, California, from 3 to 2 million years before present. U.S. Geol. Surv. Bull. 1933: 1–13.

Ager, T. A., J. V. Matthews, Jr., and W. Yeend. 1994. Pliocene terrace gravels of the ancestral Yukon River near Circle, Alaska: palynology, paleobotany, paleoenvironmental reconstruction and regional correlation. Quatern. Int. 22/23: 185–206.

Armstrong, R. L., W. P. Leeman, and H. E. Malde. 1975. K-Ar dating, Quaternary and Neogene volcanic rocks of the Snake River Plain, Idaho. Am. J. Sci. 275: 225–251.

Arroyo, M. T. K., P. H. Zedler, and M. D. Fox (eds.). 1995. Ecology and Biogeography of Mediterranean Ecosystems in Chile, California, and Australia. Ecological Studies 108. Springer-Verlag, Berlin.

Austin, J. J., A. J. Ross, A. B. Smith, R. A. Fortey, and R. H. Thomas. 1997. Problems of reproducibility—does geologically ancient DNA survive in amber-preserved insects? Proc. Roy. Soc. London B . 264: 467–474.

Axelrod, D. I. 1937. A Pliocene flora from the Mount Eden beds, southern California. Carnegie Inst. Wash. Contrib. Paleontol. 476: 125–183.

———. 1940. A record of *Lyonothamnus* in Death Valley, California. J. Geol. 48: 526–531.

———. 1944. The Sonoma flora (California). Carnegie Inst. Wash. Contrib. Paleontol. 533: 167–206.

———. 1950a. The Anaverde flora of southern California. Carnegie Inst. Wash. Contrib. Paleontol. 590: 119–158.

———. 1950b. The Piru Gorge flora of southern California. Carnegie Inst. Wash. Contrib. Paleontol. 590: 159–214.

———. 1950c. Further studies of the Mount Eden flora, southern California. Carnegie Inst. Wash. Contrib. Paleontol. 590: 73–117.

———. 1950d. A Sonoma florule from Napa, California. Carnegie Inst. Wash. Contrib. Paleontol. 590: 23–71.

———. 1956. Mio-Pliocene floras from west-central Nevada. Univ. Calif. Publ. Geol. Sci. 33: 1–322.

———. 1958. The Pliocene Verdi flora of western Nevada. Univ. Calif. Publ. Geol. Sci. 34: 91–160.

———. 1964. The Miocene Trapper Creek flora of southern Idaho. Univ. Calif. Publ. Geol. Sci. 51: 1–180.

———. 1978. Origin of coastal sage vegetation, Alta and Baja California. Am. J. Bot. 65: 1117–1131.

———. 1980. Contributions to the Neogene paleobotany of central California. Univ. Calif. Publ. Geol. Sci. 121: 1–212.

———. 1985. Rise of the grassland biome, central North America. Bot. Rev. 51: 163–201.

———. 1987. The Late Oligcene Creede flora, Colorado. Univ. Calif. Publ. Geol. Sci. 130: 1–235.

———. 1992a. The Middle Miocene Pyramid flora of western Nevada. Univ. Calif. Publ. Geol. Sci. 137: 1–50.

———. 1992b. Miocene floristic change at 15 Ma, Nevada to Washington, U.S.A. Palaeobotanist 41: 234–239.

———. 1995. The Miocene Purple Mountain flora of western Nevada. Univ. Calif. Publ. Geol. Sci. 139: 1–62.

——— and J. Cota. 1993. A further contribution to the closed-cone pine (Oocarpae) history. Am. J. Bot. 80: 743–751.

Axelrod, D. I. and T. A. Deméré. 1984. A Pliocene flora from Chula Vista, San Diego County, California. Trans. San Diego Soc. Nat. Hist. 20: 277–300.

Axelrod, D. I., M. T. Kalin Arroyo, and P. H. Raven. 1991. Historical development of temperate vegetation in the Americas. Rev. Chil. Hist. Nat. 64: 413–446.

Axelrod, D. I. and H. E. Schorn. 1994. The 15 Ma floristic crisis at Gillam Spring, Washoe County, northwestern Nevada. PaleoBios 16: 1–10.

Barnowsky, C. W. 1984. Late Miocene vegetational and climatic variations inferred from a pollen record in northwest Wyoming. Science 223: 49–51.

Barrett, P. J., C. J. Adams, W. C. McIntosh, C. C. Swisher III, and G. S. Wilson. 1992. Geochronological evidence supporting Antarctic deglaciation three million years ago. Nature 359: 816–818.

Bennike, O. and J. Böcher. 1990. Forest–tundra neighbouring the North Pole: plant and insect remains from the Plio–Pleistocene Kap København Formation, north Greenland. Arctic 43: 331–338.

Berggren, W. A., D. V. Kent, M.-P. Aubry, and J. Hardenbol (eds.). 1995. Geochronology, Time Scales and Global Stratigraphic Correlations. Society for Sedimentary Geology, Tulsa, OK. Special Publication 54.

Brooks, B. W. 1935. Fossil plants from Sucker Creek, Idaho. Ann. Carnegie Mus. 24: 275–337.

Call, V. B. and D. L. Dilcher. 1995. Fossil *Ptelea* samaras (Rutaceae) in North America. Am. J. Bot. 82: 1069–1073.

Cano, R. J., H. N. Poinar, D. W. Roubik, and G. O. Poinar, Jr. 1992. Enzymatic amplification and nucleotide sequencing of portions of the 18S rDNA gene of the bee *Prolebeia dominicana* (Apidae: Hymenoptera) isolated from 25–40 million year old Dominican amber. Med. Sci. Res. 20: 619–622.

Chamley, H. 1979. North Atlantic clay sedimentation and paleoenvironment since the Late Jurassic. In: M. Talwani, W. Hay, and W. B. F. Ryan (eds.), Deep Drilling Results in the Atlantic Ocean: Continental Margins and Paleoenvironment. Amer. Geophysical Union, Washington, D.C. pp. 342–361.

Cerling, T. E. et al. (+ 6 authors). 1997. Global vegetation change through the Miocene/Pliocene boundary. Nature 389: 153–158.

Chaney, R. W. 1938. The Deschutes flora of eastern Oregon. Carnegie Inst. Wash. Contrib. Paleontol. 476: 185–216.

———. 1959. Miocene floras of the Columbia Plateau. Part I. Composition and interpretation. Carnegie Inst. Wash. Contrib. Paleontol. 617: 1–134.

——— and D. I. Axelrod. 1959. Miocene floras of the Columbia Plateau. Part II. Systematic considerations. Carnegie Inst. Wash. Contrib. Paleontol. 617: 135–237.

Chaney, R. W. and M. K. Elias. 1938. Late Tertiary floras from the High Plains. Carnegie Inst. Wash. Contrib. Paleontol. 476: 1–72.

Coates, A. G. et al. (+ 7 authors). 1992. Closure of the Isthmus of Panama: the near-shore marine record of Costa Rica and western Panama. Geol. Soc. Am. Bull. 104: 814–828.

Condit, C. 1938. The San Pablo flora of west central California. Carnegie Inst. Wash. Contrib. Paleontol. 476: 217–268.

Cronin, T. M. 1991. Pliocene shallow water paleoceanography of the North Atlantic Ocean based on marine ostracodes. Quat. Sci. Rev. 10: 175–188.

——— and H. J. Dowsett (eds.). 1991. Pliocene Climates. Quat. Sci. Rev. 10: 1–296 [special issue].

Cross, A. T. and R. E. Taggart. 1982. Causes of short-term sequential changes in fossil plant assemblages: some considerations based on a Miocene flora of the northwest United States. Ann. Missouri Bot. Garden 69: 676–734.

Crowley, T. J. 1991. Modeling Pliocene warmth. Quat. Sci. Rev. 10: 275–282.

——— and G. R. North. 1991. Paleoclimatology. Oxford Monographs in Geology and Geophysics.Oxford University Press, Oxford, UK. No. 16.

Curry, W. B. and K. G. Miller. 1989. Oxygen and carbon isotopic variation in Pliocene benthic foraminifers of the equatorial Atlantic. Ocean Drill. Program Sci. Results 108: 157–166.

Davis, O. K. In press. Preliminary pollen analysis of Cenozoic sediments of the Great Salt Lake, U.S.A. Ameican. Association of Stratigraphic Palynologists Pliocene Symposium, Baton Rouge, LA, October 1993.

DeSalle, R., J. Gatesy, W. Wheeler, and D. Grimaldi. 1992. DNA sequences from a fossil termite in Oligo–Miocene amber and their phylogenetic implications. Science 257: 1933–1936.

Dorf, E. 1930. Pliocene floras of California. Carnegie Inst. Wash. Contrib. Paleontol. 412: 1–112.

Downing, K. F. and C. C. Swisher III. 1993. New $^{40}Ar/^{39}Ar$ dates and refined geochonology of the Sucker Creek Formation, Oregon. J. Vert. Paleontol. 13: 33A.

Dowsett, H. J. and T. M. Cronin. 1990. High eustatic sea level during the Middle Pliocene: evidence from the southeastern U.S. Atlantic Coastal Plain. Geology 18: 435–438.

Dowsett, H. J. and P. Loubere. 1992. High resolution Late Pliocene sea-surface temperature record from the northeast Atlantic Ocean. Mar. Micropaleontol. 20: 91–105.

Elias, M. K. 1932. Grasses and other plants from the Tertiary rocks of Kansas and Colorado. Univ. Kansas Sci. Bull. 20: 333– 367.

———. 1935. Tertiary grasses and other prairie vegetation from the High Plains of North America. Am. J. Sci., Ser. 5, 29: 24–33.

———. 1942. Tertiary Prairie Grasses and Other Herbs from the High Plains. Geology Society of America, Boulder, CO. Special Paper 41.

Emslie, S. D. and G. S. Morgan. 1994. A catastrophic death assemblage and paleoclimatic implications of Pliocene seabirds of Florida. Science 264: 684–685.

Essay Collection. 1996. Marine Geology 27.

Evernden, J. F. and G. T. James. 1964. Potassium-argon dates and the Tertiary floras of North america. Am. J. Sci. 262: 945–974.

Eyde, R. H. and E. S. Barghoorn. 1963. Morphological and paleobotanical studies of the Nyssaceae, II the fossil record. J. Arnold Arb. 44: 328–370.

Fields, P. F. 1983. A Review of the Miocene Stratigraphy of Southwestern Idaho, with Emphasis on the Payette Formation and Associated Floras. Masters thesis. University of California, Berkeley, CA.

———. 1990. Succor Creek floral taxonomic diversity: a continuing work in progress. Maps Digest, Expo XII Ed. 13: 50–59.

———. 1992. Floras of Idaho's past: geographic and paleobotanic overview of the Middle Miocene Succor Creek flora as an example. J. Idaho Acad. Sci. 28: 44–56.

———. 1996. The Succor Creek Flora of the Middle Eocene Sucker Creek Formation, Southwestern Idaho and Eastern Oregon: Systematics and Paleoecology. Ph.D. dissertation. Michigan State University, East Lansing, MI.

Frederiksen, N. O. 1984. Stratigraphic, paleoclimatic, and

paleobiogeographic significance of Tertiary sporomorphs from Massachusetts. U.S. Geological Survey, Reston, VA. Prof. Paper 1308, pp. 1–25.

———. 1989. Late Cretaceous and Tertiary floras, vegetation, and paleoclimates of New England. Rhodora 91: 25–48.

Funder, S., N. Abrahamsen, O. Bennike, and R. W. Feyling-Hanssen. 1985. Forested Arctic: evidence from north Greenland. Geology 13: 542–546.

Fyles, J. G. et al. (+ 6 authors). 1994. Ballast Brook and Beaufort formations (Late Tertiary) on northern Banks Island, Arctic Canada. Quatern. Int. 22/23: 141–171.

Golenberg, E. M. 1991. Amplification and analysis of Miocene plant fossil DNA. Phil. Trans. R. Soc. Lond. B 333: 419–427.

——— et al. (+ 6 authors). 1990. Chloroplast DNA sequence from a Miocene *Magnolia* species. Nature 344: 656–658.

Graham, A. 1963. Systematic revision of the Sucker Creek and Trout Creek Miocene floras of southeastern Oregon. Am. J. Bot. 50: 921–936.

———. 1965. The Sucker Creek and Trout Creek Miocene floras of southeastern Oregon. Kent State Univ. Bull., Res. Ser. IX.

———. 1976a. Late Cenozoic evolution of tropical lowland vegetation in Veracruz, Mexico. Evolution 29: 723–735.

———. 1976b. Studies in neotropical paleobotany. II. The Miocene communities of Veracruz, Mexico. Ann. Missouri Bot. Garden 63: 787–842.

———. 1992. Utilization of the isthmian land bridge during the Cenozoic—paleobotanical evidence for timing, and the selective influence of altitudes and climate. Rev. Palaeobot. Palynol. 72: 119–128.

Greller, A. M. and L. D. Rachele. 1983. Climatic limits of exotic genera in the Legler palynoflora, Miocene, New Jersey, U.S.A. Rev. Palaeobot. Palynol. 40: 149–163.

Groot, J. J. 1991. Palynological evidence for Late Miocene, Pliocene, and Early Pleistocene climate changes in the middle U.S. Atlantic coastal plain. Quat. Sci. Rev. 10: 147–162.

Haggard, K. K. and B. H. Tiffney. 1997. The flora of the Early Miocene Brandon lignite, Vermont, U.S.A. VII. *Caldesia* (Alismataceae). Am. J. Bot. 84: 239–252.

Hesse, C. J. 1936. Lower Pliocene vertebrate fossils from the Ogallala Formation (Laverne zone) of Beaver County, Oklahoma. Carnegie Inst. Wash. Contrib. Paleontol. 467: 47–72.

Hills, L. V., J. E. Klovan, and A. R. Sweet. 1974. *Juglans eocinerea* n. sp., Beaufort Formation (Tertiary), southwestern Banks Island, Arctic Canada. Can. J. Bot. 52: 65–90.

Hopkins, D. M., J. V. Matthews, Jr., J. A. Wolfe, and M. L. Silberman. 1971. A Pliocene flora and insect fauna from the Bering Strait region. Palaeogeogr., Palaeoclimatol., Palaeoecol. 9: 211–231.

Huber, N. K. 1981. Amount and timing of Late Cenozoic uplift and tilt of the central Sierra Nevada, California—evidence from the upper San Joaquin River Basin. U.S. Geological Survey, Reston, VA. Professional Paper 1197, pp. 1–28.

Hutchison, J. H. 1982. Turtle, crocodilian, and champsosaur diversity changes in the Cenozoic of the north-central region of western United States. Palaeogeogr., Palaeoclimatol., Palaeoecol. 37: 149–164.

Knowlton, F. H. 1926. Flora of the Latah Formation of Spokane, Washington, and Coeur d'Alene, Idaho. U.S. Geological Survey, Reston, VA. Professional Paper 140A, pp. 1–81.

Krantz, D. E. 1991. A chronology of Pliocene sea-level fluctuations: the U.S. Middle Atlantic Coastal Plain record. Quatern. Sci. Rev. 10: 163–174.

Larsen, H. C. et al. (+ 6 authors). 1994. Seven million years of glaciation in Greenland. Science 264: 952–955.

Leg 105 Shipboard Scientific Party. 1986. High-latitude palaeoceanography. Nature 230: 17–18.

Leopold, E. B. and M. F. Denton. 1987. Comparative age of grassland and steppe east and west of the northern Rocky Mountains. Ann. Missouri Bot. Garden 74: 841–867.

Leopold, E. B. and G. Liu. 1994. A long pollen sequence of Neogene age, Alaska Range. Quatern. Int. 22/23: 103–140.

Leopold, E. B. and V. C. Wright. 1985. Pollen profiles of the Plio–Pleistocene transition in the Snake River Plain, Idaho. In: C. J. Smiley (ed.), Late Cenozoic History of the Pacific Northwest. Pacific Division, American Association for the Advancement of Science, San Francisco, CA. pp. 323–348.

MacGinitie, H. D. 1933. The Trout Creek flora of southeastern Oregon. Carnegie Inst. Wash. Contrib. Paleontol. 416: 21–68.

———. 1962. The Kilgore flora, a Late Miocene flora from northern Nebraska. Univ. Calif. Publ. Geol. Sci. 35: 67–158.

Manchester, S. R., P. R. Crane, and D. L. Dilcher. 1991. *Nordenskioldia* and *Trochodendron* (Trochodendraceae) from the Miocene of northwestern North America. Bot. Gaz. 152: 357–368.

Marchant, D. R., C. C. Swisher III, D. R. Lux, D. P. West, Jr., and G. H. Denton. 1993. Pliocene paleoclimate and East Antarctic ice-sheet history from surficial ash deposits. Science 260: 667–670.

Masterson, J. 1994. Stomatal size in fossil plants: evidence for polyploidy in majority of angiosperms. Science 264: 421–424.

Matthews, J. V., Jr. 1987. Plant Macrofossils from the Neogene Beaufort Formation on Banks and Meighen Islands, District of Franklin. Current Research, Part A, Geologic Survey of Canada, Ottawa, Paper 87–1A, pp. 73–87.

——— and L. E. Ovenden. 1990. Late Tertiary plant macrofossils from localities in Arctic/Subarctic North America: a review of the data. Arctic 43: 364–392.

McCartan, L. et al. (+ 7 authors). 1990. Late Tertiary floral assemblage from upland gravel deposits of the southern Maryland coastal plain. Geology 18: 311–314.

Miller, C. N., Jr. and L. Ping. 1994. Structurally preserved larch and spruce cones from the Pliocene of Alaska. Quatern. Int. 22/23: 207–214.

Mudie, P. J. and J. Helgason. 1983. Palynological evidence for Miocene climatic cooling in eastern Iceland about 9.8 Myr ago. Nature 303:689–692.

Nelson, R. E. and L. D. Carter. 1985. Pollen analysis of a Late Pliocene and early Pleistocene section from the Gubik Formation of Arctic Alaska. Quatern. Res. 24: 295–306.

———. 1992. Preliminary interpretation of vegetation and paleoclimate in northern Alaska during the Late

Color illustration of fossil leaf of *Zelkova oregoneniana* showing preservation of flavonoid pigments for ~14–15 Ma (Niklas and Giannasi, 1977; Niklas, 1981). This was an early demonstration of the potential preservation of biochemical compounds in megafossils of Tertiary age. Other examples of paleobiochemical fossils include pigment compounds from lake sediments, alkenones (p. 32), ancient DNA (pp. 255–256), and optically active compounds from more ancient deposits. The gradual accumulation of such records may eventually make possible a more complete reading of the Earth's biotic history in terms of molecular evolution. Photograph courtesy of Karl J. Niklas.

Pliocene Colvillian marine transgression. U.S. Geol. Surv. Bull. 1999: 219–222.

Niklas, K. J. 1981. The chemistry of fossil plants. BioScience 31: 820–825.

——— and D. E. Giannasi. 1977. Flavonoides and other chemical constituents of fossil Miocene *Zelkova* (Ulmaceae). Science 196: 877–878.

Ovenden, L. 1993. Late Tertiary mosses of Ellesmere Island. Rev. Palaeobot. Palynol. 79: 121–131.

Owens, J. P. et al. (+ 5 authors). 1988. Stratigraphy of the Tertiary Sediments in a 945-Foot-Deep Corehole Near Mays Landing in the Southeastern New Jersey Coastal Plain. U.S. Geological Survey Professional Paper 1484, pp. 1–39.

Pazzaglia, F. J., R. A. J. Robinson, and A. Traverse. 1997. Palynology of the Bryn Mawr Formation (Miocene): insights on the age and genesis of middle Atlantic margin fluvial deposits. Sedimentary Geology 108: 19–44.

Poinar, H. N., M. Höss, J. L. Bada, and S. Pääbo. 1996. Amino acid racemization and the preservation of ancient DNA. Science 272: 864–866.

Rachele, L. D. 1976. Palynology of the Legler lignite: a deposit in the Tertiary Cohansey Formation of New Jersey, U.S.A. Rev. Palaeobot. Palynol. 22: 225–252.

Raymo, M. E., B. Grant, M. Horowitz, and G. H. Rau. 1996. Mid-Pliocene warmth: stronger greenhouse and stronger conveyor. Mar. Micropaleontol. 27: 313–326.

Raymo, M. E. and G. Rau. 1992. Plio–Pleistocene atmospheric CO_2 levels inferred from POM $\partial^{13}C$ at DSDP site 607. Trans. Am. Geophys. Union 73: 95–96.

———. 1993. Mid-Pliocene warmth: higher CO_2 or stronger ocean heat transport? Trans. Am. Geophys. Union 74: 49.

Rea, D. K., M. Leinen, and T. R. Janecek. 1985. Geologic approach to the long-term history of atmospheric circulation. Science 227: 721–725.

Renney, K. M. 1969. The Miocene Temblor Flora of West Central California. Masters Thesis. University of California, Davis, CA.

Repenning, C. A. and E. M. Brouwers. 1992. Late Pliocene–Early Pleistocene ecologic changes in the Arctic Ocean borderland. U.S. Geol. Surv. Bull. 2036: 1–37.

Rogers, S. O. and A. J. Bendich. 1985. Extraction of DNA from milligram amounts of fresh, herbarium, and mummified plant tissues. Plant Mol. Biol. 5: 69–76.

Schoell, M., S. Schouten, J. S. Sinninghe Damsté, J. W. de Leeuw, and R. E. Summons. 1994. A molecular organic carbon isotope record of Miocene climate changes. Science 263: 1122–1125.

Schorn, H. E. 1994. A preliminary discussion of fossil larches (*Larix*, Pinaceae) from the Arctic. Quatern. Int. 22/23: 173–183.

——— and N. L. Gooch. 1994. *Amelanchier hawkinsae* sp. nov. (Rosaceae, *Maloideae*) from the Middle Miocene of Stewart Valley, Nevada, and a review of the genus in the Nevada Neogene. PaleoBios 16: 1–17.

Smiley, C. J. 1963. The Ellensburg flora of Washington. Univ. Calif. Publ. Geol. Sci. 35: 159–276.

———, J. Gray, and L. M. Huggins. 1975. Preservation of Miocene fossils in unoxidized lake deposits, Clarkia, Idaho. J. Paleontol. 49: 833–844.

Smiley, C. J. and W. C. Rember. 1981. Paleoecology of the Miocene Clarkia Lake (northern Idaho) and its environs. In: J. Gray, A. J. Boucot, and W. B. N. Berry (eds.), Communities of the Past. Hutchinson Ross, Stroudsburg, PA. pp. 551–590.

———. 1985. Physical setting of the Miocene Clarkia fossil beds, northern Idaho. In: C. J. Smiley (ed.), Late Cenozoic History of the Pacific Northwest (Interdisciplinary Studies on the Clarkia Fossil Beds of Northern Idaho). Pacific Division, American Association for the Advancement of Science, San Francisco, CA.

Smith, C. A. S., C. A. Fox, and H. Kodama. 1994. Paleosols associated with Miocene basalts, Porcupine River, northeastern Alaska: implications for regional paleoclimates. Quatern. Int. 22/23: 79–90.

Smith, G. R. and W. P. Patterson. 1994. Mio–Pliocene seasonality on the Snake River Plain: comparison of faunal and oxygen isotopic evidence. Palaeogeogr., Palaeoclimatol., Palaeoecol. 107: 291–302.

Smith, H. V. 1938. Some new and interesting late Tertiary plants from Sucker Creek, Idaho–Oregon boundary. Bull. Torrey Bot. Club 65: 557–564.

———. 1939. Additions to the fossil flora of Sucker Creek, Oregon. Papers Mich. Acad. Sci., Arts, Lett. 24: 107–120.

Soltis, P. S. and D. E. Soltis. 1993. Ancient DNA: prospects and limitations. N.Z. J. Bot. 31: 203–209.

——— and C. J. Smiley. 1992. An *rbc*L sequence from a Miocene *Taxodium* (bald cypress). Proc. Natl. Acad. Sci. USA 89: 449–451.

Stanton, R. J., Jr. and J. R. Dodd. 1970. Paleoecologic techniques—comparison of faunal and geochemical analyses of Pliocene paleoenvironments, Kettleman Hills, California. J. Paleontol. 44: 1092–1121.

Taggart, R. E. and A. T. Cross. 1990. Plant successions and interruptions in Miocene volcanic deposits, Pacific Northwest. In: M. G. Lockley and A. Rice (eds.), Volcanism and Fossil Biotas. Geological Society of America, Boulder, CO. Special Paper 244, pp. 57–68.

——— and L. Satchell. 1982. Effects of periodic volcanism on Miocene vegetation distribution in eastern Oregon and western Idaho. Proc. Third North Am. Paleontol. Convention 2: 535–540.

Thelin, G. P. and R. J. Pike. 1991. Landforms of the conterminous United States—A Digital Shaded-Relief Portrayal. U.S. Geological Survey, Reston, VA. Miscell. Invest. Ser. Map I–2206.

Thomasson, J. R. 1978a. Epidermal patterns of the lemma in some fossil and living grasses and their phylogenetic significance. Science 199: 975–977.

———. 1978b. Observations on the characteristics of the lemma and palea of the Late Cenozoic grass *Panicum elegans*. Am. J. Bot. 65: 34–39.

———. 1979. Angiosperms from the Late Tertiary Keller local fauna of Ellis County, Kansas. Contrib. Geol. Univ. Wyoming 17: 59–63.

———. 1980a. A fossil *Equisetum* sp. (family Equisetaceae, subgenus *Hippochaetae*) from the Late Tertiary Ash Hollow Formation of Nebraska. Am. J. Bot. 67: 125–127.

———. 1980b. *Archaeoleersia nebraskensis* gen. et sp. nov. (Gramineae–Oryzeae), a new fossil grass from the Late Tertiary of Nebraska. Am. J. Bot. 67: 876–882.

———. 1980c. *Paleoeriocoma* (Gramineae, Stipeae) from the Miocene of Nebraska: taxonomic and phylogenetic significance. Syst. Bot. 5: 233–240.

———. 1982. Fossil grass anthoecia and other plant fossils from anthropod burrows in the Miocene of western Nebraska. J. Paleontol. 56: 1011–1017.

———. 1983. *Carex graceii* sp.n., *Cyperocarpus eliasii* sp.n., *Cyperocarpus terrestris* sp.n., and *Cyperocarpus pulcherrima* sp.n. (Cyperaceae) from the Miocene of Nebraska. Am. J. Bot. 70: 435–449.

———. 1984. Miocene grass (Gramineae: Arundinoideae) leaves showing external micromorphological and internal anatomical features. Bot. Gaz. 145: 204–209.

———. 1985. Tertiary fossil plants found in Nebraska. Natl. Geogr. Soc. Res. Rep. 19: 553–564.

———, R. J. Zakrzewski, H. A. Lagarry, and D. E. Mergen. 1990. A Late Miocene (late Early Hemphillian) biota from northwestern Kansas. Natl. Geogr. Res. 6: 231–244.

Thompson, R. S. 1991. Pliocene environments and climates in the western United States. Quatern. Sci. Rev. 10: 115–132.

——— (ed.). 1994. Pliocene terrestrial environments and data/model comparisons. U.S. Geological Survey, Reston, VA. Open-file Rep. 94–23, pp. 1–93.

———. 1996. Pliocene and Early Pleistocene environments and climates of the western Snake River Plain, Idaho. Mar. Micropaleontol. 27: 141–156.

Ward, L. W. and D. S. Powars. 1989. Tertiary Stratigraphy and Paleontology, Chesapeake Bay Region, Virginia and Maryland. 28th International Geology Congress Field Trip Guidebook T–216. American Geophysical Union, Washington, D.C.

——— and G. L. Strickland. 1985. Outline of Tertiary stratigraphy and depositional history of the U.S. Atlantic coastal plain. In: C. W. Poag (ed.), Geologic Evolution of the United States Atlantic Margin. Van Nostrand Reinhold, New York. pp. 87–123.

Webb, S. D. 1983. The rise and fall of the Late Miocene ungulate fauna in North America. In: M. H. Nitecki (ed.), Coevolution. University of Chicago Press, Chicago. pp. 267–306.

Webber, I. E. 1933. Woods from the Ricardo Pliocene of Last Chance Gulch, California. Carnegie Inst. Wash. Contrib. Paleontol. 412: 113–134.

Wen, J. and T. F. Stuessy. 1993. The phylogeny and biogeography of *Nyssa* (Cornaceae). Syst. Bot. 18: 68–79.

Wheeler, E. A. and C. G. Arnette, Jr. 1994. Identification of Neogene woods from Alaska–Yukon. Quatern. Int. 22/23: 91–102.

White, J. M. and T. A. Ager. 1994. Palynology, paleoclimatology and correlation of Middle Miocene beds from Porcupine River (Locality 90–1), Alaska. Quatern. Int. 22/23: 43– 77.

White, J. M. et al. (+ 6 authors). 1997. An 18 million year record of vegetation and climate change in northwestern Canada and Alaska: tectonic and global climatic correlates. Palaeogeogr., Palaeoclimatol., Palaeocol. 130: 293–306.

Willard, D. A. 1993. Pliocene Palynomorph Census Data from the Pinecrest Beds (Tamiami Formation) in Quality Aggregates Phase 6 Pit, Southwestern Florida. U.S. Geological Survey, Reston, VA. Open-file Rep. 93-563, pp. 1–9.

———, T. M. Cronin, S. E. Ishman, and R. J. Litwin. 1993. Terrestrial and marine records of climatic and environmental changes during the Pliocene in subtropical Florida. Geology 21: 679–682.

Wolfe, J. A. 1966. Tertiary Plants from the Cook Inlet Region, Alaska. U.S. Geological Survey, Reston, VA. Professional Paper 398–B, pp. 1–32.

———. 1972. An interpretation of Alaskan Tertiary floras. In: A. Graham (ed.), Floristics and Paleofloristics of Asia and Eastern North America. Elsevier Science Publishers, Amsterdam. pp. 201–233.

———. 1978. A paleobotanical interpretation of Tertiary climates in the northern hemisphere. Am. Sci. 66: 694–703.

———. 1981. A chronologic framework for Cenozoic megafossil floras of northwestern North America and its relation to marine geochronology. In: J. M. Armentrout (ed.), Pacific Northwest Cenozoic Biostratigraphy. Geological Society of America, Boulder, CO. Special Paper 184, pp 39–47.

———. 1985. Distribution of major vegetational types during the Tertiary. In: E. T. Sundquist and W. S. Broecker (eds.), The Carbon Cycle and Atmospheric CO_2: Natural Variations Archean to Present American Geophysical Union, Washington, D.C. Monograph 32, pp. 357–375.

———. 1990. Estimates of Pliocene precipitation and temperature based on multivariate analysis of leaf physiognomy. In: L. B. Gosnell and R. Z. Poore (eds.), Pliocene Climates: Scenario for Global Warming. U.S. Geological Survey, Reston, VA. Open-File Rep. 90–64, pp. 39–42.

——— and D. M. Hopkins. 1967. Climatic changes recorded by Tertiary land floras in northwestern North America. In: K. Hatai (ed.), Tertiary Correlations and Climatic Changes in the Pacific. 11th Pacific Science Congress, Tokyo, 1966 Symposium 25, Sendai, Sasaki, pp. 67–76.

——— and E. B. Leopold. 1966. Tertiary stratigraphy and paleobotany of the Cook Inlet region, Alaska. U.S. Geological Survey, Reston, VA. Professional Paper 398–A, pp. 1–29.

Wolfe, J. A. and E. B. Leopold. 1967. Neogene and Early Quaternary vegetation of northwestern North America and northeastern Asia. In: D. M. Hopkins (ed.), The Bering Land Bridge. Stanford University Press, Stanford, CA. pp. 193–206.

Wolfe, J. A., H. E. Schorn, C. E. Forest, and P. Molnar. 1997. Paleobotanical evidence for high altitudes in Nevada during the Miocene. Science 276: 1672–1675.

Wolfe, J. A. and T. Tanai. 1980. The Miocene Seldovia Point flora from the Kenai Group, Alaska. U.S. Geological Survey, Reston, VA. Professional Paper 1105, pp. 1–52.

———. 1987. Systematics, phylogeny, and distribution of *Acer* (maples) in the Cenozoic of western North America. J. Fac. Sci., Hokkaido Univ. 22: 1–246.

Zubakov, V. A. and I. I. Borzenkova. 1990. Global Palaeoclimate of the Late Cenozoic. Developments in Palaeontology and Stratigraphy, 12. Elsevier Science Publishers, Amsterdam.

General Readings

Fischman, J. 1993. Forever (in) amber. Science 262: 655–656.

Golenberg, E. M. 1994. Antediluvian DNA research. Nature 367: 692.

Holden, C. 1997. "No go" for Jurassic Park-style dinos. Science 276: 361.

Lindahl, T. 1993. Recovery of antediluvian DNA. Nature 365: 700.

Poinar, G. O., Jr. 1993. Recovery of antediluvian DNA—reply. Nature 365: 700.

Poinar, H. N., R. J. Cano, and G. O. Poinar, Jr. 1993. DNA from an extinct plant. Nature 363: 677.

Service, R. F. 1996. Just how old is that DNA, anyway? Science 272: 810.

Sykes, B. 1997. Lights turning red on amber. Nature 386: 764–765.

EIGHT

Quaternary North American Vegetational History

1.6 Ma to the Present

CONTEXT SUMMARY

The Quaternary Period encompasses the Pleistocene and the Holocene or Recent Epochs (Fig. 7.2). The date used for the beginning of the Pleistocene depends upon which globally recognizable event is selected as representing a significant break with the preceding Pliocene Epoch. Candidates include the Gauss–Matuyama magnetopolarity boundary (~2.8 Ma; see Quaternary International, 1997); the initiation of widespread permafrost, a frigid Arctic Ocean, and rapid glaciation in the high northern latitudes (~2.4 Ma; Shackleton and Opdyke, 1977; Shackleton et al., 1984); or the African Olduvai paleomagnetic event between 1.87 and 1.67 Ma. The transition from hothouse to icehouse conditions was gradual, but the Pleistocene is typified at Vrica, Italy, as beginning at ~1.67 Ma (Aguirre and Pasini, 1985; Richmond and Fullerton, 1986; oxygen isotope stage 62), and that is the date used here. In the conterminous United States the Elk Creek till of Nebraska is 2.14 m.y. in age (Hallberg, 1986), and the onset of the full ice age is represented by the onset of repeated glaciations at ~850 Kya when glaciers extended down the Mississippi River Valley (Fig. 8.1). Subsequently, glacial–interglacial conditions fluctuated until the latest retreat at ~11 Kya that began the Holocene or Recent Epoch.

The chronology of ice age events began with the publication of Louis Agassiz's (1840) *Etudes sur les Glaciers*. In the absence of evidence to the contrary, a single glacial advance was envisioned as blanketing the high latitudes. In the 1940s Willard F. Libby at the University of Chicago perfected the technique of radiocarbon dating, and Flint and Rubin (1955) applied this methodology of "isotopic clocks" to establishing the absolute chronology of drift deposits from the eastern and midwestern United States. Their radiocarbon dates showed evidence of two or more times of continental-scale glaciations; older organic material was "radiocarbon inert" and beyond the ~40-Ky range of the technique. A standard chronology eventually became established for North America that included four major glacial stages (Nebraskan, oldest; Kansan; Illinoian; and Wisconsin) separated by four interglacials (Aftonian, oldest; Yarmouth, Sangamon, and the present Holocene).[1] Then, long cores from the Atlantic Ocean documented at least nine glacial stages (Phleger et al., 1953), and Emiliani (1955) discovered seven complete glacial–interglacial cycles in a core from the Caribbean based on $^{18}O/^{16}O$ measurements. A total of 18–20 stages is presently recognized for the past 1.6 m.y. Study of another Caribbean core showed an eccentricity rhythm of 100 Ky (Broecker and van Donk, 1970; Fig. 8.2), while a core from the Indian Ocean revealed tilt and precession cycles (Hays et al., 1976; Fig. 8.3). It is now known that between 2.5 and 0.735 Ma (Matuyama magnetic chron) glacial cycles in the northern latitudes followed the tilt variation of ~41 Ky; since that time they have followed the 100-Ky eccentricity variation. Toward the middle latitudes it is the 41-Ky cycle that has been primary since ~735 Kya. A close correspondence has been shown between high sea levels and periods of maximum insolation in the Milankovitch cycles (Shackleton, 1987). The climatic pattern involves slow buildup to glacial conditions (stadials), followed by abrupt switches (terminations) to warmer conditions at 100-Kya intervals (interstadials or interglacials; Fig. 8.4). The Greenland and Antarctic ice cores subsequently provided evidence of small rapid fluctuations of 7–12 Ky (Heinrich events; Fig. 8.4) and 1 Ky to ~100 years (D–O events; Fig. 8.5). Climatic instability is particularly pronounced at the onset of deglaciations. Correlation of terrestrial vegetational changes on the scale of D–O events is complicated by distance from the coast (continentality), topographic diversity, and individual lag times for migrating populations of various species that can range from decades to centuries (Davis, 1989).

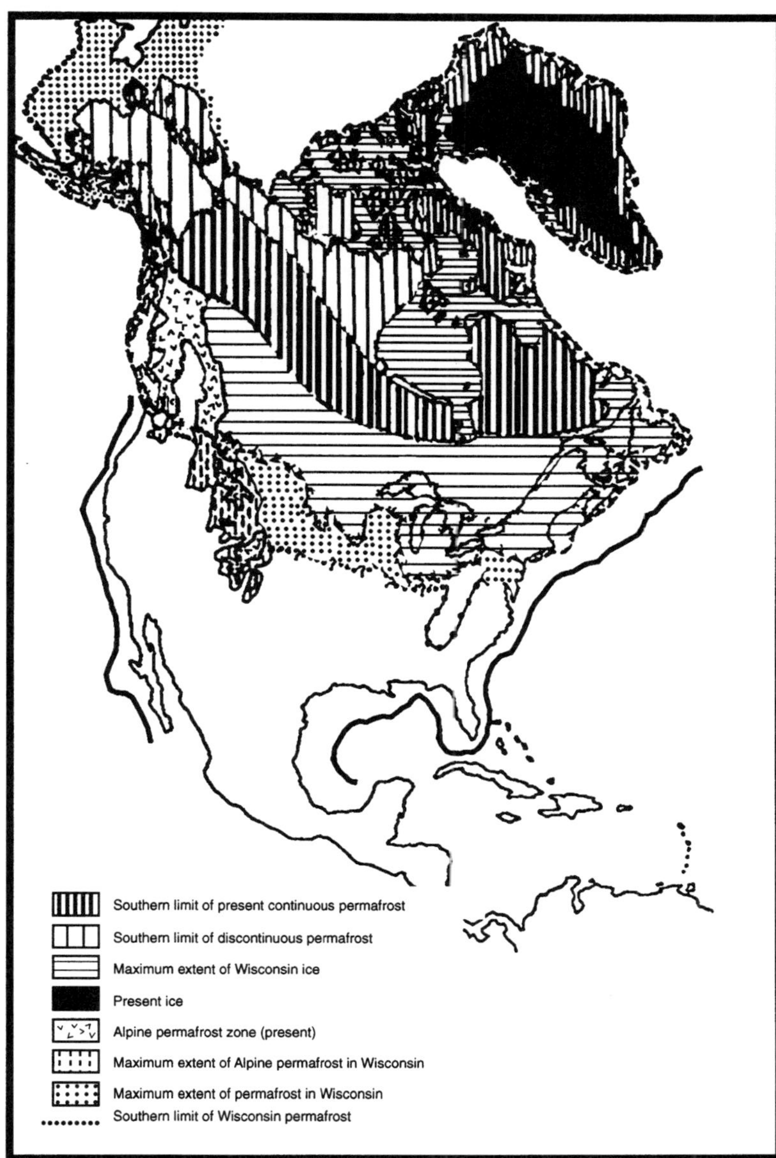

Figure 8.1. Extent of glacial and periglacial zones of North America at present and at the last glacial maximum. The solid line bordering the continent represents the approximate position of the shoreline during the full Wisconsin glaciation. The southern extension of discontinuous permafrost along the Appalachian high summits to the Great Smoky Mountains in the Late Wisconsin is based on Péwé (1983). Reprinted from Williams et al. (1993, fig. 5, and references cited therein) with the permission of M. A. J. Williams.

On the basis of marine oxygen isotope, Greenland ice core, and paleomagnetic data, the Quaternary may be subdivided into early, middle, and late epochs. The Early Quaternary extends from 1.6 Ma to the Matuyama–Brunhes geomagnetic polarity reversal at 780 Kya; the Middle Quaternary extends from 780 Kya to oxygen isotope substage 5e, corresponding to an age of 122 Kya; and the Late Quaternary extends from 122 Kya to the present. The latter is divided into the oxygen isotope stages shown in Fig. 8.5. Other time units encountered in the literature and their approximate duration are:

Sangamon interglacial: 122 Kya (132 Kya; Richmond and Fullerton, 1986) to 80 Kya, oxygen isotope stage 5e-a
Early Wisconsin: 80–28 Kya, stage 4
Middle Wisconsin: 28–23 Kya, stage 3
full-glacial Late Wisconsin: 23–16.5 Kya [18 Kya], stage 2
late-glacial Late Wisconsin: 16.5–12 Kya
Early Holocene: 12 Kya to 8.9–8.5 Kya, stage 1
Middle Holocene: 8.9–8.5 Kya to 4 Kya
Late Holocene: 4 Kya to the present

In North America there were two main sites of ice accumulation. The Laurentide ice sheet was the largest and it spread from a central dome located over Hudson Bay; the Cordilleran ice sheet formed in the northern Rocky Mountains of British Columbia. The first estimate of maximum ice thickness (by Cleveland, Ohio, geologist Charles Whittlesey, 1868 was ~1.8 km (~1 mi) for the Laurentide continental glacier. More recently a value of 3.5 km has been suggested, but Peltier (1994) interprets sea-level and crustal rebound data as indicating a thickness of ~2 km (see technical comment by Edwards, 1995; response by Peltier, 1995). The most areally extensive glaciation was the Nebraskan because till and outwash sediments of that age ex-

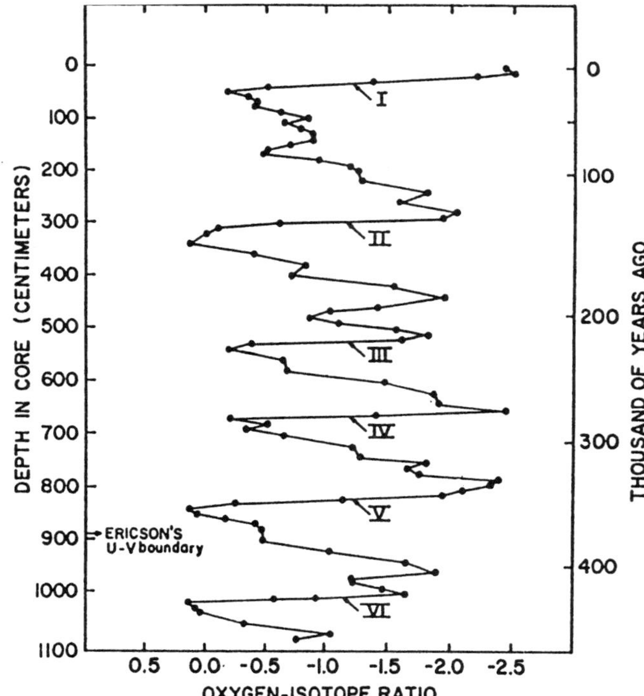

Figure 8.2. Variation in oxygen-isotope ratios from Caribbean core V12–122 showing 100-ka eccentricity cycle. Roman numerals I–VI indicate the six intervals of rapid deglaciation (terminations). Reprinted from Imbrie and Imbrie) (1979, fig. 38, and references cited therein) with the permission of Enslow Publishers, Inc.

tend beyond the cover of later advances (Bowen, 1978). The two centers were frequently confluent, and at the latest glacial maximum at 18 Kya ice covered ~40 million km^2 (30% of the Earth's surface) or ~3 times more than the present 15 million km^2. During the last or Late Wisconsin episode of glaciation, the Laurentide ice sheet reached its farthest southern extension as the Des Moines lobe in central Iowa (41° N latitude) at ~14 Kya; the maximum of the Cordilleran ice sheet was at ~15 Kya. A zone of permafrost 80–200 km wide extended along the southern edge of the Laurentide ice sheet from the western cordilleras to the Atlantic coast, as shown by the distribution of ice-wedge casts and pingos (hills that were ice covered and preserve evidence of glacial action; Péwé, 1983). Approximately 77 million km^3 or 10% of the world's water was locked up in the ice, compared to 30 million km^3 or 1.7% today, and the full-glacial position of sea level was lower than at present by 121 ± /5 m. This increased ocean salinity by ~3%, in-

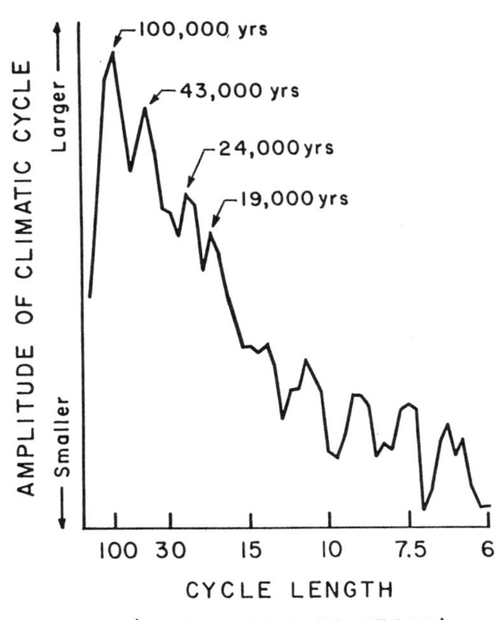

Figure 8.3. Variation in oxygen-isotope ratios from Indian Ocean cores showing temperature changes consistent with the Milankovitch variations. Reprinted from Imbrie and Imbrie (1979, fig. 42, based on Hays et al., 1976) with the permission of Enslow Publishers, Inc.

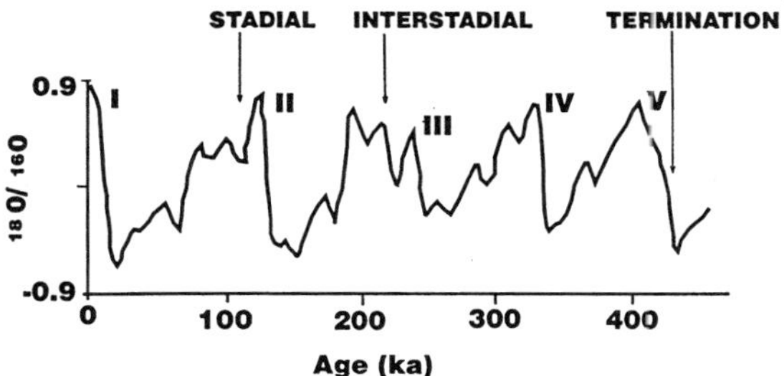

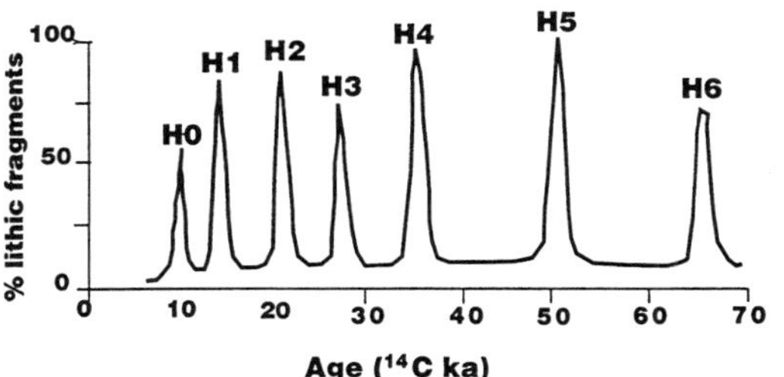

Figure 8.4. (Top) Oxygen isotope record of global ice volume for the past 450 Kya. (Bottom) Percent lithic fragments in North Atlantic deep sea sediments for the past 70 ka showing influx of debris from icebergs during Heinrich events. Reprinted from Clark (1995) with the permission of Peter Clark and the American Association for the Advancement of Science.

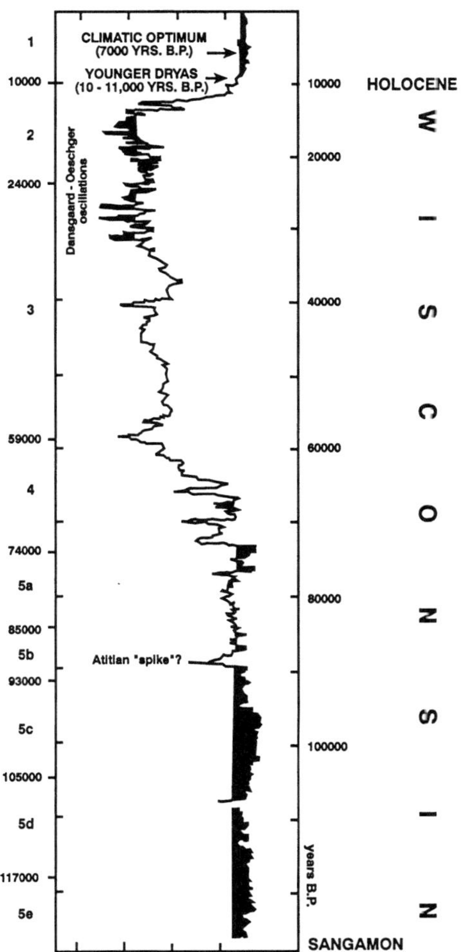

Figure 8.5. Oxygen isotope record from the Camp Century Greenland ice core showing climatic optimum, Younger Dryas, and D-O oscillations. The Atitlan spike may be the eruption of the Atitlan volcano (Los Chocoyos ash), Guatemala, 84 Kya. The spike at ~75 ka may be an eruption of Mount Pinatubo. Reprinted from Dawson (1992, fig. 2.6, based on Johnsen et al., 1972) with the permission of Alastair Dawson and *Nature*.

creased global land surfaces by ~15%, and doubled the size of Florida, primarily by exposing the shallower western coastal plain (Fig. 8.1).

The growth of ice sheets and the climatic cycles they imply are determined by a complex of interacting factors and feedbacks. The closure of the Isthmus of Panama by ~3.5 Ma strengthened the Gulf Stream and brought increasing amounts of warm water and moist air into the North Atlantic beginning in the Late Pliocene. Atmospheric CO_2 concentration was decreasing, and samples of past atmosphere from air bubbles trapped in the Greenland ice cores show a relatively low ~190–200 parts per million by volume (ppmv) in the last glacial compared to the present ~351 ppmv. As noted previously, the mechanism(s) that control the input of CO_2 into the atmosphere, beyond periods of major plate reorganization, are immensely complex and not well understood. They include regulation of CO_2 release from the ocean surface through the formation and removal of overlying sea ice, thermocline and saline stratification differences, residence time associated with oceanwater circulation, and fluctuations in the abundance and CO_2 production of marine organisms. Processes in the terrestrial realm include expansion and contraction of carbon-storing tundra (Gallimore and Kutzbach, 1996), boreal, temperate, and tropical forests, tropical grassland, and other vegetation types; CO_2 and other reactive gas storage and release from terrestrial vegetation colonizing the continental shelf, alternately exposed or inundated with changes in sea level; erosion rates, which also vary with sea level along the continental shelf; precipitation of shallow-marine carbonate sediments; and, to a lesser extent, diminished episodes of volcanic activity. Periodically these factors reduced atmospheric CO_2 to a level where low insolation phases of the Milankovitch cycles were expressed as the increasing accumulation of winter snow surviving through intervals of diminished summer abolation.

As the Laurentide ice sheet developed, the mountain of glacial ice divided the westerly jet stream into two components (Broccoli and Manabe, 1987; Kutzbach and Guetter, 1986; Kutzbach and Wright, 1985; Manabe and Broccoli, 1985). One flowed north across Alaska and the Canadian Arctic, reducing the strength of the easterlies that prevail there at present. The other crossed North America between ~33° N and the glacial margin at ~40° N. The present winter flow is centered across the northern Rocky Mountains at ~45°–47° N and shifts northward in the summer to 65°–70° N. As the air flowed across the accumulating ice it was cooled, carried into the North Atlantic by the westerlies, chilled the ocean surface, and contributed to the formation of sea ice. This provided a positive feedback to colder climates, as did increased albedo from expanding ice and decreasing soil and vegetation cover. Using either minimum or maximum estimates for ice thickness (1.8–3.5 km), a split in the jet stream is still produced according to simulations generated by computer models.

In these simulations, winters at glacial maxima in Greenland were ~11–15°C cooler (Cuffey et al., 1995) than in the Early Holocene, and summers were 2°C cooler. Adjacent areas to the south were ~6–8°C cooler, and toward the lower latitudes temperatures were 2°C (CLIMAP Project Members, 1976) to 5–6°C cooler than at present (Guilderson et al. 1994; Thompson et al., 1995). At elevations between 3 and 5 km in the mid-latitudes temperatures were ~5°C lower at glacial maximum than at present.

Within and marginal to the ice sheets, periods of glacial advance correspond to times of aridity and increased levels of dust appear in the marine and terrestrial record (dry, windy, and cold). Precipitation in these regions during glacial advances was ~30% of the present; also, dust in cores from nearby dry continental interiors, glacial outwash, and the exposed continental shelves was 70 times higher than during the interglacials. The semipermanent high-pressure anticyclone system fixed over the Laurentide dome displaced low-pressure cyclone belts to the south. This produced pluvial or wet phases in parts of periglacial North America (i.e., beyond the glacial boundary), especially in the southwest under the ascending arm of the Hadley regime and in the southeastern United States north of ~33° N, which marked the position of the jet stream or polar front (Delcourt and Delcourt, 1984). During glacial intervals south of ~33° N, the region across the southeastern United States was somewhat drier than areas to the west (Barry, 1983; Kutzbach and Guetter, 1986) and the flow of the easterlies brought dry winds to the Pacific Northwest.

The inland position of the Early Laurentide and Cordilleran centers allowed ice to accumulate without extensive calving that would have diluted the northern oceans and created a "drop dead" mode of thermohaline ocean circulation. As the glacier margins neared the coasts, such a mode did develop and this is one mechanism proposed for the rapid reversals of climate expressed in the Greenland ice cores.

The ice cores show the transition from the Late Pleistocene to the Holocene at a depth of 1150 m, corresponding to an age of ~11 Kya (10,750 years B.P.; the end of the Younger Dryas stadial). At that time glaciers worldwide began to retreat, temperatures rose, dust declined, and CO_2 and CH_4 concentrations increased; Termination I had begun. [For a series of papers on the Younger Dryas, see Quaternary Science Reviews, (1993, 1995).] As the Milankovitch cycles entered a phase of increased summer insolation and enhanced seasonal contrast glaciers began to rapidly fluctuate (Levesque et al., 1997), but generally retreat; melting and calving poured freshwater into the North Atlantic Ocean and down the Mississippi River into the Gulf of Mexico until the ice margin retreated beyond the St. Lawrence drainage at ~11 Kya. Early retreat led to accelerating feedbacks such as decreased erosion of the continental shelf as sea levels rose (less drawdown of CO_2), decreased albedo from reduced ice and expanding vegetation, increased biological activity and CO_2 production in the

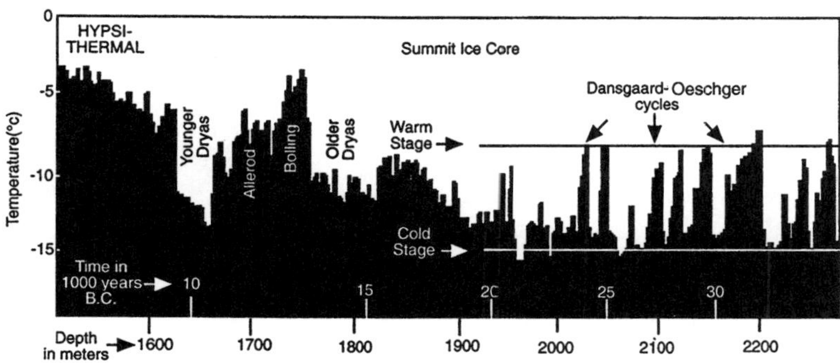

Figure 8.6. Air temperature over Greenland for the past 40 Kya based on oxygen isotope ratios in the Summit ice core. Reprinted from Kerr (1993) with the permission of the American Association for the Advancement of Science and the Greenland Ice Core Project.

warming northern waters, reduced sea ice, and greater release of CO_2 from the ocean surface. Some models indicate that increased dust over high-albedo ice and snow surfaces may have contributed to the abrupt periods of warmth in the last glacial cycle (Overpeck et al., 1996). By the Holocene, CO_2 concentration in gas bubbles from the Greenland cores had risen from 190 to 200 to 260 to 280 ppmv. Marine isotope records from the North Atlantic reveal three episodes of glacial melting at 14–12 Kya, 10-9 Kya, and at the postglacial climatic optimum at 8–6 Kya (Jansen and Veum, 1990; Mix, 1987; Fig. 8.6). During this mid-Holocene warm period, also known as the hypsithermal (eastern North America) or altithermal (western North America), summer temperatures averaged 2-4°C warmer than at present. About one-third of total ice volume was lost during the first two episodes, and the associated rise in sea level is documented by erosion terraces on the Barbados coral reefs (Fairbanks, 1989). At ~12 Kya the rise there was ~24 m in ~1 Kya, and at 9 Kya it rose 28 m over a slightly longer time. The intervening pause corresponds to the Younger Dryas stadial. Retreat of the Wisconsin glacier along its southern margin was a rapid few hundred meters per year. It had reached Hudson Bay by ~8 Kya and was essentially at its present extent by ~6 Kya. Ice that had accumulated over the previous ~100 Kya melted in only 8 Kya, and southern Greenland warmed by 7°C at the end of the Younger Dryas stadial.

Between 1450 and 1850 A.D. there was a pause in the warming trend called the Little Ice Age when MAT in the middle to high latitides declined by 1–2°C. Severe winters are verified by a variety of historical records, including poor grain and grape harvests, livestock losses, navigational difficulties, and paintings of skaters on frozen Dutch canals that rarely freeze today (15 times this century, the last in January, 1997). Imbrie and Imbrie cite another interesting consequence of this short-term fluctuation:

> The legendary winter during which Washington's troops were bivouacked at Valley Forge was actually regarded by contemporary observers as "notably mild." In fact, if Washington had camped at Valley Forge two years later, during the winter of 1779–80, the suffering of his troops would have been much greater. Even by Little Ice Age standards, that winter was "The most hard difficult winter . . . that ever was known by any person living." (1979, p. 183)

To the north, New York harbor was frozen solid. Ludlum writes:

> Though both the Hudson and the East Rivers were accustomed to freeze solidly from time to time in the olden days, there was no record of the entire Upper Bay congealing for a number of days. . . . [In late January] people walked on the ice all the way from Staten Island to Manhattan Island, a distance of five miles. . . . Heavy loads and even large cannon were dragged across the iceways to fortify the British position on Staten Island which had been subject to cross-the-ice forays from Washington's outposts in New Jersey. (1966, p. 115)

Regions within and along the margins of the glacial boundary in North America developed wetter climates during the interglacials as the jet stream and Hadley atmospheric convection cell approached modern positions. The northward shift of the low-pressure systems brought the southwestern United States back under the descending arm of the Hadley regime that, together with its inland position and rainshadow effects, restored warm dry climates to the region during each interglacial. This high-pressure system during interglacial times in the southwest also prevented the inland penetration of lows from the Gulf of Mexico and the Pacific Ocean. Climatic trends in the far southeastern and southwestern United States were complicated by the close proximity to the ocean, responsiveness to variations in the flow of ocean currents, and upwelling.

Areas are still rebounding from removal of the ice; for example, the shores around Lake Superior are rising ~38 cm (15 in.) per century. This postglacial adjustment to load (isostasy) from the ice-age cryosphere is a delayed result of the viscous and deformable nature of the underlying mantle (Peltier, 1996).

The Pleistocene land-mammal faunas of North America group into two NALMAs. The older Irvingtonian has been characterized as *Mammuthus*-pre-*Bison* (Savage and Russell, 1983). There are ~120 genera from this age (1.6–0.5 Ma) and 12 new immigrants from Asia and eight from South America (Webb, 1985). Steppe species from Asia include mammoths, musk-oxen, and microtid rodents (lemmings, mice, voles, and related types); introductions from South America include sloths (*Notrotheriops*, *Eremotherium*), anteaters, opossum, and hystricognath rodents (porcupines). By Rancholabrean time (~0.5 Ma) there were ~130 genera of land mammals in North America and 16 genera newly emigrated from Asia and nine from South America. Of special importance were the Asian microtids, goats, and bison that had significant ecological impact on the vegetation (e.g., see exclusion experiments discussed in Chapter 3, Faunas). Fluctuations in composition characterized North American mammal faunas during the Quaternary, paralleling that described for the flora. Although suggesting a degree of randomness not always evident in the fossil assemblages, the assessment by Graham and Mead is similar to that expressed by several students of the Pleistocene flora:

> [V]ertebrate communities are not tightly bound aggregates of species; rather, they are collections of species randomly distributed along environmental gradients (Whittaker, 1970). Consequently, environmental change will not illicit a response from the entire community, but instead each species will respond according to its own tolerances. The species composition of a community may therefore be constantly in flux, and significant environmental fluctuations may cause major biotic reorganizations. Modern communities are not direct analogues for past ones, but changes in the distribution, abundance, and clinal variation of individual species can provide invaluable information about past environments. (1987, p. 371)

The extinction of most large land mammals in North America is attributed to the rapid climatic changes of the Late Pleistocene, particularly during the transition from Late-glacial to Holocene times (Termination I), and to pressures from newly arrived Asian hunters following the "ice-free corridor" of western Canada. [For a series of papers on the corridor, see Quaternary International (1996).]

The immigration of humans that were tracking herds of large mammals from Asia across the Bering land bridge is a new factor that later affected North American biotas through agricultural disturbance as shown in pollen and spore diagrams. Hunters and gatherers may have been in Siberia by 30–35 Kya (Klein, 1975) and although the dates are unsettled, they possibly reached North America by 30 Kya, corresponding to a glacial interval with lower sea level. At this time the Bering land bridge was 1500 km wide and unglaciated portions supported various graminoid–shrub tundra associations described in Chapter 1. At ~12 Kya the MacKenzie River Valley was mostly ice free and there was another wave of immigration (Martin, 1973). Human populations in Beringia and elsewhere in North America were definitely established by ~12 Kya (Josenhans et al., 1997) and the spread of PaleoIndian culture correlates temporally with extinction of the North American megafaunas at 12-10 Kya.

In the older four-part chronology, glacials and interglacials were considered about equal in duration (~175 Ky) or the interglacials longer, but the ^{18}O and ice-core evidence now document that the principal interglacials lasted only ~12 Ky. There have been few periods within the past 130 Ky when ice volume was as low as it is at present. Holocene temperatures reached their maximum between 8 and (~6) 4 Kya, and since that time they have shown an overall decline. As noted by Davis for North America, "The Pleistocene must be viewed as a cold, glaciated epoch, interrupted periodically by catastrophic warm events—the brief interglacials with climate similar to that of today." (1976, p. 14). This raises the important and complex possibility that the Earth entered another cold interval ~4-6 Kya and that, consistent with previous history, a new series of glacial advances may be imminent. For example, during the past 3 Ky there has been a southward displacement of the ecotones between the tundra and boreal forest and between the boreal and lake states forest. The unresolved question is the extent to which human-induced global warming will delay or offset these solar–ocean circulation-induced climatic trends. Efforts are under way to predict future biome distributions under various future-climate scenarios (VEMAP, 1995)

PLEISTOCENE AND HOLOCENE VEGETATION

Information on North American Quaternary vegetation is not uniformly available geographically or stratigraphically. Within the glacial boundary the advance of the Laurentide ice sheet eroded underlying bedrock to a depth of ~120 m (Bell and Laine, 1985). With subsequent retreat of the ice an irregular topography was created, augmented by numerous kettleholes, which are steep-sided and round to flat bottom depressions formed by remnants of stagnant ice. Melting ice blocks in this poorly drained landscape left numerous lakes and bogs. The depth of some lakes and the rapid accumulation of sediments (typically 1 m/1000 years) reduced oxidation of organic material and allowed preservation of plant and animal fossils, especially pollen, spores (algal, fungal, bryophyte, fern and other vascular nonseed plants), phytoliths, cuticles, epidermal fragments, wood, diatoms, unicellular (e.g., *Pediastrum*) and colonial (e.g., *Botryococcus*) algae, insects [Coleoptera (beetles), Morgan, 1987; midges, Walker et al., 1991], molluscs, and even small mammals. Vegetation developing around the lake margins eventually filled some basins to form peat bogs. The acidity of these bogs reduced bacterial and fungal activity, further contributing to the preservation of the fossils. The result is that much information on Late Quaternary vegetational history is available from lake and bog sequences within and marginal to glaciated North America. In periglacial regions suitable sites are fewer, although in the southeast there are karst lakes and sinkholes, bays, buried soils, lagoon deposits, and some bogs, glades, and fens (Jacobson et al., 1987, fig. 1). In the southwest, playa

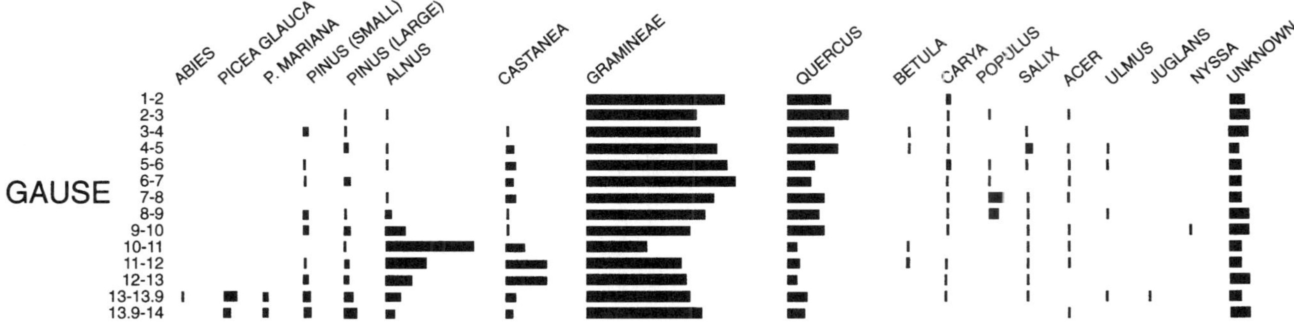

Figure 8.7. Pollen diagram from the Gause Bog, Milan County, Texas. Note the reported presence of ~11% *Abies* and *Picea* pollen at the ~13–13.7 ft level and ~20% *Castanea* pollen between the 11- and 13-ft levels Reprinted from Potzger and Tharp (1954, fig. 1) with the permission of The Ecological Society of America.

lakes, packrat middens, and dendrochronology provide information on biotic and environmental history, especially when interpreted within context information discussed in previous chapters.

Sediments were mostly eroded away with each new glacial advance, so less is known about the Nebraskan, Aftonian, Kansan, Yarmouth, Illinoian, and Sangamon vegetation and terrestrial environments (Early and Middle Quaternary) than for the Wisconsin and Holocene (Late Quaternary). Also, improvements in the absolute dating technique of ^{14}C have extended the dating range back from ~40 Kya to as much as 75 Kya. Seasonal layers in lake sediments called varves are recognizable from the rapid accumulation and larger particles deposited during the spring than in the summer and winter. This allows temporal resolution to 1 year in some Late Quaternary sequences, but annually laminated varves are not widely available in older Quaternary deposits. Layers of tephra (volcanic ejecta) occasionally can be isotopically dated or tied to times of known eruptions such as Mount Mazama and early Mount St. Helens. Generally, however, the chronology of events is less precise for the Early and Middle Quaternary than for later times.

Methodologies

A considerable amount of information on terrestrial vegetational history for the Quaternary has come from pollen and spore studies. The procedures for retrieving and processing samples are described in numerous publications (Faegri et al., 1989; Gray, 1965; Moore et al. 1991; Traverse, 1998; Chapter 4, Paleopalynology). Microfossils are identified by comparison with types in a reference collection, and their representations are presented in the form of diagrams constructed by counting the microfossils present at various levels. The sampling interval is commonly every 10 or 15 cm along the sediment sequence, and a constant number (ranging from 200 to 1000 microfossils) are tabulated for each level. Percentages for all types at a given depth level constitute a spectrum, which gives an inventory mostly of the prominent wind-pollinated plants present at the time—the "pollen and spore rain" that accumulated on the ancient bog or mud surface. From this information the paleovegetation for that interval is characterized. The spectra may be superimposed into a vertical sequence to provide profiles of the abundance of each type of microfossil through time. The collective spectra and profiles constitute a diagram, and from it changes in vegetation over time can be traced.

The early pollen and spore diagrams were presented as percentage histograms (see e.g., Fig. 8.7) that usually also included sediment depth and occasionally sediment type. More recent diagrams may give ^{14}C ages and separate calculations for total tree pollen (arboreal pollen or AP) versus herb and shrub pollen (nonarboreal pollen or NAP). This ratio reveals changes in vegetation density of forest relative to nonforest communities (e.g., from closed forest to savanna or nonforested tundra vegetation). Pollen zones are drawn to delineate major vegetation types such as tundra, boreal forest, deciduous forest, or to characterize regional climatic events such as the climatic optimum and the Little Ice Age. When ^{14}C dates are available, the age and duration of each zone can be determined, the zones can be correlated between sites to reconstruct regional vegetation, and the migratory history of individual taxa can be traced via isopols (e.g., Delcourt and Delcourt, 1987, 1993). Isopolls are lines on a map that connect areas of equal percentages of a given palynomorph. These denote first appearances and similar percentages of pollen and spores at different sites and can reveal the source, direction, and pace of plant migrations (Delcourt et al., 1984; Jacobson et al., 1987; Webb, 1988). This same information can be used to detect the movement of human populations through their impact on vegetation. Such evidence includes the simultaneous reduction of varied forest types, the correlated appearance of charcoal layers, and the appearance of introduced cultivars and weeds associated with forest clearance. The classic studies of Iversen (1941) demonstrated the migration of early agriculturalists from the Fertile Crescent into western Europe near the end of the last ice age. Archeological zones and cultural periods used in anthropology [Paleolithic or Early Stone Age (~2 Ma–10 Kya); Mesolithic or Middle Stone Age (10–8 Kya); Neolithic or Late Stone Age (8–3.5 [5.5] Kya); Bronze Age (3.5–1 [5.5] Kya); Iron Age beginning at 1 Kya; and Historic] establish the temporal relationship between vegetation, climate, and human history (Dimbleby, 1985; Gray and Smith, 1962; Williams et al., 1993, chapter 9). Accessory diagrams may include fungal spores, diatoms, ethnobotanical remains, and macrofossils (fruits, seeds, leaves, twigs, wood; Jackson et al., 1997). Chemical analyses provide data on lake history (Hedges et al., 1982; Leopold et al., 1982).

Significant advances have been made in reconstructing the composition, arrangement, and dynamics of paleocommunities from pollen and spore diagrams (Grimm, 1988). The sources of error discussed in Chapter 4 have mostly been taken into account, but Davis (1963) has emphasized the need for continued improvement in the way vegetation is inferred from fossil pollen and spore assemblages. A problem with percentage counts is that fluctuations in the abundance of any one entity affects the representation of others, irrespective of their real history in the vegetation. Also, the concentration of microfossils in sediments varies with compaction and sedimentation rate, as well as with actual changes in the vegetation (e.g., from dense forest to open-ground tundra or with cultural burning of forests and land clearing for cropland). Several methods have been developed that allow a more accurate reconstruction of the composition and the abundance and arrangement of component associations from pollen and spore diagrams. One is to compare the modern pollen rain on moss polsters, bogs, and other surfaces with the surrounding vegetation that has been inventoried and mapped. This gives an indication of the kind of pollen spectrum produced by known vegetation types and helps in identifying most similar modern-day analogs for the ancient fossil assemblages (Delcourt and Delcourt, 1985a).

By counting either a standard volume of material or introducing a known concentration of exotic pollen, rather than tabulating a fixed number of microfossils from each level, a calculation can be made of pollen influx or microfossils/unit of sediment (Benninghoff, 1962; Davis, 1965). This removes the variable of sedimentation rate from estimates of vegetation density. When several ^{14}C dates are obtained from a single section, a further calculation can be made of absolute pollen frequency (or pollen accumulation rates, PAR) in microfossils/unit of sediment/unit of time (Davis, 1967; Davis et al., 1973; see Williams et al. 1993, fig. 8.9, for a comparison of percentage and pollen influx diagrams from Rogers Lake, Connecticut). The results are then combined with data from studies of modern pollen rain to better establish the qualitative and quantitative relationships between various plant populations, modern vegetation types, and their pollen and spore signatures. These innovations have significantly improved the accuracy of paleovegetation reconstructions from pollen and spore assemblages, but each approach has its own set of limitations that still provide only a general picture of ancient plant communities (Peteet, 1991; Prentice, 1988).

A knowledge of basic ecology and a familarity with vegetation similar to that reflected by the various spectra are also important in reconstructing paleocommunities from pollen assemblages. Improved numerical methods such as transfer functions are standardizing data analysis and are yielding more accurate and reproducible results (Birks and Gordon, 1985). Collectively, accessory diagrams, modern pollen rain studies, pollen frequency, pollen influx, increased understanding of ecological processes and vegetation dynamics, and improved statistical analyses are revealing the history of Quaternary vegetation and environments in greater detail than for any other segment of geologic time (Bartlein, 1988; Birks, 1993; Gajewski, 1993; Schoonmaker and Foster, 1991). Understanding this history is enhanced by a knowledge of cause and effect mechanisms discussed in Chapter 2 and context information discussed in Chapter 3. The literature is vast and numerous summaries are available for the North American continent (e.g., Bryant and Holloway, 1985; Delcourt and Delcourt, 1987, 1993; Heusser and King, 1988; Porter, 1983; Ruddiman and Wright, 1987; Wright, 1983; Wright et al., 1993; see also papers in Huntley and Webb, 1988). The intent of this chapter is to summarize broad-scale changes in regional vegetation that produced the current version of North American plant formations and associations from Late Pliocene predecessors. Tracing this history reveals the dynamic and complex nature of biotic–environmental interactions. This is done within the constraint of a major gap in the record between the latest Pliocene and the Wisconsin.

Southeast

Little is known about Early and Middle Quaternary vegetation in the southeastern United States (Elsik, 1969; Groot, 1991). It may be assumed on the basis of events in the Late Glacial and Holocene that during earlier cold intervals, upland (mostly coniferous) elements moved downward and northern temperate elements expanded their range southward. At present, *Tsuga canadensis*, *Acer saccharum*, *Fagus grandifolia*, and *Betula allegheniensis* grow at middle to high elevations in the central and southern Appalachians; *Abies* and *Picea* occur at high elevations in the central Appalachians and sporadically only above ~1500 m to the south. *Pinus banksiana* (jack pine) is presently a northern species that grew in the south in glacial times. With the onset of warmer interglacial climates, cool- to cold-temperate elements either persisted in refugia at higher altitudes and along river valleys with cold-air drainage, evolved new ecotypes, disappeared from the re-

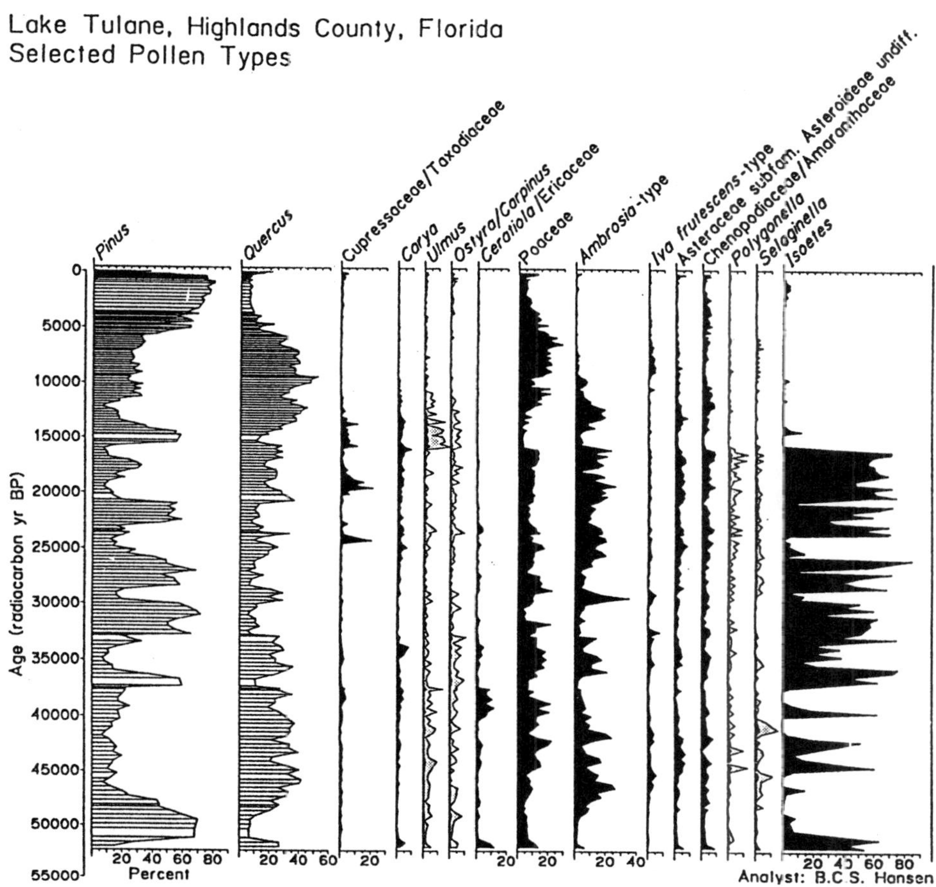

Figure 8.8. Pollen diagram from Lake Tulane, Highlands County, Florida. Reprinted from Watts and Hansen (1994, fig.) with the permission of Elsevier Science-NL.

gion, or became extinct at the species level. At Wilcox Bluff in Louisiana, a fluvial terrace assemblage probably dates from the Sangamon Interglacial or Early Wisconsin (Delcourt and Delcourt, 1977, in press). The vegetation is warm-temperate and includes *Juniperus, Pinus, Alnus, Carya, Cephalanthus* (buttonbush), *Fagus, Fraxinus, Ilex, Itea* (Virginia willow), *Leitneria* (corkwood), *Liquidambar, Liriodendron, Nyssa, Ostrya–Carpinus, Platanus, Quercus, Salix, Tilia,* and *Vitis.* Warm-temperate to subtropical elements (e.g., palms, *Bursera, Ficus, Rhizophora*) were likely introduced elsewhere in the southeast and/or expanded their range northward and into higher elevations during the interglacials and migrated southward, downward, or disappeared from the region in glacial times. The most recent introduction of tropical elements into peninsular Florida probably occurred 5–6 Kya. The north–south alignment of the Appalachian Mountains, climatic changes, and continuous land connections to boreal zones to the north and tropical zones to the south made the region both a pathway and a refugium for plants and animals throughout Cenozoic time. The result is the most species-rich biotic province in North America north of Mexico.

The Late Quaternary history of vegetation in the Florida Peninsula is preserved in numerous lakes formed in limestone sinkholes. One of these is Lake Tulane in Avon Park, south-central Florida. An 18.5-m core extends from the present back to ~50 Kya (Watts and Hansen, 1994; Fig. 8.8). The Wisconsin portion records a series of peaks in *Pinus* pollen at ~14–16, 21–23, 26–28, 30–33, 36–37, and 48–51 Kya. All but the 30–33 Kya interval correlate with Heinrich events H1–H5 (Bond et al., 1992; Grimm et al., 1993; Fig. 8.9). Alternating with pine are zones of abundant *Quercus* pollen and associated herbs (Gramineae, Chenopodiaceae, *Iva, Ambrosia* types). Pollen resembling *Taxodium* is present throughout the profile and it becomes more abundant during high stands of sea level. The probable scrub nature of the pine association during cold intervals was due in part to sea levels (and water tables) that were lower by ~20 m. At glacial maximum the MAT is estimated to have been ~2°C cooler (Brunner, 1982); but if the oxygen–strontium readings from the Barbados corals are correct (5°C lower; Guilderson et al., 1994), it may have been cooler.

In the Late Holocene after ~6 Kya, pine woods (*Pinus clausa*, sand pine; *P. elliottii*, slash pine), oak woods (*Quercus laevis*, turkey oak), and a flood-plain vegetation of bald cypress (*Taxodium distichum*) were reestablished, but with a herbaceous understory vegetation typical of the modern associations. According to the Lake Tulane profile, the re-

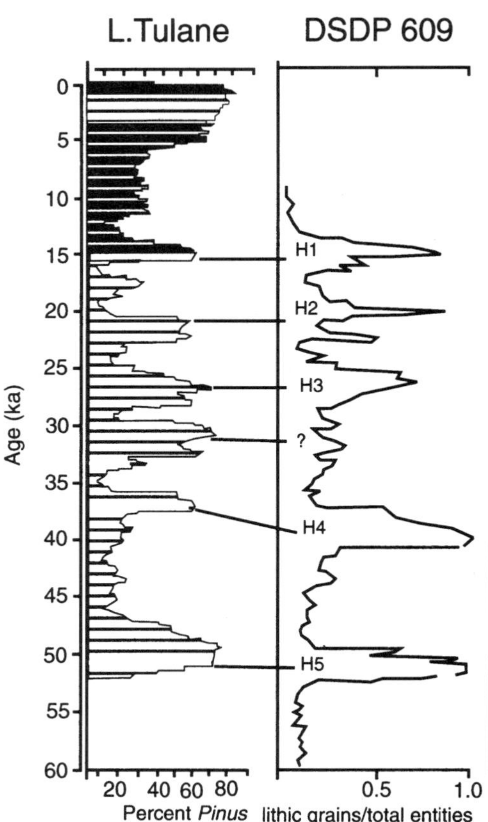

Figure 8.9. Lake Tulane pine pollen peaks and Heinrich events 1–5. Reprinted from Watts and Hansen (1994 fig. 5) with the permission of Elsevier Science-NL.

cent history of interior peninsular Florida vegetation has been pine–oak–grassland (warm intervals) alternating with open pine woods (cold intervals) within a mosaic of sand-dune scrub and with the present interglacial version dating back ~6 Ky.

In northwest Florida, Camel Lake shows a similar alternation between *Pinus* and *Quercus*, but there is better representation of *Carya* and *Fagus* pollen than at Lake Tulane to the south and at one brief interval between 12 and 14 Kya *Picea* is present. This is the only locality in Florida where prehistoric spruce pollen occurs in even modest quantity (typically 2%, one spectrum up to 8%). While *Pinus banksiana* (jack pine) decreased in the southern Appalachian Mountains beginning ~15 Kya and was replaced mostly by *Quercus*, *Picea* remained abundant until ~12 Kya to form a spruce-oak forest as in the lower midwestern United States. Both *Fagus* and *Picea* grow north of Florida in the Appalachian Mountains, which suggests temperatures in northern Florida were lower than at present (possibly to a -5°C January mean; Watts and Hansen, 1994).

To the east, Sheelar Lake is closer to the southeastern terminus of the Appalachian Mountains, and the fossil assemblage includes more mesic forest elements. At 12–14 Kya a deciduous forest was present (*Carya*, *Celtis*, *Fagus*, *Fraxinus*, *Ostrya-Carpinus*, *Platanus*, *Ulmus*). After ~12.7 Kya a warmer, drier climate developed and grassy openings appeared as shown by increasing Gramineae, chenoam, and Compositae–Asteraceae pollen. This was followed by an increase in *Quercus* then *Pinus* pollen at ~8 Kya to form the most recent version of the modern pine (*Pinus palustris*, long-leaf pine)–oak (*Quercus laevis*, turkey oak) vegetation. The Sheelar Lake region served as one refugium for mesic deciduous elements during the cold intervals of the Quaternary.

In the Tunica Hills region of Louisiana–Mississippi, sediments dated between 25,250 and 12,500 years B.P. (Delcourt and Delcourt, in press) contain megafossils and abundant pollen of *Picea* (40–70%). *Abies* and *Larix* are absent, but they are found to the north in the central Mississippi River Valley (Royall et al., 1991). Deciduous hardwoods are rare, and the climate during the Late Wisconsin was cooler than at present but not "boreal" (Jackson and Givens, 1994).

In the western part of southeastern United States an early study was made on two bogs in central Texas (Potzger and Tharp, 1943, 1947, 1954). The most significant result was the report of *Abies* and *Picea* pollen totaling 11% in the lowermost level of the Gause bog (Fig. 8.7). The present vegetation is an oak (*Quercus stellata*, post oak)–hickory (*Carya texana*) association, or post-oak savanna, with abundant grasses in the understory. *Juniperus* and *Prosopis* are also abundant, but they have increased in modern times from overgrazing. The presence of two northern conifers contributing 11% pollen implies a substantial boreal forest in the region. The nearest populations of *Abies* and *Picea* today are several hundred miles to the northeast in the

Appalachian Mountains and to the west in the southern Rocky Mountains. Brown (1938) reported cones, twigs, and wood of the boreal gymnosperms *Larix*, *Picea*, and *Thuja* from Pleistocene deposits in Louisiana; Wilson (in Davis, 1946) found pollen of *Picea* in peats from northern Florida. Collectively, these studies suggested that during glacial periods, climates had changed so dramatically in the south that a boreal forest extended from the glacial margin south to the Gulf Coast from Texas to Florida (Deevey, 1949).

In the absence of further information from the fossil record of the Gulf Coast, it was difficult to evaluate this model. The first direct evidence that revisions were necessary came from a restudy of the Texas material (Graham and Heimsch, 1960).[2] *Abies* pollen was not encountered, *Picea* pollen ranged from absent to 1.5% to a maximum of 3% (6 grains/200 count at one level), and pollen of *Castanea* was not recovered. The closest stands of *Castanea* (*C. pumila*, Allegheny chinquapin) are in northeast Texas where numerous eastern deciduous trees, shrubs, and vines have their western limits before encountering the warm-temperate to arid habitats and deep sand to rocky soils of the south and west. This is an important floristic boundary for many elements that shifted their range with climatic changes of the Quaternary. Eastern deciduous forest representatives here include *Alnus serrulata*, *Aralia spinosa* (devil's walking stick), *Asimina triloba* (pawpaw), *Betula nigra*, *Carpinus caroliniana* (American hornbeam), *Cercis canadensis* (redbud), *Chionanthus virginicus* (fringetree), *Diospyros virginiana* (persimmon), *Euonymus atropurpureus* (eastern wahoo), *Fagus grandifolia*, *Gymnocladus dioicus* (Kentucky coffee tree), *Juglans nigra*, *Liquidambar styraciflua*, *Maclura pomifera* (osage orange), *Magnolia grandiflora*, *M. virginiana*, *Myrica cerifera* (wax myrtle), *Ostrya virginiana* (eastern hop hornbeam), *Oxydendrum arboreum* (sourwood), *Persea borbonia* (red bay), *P. palustris* (swamp bay), *Platanus occidentalis*, *Populus deltoides* (eastern cottonwood), *Pyrus angustifolia* (crabapple), *Rhamnus caroliniana* (Carolina buckthorn), *Sabal minor* (bush palmetto), *Sambucus canadensis* (American elder), *Sassafras albidum*, *Stewartia malacodendron* (Virginia stewartia), *Styrax grandifolius* (snowbell), *Toxicodendron vernix* (poison sumac), *Vaccinium arboreum*, and species of *Acer*, *Aesculus*, *Bumelia*, *Carya*, *Celtis*, *Cornus*, *Crataegus*, *Fraxinus*, *Gleditsia*, *Halesia*, *Ilex*, *Morus*, *Nyssa*, *Prunus*, *Rhus*, *Salix*, *Tilia*, *Ulmus*, *Viburnum*, and *Zanthoxylum*.

Megafossils identified as *Thuja occidentalis* (northern white cedar) in the Louisiana deposits were later found to be *Chamaecyparis thyoides* (southern white cedar), while *Picea* and *Larix* were confirmed in sediments dated at 12,740 ^{14}C years (Delcourt and Delcourt, 1977, in press). These data indicate significant but less profound change in the vegetation of the region than first described. Boreal elements migrated southward partly along the Mississippi River Valley system with cold-air drainage and increased moisture from fog. This brought stands of *Picea* within pollen dispersal range of central Texas but not extensive boreal forests along the entire Gulf Coast.

Letters from John Potzger to B. C. Tharp provide some interesting insight into the Texas studies. In a letter dated May 5, 1942, Potzger notes, "My one fond hope is that the lowest levels will show pollen of northern species, if possible spruce and fir," and upon receiving word that the samples had been sent, he writes (November 19, 1942), "I can hardly wait for the arrival of the samples. I hope we find *Picea* in the bottom layers, that would be one of the most outstanding finds of the decade." Only spruce was initially identified, but in another letter (May 12, 1943) he "Had word from my friend today with respect to the pollen from Patschke bog which I had interpreted as *Picea*. He agrees with me on some but thinks that *Abies* is also represented." After that Potzger also began recognizing *Abies* in the material. In these early days the morphologic characteristics that distinguish the winged pollen of *Abies*, *Picea*, and *Pinus* were not well known and, indeed, confusion about them led to misidentifications in other material (Potzger, 1944).[3]

Recent studies have provided additional information on the late-glacial and postglacial history in the southwestern part of the southeastern United States (Bryant, 1977; Bryant and Holloway, 1985; Bryant and Shafer, 1977; Holloway et al., 1987; Larson et al., 1972; Fig. 8.10). Peats dated as older than 15,590 ± 250 years B.P. (late full glacial) confirm the presence of *Picea* within pollen range of the Texas sites. Deciduous elements from northeast Texas extended to the southwest along rivers, lakes, and in swampy areas (*Acer*, *Alnus*, *Betula*, *Carya*, *Cornus*, *Corylus*, *Fraxinus*, *Populus*, *Tilia*). Drier areas supported oak savanna (*Quercus* and Gramineae) and grassland during the full glacial in the southern High Plains (Hall and Valastro, 1995). The mean July temperature at the end of full-glacial times is estimated at ~2–3°C cooler than the present 27°C, and rainfall as about the same (Dillon, 1956) or slightly greater than the present 810 mm (Bryant, 1977). Vertebrate records that extend back ~20 Ky on the adjacent Edwards Plateau to the west indicate mean summer temperatures ~6°C lower than the present 27°C at glacial maximum and within 2–3°C of present values at ~13 Kya (Toomey et al., 1993). Noble gases (Ar, Kr, Xe) dissolved in groundwater from the Carizzo Aquifer in south Texas indicate a MAT ~5°C lower than at present for the last glacial maximum (Stute et al., 1992). Precipitation at full glacial was less than the present 810–510 mm (east to west gradient). In the late glacial many of the deciduous elements disappeared, and oak savanna with juniper (*Juniperus ashei*) and mesquite (*Prosopis glandulosa*) was established.

The history of this ecotonal region involved introduction of *Picea* and mesic deciduous vegetation from the north during cool moist intervals, the establishment of oak savanna or an oak–hickory (*Carya illinoiensis*) association in warmer drier times, and possibly the incursion of arid elements from the west during periods of maximum tem-

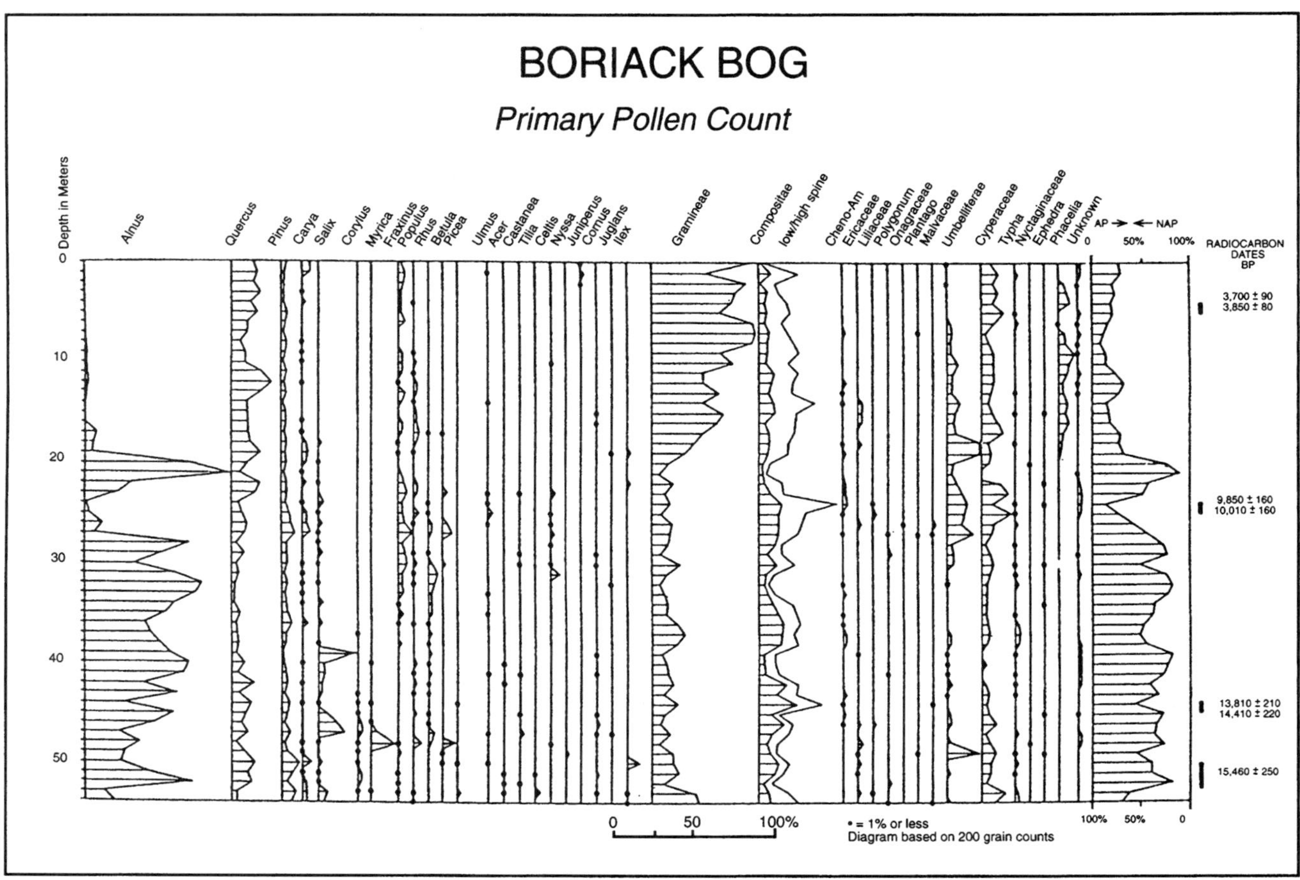

Figure 8.10. Pollen diagram from the Boriack Bog, Lee County, Texas. Reprinted from Bryant (1977, fig. 2) with the permission of the American Association of Stratigraphic Palynologists Foundation.

perature and dryness. The area is presently bordered to the west and southwest by *Juniperus* and *Prosopis* woodland, shortgrass prairie, and the Chihuahuan Desert. The present high-pressure system over the southwest mostly prevents moist air from the Gulf of Mexico from penetrating into central Texas and westward, but during glacial times the presence of the ascending low-pressure arm of the Hadley convection cell allowed moisture from the Gulf of Mexico to reach further inland.

In the central Appalachians more sites are available and the vegetational history is known in greater detail. The area is further north, closer to the Wisconsin glacial boundary, and many of the communities are arranged along relatively steep altitudinal gradients. As a result, changes in climate and other factors produce complex alterations in vegetation that are reflected in the pollen profiles. At the last full-glacial maximum, a form of alpine tundra was present at the highest elevations in the central Appalachian Mountains and to the north (Chester County, Pennsylvania, Martin, 1958; Buckle's Bog, Allegheny Plateau of Maryland, Maxwell and Davis, 1972; Potts Mountain Pond, Virginia, Watts, 1979). This is inferred from low percentages of tree pollen, abundant Cyperaceae and other herbaceous plants, and a high influx of minerals suggesting open ground (Delcourt and Delcourt, 1986; Mills and Delcourt, 1991). The present occurrence of exposed unglaciated granitic outcrops and invasion of sparsely vegetated surfaces by pioneer species of Cyperaceae, Gramineae, and other herbs and shrubs with pollen indistinguishable from tundra species (e.g., tree vs. dwarf species of *Betula*; Edwards et al., 1991) illustrates the difficulty in characterizing the ancient community. The characterization of this paleovegetation as equivalent to high Arctic tundra is speculative and has unlikely implications (e.g., for light regimes; Chapter 1, Tundra). The composition and prevailing physical conditions are more suggestive of an alpine version of low Arctic tundra or tundra–spruce parkland. At lower elevations there was a version of the Appalachian coniferous forest (*Abies*, *Picea*), including additional elements of the boreal forest growing south of their present distribution (e.g., *Pinus banksiana*, jack pine; Delcourt and Delcourt, 1985b; Watts, 1970; Whitehead, 1981; see also distribution in Elias, 1980). Fossil pollen of *P. banksiana* probably can be recognized to species level based on size and morphology (e.g., Ammann, 1977), but megafossils at several sites further confirm the presence of *P. banksiana* in periglacial southeastern and midwestern United States during the latest full glacial. The record implies that the

Appalachian coniferous forest expanded and contracted, and that boreal forest elements appeared and disappeared repeatedly from the region during the glacial–interglacial cycles of the Quaternary. The disjunct distribution of other northern species in the central and southern Appalachian highlands, which cannot be identified to species level from pollen (e.g., *Betula populifolia*; Elias, 1980), could be the result of long-distance dispersal any time during the Late Cenozoic; but target areas were certainly increased during the numerous cold intervals.

The dynamic nature of the vegetation in the southeastern United States is documented by studies at Anderson Pond (elevation 300 m, Middle Tennessee; Delcourt, 1979; Delcourt and Delcourt, 1985b). This site presently supports mixed mesophytic forest, but at 19 Kya it was Appalachian coniferous forest. If interglacials are atypical "catastrophic warm events" occurring regularly every 100 Ky during the predominantly glacial climates of the past 1.6 Ma (Davis, 1976), then the present extensive deciduous forest is exceptional in eastern North America, which was covered most of the time by versions of an Appalachian coniferous–boreal forest.

A pollen assemblage from the Piedmont region of northeastern Georgia, dated at ~26–22 Kya (just before the Wisconsin glacial maximum), was rich in *Pinus* and *Quercus* and had lesser representation of *Abies*, *Picea*, and *Carya* (Jackson and Whitehead, 1993). At glacial maximum the Carolina Bays on the coastal plain were surrounded by spruce and northern pine and had few representatives of the deciduous forest (Whitehead, 1981). During the interglacials southern yellow pine was dominant. Thus, the Atlantic Coastal Plain was not a refugium for the deciduous forest in glacial times.

During these cold periods the mixed mesophytic and other associations of the deciduous forest formation did persist in at least three refugia identified from the pollen and associated megafossil record: Nonconnah Creek in the blufflands of southwestern Tennessee (Delcourt et al., 1980); Goshen Springs in south-central Alabama (Delcourt, 1980); and, as previously noted, Sheelar Lake in northern Florida (Watts and Hansen, 1994; Watts and Stuiver, 1980). The deciduous forest spread from these Gulf Coast refugia during warm intervals in patterns that seem clear for the Holocene (Davis 1981b, 1983; Delcourt and Delcourt, 1975). However, the few number of refugia identified to date make provisional any detailed reconstruction of the direction and chronology of migrations during full glacial and earlier times (Ritchie, 1987).

In the late-glacial (16.5–12.5 Kya) pollen of the *Pinus banksiana* type began to disappear, the southern limit of boreal elements moved northward, and there was an increase in pollen of mesic deciduous species. TAMS ^{14}C dates from Virginia indicate warming in the south may have begun by 17 Kya (Kneller and Peteet, 1993). During the climatic optimum or hypsithermal interval (8500–4 Kya), pollen of drier habitat species of *Quercus* and *Carya* increased and these trees were probably present on the coastal plain, but by 5 Kya they were replaced there by the modern pine woods association. Such interchanges must have occurred frequently during the Quaternary, but the latest version of the southern yellow pine woods association along the coastal plain dates from ~5 Kya. *Castanea dentata* extended its range into the central Appalachian Mountains to form an oak–chestnut forest, and *Pinus echinata* (short-leaf pine) moved from the south and west into the Ozark region of Oklahoma and Missouri (Delcourt and Delcourt, 1991).

Evidence for early human occupation in eastern North America comes from Meadowcroft Rockshelter in Pennsylvania. Ages ranging from 12 to 20 Kya have been suggested, but estimates from well-dated sites favor occupation after ~12 Kya; there was little impact on regional vegetation until the last few hundred years (McAndrews, 1988; Fig. 8.11). The influence of AmerIndian populations on the vegetation begins in the Late Holocene with the recovery of pollen of *Zea mays* (2 Kya; Whitehead, 1965; Whitehead and Oaks, 1979; Whitehead and Sheehan, 1985) and remains of *Lagenaria siceraria* (gourd), *Cucurbita pepo* (squash), and *Phaseolus vulgaris* (bean; Chapman et al., 1982; Cridlebaugh, 1984). Also present are pollen of plants indicating open habitats associated with cultivation (*Ambrosia*, *Chenopodium* type, *Iva*, *Plantago*, *Rumex*) and increased charcoal resulting from land clearing and food preparation (Delcourt, 1987; Delcourt et al., 1986).

Northeast

Glaciers extended south to ~40° N at 18 Kya, then melting began and intensified between 12 and 10 Kya. The general sequence of vegetation change near the glacial boundary was outlined early from bog and lake sequences in Connecticut (Deevey, 1939; see also Leopold, 1956). The presence of tundra along the margin of the Laurentide ice sheet was considered likely, but percentage tabulations were not adequate to detect treeless vegetation receiving pollen input from adjacent forests. Either tundra or park tundra (herbs with pine and spruce pollen, Peteet et al., 1993; spruce needles, Watts, 1979) has been documented during the full glacial in Maryland (Maxwell and Davis, 1972), Pennsylvania (Martin, 1958), and New England (Gaudreau and Webb, 1985); the presence of an 80–200 km wide zone of permafrost fringing the ice suggests this vegetation was typical of the ice margins (Miller, 1992). A bryophyte flora from Connecticut dated at 13.5 Kya reveals tundra similar to that now found along the northern margin of the deciduous forest (Miller, 1993). By 11.5 Kya the region was boreal woodland. In several places megafossils of tree species were recovered, and whether this vegetation should be characterized as tundra typical of Arctic regions is open to question (Birks, 1976).

After the tundra–park tundra phase, Deevey (1939) recognized three zones (A, B, C) with several subzones (e.g., A-1, A-2) to describe the sequence of vegetation and implied

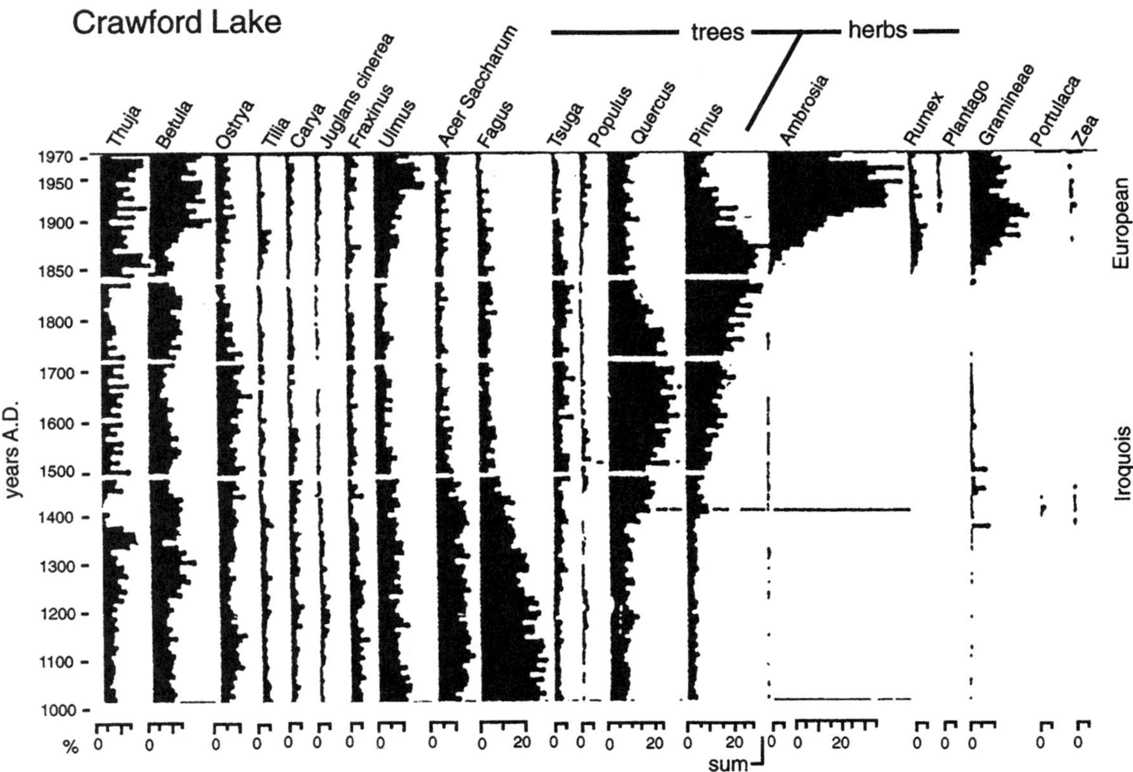

Figure 8.11. Pollen diagram from Crawford Lake, Ontario, Canada. Reprinted from McAndrews (1988, fig. 5) with the permission of J. H. McAndrews, Kluwer Academic Publishers, and the Geological Survey of Canada.

climates for the region. Although details vary at different places in the northeast and adjacent regions (e.g., Adirondacks, Jackson and Whitehead, 1991; New Jersey, Peteet et al., 1990, 1993; Pennsylvania, Watts, 1979, Fig. 8.12; and Whitehead et al., 1989), the sequences are similar and begin with a spruce–fir forest with birch locally abundant, invading the tundra as climates warmed and the glacial margin retreated (zone A-1). Spruce and fir reached a maximum (A-2) then declined. As climate warmed to the climatic optimum, pine increased (B); then with amelioration the deciduous forest became established through oak and hemlock (C-1), hickory (C-2; oak and hemlock decline), and chestnut–hemlock–oak phases (C-3). Evidence of human disturbance appears late in C. These phases were later modified (Davis, 1969; Peteet et al., 1993) and are summarized in Table 8.1.

At the western limits of the deciduous forest in eastern and central Ohio there is a similar sequence from spruce, to pine, and then deciduous trees (Shane, 1975, 1987, 1989a; Shane and Anderson, 1993). Slightly further to the west on the till plains of Ohio and Indiana the history begins with an open spruce forest tundra between 15.5 and 13.5 Kya. There follows a deciduous open woodland (13.5-11 Kya), reversal to a cooler period (increase in spruce) correlated with the Younger Dryas event (11–10.3 Kya), and rapid warming between ~10.3 and 4 Kya (decline of hemlock, jack pine, red pine, white pine). After 10 Kya a mixed deciduous forest was established; continued warming led to the development of an open oak forest between 8 and 4 Kya (King, 1981; Shane, 1987; Webb et al., 1983). Transfer function equations give an estimate of January mean temperatures rising from −14 to −2°C and July mean temperatures from 19.4 to 23.5°C between 13 Kya and 4500 years B.P. In the latter part of this period, the June mean temperature cooled to the present 22.5°C. Annual precipitation varied from ~660 to ~750 mm (Shane, 1989b).

It is tempting to attribute all these regional and local changes in vegetation to climate, but the history is more complex. At ~4800 years B.P. hemlock declined rapidly, then recovered beginning ~3400 years B.P. Other associates of hemlock with comparable ecological requirements did not show a similar pattern, and it is probable that the decline was due to epidemic disease (Allison et al., 1986; Davis, 1981a). Considering the recent history of species of *Castanea* (chestnut) and *Ulmus* (elm), it is not surprising that pollen diagrams record similar events in the past (Patterson and Backman, 1988). Also, the individual migration rates of organisms cause them to enter habitats at various times after displacement and to form different alliances with ecologically compatible species through time (Davis, 1981b, 1983; Webb, 1987). Past changes in the distribution of plant formations and associations are most frequently the consequence of climate; but if individual elements de-

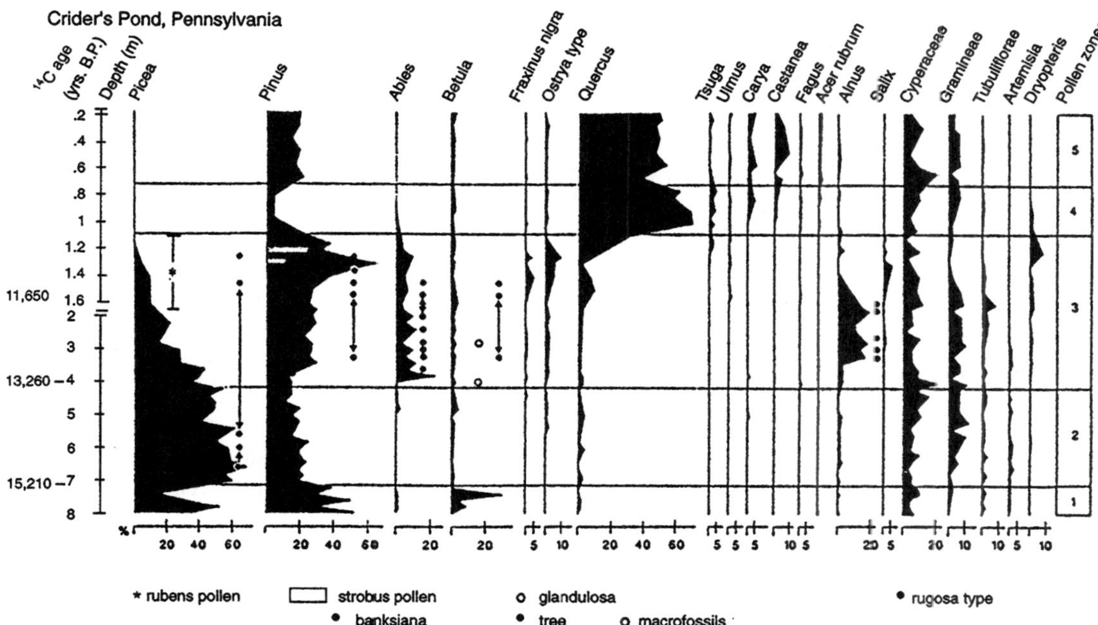

Figure 8.12. Pollen diagram from Crider's Pond, Pennsylvania. Reprinted from Ritchie (1987, fig. 4.5) with the permission of Cambridge University Press.

viate from the pattern, several explanations are available. These include disease, differential migration rates, individual accommodation to stages of soil formation, and selective anthropogenic impact.

Further to the north, arid permafrost conditions existed in Greenland, Spitsbergen, and the Arctic Islands throughout the Quaternary, and persist today as the High Arctic polar desert. Pollen from Sangamon deposits gives an insight into the vegetation and climate at the end of the last interglacial into the Early Wisconsin. At East Milford Quarry in Nova Scotia (>50 Kya) the sequence is hardwood forest (*Acer, Carya, Fagus, Quercus, Tilia, Ulmus*; climates at least as warm as at present), a mixed forest (*Abies, Picea, Betula*; cooling), and coniferous forest (*Abies, Picea*; cold; Mott et al., 1982).

Very little of present-day Canada was ice free at the last glacial maximum, so records begin progressively later toward the glacial center and preserve events primarily from the late glacial and Holocene (Ritchie, 1987). The Laurentide ice sheet separated from the northern Appalachian glaciers at ~11 Kya and northern Maine was subsequently ice free. Beginning ~14 Kya melting in the Gulf of St. Lawrence region was calving ice into the St. Lawrence Estuary, and by ~13–11 Kya vegetation was recolonizing the coastal areas of the Gaspé Peninsula and Anticosti Island (Richard, 1994a). Blocks of stagnant ice remained in central Quebec until ~6 Kya. As the conifer-dominated Appalachian forest spread northward with the retreating ice sheet and as mesic deciduous elements expanded from refugia, a variably defined lake states northern deciduous–boreal coniferous forest became widespread, especially in extensive areas of low to moderate physiographic relief. This transitional vegetation shifted northward during the climatic optima in all of the various interglacials, most recently at ~7 Kya.

Table 8.1. General sequence of pollen zones and inferred vegetation and climate for New England region.

Pollen Zones	Years B.P.	Vegetation	Climate
C-3	Present	Oak–Chestnut–Hemlock (*Ambrosia, Rumex*, other weeds)	Moister, Cooler
C–2		Oak, with hickory	Warm–dry
C-1	7,900	Oak–hemlock (*Ambrosia*)	Warm–moist
B	7,900–8,100	Pine maximum, with oak	Warm–dry
A-4	9,100–10,200	Spruce–fir with larch, birch, alder	Cooler–moister (Younger Dryas)
A-3	10,200	Spruce–oak, with pine	Warmer–drier
A-2	11,700	Spruce, with oak	
A-1	11,700–12,150	Spruce–fir–birch	Cool, moist
"T"	12,150–14,300	Tundra	Cold, dry

Adapted from Davis (1969), Deevey (1939), Leopold (1956), and Peteet et al. (1993). T = tundra.

Early studies on the late-glacial and postglacial vegetation of eastern Canada were made by Terasmae (e.g., 1958, 1960, 1967) and more recently by Anderson (1985), Gajewski et al. (1993), Payette (1993), Richard (1993, 1994a,b), Ritchie (1987), and others. The sequences show an early nonarboreal stage prior to ~10 Kya of Cyperaceae, Gramineae, dwarf birch (*Betula glandulosa*), and *Salix*, which was interpreted as a desert tundra that changed to an herb-rich tundra and then to a shrubby (dwarf birch) tundra. There followed an afforestation stage with increasing AP (*Abies*, *Picea*, *Pinus banksiana*, *Betula*, *Populus*) between ~11 and 8.5 Kya (beginning later in the highlands at ~8 Kya and lasting until 5 Kya). There were brief reversals in the trend toward greater forest vegetation as shown by pollen and midge evidence from the Canadian Maritime provinces (Levesque et al., 1993). There a cold interval occurred between 11,160 and 10,910 years B.P., just before the Younger Dryas event. After that the canopy gradually closed with the addition or increase in *Larix*, *Picea*, *Fraxinus*, *Ostrya*, *Quercus*, and *Ulmus*; the modern boreal forest was established across broad regions of central Canada. It extended farthest north most recently at ~6 Kya (Webb, 1987).

The chronology of these events in eastern to central Canada is generally time-transgressive from south to north and from east (coastal) to west (interior). The present cline in vegetation from coastal (and lake margin) cold-temperate deciduous forest to boreal forest westward to tundra northward in east central Canada was established by ~3 Kya (Richard, 1994b). High-resolution pollen studies anchored by ^{14}C dates further demonstrate that the invasion of these vegetation types was rapid. The change from tundra to closed canopy *Picea mariana* (black spruce) forest in central Canada occurred in 150 years between 5 and 4 Kya at rates estimated at 200 km/century (MacDonald et al., 1993). In the past 3 Ky, tundra in northern Québec has increased at the expense of forest (Gajewski et al., 1993). In the far north around Ungava Bay in northern Québec, tundra persisted throughout the Holocene.

The spread of tree species from the south followed two principal routes along the sides of the Laurentide ice sheet: the Labradorean pathway (boreal deciduous trees favored by humid environments) and the Hudsonian pathway (*Picea mariana*, black spruce). Many of these elements illustrate again one of the important results of recent research on Quaternary vegetational history: the associations are temporal in nature. *Pinus banksiana* is in equilibrium with the present climate and is not expanding its range, *Picea glauca* is in equilibrium in eastern Québec but it is spreading along Hudson Bay, and *Abies balsamea* is expanding in the northern James Bay region (Payette, 1993).

Plains and Upper Midwest

To the west in the Plains area there are few plant assemblages of Early to Middle Quaternary age (Baker and Waln, 1985; Watts, 1983). In southwestern Kansas and adjacent Oklahoma pollen from possibly Illinoian full-glacial deposits are rich in *Pinus* and have smaller amounts of *Picea*, Gramineae, Compositae, *Artemisia*, and rarer *Alnus*, *Betula*, *Carya*, *Fraxinus*, *Juglans*, *Quercus*, and *Salix*. The vegetation is interpreted as a pine savanna with spruce in local moist habitats and deciduous trees confined mostly to streamsides (Kapp, 1970). Toward the warmer Sangamon interglacial *Pinus* decreases, *Picea* disappears, and grasses and other herbs increase (drier and/or warmer). Near the beginning of the Wisconsin, *Pinus* and *Picea* return (moister and/or cooler).

In the Wisconsin stage of northeastern Kansas *Picea* was prominent between 24 and 14 Kya, followed by an increase in mesic deciduous elements (*Carpinus*, *Fraxinus*, *Quercus*, *Ulmus*) and establishment of the modern grassland by ~10 Kya (Grüger, 1973). Megafossils from the Middle Wisconsin of Missouri confirm at least some of the pine as *Pinus banksiana* (King, 1973). By the Late Wisconsin and near the glacial maximum (~20 Kya) the Ozark highlands in west-central Missouri were covered by spruce forest until ~14 Kya, followed by mesic deciduous elements (*Fraxinus*, *Ostrya*, *Quercus*, *Ulmus*). In adjacent northeastern Kansas *Picea* had disappeared by ~11.5 Kya, and at ~5 Kya the modern open woodland of *Quercus* and *Carya* with understory grasses and other herbs was established. After ~5 Kya in southern Oklahoma the vegetation fluctuated between *Quercus* savanna, the present-day *Quercus*–*Carya* woodland (extending into central Texas), and pine woodland probably extending inland from the east Texas–Gulf Coast region (*P. echinata*, short leaf; *P. taeda*, loblolly; *P. palustris*, long leaf). The latest *Quercus*–*Carya* woodland has existed for the past ~2 Ky. The broad history of vegetation in this region has been repeated shifts between grassland and savanna–woodland (*Pinus*, *Artemisia*, *Carya*, *Quercus*) with incursions of boreal conifers during the coldest periods.

The northern grasslands are bordered by boreal forest to the north and deciduous forest to the east and, as expected, the ecotones shifted with the climatic changes of the Quaternary. Among the first studies demonstrating the dynamic aspect of these Pleistocene biotas were those of Wright et al. (1963) and McAndrews (1966) that showed shifts in the grassland–forest boundary in the upper Midwest. With these shifts temporary associations are inferred that have no modern counterparts. Spruce woodland, spruce–oak woodland, and black ash tundra are examples (Chapter 3; Fig. 8.13). In eastern Iowa and Missouri the vegetation between 34 and 22 Kya was an open *Picea*–*Pinus* forest similar to tundra–tree-line vegetation (Baker et al., 1986). Pollen of *Pinus* then declined, *Picea* remained about the same or increased, and the pollen of herbs increased, probably reflecting one of the many stadials of the Wisconsin ice sheet. One of the differences between ice margin vegetation in the upper Midwest and in eastern North America at glacial maxima was that the former was tundra–parkland (with spruce), while the latter had fewer trees. A com-

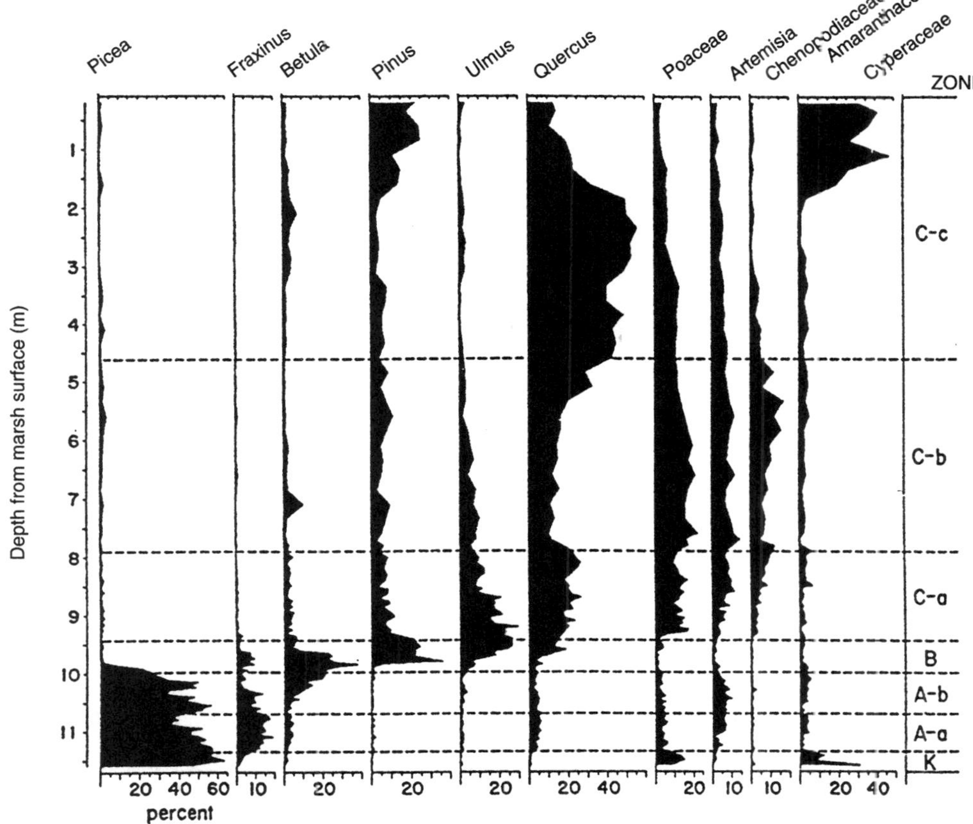

Figure 8.13. Pollen diagram from Kirchner Marsh, Dakota County, Minnesota. Reprinted from Grimm (1988, fig. 3, based on Wright et al., 1963) with the permission of Eric Grimm, the Geological Society of America, the Geological Survey of Canada, and Kluwer Academic Publishers.

bination of *Picea* and *Larix*, with some *Fraxinus*, was present between 14 and 12.5 Kya. This was followed by incursions of *Acer negundo*, *Corylus cornuta*, and *Quercus rubra* into the *Picea–Larix* forests; disappearance of *Picea–Larix* and an increase of deciduous trees between ~11 and 9 Kya; and decline of deciduous trees with an increase of grassland elements at 9 Kya. By 7 Kya grasslands were established in western Iowa and deciduous forest grew to the east. At the height of postglacial warming (middle hypsithermal interval, ~6 Kya), the grassland reached its greatest eastern extent and grew beyond the present prairie peninsula of Illinois. At Wolf Creek in Minnesota the sequence is tundra (*Antennaria*, *Dryas*, *Silene*, *Vaccinium*; 20.5–14.7 Kya), a transitional shrub tundra (*Alnus*, *Betula*, *Empetrum*, *Salix*, *Shepherdia*; 14.6–13.6 Kya); *Picea* woodland (13.6–10 Kya); and shifts between the limits of the mixed conifer (*Pinus banksiana*)–deciduous forest of the present (*Acer*, *Carya*, *Castanea*, *Celtis*, *Fagus*, *Ostrya–Carpinus*, *Tilia*, *Ulmus*) and grasslands to the south (Birks, 1976). At Wolsfeld and French lakes in southern Minnesota grassland was present until ~5 Kya, then an oak savanna developed and persisted until ~300 years ago when during the Little Ice Age *Acer saccharum*, *Ostrya virginiana*, *Tilia americana*, and *Ulmus* arrived and formed the present Big Woods of that region (Grimm, 1983). A detailed history of Elk Lake Minnesota, for the past 10 Ky is presented by Bradbury and Dean (1993). In southern Ontario oak savanna was also present after ~6 Kya (Szeicz and MacDonald, 1991). Prior to that study it was thought that the oak savanna was a result of burning by AmerIndians. The history of *Tsuga* in the forest of the Great Lakes region provides an example of factors that can delay the introduction of a species even though climates are suitable. In this case the delay was due to the physical–biotic barrier presented by Lake Michigan surrounded along its southern boundary by dry grassland (Davis et al., 1986).

Mean July temperatures at glacial maximum are estimated at 10–12°C cooler than at present. The isotopic composition of the waters of glacial Lake Agassiz in North Dakota and southern Manitoba provides an estimate of average air temperature during the Late Wisconsin–Early Holocene (11.7 Ka–9500 years B.P.) of –16°C compared to the present 0°C (Remenda et al., 1994). Rainfall during cold periods was 20% below the present in some areas (Bartlein et al., 1984).

Model simulations (Kutzbach, 1987) generally parallel

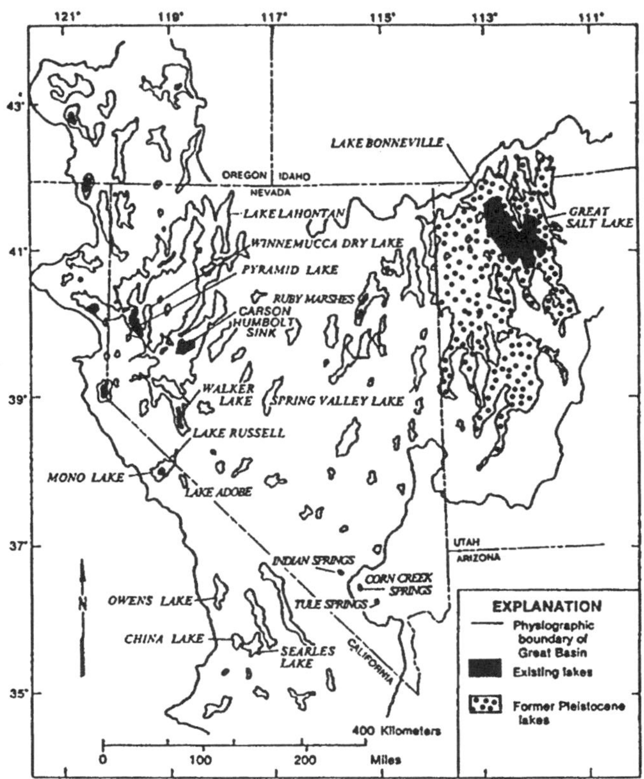

Figure 8.14. Late Pleistocene lakes and marshes in the Great Basin. Reprinted from Benson and Thompson (1987, fig. 1, after Spaulding et al., 1983) with the permission of the Geological Society of America.

results from the pollen and spore studies (Webb et al., 1987). At 18 Kya temperatures in the central and eastern sections of North America were an estimated 5°C colder than at present, while in the southeast they were 1–2°C colder. At 12 Kya the jet stream switched to a single flow as the Laurentide ice sheet was reduced in area and elevation (~50%). Temperatures in the north-central and northeast regions warmed to 2–4°C cooler than at present and were ~1°C warmer in the southeast. At 9 Kya the ice sheet was reduced to ~800 m in elevation and July insolation was ~8% greater than now, producing the early stages of the climatic optimum wherein temperatures across North America averaged 1–2°C warmer than at present. By 6 Kya the Laurentide ice sheet had essentially disappeared and conditions were at or near present values.

Southwest and West

One effect of the high-pressure anticyclone systems located over the Laurentide and Cordilleran ice sheets during glacial maxima was the displacement of low-pressure cyclone belts southward. During these times there was increased precipitation and lower temperatures, and over 100 pluvial lakes formed in the southwest (Fig. 8.14). These lakes muted seasonal temperature variations and served as a positive feedback to increased precipitation. Tree lines lowered by 500–1000 m as determined by *Picea–Pinus* ratios from bogs compared to the ratios in surface samples collected along altitudinal and vegetation gradients (e.g., Maher, 1963). Pine–juniper woodland moved into the plains and basins, displacing desert shrubland and grassland, although recent studies indicate that *Pinus* may be overrepresented at several sites and that desert grassland was more prominent than thought earlier (Hall, 1985). These conditions were reversed in each of the 18–20 interglacial periods of the past 1.6 m.y. The fossil record from the arid southwest shows that plant species there also responded to climatic change individually, forging combinations that are now rare or absent (Thompson, 1988; Van Devender et al., 1987). Physiographic diversity is greater in the western cordilleran region of North America than in the east or in the Plains. As a result, biotic history is more reflective of localized basins, plateaus, slopes, and peaks. Although correlated global and hemispheric events, such as the Little Ice Age, the Younger Dryas, and the altithermal interval, are broadly evident throughout western North America, the regional histories tend to be more individualized.

Pollen-bearing sediments are not extensive in this region, and diversity in assemblages representing low to middle elevation habitats may be restricted to 40 or fewer pollen types. *Juniperus*, *Pinus*, Gramineae, *Artemisia*, chenoams, and Compositae often account for 90% or more of the plant microfossils. Continuous and well-dated Holocene sequences come mostly from montane bogs where pollen of *Abies*, *Picea*, *Pinus*, and herbaceous pollen, usually undifferentiated to genus, are prominent. Consequently, information on Quaternary biotas and environ-

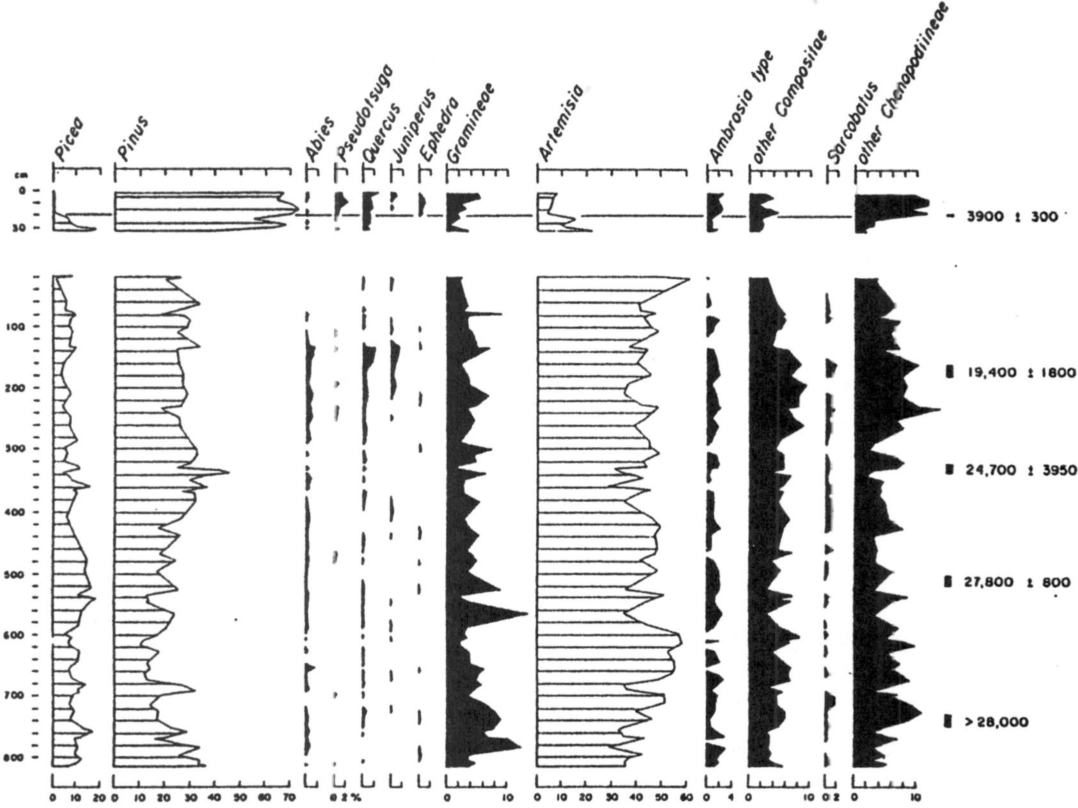

Figure 8.15. Pollen diagram from Dead Man Lake, Chuska Mountains, New Mexico. Reprinted from Hall (1985, fig. 7, after Wright et al., 1973) with the permission of the American Association of Stratigraphic Palynologists Foundation.

ments must be assembled from fossil faunas, packrat middens, dendrochronology, megafossil assemblages, and pollen from bogs, playa lakes, and cave and rock shelter deposits (see papers in Jacobs et al., 1985). Also, the correlation of vegetation change with climatic events is complicated by topographic diversity, varied distances from sources of oceanic moisture, and possible lag time or "vegetation inertia" (Cole, 1985; Spaulding, 1990a). The latter concept is based on midden data from the Grand Canyon region of northern Arizona. In the early stages of a new climatic cycle the first organisms to disappear are those near the limits of their ecological tolerance, which is recorded in the diagrams as a decline in species diversity. However, the dominants to some extent sustain their own microenvironment (sunlight, moisture, soil chemistry) and persist longer into the cycle. Although some correspondence is now emerging between high-resolution pollen diagrams and the oxygen isotope record (Williams et al., 1993, fig. 8.22), if climatic changes are rapid, as in the Wisconsin–Holocene transition, the plant record may not vary precisely in concert because of lag time. Another early source of confusion in reconstructing regional climate and vegetational history, and comparing these with events in adjacent regions, was pollen evidence suggesting that the interval between ~7.5 and 4 Kya years in the southwest was moist (Martin, 1963a,b). Geomorphic and other data indicated it was arid (the "long drought"; Antevs, 1938). The problem was resolved with the discovery that the sections representing the altithermal at several sites (e.g., Whitewater Draw, Arizona) had been removed by erosion.

Little is known about pre-Wisconsin vegetation in the arid southwest, but in the far northwestern corner of Texas the small Rita Blanca megafossil flora provides some insight into Nebraskan age vegetation (Tucker, 1969). The prominent member is *Quercus* that is similar to the *Q. gambelii* that presently grows along the western forest–grassland border. Associates suggest higher water tables and more mesic conditions than at present (*Crataegus, Maclura, Populus, Ribes, Salix, Sapindus*). Drier elements include *Amorpha* (mock lotus) and *Baccharis* (buckbrush); the pollen flora is rich in *Artemisia*, other composites, and grasses.

For Wisconsin time more information is available from the southwest. Many of the full-glacial records (24–14 Kya) are rich in *Artemisia* and *Pinus*, have smaller amounts of *Abies* and *Picea* (Fig. 8.15), and represent a cool sagebrush steppe or sagebrush–pine woodland. The latter is an association of genera and a vegetation type that is not widespread today. Alpine tundra expanded at the highest elevations as tree lines shifted downward. The transition from glacial to postglacial conditions occurred between ~14 and

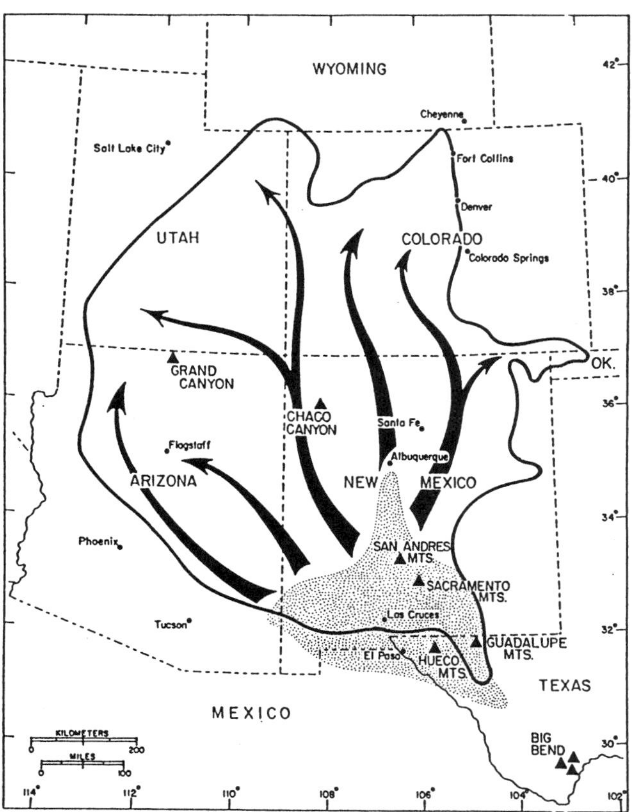

Figure 8.16. Modern distribution (solid line) and Late Wisconsin distribution (stippled) of *Pinus edulis* (piñon pine). Arrows represent suggested migration routes of *P. edulis* into its present range. Triangles are packrat midden sites. Reprinted from Van Devender et al. (1984, fig. 3) with the permission of Quaternary Research.

12 Kya in the southwest; after an Early Holocene continuation of cool conditions, climates warmed in the Middle Holocene and vegetation trended toward drier communities and an upward shift in ecotones.

Packrat middens from the Chihuahuan Desert of northeastern Mexico, Texas, and New Mexico show a sequence from piñon pine (*P. edulis*)–juniper (*J. scopulorum*) woodland (mesic, Late Wisconsin), to oak–juniper woodland (Early Holocene transition, ~11 Kya), to the present desert grassland–shrubland that developed between 8 and 4 Kya (Van Devender, 1985, 1990a; Van Devender et al., 1984). A similar sequence was likely repeated throughout the Quaternary as the low-pressure systems of the glacial intervals contributed moisture and allowed deeper penetration of cyclones from the Gulf of Mexico and the Pacific Ocean, followed by the return of high-pressure systems in interglacial times. Playa lakes were numerous and some of these provided the basis for early pollen studies in the southwest from the San Agustin Plains of New Mexico (Clisby and Sears, 1956; see also Markgraf et al., 1984) and the Wilcox Playa of Arizona (Martin, 1963c). Forests and woodland grew 1000 m lower than at present and mountain glaciers formed in the Sacramento Mountains of New Mexico. These and other highlands likely served as refugia for mesic elements that spread during glacial (pluvial) times (Fig. 8.16). In each of the 18–20 principal interglacials of the Quaternary, arid conditions returned and vegetation similar to that of the present was established.

In the Trans Pecos region of west Texas the present vegetation is a desert scrubland of *Flourensia*, *Larrea*, grasses, cacti, and composites. At higher elevations there is an oak–juniper woodland. During the latest full glacial (20–14 Kya) the vegetation was a piñon pine–juniper woodland (Spaulding et al., 1983), reflecting cooler and more moist conditions. The pine–spruce, ponderosa pine, and piñon pine–juniper zones shifted downward and the latter spread onto the lowland plains now occupied by desert grassland and shrubland. By Holocene times conditions at Bonfire Shelter in south-central Texas at the Mexican border were warmer and drier than at present. Pollen of grasses and composites increased after 10.5 Kya (Bryant and Holloway, 1985). Later (8.7–6 Kya) pollen of the xerophytes *Agave* and *Dasylirion* increased, corresponding to the warm-dry climate of the climatic optimum (Hall, 1985). Rainfall continued to decrease until ~2.5 Kya, based on faunal records from the Edwards Plateau region (disappearance of high moisture-requiring taxa *Pipistrellus subflavus*, pipistrelle bat; *Microtus pinetorum*, woodland vole; Toomey et al., 1993). Arid late-glacial and interglacial conditions are also suggested by landscape morphology and pollen assemblages from the San Juan Basin of northwestern New Mexico (Hall, 1977, 1985, 1990) and by low water tables in the southern High Plains of Texas (Meltzer, 1991). After ~2.5 Kya slightly more mesic conditions returned to the Edwards Plateau region, and essentially modern biotas and environments have prevailed.

To the northwest along the Texas–New Mexico border, early pollen studies were made at several sites on the Llano Estacado (southern High Plains; Hafsten, 1964). The present rainfall there is 400–500 mm, average January temperature is ~5°C, and average July temperature is ~27°C. The vegetation is a shrubland (juniper, oak)–desert grassland (*Bouteloua*, grama grass; *Buchloë*, buffalo grass; chenopods; composites; *Artemisia*, sagebrush; *Ephedra*; Küchler, 1964). At about 33.5 Kya the vegetation was an open woodland of pine and spruce, summer temperatures were an estimated ~5°C cooler than at present, and moisture was more plentiful. The principal fluctuations in vegetation involved decreases in trees with more *Artemisia* (warmer and drier), then a return to pine and spruce parkland. At the Wisconsin glacial maximum at 22.5–14 Kya, corresponding to an 8–10°C cooler interval, more closed woodlands of pine and spruce prevailed. Near the end of the Wisconsin stage the *Artemisia* steppe conditions of the present were attained. It is assumed that a similar history of woodland (with boreal and western montane conifers), to less dense parkland, to open shrubland with abundant grasses characterized the region from the Nebraskan glacial stage throughout the Quaternary. In the nearby Hueco Mountains (El Paso and Hudspeth Counties, Texas) packrat middens preserve a piñon pine–juniper–oak woodland at ~13.5–12 Kya, changing to drier oak–juniper woodland by ~9.4 Kya, to desert grassland (8.9–4 Kya), to modern Chihuahuan Desert scrub after 4 Kya (Van Devender and Riskind, 1979; Van Devender et al., 1987). At Chaco Canyon in northwestern New Mexico (Betancourt and Van Devender, 1981), pollen evidence suggests less piñon pine–juniper woodland (midden data) and more shrubby grassland (Hall–1981, 1982, 1985). If the pine and juniper were localized stands in canyons, they may be overrepresented in the middens; and the vegetation may have been mostly an arid shrub grassland for the past 7 Kya (Hall, 1977).

The Puerto Blanco Mountains of southwestern Arizona are located in one of the most arid regions of North America (Organ Pipe Cactus National Monument, Sonoran Desert). A Holocene history has been reconstructed from middens (Van Devender, 1987). The sequence is from juniper–joshua tree woodland (~14 Kya) to desert shrubland with *Carnegiea gigantea* (saguaro) and *Encelia farinosa* (brittle bush) as dominants and *Acacia greggii*, *Prosopis velutina*, and *Cercidium floridum* as associates (~10.5 Kya). There followed Sonoran Desert vegetation with *Sapium biloculare* (Mexican jumping bean), *Olneya tesota* (ironwood), and *Stenocereus thurberi* (organ pipe cactus) at ~4 Kya. In the Waterman Mountains just to the east in southern Arizona, the Middle to Late Wisconsin vegetation was piñon pine–juniper woodland with *Artemisia* (sagebrush), *Vauquelinia californica* (Arizona rosewood), and *Yucca* (Joshua tree) at 22.5–11.5 Kya, followed by communities similar to those recorded at Puerto Blanco (Anderson and Van Devender, 1991; Van Devender, 1990b). The Clovis complex at Murray Hills, Arizona, is the oldest documented PaleoIndian tradition in the southwest. It includes remains of browsing mammoth and grazing bison and dates from ~11.2 to 10.9 Kya (Hoffecker et al., 1993).

Middens from the Scodie Mountains in the western Mojave Desert are from 13,330 to 12,870 years old. The site is at 1215 m elevation and the present vegetation is a *Larrea tridentata* (creosote bush)–*Ambrosia dumosa* (bur sage) desert. In the uplands there is a Joshua tree (*Yucca brevifolia*) desert at 1300–1900 m and a piñon pine (*P. monophylla*)–oak (*Q. chrysolepis*) woodland at 1900–2200 m. At ~13 Kya Death Valley (elevation −86 m) in the Mojave Desert was a ~173-m deep lake (90 m from 10 to 35 Kya; Li et al., 1996). In the Late glacial there was a piñon pine–juniper (*J. californica*) woodland (Thompson, 1990). Lower ecotones, plants no longer in the immediate vicinity (54.5%; e.g., *Ceanothus greggii*), and modern analogs elsewhere suggest a January mean temperature of −0.75°C (presently 2.5°C), a July mean of 21.3°C (25.3°C), and precipitation of 218 mm/year (168 mm/year; McCarten and Van Devender, 1988). The MAT at glacial maximum (~18 Kya) has been estimated to have been ~6°C colder than at present. A second set of middens from the McCollough Range in the southeastern Mojave Desert of Nevada are 6800–5060 and 1250 years old (Spaulding, 1991). The present vegetation is dominated by *Larrea tridentata* and *Acacia greggii*, with *Ephedra torreyana*, *Chrysothamnus teretifolius* (rabbit brush), *Encelia* cf. *virginensis* (brittlebush), *Krameria parvifolia*, *Opuntia basilaris*, *Peucephyllum schottii* (desert spruce), and *Viguiera reticulata*. Cooperative Holocene Mapping Project (COHMAP) reconstructions (COHMAP Members, 1988) depict Middle Holocene aridity (climatic optimum) for the Pacific Northwest and the Great Basin, which may have extended into the Mojave Desert and parts of the Sonoran Desert (Hall, 1985). The 6800–5060 year old McCollough Range samples (Middle Holocene) do reflect drier conditions. The most xeric members (e.g., *Larrea*, *Peucephyllum*) are consistently more abundant than in the Late Holocene or modern assemblages, while more mesic species are less abundant (*Chrysothamnus*, *Encelia*). Studies of middens in the Mojave Desert generally indicate that in the last glaciation juniper and piñon pine woodlands were widespread (Spaulding, 1990b). The woodlands gave way to desert scrub in the latest Wisconsin and Early Holocene, with possible reversals to brief moist conditions at ~10 and 8 Kya. Early Wisconsin (pre-24 Kya) pollen sequences from Tule Springs, Nevada (Mehringer, 1965, 1967), and elsewhere show that *Artemisia* was in the lowlands and that zones of *Abies*, *Picea*, and *Pinus* were lower by 900–1400 m. The most recent version of the Mojave Desert vegetation was established by 7 Kya, although the meager records from the area do not preclude later fluctuations in vegetation. In the White Mountains along the California–Nevada border, southeast of Yosemite National Park, the hydrogen isotope composition of bristlecone pine tree rings provides a temperature chronology for the past 8 Ky (Feng and Epstein, 1994). The postglacial climatic optimum is recorded at ~6.8 Kya followed by gradual cooling then rapid cooling at ~1700 A.D. (Little Ice Age). Both events are also reflected in tree-ring analyses in the southern Sierra Nevada (Scuderi, 1993).

A pollen assemblage from the Santa Barbara Basin in southern coastal California shows a change from upland coniferous forest (*Pinus*) to lower montane and lowland forest (*Quercus* with *Artemisia*) between ~12 and 7.8 Kya, then to modern chaparral (*Adenostoma*, *Cercocarpus*, *Rhamnus*, *Rhus*, *Ceanothus*) and coastal sage scrub after 5.7 Kya (*Artemisia*, *Erigonum*, *Salvia*; Heusser, 1978). Relative frequency and pollen influx diagrams reflect the reduction in vegetation density. The inferred climate is cool (1–2°C cooler) and wet (150–250 mm more rainfall) with a 300-m lowering of tree line in the late glacial, trending to warm and dry by the climatic optimum. Treeline studies in the White Mountains of California suggest temperatures ~2°C lower at 6 Kya and general estimates of temperature for the U.S. Pacific Northwest are 6°C in the early postglacial and 4°C in the late postglacial. Middle Holocene warmth is evident in profiles from the Steens Mountains of southern Oregon (Mehringer, 1985).

Pollen and plant megafossil profiles from three high-altitude lakes in the central Sierra Nevada (~2400–3000 m), near the western edge of Yosemite National Park, record vegetation for the past 12.5 Ky (Anderson, 1990). Prior to 10 Kya trees were sparse to absent, probably restricted by poor soils following deglaciation. Between 10 and 6 Kya *Pinus flexilis* (limber pine), *P. monticola* (western white pine), and *P. murrayana* (lodge–pole pine) become established at various sites. Montane chaparral shrubs of *Arctostaphylos* sp., *Artemisia* sp., *Cercocarpus betuloides* (mountain mahogany), and *Chrysolepis sempervirens* (bush chinquapin) were abundant, suggesting the forest was open. *Abies magnifica* (red fir), *Pinus albicaulis* (whitebark pine), and *Tsuga mertensiana* (mountain hemlock) were confined to mesic habitats. The inferred climate during this interval was warm and dry. After ~6 Kya precipitation increased, which is shown by a greater abundance of the mesic *Abies magnifica*, *Pinus murrayana*, and *Tsuga mertensiana* and fewer dry shrubs and herbs. Cooling is evident by 3– 2.5 Kya with downward shift in the upper altitudinal limits of *Abies magnifica* and *Tsuga mertensiana* and in the lower limits of *Pinus albicaulis*. A similar lowering of ecotones occurred elsewhere in the Sierra Nevada region (Koehler and Anderson, 1994): the montane conifer forest was lower by 300–700 m in Kings Canyon (Cole, 1983), and *Artemisia* and *Sarcobatus* move into the present grasslands of the San Joaquin Valley (Atwater et al., 1986).

Long pollen profiles from Clear Lake near the southern terminus of the northern California Coast Ranges provide a record of climates and vegetation at an elevation of 404 m (Adam, 1988). The cores cover the interval from the Late Illinoian to the present. The modern vegetation is mostly an oak woodland (*Q. douglasii*, blue oak) with digger pine (*Pinus sabiniana*). Other smaller and/or more distant formations include mixed hardwood forest, chaparral, and Coast Range montane coniferous forest. Despite the lengthy time span represented by the core, the vegetation and climatic changes throughout are reflected by just two components. There are numerous rapid shifts between associations dominated by oak (warm and dry) and those dominated by pine and TCT-type pollen (cool and moist). An interesting feature is that the patterns correlate well with those for the Eemian and Würm (Weichsel–Devensian) of Europe. The implication is that if cores are long, continuous, well-dated, and sufficiently near the ocean, they may preserve responses to Milankovitch (long term) and thermohaline–ocean circulation (shorter term) forcing.

Additional evidence is accumulating that the abrupt climatic changes that characterized the North Atlantic also affected the North Pacific region. A 196-m core from the Santa Barbara Basin records rapid shifts in climate during the past 20 Ky that parallel those from the Greenland ice cores and adjacent marine sediments (Kennett and Ingram, 1995). Another core from the subarctic Pacific Ocean extends the pattern back to 95 Kya (Kotilainen and Shackleton, 1995), as do results from the eastern North Pacific Ocean (Thunell and Mortyn, 1995). Hydrological cycles of the Owens Basin in the Great Basin of northwestern California during the last glacial termination parallel climatic changes in the North Atlantic (Benson et al., 1997). Correlation between cosmogenic chlorine-36 dates on moranes in the Sierra Nevada and radiocarbon chronologies from Owens Lake reveal that glacial advances correspond to Heinrich events 5, 3, 2, and 1 (Phillips et al., 1996). Diatom sequences and fluctuations in percentages of *Juniperus* pollen reflect cooler and more moist climates during glacial intervals by displacement of storm tracts from the Alutian low pressure system (Bradbury, 1997a,b; Smith et al., 1997). In sediments from the Olympic Peninsula of Washington, and continuing north to the Alaska Panhandle, there are peaks of *Tsuga mertensiana* pollen, reversals from forested to nonforest vegetation, and other evidence for a cold interval between 11 and 10 Kya suggesting a Younger Dryas climatic reversal (Mathewes, 1993). These new data provide a context for interpreting terrestrial environmental history of the Pacific Northwest comparable to that provided by the Greenland data for the North Atlantic. They are also of theoretical importance because they show that climate forcing mechanisms in the Quaternary may have been circumpolar in extent, which would require that developing models account for such correlated hemispheric events. One implication is that freshwater influx into the North Pacific Ocean from the Cordilleran ice sheet may have interfered with some degree of thermohaline circulation.

Along the coast an indication of Late Pleistocene environments is provided by the remarkable fossil fauna at Rancho La Brea (the La Brea tar pits; Stock, 1992). About 600 different kinds of organisms are known and include

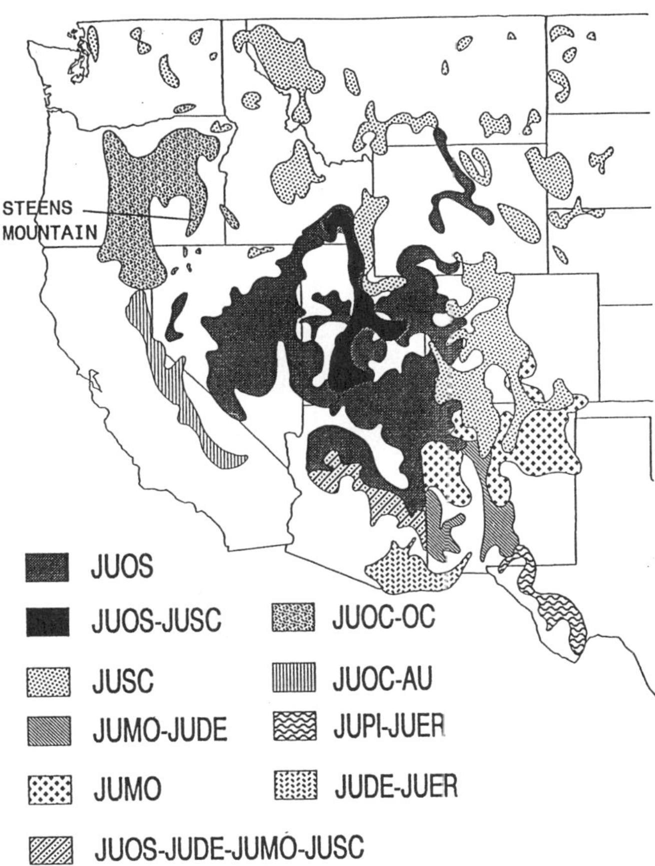

Figure 8.17. General distribution of common species of *Juniperus* in the western United States. JUOS, *J. osteosperma* (Utah juniper); JUSC, *J. scopulorum* (Rocky Mountain juniper); JUMO, *J. monosperma* (one-seeded juniper); JUDE, *J. deppeana* (alligator juniper); OC, *J. occidentalis* (western juniper var. occidentalis); JUOC-AU, *J. occidentalis* (western juniper var. australis); JUPI, *J. pinchotii* (Pinchot's juniper); JUER, *J. erythrocarpa* (redberry juniper). Reprinted from Miller and Wigand (1994, fig. 2, based on Critchfield and Little, 1966) with the permission of Richard Miller, Peter Wigand, and the American Institute of Biological Sciences.

bats, imperial mammoth, mastodon, saber-toothed cat (abundant), puma, lynx, dire wolf (abundant), coyote (abundant), bear (short-faced, black, grizzly), weasel, badger, skunk, rodents (gopher, mice, kangaroo rat, voles, ground squirrel, rat, chipmunk, jackrabbit, brush rabbit, cottontail), antelope, pronghorns, peccary, bison, camel, tapir, horse, deer, ground sloth, and giant ground sloth. The plants have not been studied but include a chaparral association of *Adenostoma* (chamise), *Arctostaphylos* (manzanita), *Ceanothus* (buckbrush, California lilac), *Juglans*, *Quercus*, *Sambucus* (elderberry), and others.

In the Great Basin region, Great Salt Lake in Utah has a surface area of 6200 km^2. Its predecessor glacial Lake Bonneville at 18 Kya had a surface area of 47,800 km^2 (Hostetler et al., 1994) to 51,300 km^2 (Benson and Thompson, 1987), a volume of 9500 km^3, and was ~300 m deeper than at present. Late Pleistocene precipitation is estimated at 1.6–1.9 m/year (present 0.4–1.5 m/year) or 33% greater than at present (Lemons et al., 1996). Later glacial Lake Lahontan in western Nevada reached a size of 14,700 km^2 (remnants presently total 2500 km^2; Benson and Thompson, 1987) during the 1400-year period between Heinrich event 1 at 14 Kya and the resumption of thermohaline circulation at 12.7 Kya. Zones of permafrost in the uplands of Idaho, Montana, Wyoming, and Colorado were lowered by 1000 m and MAT is estimated at 9–11°C below that of the present (Péwé, 1983). In the southern Great Basin MAT is estimated to be 6°C cooler and precipitation 30–40% greater (Spaulding, 1990b). The Middle Holocene period of aridity between ~7.5 and 4.5 Kya, as proposed early by Antevs (1938), has generally been confirmed, although it may have been expressed at slightly different times in various parts of the west.

The montane coniferous forest shifted downward by ~600 m and its range expanded during the cool moist intervals of the glacials and Early Holocene (*Picea*, *Pinus flexilis*; *Abies* locally). The lower limit of the forest was at ~1000 m, compared to the present 1500–2000 m limit. Below the montane coniferous forest, the mountain slopes between ~1500 and 2250 m are presently occupied by piñon–juniper woodland (Fig. 8.17). Farther down on the lower slopes and basin floor there is a woodland (steppe) of *Artemisia*. It is obvious from Fig. 8.17 that the regional vegetation is a complex mosaic of communities and that even slight environmental change would have produced considerable response from biotas arranged along the steep altitudinal gradients. Changes did occur throughout the Quaternary and they are particularly evident in the turbulent transitions between glacial and interglacial stages. During cool moist glacial intervals the drier habitat species of *Juniperus* were 500–600 km further south and up to 1500 m lower in elevation (Miller and Wigand, 1994). Mesic species (e.g., *J. osteosperma*, Utah juniper) extended into the Mojave and Sonoran deserts (Thompson, 1990). In the

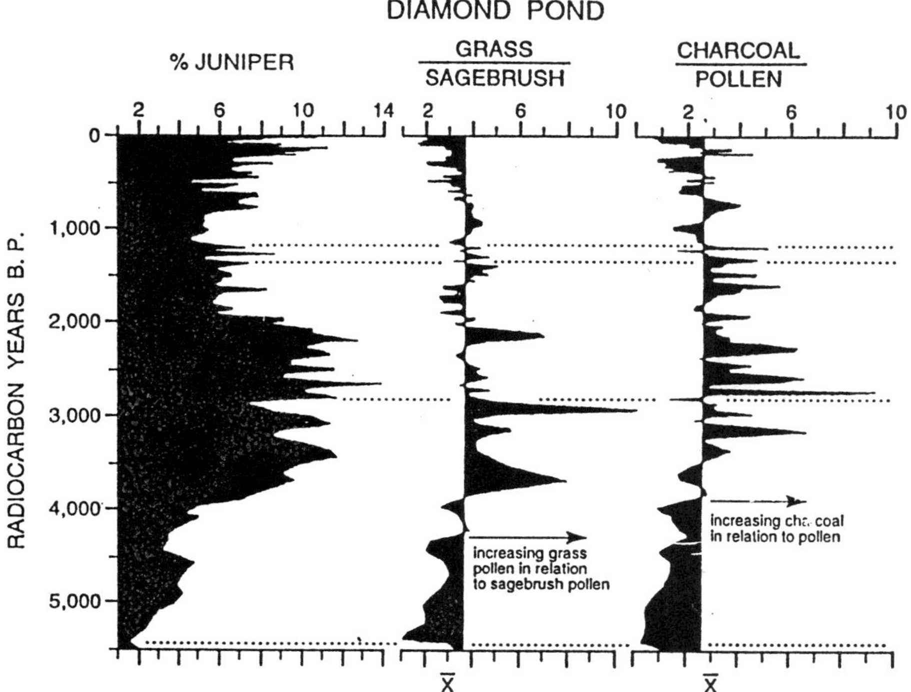

Figure 8.18. Diagram of relative abundance of *Juniperus* pollen, and ratios of Gramineae to *Artemisia* and charcoal to pollen, Diamond Pond, southeastern Oregon. Reprinted from Mehringer and Wigand (1990) with the permission of the University of Arizona Press.

Middle Holocene climatic optimum (~7–5.4 Kya) there was widespread shrubland (steppe) of *Artemisia*, *Atriplex confertifolia* (shadscale), and *Sarcobatus* (greasewood) in the lowlands, followed after ~4.5 Kya by slight expansion of the coniferous forest at high altitudes and more mesic species of juniper (downslope ~150 m; Mehringer, 1985; Figs. 8.18–8.20). Similar fluctuations probably occurred in earlier glacial times (Coplen et al., 1994). Throughout the past 1.6 m.y., however, the seasonally cold dry shrubland–steppe has been the dominant vegetation at low to midelevations of the Great Basin; at times it was even more extensive than at present.

In Grand Teton National Park, Wyoming, the vegetation near the end of the Pinedale Glaciation at ~14–11 Kya was alpine meadow with *Betula* and *Juniperus* (Whitlock, 1993). Treeline was lower by ~600 m and temperatures were an estimated 5–6°C cooler. Temperatures and winter precipitation increased between 11.5 and 10.5 Kya, allowing *Picea*, *Abies*, then *Pinus* cf. *albicaulis* to invade. By ~10.5 Kya modern subalpine forest was established. At Yellowstone National Park, Wyoming, highland sites prior to 10 Kya were treeless (Whitlock, 1993); at slightly lower altitudes pollen from a pond core included *Juniperus*, *Picea*, Cyperaceae, Gramineae, *Artemisia*, *Betula*, and *Salix*, indicating a treeline lower by ~500 m (Waddington and Wright, 1974).

Glacial history is known in considerable detail for Glacier National Park in Montana (Carrara, 1989); along with dated ash sequences, a high-resolution context is available for reconstructing the vegetational history (Barnosky et al., 1987a; Mack et al., 1983). Highland areas were mostly deglaciated by 11.2–11.4 Kya and long before that in the adjacent plains, as shown by fossils below the Glacier Peak (11.2 Kya) and early Mount St. Helens (11.5 Kya) ash layers. Pollen, wood, needles, and insects below the Mount St. Helens ash show that treeline was ~500–700 m lower than at present. At ~11.2 Kya in the Kootenai River Valley, 150 km west of Glacier National Park, basal sediments from a lake at 1270-m elevation contain pollen of pine, spruce, fir, alder, and willow, suggesting a cool moist climate (Mack et al., 1983). At Sheep Mountain bog 150 km south of the park, douglas fir has occupied the site since at least 10 Kya (Mehringer, 1985). Reforestation in the highlands at Glacier National Park began ~10 Kya as an open conifer forest of *Pinus contorta* (lodgepole pine) with spruce (pollen) and spruce or larch (wood). This type of vegetation is characteristic of the present alpine tundra–subalpine forest boundary at treeline. By 10 Kya the glaciers were occupying the cirques and depressions of the present; similar chronologies are known from the North Cascades, Banff, Jasper, and Yoho National Parks (pine–fir forests at 10 Kya). At ~9.5 Kya the treeline was near its present position in the Front Range of Colorado (Benedict, 1973) and ~80 m higher in the San Juan Mountains of southwest Colorado (Carrara et al., 1984). At lower elevations the sequence has been from mostly pine-dominated (warm) to *Artemisia*-dominated (colder) steppe tundralike vegetation. Longer records from Yellowstone National Park document that similar shifts in treeline, involving essentially modern regional communities, probably occurred throughout the Quaternary (Fig. 8.21).

After the late-glacial to Early Holocene cool period, cli-

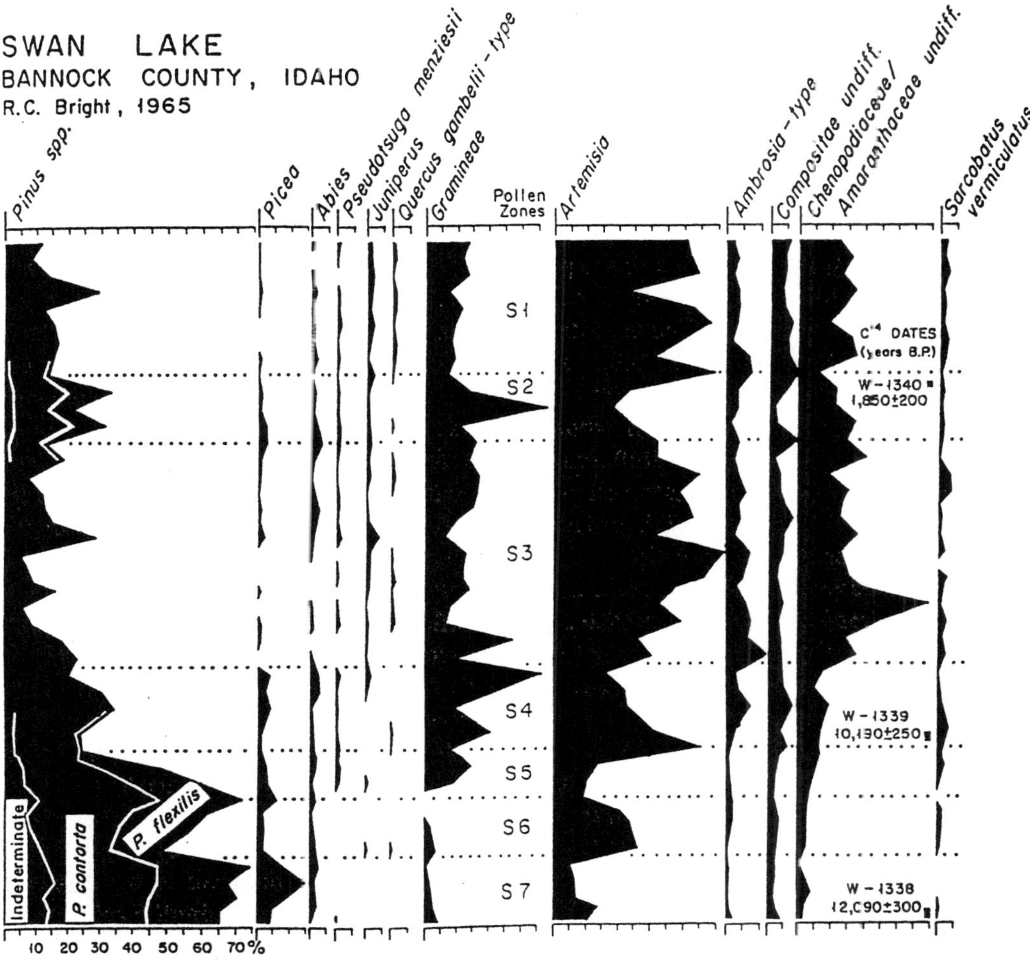

Figure 8.19. Pollen diagram from Swan Lake, Bannock County, Idaho. Reprinted from Mehringer (1985, fig. 5, and references cited therein) with the permission of the American Association of Stratigraphic Palynologists Foundation.

mates in the northern Rocky Mountains warmed to the Middle Holocene climatic optimum. MAT at 5900-m elevation in Jasper National Park was ~1.9°C warmer than at present (Osborn, 1982). In the mid-1700s there was a brief period of glacial advance correlated with the Little Ice Age, and records of fir pollen indicate that treeline reached its lowest altitudinal limit in the last 500 years (Kearney and Luckman, 1983).

Northwest

The maximum extension of continental ice in the Pacific Northwest was near Olympia, Washington, in the Puget Lowland, and in the northern Columbia Basin at ~14.5 Kya. Early pollen studies in this region were made by Hansen (1939 et seq., see summary, 1947), Heusser (1952, 1977 et seq.), and Mathewes (1973, 1979), and more recently by Barrie et al. (1994), Mann and Hamilton (1995), Mathewes (1985, 1991), Mathewes and Clague (1994), McLachlan and Brubaker (1995), Mock and Bartlein (1995), Thompson et al. (1993), Whitlock (1992, 1993), and Whitlock and Bartlein (1997). The Quaternary history of coastal communities from southern Oregon to British Columbia during glacial times is generally one of lowered forest ecotones, cold dry steppe in inland areas, and subalpine-type forests along the coast. The downward shifts did not involve entire vegetation zones, only differential movement of species according to their individual ecological requirements. Thus, mixtures of montane and lowland elements developed that do not occur today. This is consistent with the individual responses of various taxa to Quaternary climatic change described for the eastern deciduous forest and the arid southwest (e.g., sagebrush–pine). Further north in Alaska, the Wisconsin and earlier glacial-stage vegetation was mostly tundra of modern aspect.

When this paleoecological information is used as validation data for CCM simulations (Kutzbach, 1987), the model results are generally supported. At glacial maxima the splitting of the jet stream by the Cordilleran and Laurentide ice sheets brought the southern arm across latitude ~45° N (Oregon–Washington border). An anticyclone system developed between this latitude and the glacial margin

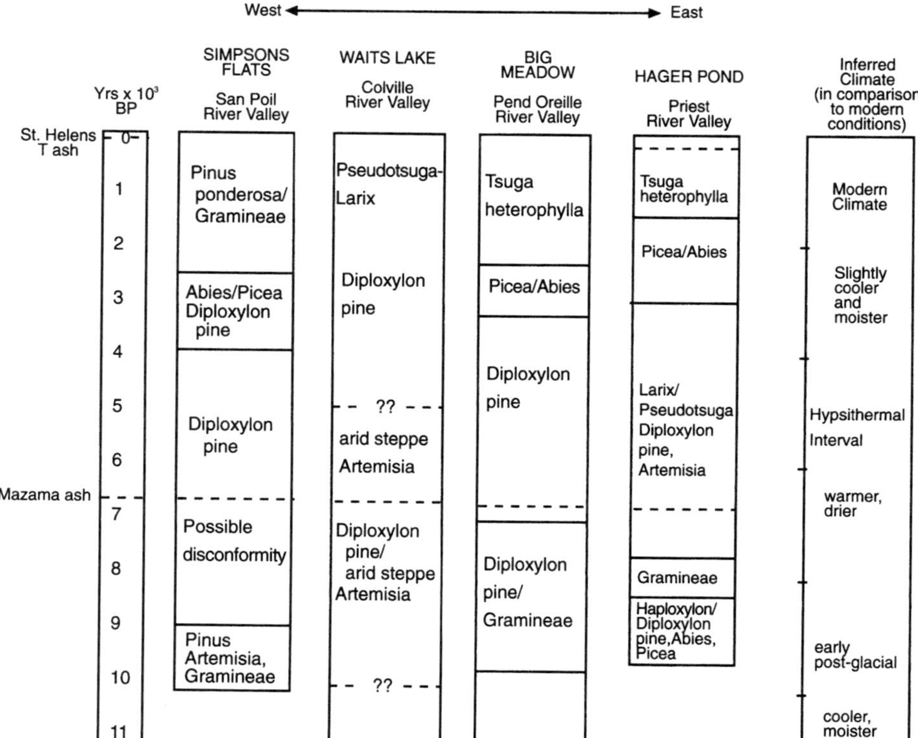

Figure 8.20. Summary of vegetation and inferred climates for northeastern Washington and northern Idaho since the Wisconsin glaciation. Reprinted from Mack et al. (1978, fig. 3) with the permission of The University of Chicago Press.

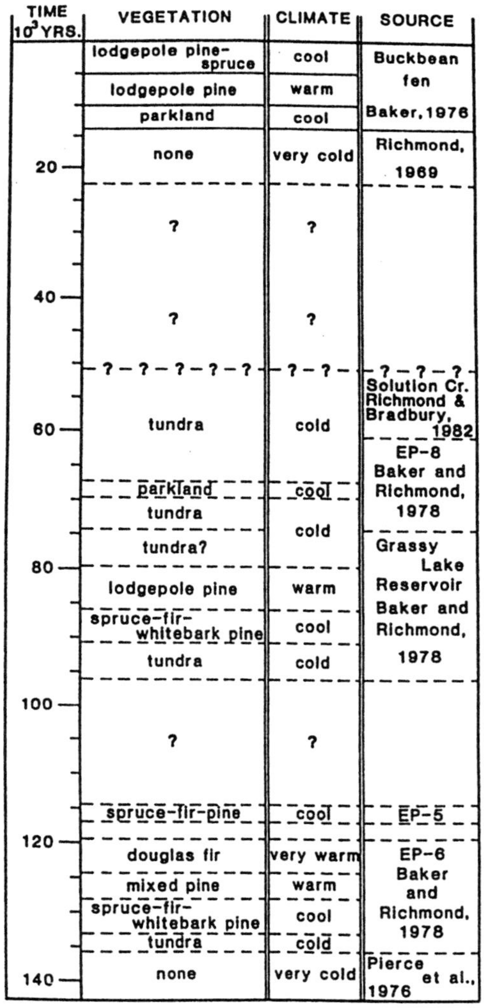

TIME 10^3 YRS.	VEGETATION	CLIMATE	SOURCE
	lodgepole pine-spruce	cool	Buckbean fen
	lodgepole pine	warm	
	parkland	cool	Baker, 1976
20	none	very cold	Richmond, 1969
	?	?	
40	?	?	
	? – ? – ? – ? – ?	? – ? – ?	? – ? – ?
60	tundra	cold	Solution Cr. Richmond & Bradbury, 1982
	parkland	cool	EP-8 Baker and Richmond, 1978
	tundra	cold	
	tundra?		Grassy Lake Reservoir Baker and Richmond, 1978
80	lodgepole pine	warm	
	spruce-fir-whitebark pine	cool	
	tundra	cold	
100	?	?	
	spruce-fir-pine	cool	EP-5
120	douglas fir	very warm	EP-6 Baker and Richmond, 1978
	mixed pine	warm	
	spruce-fir-whitebark pine	cool	
	tundra	cold	
140	none	very cold	Pierce et al., 1976

Figure 8.21. Estimated vegetational changes for the past 135 Kya, Yellowstone National Park. Reprinted from Baker and Waln (1985, fig. 3, and references cited therein) with the permission of the American Association of Stratigraphic Palynologists Foundation.

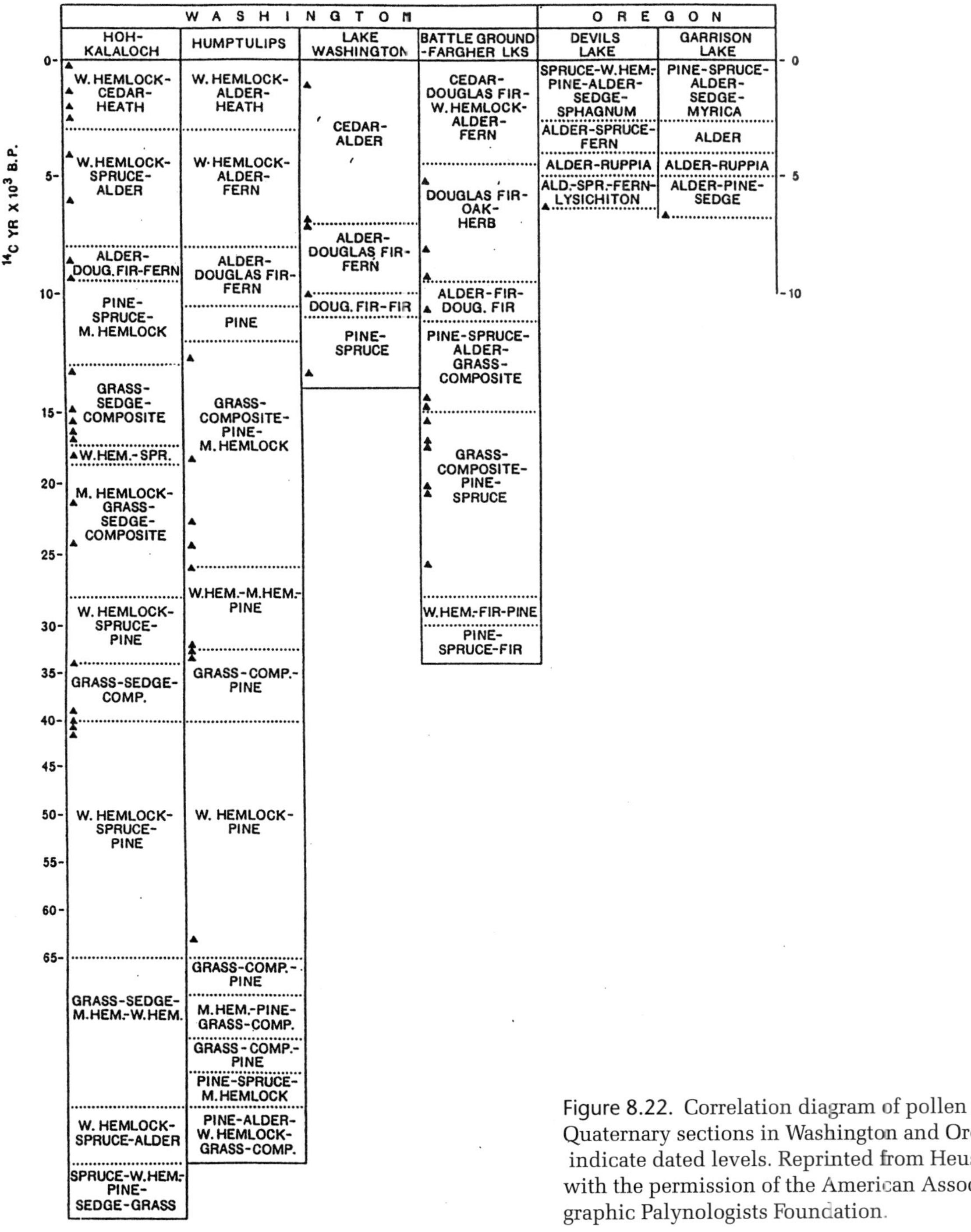

Figure 8.22. Correlation diagram of pollen zones from Quaternary sections in Washington and Oregon. Triangles indicate dated levels. Reprinted from Heusser (1985, fig. 11) with the permission of the American Association of Stratigraphic Palynologists Foundation.

to the north, resulting in increased easterly winds off the Laurentide and Cordilleran ice sheets, cooler temperatures, and drier conditions east of the Cascade Mountains. The simulation is paralleled in the fossil record by steppe and tundra characteristic of cold dry climates in rainshadow areas. By 12 Kya July temperatures were warming as indicated by model simulations (Kutzbach, 1987) and validation data (Barnosky et al., 1987b). This low precipitation signal is in contrast to pluvial conditions in the southwest during glacial maxima.

Only maritime areas to the west of the Coast Ranges remained warm and wet. During interglacial times, the Pacific coast coniferous forest (*Pinus contorta*, lodgepole pine) and boreal forest (*Picea sitchensis*, sitka spruce) migrated from refugia (e.g., western Olympic Peninsula, Queen Charlotte Islands, interior Alaska) and established a distribution and composition similar to that of the present (Heusser, 1985); Fig. 8.22). Pollen diagrams from Hoh-Kalaloch, Washington, show tundra or park tundra at glacial maxima and elsewhere the later development of spruce parkland, pine woodland, and forest communities after 13 Kya. At 6.3 Kya the vegetation at Devils Lake in coastal Oregon included *Picea sitchensis*, *Alnus*, *Lysichiton* (Araceae), and ferns, with later introduction of *Tsuga heterophylla* (western hemlock), *Thuja plicata* (western red cedar), *Pseudotsuga menziesii* (douglas fir), *Pinus*, and *Alnus*. At Carp Lake in the Columbia Basin of south-central Washington the full-glacial vegetation was a cool, dry, nonarboreal steppe of *Artemisia* and grasses with alpine elements in the highlands (Barnosky, 1985; Whitlock and Bartlein, 1997). MAT is estimated at ~10°C (presently ~15°C) and precipitation at 1300 mm along coastal Oregon during the Wisconsin glacial maximum (presently ~2600 mm). At 8 Kya temperatures increased to ~13°C and precipitation to ~2400 mm, representing the period of maximum warmth in the region. After that time the climate became slightly warmer and wetter.

At glacial maximum ~14,400 years B.P. in coastal British Columbia, this cold interval corresponds to Heinrich event H1 as represented by coarse sediments in an otherwise fine sediment sequence from the continental slope (Blaise et al., 1990; Mathewes, 1996). As noted in the discussions of Clear Lake and glacial chronologies in the Sierra Nevada, detailed analyses of pollen cores near coastal and other areas in the west are revealing fluctuations in closer accord with orbital forcing, ocean circulation, and disruption of some level of thermohaline circulation. Although such correlations are not yet numerous and exact, there is growing evidence to suggest that Quaternary climatic histories in the North Pacific and the North Atlantic were more similar than previously thought (see Mathewes, 1993; Mikolajewicz et al., 1997). It should be noted, however, that lower salt content of northern Pacific Ocean surface waters prevents the same degree of sinking as in the northern Atlantic Ocean. After glacial retreat, the forest sequence in the lowlands of British Columbia during the Early Holocene was *Pinus contorta* (cool, probably dry); *Abies*, *Picea*, *Pinus*, and *Tsuga* (cool and moist); and modern vegetation at 4–2 Kya. At the same time the highlands of the southern interior supported *Artemisia* and grasses. Other fluctuations in various parts of British Columbia are detailed by Hebda (1995).

The region of eastern Washington (Johnson et al., 1994) experienced catastrophic flooding between ~15.3 and 12 Kya that formed the scablands (Chapter 2, Missoula Floods). At nearly the same time (11,250 years B.P.) two falls of the Glacier Peak ash, about 25 years apart, blanketed parts of the Pacific Northwest to a depth of ~6 m (Mehringer et al., 1977). The ash is an important marker for dating events in the region, and a layer is present in the flood sediments. Undoubtedly both the floods and ash falls affected the vegetation, but their impact is difficult to identify from pollen and spore diagrams presently available. In the case of the ash fall, the recovery time from colonizing vegetation to woodland (steppe) and forest was brief and is not evident on pollen diagrams from adjacent regions when standard sampling intervals are utilized. Scouring by the floods removed all sediments, so records in the immediate vicinity begin after the event. Profiles from adjacent regions might reflect some change in the vegetation due to catastrophy, but these would likely be subtle and blurred by local pollen and spore input. Fine-resolution sampling and appropriate statistical tests (e.g., analysis of variance) may eventually allow the effects of the ash fall and flooding to be identified and isolated from those caused by fluctuating climates at the Pleistocene–Holocene transition.

In interior and western Canada a rich array of localities are available for interpreting Quaternary vegetational history and paleoenvironments (e.g., see Ritchie, 1984, 1985, table 1 and fig. 2). A site along the Roaring River in Manitoba preserves a pollen record for the Middle Pleistocene. The sequence is boreal forest (*Picea*; cool climate), grasslands (Gramineae, *Artemisia*, chenoams), and oak savanna (*Quercus*; warm), and a return back to boreal forest (Klassen et al., 1967). A site on Baffin Island (Miller et al., 1977; Terasmae et al., 1966) is tentatively assigned to Isotope Stage 5e (Sangamon interglacial) and the vegetation is a low Arctic shrub tundra (*Alnus*, *Betula*, Ericaceae, *Salix*). Megafossils indicate that the birches were of the dwarf-shrub type (*B. glandulosa*, *B. nana*). Mean July temperature is estimated at ~3° C warmer than at present. In east-central Alberta, Canada, the sequence is birch–spruce-dominated boreal forest (~11.4 Kya), giving way to open parkland–grassland by ~10 Kya in response to Holocene warmth (Hickman and Schweger, 1996).

Late glacial and Holocene records from British Columbia, Alberta, and the Yukon Territory were summarized by Ritchie (1985) and show a general sequence from late glacial to Early Holocene of tundra giving rise to boreal forest (*Picea*) and a Middle Holocene period of warmth characterized by increased *Artemisia* (southern Canada), grasses (central), or mixed coniferous forest (northern).

Vegetational history in the western Arctic region is reviewed by Lamb and Edwards (1988). In southeastern Alaska winds off the Gulf of Alaska bring moisture inland, and the area was extensively glaciated during the Quaternary. Hence, most pollen records begin in the Late Pleistocene and Holocene. In the Aleutian Islands tundra vegetation is evident throughout sections that date from the last glacial retreat at 12–10 Kya (Heusser, 1985, 1990; Fig. 8.23). Inland, an early sedge tundra (grasses, willow, sage; 25–14 Kya) is supplemented by dwarf birch (14–12 Kya), followed by an *Alnus* shrub tundra with aspen and balsam poplar between ~10 and 7 Kya. Spruce first became established at 9.5 Kya, western hemlock came in at ~3.5 Kya, and *Tsuga mertensiana* (mountain hemlock) arrived at 3 Kya. There is evidence that the cold Younger Dryas interval (11–10 Kya), well documented in the North Atlantic and western Europe, may be evident in the Alaskan sequences. Between 10.8 and 9.8 Kya forest parkland was replaced by shrub tundra followed by spruce and hemlock. If the tundra developed in response to Younger Dryas cooling, it

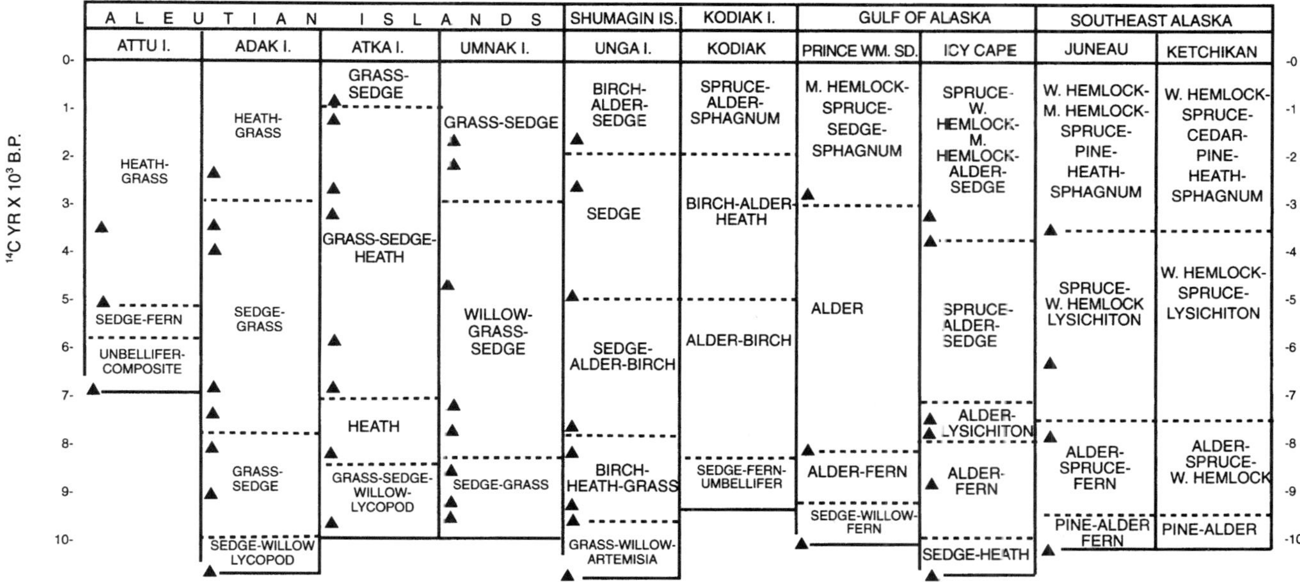

Figure 8.23. Correlation diagram of pollen assemblage zones of Quaternary sections in Alaska. Reprinted from Heusser (1985, fig. 9) with the permission of the American Association of Stratigraphic Palynologists Foundation.

would mean that this event, previously ascribed to changes in North Atlantic ocean circulation, may have been more global in scope (Engstrom et al., 1990).

Winds and moisture were prevented from penetrating into the interior by the Alaska Range, and much of this region remained ice free during the Quaternary. Thus, it served as an important refugium for biotas that spread outward during the warm interglacials onto newly exposed surfaces (Ager and Brubaker, 1985; Edwards and Barker, 1994, northeastern Alaska). Examples of the interactions between herb tundra, shrub tundra, and boreal coniferous forest are shown in the composite diagram for Tanana Valley Lakes near Fairbanks (Fig. 8.24), and the Late Wisconsin and Holocene history for south-central and interior Alaska is summarized in Fig. 8.25. In the central Mackenzie Mountains of Canada an *Artemisia* and *Salix* herb tundra was present from at least 12–10.2 Kya, followed by a shrub tundra of *Betula glandulosa* and *Populus balsamifera*. *Picea* arrived by 8.5 Kya and, with the expansion of *Alnus*, the present vegetation was essentially established by 6 Kya.

The presence of *Artemisia* pollen opens the possibility that regions within the Arctic Circle may have supported sagebrush steppe during glacial maxima (most recently between 22 and 14 Kya). This interpretation is consistent with abundant bones of mammoths, horses, bison, and other ungulates in Beringian sediments—the mammoth steppe. Pollen analysis suggests mostly tundra, and the relative prominence of the two ecosystems is unsettled. New ocean cores from submerged portions of Beringia in the Bering and Chukchi Seas have yielded meager amounts of *Artemisia* pollen, and the profiles are interpreted as reflecting tundra (Elias et al., 1996). The attractiveness of a steppe is that it is better suited to support large herds of ungulates and, in turn, human hunters following their migrations into North America. As noted by Colinvaux (1996), a spirited response from the big-mammal community doubtless will be forthcoming.

Early pollen studies on Alaskan Quaternary environments were made by Livingstone (1955, 1957) and Colinvaux (1963, 1964a,b). This work is important because it documented that the Arctic was cold during times of glaciation, in contrast to a theory proposed by Ewing and Donn (1956) that the region must have been warm to provide the moisture necessary for the growth of glaciers. Ice-wedge evidence indicates that the MAT was −6°C during the Wisconsin and similar conditions probably prevailed during previous glacial intervals. As noted in Chapter 2 (Orogeny), model simulations show warm winters in unglaciated interior Alaska while the ground evidence shows that the vegetation was tundra and the climate cold and dry. Approximately all of northwestern Canada and 82% of Alaska is now underlain by permafrost (Péwé, 1983), and on the North Slope in the vicinity of Barrow and Prudhoe Bay it is presently 400 m thick (Walker, 1973).

VEGETATION SUMMARY

In eastern North America the deciduous forest at the formation level was modernized in the Pliocene with the disappearance of present-day Asian genera. Beginning ~2.4 Ma Tertiary predecessors of the Appalachian montane coniferous forest coalesced and spread from high-altitude sites. *Picea* was present in Jackson (Late Eocene) time, *Tsuga* is known from the Middle Miocene, and *Abies* appears in the Pliocene in the southern Appalachian region.

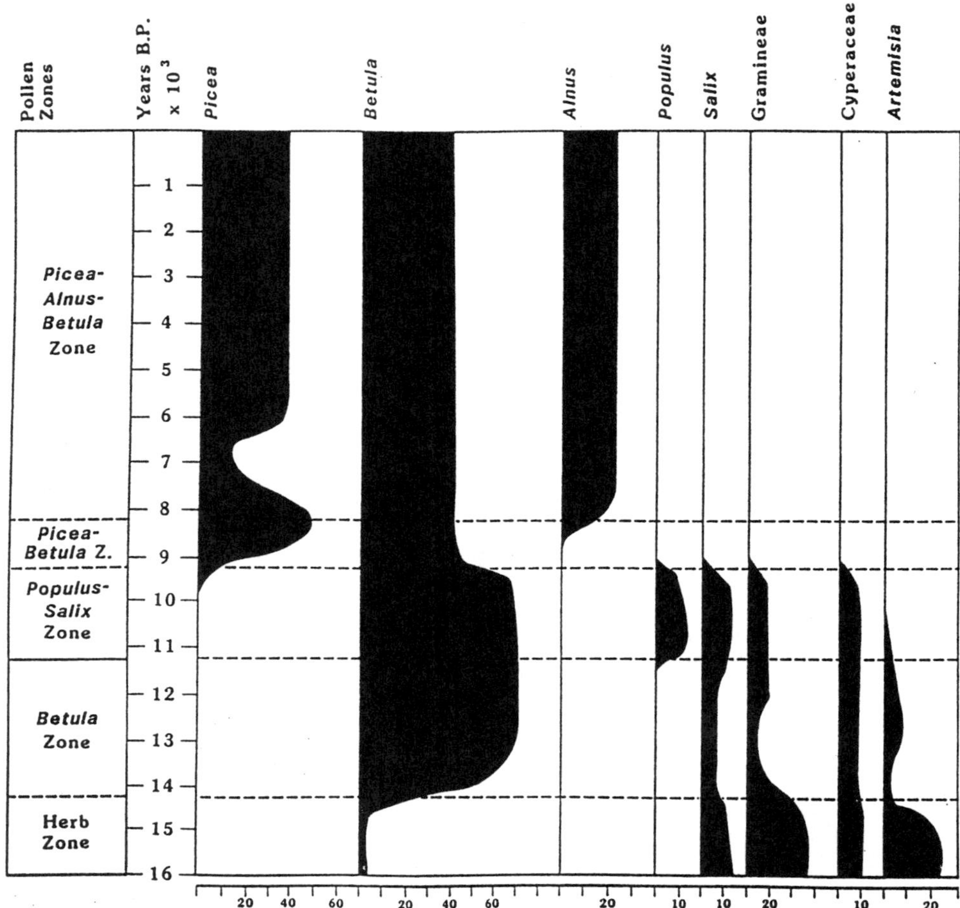

Figure 8.24. Composite pollen diagram from several sites in the Tanana Valley, Alaska. Reprinted from Ager and Brubaker (1985, fig. 8) with the permission of the American Association of Stratigraphic Palynologists Foundation.

This association was prominent during the subsequent cool to cold glacial intervals when present-day boreal elements were also periodically introduced and eliminated from the eastern forest (e.g., *Pinus banksiana*). A form of alpine tundra grew at the highest elevations in the central and northern Appalachian Mountains and perhaps at more restricted sites to the south. The cold climates and Appalachian coniferous forest are now recognized as typical for ~90% of Quaternary time. The deciduous forest mostly occupied refugia such as those identified from southwestern Tennessee, south-central Alabama, and northern Florida. During the brief warm intervals it spread from these refugia to form the extensive, and exceptional, modern formation while the coniferous forest moved into the highlands. This scenario was probably repeated in each of the 18–20 cold–warm cycles of the Quaternary and reveals the temporal and dynamic nature of the Appalachian montane coniferous forest and the deciduous forest.

Much of Florida and the southern United States coastal plain was inundated during interglacial times, and the flood-plain forest association expanded. With cooling climates, glaciers advanced, sea levels lowered, and the coastal plains were vegetated by forests capable of growing under cold conditions and in the physiologically arid deep sand. In Florida the plant cover alternated between open pine–oak forest with understory grasses in the warm intervals and pine forest during colder times. These changes correlate with Heinrich events and suggest that alterations in ocean circulation were a primary forcing mechanism. To the north in the Carolina Bays region it was *Picea* rather than extensive deciduous elements that grew with *Pinus* along the coastal plain during the glacial intervals. The latest version of the pine woods association begins at ~5 Kya. In the far southwestern portion of the deciduous forest region (to central Texas), deciduous elements with some spruce moved into mesic sites from the east and north during cool periods; drier areas supported oak–hickory savannas. The latter expanded in the warmer interglacial climates, and the pattern repeated throughout the Quaternary. In the northwestern part of the deciduous forest the interplay was between forest, grassland, and spruce (with ash) savanna.

To the northeast, a four-part chronology describes the vegetation from the last glaciation to the present; by infer-

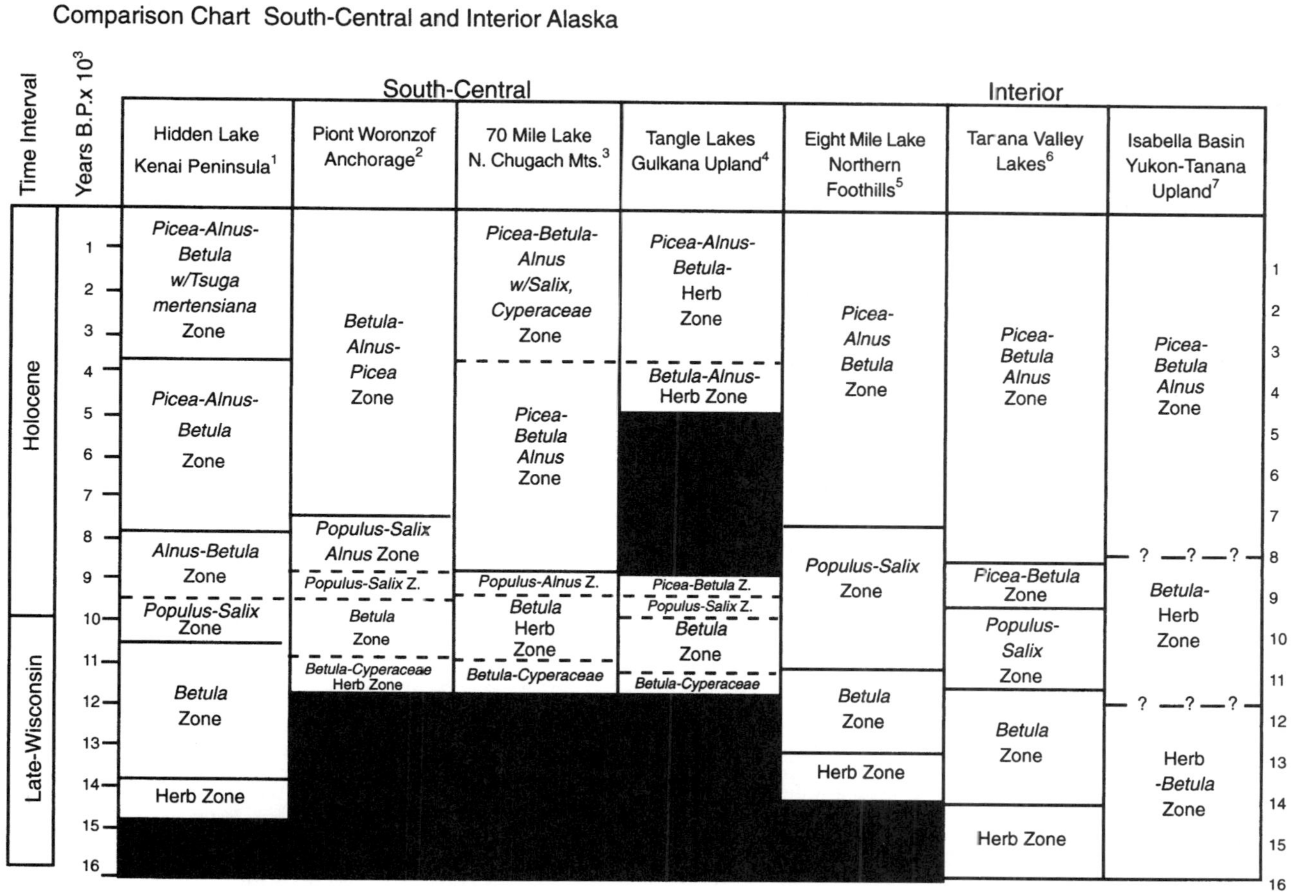

Comparison Chart South-Central and Interior Alaska

Region	Site	Pollen zones (youngest to oldest)
South-Central	Hidden Lake Kenai Peninsula[1]	*Picea-Alnus-Betula w/Tsuga mertensiana* Zone; *Picea-Alnus-Betula* Zone; *Alnus-Betula* Zone; *Populus-Salix* Zone; *Betula* Zone; Herb Zone
South-Central	Piont Woronzof Anchorage[2]	*Betula-Alnus-Picea* Zone; *Populus-Salix Alnus* Zone; *Populus-Salix* Z.; *Betula* Zone; *Betula-Cyperaceae* Herb Zone
South-Central	70 Mile Lake N. Chugach Mts.[3]	*Picea-Betula-Alnus w/Salix, Cyperaceae* Zone; *Picea-Betula Alnus* Zone; *Populus-Alnus* Z.; *Betula* Herb Zone; *Betula-Cyperaceae*
South-Central	Tangle Lakes Gulkana Upland[4]	*Picea-Alnus-Betula*-Herb Zone; *Betula-Alnus*-Herb Zone; *Picea-Betula* Z.; *Populus-Salix* Z.; *Betula* Zone; *Betula-Cyperaceae*
Interior	Eight Mile Lake Northern Foothills[5]	*Picea-Alnus Betula* Zone; *Populus-Salix* Zone; *Betula* Zone; Herb Zone
Interior	Tarana Valley Lakes[6]	*Picea-Betula Alnus* Zone; *Picea-Betula* Zone; *Populus-Salix* Zone; *Betula* Zone; Herb Zone
Interior	Isabella Basin Yukon-Tanana Upland[7]	*Picea-Betula Alnus* Zone; ? —? —?; *Betula*-Herb Zone; ? —? —?; Herb -*Betula* Zone

Time Interval: Holocene; Late-Wisconsin. Years B.P. x 10^3: 1–16.

Figure 8.25. Comparison chart of Late Quaternary pollen zones in south-central and interior Alaska. Dashed lines represent uncertainty about boundary positions. Reprinted from Ager and Brubaker (1985, fig. 5) with the permission of the American Association of Stratigraphic Palynologists Foundation.

ence a similar pattern was likely followed during each of the 18–20 glacial–interglacial transitions of the past 1.6 m.y. The sequence was a herb zone bordering the retreating glacial margin (tundra to park tundra); spruce, fir, and other boreal species; pine (climatic optimum); and dedicuous forest (oak with hemlock, birch, and beech in the northern region).

The boreal coniferous forest had its primary center of origin in the highlands of the northern Rocky Mountains in the Middle Eocene (*Abies, Chamaecyparis, Picea, Pinus, Pseudolarix, Thuja, Tsuga, Betula*). Beginning at ~3.4 Ma it began expanding extensively into the lowlands and subsequently alternated between numerous regional eradications by glacial advances and reintroductions during interglacial times. The current version dates from the retreat of the last glaciers at ~6 Kya, and some areas have supported boreal coniferous forest for only 3 Ky.

In the Plains area, trees diminished to the point where the vegetation could be designated a grassland near the onset of extensive Arctic glaciation at ~2.4 Ma. After that time the southern region broadly supported oak–hickory savanna alternating with increased deciduous species mixed with some spruce. Southern coastal pines also extended inland with cooler climates. Along the western margin pine–juniper and *Artemisia* periodically extended onto the grassland. In the north, the ecotone was with the boreal forest (especially *Picea*); along the eastern margin it was with the deciduous forest.

In one respect vegetational history in the west and southwestern United States parallels the observations of Davis (1976) for the deciduous forest of the east: the present-day widespread warm deserts (Chihuahuan, Sonoran, Mojave) and desert grasslands are the exceptions for the past 1.6 m.y. For most of this time the region was characterized by low-pressure systems, higher rainfall, moderate temperatures, numerous lakes, lower forest ecotones, and savanna–woodlands (e.g., juniper–piñon pine; evergreen oak–laurel–madrone; *Artemisia* steppe) in the presently desert lowlands. Desert elements, in turn, were mostly restricted to edaphically or topographically favorable habitats. California chaparral and coastal sage communities developed from Pliocene predecessors that grew as sclerophyllous shrubs in the understory of oak–laurel–madrone woodlands (Axelrod, 1978, 1989). Some of these were derived from plants of the Sonoran Desert of northern Mexico that grew in marginal areas of greater summer rainfall. Others came from temperate forests to the north: *Aesculus palmeri*, *Baccharis sergiloides*, *Fraxinus trifoliata*, *Malosma laurina*, *Rhus integrifolia*, *Sanicula deserticola*, and *Satureja chandleri*. Their coalescence into a distinct recognizable community was favored by the Mediterranean (summer dry) climates and fires that increased with the first Aftonian interglacial period of aridity and by subsequent hypsithermal intervals, including the last one between 8 and 4 Kya. Some communities had different combinations of ecologically compatible taxa after each reshuffling, and this fluidity has probably characterized western North American vegetation at the association level throughout its history.

At about 3.4 Ma the Arctic tundra included an estimated 1500 species from alpine and cool-temperate predecessors that had gradually become adapted to the light, temperature, and precipitation regimes of the far north. With the rigors of subsequent climatic changes the number has been reduced to less than 1100 species, of which about 700 occur in the North American Arctic (Löve and Löve, 1974).

Little is known about the Early Quaternary history of tropical elements in peninsular Florida. It is assumed that *Rhizophora* and associated taxa increased in each of the warm intervals and declined or disappeared during cold times.

The temperature difference in MAT in much of North America between the last glacial maximum at 18 Kya and the climatic optimum at ~7 Kya was ~10°C. Recall from the paleotemperature curve (Fig. 3.1) that this is approximately the average change in benthic temperatures during the 70-m.y. span between the Late Cretaceous and the beginning of the Pleistocene. A direct comparison is misleading, but it is clear that the pace and intensity of climatic change increased in Quaternary times. Three of the most extensive North American plant formations (desert, grassland, tundra) appear in this period and are among the most recent to develop in modern form. The current versions date back mostly to the end of the last glaciation. At ~6 Kya (the midpoint of the current 10–12 Kya interglacial) they began responding to a new cooling phase, but this trend has been countered in the past 200 years by global warming.

The Quaternary is the interval in which paleoecology interfaces most closely with the environments and processes operating in the modern biota. Thus, our reconstruction of Quaternary vegetational history is more complete than for any previous time. In turn, this history provides a better understanding of the modern flora and constitutes the only basis available for estimating its future. Although the information is comparatively detailed, the complexities of environmental change and biotic response still exceed our capacity to describe them precisely. In particular, even though some components of paleovegetation analysis have been improved and can be applied with great precision, others allow considerable latitude in interpretation. For example, the ability to identify plant microfossils mostly to genus or family is especially limiting when considering vegetation types often characterized by different species of *Pinus*, Cyperaceae, Gramineae, *Acer*, Compositae, *Betula*, Ericaceae, *Quercus*, and others. In this connection megafossils are valuable and context information is essential.

Even with these limitations, however, the last decade has produced results that are both revolutionary and fundamental to our understanding of biotic history:

1. climates switch modes much faster than previously realized;
2. changes in climate first evident in ice cores and marine sediments from the North Atlantic may be paralleled in the North Pacific and hemispheric in scope;
3. the present warm interglacial is by far the exceptional climate during the past 1.6 m.y.;
4. vegetation characteristic of extensive regions at present (deciduous forest in the east, desert in the southwest) is not typical of the past 1.6 m.y. when montane coniferous forest and juniper–piñon pine woodland and steppe occupied much of these areas; and
5. organisms respond to environmental change as individuals.

Many of these generalizations are discernable in the Tertiary record, but they are more evident in the Quaternary where the pace was faster and the record is clearer. The average migratory rate for *Picea* in the Holocene of eastern North America was 14.1 km/century; *Quercus* 12.6; northern species of *Pinus* 13.5; and southern pines 8.1 (Delcourt and Delcourt, 1987). Where genetically controlled physiological processes are similar among taxa, the organisms potentially may have similar past histories; but even this potential is subject to the vicissitudes of barriers and seed dispersal, edaphic factors, anthropogenic influence, and epidemic disease.

Realization of the temporal nature of communities has implications for such time-honored concepts as succession, climax, and geofloras. If climates change on a scale more rapidly than implied by the older four-part glacial chronology, then designating *the* vegetation type characteristic of a given region loses much of its meaning, especially for the turbulent times of glacial–interglacial transitions. It also reduces the value of a concept that envisions the movements of vegetation as large intact blocks. On the other hand, it is worth noting that the number of taxa having similar ecological and dispersal potentials increases with higher units of vegetation. Thus, at the association level there are communities that are rare or absent in the modern biota (spruce–ash parkland in the upper midwest, sagebrush–pine woodland in the southwest). At the same time there are groupings at the formation level, which are frequently recognized by Tertiary paleontologists, that have persisted for long periods of time. Discussion of such conceptual matters improves with better definition of the time span and the hierarchical level being considered.

The actual association of elements in a community, especially when they represent novel combinations based on a mixture of wind-blown pollen types, usually can only be assummed. The diagrams presented are a general reflection of major trends in the history of a limited number of components, and interpretations can easily go beyond the sensitivity of the method. One approach to Late Neogene and Quaternary vegetational history is to begin with the assumption that the modern vegetation has persisted relatively unchanged back through time. A change is accepted only when the data actually force that conclusion. This is in contrast to a mind-set that anticipates the dramatic and leads to quick acceptance of evidence in apparent support of a dynamicism that approaches chaos and randomness, be it sizable amounts of spruce and fir pollen in Texas bogs or Arctic tundra and glacial striations on rocks in the southern Appalachian Mountains. As noted previously, the first makes for sound science but it can delay recognition and acceptance of the unexpected. The second makes for good theater, but it is more receptive to new ideas and innovative interpretations. Even though a conservative route has been followed in this summary, the vegetation of North America still emerges as dynamic and immensely complex. Its history is intriguing to study and offers a great deal to contemplate about future biotic changes.

Notes

1. Comparable glacial stages in the European Alps are named for four tributaries of the Danube west of Munich: the Günz, Mindel, Riss, and Würm. The interglacials are conventionally designated as G/M (Günz/Mindel), M/R, and R/W or Eemian (the Sangamon interglacial in North America).

2. On the surface of several of the Texas bogs there is an alga (*Zygogonium ericetorum*) belonging to the Chlorophyta that is uniquely purple. Ralph Alston, coauthor with Billie Turner of the classic text on Biochemical Systematics, was interested in these pigments and in 1957 he accompanied me on a collecting trip to the Soefje bog near Palmetto State Park. While I was coring and he was collecting algae, a bull approached, making all the signs of defending his territory. I told Dr. Alston I thought we should leave, but he said it was a bluff. The bull charged and we barely made it out by throwing corer, rods, vials, and plant presses over a barbed wire fence and diving over ourselves. Later I received a reprint with the inscription: "When you see this, you will remember that in the tradition of great scientists the world over *I* was willing to risk disfigurement, even death, to collect this material while *you* cringed behind the protection of a barbed wire fence clutching a rare orchid which, as I suspected all along, turned out to be an *Iris*." Ralph Alston died early in his career shortly afterward, and the reprint is a treasured memento of a respected scientist and generous mentor.

3. The question of "significant" versus "moderate" changes in the vegetation of southeastern United States has been long debated because the terms are relative and depend, in part, on perspective. Specialists may see events as profound and dramatic, while generalists see them in a broader context. This opens the opportunity for endless discussions and, in hindsight, amusing debates. As a recent Master's student from the University of Texas, I made my first foray into the national meetings scene at the American Institute of Biological Sciences conference at Indiana University in 1958. My paper was on a comparison of Potzger and Tharp's results and my own findings from the Texas bogs. When the session was opened for questions someone, whom I later found to be Dan Livingstone, provided a baptism of fire with a critique that covered methodology, results, interpretation, and conclusions. Another member of the audience, Paul Martin, waded in with a defense. As I watched these titans shred and reassemble my paper, I recall concluding that research in paleoecology was going to be anything but dull. Afterward, Livingstone said, "Nice paper."

In 1964 I was asked by the Society for the Study of Evolution to present a summary of the vegetational history of the southeastern United States at its meetings in Chapel Hill. Coming from the perspective of a specialist in the Tertiary, where the vegetation included extinct genera and elements from tropical America, the Old World tropics, and eastern Asia, I noted that, within this context, the changes in the biota of the southeastern United States were not extensive (Graham, 1965). Those specializing in the Quaternary correctly regarded this as an understatement. Following the Chapel Hill meeting I was invited to present the paper at a midwestern university where two colleagues had orchestrated a response. They concluded their critique with the revelation that rocks with purportedly glacial striations had just been discovered in the southern Appalachian Mountains. However, these striations were unusual in being all about the same diameter, U shaped in outline, and coated with iron oxide. They turned out to be cable grooves made during old lumbering operations as logs were dragged over the granite rocks. Periglacially modified strata do occur locally at high elevations in the northern and central Appalachian Mountains (Clark, 1968; Péwé, 1983).

References

Adam, D. P. 1988. Palynology of two Upper Quaternary cores from Clear Lake, Lake County, California. U.S. Geological Survey, Reston, VA. Professional Paper 1363, pp. 1–86.

Agassiz, L. 1840. Etudes sur les Glaciers. Privately published essay, Neuchâtel, Switzerland.

Ager, T. A. and L. B. Brubaker. 1985. Quaternary palynology and vegetational history of Alaska. In: V. M. Bryant, Jr. and R. G. Holloway (eds.), Pollen Records of Late-Quaternary North American Sediments. American Association of Stratigraphic Palynologists Foundation, Dallas, TX. pp. 353–384.

Aguirre, E. and G. Pasini. 1985. The Pliocene–Pleistocene boundary. Episodes 8: 116–120.

Allison, T. D., R. E. Moeller, and M. B. Davis. 1986. Pollen in laminated sediments provides evidence for a mid-Holocene forest pathogen outbreak. Ecology 67: 1101–1105.

Ammann, B. R. 1977. A pollen morphological distinction between *Pinus banksiana* lamb. and *P. resinosa* ait. Pollen Spores 19: 521–529.

Anderson, R. S. 1990. Holocene forest development and

paleoclimates within the central Sierra Nevada, California. J. Ecol. 78: 470–489.

——— and T. R. Van Devender. 1991. Comparison of pollen and macrofossils in packrat (*Neotoma*) middens: a chronological sequence from the Waterman Mountains of southern Arizona, U.S.A. Rev. Palaeobot. Palynol. 68: 1–28.

Anderson, T. W. 1985. Late-Quaternary pollen records from eastern Ontario, Québec, and Atlantic Canada. In: V. M. Bryant, Jr. and R. G. Holloway (eds.), Pollen Records of Late-Quaternary North American Sediments. American Association of Stratigraphic Palynologists Foundation, Dallas, TX. pp. 281–326.

Antevs, E. 1938. Postpluvial climatic variation in the southwest. Am. Meteorl. Soc. Bull. 19: 190–193.

Atwater, B. F. et al. (+ 8 authors). 1986. A fan dam for Tulare Lake, California, and implications for the Wisconsin glacial history of the Sierra Nevada. Geol. Soc. Am. Bull. 97: 97–109.

Axelrod, D. I. 1978. The origin of coastal sage vegetation, Alta and Baja California. Am. J. Bot. 65: 1117–1131.

———. 1989. Age and origin of chaparral. Nat. Hist. Mus. LA County, Sci. Ser. 34: 7–19.

Baker, R. G. and K. A. Waln. 1985. Quaternary pollen records from the Great Plains and central United States. In: V. M. Bryant, Jr. and R. G. Holloway (eds.), Pollen Records of Late-Quaternary North American Sediments. American Association of Stratigraphic Palynologists Foundation, Dallas, TX. pp. 191–203.

Baker, R. G. et al. (+ 6 authors). 1986. A full-glacial biota from southeastern Iowa. J. Quatern. Sci. 1: 91–107.

Barnosky, C. W. 1985. Late Quaternary vegetation in the southwestern Columbia Basin, Washington. Quatern. Res. 23: 109–122.

———, P. M. Anderson, and P. J. Bartlein. 1987a. The northwestern U.S. during deglaciation; vegetational history and paleoclimatic implications. In: W. F. Ruddiman and H. E. Wright, Jr. (eds.), North America and Adjacent Oceans during the Last Deglaciation. The Geology of North America, Decade of North American Geology. Geological Society of America, Boulder, CO. Volume K-3, pp. 289–321.

Barnowsky, C. W., E. C. Grimm, and H. E. Wright, Jr. 1987b. Towards a postglacial history of the northern Great Plains: a review of paleoecologic problems. Ann. Carnegie Mus. 56: 259–273.

Barrie, J. V., K. W. Conway, R. W. Mathewes, H. W. Josenhans, and M. J. Johns. 1994. Submerged Late Quaternary terrestrial deposits and paleoenvironment of northern Hecate Strait, British Columbia continental shelf, Canada. Quatern. Int. 20: 123–129.

Barry, R. G. 1983. Late-Pleistocene climatology. In: S. C. Porter (ed.), Late-Quaternary Environments of the United States. The Late Pleistocene. University of Minnesota Press, Minneapolis, MN. Vol. 1, pp. 390–407.

Bartlein, P. J. 1988. Late-Tertiary and Quaternary paleoenvironments. In: B. Huntley and T. Webb III (eds.), Vegetation History. Kluwer Academic Publishers, Dordrecht. pp. 113–152.

———, T. Webb III, and E. Fleri. 1984. Holocene climatic change in the northern midwest: pollen-derived estimates. Quatern. Res. 22: 361–374.

Bell, M. and E. P. Laine. 1985. Erosion of the Laurentide region of North America by glacial and glaciofluvial processes. Quatern. Res. 23: 154–174.

Benedict, J. B. 1973. Chronology of cirque glaciation, Colorado Front Range. Quatern. Res. 3: 584–599.

Benninghoff, W. S. 1962. Calculation of pollen and spore density in sediments by addition of exotic pollen in known quantities. Pollen Spores 4: 332–333.

Benson, L. J. Burdett, S. Lund, M. Kashgarian, and S. Mensing. 1997. Nearly synchronous climate change in the Northern Hemisphere during the last glacial termination. Nature 388: 263–265.

Benson, L. and R. S. Thompson. 1987. The physical record of lakes in the Great Basin. In: W. F. Ruddiman and H. E. Wright, Jr. (eds.), North America and Adjacent Oceans during the Last Deglaciation. The Geology of North America, Decade of North American Geology, Geological Society of America, Boulder, CO. Vol. K-3, pp. 241–260.

Betancourt, J. L. and T. R. Van Devender. 1981. Holocene vegetation in Chaco Canyon, New Mexico. Science 214: 656–658.

Birks, H. J. B. 1976. Late-Wisconsinan vegetational history at Wolf Creek, central Minnesota. Ecol. Monogr. 46: 395–429.

———. 1993. Quaternary palaeoecology and vegetation science—current contributions and possible future developments. Rev. Palaeobot. Palynol. 79: 153–177.

——— and A. D. Gordon. 1985. Numerical Methods in Quaternary Pollen Analysis. Academic Press, New York.

Blaise, B., J. J. Clague, and R. W. Mathewes. 1990. Time of maximum Late Wisconsin glaciation, west coast of Canada. Quatern. Res. 34: 282–295.

Bond, G. et al. (+ 13 authors). 1992. Evidence for massive discharges of icebergs into the North Atlantic Ocean during the last glacial period. Nature 360: 245–249.

Bowen, D. Q. 1978. Quaternary Geology, A Stratigraphic Framework for Multidisciplinary Work. Pergamon Press, Oxford, U.K. [See also later reprints and additions to 1988.]

Bradbury, J. P. 1997a. A Diatom record of climate and hydrology for the past 200ka from Owens Lake, California with comparison to other Great Basin records. Quaternary Sci. Rev. 16: 203–219.

———. 1997b. A diatom-based paleohydrologic record of climate change for the past 800 k.y. from Owens Lake, California. Geol. Soc. Am. Sp. Paper 317: 99–112.

Bradbury, J. P. and W. E. Dean (eds.). 1993. Elk Lake, Minnesota: Evidence for Rapid Climate Change in the North-Central United States. Geological Society of America, Boulder, CO. Special Paper 276.

Broccoli, A. J. and S. Manabe. 1987. The effects of the Laurentide ice sheet on North American climate during the last glacial maximum. Géograph. Phys. Quatern. 41: 291–299.

Broecker, W. S. and J. van Donk. 1970. Insolation changes, ice volumes, and the O^{18} record in deep-sea cores. Rev. Geophys. Space Phys. 8: 169–197.

Brown, C. A. 1938. The flora of Pleistocene deposits in the western Florida Parishes, west Feliciana Parish, and east Baton Rouge Parish, Louisiana. In: H. N. Fisk, H. G. Richards, C. A. Brown, and W. C. Steere (eds.). Contributions to the Pleistocene History of the Florida Parishes of Louisiana. Department of Conservation,

Louisiana Geological Survey, New Orleans, LA. Geological Bull. 12, pp. 59–96.

Brunner, C. A. 1982. Paleoceanography of surface waters in the Gulf of Mexico during the Late Quaternary. Quatern. Res. 17: 105–119.

Bryant, V. M., Jr., 1977. A 16,000 year pollen record of vegetational change in central Texas. Palynology 1: 143–156.

——— and R. G. Holloway. 1985. A Late Quaternary paleoenvironmental record of Texas: an overview of the pollen evidence. In: V. M. Bryant, Jr. and R. G. Holloway (eds.), Pollen Records of Late Quaternary North American Sediments. American Association of Stratigraphic Palynologists Foundation, Dallas, TX. pp. 39–70.

Bryant, V. M., Jr. and H. J. Shafer. 1977. The Late Quaternary paleoenvironment of Texas: a model for the archeologist. Bull. Texas Archeol. Soc. 48:1–25.

Carrara, P. E. 1989. Late Quaternary glacial and vegetative history of the Glacier National Park region, Montana. U.S. Geol. Surv. Bull. 1902: 1–64.

———, W. N. Mode, M. Rubin, and S. W. Robinson. 1984. Deglaciation and postglacial timberline in the San Juan Mountains, Colorado. Quatern. Res. 21: 42–55.

Chapman, J., P. A. Delcourt, P. A. Cridlebaugh, A. B. Shea, and H. R. Delcourt. 1982. Man–land interaction: 10,000 years of American Indian impact on native ecosystems in the Lower Little Tennessee River Valley, eastern Tennessee. Southeast. Archaeol. 1: 115–121.

Clark, G. M. 1968. Sorted patterned ground: new Appalachian localities south of the glacial border. Science 161: 355–356.

Clark, P. U. 1995. Fast glacier flow over soft beds. Science 267: 43–45.

CLIMAP Project Members. 1976. The surface of the ice-age earth. Science 191: 1131–1137.

Clisby, K. H. and P. B. Sears. 1956. San Agustin Plains—Pleistocene climatic changes. Science 124: 537–538.

COHMAP Members. 1988. Climatic changes of the last 18,000 years: observations and model simulations. Science 241: 1043–1052.

Cole, K. 1983. Late Pleistocene vegetation of Kings Canyon, Sierra Nevada, California. Quatern. Res. 19: 117–129.

———. 1985. Past rates of change, species richness, and a model of vegetational inertia in the Grand Canyon, Arizona. Am. Nat. 125: 289–303.

Colinvaux, P. A. 1963. A pollen record from Arctic Alaska reaching glacial and Bering land bridge times. Nature 198: 609–610.

———. 1964a. Origin of ice ages: pollen evidence from Arctic Alaska. Science 145: 707–708.

———. 1964b. The environment of the Bering land bridge. Ecol. Monogr. 34: 297–329.

———. 1996. Low-down on a land bridge. Nature 382: 21–23.

Coplen, T. B., I. J. Winograd, J. M. Landwehr, and A. C. Riggs. 1994. 500,000-year stable carbon isotopic record from Devils Hole, Nevada. Science 263: 361–365.

Corridor essays. 1996. Quatern. Int. 32.

Cridlebaugh, P. A. 1984. American Indian and Euro-American impact on Holocene vegetation in the Lower Little Tennessee River Valley, east Tennessee. Ph.D. dissertation, University of Tennessee, Knoxville, TN.

Critchfield, W. B. and E. L. Little, Jr. 1966. Geographic distribution of the pines of the world. U.S. Forest Service, Washington, D.C. Miscellaneous Publ. 991.

Cuffey, K. M. et al. (+ 5 authors). 1995. Large Arctic temperature change at the Wisconsin–Holocene transition. Science 270: 455–458.

Davis, J. H., Jr. 1946. The peat deposits of Florida, their occurrence, development, and uses. Fla. Dept. Conserv., Geol. Surv., Geol. Bull. 30: 1–247.

Davis, M. B. 1963. On the theory of pollen analysis. Am. J. Sci. 261: 897–912.

———. 1965. A method for determination of absolute pollen frequency. In: B. Kummel and D. Raup (eds.), Handbook of Paleontological Techniques. W. H. Freeman and Co., San Francisco, CA. pp. 674–686.

———. 1967. Pollen accumulation rates at Rogers Lake, Connecticut, during late- and postglacial time. Rev. Palaeobot. Palynol. 2: 219–230.

———. 1969. Climatic changes in southern Connecticut recorded by pollen deposition at Rogers Lake. Ecology 50: 409–422.

———. 1976. Pleistocene biogeography of temperate deciduous forests. Geosci. Man 13: 13–26.

———. 1981a. Outbreaks of forest pathogens in Quaternary history. In: Proceedings of the IVth International Palynologists Conference, Lucknow, 1976-1977, Vol. 3, pp. 216–227.

——— 1981b. Quaternary history and the stability of forest communities. In: D. C. West, H. H. Shugart, and D. B. Botkin (eds.), Forest Succession: Concepts and Applications. Springer-Verlag, Berlin. pp. 132–153.

———. 1983. Holocene vegetational history of eastern United States. In: H. E. Wright, Jr. (ed.), Late-Quaternary Environments of the United States, The Holocene. University of Minnesota Press, Minneapolis, MN. Vol. 2, pp. 166–181.

———. 1989. Insights from paleoecology on global change. Ecol. Soc. Am. Bull. 70: 222–228.

———, L. B. Brubaker, and T. Webb III. 1973. Calibration of absolute pollen influx. In: H. J. B. Birks and R. G. West (eds.), Quaternary Plant Ecology. Blackwell Science Publishers, Oxford, U.K. pp. 9–25.

Davis, M B., K. D. Woods, S. L. Webb, and R. P. Futyma. 1986. Dispersal versus climate: expansion of *Fagus* and *Tsuga* into the upper Great Lakes region. Vegetatio 67: 93–103.

Dawson, A. G. 1992. Ice Age Earth, Late Quaternary Geology and Climate. Routledge, London.

Deevey, E. S., Jr. 1939. Studies on Connecticut Lake Sediments. I. A postglacial climatic chronology for southern New England. Am. J. Sci. 237: 691–724.

———. 1949. Biogeography of the Pleistocene. Bull. Geol. Soc. Amer. 60: 1315–1416.

Delcourt, H. R. 1979. Late Quaternary vegetation history of the eastern Highland Rim and adjacent Cumberland Plateau of Tennessee. Ecol. Monogr. 49: 255–280.

———. 1987. The impact of prehistoric agriculture and land occupation on natural vegetation. Trends Ecol. Evol. 2: 39–44.

——— and P. A. Delcourt. 1975. The Blufflands: Pleistocene pathway into the Tunica Hills. Am. Midland Nat. 94: 385–400.

———. 1985a. Comparison of taxon calibrations, modern analogue techniques, and forest-stand simulation models for the quantitative reconstruction of past vegetation. Earth Surf. Processes Landforms 10: 293–304.

———. 1985b. Quaternary palynology and vegetational history of the southeastern United States. In: V. M. Bryant, Jr. and R. G. Holloway (eds.), Pollen Records of Late-Quaternary North American Sediments. American Association of Stratigraphic Palynologists Foundation, Dallas, TX. pp. 1–37.

———. 1986. Late Quaternary vegetational change in the central Atlantic States. In: J. N. McDonald and S. O. Bird (eds.), The Quaternary of Virginia. Virginia Division of Mineral Resources, Charlottesville, VA. Publication 75, pp. 23–35.

———. 1991. Late-Quaternary vegetation history of the Interior Highlands of Missouri, Arkansas, and Oklahoma. In: D. Henderson and L. D. Hedrick (eds.), Restoration of Old Growth Forests in the Interior Highlands of Arkansas and Oklahoma. Ouachita National Forest and Winrock International Institute for Agricultural Development, Morrilton, AR. pp. 15–30.

Delcourt, P. A. 1980. Goshen Springs: Late Quaternary vegetation record for southern Alabama. Ecology 61: 371–386.

——— and H. R. Delcourt. 1977. The Tunica Hills, Louisiana–Mississippi: late glacial locality for spruce and deciduous forest species. Quatern. Res. 7: 218–237.

———. 1984. Late Quaternary paleoclimates and biotic responses in eastern North America and the western North Atlantic Ocean. Palaeogeogr., Palaeoclimatol., Palaeoecol. 48: 263–284.

———. 1987. Long-Term Forest Dynamics of the Temperate Zone: A Case Study of Late-Quaternary Forests in Eastern North America, Ecological Studies 63. Springer-Verlag, Berlin.

———. 1993. Paleoclimates, paleovegetation, and paleofloras during the Late Quaternary. In: Flora of North America Editorial Committee (eds.), Flora of North America North of Mexico. Oxford University Press, Oxford, U.K. Vol. 1, pp. 71–94.

———. In press. Quaternary paleoecology of the lower Mississippi Valley. Eng. Geol.

———, R. C. Brister, and L. E. Lackey. 1980. Quaternary vegetation history of the Mississippi Embayment. Quatern. Res. 13: 111–132.

———, P. A. Cridlebaugh, and J. Chapman. 1986. Holocene ethnobotanical and paleoecological record of human impact on vegetation in the Little Tennessee River Valley, Tennessee. Quatern. Res. 25: 330–349.

———, and T. Webb III. 1984. Atlas of Mapped Distributions of Dominance and Modern Pollen Percentages for Important Tree Taxa of Eastern North America. American Association of Stratigraphic Palynologists Foundation, Dallas, TX. Contrib. Ser. 14.

Dillon, L. S. 1956. Wisconsin climate and life zones in North America. Science 123: 167–176.

Dimbleby, G. W. 1985. The Palynology of Archaeological Sites. Academic Press, New York.

Edwards, M. E. and E. D. Barker, Jr. 1994. Climate and vegetation in northeastern Alaska. Palaeogeogr., Palaeoclimatol., Palaeoecol. 109: 127–135.

Edwards, M. E., J. C. Dawe, and W. S. Armbruster. 1991. Pollen size of *Betula* in northern Alaska and the interpretation of Late Quaternary vegetation records. Can. J. Bot. 69: 1666–1672.

Edwards, R. L. 1995. Technical Comments—paleotopography of glacial-age ice sheets. Science 267: 536 [see also Response, Peltier, 1995].

Elias, S. A., S. K. Short, C. H. Nelson, and H. H. Birks. 1996. Life and times of the Bering land bridge. Nature 382: 60–63.

Elias, T. S. 1980. The Complete Trees of North America, Field Guide and Natural History. Van Nostrand Reinhold, New York.

Elsik, W. C. 1969. Late Neogene palynomorph diagrams, northern Gulf of Mexico. Trans. Gulf Coast Assoc. Geol. Soc. 19: 509–528.

Emiliani, C. 1955. Pleistocene temperatures. J. Geol. 63: 538–578.

Engstrom, D. R., B. C. S. Hansen, and H. E. Wright, Jr. 1990. A possible Younger Dryas record in southeastern Alaska. Science 250: 1383–1385.

Ewing, M. and W. L. Donn. 1956. A theory of ice ages. Science 123: 1061–1066.

Faegri, K., P. Kaland, and K. Krzywinski. 1989. Textbook of Pollen Analysis. Wiley, New York. 4th edition.

Fairbanks, R. G. 1989. A 17,000-year glacio-eustatic sea level record: influence of glacial melting rates on the Younger Dryas event and deep-ocean circulation. Nature 342: 637–642.

Feng, X. and S. Epstein. 1994. Climatic implications of an 8000-year hydrogen isotope time series from bristlecone pine trees. Science 265: 1079–1081.

Flint, R. F. and M. Rubin. 1955. Radiocarbon dates of pre-Mankato events in eastern and central North America. Science 121: 649–658.

Gajewski, K. 1993. The role of paleoecology in the study of global climate change. Rev. Palaeobot. Palynol. 79: 141–151.

———, S. Payette, and J. C. Ritchie. 1993. Holocene vegetation history at the boreal-forest–shrub-tundra transition in north-western Québec. J. Ecol. 81: 433–443.

Gallimore, R. G. and J. E. Kutzbach. 1996. Role of orbitally induced changes in tundra area in the onset of glaciation. Nature 381: 503–505.

Gaudreau, D. C. and T. Webb III. 1985. Late-Quaternary pollen stratigraphy and isochrone maps for the northeastern United States. In: V. M. Bryant, Jr. and R. G. Holloway (eds.), Pollen Records of Late-Quaternary North American Sediments. American Association of Stratigraphic Palynologists Foundation, Dallas, TX. pp. 247–280.

Graham, A. 1965. Origin and evolution of the biota of southeastern North America: evidence from the fossil plant record. Evolution 18: 571–585.

——— and C. Heimsch. 1960. Pollen studies of some Texas peat deposits. Ecology 41: 785–790.

Graham, R. W. and J. I. Mead. 1987. Environmental fluctuations and evolution of mammalian faunas during the last deglaciation in North America. In: W. F. Ruddiman and H. E. Wright, Jr. (eds.), North America and Adjacent Oceans during the Last Deglaciation. The Geology of North America, Decade of North American Geology. Geological Society of America, Boulder, CO. Vol. K-3, pp. 371–402.

Gray, J. (Coordinator). 1965. Techniques in palynology. In: B. Kummel and D. Raup (eds.), Handbook of Paleontological Techniques. W. H. Freeman and Co., San Francisco, CA. pp. 471–706.

——— and W. Smith. 1962. Fossil pollen and archaeology. Archaeology 15: 16–26.

Grimm, E. C. 1983. Chronology and dynamics of vegetation change in the prairie–woodland region of southern Minnesota, U.S.A. New Phytol. 93: 311–350.

———. 1988. Data analysis and display. In: B. Huntley and T. Webb III (eds.), Vegetation History. Kluwer Academic Publishers, Dordrecht. pp. 43–76.

———, G. L. Jacobson, Jr., W. A. Watts, B. C. S. Hansen, and K. A. Maasch. 1993. A 50,000-year record of climate oscillations from Florida and its temporal correlation with the Heinrich events. Science 261: 198–200.

Groot, J. J. 1991. Palynological evidence for Late Miocene, Pliocene and Early Pleistocene climate changes in the middle U.S. Atlantic Coastal Plain. Quatern. Sci. Rev. 10: 147–162.

Grüger, J. 1973. Studies on the Late Quaternary vegetation history of northeastern Kansas. Geol. Soc. Am. Bull. 84: 239–250.

Guilderson, T. P., R. G. Fairbanks, and J. L. Rubenstone. 1994. Tropical temperature variations since 20,000 years ago: modulating interhemispheric climatic change. Science 263: 663–665.

Hafsten, U. 1964. A standard pollen diagram for the southern High Plains, USA, covering the period back to the Early Wisconsin glaciation. In: Report of the VIth International Congress on the Quaternary, Warsaw, 1961, Vol. II, Palaeobotanical Sect., pp. 407–420.

Hall, S. A. 1977. Late Quaternary sedimentation and paleoecologic history of Chaco Canyon, New Mexico. Geol. Soc. Am. Bull. 88: 1593–1618.

———. 1981. Holocene vegetation at Chaco Canyon: pollen evidence from alluvium and pack rat middens. In: Society of American Archeology. 46th Annual Meeting, San Diego, 1981. Program and Abstracts, p. 61.

———. 1982. Reconstruction of local and regional Holocene vegetation in the arid southwestern United States based on combined pollen analytical results from *Neotoma* middens and alluvium. In: International Union for Quaternary Research, XI Congress, Moscow, Vol. 1, p. 130.

———. 1985. Quaternary pollen analysis and vegetational history of the southwest. In: V. M. Bryant, Jr. and R. G. Holloway (eds.), Pollen Records of Late-Quaternary North American Sediments. American Association of Stratigraphic Palynologists Foundation, Dallas, TX. pp. 95–123.

———. 1990. Holocene landscapes of the San Juan Basin, New Mexico; geomorphic, climatic, and cultural dynamics. In: N. P. Lasca and J. Donahue (eds.), Archaeological Geology of North America. Geological Society of America, Boulder, CO. Centennial Special Vol. 4. pp. 323–334.

——— and S. Valastro, Jr. 1995. Grassland vegetation in the southern Great Plains during the last glacial maximum. Quatern. Res. 44: 237–245.

Hallberg, G. R. 1986. Pre-Wisconsin glacial stratigraphy of the central plains region in Iowa, Nebraska, Kansas, and Missouri. In: V. Sibrava, D. Q. Bowen, and G. M. Richmond (eds.), Quaternary Glaciations in the Northern Hemisphere. Report of the International Geological Correlation Programme Project 24. Pergamon Press, Oxford, U.K. pp. 11–15.

Hansen, H. P. 1939. Pollen analysis of a bog near Spokane, Washington. Bull. Torrey Bot. Club. 66: 215–220.

———. 1947. Postglacial forest succession, climate and chronology in the Pacific Northwest. Am. Philos. Soc. 37: 1–130.

Hays, J. D., J. Imbrie, and N. J. Shackleton. 1976. Variations in the earth's orbit: pacemaker of the ice ages. Science 194: 1121–1132.

Hebda, R. J. 1995. British Columbia vegetation and climate history with focus on 6 Kya BP. Géograph. Phys. Quatern. 49: 55–79.

Hedges, J. I., J. R. Ertel, and E. B. Leopold. 1982. Lignin geochemistry of a Late Quaternary sediment core from Lake Washington. Geochim. Cosmochim. Acta 46: 1869–1877.

Heusser, C. J. 1952. Pollen profiles from southeastern Alaska. Ecol. Monogr. 22: 331–352.

———. 1977. Quaternary palynology of the Pacific slope of Washington. Quatern. Res. 8: 282–306.

———. 1985. Quaternary pollen records from the interior Pacific Northwest coast: Aleutians to the Oregon–California boundary. In: V. M. Bryant, Jr. and R. G. Holloway (eds.), Pollen Records of Late-Quaternary North American Sediments. American Association of Stratigraphic Palynologists Foundation, Dallas, TX. pp. 141–165.

———. 1990. Late Quaternary vegetation of the Aleutian Islands, southwestern Alaska. Can. J. Bot. 68: 1320–1326.

Heusser, L. E. 1978. Pollen in Santa Barbara Basin, California: a 12,000-yr record. Geol. Soc. Am. Bull. 89: 673–678.

——— and J. E. King. 1988. North America, with special emphasis on the development of the Pacific coastal forest and prairie/forest boundary prior to the last glacial maximum. In: B. Huntley and T. Webb III (eds.), Vegetation History. Kluwer Academic Publishers, Dordrecht. pp. 193–236.

Hickman, M. and C. E. Schweger. 1996. The Late Quaternary palaeoenvironmental history of a presently deep freshwater lake in east-central Alberta, Canada and palaeoclimate implications. Palaeogeogr., Palaeoclimatol., Palaeoecol. 123: 161–178.

Hoffecker, J. F., W. R. Powers, and T. Goebel. 1993. The colonization of Beringia and the peopling of the New World. Science 259: 46–53.

Holloway, R. G., L. M. Raab, and R. Stuckenrath. 1987. Pollen analysis of Late Holocene sediments from a central Texas bog. Texas J. Sci. 39: 71–80.

Hostetler, S. W., F. Giorgi, G. T. Bates, and P. J. Bartlein. 1994. Lake-atmosphere feedbacks associated with paleolakes Bonneville and Lahontan. Science 263: 665–668.

Huntley, B. and T. Webb III (eds.). 1988. Vegetation history. In: H. Lieth, (ed.). Handbook of Vegetation Science. Kluwer Academic Publishers, Dordrecht. Vol. 7.

Imbrie, J. and K. P. Imbrie. 1979. Ice Ages, Solving the Mystery. Harvard University Press, Cambridge, MA.

Iversen, J. 1941. Land occupation in Denmark's Stone Age. A pollen-analytical study of the influence of farmer culture on the vegetational development. Danmarks Geol. Unders. 66: 20–65.

Jackson, S. T. and C. R. Givens. 1994. Late Wisconsinan vegetation and environment of the Tunica Hills region, Louisiana/Mississippi. Quatern. Res. 41: 316–325.

Jackson, S. T., J. T. Overpeck, T. Webb III, S. E. Keattch, and K. H. Anderson. 1997. Mapped plant-macrofossil and pollen records of Late Quaternary vegetation change in eastern North America. Quaternary Sci. Rev. 16: 1–70.

Jackson, S. T. and D. R. Whitehead. 1991. Holocene vegetation patterns in the Adirondack Mountains. Ecology 72: 641–653.

———. 1993. Pollen and macrofossils from Wisconsinan interstadial sediments in northeastern Georgia. Quatern. Res. 39: 99–106.

Jacobs, B. F., P. L. Fall, and O. K. Davis (eds.). 1985. Late Quaternary vegetation and climates of the American southwest. Am. Asso. Stratigr. Palynol. Contrib. Ser. 16: 1–185.

Jacobson, G. L., Jr., T. Webb III, and E. C. Grimm. 1987. Patterns and rates of vegetation change during the deglaciation of eastern North America. In: W. F. Ruddiman and H. E. Wright, Jr. (eds.), North America and Adjacent Oceans During the Last Deglaciation. The Geology of North America, Decade of North American Geology. Geological Society of America, Boulder, CO. Vol. K-3, pp. 277–288.

Jansen, E. and T. Veum. 1990. Evidence for two-step deglaciation and its impact on North Atlantic deep-water circulation. Nature 343: 612–616.

Johnson, C. G., Jr., R. R. Clausmitzer, P. J. Mehringer, and C. D. Oliver. 1994. Biotic and Abiotic Processes of East-side Ecosystems: The Effects of Management on Plant and Community Ecology, and on Stand and Landscape Vegetation Dynamics. U.S. Dept. Agriculture, Forest Service, Pacific Northwest Station, Portland, OR. General Tech. Rep. PNW-GTR-322.

Josenhans, H., D. Fedje, R. Pienitz, and J. Southon. 1997. Early humans and rapidly changing Holocene sea levels in the Queen Charlotte Islands–Hecate Strait, British Columbia, Canada. Science 277: 71–74.

Kapp, R. O. 1970. Pollen analysis of pre-Wisconsin sediments from the Great Plains In: W. Dort and J. K. Jones (eds.), Pleistocene and Recent Environments of the Central Great Plains. University of Kansas Press, Lawrence, KS. pp. 143–155.

Kearney, M. S. and B. H. Luckman. 1983. Holocene timberline fluctuations in Jasper National Park, Alberta. Science 221: 261–263.

Kennett, J. P. and B. L. Ingram. 1995. A 20,000-year record of ocean circulation and climate change from the Santa Barbara Basin. Nature 377: 510–514.

Kerr, R.A. 1993. How ice age climate got the shakes. Science 260: 890–892.

King, J. E. 1973. Late Pleistocene palynology and biogeography of the western Missouri Ozarks. Ecol. Monogr. 43: 539–565.

———. 1981. Late Quaternary vegetational history of Illinois. Ecol. Monogr. 51: 43–62.

Klassen, R. W., L. D. Delorme, and R. J. Mott. 1967. Geology and paleontology of Pleistocene deposits in southwestern Manitoba. Can. J. Earth Sci. 4: 433–477.

Klein, R. G. 1975. The relevance of Old World archeology to the first entry of man into the New World. Quatern. Res. 5: 391–394.

Kneller, M. and D. Peteet. 1993. Late-Quaternary climate in the Ridge and Valley of Virginia, U.S.: changes in vegetation and depositional environment. Quatern. Sci. Rev. 12: 613–628.

Koehler, P. A. and R. S. Anderson. 1994. The paleoecology and stratigraphy of Nichols Meadow, Sierra National Forest, California, USA. Palaeogeogr., Palaeoclimatol., Palaeoecol. 112: 1–17.

Kotilainen, A. T. and N. J. Shackleton. 1995. Rapid climatic variability in the North Pacific Ocean during the past 95,000 years. Nature 377: 323–326.

Küchler, A. W. 1964. Manual to Accompany the Map Potential Natural Vegetation of the Conterminous United States. American Geographical Society, New York, Special Publ.

Kutzbach, J. E. 1987. Model simulations of the climatic patterns during the deglaciation of North America. In: W. F. Ruddiman and H. E Wright, Jr. (eds.), North America and Adjacent Oceans during the Last Deglaciation. The Geology of North America, Decade of North American Geology. Geological Society of America, Boulder, CO. Vol. K-3, pp. 425–446.

——— and P. J. Guetter. 1986. The influence of changing orbital parameters and surface boundary conditions on climate simulations for the past 18,000 years. J. Atmos. Sci. 43: 1726–1759.

Kutzbach, J. E. and H. E. Wright, Jr. 1985. Simulation of the climate of 18,000 years BP; results for the North American/North Atlantic/European sector and comparison with the geological record of North America. Quatern. Sci. Rev. 4: 147–187.

Lamb, H. F. and M. E. Edwards. 1988. The Arctic. In: B. Huntley and T. Webb III (eds.), Vegetation History. Kluwer Academic Publishers, Dordrecht. pp. 519–555.

Larson, D. A., V. M. Bryant, Jr., and T. S. Patty. 1972. Pollen analysis of a central Texas bog. Am. Midland Nat. 88: 358–367.

Lemons, D. R., M. R. Milligan, and M. A. Chan. 1996. Paleoclimatic implications of Late Pleistocene sediment yield rates for the Bonneville Basin, northern Utah. Palaeogeogr., Palaeoclimatol., Palaeoecol. 123: 147–159.

Leopold, E. B. 1956. Two late-glacial deposits in southern Connecticut. Proc. Natl. Acad. Sci. 52: 863–867.

———, R. Nickmann, J. I. Hedges, and J. R. Ertel. 1982. Pollen and lignin records of Late Quaternary vegetation, Lake Washington. Science 218: 1305–1307.

Levesque, A. J., L. C. Cwynar, and I. R. Walker. 1997. Exceptionally steep north-south gradients in lake temperatures during the last deglaciation. Nature 385: 423–426.

Levesque, A. J., F. E. Mayle, I. R. Walker, and L. C. Cwynar. 1993. A previously unrecognized late-glacial cold event in eastern North America. Nature 361: 623–626. [See Nature 363: 188, for a corrected version of the pollen diagrams.]

Li, J., T. K. Lowenstein, C. B. Brown, T.-L. Ku, and S. Luo. 1996. A 100 Kya record of water tables and paleoclimates from salt cores, Death Valley, California. Palaeogeogr., Palaeoclimatol., Palaeoecol. 123: 147–159.

Livingstone, D. A. 1955. Some pollen profiles from Arctic Alaska. Ecology 36: 587–600.

———. 1957. Pollen analysis of a valley fill near Umiat, Alaska. Am. J. Sci. 255: 254–260.

Löve, D. and A. Löve. 1974. Origin and evolution of Arctic and alpine floras. In: J. D. Ives and R. G. Barry (eds.), Arctic and Alpine Environments. Methuen, London. pp. 571–603.

Ludlum, D. 1966. Early American Winters: 1604-1820. American Meterological Society, Boston.

MacDonald, G. M., T. W. D. Edwards, K. A. Moser, R. Pienitz, and J. P. Smol. 1993. Rapid response of treeline vegetation and lakes to past climate warming. Nature 361: 243–246.

Mack, R. N., N. W. Rutter, and S. Valastro. 1983. Holocene vegetational history of the Kootenai River Valley, Montana. Quatern. Res. 20: 177–193.

———, and V. M. Bryant, Jr. 1978. Late Quaternary vegetation history at Waits Lake, Colville River Valley, Washington. Bot. Gaz. 139: 499–506.

Maher, L. J., Jr. 1963. Pollen analyses of surface materials from the southern San Juan Mountains, Colorado. Geol. Soc. Am. Bull. 74: 1485–1504.

Manabe, S. and A. J. Broccoli. 1985. The influence of continental ice sheets on the climate of an ice age. J. Geophys. Res. 90: 2167–2190.

Mann, D. H. and T. D. Hamilton. 1995. Late Pleistocene and Holocene paleoenvironments of the North Pacific coast. Quatern. Sci. Rev. 14: 449–471.

Markgraf, V., J. P. Bradbury, R. M. Forester, G. Singh, and R. S. Sternberg. 1984. San Agustin Plains, New Mexico: age and paleoenvironmental potential reassessed. Quatern. Res. 22: 336–343.

Martin, P. S. 1958. Taiga-tundra and the full-glacial period in Chester County, Pennsylvania. Am. J. Sci. 256: 470–502.

———. 1963a. The Last 10,000 Years. A Fossil Pollen Record of the American Southwest. University of Arizona Press, Tucson, AZ.

———. 1963b. Early man in Arizona: the pollen evidence. Am. Antiq. 29: 67–73.

———. 1963c. Geochronology of pluvial Lake Cochise, southern Arizona. II. Pollen analysis of a 42-meter core. Ecology 44: 436–444.

———. 1973. The discovery of America. Science 179: 969–974.

Martin, W. C. 1969. Fossil flora of the Rita Blanca lake deposits. Geol. Soc. Am. Memoir 113: 101–106.

Mathewes, R. W. 1973. A palynological study of postglacial vegetation changes in the University Research Forest, southwestern British Columbia. Can. J. Bot. 51: 2085–2103.

———. 1979. A paleoecological analysis of Quadra Sand at Point Grey, British Columbia, based on indicator pollen. Can. J. Earth Sci. 16: 847–858.

———. 1985. Paleobotanical evidence for climatic change in southern British Columbia during late-glacial and Holocene time. Syllogeus 55: 397–422.

———. 1991. Climatic conditions in the western and northern cordillera during the last glaciation: paleoecological evidence. Géograp. Phys. Quatern. 45: 333–339.

———. 1993. Evidence for Younger Dryas-age cooling on the North Pacific coast of America. Quatern. Sci. Rev. 12: 321–331.

———. 1996. Paleoecology of Fraser (Late Wisconsin) glaciation in coastal British Columbia, Canada. In: IXth International Palynological Congress, Houston, TX, June 1996, Program and Abstracts, p. 99.

——— and J. C. Clague. 1994. Detection of large prehistoric earthquakes in the Pacific Northwest by microfossil analysis. Science 264: 688–691.

Maxwell, J. A. and M. B. Davis. 1972. Pollen evidence of Pleistocene and Holocene vegetation on the Allegheny Plateau, Maryland. Quatern. Res. 2: 506–530.

McAndrews, J. H. 1966. Postglacial history of prairie, savanna, and forest in northwestern Minnesota. Mem. Torrey Bot. Club. 22: 1–72.

———. 1988. Human disturbance of North American forests and grasslands: the fossil pollen record. In: B. Huntley and T. Webb III (eds.), Vegetation History. Kluwer Academic Publishers, Dordrecht. pp. 673–697.

McCarten, N. and T. R. Van Devender. 1988. Late Wisconsin vegetation of Robber's Roost in the western Mojave Desert, California. Madroño 35: 226–237.

McLachlan, J. S. and L. B. Brubaker. 1995. Local and regional vegetation change on the northeastern Olympic Peninsula during the Holocene. Can. J. Bot. 73: 1618–1627.

Mehringer, P. J., Jr. 1965. Late Pleistocene vegetation in the Mohave Desert of southern Nevada. J. Ariz. Acad. Sci. 3: 172–188.

———. 1967. Pollen analysis of the Tule Springs area, Nevada. In: H. M. Wormington and D. Ellis (eds.), Nevada State Museum Anthropological Papers. Carson City, NV. No. 13. pp. 129–200.

———. 1985. Late-Quaternary pollen records from the interior Pacific Northwest and northern Great Basin of the United States. In: V. M. Bryant, Jr. and R. G. Holloway (eds.), Pollen Records of Late-Quaternary North American Sediments. American Association of Stratigraphic Palynologists Foundation, Dallas, TX. pp. 167–189.

———, E. Blinman, and K. L. Petersen. 1977. Pollen influx and volcanic ash. Science 198: 257–261.

Meltzer, D. J. 1991. Altithermal archaeology and paleoecology at Mustang Springs, on the Southern High Plains of Texas. Am. Antiq. 56: 236–267.

Mikolajewicz, U., T. J. Crowley, A. Schiller, and R. Voss. 1997. Modeling teleconnections between the North Atlantic and North Pacific during the Younger Dryas. Nature 387: 384–387.

Miller, G. H., J. T. Andrews, and S. K. Short. 1977. The last interglacial–glacial cycle, Clyde Foreland, Baffin Island, N.W.T.: stratigraphy, biostratigraphy, and chronology. Can. J. Earth Sci. 14: 2824–2857.

Miller, N. G. 1992. A contribution toward a history of the Arctic moss flora. Contrib. University of Mich. Herb. 18: 73–86.

———. 1993. New Late-Pleistocene moss assemblages from New England, U.S.A., and their bearing on the migrational history of the North American moss flora. J. Hattori Bot. Lab. 74: 235–248.

Miller, R. F. and P. E. Wigand. 1994. Holocene changes in semiarid pinyon–juniper woodlands. BioScience 44: 465–474.

Mills, H. H. and P. A. Delcourt. 1991. Quaternary geology of the Appalachian Highlands and interior low plateaus. In: R. B. Morrison (ed.), Quaternary Nonglacial Geology: Conterminous U.S. The Geology of North America, Decade of North American Geology. Geological Society of America, Boulder, CO. Vol. K-2, pp. 611–628.

Mix, A. C. 1987. The oxygen-isotope record of glaciation. In: W. F. Ruddiman and H. E. Wright, Jr. (eds.), North America and Adjacent Oceans during the Last Deglaciation. The Geology of North America, Decade of North American Geology. Geological Society of America, Boulder, CO. Vol. K-3, pp. 111–136.

Mock, C. J. and P. J. Bartlein. 1995. Spatial variability of Late-Quaternary paleoclimates in the western United States. Quatern. Res. 44: 425–433.

Moore, P. D., J. A. Webb, and M. E. Collinson. 1991. Pollen Analysis. Blackwell Science Publishers, Oxford, U.K. 2nd edition.

Morgan, A. V. 1987. Late Wisconsin and Early Holocene paleoenvironments of east-central North America based on assemblages of fossil Coleoptera. In: W. F. Ruddiman and H. E. Wright, Jr. (eds.), North America and Adjacent Oceans During the Last Deglaciation. The Geology of North America, Decade of North American Geology. Geological Society of America, Boulder, CO. Vol. K-3, pp. 353–370.

Mott, R. J., T. W. Anderson, and J. V. Matthews, Jr. 1982. Pollen and macrofossil study of an interglacial deposit in Nova Scotia. Géograp. Phys. Quatern. 36: 197–208.

Osborn, G. D. 1982. Holocene glacier and climate fluctuations in the southern Canadian Rocky Mountains, a review. Striae 18: 15–25.

Overpeck, J., D. Rind, A. Lacis, and R. Healy. 1996. Possible rate of dust-induced regional warming in abrupt climate change during the last glacial period. Nature 384: 447–449.

Patterson, W. A. III and A. E. Backman. 1988. Fire and disease history of forests. In: B. Huntley and T. Webb III (eds.), Vegetation History. Kluwer Academic Publishers, Dordrecht. pp. 603–632.

Payette, S. 1993. The range limit of boreal tree species in Québec–Labrador: an ecological and palaeoecological interpretation. Rev. Palaeobot. Palynol. 79: 7–30.

Peltier, W. R. 1994. Ice age paleotopography. Science 265: 195–201.

———. 1995. Response—Paleotopography of glacial-age ice sheets. Science 267: 536–538. [See also Technical Comments—Edwards, 1995.]

———. 1996. Mantle viscosity and ice-age ice sheet topography. Science 273: 1359–1364.

Peteet, D. M. 1991. Postglacial migration history of lodgepole pine near Yakutat, Alaska. Can. J. Bot. 69: 786–796.

Peteet, D. M. et al. (+ 5 authors). 1990. Younger Dryas climatic reversal in northeastern USA? AMS ages for an old problem. Quatern. Res. 33: 219–230.

——— et al. (+ 5 authors). 1993. Late-glacial pollen, macrofossils and fish remains in northeastern U.S.A.—the Younger Dryas oscillation. Quatern. Sci. Rev. 12: 597–612.

Péwé, T. L. 1983. The periglacial environment in North America during Wisconsin time. In: S. C. Porter (ed.), Late-Quaternary Environments of the United States, The Late Pleistocene. University of Minnesota Press, Minneapolis, MN. Vol. 1, pp. 157–189.

Phillips, F. M. et al (+ 5 authors). 1996. Chronology for fluctuations in late Pleistocene Sierra Nevada glaciers and lakes. Science 274: 749–751.

Phleger, F. B., F. L. Parker, and J. F. Peirson. 1953. North Atlantic foraminifera. Rept. Swed. Deep-Sea Expedit. 1947–1948 7: 1–122.

Porter, S. C. (ed.). 1983. Late-Quaternary Environments of the United States, The Late Pleistocene. University of Minnesota Press, Minneapolis, MN. Vol. 1.

Potzger, J. E. 1944. Pollen frequency of *Abies* and *Picea* in peat: a correction on some published records from Indiana bogs and lakes. Butler University of Bot. Stud. 6: 123–130.

——— and B. C. Tharp. 1943. Pollen record of Canadian spruce and fir from Texas bog. Science 98: 584–585.

———. 1947. Pollen profile from a Texas bog. Ecology 28: 274–280.

———. 1954. Pollen study of two bogs in Texas. Ecology 35: 462–466.

Prentice, I. C. 1988. Records of vegetation in time and space: the principles of pollen analysis. In: B. Huntley and T. Webb III (eds.), Vegetation History. Kluwer Academic Publishers, Dordrecht. pp. 17–42.

Remenda, V. H., J. A. Cherry, and T. W. D. Edwards. 1994. Isotopic composition of old ground water from Lake Agassiz: implications for Late Pleistocene climate. Science 266: 1975–1978.

Richard, P. J. H. 1993. Origine et dynamique postglaciaire de a forêt mixte au Québec. Rev. Palaeobot. Palynol. 79: 31–68.

———. 1994a. Wisconsinan Late-Glacial environmental change in Québec: a regional synthesis. J. Quatern. Sci. 9: 165–170.

———. 1994b. Postglacial palaeophytogeography of the eastern St. Lawrence River watershed and the climatic signal of the pollen record. Palaeogeogr., Palaeoclimatol., Palaeoecol. 109: 137–161.

Richmond, G. M. and D. S. Fullerton. 1986. Quaternary glaciations in the United States of America. In: V. Sibrava, D. Q. Bowen, and G. M. Richmond (eds.), Quaternary Glaciations in the Northern Hemisphere. Report of the International Geological Correlation Programme Project 24. Pergamon Press, Oxford, U.K. pp. 3–10.

Ritchie, J. C. 1984. Past and Present Vegetation of the Far Northwest of Canada. University of Toronto Press, Toronto.

———. 1985. Quaternary pollen records from the western interior and the Arctic of Canada. In: V. M. Bryant, Jr. and R. G. Holloway (eds.), Pollen Records of Late-Quaternary North American Sediments. American Association of Stratigraphic Palynologists Foundation, Dallas, TX. pp. 327–352.

———. 1987. Postglacial Vegetation of Canada. Cambridge University Press, Cambridge, U.K.

Royall, P. D., P. A. Delcourt, and H. R. Delcourt. 1991. Late Quaternary paleoecology and paleoenvironments of the central Mississippi alluvial valley. Geol. Soc. Am. Bull. 103: 157–170.

Ruddiman, W. F. and H. E. Wright, Jr. (eds.) 1987. North America and Adjacent Oceans during the Last Deglaciation. The Geology of North America, Decade of North American Geology. Geological Society of America, Boulder, CO. Vol. K-3.

Savage, D. E. and D. E. Russell. 1983. Mammalian Paleofaunas of the World. Addison-Wesley, London.

Schoonmaker, P. K. and D. R. Foster. 1991. Some implications of paleoecology for contemporary ecology. Bot. Rev. 57: 204–245.

Scuderi, L. A. 1993. A 2000-year tree ring record of annual temperatures in the Sierra Nevada Mountains. Science 259: 1433–1436.

Shackleton, N. J. 1987. Oxygen isotopes, ice volume and sea level. Quatern. Sci. Rev. 6: 183–190.

——— and N. D. Opdyke. 1977. Oxygen isotope and palaeomagnetic evidence for early Northern Hemisphere glaciation. Nature 270: 216–219.

Shackleton, N. J. et al. (+ 16 authors). 1984. Oxygen isotope calibration of the onset of ice-rafting and history of glaciation in the North Atlantic region. Nature 307: 620–623.

Shane, L. C. K. 1975. Palynology and radiocarbon chronology of Battaglia Bog, Portage County, Ohio. Ohio J. Sci. 75: 96–102.

———. 1987. Late-glacial vegetational and climatic history of the Alleghney Plateau and the till plains of Ohio and Indiana, U.S.A. Boreas 16: 1–20.

———. 1989a. Changing palynological methods and their role in three successive interpretations of the late-glacial environments at Bucyrus Bog, Ohio, USA. Boreas 18: 297–309.

———. 1989b. The History of Vegetation, Climate, and Relict Plant Taxa: Palynology of the Gott Fen/Frame Lake Area, Portage Co., Ohio. Final Report. Ohio Department of Natural Resources, Columbus, OH. pp. 1–20.

——— and K. H. Anderson. 1993. Intensity, gradients and reversals in late glacial environmental change in east-central North America. Quatern. Sci. Rev. 12: 307–320.

Smith, G. L., J. L. Bischoff, and J. P. Bradbury. 1997. Synthesis of the paleoclimatic record from Owens Lake core OL-92. Geol. Soc. Am. Sp. Paper 317: 143–160.

Spaulding, W. G. 1990a. Vegetation dynamics during the last deglaciation, southeastern Great Basin, U.S.A. Quatern. Res. 33: 188–203.

———. 1990b. Vegetational and climatic development of the Mojave Desert: the last glacial maximum to the present. In: J. L. Betancourt, T. R. Van Devender, and P. S. Martin (eds.), Packrat Middens, The Last 40,000 Years of Biotic Change. University of Arizona Press, Tucson, AZ. pp. 166–199.

———. 1991. Pluvial climatic episodes in North America and North Africa: types and correlation with global climate. Palaeogeogr., Palaeoclimatol., Palaeoecol. 84: 217–227.

———, E. B. Leopold, and T. R. Van Devender. 1983. Late Wisconsin paleoecology of the American southwest. In: S. C. Porter (ed.), Late-Quaternary Environments of the United States. The Late Pleistocene. University of Minnesota Press, Minneapolis, MN. Vol. 1, pp. 259–293.

Stock, C. (revised by J. M. Harris). 1992. Rancho La Brea, a Record of Pleistocene Life in California. Natural History Museum of Los Angeles County, Los Angles. 7th edition, Science Ser. 37.

Stute, M., P. Schlosser, J. F. Clark, and W. S. Broecker. 1992. Paleotemperatures in the southwestern United States derived from noble gases in ground water. Science 256: 1000–1003.

Szeicz, J. M. and G. M. MacDonald. 1991. Postglacial vegetation history of oak savanna in southern Ontario. Can. J. Bot. 69: 1507–1519.

Terasmae, J. 1958. Contributions to Canadian palynology. Geol. Surv. Can. Bull. 46: 1–35.

———. 1960. Contributions to Canadian palynology no. 2. Geol. Surv. Can. Bull. 56: 1–41.

———. 1967. A Review of Quaternary Palaeobotany and Palynology in Canada. Geological Survey of Canada, Ottawa, Paper 67–13, pp. 1–12.

———, P. J. Webber, and J. T. Andrews. 1966. A study of Late-Quaternary plant-bearing beds in north-central Baffin Island, Canada. Arctic 19: 296–318.

Thompson, L. G. et al. (+ 7 authors). 1995. Late glacial stage and Holocene tropical ice core records from Huascarán, Peru. Science 269: 46–51.

Thompson, R. S. 1988. Western North America, vegetation dynamics in the western United States: modes of response to climatic fluctuations. In: B. Huntley and T. Webb III (eds.), Vegetation History. Kluwer Academic Publishers, Dordrecht. pp. 415–458.

———. 1990. Late Quaternary vegetation and climate in the Great Basin. In: J. L. Betancourt, T. R. Van Devender, and P. S. Martin (eds.), Packrat Middens: The Last 40,000 Years of Biotic Change. University of Arizona Press, Tucson, AZ. pp. 200–239.

———, C. Whitlock, P. J. Bartlein, S. P. Harrison, and W. G. Spaulding. 1993. Climatic changes in the western United States since 18,000 yr B.P. In: H. E. Wright, Jr. (ed.), Global Climates Since the Last Glacial Maximum. University of Minnesota Press, Minneapolis, MN. pp. 468–513.

Thunell, R. C. and P. G. Mortyn. 1995. Glacial climate instability in the northeast Pacific Ocean. Nature 376: 504–506.

Toomey, R. S. III, M. D. Blum, and S. Valastro, Jr. 1993. Late Quaternary climates and environments of the Edwards Plateau, Texas. Global Planet. Change 7: 299–320.

Traverse, A. 1988. Paleopalynology. Unwin Hyman, Boston.

Tucker, J. M. 1969. Oak leaves of the Rita Blanca lake deposits. Memoir Geol. Soc. Am. 113: 97–99.

Van Devender, T. R. 1985. Climatic cadences and the composition of Chihuahuan Desert communities: the Late Pleistocene packrat midden record. In: J. Diamond and T. J. Case (eds.), Community Ecology. Harper & Row, New York. pp. 285–299.

———. 1987. Holocene vegetation and climate in the Puerto Blanco Mountains, southwestern Arizona. Quatern. Res. 27: 51–72.

———. 1990a. Late Quaternary vegetation and climate of the Chihuahuan Desert, United States and Mexico. In: J. L. Betancourt, T. R. Van Devender, and P. S. Martin (eds.), Packrat Middens, The Last 40,000 Years of Biotic Change. University of Arizona Press, Tucson, AZ. pp. 104–133.

———. 1990b. Late Quaternary vegetation and climate of the Sonoran Desert, United States and Mexico. In: J. L. Betancourt, T. R. Van Devender, and P. S. Martin (eds.), Packrat Middens, The Last 40,000 years of Biotic Change. University of Arizona Press, Tucson, AZ. pp. 134–165.

———, J. L. Betancourt, and M. Wimberly. 1984. Biogeographic implications of a packrat midden sequence from the Sacramento Mountains, south-central New Mexico. Quatern. Res. 22: 344–360.

Van Devender, T. R. and D. H. Riskind. 1979. Late Pleistocene and early Holocene plant remains from Hueco Tanks State Historical Park: the development of a refugium. Southwest. Nat. 24: 127–140.

Van Devender, T R., R. S. Thompson, and J. L. Betancourt. 1987. Vegetation history of the deserts of southwestern North America; the nature and timing of the Late Wisconsin–Holocene transition. In: W. F. Ruddiman and

H. E. Wright, Jr. (eds.), North America and Adjacent Oceans during the Last Deglaciation. The Geology of North America, Decade of North American Geology. Geological Society of America, Boulder, CO. Vol. K-3, pp. 323–352.

VEMAP (Vegetation/Ecosystem Modeling and Analysis Project). 1995. A comparison of biogeography and biogeochemistry models in the context of global climate change. Global Biogeochemistry Cycles 9: 407–437.

Waddington, J. C. B. and H. E. Wright, Jr. 1974. Late Quaternary vegetational changes on the east side of Yellowstone Park, Wyoming. Quatern. Res. 4: 175–184.

Walker, H. J. 1973. Morphology of the North Slope. In: M. E. Britton (ed.), Alaskan Arctic Tundra. Arctic Institute of North America, Fairbanks, AK, Tech. Paper 25, pp. 49–92.

Walker, I. R., R. J. Mott, and J. P. Smol. 1991. Allerød–Younger Dryas lake temperatures from midge fossils in Atlantic Canada. Science 253: 1010–1012.

Watts, W. A. 1970. The full-glacial vegetation of northwestern Georgia. Ecology 51: 17–33.

———. 1979. Late Quaternary vegetation of central Appalachia and the New Jersey Coastal Plain. Ecol. Monogr. 49: 427–469.

———. 1983. Vegetational history of the eastern United States 25,000 to 10,000 years ago. In: S. C. Porter (ed.), Late-Quaternary Environments of the United States, The Late Pleistocene. University of Minnesota Press, Minneapolis, MN. Vol. 1, pp. 294–310.

——— and B. C. S. Hansen. 1994. Pre-Holocene and Holocene pollen records of vegetation history from the Florida Peninsula and their climatic implications. Palaeogeogr., Palaeoclimatol., Palaeoecol. 109: 163–176.

Watts, W. A. and M. Stuiver. 1980. Late Wisconsin climate of northern Florida and the origin of species-rich deciduous forest. Science 210: 325–327.

Webb, S. D. 1985. Main pathways of mammalian diversification in North America. In: F. G. Stehli and S. D. Webb (eds.), The Great American Biotic Interchange. Plenum, New York. pp. 201–217.

Webb, T. III. 1987. The appearance and disappearance of major vegetational assemblages: long-term vegetational dynamics in eastern North America. Vegetatio 69: 177–187.

———. 1988. Eastern North America. In: B. Huntley and T. Webb III (eds.), Vegetation History. Kluwer Academic Publishers, Dordrecht. pp. 385–414.

———, P. J. Bartlein, and J. E. Kutzbach. 1987. Climatic change in eastern North America during the past 18,000 years; comparisons of pollen data with model results. In: W. F. Ruddiman and H. E. Wright, Jr. (eds.), North America and Adjacent Oceans during the Last Deglaciation. The Geology of North America, Decade of North American Geology. Geological Society of America, Boulder, CO. Vol. K-3, pp. 447–462.

Webb, T. III, E. J. Cushing, and H. E. Wright, Jr. 1983. Holocene changes in the vegetation of the Midwest. In: H. E. Wright, Jr. (ed.), Late Quaternary Environments of the United States, The Holocene. University of Minnesota Press, Minneapolis, MN. Vol. 2, pp. 142–165.

Whitehead, D. R. 1965. Prehistoric maize in southeastern Virginia. Science 150: 881–883.

———. 1981. Late-Pleistocene vegetational changes in northeastern North Carolina. Ecol. Monogr. 51: 451–471.

———, D. F. Charles, S. T. Jackson, J. P. Smol, and D. R. Engstrom. 1989. The developmental history of Adirondack (N.Y.) lakes. J. Paleolimnol. 2: 185–206.

Whitehead, D. R. and R. Q. Oaks, Jr. 1979. Developmental history of the Dismal Swamp. In: P. W. Kirk, Jr. (ed.), The Great Dismal Swamp. University of Virginia Press, Charlottesville, VA. pp. 25–43.

Whitehead, D. R. and M. C. Sheehan. 1985. Holocene vegetational changes in the Tombigbee River Valley, eastern Mississippi. Am. Midland Nat. 113: 122–137.

Whitlock, C. 1992. Vegetational and climatic history of the Pacific Northwest during the last 20,000 years: implications for understanding present-day biodiversity. Northwest Environ. J. 8: 5–28.

———. 1993. Postglacial vegetation and climate of Grand Teton and southern Yellowstone National Parks. Ecol. Monogr. 63: 173–198.

Whitlock, C. and P. J. Bartlein. 1997. Vegetation and climate change in northwest America during the past 125kyr. Nature 388: 57–61.

Whittaker, R. H. 1970. Communities and Ecosystems. MacMillan Co., New York.

Whittlesey, C. 1868. Depression of the ocean during the ice period. Proc. Am. Assoc. Adv. Sci. 16: 92–97.

Williams, M. A. J., D. L. Dunkerley, P. De Deckker, A. P. Kershaw, and T. Stokes. 1993. Quaternary Environments. Edward Arnold, London.

Wright, H. E., Jr. (ed.). 1983. Late-Quaternary Environments of the United States, The Holocene. University of Minnesota Press, Minneapolis, MN. Vol. 2.

H. E. Wright Jr., J. E. Kutzbach, T. Webb III, W. F. Ruddiman, S. A. Street-Perrott, and P. J. Bartlein (eds.). 1993. Global Climates Since the Last Glacial Maximum. University of Minnesota Press, Minneapolis, MN.

Wright, H. E. Jr., A. M. Bent, B. S. Hansen, and L. J. Maher, Jr. 1973. Present and past vegetation of the Chuska Mountains, northwestern New Mexico. Geol. Soc. Am. Bull. 84: 1155–1180.

Wright, H. E. Jr., T. C. Winter, and H. L. Patten. 1963. Two pollen diagrams from southeastern Minnesota: problems in the regional late glacial and postglacial vegetational history. Geol. Soc. of Am. Bull. 74: 1371–1396.

Younger Dryas essays. 1993/1995. Quatern. Sci. Rev. 12/14.

General Readings

Bradley, R. S. (ed.). 1991. Global Changes of the Past. UCARC, Interdisciplinary Earth Studies, Boulder, CO.

Broecker, W. S. 1994. Massive iceberg discharges as triggers for global climate change. Nature 372: 421–424.

———. 1995. The Glacial World According to Wally. Lamont-Doherty Earth Observatory, Palisades, N.Y.

Chandler, M. 1996. Trees retreat and ice advances. Nature 381: 477–478.

Clark, J. S. et al. (+ 12 authors). 1998. Reid's paradox of rapid plant migration. BioScience 48: 13–24.

Delcourt, H. R. and P. A. Delcourt. 1991. Quaternary Ecology, A Paleoecological Perspective. Chapman & Hall, New York.

Gibbons, A. 1993. Pleistocene population explosions. Science 262: 27–28.

Gibbons, A. 1996. The peopling of the Americas. Science 274: 31–33.

Figure 9.2. *Larrea tridentata*, Arizona. (Left) Habit. (Right) Close-up showing flowers. The creosote bush also occurs in the dry regions of Argentina and Chile.

South America, and the Antilles and there is presently no evidence for a belt of arid vegetation or subhumid environments (e.g., Graham, 1976, 1985, 1987, 1988a,b, 1989, 1990, 1991a–c; Graham and Jarzen, 1969). Arid habitats were more extensive in the Late Tertiary than in the Cretaceous when maximum elevations were attained in the coastal mountains, and in the Quaternary when aridity developed in association with cooling climates. Indeed, dry areas are more extensive now than at anytime in the Cenozoic. However, in this connection it should be noted that earlier references to the origin of Mediterranean climates as dating to the Late Pleistocene (e.g., Raven, 1972; Raven and Axelrod, 1978) require revision in light of the recent Greenland ice core and marine oxygen isotope data. Conditions similar to the latest glacial–interglacial cycle have likely prevailed numerous times since the onset of Late Cenozoic cold climates in the Late Pliocene (~2.4 Ma) and Quaternary (Chapter 2, The Greenland Ice Core Record, The Thermohaline Cell; Chapter 8). In Europe Mediterranean climates date from the Pliocene (Suc, 1984). After separation of North and South America in the Cretaceous, a marine barrier extended across much of Central America until ~3.5 Ma; and it was not until ~2.4 Ma that land surfaces existed for the exchange of large upland terrestrial animals and plants lacking means of long-distance dispersal (Coates et al., 1992; Graham, 1992). It is likely that the affinity is of recent origin and is a result of long-distance dispersal. Many of the plant examples are self-compatible species, occupy open habitats convenient for introduction and establishment, and have edible and/or small seeds. Each year millions of birds migrate between the regions (e.g., *Charadrius vulgaris*, the semipalmated plover; see Raven, 1972) affording one mode of transport for plant propagules.

FLORISTIC AFFINITIES BETWEEN EASTERN NORTH AMERICA AND EASTERN ASIA

The deciduous forest of eastern North America merges northward through the lake states forest into the boreal forest and is bordered on the west by the midcontinent grasslands (Fig. 9.3). Further west and south there is the low-altitude desert and mid- to high-altitude montane coniferous forest where deciduous species are confined mostly to gallery and mixed gymnosperm–angiosperm transition vegetation. In western Europe natural arborescent vegetation is depauperate due to a combination of topography, climatic events in the Late Cenozoic and Quaternary, and human history (see later section). In the Middle East, arid conditions developing in the Late Tertiary reduced and currently prevent the widespread occurrence of deciduous forest. The community reappears in climatically and topographically suitable areas of the Far East, especially in Japan and in the central provinces of China, where it shows a remarkable similarity to the vegetation of eastern North America (Graham, 1972a).

The first recognition that plants similar to those of

Figure 9.3. Eastern North America–eastern Asia distribution of *Carya*. Reprinted from Wu (1983) with the permission of Zhengyi Wu and the Missouri Botanical Garden.

North America also occur in eastern Asia is recorded in the dissertation of Jonas P. Halenius (1750; Fig. 9.4), a student of Linnaeus, entitled Plantae Rariores Camschatcenses (Graham, 1966, 1972b). Linnaeus sponsored 186 of these student theses that constitute a series known as the *Amoenitates Academicae*. In those days, the procedures for obtaining advanced degrees differed considerably from those of today. In Sweden until about 1885 and later in other European universities, it was the responsibility of the student to defend in public debate in Latin a thesis for which the professor was largely responsible. The purposes of the examination were to demonstrate the student's familiarity with the rules of formal disputation and fluency in Latin. The content of the thesis and its originality did not matter greatly. Writing about another student's thesis, Linnaeus stated, "Mr. Kalm ought to dispute pro gradu *de Oeconomiae patriae augmento opera L. B. Bjelke.* This he ought to do out of gratitude . . . I will dictate the first version." This system differs considerably from the modern one in which the student is primarily responsible for the research but is mercifully spared the ordeal of publicly defending it in Latin within the rules of formal debate. It is an interesting concept, however.[1]

In paragraph IV of the dissertation, Halenius (Linnaeus) notes that

> Last summer Gregory Demidoff, a distinguished Ukrainian and excellent judge of plants, submitted to the examination of our President, an enormous collection of very rare plants which Lerche, a botanist of great insight, has collected toward the end of last year in Kamchatka, the remotest part of Asia and that closest to the American continent. With some astonishment, I saw among them not only a great number which are found in Lapland, but others also—some of which were altogether unknown, some only slightly treated before. Finally I saw some which are also found in Canada, the reason being that Canada is not far distant from Kamchatka. The following are examples of species previously seen only in North America but now found also at the farthest limits of Siberia. (1750)

The list is in the polynomial form used prior to Linnaeus' consistent application of his binomial system and includes *Claytonia*, *Anemone*, *Paris*, *Kleinia*, *Heuchera*, *Uvularia*, *Spiraea*, *Asplenium*, and *Lycopodium*.

In a 1787 reprint of Linnaeus' Critica Botanica the list is given according to the binomial system and Hultén (1927–1930) cites the names used in Plantae Rariores Camschatcenses in his synonymmies for the Flora of Kamtchatka. This makes it possible to relate the names in Halenius's thesis to those now in use. It is evident from the combined distribution of the plants that Linnaeus was referring to eastern North America because this is the region where they grow together. On the other hand, it is unlikely that Linnaeus was aware of the major disjunction that the plants represented because he knew virtually nothing about the unexplored vegetation of western North America. His knowledge of the North American flora came from the collections of Clayton in Virginia and those of Sarrazin and Gaulthier in eastern Canada and he was probably commenting more on the widespread distribution of the plants than on disjunct relationships with the vegetation of eastern Asia. That analysis was made by Asa Gray (1840, 1846, 1859) and subsequently documented in greater detail by others (Boufford and Spongberg, 1983; Graham, 1972a; Li, 1952; Tiffney, 1985; Wood, 1970, 1972). Among the tree and shrub genera illustrating the pattern are *Acer*, *Aesculus*, *Buckleya*, *Carya* (Fig. 9.3), *Celastrus*, *Cornus*, *Gleditsia*, *Diospyros*, *Gordonia*, *Gymnocladus*, *Hamamelis*, *Ilex*, *Liriodendron*, *Magnolia*, *Mahonia*, *Nyssa*, *Pieris*, *Rhus*, *Robinia*, *Sassafras*, *Symplocarpus*, *Symplocos*, *Vitis*, and *Zanthoxylum*.

The forests of eastern North America and eastern Asia have been viewed as remnants of the Arcto–Tertiary geoflora that extended around the temperate latitudes of the northern hemisphere primarily from the Late Eocene through about the Middle Miocene and in Europe until the Early Quaternary. To the extent that the concept implies a relatively uniform broad-leaved deciduous forest throughout this zone, it requires modification to include a prominent evergreen gymnosperm element (e.g., *Sequoia*), to

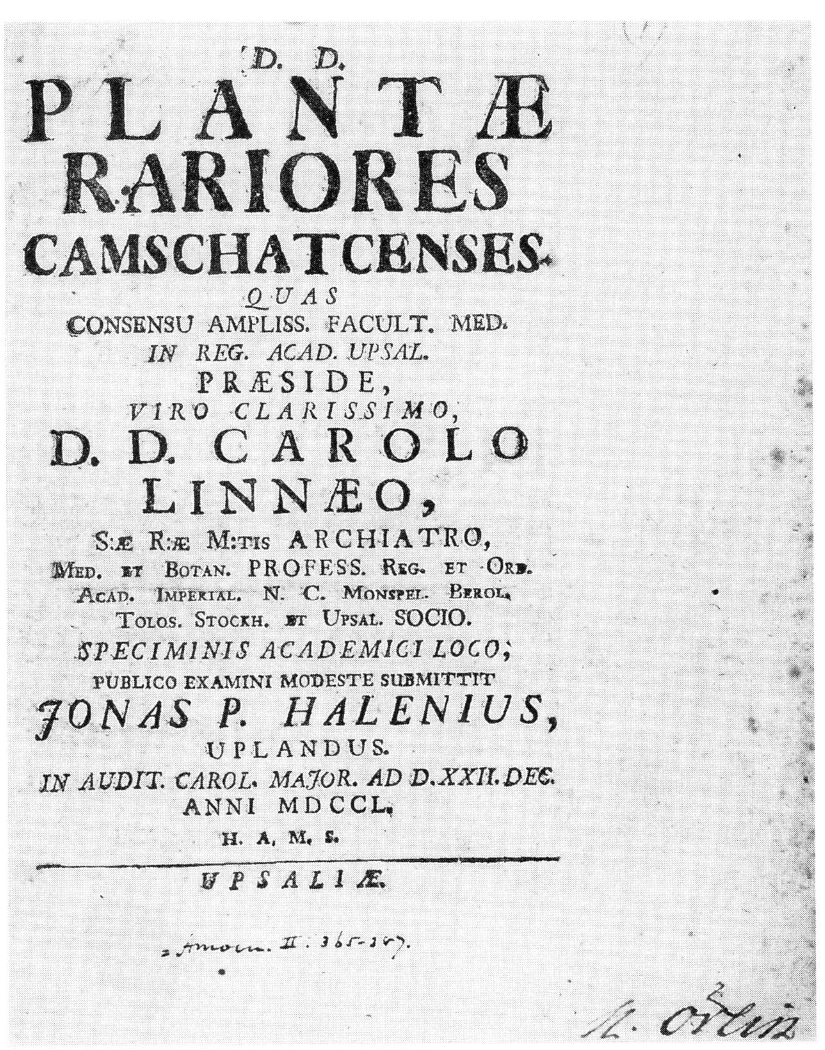

D. D.

PLANTÆ
RARIORES
CAMSCHATCENSES.

QUAS
CONSENSU AMPLISS. FACULT. MED.
IN REG. ACAD. UPSAL.
PRÆSIDE,
VIRO CLARISSIMO,
D. D. CAROLO
LINNÆO,
S:Æ R:Æ M:TIS ARCHIATRO,
MED. ET BOTAN. PROFESS. REG. ET ORD.
ACAD. IMPERIAL. N. C. MONSPEL. BEROL.
TOLOS. STOCKH. ET UPSAL. SOCIO.
SPECIMINIS ACADEMICI LOCO;
PUBLICO EXAMINI MODESTE SUBMITTIT
JONAS P. HALENIUS,
UPLANDUS.
IN AUDIT. CAROL. MAJOR. AD D. XXII. DEC.
ANNI MDCCL.
H. A. M. S.

UPSALIÆ.

Figure 9.4. Title page from the 1750 dissertation of Jonas P. Halenius, a student of Linnaeus, where first reference is made to the similarities in vegetation between North America and Asia.

avoid the impression that its demise in Europe was exclusively a Quaternary event (Chaney, 1944), and that it was a monolithic assemblage responding to climatic change as a unified block of vegetation (see Davis, 1983; Wolfe, 1978, 1979). In North America, and especially in Europe, many temperate deciduous trees were already disappearing in the Tertiary. The essential point, however, is that even though the structure, composition, and history of the northern temperate biota have been oversimplified by the geoflora concept, there was a belt of vegetation similar in structure and composition extending around much of the northern hemisphere during the Tertiary. Present-day Asian genera that occur in the fossil record of North America include *Ailanthus* (also widespread in the Tertiary of Europe), *Amentotaxus, Cunninghamia, Ginkgo, Glyptostrobus, Metasequoia, Castanopsis, Cercidiphyllum, Craigia, Cyclocarya, Diplopanax, Dipteronia, Engelhardia, Koelreuteria, Mastixia, Pachysandra, Platycarya, Tapiscia, Trapa, Trochodendron*, and others. Fossil taxa shared between the Paleocene Fort Union flora and the Altai flora of Xinjiang in northwestern China include *Corylites, Ditaxocladus, Nordenskioldia, Nyssidium, Paleocarpinus, Platanus, Trochodendroides*, and *Ziziphoides* (Manchester, 1995; Manchester and Guo, 1996). The present-day American genera *Chamaecyparis, Taxodium, Thuja, Torreya, Tsuga, Asimina, Carya, Diospyros, Lindera, Liquidambar, Liriodendron, Magnolia, Morus, Nyssa, Persia, Robinia, Sabal, Sapindus*, and *Sassafras* were in Europe during the late Cenozoic (Niemelä and Mattson, 1966). It was the disruption of various sectors of this belt at different times during the Cenozoic that produced the residual floristic similarities evident between eastern North America and eastern Asia.

An early disruption occurred in the midcontinent region of North America. Recall that after the retreat of the Cretaceous sea there was an initial lowering of temperature and a diversification of Juglandaceae (*Carya*), along with the appearance of *Ginkgo, Glyptostrobus, Metasequoia*, and genera identified as *Alangium, Betula, Carya, Castanea, Celtis, Cornus, Corylus, Juglans, Magnolia, Pterocarya, Quercus, Symplocos, Ulmus*, and *Zelkova* in the Fort Union flora. In eastern North America pollen of *Alnus, Betula, Carya*, and *Platycarya* is preserved in the Paleocene Oak Grove core of northern Virginia; in addition, megafossils of *Metasequoia*, platanoids, and trochodendroids are

Figure 9.5. Mixed broad-leaved deciduous and needle-leaved evergreen forest, Changpai Mountains, China (elevation 500–1000 m). Trees include *Acer*, *Carpinus*, *Fraxinus*, *Juglans*, *Tilia*, and *Ulmus*. Reprinted from Hou (1983) with the permission of Hsioh-Yu Hou and the Missouri Botanical Garden.

known from the Atanikerdluk flora of Greenland. The Late Paleocene–Early Eocene Thyra Ø flora of Greenland includes *Cephalotaxus*, *Fokienia*, *Ginkgo*, *Metasequoia*, Betulaceae, *Cercidiphyllum*, cf. *Corylus*, *Platanus*, and others. In the midcontinent of North America this vegetation was replaced by a seasonally dry broad-leaved evergreen forest in the Late Paleocene and Eocene (Willwood flora). Paleosol evidence then suggests dry deciduous forest (34 Ma), dry woodland (33 Ma), wooded grassland with gallery forest (32 Ma), and more open savanna with increasing grasses (30 Ma). The Kilgore flora of northern Nebraska (13–14 Ma) is a streamside assemblage of broad-leaved deciduous plants with *Pinus*, *Artemisia*, and grasses in the surrounding uplands. Continued lowering of winter temperatures and more seasonal distribution of rainfall further disrupted the deciduous forest and favored savanna and grassland in the Plains. Further to the west, the rich Miocene forests of the Columbia Plateau were replaced by expanding montane coniferous forest, steppe, and desert in the Late Miocene and especially in the Pliocene with the rise of the Sierra Nevada and Coast Ranges. Asian exotics (*Alangium*, *Cyrilla*, *Trapa*, *Pterocarya*, *Zelkova*) had mostly disappeared from the eastern North American forest by the Pliocene, persisting somewhat later west of the Rocky Mountains. The trend toward grassland in the Plains and desert, steppe, and montane coniferous forest in the west culminated in the Quaternary with virtual elimination there of remnants of the deciduous forest. For North America this left a modified version of the earlier formation only in the eastern part of the continent.

During the Late Tertiary and Quaternary, the diverse physiography and the north–south orientation of the American mountains allowed the deciduous forest to persist in local refugia and to migrate in response to changes in climate. In Europe, however, the vegetation was caught between the advancing Fennoscandian ice sheet from the north and the alpine glaciers of the east–west trending Pyrennes–Alp mountain system to the south. The result was the disappearance of the gymnosperm–broad-leaved deciduous forest during glacial intervals. Some elements probably persisted in refugia on the southern slopes of the Alps and in the Middle East (van der Hammen et al., 1971); but like glacial refugia in the southeastern United States, subsequent reforestation was a complex process with no single center serving as a central refuge and with various deciduous forest elements migrating as individuals. Later, as Holocene forest regeneration was underway, agriculturalists began moving in from the Fertile Crescent between the Tigris and Euphrates Rivers, clearing the land with fire and ax and planting crops, as reflected in the dramatic pollen diagrams of Iversen (1941). With the development of arid conditions in much of the Middle East in the Late Tertiary and Quaternary, remnants of the once widespread temperate gymnosperm–broad-leaved angiosperm forest remain primarily in eastern North America (Fig. 1.9) and eastern Asia (Fig. 9.5).

FLORISTIC AFFINITIES BETWEEN EASTERN NORTH AMERICA AND EASTERN MEXICO

The deciduous forest formation extends into northeastern Texas where a number of temperate genera reach their southern limits in the conterminous United States (Fig. 9.6; Chapter 8, Southeast). Farther south are the coastal grasslands and interior Chihuahuan Desert of Texas and north-

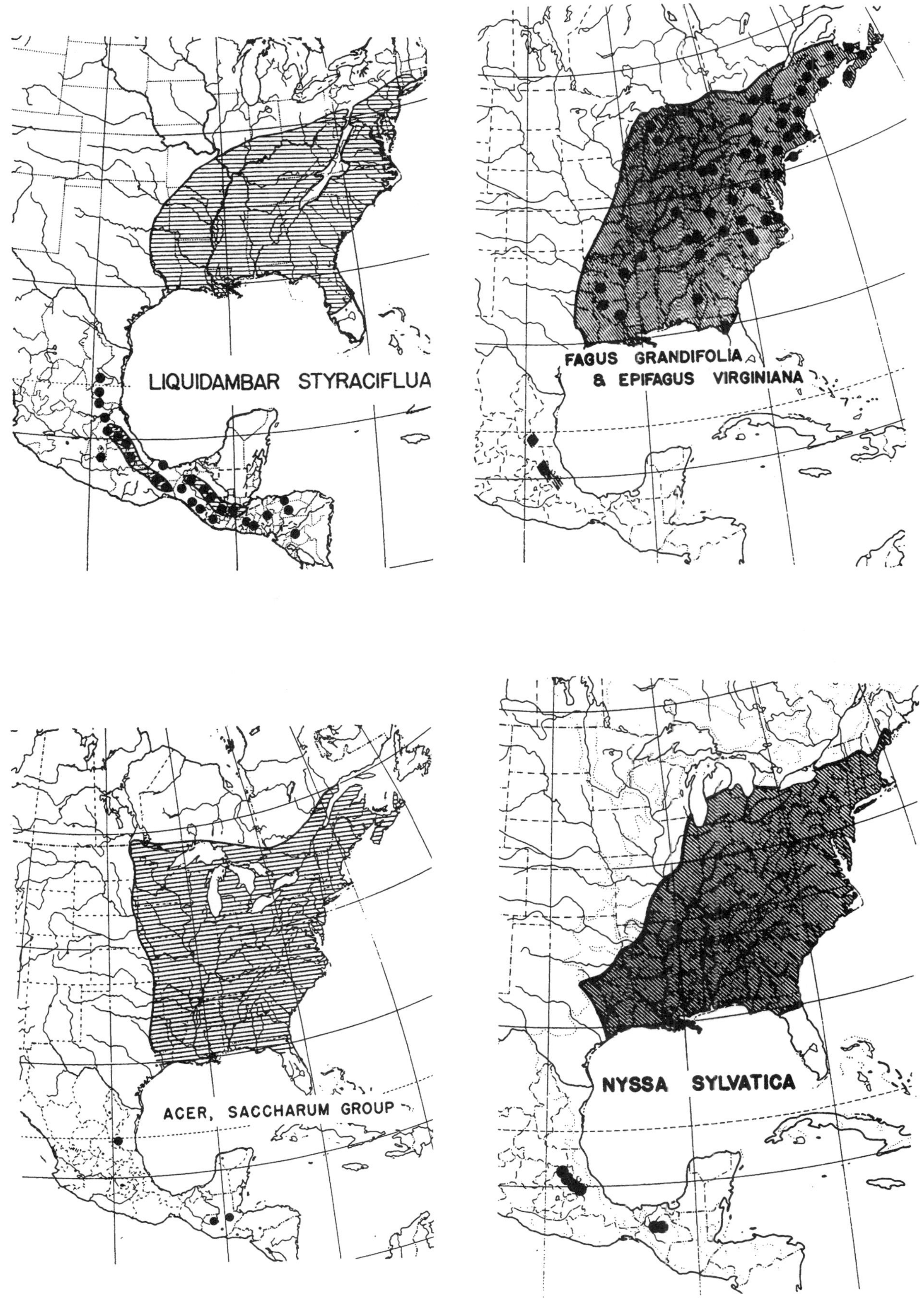

Figure 9.6. Eastern North America–eastern Mexico distribution of four forest taxa. Reprinted from Martin and Harrell (1957) with the permission of The Ecological Society of America.

Figure 9.7. Remnants of the deciduous forest, Jalapa, Veracruz, Mexico. The principal trees are *Liquidambar* and *Quercus*.

ern Mexico. Elements of the temperate broad-leaved deciduous forest reappear in a zone between ~1000 and 2000 m along the eastern escarpment of the Mexican Plateau and the Sierra Madre Oriental (Gómez–Pompa, 1973; Fig. 9.7). They extend in diminishing numbers through the state of Chiapas in southern Mexico (Breedlove, 1973) and into Guatemala. *Pinus* reaches its southern limit in the Cordillera Dariense of Nicaragua (*P. oocarpa, P. patula*; Perry, 1991; Styles, 1993); *Alnus, Ilex, Myrica*, and *Quercus* occur scattered through Central America into northern South America.

Among the plants disjunct between the eastern United States and eastern Mexico are species of *Acer, Alnus, Carpinus, Carya, Cercis, Cornus, Fagus, Fraxinus, Juglans, Liquidambar, Magnolia, Myrica, Nyssa, Ostrya, Platanus, Prunus, Rhus, Smilax, Tilia*, and *Ulmus*. The floristic affinity between the eastern United States and eastern Mexico has been noted by Fernald (1931), Graham (1973a), McVaugh (1943), and especially by Miranda and Sharp (1950). These and other authors have viewed the plants as Tertiary relicts, as ultimately they have proved to be. However, this was initially an intuitive assessment because there were no paleobotanical data to indicate when northern temperate elements first became established in Mexico. Recall that Potzger and Tharp (1943, 1947, 1954) had reported up to 11% pollen of *Abies* and *Picea* from central Texas; Brown (1938) identified megafossils from Quaternary deposits in Louisiana as northern species of *Larix, Picea*, and *Thuja*; and Wilson (in Davis, 1946) found pollen of *Picea* in peats from Florida. These reports seemed to suggest that climates in the southeastern United States were so cold that the boreal coniferous forest extended to the Gulf Coast. The implication was that elements of the deciduous forest were displaced even farther south into Mexico and that with the return of warmer conditions in the Holocene, remnants were left stranded in the midaltitude temperate zones of eastern Mexico (Deevey, 1949). By this model the origin of biotic affinities between the two regions dates primarily from the Pleistocene. Another implication of the model is that if these climatic changes and migrations characterize the most recent glacial–interglacial cycle, they must have happened repeatedly throughout the Quaternary. Earlier it was thought that there were only four of these cycles within the last 1.6 m.y. at intervals of ~175,000 years. It is now known from oxygen isotope, Greenland ice core, and other data that there were 18–20 such cycles; this conjures up a tempo of disruption and reunion reminiscent of Joleaud's opening and closing of the Atlantic Ocean.

As noted in Chapter 8, the absence of sufficient information from the fossil record of the Gulf Coast and from eastern Mexico made it difficult to evaluate the model. However, other studies on the biogeography and taxonomy of modern assemblages were creating an intuitive feeling that modifications would be forthcoming as new data emerged. For example, the biotic affinities that exist between the eastern United States and eastern Mexico are mostly at the generic level, implying a substantial period of geographic isolation. This is evident in the vegetation (Braun, 1955; Miranda and Sharp, 1950), reptiles and amphibians (Martin, 1958), and vertebrates (Martin and Harrell, 1957). If climatic events and the displacement of biotic zones were as extensive and frequent as suggested by the early paleobotanical data, there would be multiple contacts between the biotas extending up to ~11 Kya, which would likely break down reproductive barriers and maintain similarity at the species level. There is no evidence in southern Texas and northeastern Mexico for repeating buried soil horizons of the podzol type required to support a rich deciduous forest, and there is no archeological evidence that AmerIndian populations changed cultural activities from those typical of present-day central Texas to ones suggesting climates of the upper midwest.

Later, several lines of evidence further documented the need to modify concepts about the time of origin of floristic affinities between the eastern United States and eastern Mexico. One was the discovery that the 11% *Abies* and *Picea* pollen reported from the central Texas bogs was incorrect and that boreal conifers were represented by ~1.5% of *Picea* only (Graham and Heimsch, 1960). Another was that megafossils of the boreal *Thuja occidentalis* reported from Louisiana were actually the southern white cedar *Chamaecyparis thyoides* (Delcourt and Delcourt, 1977). These revisions showed that although some *Picea* filtered southward, closer to the bogs in central Texas from more extensive populations in the upper and central midwest, there was no evidence of a boreal coniferous forest extending all the way to the Gulf Coast. This further reduced the likelihood that the eastern Mexican temperate elements were exclusively a result of repeated and frequent interchange during the Quaternary.

Evidence that north temperate deciduous forest species were already established in eastern Mexico in the Tertiary came from a study of pollen and spores from the Paraje Solo Formation near Coatzacoalcos, Veracruz, Mexico (Graham, 1976). These deposits were initially estimated to be Miocene in age, but ostracode and nannofossil assemblages now show they are of Middle Pliocene age (~3.4 Ma). Among the plant microfossils recovered were *Abies*, *Picea*, *Alnus*, *Celtis*, *Ilex*, *Juglans*, *Liquidambar*, *Populus*, *Quercus*, and *Ulmus*. It has been suggested that these representatives of the deciduous forest were not introduced as a result of Late Cenozoic cooling (Graham, 1973b) but were remnants of a more ancient widespread vegetation existing from Paleogene times (Axelrod, 1975). Initially the paleobotanical data showed only that none of these elements were present in Eocene or Miocene deposits of northern Latin America (Graham, 1985 et seq.) but that they were present in the Eocene and later epochs of the eastern United States (Frederiksen, 1981; Gray, 1960), suggesting a southward introduction with cooling Cenozoic climates. A new study of plant microfossils from the Simojovel Group of Chiapas, Mexico (A. Graham, unpublished data), further shows that northern temperate elements were not yet established in that region in the Oligo–Miocene. Although the data base for northern Latin America is still meager, if elements of the eastern deciduous forest were present there in Early and Middle Tertiary times, these phantom forests left no evidence in the fossil floras known to date. As noted long ago by Arnold with reference to Arber and Parkin's (1907) hypothetical Hemiangiospermae, "workable phylogenies [and paleocommunities] cannot be built upon forms that exist only in the mind." (1947). The paleotemperature curve (Fig. 3.1) and the existing paleobotanical evidence suggests that the initial affinities probably date from the Middle Miocene when global temperatures reached new lows, allowing the deciduous forest to reach its southernmost extent. Later Cenozoic and glacial–interglacial aridity left elements isolated in the midaltitude temperate environments of eastern Mexico. The question is still open as to whether some elements were reintroduced after the Miocene through climatic change and by long-distance transport. This possibility is currently being addressed through new and innovative biogeograpic analyses.

EMERGING METHODS IN BIOGEOGRAPHY

An essential prerequisite for determining the origin of biogeographic affinities is an accurate phylogeny. Taxonomic treatments that do not closely reflect phylogenetic relationships may suggest false patterns and affinities by unifying taxonomically heterogeneous assemblages or dividing natural assemblages into taxa distinguished primarily by geography. As noted by Raven:

> The moss *Macromitrium sullivantii* C. Müll., thought to be an endemic of a small area of the southeastern Blue Ridge escarpment of the United States, has recently been shown to be identical with *Macrocoma hymenostomum* (Mont.) Grout, a well known species that occurs throughout the American tropics and South America. The range as now understood becomes a striking example of a disjunct distribution. On the other hand, the genus *Boisduvalia* (Onagraceae), comprising six species of semiarid western North and South America, was until recently thought to include a species of the mountains of southeastern Australia and Tasmania, *B. tasmanica* (Hook. f.) Munz. With the demonstration that this species is actually an *Epilobium*, now known as *E. curtisiae* Raven, which forms natural hybrid populations with closely related Australian species, the situation demands a very different interpretation. (1972, p. 234)

The realization has emerged, especially within the past decade, that morphological features alone are often insufficient to reveal phylogenetic relationships. Powerful new data sources and analytical tools are now available to complement traditional morphological approaches. In addition to clarifying natural relationships, these techniques are being applied to questions of biogeography. As noted by Bremer, "It is the previous methodological approaches to the search for centers of origin [and patterns of subsequent dispersal], not the search per se, that are spurious." (1992). The new methods are being refined and are yielding results of potentially great value to biogeography.

Allozyme–Isozyme Analysis and Genetic Distance

It has been known for some time that within individual organisms there are multiple forms of enzymes that catalyze the same reaction (Hunter and Markert, 1957; Markert and Møller, 1959). Earlier, when the genetic basis for these different forms was unknown, they were collectively referred to as isozymes. Now that genetic factors determining some of them are better understood, a more precise terminology is emerging. Different forms of an enzyme encoded by different alleles at the same gene locus are called allozymes. Forms catalyzing the same reaction encoded at different loci, or when their genetic basis is unknown, are called isozymes (Crawford, 1990). The enzymes may be separated from leaves, seeds, and other plant tissue by gel electrophoresis and bands on the gel can be compared within populations of the same species and between different species. The data may be analyzed via genetic identity or genetic distance calculations (Nei, 1972), and a value is obtained that may reflect the degree of relationship and relative times of divergence. The method has been used to assess the relationship between populations and species of *Liquidambar* disjunct in eastern North America, eastern Asia, and eastern Mexico.

Although the species of *Liquidambar* are morphologically similar, the level of isozyme divergence is comparatively high (Hoey and Parks, 1991, 1994). *Liquidambar styraciflua* in the eastern United States shows little divergence among different populations, while those in eastern Mexico exhibit greater divergence. This may be the result of range restriction into the various refugia described for eastern North American during glacial intervals so that most of its current range is of recent origin. If populations in eastern Mexico were not subjected to the same homogenizing effect of Quaternary climatic change, they may have experienced less alteration in range and persisted as a more continuous community for longer periods of time.

Another intriguing possibility raised by the isozyme studies is that there may have been contact between the populations of *L. styraciflua* in the eastern United States and eastern Mexico after their initial introduction into Mexico in the Tertiary. There is high genetic identity among three populations from Mexico and those from the eastern United States (0.985, 0.930, 0.940). These values are greater than would be expected if the populations had been isolated since the Middle to Late Miocene (see discussion below). This supports the possibility of some interchange through climatic fluctuations during the late Cenozoic and Quaternary and throughout the Cenozoic by long-distance dispersal.

A further example of the potential value of isozyme studies for better understanding North American floristic affinities is provided by comparison of populations of *Liquidambar* in eastern North America and eastern Asia. Genetic identity calculations between the eastern American *L. styraciflua* and the Asian *L. orientalis*, *L. formosana*, and *L. acalcina* gave values of 0.512, 0.431, and 0.485. This suggests a longer period of isolation than between the eastern North American–eastern Mexican populations, dating perhaps also from the Miocene but without subsequent interchange. Similar molecular divergence analyses have been made on the *Magnolia* section *Rytidospermum* (Qiu et al., 1995a). When molecular data for the taxa were compared, several levels of divergence were evident among the eastern Asia–eastern North America disjuncts. This is consistent with the emerging view that organisms migrate as individuals rather than as unified blocks of vegetation; that is, various genera and species within a genus may respond differently and follow individual routes of migration in response to changes in climate and the physical environment, encounter different barriers over time, and be selectively affected by epidemic disease. All of these data are preliminary, but they clearly demonstrate the usefulness of molecular information in better quantifying taxonomic relationships and in establishing the time of origin of floristic affinities.

Molecular Clocks

Another early observation that has great potential for phylogenetic and biogeographic studies was that the rate of amino acid substitution in hemoglobin and cytochrome C proteins is about the same among different mammalian groups (Margoliash, 1963; Zuckerkandl, 1987; Zuckerkandl and Pauling, 1962, 1965a). Later Zuckerkandl and Pauling (1965b) proposed that the rate of protein change (molecular evolution) is constant over time in all lineages—that is, there is a molecular clock (Li and Graur, 1991). The method is relevant to biogeographic problems of the North American flora because, for example, it has been suggested that the Mexican populations of *Liquidambar* (*L. macrophylla*, Gómez–Pompa, 1973) represent a separate species from that in eastern North America (*L. styraciflua*). If so, the protein differences could be used to date the time of species divergence that would also reflect the time of geographic isolation.

Unfortunately, there are uncertainties in the method that need to be resolved. One is that the uneven pace of speciation reflected by the fossil record is not in accord with the constant rate implied by the molecular clock.

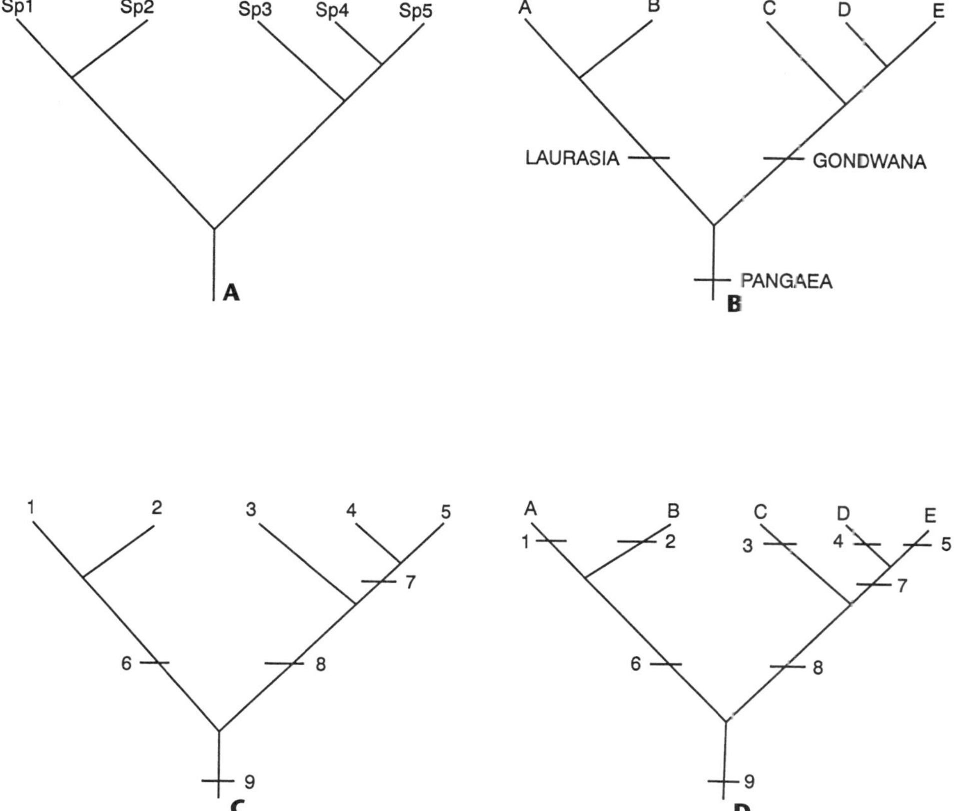

Figure 9.8. Steps in the preparation of area cladograms from phylogenetic and historical geologic data. (A,D) A, Eurasia; B, North America; C, Australia; D, South America; and E, Africa. Adapted from Brooks and McLennan (1991, chapter 7).

Modern genetic theory also suggests that the tempo of evolution varies among different groups of organisms and that it is much more rapid during times of adaptive radiation, at the margin of ranges where the taxon is at the limits of its ecological tolerance, and during times of increased environmental instability. Other problems include different generation times for various lineages and the differing efficiency of DNA repair systems. Thus, it has yet to be established that there is a molecular clock and that it runs at a constant pace. Even so, it is worthwhile for biogeographers to follow developments in this field because refinements could make available an additional technique for investigating the temporal history and relationships of elements in the North American flora.

Cladistics, Area Cladograms, and Vicariance Biogeography

Another powerful tool for estimating natural relationships among taxa is cladistic analysis (Cronquist, 1987; Donoghue and Cantino, 1988; Humphries and Chappill, 1988; Ronquist, 1997 Wiley et al., 1991). Morphological, molecular, and cytological characteristics of taxa may be entered into the rows and columns of a data matrix. The character states are coded in multistate or binary format (e.g., 0 = feature present, 1 = feature absent). The information is subjected to cladistic analysis via software such as PAUP or Hennig86. These programs include facilities for polarizing characters by outgroup comparison and weighting of characters when one (e.g., fusion of petals) is considered more important than others (e.g., petal color). The programs generate a series of trees or cladograms showing branching patterns from which ancestors, divergence, and evolutionary patterns can be inferred and which cluster taxa on the basis of shared derived characters into groups or "clades" (Fig. 9.8A). Several to many trees are usually produced and the selection of the one(s) most likely reflecting the "true" phylogeny is based on the principle of parsimony: in the absence of evidence to the contrary, the shortest tree is favored (i.e., the one invoking the fewest number of steps in the development of each character). Another option is the generation of a consensus tree when there are several equally parsimonious trees. From these cladograms phylogenies can be inferred and monophyletic or "natural" groups (clades) can be identified.

Cladograms are important in biogeography for two principal reasons. One is that they can provide an improved understanding of natural relationships among organisms that allows real rather than imaginary geographic affinities to be deduced (Qiu et al., 1995b). The other is that with the taxon cladogram as a basis, another type of cladogram can be generated by using distributions and events in historical geology from which centers of origin can be inferred (e.g., Bremer, 1992) and which can suggest the origin of some modern patterns of distribution. These are called area cladograms, and this emerging method is based on a synthesis of primarily plate tectonic data and phylogenetic systematics (Nelson and Platnick, 1981). Their construction is described by Brooks and McLennan (1991) and Page (1988).

Assume that a phylogenetic cladogram has been generated for five taxa (Fig. 9.8A). Vicariance events within the present range of the collective taxa are identified. In this example, it is the breakup of Pangaea into Gondwana and Laurasia (Fig. 9.8B). Other possibilities could be the separation of India from Africa, the periodic isolation of England from continental Europe as a result of Quaternary sea-level changes, or the rise of the Rocky Mountains dividing once continuous ranges into eastern and western populations. The taxa are assigned numbers and a list is made of all areas in which they are found:

A = Eurasia	sp. 1 = 1
B = North America	sp. 2 = 2
C = Australia	sp. 3 = 3
D = South America	sp. 4 = 4
E = Africa	sp. 5 = 5

The taxa numbers are placed on the cladogram and each internal branch is numbered (Fig. 9.8C). These numerical codes for the taxa and areas are then entered into a data matrix using 1 as present and 0 as absent:

Taxon	Branches 1 2 3 4 5 6 7 8 9 Binary Code
sp. 1	1 0 0 0 0 1 0 0 1
sp. 2	0 1 0 0 0 1 0 0 1
sp. 3	0 0 1 0 0 0 0 1 1
sp. 4	0 0 0 1 0 0 1 1 1
sp. 5	0 0 0 0 1 0 1 1 1

The species names are then replaced with their geographic distribution:

A Eurasia	1 0 0 0 0 1 0 0 1
B North America	0 1 0 0 0 1 0 0 1
C Australia	0 0 1 0 0 0 0 1 1
D South America	0 0 0 1 0 0 1 1 1
E Africa	0 0 0 0 1 0 1 1 1

An area cladogram can now be generated that shows the evolution of the group in relation to the history of the geographic areas it occupies (Fig. 9.8D). A comparison of Fig. 9.8A (the phylogenetic cladogram) and Fig. 9.8D (the area cladogram) shows complete correspondence between clades and geography. In this simplified example, speciation events correspond perfectly to past geologic history and the inference is that speciation in this group was a consequence of the breakup of Pangaea into Laurasia (spp. 1, 2) and Gondwana (spp. 3–5). The relative chronology of these vicariance events can also be plotted on the area cladogram. There are more direct ways to construct area cladograms (Forey et al., 1994; Humphries and Parenti, 1986), but the example used here provides a more stepwise description.

In actual cases there is usually less than 100% correspondence between phylogenetic and area cladograms, indicating that some process(es) other than vicariance (movement of land) is operating (e.g., movement of organisms or dispersal, and extinctions). However, in ideal situations area cladograms may distinguish between vicariance and dispersal or other events. When different taxa show the same repeated patterns of geographical speciation, it strengthens the proposition that the vicariance event (e.g., the breakup of Pangaea), rather than dispersal, was primary in the phylogeny and in the modern distribution of the clade. In many regions, such as the Caribbean, the geologic history is problematic. Here area cladograms eventually may prove useful in suggesting previously unrecognized tectonic events made evident by biogeographic analysis. There is no consensus on this point, and indeed the whole subject of vicariance biogeography and the practical (as opposed to theoretical) value of area cladograms is still being debated (Hedges et al., 1994; Page and Lydeard, 1994).

For Canada and the conterminous United States the vicariance approach has been used to study the distribution and speciation of fish faunas, correlation of salamander ranges and suspect terranes in the western United States, and mammal occurrences in the montane regions of western North America. A number of animal groups have also been analyzed from the southeastern United States (pocket gopher, deer mouse, sunfish, salamander; Avise et al., 1987). Area cladograms for the grasshopper mouse genus *Onychomys* in arid regions of western North America suggest that its distribution reflects the pluvial–interpluvial phases that fragmented arid habitats (Riddle and Honeycutt, 1990). Among plants, analyses of phylogenies and vicariance events have been made on ferns and angiosperms in Europe (see Wiley, 1988). However, for North America examples from plants are rare, mostly dealing with geographic affinities with distant regions. Vicariance cladograms have been used to suggest that warm temperate species of *Crataegus* crossed Beringia from North America in the Tertiary and migrated southward into eastern Asia (Phipps, 1983). Modern biogeographic studies have been made on the biota of the Hawaiian Islands (Wagner and Funk, 1995), which provides an ideal setting for vicariance biogeographic analyses. In the conterminous United States, however, separation of land fragments by plate tectonic or orogenic mechanisms were either minor, gradual, occurred so long ago, and/or have been overprinted by dispersal, extinctions, and other factors that area cladograms have not been used for explaining distribution–speciation patterns for plant taxa in this continental setting.

Cladistic methods assume hierarchical rather than reticulate patterns of evolution (Rieseberg, 1995). In other words, cladistics does not handle reticulation (viz., hybridization) well. As an extrapolation, it is also true that vicariance biogeography does not handle the collision of land fragments bearing separate biotas well. This has occurred among the islands of the Caribbean. For example, in the Late Miocene Haiti and the Dominican Republic fused into the present-day island of the Dominican Republic. The suture is the Cul-de-Sac and Enriquillo Basin north of Port-au-Prince (Graham, 1990). Such events have not extensively affected the biota of North America, although there are exceptions like the transported Mint Canyon and Carmel floras of California and the Yakutat block of Alaska.

Although there are obvious limitations to the use of vicariance biogeography and area cladograms, the potential is there for important contributions to our understanding of distribution patterns. Through continued reexamination of assumptions and refinement in procedures, area cladograms that have proven useful in identifying centers of origin and pockets of speciation and endemism may suggest probable tectonic–orogenic events, geographic affinities, and pathways of migration for elements of the North American flora. In addition, a by-product of these emerging methods is that they have revived interest in biogeography and are attracting a new generation of investigators intrigued by molecular, mathematical, and theoretical considerations; this is to the benefit of the science.

EPILOGUE

This survey of North American vegetation and paleoenvironments began with a Late Cretaceous continent having rugged mountains in the east and modest topographic relief in the west. It was divided into eastern and western provinces by an epicontinental sea that extended from the Gulf of Mexico to the Arctic Ocean. Sea level was higher by 300 m, land area was 40% less, and temperatures were 10–12° C higher than at the last glacial maximum. The vegetation consisted of four formations: tropical forest, paratropical rain forest, notophyllous broad-leaved evergreen forest, and polar broad-leaved evergreen forest. Initially gymnosperms formed the canopy in many areas; various herbaceous–shrubby dicots, palms, ferns, and bryophytes formed the ground cover. Angiosperms were more prominent in disturbed areas, as along streams, where the unpredictability of the habitat contributed to the development of deciduousness.

Increased warmth and precipitation in the Paleocene and Early Eocene resulted in a tropical rain forest, paratropical rain forest, notophyllous broad-leaved evergreen forest, and a much restricted polar broad-leaved deciduous forest in the uplands of the high latitudes.

Cooling in the Middle Eocene produced intermittent glaciers on Antarctica and the beginning of a montane coniferous forest in the volcanic highlands west of the Rocky Mountains. In the Oligocene global climates fluctuated between cool and warm, modified locally by topography. Glaciers gradually became a more widespread and consistent feature of the Antarctic landscape and increasingly contributed cold water to circulating ocean currents. Elements of the broad-leaved deciduous forest and the montane coniferous forest consolidated into communities resembling the modern formations. The montane coniferous forest moved into lower elevations to form the boreal coniferous forest. The broad-leaved deciduous forest spread southward and joined elements from other locales in a complex pattern of evolution and migration that has been oversimplified by the geoflora concept. Another drop in temperature in the Middle Miocene ushered in Arctic glaciation, cooler temperatures, and more seasonal rainfall. Preadapted local Eocene progenetors of shrubland/chaparral–woodland–savanna consolidated, the broad-leaved deciduous forest extended as far south as the uplands of eastern Mexico, and trees began to decrease in the central and northern plains and in the high latitudes. By the Late Pliocene temperatures were even cooler and rainfall was less and more seasonal. North American vegetation then consisted of near tundra, coniferous forest, deciduous forest, near grasslands, near deserts, shrubland/chapparal–woodland–savanna, and a few residual tropical elements in the southeastern United States. After 1.6 Ma the modern plant formations had developed, but they were characterized by associations of a composition and range that fluctuated with the rapid environmental changes of the Quaternary. For most of the past 1.6 m.y., which was glacial in nature, the prominent vegetation formations of North America were widespread coniferous forest (boreal and montane); a deciduous forest restricted to local refugia; grassland; woodland (oak–hickory, juniper–piñon pine) in warmer regions, spruce savanna in the upper midwest, and *Artemisia* steppe in cooler western regions; elements of a desert vegetation restricted to favorable slope and edaphic habitats in the pluvial southwestern United States; and expanded Arctic and alpine tundra. The present interglacial represents conditions of anomalous warmth with widespread deciduous forest in the east, desert in the west, and tropical elements in the southeastern United States.

The recent decade has witnessed many new and improved methodologies and numerous contributions of fundamental importance to our understanding of North American Late Cretaceous and Cenozoic environments. There has been recognition from the study of ocean and Greenland–Antarctic ice cores that great and sudden climatic changes characterize recent geologic time, especially at glacial–interglacial transitions. It seems increasingly likely that the forcing mechanisms for these changes are more global in extent than suggested by earlier models based primarily on data from the North Atlantic. The new data include evidence for cool intervals in Florida and the Pacific Northwest, which were identified from pollen and spore diagrams, that correlate with Heinrich events. Some diagrams from the Pacific Northwest also provide preliminary evidence for a cold period corresponding to the Younger Dryas. The view is emerging that fluctuations and cooling in the tropics (Barbados, Peru) were greater than suggested by the earlier CLIMAP studies. Glaciers were present on Antarctica in the Eocene, rather than first appearing in the Miocene; the real extent of exotic terranes in the American west is being documented; and a greater role for solar variablity is emerging as a potential forcing mechanism for climatic change.

Paleopalynological and paleobotanical studies are also contributing other new information that is modernizing our concepts of North American vegetational history. Detailed paleofloristic studies and revisions are underway on floras from the Eocene of the southeastern United States; the Paleocene and Eocene of the Plains states and adjacent northern Rocky Mountains; and in Oregon, Washington, and Alberta. Additions continue to be made to Miocene floras from Vermont and Oregon, and a better understanding is emerging of megafossil and microfossil floras from the high latitudes. These revisions allow paleomonographic studies of lineages, which are facilitated by increasing international cooperative efforts. Such efforts are witnessed by the recent Workshop on US–Eastern European Collaboration in Paleobotany organized by Steven Manchester and sponsored by the National Science Foundation as part of the International Organization of Paleobotany meetings in Santa Barbara, California, in 1996. Other cooperative projects were initiated or further developed as part of the International Conference on Diversification and Evolution of Terrestrial Plants in Geological Time in Nanjing in 1995, and there are numerous other examples. This trend toward globalization of paleobiological research, and its integration into other types of large-scale environmental studies, is an exciting development and it is essential for achieving a full and accurate understanding of biotic history.

As large amounts of data accumulate, it has become necessary to communicate the results electronically and to develop data bases to store and manipulate information. Examples can be located on the Worldwide Web via search engines under such topics as paleobotany, palynology, paleoclimatology, and related subjects. They include the World Data Center-A (http://www.ngdc.noaa.gov/paleo/paleo.html), the bulletin board POLPAL-L (listserv@uoguelph.ca), the commercial Palynodata Inc. (CD-ROM disks by subscription), the home page of the American Association of Stratigraphic Palynologists, the list service PALEOLIM (paleolim@nervm.nerdc.ufl.edu), the Plant Fossil Record Data Base, Electronica Paleontologica (http://www.nhm.ac.uk/paleonet/pe/glines.html) and many others.

Detailed and high-resolution studies of individual fossil floras are revealing the effects of volcanism on floras, faunas, and their representation in the fossil record. These studies further reveal that abrupt threshold changes probably characterized climates and biotas throughout Late Cretaceous and Cenozoic time, as shown by the recently identified cool interval within the Paleogene period of maximum warmth. Paleosols, otoliths, phytoliths, packrat middens, dendrochronology, isotopes, seismic profiles, and stomatal analyses are further expanding the arsenal of investigative techniques.

There has been a gradual, hard-fought acceptance that the idea of geofloras requires at least substantial renovation, with the concomitant emergence of the boreotropical concept as an alternative for envisioning the development and movement of high-latitide biotas through time. The temporal nature of associations and the increasing documentation that organisms migrate as individuals requires rethinking not only of geofloras, but also of time-honored views about climax communities. History has repeatedly shown, however, that new ideas can generate exuberant claims of success, as well as entrenched attachment to old ways and previous positions. Eventually a rational balance is usually achieved. In the case of community stability versus randomness, climax, and geofloras, it must also be acknowledged that some assemblages do exist in the Tertiary that resemble modern associations and that more and longer lasting ones are evident at the level of formations.

Equally vigorous debates have centered around the use of leaf physiognomy as a means of reconstructing past climates, but continued refinements and pioneering innovations such as CLAMP and associated statistical analyses are providing convincing successes for the method. In turn, this has generated new interest and opportunities for estimating paleoelevations. If local paleoelevations near modern values were present in western North America by the Paleogene, this would be of great importance for models attempting to simulate the effects of uplift on atmospheric circulation and climate. The challenges of incorporating molecular technologies and cladistics into biogeography and phylogenetic reconstructions are attracting new investigators into this rejuvenated field. Alkenones, hopanes, and sterenes in sediments, oxygen and hydrogen isotopes in wood, flavonoids in fossil leaves of *Zelkova*, and ancient DNA in fossil tissues are related discoveries. As noted in the Prologue, there is an excitement attendant to the realization that with the development and application of new technology, we are on the threshold of attaining a much improved understanding of the dynamics of past environmental change and biotic response. The importance of this understanding for predicting future changes is just being realized.

> Prediction is very difficult—especially when it involves the future. (Nils Bohr, quoted from Wehr, 1995)

The future of North American plant formations is unclear because of the novel forcing mechanisms of greenhouse warming, ozone depletion, destruction of the rain forest, and unprecedented environmental disruption through intensifying agricultural and industrial activities. The next generation of paleoecologists will witness clearer signals about trends in climate, biotic response, relevancy of past patterns to future events, and the degree to which computer modeling can predict these events. The task will be complicated, however, by continuing rapid modification of the modern analogs used to interpret the ancient communities. After all, as Derek Ager notes, "It's the only present we've got." (1993, chapter 12).

Beyond theoretical considerations, the world's plant resources are of immense economic, as well as strategic, importance: witness the increasing use of food as an inducement for political conformity. Clearly shifts in climate affect the geography of agricultural surplus and deficit (Nicholls, 1997) and some could have serious political implications. Computer models predict that greenhouse warming by enhanced CO_2 concentration will benefit agriculture in North America and adversely affect it in the lower latitudes that include many developing countries that can least afford a reduction in agricultural productivity (Rosenzweig and Hillel, 1995). Industrialized countries will certainly not be immune to the unrest associated with this reduction.

Predicting the impact of future climatic changes on sea levels is critical for coastal centers of population and for determining suitable sites for the storage of nuclear waste with a half-life up to 10–30 ka (Leopold, 1987). The recent collapse of an underground salt dome in Louisiana illustrates the vulnerability of such potential sites to the inevitable ebb and flow of groundwater and dissolution resulting from sea-level changes. The recent revelations that glacial–interglacial stages, rather than proceeding at a steady pace of 175 Ky, consist of interglacials lasting only 10–12 Ky and that the present one began ~11 Kya, underscores the importance of paleoclimatic data in formulating rational long-term strategies and resource management practices.

Vegetational history provides important proxy data for determining past patterns of environmental change. It is also our principal measure of how far we have gone in creating a future ecological catastrophy through the destruction of rain forest and other habitats and through unprecedented reduction in biodiversity. This history can serve to place in perspective the nature of modern pertubations on the Earth's environment and its biota. It is worthwhile to contemplate that the current rate of extinction equals or exceeds that of the terminal Cretaceous event and constitutes the greatest catastrophy ever imposed on the planet. As Kenneth Hsü noted, "What should we be afraid of? Our neighbors are not our enemy, nor or we likely to be exterminated by our nearest relatives. Our enemy is the disastrous disregard of the manmade catastrophe that threatens to exterminate many species on earth, including, eventually, *Homo sapiens*." (1992). It is an exciting time for studying, interpreting, and assessing the implications of vegetational, faunal, and environmental history. The broadest scope of talent, methodologies, and context information will be needed. A lot is riding on the outcome.

References

Ager, D. 1993. The New Catastrophism. Cambridge University Press, Cambridge, U.K.

Arber, E. A. N. and J. Parkin.1907. The origin of the angiosperms. Bot. J. Linn. Soc. 38: 29–80.

Arnold, C. A. 1947. An Introduction to Paleobotany. McGraw-Hill, New York.

Avise, J. C. et al. (+ 7 authors). 1987. Intraspecific phytogeography: the mitochondrial DNA bridge between population genetics and systematics. Annu. Rev. Ecol. Syst. 18: 489–522.

Axelrod, D. I. 1975. Evolution and biogeography of the Madrean–Tethyan sclerophyll vegetation. Ann. Missouri Bot. Garden 62: 280–334.

Boufford, D. E. and S. A. Spongberg. 1983. Eastern Asian–eastern North American phytogeographical relationships—a history from the time of Linnaeus to the twentieth century. Ann. Missouri Bot. Garden 70: 423–439.

Braun, E. L. 1955. The phytogeography of unglaciated eastern United States and its interpretation. Bot. Rev. 21: 297–375.

Breedlove, D. E. 1973. The phytogeography and vegetation of Chiapas (Mexico). In: A. Graham (ed.), Vegetation and Vegetational History of Northern Latin America. Elsevier Science Publishers, Amsterdam. pp. 149–165.

Bremer, K. 1992. Ancestral areas: a cladistic reinterpretation of the center of origin concept. Syst. Biol. 41: 436–445.

Brooks, D. R. and D. A. McLennan. 1991. Phylogeny, Ecology, and Behavior, a Research Program in Comparative Biology. University of Chicago Press, Chicago.

Brown, C. A. 1938. The flora of Pleistocene deposits in the western Florida Parishes, west Feliciana Parish, and east Baton Rouge Parish, Louisiana. In: H. N. Fisk, H. G. Richards, C. A. Brown, and W. C. Steere (eds.), Contributions to the Pleistocene History of the Florida Parishes of Louisiana. Department of Conservation, Louisiana Geological Survey, New Orleans, LA. Geol. Bull. 12, pp. 59–96.

Chaney, R. W. 1944. Pliocene floras of California and Oregon. (Introduction and subsequent papers). Carnegie Inst. Wash. Contrib. Paleontol. 553.

Coates, A. G. et al. (+ 7 authors). 1992. Closure of the Isthmus of Panama: the near-shore marine record of Costa Rica and western Panama. Geol. Soc. Am. Bull. 104: 814–828.

Crawford, D. J. 1990. Plant Molecular Systematics, Macromolecular Approaches. Wiley-Interscience, New York.

Cronquist, A. 1987. A botanical critique of cladism. Bot. Rev. 53: 1–52.

Davis, J. H., Jr. 1946. The peat deposits of Florida, their occurrence, development, and uses. Fla. Dept. Conserv., Geol. Surv., Geol. Bull. 30: 1–247.

Davis, M. B. 1983. Quaternary history of deciduous forests of eastern North America and Europe. Ann. Missouri Bot. Garden 70: 550–563.

Deevey, E. S., Jr. 1949. Biogeography of the Pleistocene. Bull. Geol. Soc. Am. 60: 1315–1416.

Delcourt, P. A. and H. R. Delcourt. 1977. The Tunica Hills, Louisiana–Mississippi: late glacial locality for spruce and deciduous forest species. Quatern. Res. 7: 218–237.

Donoghue, M. J. and P. D. Cantino. 1988. Paraphyly, ancestors, and the goals of taxonomy: a botanical defense of cladism. Bot. Rev. 54: 107–128.

Fernald, M. L. 1925. Persistence of plants in unglaciated areas of boreal America. Mem. Am. Acad. Arts Sci. 15: 241–342.

———. 1931. Specific segregations and identities in some floras of eastern North America and the Old World. Rhodora 33: 25–63 [reprinted as Contrib. Gray Herb. 93].

Forey, P. L. et al. (+ 5 authors). 1994. Cladistics, A Practical Course in Systematics. Oxford Science Publications, Claredon Press, Oxford, U.K. Systematics Association Publ. 10.

Frederiksen, N. O. 1981. Middle Eocene to Early Oligocene plant communities of the Gulf Coast. In: J. Gray, A. J. Boucot, and W. B. Berry (eds.), Communities of the Past. Hutchinson Ross Publishing Co., Stroudsburg, PA. pp. 493–549.

Fritsch, P. 1996. Isozyme analysis of intercontinental disjuncts within *Styrax* (Styracaceae): implications for the Madrean–Tethyan hypothesis. Am. J. Bot. 83: 342–355.

Gómez-Pompa, A. 1973. Ecology of the vegetation of Veracruz. In: A. Graham (ed), Vegetation and Vegetational History of Northern Latin America. Elsevier Science Publishers, Amsterdam. pp. 73–148.

Graham, A. 1966. Plantae rariores Camschatcenses: a translation of the dissertation of Jonas P. Halenius, 1750. Brittonia 18: 131–139.

——— (ed.). 1972a. Floristics and Paleofloristics of Asia and Eastern North America. Elsevier Science Publishers, Amsterdam.

———. 1972b. Outline of the origin and historical recognition of floristic affinities between Asia and eastern North America. In: A. Graham (ed.), Floristics and Paleofloristics of Asia and Eastern North America. Elsevier Science Publishers, Amsterdam. pp. 1–18.

——— (ed.). 1973a. Vegetation and Vegetational History of Northern Latin America. Elsevier Science Publishers, Amsterdam.

———. 1973b. History of the arborescent temperate element in the Latin American biota. In: A. Graham (ed.), Vegetation and Vegetational History of Northern Latin America. Elsevier Science Publishers, Amsterdam. pp. 301–314.

———. 1976. Studies in neotropical paleobotany. II. The Miocene communities of Veracruz, Mexico. Ann. Missouri Bot. Garden 63: 787–842.

———. 1982. Diversification beyond the Amazon Basin. In: G. T. Prance (ed.), Biological Diversification in the Tropics. Columbia University Press, New York. pp. 78–90.

———. 1985. Studies in neotropical paleobotany. IV. The Eocene communities of Panama. Ann. Missouri Bot. Garden 72: 504–534.

———. 1987. Miocene communities and paleoenvironments of southern Costa Rica. Am. J. Bot. 74: 1501–1518.

———. 1988a. Studies in neotropical paleobotany. V. The lower Miocene communities of Panama—the Culebra Formation. Ann. Missouri Bot. Garden 75: 1440–1466.

———. 1988b. Studies in neotropical paleobotany. VI. The lower Miocene communities of Panama—the Cucaracha Formation. Ann. Missouri Bot. Garden 75: 1467–1479.

———. 1989. Studies in neotropical paleobotany. VII. The lower Miocene communities of Panama—the La Boca Formation. Ann. Missouri Bot. Garden 76: 50–66.

———. 1990. Late Tertiary microfossil flora from the Republic of Haiti. Am. J. Bot. 77: 911–926.

———. 1991a. Studies in neotropical paleobotany. VIII. The Pliocene communities of Panama—introduction and ferns, gymnosperms, angiosperms (monocots). Ann. Missouri Bot. Garden 78: 190–200.

———. 1991b. Studies in neotropical paleobotany. IX. The Pliocene communities of Panama—angiosperms (dicots). Ann. Missouri Bot. Garden 78: 201–223.

———. 1991c. Studies in neotropical paleobotany. X. The Pliocene communities of Panama—composition, numerical representations, and paleocommunity paleoenvironmental reconstructions. Ann. Missouri Bot. Garden 78: 465–475.

———. 1992. Utilization of the isthmian land bridge during the Cenozoic—paleobotanical evidence for timing, and the selective influence of altitudes and climate. Rev. Palaeobot. Palynol. 72: 119–128.

——— and C. Heimsch. 1960. Pollen studies of some Texas peat deposits. Ecology 41: 785–790.

Graham, A. and D. M. Jarzen. 1969. Studies in neotropical paleobotany. I. The Oligocene communities of Puerto Rico. Ann. Missouri Bot. Garden 56: 308–357.

Gray, A. 1840. Dr. Siebold, Flora Japonica. Am. J. Sci. Arts 39: 175–176.

———. 1846. Analogy between the flora of Japan and that of the United States. Am. J. Sci. Arts II, 2: 135–136.

———. 1859. Diagnostic characters of new species of phanerogamous plants collected in Japan by Charles Wright, botanist of the U.S. North Pacific Exploring Expedition. (Published by request of Captain James Rodgers, Commander of the Expedition). With observations upon the relations of the Japanese flora to that of North America, and of other parts of the northern temperate zone. Mem. Am. Acad. Arts 6: 377–452.

Gray, J. 1960. Temperate pollen genera in the Eocene (Claiborne) flora, Alabama. Science 132: 808–810.

Halenius, J. P. 1750. Plantae Rariores Camschatcenses. Thesis, University of Uppsala, Uppsala, Sweden.

Hedges, S. B., C. A. Hass, and L. R. Maxson. 1994. Reply: Towards a biogeography of the Caribbean. Cladistics 10: 43–55.

Hoey, M. T. and C. R. Parks. 1991. Isozyme divergence between eastern Asian, North American, and Turkish species of *Liquidambar* (Hamamelidaceae). Am. J. Bot. 78: 938–947.

——— and ———. 1994. Genetic divergence in *Liquidambar styraciflua*, *L. formosana*, and *L. acalycina* (Hamamelidaceae). Syst. Bot. 19: 308–316.

Hou, H.-Yu. 1983. Vegetation of China with reference to its geographical distribution. Ann. Missouri Bot. Garden 70: 509–548.

Hsü, K. J. 1992. Challenger at Sea, a Ship that Revolutionized Earth Science. Princeton University Press, Princeton, NJ.

Hultén, E. 1927–1930. Flora of Kamtchatka and the Adjacent Islands, Parts I–IV. Almquist & Wiksell, Stockholm.

———. 1937. Outline of the History of Arctic and Boreal Biota During the Quaternary Period. Thule, Stockholm.

Humphries, C. J. and J. A. Chappill. 1988. Systematics as science: a response to Cronquist. Bot. Rev. 54: 129–144.

——— and L. R. Parenti. 1986. Cladistic Biogeography. Oxford Monographs on Biogeography. Clarendon Press, Oxford, U.K. No. 2.

Hunter, R. L. and C. L. Markert. 1957. Histochemical demonstration of enzymes separated by zone electrophoresis in starch gels. Science 125: 1294–1295.

Hunziker, J. H., R. A. Palacios, A. de Valesi, and L. Poggio. 1972. Species disjunctions in *Larrea*: evidence from morphology, cytogenetics, phenolic compounds, and seed albumins. Ann. Missouri Bot. Garden 59: 224–233.

Iversen, J. 1941. Land occupation in Denmark's Stone Age. A pollen-analytical study of the influence of farmer culture on the vegetational development. Danmarks Geol. Unders. 66: 20–65.

Johnston, I. M. 1940. The floristic significance of shrubs common to North and South American areas. J. Arnold Arb. 21: 356–363.

Kalin Arroyo, M. T., P. H. Zedler, and M. D. Fox (eds.). 1995. Ecology and Biogeography of Mediterranean Ecosystems in Chile, California, and Australia. Ecological Studies 108. Springer-Verlag, Berlin.

Leopold, E. B. 1987. Past and future climatic and hydrologic patterns at Hanford, Washington. Northwest Environ. J. 3: 75–96.

Li, H.-L. 1952. Floristic relationships between eastern Asia and eastern North America. Trans. Am. Philos. Soc. 42: 371–429.

Li, W.-H. and D. Graur. 1991. Fundamentals of Molecular Evolution. Sinauer Associates, Sunderland, MA.

Manchester, S. R. 1995. Paleophytogeography of extant eastern Asian "endemic" seed plant genera with paleobotanical records in the Tertiary of North America. In: International Conference of Diversification and Evolution of Terrestrial Plants in Geological Time (ICTPG), Nanjing, 1995. p. 82.

——— and S.-X. Guo. 1996. *Palaeocarpinus* (extinct Betulaceae) from northwestern China: new evidence for Paleocene floristic continuity between Asia, North America, and Europe. Int. J. Plant Sci. 157: 240–246.

Margoliash, E. 1963. Primary structure and evolution of cytochrome C. Proc. Natl. Acad. Sci. USA 50: 672–679.

Marie-Victorin, F. 1938. Phytogeographical problems in eastern Canada. Am. Midland Nat. 19: 489–558.

Markert, C. L. and F. Møller. 1959. Multiple forms of enzymes: tissue, ontogenetic and species-specific patterns. Proc. Natl. Acad. Sci. USA 45: 753–763.

Martin, P. S. 1958. A biogeography of reptiles and amphibians in the Gomez Farias region, Tamaulipas, Mexico. Miscellaneous Publs. 101. Museum of Zoology University of Michigan, Ann Arbor, MI.

——— and B. E. Harrell. 1957. The Pleistocene history of temperate biotas in Mexico and eastern United States. Ecology 38: 468–480.

Marvin, U. B. 1973. Continental Drift, the Evolution of a Concept. Smithsonian Institution Press, Washington, D.C.

McVaugh, R. 1943. The vegetation of the granitic flat-rocks of southeastern United States. Ecol. Monogr. 13: 120–166.

Meusel, H. Von. 1969. Beziehungen in der florendifferenzierung von Eurasien und Nordamerika. Flora 158: 537–564.

Miranda, F. and A. J. Sharp. 1950. Characteristics of the vegetation in certain temperate regions of eastern Mexico. Ecology 31: 313–333.

Morrone, J. J. and J. V. Crisci. 1995. Historical biogeography: introduction to methods. Annu. Rev. Ecol. Syst. 26: 373–401.

Nei, M. 1972. Genetic distance between populations. Am. Nat. 106: 283–293.

———. 1987. Molecular Evolutionary Genetics. Columbia University Press, New York.

Nelson, G. 1978. From Candolle to Croizat: comments on the history of biogeography. J. Hist. Biol. 11: 269–305.

——— and N. Platnick. 1981. Systematics and Biogeography, Cladistics and Vicariance. Columbia University Press, New York.

Nichols, N. 1997. Increased Australian wheat yield due to recent climate trends. Nature 387:484–485.

Niemelä, P. and W. J. Mattson. 1996. Invasion of North American forests by European phytophagous insects. BioScience 46: 741–753.

Orians, G. H. and O. T. Solbrig (eds.). 1977. Convergent Evolution in Warm Deserts. Dowden, Hutchinson & Ross, Stroudsburg, PA.

Page, R. D. M. 1988. Quantitative cladistic biogeography: constructing and comparing area cladograms. Syst. Zool. 37: 254–270.

——— and C. Lydeard. 1994. Towards a cladistic biogeography of the Caribbean. Cladistics 10: 21–41.

Perry, J. P., Jr. 1991. The Pines of Mexico and Central America. Timber Press, Portland, OR.

Phipps, J. B. 1983. Biogeographic, taxonomic, and cladistic relationships between east Asiatic and North American *Crataegus*. Ann. Missouri Bot. Garden 70: 667–700.

Potzger, J. E. and B. C. Tharp. 1943. Pollen record of Canadian spruce and fir from Texas bog. Science 98: 584–585.

———. 1947. Pollen profile from a Texas bog. Ecology 28: 274–280.

———. 1954. Pollen study of two bogs in Texas. Ecology 35: 462–466.

Qiu, Y.-L., C. R. Parks, and M. W. Chase. 1995a. Molecular divergence in the eastern Asia–eastern North America disjunct Section *Rytidospermum* of *Magnolia* (Magnoliaceae). Am. J. Bot. 82: 1589–1598.

Qiu, Y.-L., M. W. Chase, and C. R. Parks. 1995b. A chloroplast DNA phylogenetic study of the eastern Asia–eastern North America disjunct section *Rytidospermum* of *Magnolia* (Magnoliaceae). Am. J. Bot. 82: 1582–1588.

Raven, P. H. 1963. Amphitropical relationships in the floras of North and South America. Q. Rev. Biol. 38: 151–177.

———. 1971. The relationships between "Mediterranean" floras. In: P. H. Davis, P. C. Harper and I. C. Hedge (eds.), Plant Life of South-West Asia. Botanical Society of Edinburgh, Edinburgh, Scotland. pp. 119–134.

———. 1972. Plant species disjunctions: a summary. Ann. Missouri Bot. Garden 59: 234–246.

———. 1973. The evolution of Mediterranean floras. In: F. di Castri and H. A. Mooney (eds.), Mediterranean Type Ecosystems, Origin and Structure. Ecological Studies. Springer-Verlag, Berlin. Vol. 7. pp. 213–224.

——— and D. I. Axelrod. 1975. History of the flora and fauna of Latin America. Am. Sci. 63: 420–429.

———. 1978. Origin and relationships of the California flora. Univ. Calif. Publ. Bot. 72: 1–134.

Riddle, B. R. and R. L. Honeycutt. 1990. Historical biogeography in North American arid regions: an approach using mitochondrial–DNA phylogeny in grasshopper mice (genus *Onychomys*). Evolution 44: 1–15.

Rieseberg, L. H. 1995. The role of hybridization in evolution: old wine in new skins. Am. J. Bot. 82: 944–953.

Ronquist, F. 1997. Dispersal-vicariance analysis: a new approach to the quantification of historical biogeography. Syst. Biol. 46: 195–203.

Rosenzweig, C. and D. Hillel. 1995. Potential impacts of climate change on agriculture and food supply. Consequences, Nature Implications Environ. Change 1: 23–32.

Sarich, V. M. 1977. Rates, sample sizes, and the neutrality hypothesis for electrophoresis in evolutionary studies. Nature 265: 24–28.

Simpson, G. G. 1943. Mammals and the nature of continents. Am. J. Sci. 241: 1–31.

Simroth, H. 1914. Die Pendulations Theorie. 2nd edition. K. Grethlein, Leipzig.

Solbrig, O. T. 1972. The floristic disjunctions between the "Monte" in Argentina and the "Sonoran Desert" in Mexico and the United States. Ann. Missouri Bot. Garden 59: 218–223.

Styles, B. T. 1993. Genus *Pinus*: a Mexican purview. In: T. P. Ramamoorthy, R. Bye, A. Lot, and J. Fa (eds.), Biological Diversity of Mexico, Origins and Distribution. Oxford University Press, Oxford, U.K. pp. 397–420.

Suc, J. P. 1984. Origin and evolution of the Mediterranean vegetation and climate in Europe. Nature 307: 429–432.

Thrower, N. J. W. and D. E. Bradbury (eds.). 1977. Chile–California Mediterranean Scrub Atlas, A Comparative Analysis. Dowden, Hutchinson & Ross, Stroudsburg, PA.

Tiffney, B. H. 1985. Perspectives on the origin of the floristic similarity between eastern Asia and eastern North America. J. Arnold Arb. 66: 73–94.

Udvardy, M. D. F. 1969. Dynamic Zoogeography, with Special Reference to Land Animals. Van Nostrand Reinhold, New York.

van der Hammen, T., T. A. Wijmstra, and W. H. Zagwijn. 1971. The floral record of the Late Cenozoic of Europe. In: K. K. Turekian (ed.), The Late Cenozoic Glacial Ages. Yale University Press, New Haven, CT. pp. 391–424.

Van Dyke, E. C. 1940. The origin and distribution of the coleopterous insect fauna of North America. Proc. Sixth Pacif. Sci. Congr. 4: 255–268.

Wagner, W. L. and V. A. Funk (eds.). 1995. Hawaiian Biogeography: Evolution on a Hot Spot Archipelago. Smithsonian Institute Press, Washington, D.C.

Wehr, W. C. 1995. Early Tertiary flowers, fruits, and seeds of Washington state and adjacent areas. Wash. Geol. 23: 3–16.

Wiley, E. O. 1988. Vicariance biogeography. Annu. Rev. Ecol. Syst. 19: 513–542.

———, D. Siegel-Causey, D. R. Brooks, and V. A. Funk. 1991. The Complete Cladist, A Primer of Phylogenetic Procedures. University of Kansas Museum of Natural History. Lawrence, KS, Special Publ. 19.

Willis, J. C. 1922. Age and Area: A Study in Geographical Distribution and Origin of Species. Cambridge University Press, Cambridge, U.K.

———. 1949. The birth and spread of plants. Boissiera 8: 1–561.

Wolfe, J. A. 1975. Some aspects of plant geography of the northern hemisphere during the Late Cretaceous and Tertiary. Ann. Missouri Bot. Garden 62: 264–279.

———. 1978. A paleobotanical interpretation of Tertiary climates in the northern hemisphere. Am. Sci. 66: 694–703.

———. 1979. Temperature Parameters of Humid to Mesic Forests of Eastern Asia and Relation of Forests of Other Regions of the Northern Hemisphere and Australasia. U.S. Geology Survey, Reston, VA. Professional Paper 1106, pp. 1–37.

Wood, C. E., Jr. 1970. Some floristic relationships between the southern Appalachians and western North America. In: P. C. Holt (ed.), The Distributional History of the Biota of the Southern Appalachians. Part II: Flora. Virginia Polytechnic Institute State University (Blacksburg). Research Div. Monogr. 2, pp. 331–404.

———. 1972. Morphology and phytogeography: the classical approach to the study of disjunctions. Ann. Missouri Bot. Garden 59: 107–124.

Wu, Z. 1983. On the significance of Pacific intercontinental discontinuity. Ann. Missouri Bot. Garden 70: 577–590.

Zuckerkandl, E. 1987. On the molecular evolutionary clock. J. Mol. Evol. 26: 34–46.

——— and L. Pauling. 1962. Molecular disease, evolution and genetic heterogeneity. In: M. Kash and B. Pullman (eds.), Horizons in Biochemistry. Academic Press, New York. pp. 189–225.

———. 1965a. Molecules as documents of evolutionary history. J. Theor. Biol. 8: 357–366.

———. 1965b. Evolutionary divergence and convergence in proteins. In: V. Bryson and H. J. Vogel (eds.), Evolving Genes and Proteins. Academic Press, New York. pp. 97–166.

General Readings

Herrera, C. M. 1995. Plant-vertebrate seed dispersal systems in the Mediterranean: ecological, evolutionary, and historical determinants. Annu. Rev. Ecol. Syst. 26: 705–727.

Nixon, K. C. 1996. Paleobotany in cladistics and cladistics in paleobotany: enlightenment and uncertainty. Rev. Palaeobot. Palynol. 90: 361–373.

Simpson, B. B. and J. Cracraft. 1995. Systematics: the science of biodiversity. BioScience 45: 670–672.

Notes

1. A notable feature of these botanical dissertations was, appropriately enough, the flowery language used in the dedications. Halenius wrote:

> To my foremost patron, the most Reverend and Renowned Lord and Master Engelbert Halenius, Professor of Sacred Theology, Distinguished member of the Ecclesiastical and Academic Consistoires, Pastor in Dannemarck.
>
> Your exceptional favor and kindness, showered abundantly and continually on me, have induced me, my distinguished patron, to enhance my modest labors with your renowned name. For as soon as it was my good fortune, after the loss of my former director of studies, to come to this temple of Minerva, you saw fit, noble sir, to receive me most graciously, undeserving though I was. You have been willing not only to look out for my fortunes in every way, but you deigned also to further all my undertakings. In view of these evident tokens of your favor, I shall always be deeply devoted to you. I therefore now offer you the first-fruits of my studies, and I earnestly urge you to accept them graciously and in the future also bestow upon me your continued favor. As long as I live I shall never cease to offer to Almighty God the most fervent prayers that happiness may come to you and yours. Your Reverence's most humble servant. (1750)

Particularly poignant for those who have provided a modern-day education to our own issue is the dedication to the

> [M]ost Reverend and Learned Peter Halenius, most vigilant Co-Minister of the Church of Hedesunda, and my most indulgent father, with all due reverence until the day of my death.
>
> The day has finally dawned [!] when I can publicly bring to an end my academic apprenticeship. The work which is, as it were, the first-fruits of that period of study—a work which I owe to your eager interest and financial assistance [!!]—I unhesitatingly dedicate to you, since in a sense it makes that dedication of its own accord. The countless benefits you have shown me from my earliest years are too great to be recompensed by so modest a work. But since I am well aware, my dear father, that you ask nothing more from your son than a grateful heart, I earnestly ask you to accept now with fatherly affection these pages which bear witness to my sentiments of respect for you. For the rest, be sure, father, that more filial piety for you will ever be deep in my heart than I can ever express in words. With devoted prayers to Almighty God for your unimpaired health and long life, I shall remain until death, Your most obedient son. (1750)

I mentioned to a class that such expressions of reverence for their mentors in theses has regrettably passed out of fashion. A student in that class later did graduate work with me, and when her thesis was submitted it contained the following dedication:

> Discipula deterrima, magna cum modestia, summissa animo, animo grata omniscienti, benignissimo, generosissimo consiliario, cui libellum dedicatum est, gratiam referre vult. Discipula nequissima quae ad pedes praeclari magistri prosternata est, colens praesentiam eius et laudans sapientiam, elegantiam, urbanitatem, humanitatemque immensam, magnum gaudium, decus magnum verbis apte exprimere non potest. Multitudo caelestis illum propriosque defendant.
>
> It is with great humility that this most worthless student wishes to convey a cornucopia of gratitude to the most gracious, all knowledgeable, most munificent of mentors, to whom this humble essay is dedicated. Mere words cannot adequately express the great joy and honor of this lowly servant, who prostrates herself at the feet of said illustrious master in homage to his presence and in praise of his infinite wisdom, gentility and refinement. May the celestial hosts glorify his name and smile upon his ancestors, as surely he is descended from the heavens.

This is a remarkably perceptive piece of work.

INDEX

Organisms are indexed to genus with fossil occurrences indicated separately. See Contents for major subject headings. Extinct plant formations are indexed; for extant plant formations see vegetation summaries (chapters 5–8) and Epilogue. Notes and Prologue are not indexed.